Chp 3 25b-f, 32, 40, 58, 66

Chp 4 4, 16

1442
1143

SPSS 6.1

INTRODUCTION TO PROBABILITY AND STATISTICS

Principles and Applications for Engineering and the Computing Sciences

McGraw-Hill Series in Probability and Statistics

Brillinger: *Time Series: Data Analysis and Theory*
De Groot: *Optimal Statistical Decisions*
Dixon and Massey: *Introduction to Statistical Analysis*
Milton and Arnold: *Introduction to Probability and Statistics: Principles and Applications for Engineering and the Computing Sciences*
Milton: *Statistical Methods in the Biological and Health Sciences*
Mood, Graybill, and Boes: *Introduction to the Theory of Statistics*
Morrison: *Multivariate Statistical Methods*
Snell: *Introduction to Probability*
Steel and Torrie: *Principles and Procedures of Statistics*

Schaum's Solved Problems Series

Each title in this series is a complete and expert source of solved problems with solutions worked out in step-by-step detail.

Titles on the Current List Include:

3000 Solved Problems in Calculus
2500 Solved Problems in College Algebra and Trigonometry
2500 Solved Problems in Differential Equations
2000 Solved Problems in Discrete Mathematics
3000 Solved Problems in Linear Algebra
2000 Solved Problems in Numerical Analysis
3000 Solved Problems in Precalculus

INTRODUCTION TO PROBABILITY AND STATISTICS

Principles and Applications for Engineering and the Computing Sciences

Third Edition

J. S. Milton

Radford University

Jesse C. Arnold

Virginia Polytechnic Institute and State University

Irwin McGraw-Hill

Boston, Massachusetts Burr Ridge, Illinois Dubuque, Iowa
Madison, Wisconsin New York, New York San Francisco, California St. Louis, Missouri

Irwin/McGraw-Hill

A Division of The ***McGraw-Hill*** *Companies*

This book was set in Times Roman by Science Typographers, Inc.
The editors were Jack Shira and James W. Bradley;
the production supervisor was Paula Keller.
The cover was designed by John Hite.
R. R. Donnelley & Sons Company was printer and binder.

INTRODUCTION TO PROBABILITY AND STATISTICS

Principles and Applications for Engineering and the Computing Sciences

This book is printed on acid-free paper.

6 7 8 9 0 DOC DOC 9 0 9 8

ISBN 0-07-042623-6

Library of Congress Cataloging-in-Publication Data

Milton, J. Susan (Janet Susan)
Introduction to probability and statistics: principles and applications for engineering and the computing sciences / J. S. Milton, Jesse C. Arnold.—3rd ed.
p. cm.
ISBN 0-07-042623-6
1. Engineering mathematics. 2. Electronic data processing—Mathematics. 3. Probabilities. 4. Statistics. I. Arnold, Jesse C. II. Title.
TA330.M486 1995
519.5—dc20 94-31189

ABOUT THE AUTHORS

J. S. Milton is Professor of Statistics at Radford University. Dr. Milton received the B.S. degree from Western Carolina University, the M.A. degree from the University of North Carolina at Chapel Hill, and the Ph.D. degree in statistics from Virginia Polytechnic Institute and State University. She is a Danforth Associate and is a recipient of the Radford University Foundation Award for Excellence in Teaching. Dr. MIlton is the author of *Statistical Methods in the Biological and Health Science* as well as *Introduction to Statistics*, *Probability with the Essential Analysis*, *A First Course in the Theory of Linear Statistical Models*, and *Applied Statistics with Probability*.

Jesse C. Arnold is Professor of Statistics at Virginia Polytechnic Institute and State University. Dr. Arnold received the B.S. degree from Southeastern State University, and the M.A. and Ph.D. degrees in statistics from Florida State University. Dr. Arnold is very active professionally. He served as head of the statistics department for ten years, is a fellow of the American Statistical Association, and an elected member of the Internationaal Statistics Institute. He has been President of the Internationl Biometric Society Eastern North American Region, and Chairman of the Statistical Educational Section of the American Statistical Association.

TO

Peggy Arnold

Enid K. Milton

AND IN LOVING MEMORY OF GEORGE A. MILTON

CONTENTS

PREFACE

It has become increasingly evident that the interpretation of much of the research in the engineering and computing sciences depends to some extent on statistical methods. Furthermore, the practicing engineer will be expected to understand and help implement statistical quality control techniques in the work place. For these reasons, it is essential that students in these fields be exposed to statistical reasoning early in their careers. This text is intended as a first course in probability and applied statistics for students in the engineering and computing sciences. It is hoped that this first course will occur on the undergraduate level. However, the text can be used to advantage by graduate students who have little or no prior experience with statistical methods.

This text is not a statistical cookbook, nor is it a manual for researchers. We attempt to find a middle road—to provide a text that gives the student an understanding of the logic behind statistical techniques as well as practice in using them. A one-year course in elementary calculus should provide an adequate background for understanding everything presented here.

We chose the examples and exercises specifically for the student in the engineering and computing sciences. Most data sets are simulated. However, the simulation was done with care, so that the results of the analysis are consistent with recently reported research. References to reports upon which the data are based are given whenever possible. In this way, the student will gain some insight into the types of engineering problems that can be handled statistically. Many exercises are left open-ended in hopes of stimulating some classroom discussion.

It is assumed that the student has access to some type of electronic calculator. Many such calculators are on the market, and most have some built-in statistical capability. The use of these calculators is encouraged, for it allows the student to concentrate on the interpretation of the analysis rather than on the arithmetic computations.

We should point out that most of the data sets are rather small so that the student will not be overwhelmed by the computational aspects of statistics. We do not intend to imply that very small data sets are routinely used in the engineering fields. In fact, most major research projects involve a tremendous investment in time and money and result in a large body of data. Such data lend themselves to analysis via the electronic computer. For this reason, we include some instruction in the use of statistical packages. The package chosen for illustrative purposes is SAS (Statistical Analysis System: SAS Institute, Inc., Raleigh, NC). This was done because of its widespread availability and ease of use. We do not intend to imply that it is superior to other well-known packages such as SPSS (Statistical Package for the Social Sciences), BMD (Biomedical Computer Programs, University of California Press), or MINITAB (Duxbury Press). The computer methods are presented as "computing supplements." The student with access to the computer is encouraged to give them a try! However, the supplements can be skipped with no loss of continuity.

Each chapter ends with a chapter summary that is intended to remind the student of the major topics presented in the chapter. This chapter summary also includes a list of important terms. A set of exercises is provided for each section of each chapter. In addition, each chapter has a set of review exercises in which the problems are presented in random order. It is hoped that this will help the student develop the ability to recognize the appropriate analysis. The appendices include statistical tables and answers to selected exercises.

A number of different courses can be taught from this book. They can vary in length from one quarter to one year. It is difficult to determine exactly what material can be covered in a given time, since this is a function of class size, academic maturity of the students, and inclination of the instructor. However, we do offer some guidelines for the use of this text. In particular, the type of course presented can vary from one whose chief aim is to familiarize the student with the computational aspects of probability and the handling of data sets to one of a more theoretical nature. In many cases we include the proof or derivation of theorems in the text labeled as such. If an instructor wants to deemphasize theory, these proofs can be skipped easily with no loss of continuity. Starred exercises in the text are either a little more difficult computationally or theoretical in nature. These can be included or deleted to help set the tone of the course. When they are included, the course tends to take on a more theoretical nature.

CHANGES IN THE THIRD EDITION

At the suggestion of users of the first two editions of the text, some new material has been added to the Third Edition. In particular, the chapter on quality control (Ch. 16) has been rewritten and expanded. A new discussion of residual analysis via graphics has been added to Chapter 11 and more graphical ideas have been incorporated throughout the text. New material on model transformations has also been included. Many sections of the text have been rewritten and subsections have been added to make it easier for the reader to follow the flow of ideas discussed. Data sets have been updated and new exercises and examples have been added throughout the text.

Supplemental teaching and learning aids now include a student solutions manual and an instructor's manual. The student manual contains solutions to the odd numbered exercises. (The answers to these exercises are in the text.) The instructor's manual contains solutions to all exercises, a computing supplement, and a data disk. The computing supplement contains suggestions on how to use the data disk and a discussion of how to use MINITAB to reproduce the results of the SAS supplements contained in the text. The computing supplement and the data disk may be copied and distributed to students.

The following is a brief summary of each chapter:

Chapter 1 This chapter provides an introduction to probability and counting.

Chapter 2 The study of probability is continued. The laws governing probability are presented and the notions of conditional probability and independence are introduced.

Chapter 3 The notion of random variables is introduced. General properties of discrete distributions are discussed. The notion of expected value is introduced and the idea of the mean and variance of a distribution is developed. The moment generating function is presented as a means of finding the first two moments of a distribution. Important discrete distributions are studied in detail. The chapter closes with an optional section on simulating discrete distributions.

Chapter 4 Parallels Chap. 3 with emphasis being on continuous distributions.

Chapter 5 Discusses joint distributions of both the discrete and continuous types. The notions of covariance, correlation, and regression are introduced in the theoretical sense.

Chapter 6 is the link between the more theoretical concepts of statistics and the methods of data analysis. Here we present an introduction to classical data-handling techniques and descriptive statistics. We also introduce some of the newer techniques of exploratory data analysis.

Chapter 7 considers the notion of point and interval estimation of population parameters. Method of moments, maximum likelihood, and unbiased estimators are considered. Some distribution theory is also discussed. In particular, the distribution of $\bar{X}$ is investigated. The moment generating function is used as a fingerprint to help pinpoint the distribution of some important random variables that will underlie the statistical methods developed in later chapters.

Chapter 8 begins the study of the classical methods of data analysis. The topic of interest is inferences on the location and variability of a distribution based on a single sample. Both estimation and hypothesis testing are discussed and the T distribution is introduced. A full discussion of significance testing is included. The methods presented assume that sampling is from a normal distribution. For this reason, the Lilliefors test, a graphical test for normality, is included here. The chapter closes with a section on nonparametric tests for location. These tests are especially useful when the normality assumption appears to be violated.

Chapter 9 In this chapter inferences on a single proportion are considered. The study of two sample problems is begun by showing how to compare two proportions based on independent random samples.

Chapter 10 is concerned with methods used to compare two variances and two means. The F distribution is introduced as a means of comparing variances. Means are compared first when variances are assumed to be equal. The Smith-Satterthwaite procedure is used to compare means when variances appear to be unequal. These procedures all assume independent sampling. A procedure for comparing means based on paired data is presented. The chapter ends with a section on nonparametric two-sample tests for location.

Chapter 11 studies simple linear regression and correlation. The least-squares method is given for estimating parameters in the regression model. Estimation and hypothesis testing is presented. Development for the simple linear regression model is quite thorough as preparation for the more general regression cases discussed in Chap. 12. The bivariate normal distribution is presented as needed for estimation and testing for product-moment correlation. A new section on the analysis of residuals is included.

Chapter 12 The simple linear regression model is extended to multiple and polynomial models. The methods of Chap. 11 are extended in matrix form. Variable selection procedures are discussed along with examples.

Chapter 13 The analysis of variance procedure is studied for various one-factor experimental designs. This chapter includes a discussion of randomized complete blocks and some results on the effectiveness of blocking are given. A new section on Latin squares is included as well as material on Bonferroni-type multiple comparisons. Variance component estimation in random effects models is discussed.

Chapter 14 This chapter discusses factorial experiments and contains material on fractional factorials.

Chapter 15 is an introduction to the study of categorical data. Chi-squared goodness of fit tests are presented. Contingency table tests for independence and homogeneity are discussed in both the 2×2 and $r \times c$ cases.

Chapter 16 discusses the basic concepts of statistical quality control. Process control is discussed using control charts, and basic ideas of acceptance sampling are presented. The relationship of acceptance sampling with usual hypothesis testing is presented. Taguchi methods are discussed briefly.

You should be aware of the fact that statistics is an art as well as a science. For this reason, there is always room for debate on how to properly analyze a given data set. We have presented in this text methods that have stood the test of time as well as some that are relatively new. In many cases we have intentionally left to you the decision of whether or not to reject a particular null hypothesis. The reason for this is simple! No one can really say how small a probability must be in order to claim that it is too small to have occurred by chance. You might disagree with our conclusions at times. Feel free to do so!

ACKNOWLEDGMENTS

We wish to thank the Chemical Rubber Company, Bell Laboratories, and the American Society for Testing and Materials for use of statistical tables. Special thanks go to SAS Institute for permission to use their package for illustrative purposes.

A particular thanks goes to Dr. Jill Alcorn for her help in developing the chapter on experimental design and her many hours spent checking answers and writing the solutions manuals.

Thanks are also due to Linda Seawell and Jo Ann Fisher for typing the manuscript. Very special thanks are offered to the following reviewers for their many helpful suggestions during the preparation of this, and the previous two editions: Lynne Billard, University of Georgia; Martha L. Bouknight, Meredith College; David C. Brooks, Seattle Pacific University; Daren B. H. Cline, Texas A & M University; Michael W. Ecker, Pennsylvania State University; Peter G. Furth, Northeastern University; Robert Lacher, South Dakota State University; H. N. Nagaraja, The Ohio State University; Roxanne Peck, California Polytechnic and State University; Larry G. Richards, University of Virginia; Paul Speckman, University of Missouri—Columbia; Larry Stephens, University of Nebraska; Harrison M. Wadsworth, Georgia Institute of Technology; and Vasant Waikar, Miami University of Ohio.

Finally, we wish to acknowledge the helpful discussions with our colleagues at Radford University and Virginia Polytechnic Institute and State University. From J.C.A., special thanks for the unfailing inspiration and support of my wife Peggy and for the love and encouragement of our children Christa and Chuck.

J. S. Milton
Jesse C. Arnold

CHAPTER 1

INTRODUCTION TO PROBABILITY AND COUNTING

What is "statistics" and why is its study important to engineers and scientists? To answer this question; let us describe an aspect of the work of a scientist known as "model building."

Basically, the job of a scientist is to describe what he or she sees, to try to explain what is observed, and to use this knowledge to predict events in the world in which we live. The explanation often takes the form of a physical model. A *model* is a theoretical explanation of the phenomenon under study and, at the outset, is usually expressed verbally. To use the model for predictive purposes, this verbal description must be translated into one or more mathematical equations. These equations can be used to determine the value of a specific variable in the model based on the knowledge of the values assumed by other model variables. For example, the Perfect Gas Law states that the pressure and volume of a gas may both vary simultaneously when the temperature of the gas is changed. This verbal model can be translated into a mathematical equation by writing

$$\text{Perfect Gas Law: } PV = RT$$

where P is the pressure of the gas, V is its volume, T is its temperature, and R is a constant, called the *gas constant*. The numerical value of the gas constant depends on the physical units chosen for the other terms in the model. Once we know the values assumed by two of the three variables P, V, or T, we can calculate the value of the third via this mathematical model. For example, under a pressure of 760 mm mercury and a temperature of 273 kelvins a mole of any gas is thought to have a volume of 22.4 liters. The gas constant in this case has a value of approximately 62.36. Based on the Perfect Gas Law, a gas with a volume of 5 liters at a temperature of 100 kelvins has pressure P given by

$$PV = RT = 62.36T$$

or

$$P(5) = 62.36(100)$$

$$P = 1247.2 \text{ mm mercury}$$

That is, our model leads us to expect the pressure to be 1247.2 mm mercury. A model such as the Perfect Gas Law is said to be "deterministic." It is deterministic in the sense that it allows us to determine an exact value for the variable of interest under specified experimental conditions. The Perfect Gas Law does describe *some* real gases at moderate temperatures and pressures. Unfortunately, many real gases cannot be described by this or any other deterministic model, especially at extreme temperatures and pressures! Under these circumstances we must find another way to predict the behavior of the gas with some degree of certainty. This can be done with the aid of statistical methods.

What do we mean by statistical methods? These are methods by which decisions are made based on the analysis of data gathered in carefully designed experiments. Since experiments cannot be designed to account for every conceivable contingency, there is always some uncertainty in experimental science. Statistical methods are designed to *allow us to assess the degree of uncertainty present in our results*. These methods can be classed roughly into three categories: descriptive statistics, inferential statistics, and model building. By descriptive statistics we mean those techniques, both analytic and graphical, that allow us to describe or picture a data set. Inferential statistics concerns methods by which conclusions can be drawn about a large group of objects, called the *population*, based on observing only a *sample*, or a portion, of the objects in the population. Model building entails the development of prediction equations from experimental data. These equations are called statistical models; they are models that allow us to predict the behavior of a complex system and to assess our probability of error. These categories are not mutually exclusive. That is, methods developed to solve problems in one area often find application in another. We shall be concerned with all three areas in this text.

A statistician or user of statistics is always working in two worlds. The ideal world is at the population level and is theoretical in nature. It is the world that we would like to see. The world of reality is the sample world. This is the level at which we really operate. We hope that the characteristics of our sample reflect well the characteristics of the population. That is, we treat our sample as a microcosm that mirrors the population. The idea is illustrated in Fig. 1.1.

The mathematics on which statistical methods rest is probability theory. For this reason, we begin the study of statistics by considering the basic concepts of probability.

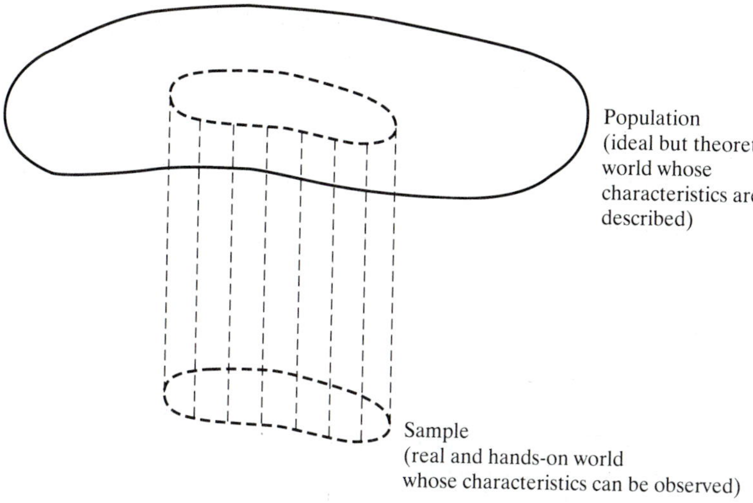

FIGURE 1.1
The sample is viewed as a miniature population. We hope that the behavior of the variable under study over the sample gives an accurate picture of its behavior in the population.

1.1 INTERPRETING PROBABILITIES

When asked, "Do you know anything about probability?" most people are quick to answer, "no!" Usually that is not the case at all. The ability to interpret probabilities is assumed in our culture. One hears the phrases "the probability of rain today is 95%" or "there is a 0% chance of rain today." It is assumed that the general public can interpret these values correctly. The interpretation of probabilities is summarized as follows:

1. Probabilities are numbers between 0 and 1, inclusive, that reflect the chances of a physical event occurring.
2. Probabilities near 1 indicate that the event is extremely likely to occur. They mean not that the event will occur, only that the event is considered to be a common occurrence.
3. Probabilities near zero indicate that the event is not very likely to occur. They do not mean that the event will fail to occur, only that the event is considered to be rare.
4. Probabilities near 1/2 indicate that the event is just as likely to occur as not.
5. Since numbers between 0 and 1 can be expressed as percentages between 0 and 100, probabilities are often expressed as percentages. This is particularly common in writings of a nontechnical nature.

These properties are guidelines for interpreting probabilities once they are available, but they do not indicate how to assign probabilities to events. Three methods are widely used: the *personal* approach, the *relative frequency* approach, and the *classical approach*. These methods are illustrated in the following examples.

Example 1.1.1. An oil spill has occurred. An environmental scientist asks, "What is the probability that this spill can be contained before it causes widespread damage to nearby beaches?" Many factors come into play, among them the type of spill, the amount of oil spilled, the wind and water conditions during the clean-up operation, and the nearness of the beaches. These factors make this spill unique. The scientist is called upon to make a value judgment; to assign a probability to the event based on informed *personal opinion*.

The main advantage of the personal approach is that it is always applicable. Anyone can have a personal opinion about anything. Its main disadvantage is, of course, that its accuracy depends on the accuracy of the information available and the ability of the scientist to assess that information correctly.

Example 1.1.2. An electrical engineer is studying the peak demand at a power plant. It is observed that on 80 of the 100 days randomly selected for study from past records, the peak demand occurred between 6 and 7 p.m. It is natural to assume that the probability of this occurring on another day is at least *approximately*

$$\frac{80}{100} = .80$$

This figure is not simply a personal opinion. It is a figure based on repeated experimentation and observation. It is a *relative frequency*.

The relative frequency approach can be used whenever the experiment can be repeated many times and the results observed. In such cases, the probability of the occurrence of event A, denoted by $P[A]$, is approximated by

$$P[A] \doteq \frac{f}{n} = \frac{\text{number of times event } A \text{ occurred}}{\text{number of times experiment was run}}$$

The disadvantage in this approach is that the experiment cannot be a one-shot situation; it must be repeatable. Remember that any probability obtained this way is an approximation. It is a value based on n trials. Further testing might result in a different approximate value. However, as the number of trials increases, the changes in the approximate values obtained tend to become slight. Thus for a large number of trials, the approximate probability obtained by using the relative frequency approach is usually quite accurate.

Example 1.1.3. What is the probability that a child born to a couple heterozygous for eye color (each with genes for both brown and blue eyes) will be brown-eyed? To answer this question, we note that since the child receives one gene from each parent, the possibilities for the child are (brown, blue), (blue, brown), (blue, blue) and (brown, brown), where the first member of each pair represents the gene received from the father. Since each parent is just as likely to contribute a gene for brown eyes as for blue eyes, all four possibilities are equally likely. Since the gene for brown eyes is dominant, three of the four possibilities lead to a brown-eyed child. Hence the probability that the child is brown-eyed is $3/4 = .75$.

The above probability is not a personal opinion, nor is it based on repeated experimentation. In fact, we found this probability by the *classical* method. This method can be used *only* when it is reasonable to assume that the possible outcomes of the experiment are equally likely. In this case, the probability of the occurrence of event A is given by

$$P[A] = \frac{n(A)}{n(S)} = \frac{\text{number of ways } A \text{ can occur}}{\text{number of ways the experiment can proceed}}$$

One advantage to this method is that it does not require experimentation. Furthermore, if the outcomes are truly equally likely, then the probability assigned to event A is not an approximation. It is an accurate description of the frequency with which event A will occur.

1.2 SAMPLE SPACES AND EVENTS

To determine what is "probable" in an experiment, we first must determine what is "possible." That is, the first step in analyzing most experiments is to make a list of possibilities for the experiment. Such a list is called a *sample space.* We define this term as follows.

Definition 1.2.1 (Sample space and sample point). A sample space for an experiment is a set S with the property that each physical outcome of the experiment corresponds to exactly one element of S. An element of S is called a sample point.

When the number of possibilities is small, an appropriate sample space usually can be found without difficulty. For instance, we have seen that when a couple heterozygous for eye color parents a child, the possible genotypes for the child are given by

$$S = \{(\text{brown}, \text{blue}), (\text{blue}, \text{brown}), (\text{blue}, \text{blue}), (\text{brown}, \text{brown})\}$$

As the number of possibilities becomes larger, it is helpful to have a system for developing a sample space. One such system is the *tree diagram*. The next example illustrates the idea.

Example 1.2.1. During a space shot the primary computer system is backed up by two secondary systems. They operate independently of one another in that the failure of one has no effect on any of the others. We are interested in the readiness of these three systems at launch time. What is an appropriate sample space for this experiment?

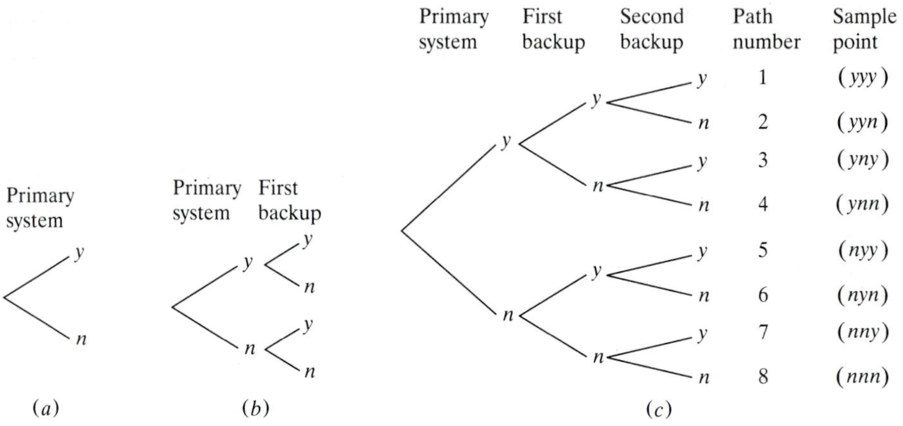

FIGURE 1.2
Constructing a tree diagram.

Since we are primarily concerned with whether each system is operable at launch, we need only find a sample space that gives that information. To generate the sample space, we use a *tree*. The primary system is either operable (yes) or not operable (no) at the time of launch. This is indicated in the tree diagram of Fig. 1.2(a), where yes = y and no = n. Likewise the first backup system either is or is not operable. This is shown in Fig. 1.2(b). Finally, the second backup system either is or is not operable. The tree is completed as shown in Fig. 1.2(c). A sample space S for the experiment can be read from the tree by following each of the eight distinct paths through the tree. Thus

$$S = \{yyy, yyn, yny, ynn, nyy, nyn, nny, nnn\}$$

Once a suitable sample space has been found, elementary set theory can be used to describe physical occurrences associated with the experiment. This is done by considering what are called *events* in the mathematical sense.

Definition 1.2.2 (Event). Any subset A of a sample space is called an event. The empty set $\varnothing$ is called the *impossible* event; the subset S is called the *certain* event.

Example 1.2.2. Consider a space shot in which a primary computer system is backed up by two secondary systems. The sample space for this experiment is

$$S = \{yyy, yyn, yny, ynn, nyy, nyn, nny, nnn\}$$

where, for example, yny denotes the fact that the primary system and second backup are operable at launch, whereas the first backup is inoperable (see Example 1.2.1). Let

A: primary system is operable

B: first backup is operable

C: second backup is operable

The mathematical event corresponding to each of these physical events is found by listing the sample points that represent the occurrence of the event. Thus we write

$$A = \{yyy, yyn, yny, ynn\}$$

$$B = \{yyy, yyn, nyy, nyn\}$$

$$C = \{yyy, yny, nyy, nny\}$$

Other events can be described using these events as building blocks. For example, the event that "the primary system *or* the first backup is operable" is given by the set $A \cup B$. Thus

$$A \cup B = \begin{matrix}\text{primary } \textit{or} \text{ first} \\ \text{backup is operable}\end{matrix} = \{yyy, yyn, yny, ynn, nyy, nyn\}$$

Note that the word "or" will denote set union. The event that "the primary system *and* the first backup is operable" is given by the set $A \cap B$. That is,

$$A \cap B = \text{primary } \textit{and} \text{ first backup operable} = \{yyy, yyn\}$$

Note that the word "and" will denote the set intersection. The event that "the primary system or the first backup is operable but the second backup is inoperable" is given by $(A \cup B) \cap C'$, where C' denotes the complement of set C. Thus

$$(A \cup B) \cap C' = \begin{matrix}\text{primary } \textit{or} \text{ first backup operable} \\ \textit{but} \text{ second backup inoperable}\end{matrix} = \{yyn, ynn, nyn\}$$

Note that the word "but" is also translated as a set intersection; the word "not" translates as a set complement.

Let us pause briefly to consider a basic difference between the sample space

$$S_1 = \{(\text{brown}, \text{blue}), (\text{blue}, \text{brown}), (\text{blue}, \text{blue}), (\text{brown}, \text{brown})\}$$

of Example 1.1.3 and

$$S_2 = \{yyy, yyn, yny, ynn, nyy, nyn, nny, nnn\}$$

of Example 1.2.1. Since each parent is just as likely to contribute a gene for brown eyes as for blue eyes, the sample points of S_1 are equally likely. This allows us to use the classical method to find the probability that a child born to a couple heterozygous for eye color will be brown-eyed. If we denote this event by A, then we can conclude that

$$P[A] = P[\{(\text{brown}, \text{blue}), (\text{blue}, \text{brown}), (\text{brown}, \text{brown})\}]$$
$$= \frac{n(A)}{n(S)} = \frac{3}{4}$$

However, it is not correct to assume that the sample points of S_2 are equally likely. This would be true if and only if each of the three computer systems is just as likely to fail as to be operable at launch time. Our technology is much better than that! The primary question to be answered is "What is the probability that at least one system will be operable at the time of the launch?" That is, what is

$$P[\{yyy, yyn, yny, ynn, nyy, nyn, nny\}]?$$

As will be shown later, this question can be answered. However, since the sample points are not equally likely, it cannot be answered using the classical method.

Not all trees are symmetric as is that pictured in Fig. 1.2. In some settings paths end at different stages of the game. Example 1.2.3 illustrates an experiment of this sort.

Example 1.2.3. Consider a production process that is known to produce defective parts at the rate of one per hundred. The process is monitored by testing randomly selected parts during the production process. Suppose that as soon as a defective part is found, the process will be stopped and all machine settings will be checked. We are interested in studying the number of parts that are tested in order to obtain the first defective part. In the tree of Fig. 1.3, c represents that the sampling continues and s represents that production is stopped. Notice that as soon as a defective item is found, the process ends and the path also ends. For this reason, some paths are much shorter than others. Notice also that theoretically this tree continues indefinitely. The sample space generated by the tree is

$$S = \{s, cs, ccs, cccs, ccccs, \dots\}$$

Since defective parts occur with probability .01, it should be evident that the paths of this tree are not equally likely.

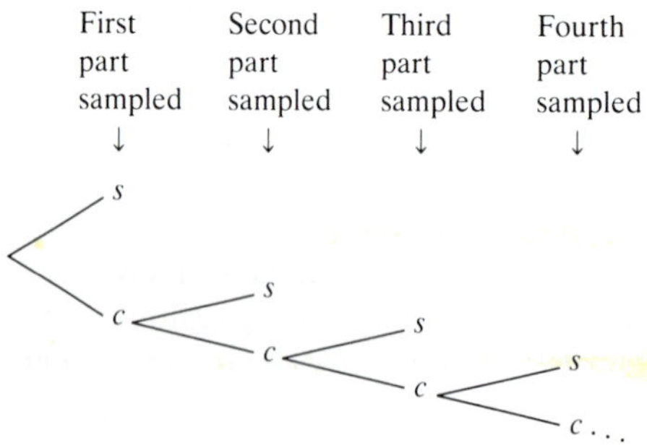

FIGURE 1.3
Sampling a production line for defective parts.

Mutually Exclusive Events

Occasionally interest centers on two or more events that cannot occur at the same time. That is, the occurrence of one event precludes the occurrence of the other. Such events are said to be *mutually exclusive*.

Example 1.2.4. Consider the sample space

$$S = \{yyy, yyn, yny, ynn, nyy, nyn, nny, nnn\}$$

of Example 1.2.1. The events

$$A_1\text{: primary system operable} = \{yyy, yyn, yny, ynn\}$$

$$A_2\text{: primary system inoperable} = \{nyy, nyn, nny, nnn\}$$

are mutually exclusive. It is impossible for the primary system to be both operable and inoperable at the same time. Mathematically, A_1 and A_2 have no sample points in common. That is, $A_1 \cap A_2 = \varnothing$.

Example 1.2.4 suggests the mathematical definition of the term "mutually exclusive events."

Definition 1.2.3 (Mutually exclusive events). Two events A_1 and A_2 are mutually exclusive if and only if $A_1 \cap A_2 = \varnothing$. Events $A_1, A_2, A_3, \ldots$ are mutually exclusive if and only if $A_i \cap A_j = \varnothing$ for $i \neq j$.

1.3 PERMUTATIONS AND COMBINATIONS

As indicated in Sec. 1.1, there are several ways to determine the probability of an event. When the physical description of the experiment leads us to believe that the possible outcomes are equally likely, then we can compute the probability of the occurrence of an event using the classical method. In this case the probability of an event A is given by

$$P[A] = \frac{n(A)}{n(S)}$$

Thus to compute a probability using the classical approach, you must be able to count two things: $n(A)$, the number of ways in which event A can occur, and $n(S)$, the number of ways in which the experiment can proceed. As the experiment becomes more complex, lists and trees become cumbersome. Alternative methods for counting must be found.

Two types of counting problems are common. The first involves *permutations* and the second, *combinations*. These terms are defined as follows:

Definition 1.3.1 (Permutation). A permutation is an arrangement of objects in a definite order.

Definition 1.3.2 (Combination). A combination is a selection of objects without regard to order.

Note that the characteristic that distinguishes a permutation from a combination is *order*. If the order in which some action is taken is important, then the problem is a permutation problem and can be solved using a technique called the multiplication principle. If order is irrelevant, then it is a combination problem and can be solved using a formula which we shall develop.

Example 1.3.1

1. Twenty different amino acids are commonly found in peptides and proteins. A pentapeptide consisting of the five amino acids

   alanine-valine-glycine-cysteine-tryptophan

   has different properties and is, in fact, a different compound from the pentapeptide

   alanine-glycine-valine-cysteine-tryptophan

   which contains the same amino acids. Peptides are permutations of amino acid units because the sequence, or order, of the amino acids in the chain is important.
2. A foundry ships engine blocks in lots of size 20. Before a lot is accepted, three blocks are selected at random and tested for hardness. Only three are tested because the testing requires that the blocks be cut in half, and is therefore destructive. The three blocks selected constitute a combination of engine blocks. We are interested only in which three are selected; we are not interested in the order in which they are chosen.

Counting Permutations

Once a problem has been identified as being one in which order is important, the next question to be answered is, "How many permutations or arrangements of the given objects are possible?" This question usually can be answered by means of the *multiplication principle*.

Multiplication principle. Consider an experiment taking place in k stages. Let n_i denote the number of ways in which stage i can occur for $i = 1, 2, 3, \ldots, k$. Altogether the experiment can occur in $\prod_{i=1}^{k} n_i = n_1 \cdot n_2 \cdot n_3 \cdots n_k$ ways.

The next example illustrates the use of this principle.

Example 1.3.2. In how many ways can the five amino acids, alanine, valine, glycine, cysteine, tryphophan, be arranged to form a pentapeptide? This is a five-stage experiment since there are five amino acids which must fall into place in the chain. This is indicated by drawing five slots and mentally noting what they represent.

____	____	____	____	____
1st acid in chain	2d acid in chain	3d acid in chain	4th acid in chain	5th acid in chain

In how many ways can the first stage of the experiment occur? Answer: Five.

There are five acids available, any one of which could fall into the first position. Indicate this by placing a 5 in the first slot.

5				
1st acid in chain	2d acid in chain	3d acid in chain	4th acid in chain	5th acid in chain

Once the first stage is complete, in how many ways can stage 2 be performed? Answer: Four. Since each pentapeptide is to contain the five amino acids mentioned, repetition of the acid first in the chain is not permitted. The second member of the chain must be one of the four acids remaining. Indicate this by placing a 4 in the second slot.

5	4			
1st acid in chain	2d acid in chain	3d acid in chain	4th acid in chain	5th acid in chain

Similar reasoning leads us to conclude that stage 3 can take place in 3 ways, stage 4 in 2 ways, and stage 5 in 1 way. By the multiplication principle there are

5 ·	4 ·	3 ·	2 ·	1 = 120
1st acid in chain	2d acid in chain	3d acid in chain	4th acid in chain	5th acid in chain

pentapeptides that can be formed from these five amino acids.

There are several guidelines to keep in mind when using the multiplication principle:

1. Watch out for repetition versus nonrepetition. Sometimes objects can be repeated; sometimes they cannot. Whether or not repetition is allowed is determined by the physical context of the problem.
2. Watch out for subtraction. Consider event A. Occasionally it will be difficult, if not impossible, to find $n(A)$ directly. However, $S = A \cup A'$. Since A and A' have no points in common, $n(S) = n(A) + n(A')$. This implies that $n(A) = n(S) - n(A')$.
3. If there is a stage in the experiment with a special restriction, then you should think about the restriction first.

These points are illustrated in the next example.

Example 1.3.3. The DNA-RNA code is a molecular code in which the sequence of molecules provides significant genetic information. Each segment of RNA is composed of "words." Each word specifies a particular amino acid and is composed of a chain of three ribonucleotides. Each of the ribonucleotides in the chain is either adenine (A), uracil (U), guanine (G), or cytosine (C).

1. How many words can be formed? Here repetition is allowed. By the multiplication principle there are $4 \cdot 4 \cdot 4 = 64$ possible RNA words.
2. How many of these words involve some repetition? To answer this question, we use subtraction. There are 64 words possible. By the multiplication principle, $4 \cdot 3 \cdot 2 = 24$ of these have no repeated nucleotides. The remaining $64 - 24 = 40$ words must involve some repetition.
3. How many of the 64 words end with the nucleotides uracil or cytosine and have no repetition? Since there is a restriction on the last position of the chain, we consider it first by placing a 2 in the third position.

		2
1st	2d	3d

Once this restriction has been taken care of, we note that repetition is not allowed. This means that the nucleotide in position 3 cannot be used again. The first position can be filled with any of the three remaining nucleotides, and the second by either of the two that will be left at that point. By the multiplication principle the number of words that end with uracil or cytosine and have no repetition is

3	·	2	·	2	= 12
1st		2d		3d	

The use of the multiplication principle often results in a product of the form $n(n-1)(n-2)\cdots 3 \cdot 2 \cdot 1$, where n is a positive integer. For example, we found that the number of pentapeptides that can be formed from the five different amino acids is $5 \cdot 4 \cdot 3 \cdot 2 \cdot 1$. This product can be denoted by using what is called *factorial* notation.

Definition 1.3.3 (Factorial notation). Let n be a positive integer. The product $n(n-1)(n-2)\cdots 3 \cdot 2 \cdot 1$ is called n factorial and is denoted by $n!$. Zero factorial, denoted by $0!$, is defined to be 1.

When we use this notation, the number of pentapeptides that can be formed from five different amino acids is $5!$. Even though the need for zero factorial is not obvious yet, its purpose will become apparent soon.

One formula for counting permutations can be derived easily from the multiplication principle. Suppose that we have n distinct objects but we are going to use only r of these objects in each arrangement. How many permutations are possible in this case? Let us denote this number by ${}_nP_r$. Note that the subscript on the left denotes the number of distinct objects available, the P denotes the fact that we are counting permutations, and the subscript on the right denotes the number of objects used per arrangement. Since each permutation is to be an arrangement of r different objects, we need r slots

$$\underset{\substack{\text{1st}\\\text{object}}}{\underline{\qquad}} \quad \underset{\substack{\text{2d}\\\text{object}}}{\underline{\qquad}} \quad \underset{\substack{\text{3d}\\\text{object}}}{\underline{\qquad}} \quad \cdots \quad \underset{\substack{r\text{th}\\\text{object}}}{\underline{\qquad}}$$

Since n distinct objects are available, we have n choices for the first slot. Repetition is not allowed, so the number of permutations is given by

$$\underset{\substack{\text{1st}\\\text{object}}}{\underline{\quad n \quad}} \cdot \underset{\substack{\text{2d}\\\text{object}}}{\underline{(n-1)}} \cdot \underset{\substack{\text{3d}\\\text{object}}}{\underline{(n-2)}} \cdots \underset{\substack{r\text{th}\\\text{object}}}{\underline{\quad (?) \quad}}$$

To find the last number in the product, note that the number subtracted from n in each factor is one less than the slot number. Thus the rth factor will be $n - (r - 1) = n - r + 1$. We now have that

$${}_nP_r = n(n-1)(n-2)\cdots(n-r+1)$$

Note that

$$\frac{n!}{(n-r)!} = \frac{n(n-1)(n-2)\cdots(n-r+1)(n-r)(n-r-1)\cdots 3\cdot 2\cdot 1}{(n-r)(n-r-1)\cdots 3\cdot 2\cdot 1}$$

$$= {}_nP_r$$

Substituting, we have shown that the formula for finding the number of permutations of n distinct objects taken r at a time is as stated in the next theorem.

Theorem 1.3.1. The number of permutations of n distinct objects used r at a time, denoted by ${}_nP_r$, is

$${}_nP_r = \frac{n!}{(n-r)!}$$

Example 1.3.4

1. $${}_9P_4 = \frac{9!}{(9-4)!} = \frac{9!}{5!} = \frac{9\cdot 8\cdot 7\cdot 6\cdot 5!}{5!} = 3024$$

2. $${}_7P_7 = \frac{7!}{(7-7)!} = \frac{7!}{0!} = \frac{7!}{1} = 5040$$

Note that in order to apply Theorem 1.3.1, the objects to be arranged must be distinct, no repetition is allowed, and there can be no restrictions on any position in the arrangement. This formula will not solve all your permutation problems! The multiplication principle should be the first thing that comes to mind once you realize that a problem involves order, either natural or imposed.

Counting Combinations

Thus far we have considered counting problems in which order is important. We now turn our attention to situations in which order is irrelevant. That is, we now consider problems involving combinations rather than permutation. One very useful formula for finding the number of combinations of n distinct objects selected r at a time can be derived. Note that arranging r objects taken from n that are available is a two-stage process. The r objects must first be selected; denote the number of ways to select these objects by ${}_nC_r$. The r objects selected must then be arranged in order; this can be done in $r!$ ways. By the multiplication principle the number of arrangements of r objects taken from n is

$${}_nP_r = {}_nC_r \cdot r!$$

Solving this equation for ${}_nC_r$ and applying Theorem 1.3.1, we see that

$${}_nC_r = \frac{{}_nP_r}{r!} = \frac{n!}{r!(n-r)!}$$

This result is summarized in the next theorem and illustrated in Example 1.3.5.

Theorem 1.3.2. The number of combinations of n distinct objects selected r at a time, denoted by ${}_nC_r$, or $\binom{n}{r}$, is given by

$${}_nC_r = \binom{n}{r} = \frac{n!}{r!(n-r)!}$$

Example 1.3.5

1. $${}_5C_3 = \frac{5!}{3!(5-3)!} = \frac{5!}{3!2!} = \frac{5 \cdot 4 \cdot 3!}{3!2 \cdot 1} = 10$$

2. $$\binom{5}{0} = {}_5C_0 = \frac{5!}{0!(5-0)!} = \frac{5!}{0!5!} = 1$$

It is usually difficult at first to distinguish combinations from permutations. Look for the key words "select" and "arrange." The former signals that the problem involves combinations; the latter, that a permutation is sought.

Example 1.3.6. A foundry ships a lot of 20 engine blocks of which five contain internal flaws. The purchaser will select three blocks at random and test them for hardness. The lot will be accepted if no flaws are found. What is the *probability* that this lot will be accepted? To answer this question, we must count two things: the number of ways to select 3 engine blocks from 20, and the number of ways to select 3 engine blocks from 20 and obtain no flawed engines. The former quantity is given by

$${}_{20}C_3 = \frac{20!}{3!17!} = \frac{20 \cdot 19 \cdot 18 \cdot 17!}{3 \cdot 2 \cdot 1 \cdot 17!} = 1140$$

In order to obtain no flawed engines, all 3 of the sampled engines must be selected from among the 15 unflawed engines in the lot. This can be done in

$$_{15}C_3 = \frac{15!}{12!3!} = \frac{15 \cdot 14 \cdot 13 \cdot 12!}{3 \cdot 2 \cdot 1 \cdot 12!} = 455$$

ways. Since the engines selected for testing are chosen at random, each of the 1140 possible samples is equally likely. Using the classical approach to probability, we have

$$P[\text{lot is accepted}] = \frac{455}{1140}$$

Permutations of Indistinguishable Objects

Thus far we have been concerned only with permutation problems that may or may not involve repetition. Now we consider situations in which repetition is inevitable. The question to be answered is, "How many distinct arrangements of n objects are possible if some of the objects are identical and therefore cannot be distinguished one from the other?" An example will show that we already have available the tools to answer this question.

Example 1.3.7. Consider a computer with 16 ports and assume that at any given time each port is either in use (u), not in use but operable (n), or inoperable (i). How many configurations are possible in which 10 ports are in use, 4 are not in use but are operable, and 2 are inoperable? A typical sequence of this sort is

$$u\ u\ i\ u\ i\ n\ n\ u\ n\ u\ u\ n\ u\ u\ u\ u$$

To determine the number of ways that these letters can be permuted to form other arrangements, consider the 16 ports. If we could control port usage, then we are faced with a three-step process. These steps are:

Step 1: Select 10 ports for usage. This can be done in $_{16}C_{10} = 8008$ ways.

Step 2: Select 4 of the remaining 6 ports to represent ports that are not in use but which are operable. This can be done in $_6C_4 = 15$ ways.

Step 3: Select the remaining 2 ports to represent inoperable ports. This can be done in $_2C_2 = 1$ way.

The multiplication principle guarantees that the entire three-step process can be done in

$$\begin{aligned} _{16}C_{10} \cdot {_6C_4} \cdot {_2C_2} &= (8008)(15)(1) \\ &= 120{,}120 \text{ ways} \end{aligned}$$

Let us use the solution to the above problem to suggest a general formula that can be used to solve other problems involving permutations of indistinguishable objects. The expression $_{16}C_{10} \cdot {_6C_4} \cdot {_2C_2}$ can be written as

$$\begin{aligned} _{16}C_{10} \cdot {_6C_4} \cdot {_2C_2} &= \frac{16!}{10!\cancel{6!}} \frac{\cancel{6!}}{4!2!} \frac{\cancel{2!}}{\cancel{2!}0!} \\ &= \frac{16!}{10!4!2!} \end{aligned}$$

Notice that the terms of this product are predictable from the original problem. The 16! appears in the numerator because there is a total of 16 ports. The three terms 10!, 4!, and 2! arise due to the fact that there are three types of letters 10 u's, 4 n's, and 2 i's being permuted. This suggests that to solve a permutation problem of this sort, we need to determine n, the total number of objects being permuted, and $n_1, n_2, \ldots n_k$, the number of each of the k types of objects involved. The general formula for the total number of distinguishable arrangements of these objects is then given by

$$\frac{n!}{n_1!n_2!\cdots n_k!} \qquad n = n_1 + n_2 + \cdots + n_k$$

A general argument similar to that given above will show that this formula does hold for any values of $n, n_1, n_2, \ldots n_k$. You are asked to verify this fact in Exercise 23.

CHAPTER SUMMARY

In this chapter we discussed how to interpret probabilities. We also presented three methods for assigning probabilities to events. These are called the personal, relative frequency, and classical approaches. We also introduced and defined important terms that you should know. These are

Sample space	Mutually exclusive events
Sample point	Permutation
Event	Combination
Impossible event	$n!$
Certain event	$0!$

In solving permutation problems, we used the multiplication principle. This principle was used to derive a formula for ${}_nP_r$, the number of permutations of n distinct objects arranged r at a time. We also derived a formula for finding ${}_nC_r$, the number of combinations of n distinct objects selected r at a time.

EXERCISES

Section 1.1

1. One environmental hazard recently identified is overexposure to airborne asbestos. In a sample of 10 public buildings over 20 years old, three were found to be insulated with materials that produced an excess number of airborne asbestos bodies. What is the approximate probability that another building of this type will have this problem? What method are you using to assign this probability?
2. A government study defines a "group 1" nuclear accident to be one involving severe core damage, melting of uranium fuel, essential failure of all safety systems, and a major breach of the reactor's containment resulting in a large release of radioactiv-

ity into the atmosphere. In 1982 officials at the Nuclear Regulatory Commission estimated the probability of such an accident occurring in the United States before the year 2000 to be .02. What approach to probability do you think was used to determine this value? (*Roanoke Times*, November 1, 1982.)

3. Hemophilia is a sex-linked hereditary blood defect of males characterized by delayed clotting of the blood which makes it difficult to control bleeding, even in the case of a minor injury. When a woman is a carrier of classical hemophilia, there is a 50% chance that a male child will inherit the disease. If a carrier gives birth to two sons, what is the probability that both boys will have the disease? What approach to probability are you using to answer this question?

4. The probability of having a fatal accident in the workplace is assessed using the fatal accident frequency rate (FAFR), defined by

$$\text{FAFR} = \text{number of fatalities per 1000 workers during a working lifetime}$$

This rate can be viewed as giving the probability of an individual having a fatal accident while at work. What approach to probability is being used? These FAFR values were reported by I. C. Clingan ("Safety at Sea—Its Risk Management," *Interdisciplinary Science Reviews*, 1981, vol. 6, no. 1, pp. 36–48).

Occupation	FAFR
Metal manufacturer	8
Coal mining	12
Construction	67
Industry overall	4

What is the approximate probability that a coal miner will suffer a fatal injury? Coal mining is an industry and is taken into account when computing the industrywide FAFR of 4. Can you explain how this rate could be so low while at least some of the components used in its computation are high?

Section 1.2

5. Fission occurs when the nucleus of an atom captures a subatomic particle called a neutron and splits into two lighter nuclei. This causes energy to be released. At the same time other neutrons are emitted, two or three on the average. If at least one of these is captured by another fissionable nucleus, then a chain reaction is possible.

(*a*) Consider a reaction in which three neutrons are emitted initially. Let c denote that a given neutron is captured by another nucleus; let n denote that the neutron is not captured by another nucleus. Construct a tree denoting the possible behavior for these three neutrons.

(*b*) List the sample points generated by the tree.

(*c*) List the sample points that constitute each of these events:

A_1: a chain reaction is possible
A_2: all three neutrons are captured
A_3: a chain reaction is not possible

(*d*) Are A_1 and A_2 mutually exclusive?
Are A_1 and A_3 mutually exclusive?
Are A_2 and A_3 mutually exclusive?
Are A_1, A_2, and A_3 mutually exclusive?

(*e*) The probability that a neutron will be captured depends on its neutron energy and is not the same for each neutron. Under these circumstances, is it correct to say that the probability that all three neutrons will be captured is 1/8 because this can occur in only one way and there are eight paths through the tree of part (*a*)? Explain.

6. In ballistics studies conducted during World War II it was found that, in ground-to-ground firing, artillery shells tended to fall in an elliptical pattern such as that of Fig. 1.4. The probability that a shell would fall in the inner ellipse is .50; the probability that it would fall in the outer ellipse is .95. ("Statistics and Probability Applied to Problems of Antiaircraft Fire in World War II," E. S. Pearson, *Statistics: A Guide to the Unknown*, Holden-Day, San Francisco, 1972, pp. 407–415.)

(*a*) A firing is considered to be a success (s) if the shell falls within the inner ellipse; otherwise it is failure (f). Construct a tree to represent the firing of four shells in succession.

(*b*) List the sample points generated by the tree.

(*c*) Let A_i $i = 1, 2, 3, 4$ denote the event that the ith firing is successful. List the sample points that constitute each of the events A_1, A_2, A_3, A_4. Are these events mutually exclusive?

(*d*) List the sample points that constitute each of these events and describe the events verbally:

$$A_1'$$
$$A_1 \cup A_2$$
$$A_1 \cap A_2$$
$$A_1 \cap A_2 \cap A_3 \cap A_4$$
$$A_1 \cap A_2 \cap A_3 \cap A_4'$$
$$(A_1 \cup A_2 \cup A_3 \cup A_4)'$$
$$A_1 \cap A_1'$$

(*e*) The probability of each of the events of part (*d*) can be found using classical probability. Why is this true? Find these probabilities.

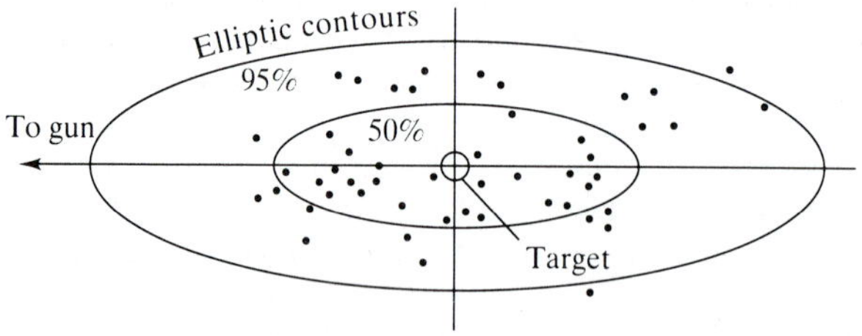

FIGURE 1.4
50% of the shells fall in the inner ellipse.

7. A home computer is tied to a mainframe computer via a telephone modem. The home computer will dial repeatedly until contact is made. Once contact has been established, the dialing process will, of course, end. Let c denote the fact that contact is made on a particular attempt and n that contact is not made.

(*a*) Construct a tree diagram to represent the dialing process.
(*b*) Are the paths through the tree equally likely?
(*c*) List the sample points generated by the tree. Can this list ever by completed?
(*d*) List the sample points that constitute event A: contact is made in at most four attempts.
(*e*) Give an example of two events that are not impossible but which are mutually exclusive.

8. A missile battery can fire five missiles in rapid succession. As soon as the target is hit, firing will cease. Let h denote a hit and m a miss.
(*a*) Draw the tree to represent the possible firing of these missiles at a single incoming target.
(*b*) Is there any difference in the tree drawn here and that illustrated in Example 1.2.3? Explain.
(*c*) List the sample points generated by the tree.
(*d*) List the sample points that constitute the events
A_1: exactly two shots are fired
A_2: at most two shots are fired
Are these events mutually exclusive? Explain.

Section 1.3

9. Evaluate each of these expressions:
(*a*) 9! (*b*) 6!
(*c*) ${}_7P_3$ (*d*) ${}_6P_2$
(*e*) ${}_5P_5$ (*f*) ${}_6P_6$

10. In investigating the Ideal Gas Law, experiments are to be run at four different pressures and three different temperatures.
(*a*) How many experimental conditions are to be studied?
(*b*) If each experimental condition is replicated (repeated) 5 times, how many experiments will be conducted on a given gas?
(*c*) How many experiments must be conducted to obtain five replications on each experimental condition for each of six different gases?

11. In setting up a computer system for his firm to use in quality control, an engineer has four choices for the main unit: IBM, VAX, Honeywell, or HP. There are six brands of CRTs that can be purchased and three types of graphics printers.
(*a*) If all equipment is compatible, in how many ways can the system be designed?
(*b*) If the engineer wants to be able to use a statistical software package that is only available on IBM and VAX equipment, in how many ways can the system be designed?

12. In Exercise 6 we considered the experiment of firing four artillery shells in succession. Each firing was classed as being either a success or a failure. Use the multiplication principle to verify that the number of paths through the tree representing this experiment is 16.

13. The Apollo mission to land men on the moon made use of a system whose basic structure is shown in Fig. 1.5. For the system to operate successfully all five components shown must function properly. Let us identify each component as being either operable (O) or inoperable (i). Thus the sequence $OOOOi$ denotes a state in which all components except the LEM engine are operable. ("Striving for Reliability," Gerald Lieberman, *Statistics: A Guide to the Unknown*, Holden-Day, San Francisco, 1972, pp. 400–406.)

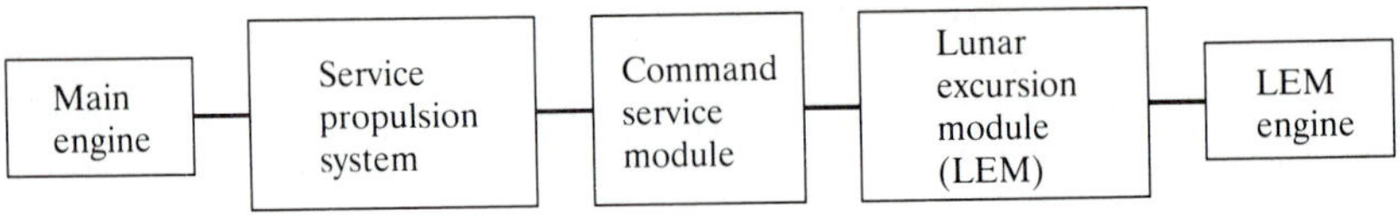

FIGURE 1.5
A simplified diagram of the Apollo system.

(*a*) How many states are possible?
(*b*) How many states are possible in which the LEM engine is inoperable?
(*c*) The mission is deemed at least partially successful if the first three components are operable. How many states represent at least a partially successful mission?
(*d*) The mission is a total success if and only if all five components are operable. How many states represent a completely successful mission?

14. The basic storage unit of a digital computer is a "bit." A bit is a storage position that can be designated as either on (1) or off (0) at any given time. In converting picture images to a form that can be transmitted electronically, a picture element, called a *pixel* is used. Each pixel is quantized into gray levels and coded using a binary code. For example, a pixel with four gray levels can be coded using two bits by designating the gray levels by 00, 01, 10, and 11.
(*a*) How many gray levels can be quantized using a four-bit code?
(*b*) How many bits are necessary to code a pixel quantized to 32 gray levels?

15. Tests will be run on five different coatings used to protect fiber optics cables from extreme cold. These tests will be conducted in random order.
(*a*) In how many orders can the tests be run?
(*b*) If two of the coatings are made by the same manufacturer, what is the probability that tests on these coatings will be run back to back?

16. The effectiveness of irradiated polymers in the removal of benzene from water is being investigated. Three polymers are to be studied. Each is to be tested at four different temperatures and three different radiation levels.
(*a*) How many different experimental conditions are under study?
(*b*) If each experimental condition is to be replicated (repeated) five times, how many experiments must be conducted?

17. Evaluate each of these expressions:
(*a*) ${}_9C_4$ (*b*) ${}_8C_3$
(*c*) $\binom{8}{5}$ (*d*) $\binom{8}{0}$

18. Prove that ${}_nC_r = {}_nC_{n-r}$.

19. BASIC is a computer language often used on home computers. Although it is easy to learn, it is also relatively slow. A study is conducted to compare the execution time for five BASIC compilers. How many pairwise comparisons can be made among the compilers? (A study of this sort is discussed in "A Comparison of Five Compilers for Apple Basic" J. H. and J. S. Taylor, *Byte*, 1982, vol. 7, no. 9, p. 440.)

20. The Delta project is a project to determine if large-scale agricultural production can succeed in Alaska. In this project, 22 persons are to be selected from a pool of 103 qualified applicants and awarded the right to purchase land parcels to be developed for agricultural purposes. ("Expanding Subarctic Agriculture," *Interdisciplinary Science Reviews*, 1982, vol. 7, no. 3, pp. 178–187.)

(*a*) In how many ways can the 22 persons be selected? (Set up only.)
(*b*) Assume that you are one of the persons in the applicant pool. In how many of the subgroups of part (*a*) will you be included? (Set up only.)
(*c*) If the selection process is done randomly, each of the subgroups of part (*a*) is equally likely. If your name is in the applicant pool, what is the probability that you will be awarded the right to purchase land?

21. A firm employs 10 programmers, 8 systems analysts, 4 computer engineers, and 3 statisticians. A "team" is to be chosen to handle a new long-term project. The team will consist of 3 programmers, 2 systems analysts, 2 computer engineers, and 1 statistician.
(*a*) In how many ways can the team be chosen?
(*b*) If the customer insists that one particular engineer with whom he or she has worked before be assigned to the project, in how many ways can the team be chosen?

22. A company receives a shipment of 20 hard drives. Before accepting the shipment, 5 of them will be randomly selected and tested. If all 5 meet specifications, then the shipment will be accepted; otherwise all 20 will be returned to the manufacturer. If, in fact, 3 of the 20 drives are defective, what is the probability that the shipment will not be accepted?

23. Consider n objects where n_1 are of type 1, n_2 of type $2, \ldots, n_k$ of type k. The number of ways in which the n objects can be arranged when objects of the same type are considered to be indistinguishable is given by

$$\frac{n!}{n_1!n_2!\cdots n_k!} \qquad n = n_1 + n_2 + \cdots + n_k$$

This theorem is proved by arguing that a k-stage process is involved. Stage 1 consists of selecting n_1 positions in which to place items of type 1. Stage 2 consists of selecting n_2 positions from the $n - n_1$ that remain in which to place items of type 2. This process continues until eventually there are n_k positions remaining in which to place the items of type k. That is,

$$\binom{n}{n_1}\binom{n-n_1}{n_2}\binom{n-n_1-n_2}{n_3}\cdots\binom{n_k}{n_k} = \frac{n!}{n_1!n_2!\cdots n_k!}$$

Verify this result.

24. The oil embargo of 1973 spurred a study of the possibility of using automatic meter readings to reduce costs to power companies. One procedure studied entailed the use of 128-bit messages. Occasionally transmission errors occur resulting in a digit reversal of one or more bits. How many messages can be sent that contain exactly two transmission errors? *Hint:* Think of a message as being a permutation of 128 objects, each of which is either correct (c) or not correct (n).

25. In studying a chemical reaction, 12 experiments will be conducted. Four different temperatures will be used 4 times, each with the temperatures run in random order. In how many orders can the series of experiments be conducted?

26. A garage door opener has six toggle switches, each with three settings: up, center, and down.
(*a*) In how many ways can these switches be set?
(*b*) If a thief knows the type of opener involved but does not know the setting, what is the probability that he or she can guess the setting on the first attempt?
(*c*) How many settings are possible in which two switches are up, two are down, and two are in the center?

27. Consider Example 1.2.1.

(*a*) Without looking at the tree diagram, how many paths through the tree will represent the fact that exactly two of the three computers are ready at the time of the launch? Verify your answer by listing these paths.

(*b*) If 10 computers were used instead of 3, the tree given in Fig. 1.1 could be expanded to answer questions posed concerning the number of computers that are ready at launch time. How many paths would such a tree entail? How many of these paths would represent the fact that exactly 7 of the 10 computers are ready at launch time?

REVIEW EXERCISES

28. Find n if $\binom{n}{2} = 21$; if $\binom{n}{2} = 105$.

29. The configuration of a particular computer terminal consists of a baud-rate setting, a duplex setting, and a parity setting. There are 11 possible baud-rate settings, two parity settings (even or odd), and two duplex settings (half or full).

(*a*) How many configurations are possible for this terminal?

(*b*) In how many of these configurations is the parity even and the duplex full?

(*c*) A line surge occurs that causes these settings to change at random. What is the probability that the resulting configuration will have even parity and be full duplex?

30. A firm offers a choice of 10 free software packages to buyers of their new home computer. There are 25 packages from which to choose. In how many ways can the selection be made? Five of the packages are computer games. How many selections are possible if exactly three computer games are selected?

31. A project manager has 10 chemical engineers on her staff. Four are women and six are men. These engineers are equally qualified. In a random selection of three workers, what is the probability that no women will be selected? Would you consider it unusual for no women to be selected under these circumstances? Explain.

32. A computer system uses passwords that consist of five letters followed by a single digit.

(*a*) How many passwords are possible?

(*b*) How many passwords consist of three As and two Bs, and end in an even digit?

(*c*) If you forget your password but remember that it has the characteristics described in part (*b*), what is the probability that you will guess the password correctly on the first attempt?

33. A mainframe computer has 16 ports. At any given time each port is either in use or not in use. How many possibilities are there for overall port usage of this computer? How many of these entail the use of at least 1 port?

34. A flashlight operates on two batteries. Eight batteries are available but three are dead. In a random selection of batteries, what is the probability that exactly one dead battery will be selected?

35. An electrical control panel has three toggle switches labeled I, II, and III, each of which can be either on (O) or off (F).

(*a*) Construct a tree to represent the possible configurations for these three switches.

(*b*) List the elements of the sample space generated by the tree.

(*c*) List the sample points that constitute the events
A: at least one switch is on
B: switch I is on
C: no switch is on
D: four switches are on

(*d*) Are events A and B mutually exclusive? Are events A and C mutually exclusive? Are events A and D mutually exclusive?

(*e*) What is the name given to an event such as D?

(*f*) If at any given time each switch is just as likely to be on as off, what is the probability that no switch is on?

36. Two items are randomly selected one at a time from an assembly line and classed as to whether they are of superior quality (+), average quality (0), or inferior quality (−).

(*a*) Construct a tree for this two-stage experiment.

(*b*) List the elements of the sample space generated by the tree.

(*c*) List the sample points that constitute the events
A: the first item selected is of inferior quality
B: the quality of each of the items is the same
C: the quality of the first item exceeds that of the second

(*d*) Are the events A and B mutually exclusive? Are the events A and C mutually exclusive?

(*e*) Give a brief verbal description of these events:
$A' \cap B$ $\quad$ $A' \cap B'$
$A \cap B'$ $\quad$ $A \cap C' \cap B$

(*f*) It is known that 90% of the items produced are of average quality, 1% are of superior quality, and the rest are of inferior quality. It is argued that since the classification experiment can proceed in nine ways with only one of these resulting in two items of average quality, the probability of obtaining two such items is 1/9. Criticize this argument.

37. An experiment consists of selecting a digit from among the digits 0 to 9 in such a way that each digit has the same chance of being selected as any other. We name the digit selected A. These lines of code are then executed:

```
IF A < 2 THEN B = 12; ELSE B = 17;
IF B = 12 THEN C = A - 1; ELSE C = 0;
```

(*a*) Construct a tree to illustrate the ways in which values can be assigned to the variables A, B, and C.

(*b*) Find the sample space generated by the tree.

(*c*) Are the 10 possible outcomes for this experiment equally likely?

(*d*) Find the probability that A is an even number.

(*e*) Find the probability that C is negative.

(*f*) Find the probability that $C = 0$.

(*g*) Find the probability that $C \leq 1$.

38. Consider Exercise 16. If experimental runs are to be done in random order, how many different sequences are possible? (Set up only!) In experiments of this sort runs are not usually done randomly. Rather, they are carefully designed so that the researcher has control of the order of experimentation. Can you think of some practical reasons for why this is necessary?

CHAPTER 2

SOME PROBABILITY LAWS

In Chap. 1 we considered how to interpret probabilities. In this chapter we consider some laws that govern their behavior. The laws that we shall present are those that will have a direct application to problem solving. These laws will be stated and illustrated numerically. Their derivations are not hard and most of them are left as exercises.

2.1 AXIOMS OF PROBABILITY

You have probably seen the development of a mathematical system in your study of high school geometry. In developing any mathematical system, one begins by stating a few basic definitions and axioms that underlie the system. The definitions are the technical terms of the system; axioms are statements that are assumed to be true and therefore require no proof. Usually one starts with as few axioms as possible and then uses these axioms and the technical definitions to develop whatever theorems follow logically. Some technical terms such as sample space, sample point, event, and mutually exclusive events have already been introduced. One can develop a useful system of theorems pertaining to probability with the aid of these definitions and three axioms, called the axioms of probability.

Axioms of probability.

1. Let S denote a sample space for an experiment:

$$P[S] = 1$$

2. $P[A] \geq 0$ for every event A.
3. Let $A_1, A_2, A_3, \ldots$ be a finite or an infinite sequence of mutually exclusive events. Then $P[A_1 \cup A_2 \cup A_3 \cdots] = P[A_1] + P[A_2] + P[A_3] + \cdots$.

Axiom 1 states a fact that most people regard as obvious; namely, the probability assigned to the certain event S is 1. Axiom 2 ensures that probabilities can never be negative. Axiom 3 guarantees that when one deals with mutually exclusive events, the probability that at least one of the events will occur can be found by adding the individual probabilities. An important consequence of this axiom is that it gives us the ability to find the probability of an event when the sample points in the same space for the experiment are not equally likely. Example 2.1.1 illustrates this point.

Example 2.1.1. The distribution of blood types in the United States is roughly 41% type A, 9% type B, 4% type AB, and 46% type O. An individual is brought into an emergency room and is to be blood-typed. What is the probability that the type will be A, B, or AB?

The sample space for this experiment is

$$S = \{\text{A}, \text{B}, \text{AB}, \text{O}\}$$

The sample points are not equally likely, so the classical approach to probability is not applicable. Let A_1, A_2, and A_3 denote the events that the patient has type A, B, and AB blood, respectively. The events A_1, A_2, and A_3 are mutually exclusive and we are looking for $P[A_1 \cup A_2 \cup A_3]$. By axiom 3,

$$\begin{aligned} P[A_1 \cup A_2 \cup A_3] &= P[A_1] + P[A_2] + P[A_3] \\ &= .41 + .09 + .04 \\ &= .54 \end{aligned}$$

An immediate consequence of these axioms is the fact that the probability assigned to the impossible event is 0, as you should suspect. The derivation of this result is outlined in Exercise 7.

Theorem 2.1.1. $P[\varnothing] = 0$.

Another consequence of the axioms is that the probability that an event will *not* occur is equal to 1 minus the probability that it will occur. For example, if the probability of a successful space shuttle mission is .99, then the probability that it will not be successful is $1 - .99 = .01$. This idea is stated in Theorem 2.1.2. Its derivation is outlined in Exercise 8.

Theorem 2.1.2. $P[A'] = 1 - P[A]$.

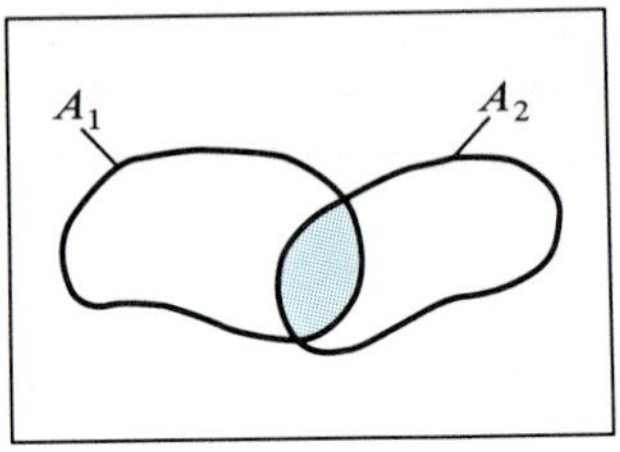

FIGURE 2.1
$A_1 \cap A_2 \neq \varnothing$.

The General Addition Rule

We have seen how to handle questions concerning the probability of one or another event occurring if those events are mutually exclusive. We now develop a more general rule that will allow us to find the probability that at least one of two events will occur when the events are not necessarily mutually exclusive. This rule is suggested by considering the Venn diagram of Fig. 2.1. Assume that the shaded region in the diagram, $A_1 \cap A_2$, is not empty so that A_1 and A_2 are not mutually exclusive. If we claim that

$$P[A_1 \cup A_2] = P[A_1] + P[A_2]$$

we have committed an obvious error. Since $A_1 \cap A_2 \subseteq A_1$ and $A_1 \cap A_2 \subseteq A_2$, $P[A_1 \cap A_2]$ has been included twice in our calculation. To correct this error, we subtract $P[A_1 \cap A_2]$ from the right-hand side of the equation to obtain the general addition rule:

General addition rule

$$P[A_1 \cup A_2] = P[A_1] + P[A_2] - P[A_1 \cap A_2]$$

This rule can be derived from the axioms of probability and the theorems that we have already developed. Its proof is outlined in Exercise 11. The key word that signals its use is the word "or."

Example 2.1.2. Components of a propulsion system can be arranged in series. However, this arrangement has a serious drawback; if one component fails, the system fails. This is obviously a risky arrangement for space travel! Consider a system in which the main engine has a backup. These engines are designed to operate independently in that the success or failure of one has no effect on the other. The engine component is operable if one *or* the other of these two engines is operable. Such a system is said to have the engine component in parallel. Assume that each engine is 90% reliable. That is, each functions correctly with probability .9. As we shall show later, it is then reasonable to assume that both engines operate correctly with probability .81. Find the probability that the engine component is operable. Let A_1: the main engine is operable, and A_2: the backup engine is operable. We are given that $P[A_1] = P[A_2] = .9$ and that $P[A_1 \cap A_2] = .81$. We want to find $P[A_1 \cup A_2]$. By the addition rule

$$\begin{aligned} P[A_1 \cup A_2] &= P[A_1] + P[A_2] - P[A_1 \cap A_2] \\ &= .9 + .9 - .81 = .99 \end{aligned}$$

The addition rule links the operations of union and intersection. If $P[A_1 \cap A_2]$ is known, the addition rule can be used to find $P[A_1 \cup A_2]$. Similarly, if $P[A_1 \cup A_2]$ is known, we can use the rule to find $P[A_1 \cap A_2]$. Venn diagrams are helpful when using this rule.

Example 2.1.3. A chemist analyzes seawater samples for two heavy metals: lead and mercury. Past experience indicates that 38% of the samples taken from near the mouth of a river on which numerous industrial plants are located contain toxic levels of lead or mercury: 32% contain toxic levels of lead and 16% contain toxic levels of mercury. What is the probability that a randomly selected sample will contain toxic levels of lead only? Let A_1 denote the event that the sample contains toxic levels of lead, and let A_2 denote that the sample contains toxic levels of mercury. We are given that $P[A_1] = .32$, $P[A_2] = .16$, and $P[A_1 \cup A_2] = .38$. By the addition rule

$$P[A_1 \cup A_2] = P[A_1] + P[A_2] - P[A_1 \cap A_2]$$

or

$$.38 = .32 + .16 - P[A_1 \cap A_2]$$

Solving this equation, we obtain $P[A_1 \cap A_2] = .10$. This is indicated in Fig. 2.2(*a*). Since $P[A_1] = .32$ and $A_1 \cap A_2 \subseteq A_1$, the probability associated with the shaded region in Fig. 2.2(*b*) is .22. Similarly, since $A_1 \cap A_2 \subseteq A_2$, a probability of .06 is associated with the shaded region of Fig. 2.2(*c*). Finally, since $P[S] = 1$, the probability assigned to the shaded area in Fig. 2.2(*d*) is .62. We are asked to find the probability that the sample will contain only lead. That is, we want to find $P[A_1 \cap A_2']$. This probability, .22, can be read from Fig. 2.2(*b*).

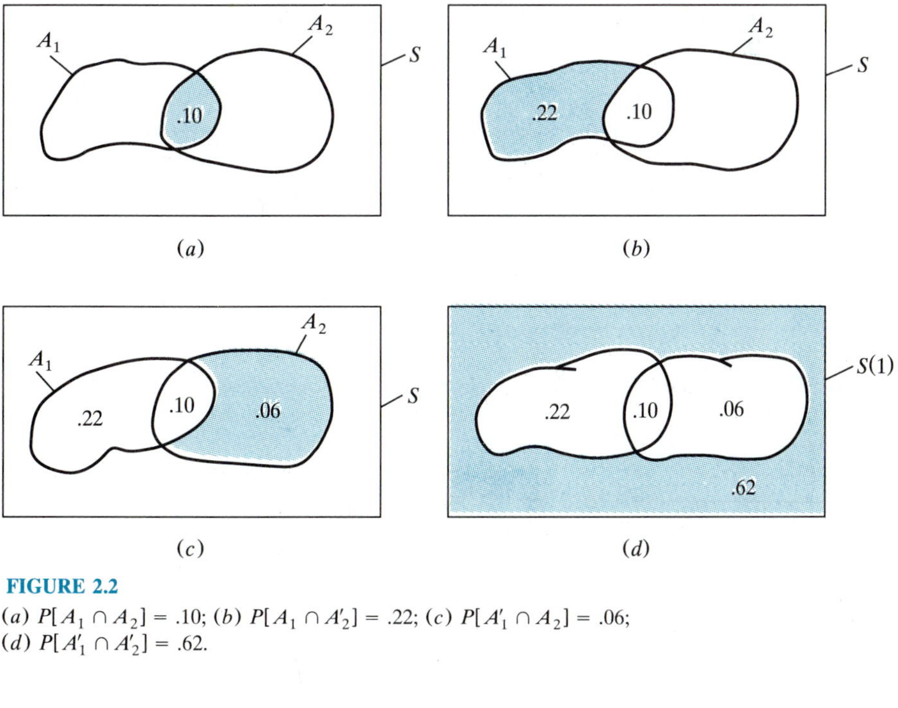

FIGURE 2.2
(*a*) $P[A_1 \cap A_2] = .10$; (*b*) $P[A_1 \cap A_2'] = .22$; (*c*) $P[A_1' \cap A_2] = .06$;
(*d*) $P[A_1' \cap A_2'] = .62$.

Notice that if the percentages reported in problems such as these are based on population data, then the probabilities calculated by use of the general addition rule are exact. However, if the percentages reported are based on samples drawn from a larger population, then the probabilities computed are relative frequency probabilities. They are *approximations* to the true probability of the occurence of the event in question. Since most percentages reported in the literature are based on samples, most of them are properly viewed as being relative frequency probabilities. We use the word "probability" with the understanding that the probabilities given and computed by using the theorems in this chapter are, in most cases, only approximations.

2.2 CONDITIONAL PROBABILITY

In this section we introduce the notion of conditional probability. The name itself is indicative of what is to be done. We wish to determine the probability that some event A_2 will occur, "conditional on" the assumption that some other event A_1 has occurred. The key words to look for in identifying a conditional question are "if" and "given that." We use the notation $P[A_2|A_1]$ to denote the conditional probability of event A_2 occurring given that event A_1 has occurred. A simple example will suggest the way to define this probability.

Example 2.2.1. In trying to determine the sex of a child a pregnancy test called "starch gel electrophoresis" is used. This test may reveal the presence of a protein zone called the pregnancy zone. This zone is present in 43% of all pregnant women. Furthermore, it is known that 51% of all children born are male. Seventeen percent of all children born are male and the pregnancy zone is present. The Venn diagram for these data is shown in Fig. 2.3. Let A_1 denote the event that the pregnancy zone is present, and A_2 that the child is male. We know that, for a randomly selected pregnant woman, $P[A_1] = .43$, $P[A_2] = .51$, $P[A_1 \cap A_2] = .17$. If asked, "What is the probability that the child is male?" the answer is .51. Suppose we are *given* the information that the pregnancy zone is present and asked, "What is the probability that the child is male?" We now have information that was not available originally. What effect, if any, does this new information have on our belief that the child is male? That is, what is $P[A_2|A_1]$? Once we know that the pregnancy zone is present, our sample space no longer includes all

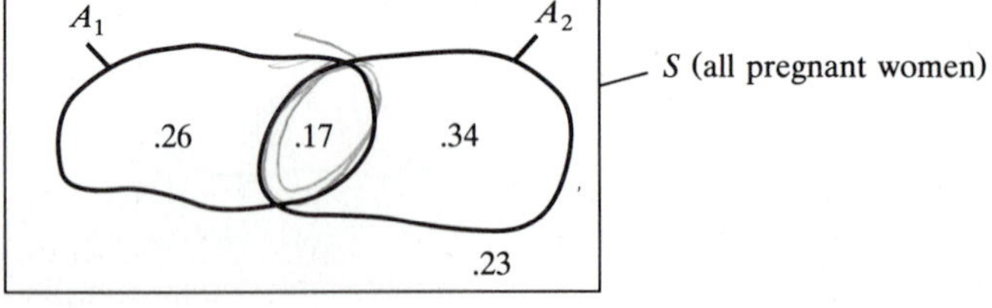

FIGURE 2.3
Partition of S.

pregnant women; it consists only of the 43% with this characteristic. Of these, $.17/.43 \doteq .395$ have male children. Logic implies that

$$P[\text{male}|\text{zone present}] = P[A_2|A_1] = .395$$

Receipt of the information that the pregnancy zone is present reduces from .51 to .395 the probability that the child is male.

To formalize the reasoning used in the previous example, note that $P[A_2|A_1]$ is found by forming a ratio whose denominator is $P[A_1]$, the probability that the *given* event will occur. The numerator is $P[A_1 \cap A_2]$, the probability that *both* the given event and the event in question will occur. That is, we define the conditional probability as follows.

Definition 2.2.1 (Conditional probability). Let A_1 and A_2 be events such that $P[A_1] \neq 0$. The conditional probability of A_2 given A_1, denoted by $P[A_2|A_1]$, is defined by

$$P[A_2|A_1] = \frac{P[A_1 \cap A_2]}{P[A_1]}$$

Sometimes receipt of the information that event A_1 has occurred has no effect on the probability assigned to event A_2. That is,

$$P[A_2|A_1] = P[A_2]$$

When this happens, A_1 and A_2 have a special relationship to one another. The nature of this relationship will be explored in the next section. In the meantime don't be surprised if you find that a particular conditional probability does not differ from the original probability assigned to the event!

2.3 INDEPENDENCE AND THE MULTIPLICATION RULE

We have used the word "independent" informally in several previous examples. Webster's dictionary defines independent objects as objects acting "irrespective of each other." Thus two events are independent if one may occur irrespective of the other. That is, the occurrence or nonoccurrence of one does not alter the likelihood of occurrence or nonoccurrence of the other. In some cases it is reasonable to assume that two events are independent from the physical description of the events themselves. For example, suppose that a couple heterozygous for eye color has two children. Since the eye color of a child is affected only by the genetic makeup of the parents and not by the eye color of the other child, it is reasonable to assume that the events A_1: the first child has brown eyes, and A_2: the second child has brown eyes, are independent. However, in most instances the issue is not clear-cut. In these cases we need a mathematical definition of the term to determine without a doubt whether two events are, in fact, independent.

To see how to characterize independence, let us consider a simple experiment that consists of rolling a single fair die once and then tossing a fair coin once. Let the first member of each ordered pair denote the number appearing on the die and the second, the face showing on the coin (H = heads, T = tails). A sample space for this experiment is

$$S = \{(1, H), (1, T), (2, H), (2, T), (3, H), (3, T),$$
$$(4, H), (4, T), (5, H), (5, T), (6, H), (6, T)\}$$

Since the die and the coin are considered to be fair, these 12 outcomes are equally likely. Consider these events:

A: the die shows one or two

B: the coin shows heads

$A \cap B$: the die shows one or two and the coin shows heads

Since knowing the result of the die roll gives us no additional information on how the coin will land, it is reasonable to assume that the events A and B are independent. Using classical probability, we easily see that

$$P[A] = P[\{(1, H), (1, T), (2, H), (2, T)\}] = 4/12 = 1/3$$
$$P[B] = P[\{(1, H), (2, H), (3, H), (4, H), (5, H), (6, H)\}]$$
$$= 6/12 = 1/2$$
$$P[A \cap B] = P[\{(1, H), (2, H)\}] = 2/12 = 1/6$$

More importantly, it is easy to see that for these physically independent events

$$P[A \cap B] = P[A] \cdot P[B]$$

Consider now an experiment that consists of drawing two coins in succession from a box containing a nickel (N), a dime (D), and a quarter (Q). The first coin is not replaced before the second is drawn. A sample space for this experiment is

$$S = \{(N, D), (N, Q), (D, N), (D, Q), (Q, N), (Q, D)\}$$

These outcomes are equally likely. Consider these events:

A: the first coin is a dime

B: the second coin is a dime

Since we do not replace the first coin before the second draw, it is evident that if event A occurs, event B cannot occur. That is, knowledge that event A has occurred does give us information on whether or not event B will occur! These events are not independent. Using classical probability, we easily see that

$$P[A] = P[\{(D, N), (D, Q)\}] = 2/6$$
$$P[B] = P[\{(N, D), (Q, D)\}] = 2/6$$
$$P[A \cap B] = P[\varnothing] = 0$$

More importantly, it is easy to see that for these events that are not independent

$$P[A \cap B] \neq P[A]P[B]$$

Thus we have noticed that when A and B are clearly independent, $P[A \cap B] = P[A]P[B]$; when they are clearly dependent, $P[A \cap B] \neq P[A]P[B]$. This is not coincidental. It is natural to use this mathematical characterization as our technical definition of the term "independent events."

Definition 2.3.1 (Independent events). Events A_1 and A_2 are independent if and only if

$$P[A_1 \cap A_2] = P[A_1]P[A_2]$$

This definition is useful in two ways. If exact probabilities are available, then it serves as a test for independence. However, since most probabilities encountered in scientific studies are approximations, it is most useful as a way to find the probability that two events will occur when the events are clearly independent. Example 2.3.1 illustrates its use as a test for independence.

Example 2.3.1. Consider the experiment of drawing a card from a well-shuffled deck of 52 cards. Let

A_1: a spade is drawn

A_2: an honor (10, J, Q, K, A) is drawn

Classical probability is used to see that $P[A_1] = 13/52$ and $P[A_2] = 20/52$. The probability that a spade and an honor, $P[A_1 \cap A_2]$, is drawn is $5/52$. Notice that these probabilities are exact. They are not approximations based on observations of card draws. Are the events A_1 and A_2 independent? To decide, note that

$$P[A_1]\,P[A_2] = (13/52)(20/52) = 5/52$$

and

$$P[A_1 \cap A_2] = 5/52$$

Since $P[A_1 \cap A_2] = P[A_1]P[A_2]$, we can conclude that these events are independent.

In Chap. 15 a test for independence will be developed that can be used when working with real data rather than with classical probabilities. Its derivation is based on the definition of independent events just discussed.

Example 2.3.2 illustrates the use of Definition 2.3.1 in finding the probability that two events will occur simultaneously when the events are clearly independent.

Example 2.3.2. In Example 1.1.3, we found that the probability that a couple heterozygous for eye color will parent a brown-eyed child is $3/4$ for each child. Genetic studies indicate that the eye color of one child is independent of that of the other. Thus if the couple has two children, then the probability that both will be brown-eyed is

$$P\begin{bmatrix}\text{first} & \text{and} & \text{second}\\ \text{brown} & & \text{brown}\end{bmatrix} = P\begin{bmatrix}\text{first}\\ \text{brown}\end{bmatrix}P\begin{bmatrix}\text{second}\\ \text{brown}\end{bmatrix}$$

$$= \frac{3}{4}\cdot\frac{3}{4}$$

$$= \frac{9}{16}$$

Definition 2.3.1 defines independence for *any* events A_1 and A_2. If at least one of the events A_1 or A_2 occurs with *nonzero* probability, then an appealing characterization of independence can be obtained. To see how this is done, assume that $P[A_1] \neq 0$. By Definition 2.3.1, A_1 and A_2 are independent if and only if

$$P[A_1 \cap A_2] = P[A_1]P[A_2]$$

Dividing by $P[A_1]$, we can conclude that A_1 and A_2 are independent if and only if

$$\frac{P[A_1 \cap A_2]}{P[A_1]} = P[A_2|A_1] = P[A_2]$$

A similar argument holds if $P[A_2] \neq 0$. We have thus derived the result given in Theorem 2.3.1

Theorem 2.3.1. Let A_1 and A_2 be events such that at least one of $P[A_1]$ or $P[A_2]$ is nonzero. A_1 and A_2 are independent if and only if

$$P[A_2|A_1] = P[A_2] \qquad \text{if } P[A_1] \neq 0$$

and

$$P[A_1|A_2] = P[A_1] \qquad \text{if } P[A_2] \neq 0$$

Since most events of real interest do occur with nonzero probability, Theorem 2.3.1 is used as a test for independence. To understand the logic behind the theorem, let us reconsider the data of Example 2.3.1.

Example 2.3.3. Consider the events A_1, a spade is drawn; and A_2, an honor is drawn. We know that $P[A_1] = 13/52$, $P[A_2] = 20/52$, and $P[A_1 \cap A_2] = 5/52$. Suppose we are asked, "What is the probability that a randomly selected card is an honor?" Our answer is $20/52$. Suppose we are now told that the card is a spade and are asked, "What is the probability that the card is an honor?" That is, "What is $P[A_2|A_1]$?" If A_1 and A_2 are independent, the new information is irrelevant and our answer should not change. That is, $P[A_2|A_1] = P[A_2]$. Otherwise our answer should change and $P[A_2|A_1] \neq P[A_2]$. In this setting, is $P[A_2|A_1] = P[A_2]$? To answer this question, note that

$$P[A_2|A_1] = \frac{P[A_1 \cap A_2]}{P[A_1]} = \frac{5/52}{13/52} = 5/13$$

and $$P[A_2] = 20/52 = 5/13$$

Since these probabilities are the same, we conclude via Theorem 2.3.1 that A_1 and A_2 are independent.

Occasionally we must deal with more than two events. Again, the question arises, "When are these events considered independent?" Definition 2.3.2 answers this question by extending our previous definition to include more than two events.

Definition 2.3.2. Let $C = \{A_i\colon i = 1, 2, \ldots, n\}$ be a finite collection of events. These events are independent if and only if, given any subcollection $A_{(1)}$, $A_{(2)}, \ldots, A_{(m)}$ of elements of C,

$$P\left[A_{(1)} \cap A_{(2)} \cap \cdots \cap A_{(m)}\right] = P\left[A_{(1)}\right]P\left[A_{(2)}\right] \cdots P\left[A_{(m)}\right]$$

Although this definition can be used to test a collection of events for independence, its main purpose is to provide a way to find the probability that a series of events that are assumed to be independent will occur. To illustrate, we reconsider a problem encountered in Chap. 1 (Example 1.2.1).

Example 2.3.4. During a space shot, the primary computer system is backed up by two secondary systems. They operate independently of one another and each is 90% reliable. What is the probability that all three systems will be operable at the time of the launch? Let

A_1: the main system is operable

A_2: the first backup is operable

A_3: the second backup is operable

We are given that $P[A_1] = P[A_2] = P[A_3] = .9$. We want $P[A_1 \cap A_2 \cap A_3]$. Since these events are assumed to be independent,

$$\begin{aligned} P[A_1 \cap A_2 \cap A_3] &= P[A_1]P[A_2]P[A_3] \\ &= (.9)(.9)(.9) \\ &= .729 \end{aligned}$$

Definition 2.3.2 must be used with care. In particular, one must be certain that it is reasonable to assume that events are independent before it is applied to compute the probability that a series of events will occur. The danger of erroneously assumed independence is illustrated in Example 2.3.5.

Example 2.3.5. An Atomic Energy Commission Study, WASH 1400, reported the probability of a nuclear accident such as that which occurred at Three Mile Island in March 1978 to be one in 10 million. Yet the accident did occur. According to Mark Stephens, "The methodology of WASH 1400 made use of event trees—sequences of actions that would be necessary for accidents to take place. These event trees did not assume any interrelation between events—that they might be caused by the same error in judgment or as part of the same mistaken action. The statisticians who assigned probabilities in the writing of WASH 1400 said, for example, that there was a one-in-a-thousand risk of one of the auxiliary feed-water control valves—the twelves—being closed. And if there is a one-in-a-thousand chance of one valve being closed, the chances of both valves being closed is one-thousandth of that, or a million to one. But both of the twelves were closed by the same man on March 26—and one had never been closed without the other." The events A_1: the first valve is closed, and A_2: the second valve is closed were not independent. However, they were treated as such when calculating the probability of an accident. This, among other things, led to an underestimate of the accident potential (from *Three Mile Island* by Mark Stephens, Random House, 1980).

Exercises 28 to 33 outline other interesting theorems concerning the idea of independence.

The Multiplication Rule

There is one further point to be made before we conclude this section. We can find $P[A_1 \cap A_2]$ if the events are assumed to be independent. Furthermore, if the proper information is given, the general addition rule can be used to find this probability. Is there any other way to find the probability of the simultaneous occurrence of two events if the events are not independent? The answer is yes, and the method is easy to derive. We know that

$$P[A_2|A_1] = \frac{P[A_1 \cap A_2]}{P[A_1]} \qquad P[A_1] \neq 0$$

regardless of whether the events are independent. Multiplying each side of this equation by $P[A_1]$, we obtain the following formula, called the *multiplication rule*:

Multiplication rule

$$P[A_1 \cap A_2] = P[A_2|A_1]P[A_1]$$

The use of this rule is illustrated in Example 2.3.6.

Example 2.3.6. Recent research indicates that approximately 49% of all infections involve anaerobic bacteria. Furthermore, 70% of all anaerobic infections are polymicrobic; that is, they involve more than one anaerobe. What is the probability that a given infection involves anaerobic bacteria *and* is polymicrobic? Let A_1 denote the event that the infection is anaerobic, and A_2 that it is polymicrobic. We are given that $P[A_1] = .49$ and that $P[A_2|A_1] = .70$. We want to find $P[A_1 \cap A_2]$. By the multiplication rule,

$$\begin{aligned} P[A_1 \cap A_2] &= P[A_2|A_1]P[A_1] \\ &= (.70)(.49) \\ &= .343 \end{aligned}$$

2.4 BAYES' THEOREM

The topic of this section is the theorem formulated by the Reverend Thomas Bayes (1761). It deals with conditional probability. Bayes' theorem is used to find $P[A|B]$ when the available information is not immediately compatible with that required to apply the definition of conditional probability directly.

Example 2.4.1 is a typical problem calling for the use of Bayes' theorem. You will find applying Bayes' rule quite natural without having seen a formal statement of the theorem!

Example 2.4.1. The blood type distribution in the United States is type A, 41%; type B, 9%; type AB, 4%; and type O, 46%. It is estimated that during World War II, 4% of inductees with type O blood were typed as having type A; 88% of those with type A were correctly typed; 4% with type B blood were typed as A; and 10% with type AB were typed as A. A soldier was wounded and brought to surgery. He was typed as having type A blood. What is the probability that this is his true blood type?

Let

A: he has type A blood

B: he has type B blood

AB: he has type AB blood

O: he has type O blood

TA: he is typed as type A

We want to find $P[A|TA]$. We are given that

$$\begin{aligned} P[A] &= .41 \qquad & P[TA|A] &= .88 \\ P[B] &= .09 & P[TA|B] &= .04 \\ P[AB] &= .04 & P[TA|AB] &= .10 \\ P[O] &= .46 & P[TA|O] &= .04 \end{aligned}$$

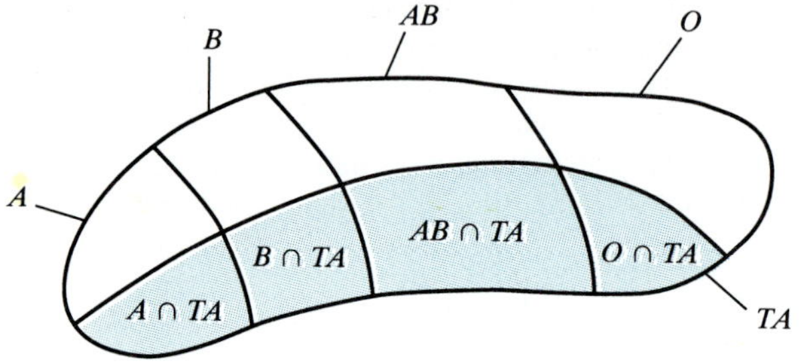

FIGURE 2.4
$TA = (A \cap TA) \cup (B \cap TA) \cup (AB \cap TA) \cup (O \cap TA)$.

Since the question asked is conditional, the first inclination is to try to apply the definition of conditional probability. Let us do so.

$$P[A|TA] = \frac{P[A \cap TA]}{P[TA]}$$

Unfortunately, neither $P[A \cap TA]$ nor $P[TA]$ is given. We must compute these quantities for ourselves. Each is easy to find. Note that by the multiplication rule,

$$\begin{aligned} P[A \cap TA] &= P[TA|A]P[A] \\ &= (.88)(.41) \\ &\doteq .36 \end{aligned}$$

Note also that the event TA can be partitioned into four mutually exclusive events, as shown in Fig. 2.4. That is,

$$TA = (A \cap TA) \cup (B \cap TA) \cup (AB \cap TA) \cup (O \cap TA)$$

By axiom 3,

$$P[TA] = P[A \cap TA] + P[B \cap TA] + P[AB \cap TA] + P[O \cap TA]$$

Applying the multiplication rule to each of the terms on the right-hand side of this equation, we obtain

$$\begin{aligned} P[TA] &= P[TA|A]P[A] + P[TA|B]P[B] \\ &\quad + P[TA|AB]P[AB] + P[TA|O]P[O] \\ &= (.88)(.41) + (.04)(.09) + (.10)(.04) + (.04)(.46) \\ &\doteq .39 \end{aligned}$$

Substituting .39 for $P[TA]$ yields

$$\begin{aligned} P[A|TA] &= \frac{P[A \cap TA]}{P[TA]} \\ &= \frac{.36}{.39} \\ &\doteq .92 \end{aligned}$$

The previous problem was solved using Bayes' rule. We now state the rule. By a partition of S we mean a collection of mutually exclusive events whose union is S.

Theorem 2.4.1 (Bayes' theorem). Let $A_1, A_2, A_3, \ldots, A_n$ be a collection of events which partition S. Let B be an event such that $P[B] \neq 0$. Then for any of the events A_j, $j = 1, 2, 3, \ldots, n$,

$$P[A_j|B] = \frac{P[B|A_j]P[A_j]}{\sum_{i=1}^{n} P[B|A_i]P[A_i]}$$

To see that Bayes' theorem was used in Example 2.4.1, make these notational changes:

$$A_1 = A \qquad A_3 = AB \qquad B = TA$$

$$A_2 = B \qquad A_4 = O$$

and use the theorem to find $P[A_1|B]$. Your answer will, of course, agree with that obtained earlier.

CHAPTER SUMMARY

In this chapter we presented some of the laws that govern the behavior of probabilities. We began with the axioms and from those we were able to derive the remaining laws. In particular, we derived the addition rule, which deals with the probability of the union of two events; the multiplication rule, which deals with the probability of the intersection of two events; and Bayes' theorem, which deals with conditional probability. We introduced and defined important terms that you should know. These are:

Conditional probability   Independent events

Care must be taken when using the concept of independence. In an applied problem, be sure that it is reasonable to assume that events A and B are independent before finding the probability of their joint occurrence via the definition $P[A \cap B] = P[A]P[B]$.

EXERCISES

Section 2.1

1. The probability that a wildcat well will produce oil is 1/13. What is the probability that it will not be productive?
2. The theft of precious metals from companies in the United States is becoming a serious problem. The estimated probability that such a theft will involve a particular

metal is given below: (Based on data reported in "Materials Theft," *Materials Engineering*, February 1982, pp. 27–31.)

tin: 1/35	platinum: 1/35	nickel: 1/35
steel: 11/35	gold: 5/35	zinc: 1/35
copper: 8/35	aluminum: 2/35	silver: 4/35
titanium: 1/35		

(Note that these events are assumed to be mutually exclusive.)

(*a*) What is the probability that a theft of precious metal will involve gold, silver, or platinum?

(*b*) What is the probability that a theft will not involve steel?

3. Assuming the blood type distribution to be A: 41%, B: 9%, AB: 4%, O: 46%, what is the probability that the blood of a randomly selected individual will contain the A antigen? That it will contain the B antigen? That it will contain neither the A nor the B antigen?

4. Assume that the engine component of a spacecraft consists of two engines in parallel. If the main engine is 95% reliable, the backup is 80% reliable, and the engine component as a whole is 99% reliable, what is the probability that both engines will be operable? Use a Venn diagram to find the probability that the main engine will fail but the backup will be operable. Find the probability that the backup engine will fail but the main engine will be operable. What is the probability that the engine component will fail?

5. When an individual is exposed to radiation, death may ensue. Factors affecting the outcome are the size of the dose, the length and intensity of the exposure, and the biological makeup of the individual. The term LD_{50} is used to denote the dose that is usually lethal for 50% of the individuals exposed to it. Assume that in a nuclear accident 30% of the workers are exposed to the LD_{50} and die; 40% of the workers die; and 68% are exposed to the LD_{50} or die. What is the probability that a randomly selected worker is exposed to the LD_{50}? Use a Venn diagram to find the probability that a randomly selected worker is exposed to the LD_{50} but does not die. Find the probability that a randomly selected worker is not exposed to the LD_{50} but dies.

6. When a computer goes down, there is a 75% chance that it is due to an overload and a 15% chance that it is due to a software problem. There is an 85% chance that it is due to an overload or a software problem. What is the probability that both of these problems are at fault? What is the probability that there is a software problem but no overload?

***7.** Derive Theorem 2.1.1.

Hint: Note that $S = S \cup \varnothing$ and that S and $\varnothing$ are mutually exclusive. Apply axioms 3 and 1.

***8.** Derive Theorem 2.1.2.

Hint: Note that $S = A \cup A'$ and that A and A' are mutually exclusive. Apply axioms 3 and 1.

***9.** Let $A \subseteq B$. Show that $P[A] \leq P[B]$.

Hint: $B = A \cup (A' \cap B)$. Apply axioms 3 and 2.

***10.** Show that the probability of any event A is at most 1.

Hint: $A \subseteq S$. Apply Exercise 9 and axiom 1.

*11. Derive the addition rule.

Hint: Note that

$$A_1 = (A_1 \cap A_2) \cup (A_1 \cap A_2')$$
$$A_2 = (A_1 \cap A_2) \cup (A_1' \cap A_2)$$
$$A_1 \cup A_2 = (A_1 \cap A_2) \cup (A_1 \cap A_2') \cup (A_1' \cap A_2)$$

Apply axiom 3 to each of these expressions.

12. Let A_1 and A_2 be mutually exclusive. By axiom 3 $P[A_1 \cup A_2] = P[A_1] + P[A_2]$. Show that the general addition rule yields the same result.

Section 2.2

13. Use the data of Exercise 5 to answer these questions.
- (*a*) What is the probability that a randomly selected worker will die given that he is exposed to the lethal dose of radiation?
- (*b*) What is the probability that a randomly selected worker will not die given that he is exposed to the lethal dose of radiation?
- (*c*) What theorem allows you to determine the answer to (*b*) from knowledge of the answer to (*a*)?
- (*d*) What is the probability that a randomly selected worker will die given that he is not exposed to the lethal dose?
- (*e*) Is $P[\text{die}] = P[\text{die}|\text{exposed to lethal dose}]$? Did you expect these to be the same? Explain.

14. Use the data of Exercise 4 to answer these questions.
- (*a*) What is the probability that in an engine system such as that described the backup engine will function given that the main engine fails?
- (*b*) Is $P[\text{backup functions}] = P[\text{backup functions}|\text{main fails}]$? Did you expect these to be the same? Explain.

15. In a study of waters near power plants and other industrial plants that release wastewater into the water system it was found that 5% showed signs of chemical and thermal pollution, 40% showed signs of chemical pollution, and 35% showed evidence of thermal pollution. Assume that the results of the study accurately reflect the general situation. What is the probability that a stream that shows some thermal pollution will also show signs of chemical pollution? What is the probability that a stream showing chemical pollution will not show signs of thermal pollution?

16. A random digit generator on an electronic calculator is activated twice to simulate a random two-digit number. Theoretically, each digit from 0 to 9 is just as likely to appear on a given trial as any other digit.
- (*a*) How many random two-digit numbers are possible?
- (*b*) How many of these numbers begin with the digit 2?
- (*c*) How many of these numbers end with the digit 9?
- (*d*) How many of these numbers begin with the digit 2 and end with the digit 9?
- (*e*) What is the probability that a randomly formed number ends with 9 given that it begins with a 2. Did you anticipate this result?

17. In studying the causes of power failures, these data have been gathered.

5% are due to transformer damage
80% are due to line damage
1% involve both problems

Based on these percentages, approximate the probability that a given power failure involves

(*a*) line damage given that there is transformer damage
(*b*) transformer damage given that there is line damage
(*c*) transformer damage but not line damage
(*d*) transformer damage given that there is no line damage
(*e*) transformer damage or line damage

Section 2.3

18. Let A_1 and A_2 be events such that $P[A_1] = .5$, $P[A_2] = .7$. What must $P[A_1 \cap A_2]$ equal for A_1 and A_2 to be independent?

19. Let A_1 and A_2 be events such that $P[A_1] = .6$, $P[A_2] = .4$ and $P[A_1 \cup A_2] = .8$. Are A_1 and A_2 independent?

20. Consider your answer to Exercise 14(*b*). Are the events A_1: the backup engine functions, and A_2: the main engine fails independent?

21. Studies in population genetics indicate that 39% of the available genes for determining the Rh blood factor are negative. Rh negative blood occurs if and only if the individual has two negative genes. One gene is inherited independently from each parent. What is the probability that a randomly selected individual will have Rh negative blood?

22. An individual's blood group (A, B, AB, O) is independent of the Rh classification. Find the probability that a randomly selected individual will have AB negative blood. *Hint:* See Example 2.1.1 and Exercise 21.

23. The use of plant appearance in prospecting for ore deposits is called geobotanical prospecting. One indicator of copper is a small mint with a mauve-colored flower. Suppose that, for a given region, there is a 30% chance that the soil has a high copper content and a 23% chance that the mint will be present there. If the copper content is high, there is a 70% chance that the mint will be present.

(*a*) Find the probability that the copper content will be high and the mint will be present.
(*b*) Find the probability that the copper content will be high given that the mint is present.

24. The most common water pollutants are organic. Since most organic materials are broken down by bacteria that require oxygen, an excess of organic matter may result in a depletion of available oxygen. In turn this can be harmful to other organisms living in the water. The demand for oxygen by the bacteria is called the biological oxygen demand (BOD). A study of streams located near an industrial complex revealed that 35% have a high BOD, 10% show high acidity, and 40% of streams with high acidity have a high BOD. Find the probability that a randomly selected stream will exhibit both characteristics.

25. A study of major flash floods that occurred over the last 15 years indicates that the probability that a flash flood warning will be issued is .5 and that the probability of dam failure during the flood is .33. The probability of dam failure given that a warning is issued is .17. Find the probability that a flash flood warning will be issued and a dam failure will occur. (Based on data reported in *McGraw-Hill Yearbook of Science and Technology*, 1980, pp. 185–186.)

***26.** The ability to observe and recall details is important in science. Unfortunately, the power of suggestion can distort memory. A study of recall is conducted as follows:

Subjects are shown a film in which a car is moving along a country road. There is no barn in the film. The subjects are then asked a series of questions concerning the film. Half the subjects are asked, "How fast was the car moving when it passed the barn?" The other half is not asked the question. Later each subject is asked; "Is there a barn in the film?" Of those asked the first question concerning the barn, 17% answer "yes"; only 3% of the others answer "yes." What is the probability that a randomly selected participant in this study claims to have seen the nonexistent barn? Is claiming to see the barn independent of being asked the first question about the barn? *Hint:*

$$P[\text{yes}] = P[\text{yes and asked about barn}] + P[\text{yes and not asked about barn}]$$

(Based on a study reported in *McGraw-Hill Yearbook of Science and Technology*, 1981, pp. 249–251.)

***27.** The probability that a unit of blood was donated by a paid donor is .67. If the donor was paid, the probability of contracting serum hepatitis from the unit is .0144. If the donor was not paid, this probability is .0012. A patient receives a unit of blood. What is the probability of the patient's contracting serum hepatitis from this source?

***28.** Show that the impossible event is independent of every other event.

***29.** Show that if A_1 and A_2 are independent, then A_1 and A'_2 are also independent. *Hint:* $A_1 = (A_1 \cap A_2) \cup (A_1 \cap A'_2)$.

***30.** Use Exercise 29 to show that if A_1 and A_2 are independent, then A'_1 and A'_2 are also independent.

***31.** It can be shown that the result of Exercise 30 holds for any collection of n independent events. That is, if $A_1, A_2, \ldots, A_n$ are independent, then $A'_1, A'_2, \ldots, A'_n$ are also independent. Use this result and the data of Example 2.3.4 to find the probability that at least one of the three computer systems will be operable at the time of the launch.

***32.** Let A_1 and A_2 be mutually exclusive events such that $P[A_1]P[A_2] > 0$. Show that these events are not independent.

***33.** Let A_1 and A_2 be independent events such that $P[A_1]P[A_2] > 0$. Show that these events are not mutually exclusive.

Section 2.4

34. Use the data of Example 2.4.1 to find the probability that an inductee who was typed as having type A blood actually had type B blood.

35. A test has been developed to detect a particular type of arthritis in individuals over 50 years old. From a national survey it is known that approximately 10% of the individuals in this age group suffer from this form of arthritis. The proposed test was given to individuals with confirmed arthritic disease, and a correct test result was obtained in 85% of the cases. When the test was administered to individuals of the same age group who were known to be free of the disease, 4% were reported to have the disease. What is the probability that an individual has this disease given that the test indicates its presence?

36. It is reported that 50% of all computer chips produced are defective. Inspection ensures that only 5% of the chips legally marketed are defective. Unfortunately, some chips are stolen before inspection. If 1% of all chips on the market are stolen, find the probability that a given chip is stolen given that it is defective.

37. As society becomes dependent on computers, data must be communicated via public communication networks such as satellites, microwave systems, and telephones. When a message is received, it must be authenticated. This is done by using a secret enciphering key. Even though the key is secret, there is always the probability that it will fall into the wrong hands, thus allowing an unauthentic message to appear to be authentic. Assume that 95% of all messages received are authentic. Furthermore, assume that only .1% of all unauthentic messages are sent using the correct key and that all authentic messages are sent using the correct key. Find the probability that a message is authentic given that the correct key is used.

REVIEW EXERCISES

38. A survey of engineering firms reveals that 80% have their own mainframe computer (M); 10% anticipate purchasing a mainframe computer in the near future (B); and 5% have a mainframe computer and anticipate buying another in the near future. Find the probability that a randomly selected firm:

(*a*) has a mainframe computer or anticipates purchasing one in the near future

(*b*) does not have a mainframe computer and does not anticipate purchasing one in the near future

(*c*) anticipates purchasing a mainframe computer given that it does not currently have one

(*d*) has a mainframe computer given that it anticipates purchasing one in the near future

39. In a simulation program, three random two-digit numbers will be generated independently of one another. These numbers assume the values $00, 01, 02, \ldots, 99$ with equal probability.

(*a*) What is the probability that a given number will be less than 50?

(*b*) What is the probability that each of the three numbers generated will be less than 50?

40. A power network involves three substations A, B, and C. Overloads at any of these substations might result in a blackout of the entire network. Past history has shown that if substation A alone experiences an overload, then there is a 1% chance of a network blackout. For stations B and C alone these percentages are 2% and 3%, respectively. Overloads at two or more substations simultaneously result in a blackout 5% of the time. During a heat wave there is a 60% chance that substation A alone will experience an overload. For stations B and C these percentages are 20 and 15%, respectively. There is a 5% chance of an overload at two or more substations simultaneously. During a particular heat wave a blackout due to an overload occurred. Find the probability that the overload occurred at substation A alone; substation B alone; substation C alone; two or more substations simultaneously.

41. A computer center has three printers A, B, and C, which print at different speeds. Programs are routed to the first available printer. The probability that a program is routed to printers A, B, and C are .6, .3, and .1, respectively. Occasionally a printer will jam and destroy a printout. The probability that printers A, B, and C will jam are .01, .05, and .04 respectively. Your program is destroyed when a printer jams. What is the probability that printer A is involved? Printer B is involved? Printer C is involved?

42. A chemical engineer is in charge of a particular process at an oil refinery. Past experience indicates that 10% of all shutdowns are due to equipment failure *alone*, 5% are due to a combination of equipment failure and operator error, and 40% involve operator error. A shutdown occurs. Find the probability that

(*a*) equipment failure or operator error is involved

(*b*) operator error alone is involved

(*c*) neither operator error nor equipment failure is involved

(*d*) operator error is involved given that equipment failure occurs

(*e*) operator error is involved given that equipment failure does not occur

43. Assume that the probability that the air brakes on large trucks will fail on a particularly long downgrade is .001. Assume also that the emergency brakes on such trucks can stop a truck on this downgrade with probability .8. These braking systems operate independently of one another. Find the probability that

(*a*) the air brakes fail but the emergency brakes can stop the truck

(*b*) the air brakes fail and the emergency brakes cannot stop the truck

(*c*) the emergency brakes cannot stop the truck given that the air brakes fail

44. Consider the problem of Example 1.2.3. Assume that sampling is independent and that at each stage the probability of obtaining a defective part when the process is working correctly is .01. If the process is working correctly, what is the probability that the first defective part will be obtained on the fourth sample? On or before the fourth sample?

CHAPTER 3

DISCRETE DISTRIBUTIONS

In the sciences one often deals with "variables." Webster's dictionary defines a variable as a "quantity that may assume any one of a set of values." In statistics we deal with *random variables*—variables whose observed value is determined by chance. Many of the examples presented in previous chapters involved random variables even though the term was not used at the time. Random variables usually fall into one of two categories; they are either discrete or continuous. We begin by learning to recognize discrete random variables. The remainder of the chapter is devoted to the study of random variables of this type.

3.1 RANDOM VARIABLES

We begin by considering three examples, each of which involves a random variable. Random variables will be denoted by uppercase letters and their observed numerical values by lowercase letters.

Example 3.1.1. Consider the random variable X, the number of brown-eyed children born to a couple heterozygous for eye color. If the couple is assumed to have two children, a priori, before the fact, the variable X can assume any one of the values 0, 1, or 2. The variable is random in that brown eyes depend on the chance inheritance of a dominant gene at conception. If for a particular couple there are two brown-eyed children, we write $x = 2$.

Example 3.1.2. The basic premise underlying the field of immunology is that an animal is immunized by injection of a suitable antigen. In one study malignant plasmacytoma cells are exposed to lymphocytes carrying a specific antigen. It is hoped that these cells will fuse since the fused cells retain the ability to grow continuously and also to retain the antibody characteristics of the antigen fused. In this way the animal is quickly immunized. Cells are exposed to the lymphocytes one at a time in the presence of polyethylene glycol, a fusion-promoting agent. It is known that the probability that such a cell will fuse is $1/2$. Let Y denote the number of cells exposed to obtain the first fusion. The variable Y is random; a priori it can assume any value in the set $\{1, 2, 3, \ldots\}$. Recall from your study of calculus that a set such as this that consists of an infinite collection of isolated points is called a countably infinite set.

Example 3.1.3. In Example 1.1.2 we considered the variable T, the time at which the peak demand for electricity occurs per day. This variable is random since its value is affected by such chance factors as time of the year, humidity, and temperature. It can conceivably assume any value in the 24-hour time span from 12 midnight one day to 12 midnight the next day.

It is easy to distinguish a discrete random variable from one that is not discrete. Just ask the question, "What are the possible values for the variable?" If the answer is a finite set or a countably infinite set, then the random variable is discrete; otherwise it is not. This idea leads to the following definition.

Definition 3.1.1 (Discrete random variable). A random variable is discrete if it can assume at most a finite or a countably infinite number of possible values.

The random variable X, the number of brown-eyed children in a two-child family, is discrete. Its set of possible values is the finite set $\{0, 1, 2\}$. The set $\{1, 2, 3, \ldots\}$ of possible values for Y, the number of cells exposed to obtain the first fusion of Example 3.1.2, is countably infinite. Thus Y is also a discrete random variable. The random variable T, the time of the peak demand for electricity at a power plant, is different from the others. Time is measured continuously and T can conceivably assume any value in the interval $[0, 24)$ where 0 denotes 12 midnight one day and 24 denotes 12 midnight the next. This set of real numbers is neither finite nor countably infinite. Any time that you ask yourself the question; "What are the possible values for the random variable?" and are forced to admit that the set of possibilities includes some interval or continuous span of real numbers, then the random variable being studied is not discrete.

3.2 DISCRETE PROBABILITY DENSITIES

When dealing with a random variable, it is not enough just to determine what values are possible. We also need to determine what is probable. We must be able to predict in some sense the values that the variable is likely to assume at any time. Since the behavior of a random variable is governed by chance, these predictions must be made in the face of a great deal of uncertainty. The best that can be done is to describe the behavior of the random variable in terms of probabilities. Two functions are used to accomplish this. We shall refer to these as the *density function* and the *cumulative distribution function*. The former is known by a variety of names in the discrete case. Some of the most commonly encountered ones being the probability function, the probability mass function, and the probability density function. In the discrete case, the density is denoted by either $p(x)$ or $f(x)$; in the continuous case it is almost always denoted by $f(x)$. For consistency we shall use $f(x)$ for the density in both cases. We begin by defining the density function for discrete random variables.

Definition 3.2.1 (Discrete density). Let X be a discrete random variable. The function f given by

$$f(x) = P[X = x]$$

for x real is called the density function for X.

There are several facts to note concerning the density in the discrete case. First, f is defined on the entire real line, and for any given real number x, $f(x)$ is the probability that the random variable X assumes the value x. For example, $f(2)$ is the probability that the random variable X assumes the numerical value of 2. Second, since $f(x)$ is a probability, $f(x) \geq 0$ regardless of the value of x. Third, if we sum f over all values of X that occur with nonzero probability, the sum must be 1. In fact, the two conditions

1. $f(x) \geq 0$
2. $\sum_{\text{all } x} f(x) = 1$

are necessary and sufficient conditions for a function f to be a discrete density. The next example illustrates these ideas.

Example 3.2.1. Consider the random variable Y, the number of cells exposed to antigen-carrying lymphocytes in the presence of polyethylene glycol to obtain the first fusion (see Example 3.1.2). We know that under these conditions the probability that a given cell will fuse is 1/2. Thus the probability that it will not fuse is also 1/2. It is reasonable to assume that the cells behave independently. The possible values for Y are $\{1, 2, 3, \ldots\}$. The probability that the first cell will fuse is 1/2. That is,

$$P[Y = 1] = f(1) = 1/2$$

The probability that the first cell will not fuse but the second one will, yielding a value of 2 for Y, is

$$\begin{aligned} P[Y = 2] = f(2) &= P[\text{first cell does not fuse}]P[\text{second cell does fuse}] \\ &= 1/2 \cdot 1/2 = 1/4 \end{aligned}$$

Similarly,

$$P[Y = 3] = f(3) = 1/2 \cdot 1/2 \cdot 1/2 = 1/8$$

TABLE 3.1

y	1	2	3	$4 \cdots$
$P[Y=y]=f(y)$	$1/2$	$1/2 \cdot 1/2$	$1/2 \cdot 1/2 \cdot 1/2$	$1/2 \cdot 1/2 \cdot 1/2 \cdot 1/2$

We can summarize the entire probability structure for Y in a density table (see Table 3.1). This is a table giving the possible values for the random variable in the first row and their corresponding probabilities in the second row. Note that there is an obvious pattern to the entries in row 2. When this occurs, we can find a closed-form expression for the density. In this case

$$f(y) = \begin{cases} (1/2)^y & y = 1, 2, 3, \ldots \\ 0 & \text{elsewhere} \end{cases}$$

Is this really a density? This function is obviously nonnegative, but does it sum to 1? To see this, note that

$$\sum_{\text{all } y} f(y) = \sum_{y=1}^{\infty} (1/2)^y$$

is a geometric series with first term $a = 1/2$ and common ratio $r = 1/2$. The properties of geometric series are well known. In particular, recall from elementary calculus that such a series can converge or diverge. The following fact will be useful in the material that follows:

Convergence of geometric series

Let $\sum_{k=1}^{\infty} ar^{k-1}$ be a geometric series.

The series converges to $\dfrac{a}{1-r}$ provided $|r| < 1$.

If we apply this result here, we see that

$$\sum_{y=1}^{\infty} (1/2)^y = \frac{a}{1-r} = \frac{1/2}{1-1/2} = 1$$

and the function f is a density.

Even though a discrete density is defined on the entire real line, it is only necessary to specify the density for those values y for which $f(y) \neq 0$. For instance, in the previous example we can write

$$f(y) = (1/2)^y \qquad y = 1, 2, 3, \ldots$$

It is understood that $f(y) = 0$ for all other real numbers.

Once it is known that a function is a density, it can be used to answer questions concerning the behavior of Y.

Example 3.2.2. What is the probability that we will need to expose four or more cells to antigen-carrying lymphocytes in the presence of polyethylene glycol to obtain the first fusion? That is, What is $P[Y \geq 4]$? The density for Y is

$$f(y) = (1/2)^y \qquad y = 1, 2, 3, \ldots$$

Although the desired probability can be found directly, it is easier to use subtraction:

$$\begin{aligned} P[Y \geq 4] &= 1 - P[Y < 4] \\ &= 1 - P[Y \leq 3] \\ &= 1 - (P[Y = 1] + P[Y = 2] + P[Y = 3]) \\ &= 1 - (f(1) + f(2) + f(3)) \\ &= 1 - \left((1/2)^1 + (1/2)^2 + (1/2)^3\right) \\ &= 1 - (1/2 + 1/4 + 1/8) \\ &= 1 - 7/8 = 1/8 \end{aligned}$$

Cumulative Distribution

The second function used to compute probabilities is the cumulative distribution function F. Most of the statistical tables used in the material that follows are tables of the cumulative distribution function for some pertinent random variable.

The word "cumulative" suggests the role of this function. It sums or accumulates the probabilities found by means of the density. This function is defined as follows:

Definition 3.2.2 (Cumulative distribution—discrete). Let X be a discrete random variable with density f. The cumulative distribution function for X, denoted by F, is defined by

$$F(x) = P[X \leq x] \qquad \text{for } x \text{ real}$$

Consider a specific real number x_0. To find $P[X \leq x_0] = F(x_0)$, we sum the density f over all values of X that occur with nonzero probability that are less than or equal to x_0. That is, computationally,

$$F(x_0) = \sum_{x \leq x_0} f(x)$$

This idea is illustrated in Example 3.2.3.

Example 3.2.3. Certain genes produce such a tremendous deviation from normal that the organism is unable to survive. Such genes are called lethal genes. An example is the gene that produces a yellow coat in mice, Y. This gene is dominant over that for gray, y. Normal genetic theory predicts that when two yellow mice heterozygous for this trait (Yy) mate, $1/4$ of the offspring will be gray and $3/4$ will be yellow. Biologists have observed that these predicted proportions do not, in

TABLE 3.2

x	0	1	2	3
$P[X = x] = f(x)$	1/27	6/27	12/27	8/27

TABLE 3.3

x	0	1	2	3
$P[X \le x] = F(x)$	1/27	7/27	19/27	27/27

TABLE 3.4

y	1	2	3	4 ···
$P[Y \le y] = F(y)$	8/16	12/16	14/16	15/16

fact, occur, but that the actual percentages produced are 1/3 gray and 2/3 yellow. It has been established that this shift is caused by the fact that 1/4 of the embryos, those homozygous for yellow (YY), do not develop. This leaves only two genotypes, Yy and yy, occurring in a ratio of 2 to 1, with the former producing a mouse with a yellow coat. For this reason, the gene Y is said to be lethal.

The density for X, the number of yellow mice in a litter of size 3, is shown in Table 3.2 and its cumulative distribution is given in Table 3.3. Notice that

$$\begin{aligned}
F(0) &= P[X \le 0] = P[X = 0] = 1/27 \\
F(1) &= P[X \le 1] = P[X = 0] + P[X = 1] = 1/27 + 6/27 \\
F(2) &= P[X \le 2] = P[X = 0] + P[X = 1] + P[X = 2] \\
&= 1/27 + 6/27 + 12/27 \\
F(3) &= P[X \le 3] = 1
\end{aligned}$$

For discrete random variables that can assume only a finite number of possible values the last entry in the bottom row of the cumulative table will always be 1.

Although cumulative probabilities are often given in table form as in the preceding example, it is sometimes possible to find express F in equation form. Example 3.2.4 illustrates this idea.

Example 3.2.4. Consider the random variable Y of Example 3.2.1 with density

$$f(y) = (1/2)^y \qquad y = 1, 2, 3, \ldots$$

A partial cumulative table for Y is shown in Table 3.4. It is formed by summing the probabilities given in the density table, Table 3.1. It is helpful to have a closed-form expression for F. In this case it is easy to obtain such an expression.

By definition,

$$F(y_0) = \sum_{y \le y_0} f(y)$$

If we let $[y_0]$ denote the greatest integer less than or equal to y_0, then in this case $F(y_0)$ can be expressed as

$$F(y_0) = \sum_{y=1}^{[y_0]} (1/2)^y$$

$$= \sum_{y=1}^{[y_0]} (1/2)(1/2)^{y-1}$$

Recall from elementary calculus that the sum of the first n terms of a geometric series is given by

Sum of first *n* terms: Geometric series

$$\sum_{k=1}^{n} ar^{k-1} = \frac{a(1-r^n)}{1-r} \qquad r \neq 1$$

where a is the first term of the series and r is the common ratio.

Apply this result with $a = 1/2$ and $r = 1/2$, to obtain

$$F(y_0) = \frac{(1/2)\left[1-(1/2)^{[y_0]}\right]}{1-1/2}$$

$$= 1 - (1/2)^{[y_0]}$$

The probability that at most seven cells must be exposed to obtain the first fusion is given by

$$P[Y \le 7] = F(y) = 1 - (1/2)^7 = \frac{127}{128}$$

Some of the mathematical properties of discrete distribution functions are outlined in Exercises 11, 12, and 13.

3.3 EXPECTATION AND DISTRIBUTION PARAMETERS

The density function of a random variable completely describes the behavior of the variable. However, associated with any random variable are constants, or "parameters," that are descriptive. Knowledge of the numerical values of these parameters gives the researcher quick insight into the nature of the variables. We consider three such parameters: the mean μ, the variance σ^2, and the standard deviation σ. If the exact density of the random variable is known, then the numerical value of each parameter can be found from mathematical considerations. That is the topic of this section. If the only thing available to the researcher is a set of observations on the random variable (a data set), then the values of these parameters cannot be found exactly. They must be approximated by using statistical techniques. That is the topic of much of the remainder of this text.

To understand the reasoning behind most statistical methods, it is necessary to become familiar with one general concept, namely, the idea of *mathematical expectation* or *expected value*. This concept is used in defining many statistical parameters and provides the logical basis for most of the methods of statistical inference presented later in this text.

A simple example will illustrate the basic idea of expectation. Consider the roll of a single fair die, and let X denote the number that is obtained. The possible values for X are $1, 2, 3, 4, 5, 6$, and since the die is fair, the probability associated with each value is $1/6$. The density for X is given by

$$f(x) = 1/6 \qquad x = 1, 2, 3, 4, 5, 6$$

When we ask for the expected value of X, we are asking for the *long-run theoretical average value of X*. If we imagine rolling the die over and over and recording the value of X for each roll, then we are asking for the theoretical average value of the rolls as the number of rolls approaches infinity. Since the density for X is symmetric and known, this average can be found intuitively. Notice that since $P[X = 1] = P[X = 6] = 1/6$, in the long run we expect to roll as many 1's as 6's. These values should counterbalance one another and their average value is $(6 + 1)/2 = 3.5$. We also expect to roll as many 2's as 4's; these numbers also average to 3.5. Likewise, the numbers 3 and 4 are expected to counterbalance one another; they average 3.5. Logic dictates that, in the long run the average or expected value of X is 3.5. We write this as $E[X] = 3.5$. Notice that this value can be calculated from the density for X as follows:

$$E[X] = 1 \cdot 1/6 + 2 \cdot 1/6 + 3 \cdot 1/6 + 4 \cdot 1/6 + 5 \cdot 1/6 + 6 \cdot 1/6 = 3.5$$

or

$$E[X] = \sum_{\text{all } x} (\text{value of } x)(\text{probability})$$

Of course, the characteristic that makes finding this expectation easy is the symmetry of the density. Can we develop a definition of expectation that will work for nonsymmetric densities and which will apply not only to X, but also to random variables that are functions of X? The answer is "yes," and the desired definition is given in Definition 3.3.1. Let us point out that in most problems interest centers first on $E[X]$. However, expectations for functions of X such as X^2, $(X - c)^2$, where c is a constant and e^{tX} are especially useful in statistical theory. For this reason, the definition of expected value is given in general terms. We now define what we mean by the expected value of some function of X which we denote by $H(X)$. The expected value of X is a special case of the definition in which $H(X) = X$.

Definition 3.3.1 (Expected value). Let X be a discrete random variable with density f. Let $H(X)$ be a random variable. The expected value of $H(X)$, denoted by $E[H(X)]$, is given by

$$E[H(X)] = \sum_{\text{all } x} H(x)f(x)$$

provided $\sum_{\text{all } x}|H(x)|f(x)$ is finite. Summation is over all values of X that occur with nonzero probability.

One thing to note concerning this definition is the fact the restriction that $\sum_{\text{all } x}|H(x)|f(x)$ exists is not particularly restrictive in practice. If the set of possible values for X is finite, it *will* be satisfied; if the set of possible values for X is countably infinite, it will *usually* be satisfied. However, it is possible to concoct a density f and a function $H(X)$ for which the series $\sum_{\text{all } x}|H(x)|f(x)$ does not converge. (See Exercise 22.) In this case we say that the expected value of the random variable $H(X)$ does not exist. An example will illustrate the use of Definition 3.3.1. Please realize that the density has been greatly oversimplified for purposes of illustration!

Example 3.3.1. A drug is used to maintain a steady heart rate in patients who have suffered a mild heart attack. Let X denote the number of heartbeats per minute obtained per patient. Consider the hypothetical density given in Table 3.5. What is the average heart rate obtained by all patients receiving this drug? That is, What is $E[X]$? By Definition 3.3.1,

$$\begin{aligned} E[X] &= \sum_{\text{all } x} H(x)f(x) \\ &= \sum_{\text{all } x} xf(x) \\ &= 40(.01) + 60(.04) + 68(.05) + \cdots + 100(.01) \\ &= 70 \end{aligned}$$

Since the number of possible values for X is finite, $\sum_{\text{all } x}|x|f(x)$ exists. Thus we can say that the *average* heart rate obtained by patients using this drug is 70 heartbeats per minute. Intuitively, we should have expected this result. Notice the symmetry of the density. In the long run we would expect as many patients with heart rates of 100 as with heart rates of 40; as many with a rate of 60 as with a rate of 80. Similarly, the rates of 68 and 72 occur with the same frequency. Each of these pairs averages to 70, the value obtained by the remaining 80% of the patients. Common sense points to 70 as the expected value for X.

TABLE 3.5

x	40	60	68	70	72	80	100
$f(x)$	.01	.04	.05	.80	.05	.04	.01

When used in a statistical context, the expected value of a random variable X is referred to as its *mean* and is denoted by μ or μ_X. That is, the terms *expected value* and *mean* are interchangeable, as are the symbols $E[X]$ and μ. The mean can be thought of as a measure of the "center of location" in the sense that it indicates where the "center" of the density lies. For this reason, the mean is often referred to as a "location" parameter.

There are three rules for handling expected values that are useful in justifying statistical procedures in later chapters. These rules hold for both continuous and discrete random variables. The rules are stated and illustrated here. We outline the proofs of the first two as exercises; the proof of rule 3 must be deferred until Chap. 5.

Theorem 3.3.1 (Rules for expectation). Let X and Y be random variables and let c be any real number.

1. $E[c] = c$ (The expected value of any constant is that constant.)
2. $E[cX] = cE[X]$ (Constants can be factored from expectations.)
3. $E[X + Y] = E[X] + E[Y]$ (The expected value of a sum is equal to the sum of the expected values.)

Example 3.3.2. Let X and Y be random variables with $E[X] = 7$ and $E[Y] = -5$. Then

$$\begin{aligned} E[4X - 2Y + 6] &= E[4X] + E[-2Y] + E[6] && \text{Rule 3} \\ &= 4E[X] + (-2)E[Y] + E[6] && \text{Rule 2} \\ &= 4E[X] - 2E[Y] + 6 && \text{Rule 1} \\ &= 4(7) - 2(-5) + 6 \\ &= 44 \end{aligned}$$

Variance and Standard Deviation

Knowledge of the mean of a random variable is important, but this knowledge *alone* can be misleading. The next example should show you the problem.

Example 3.3.3. Suppose that we wish to compare a new drug to that of Example 3.3.1. Let X denote the number of heartbeats per minute obtained using the old drug and Y the number per minute obtained with the new drug. The hypothetical density of each of these variables is given in Table 3.6. Since each of the densities is symmetric, inspection shows that $\mu_X = \mu_Y = 70$. Each drug produces *on the average* the same number of heartbeats per minute. However, there is obviously a drastic difference between the two drugs that is not being detected by the mean. The old drug produces fairly consistent reactions in patients, with 90% differing from the mean by at most 2; very few (2%) have an extreme reaction to the drug. However, the new drug produces highly diverse responses. Only 10% of the patients have heart rates within 2 units of the mean, whereas 80% show an extreme reaction. If we examined only the mean, we would conclude that the two drugs had identical effects—but nothing could be further from the truth!

TABLE 3.6

x	40	60	68	70	72	80	100
$f(x)$	.01	.04	.05	.80	.05	.04	.01

y	40	60	68	70	72	80	100
$f(y)$	.40	.05	.04	.02	.04	.05	.40

It is obvious from Example 3.3.3 that something is not being measured by the mean. That something is *variability*. We must find a parameter that reflects consistency or the lack of it. We want the measure to assume a large positive value if the random variable fluctuates in the sense that it often assumes values far from its mean; the measure should assume a small positive value if the values of X tend to cluster closely about the mean. There are several ways to define such a measure. The most widely used is the *variance*.

Definition 3.3.2 (Variance). Let X be a random variable with mean μ. The variance of X, denoted by Var X, or σ^2, is given by

$$\text{Var } X = \sigma^2 = E\left[(X - \mu)^2\right]$$

Note that the variance measures variability by considering $X - \mu$, the difference between the variable and its mean. The difference is squared so that negative values will not cancel positive ones in the process of finding the expected value. When expressed in the form $E[(X - \mu)^2]$, it is easy to see that σ^2 has the properties that we want. When the variable X often assumes values far from μ, σ^2 will be a large positive number; when the values of X tend to fall close to μ, σ^2 will assume a small positive value. Figure 3.1 illustrates the idea.

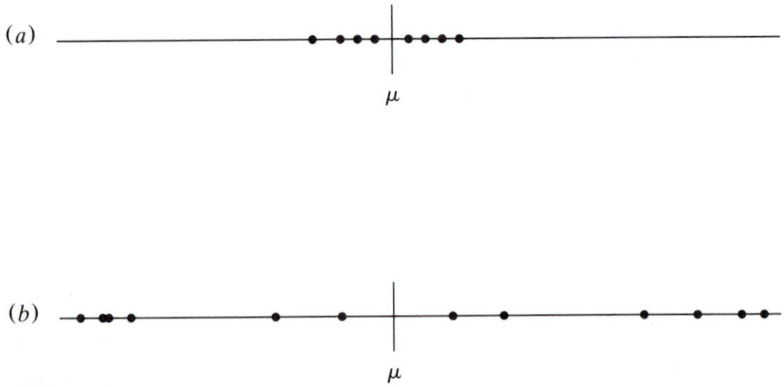

FIGURE 3.1
(*a*) A distribution with a small variance. Most of the data points, denoted by dots, lie fairly close to the average value, μ. Hence most of the differences, $x - \mu$, will be small; (*b*) a distribution with a large variance. Many of the data points lie far from the average value, μ.

Usually, the definition of σ^2 is not used to compute the variance. Rather, we use an alternative form which is given in the following theorem.

Theorem 3.3.2 (Computational formula for σ^2)

$$\sigma^2 = \operatorname{Var} X = E[X^2] - (E[X])^2$$

Proof. By definition

$$\begin{aligned}\operatorname{Var} X &= E\left[(X-\mu)^2\right] \\ &= E[X^2 - 2\mu X + \mu^2]\end{aligned}$$

Using the rules of expectation, Theorem 3.3.1, we obtain

$$\operatorname{Var} X = E[X^2] - 2\mu E[X] + \mu^2$$

Since the symbols μ and $E[X]$ are interchangeable,

$$\begin{aligned}\operatorname{Var} X &= E[X^2] - 2(E[X])^2 + (E[X])^2 \\ &= E[X^2] - (E[X])^2\end{aligned}$$

We illustrate the theorem by computing the variance of each of the random variables of Example 3.3.3.

Example 3.3.4. To find σ_X^2 and σ_Y^2 for the variables of Example 3.3.3, we first use Table 3.6 to find $E[X^2]$ and $E[Y^2]$. We know that $E[X] = E[Y] = 70$.

$$\begin{aligned}E[X^2] &= \sum_{\text{all } x} x^2 f(x) \\ &= (40^2)(.01) + (60^2)(.04) + \cdots + (100^2)(.01) \\ &= 4926.4 \\ E[Y^2] &= \sum_{\text{all } y} y^2 f(y) \\ &= (40^2)(.40) + (60^2)(.05) + \cdots + (100^2)(.40) \\ &= 5630.32\end{aligned}$$

By Theorem 3.3.2,

$$\begin{aligned}\operatorname{Var} X &= E[X^2] - (E[X])^2 \\ &= 4926.4 - 70^2 = 26.4 \\ \operatorname{Var} Y &= E[Y^2] - (E[Y])^2 \\ &= 5630.32 - 70^2 = 730.32\end{aligned}$$

As expected, $\operatorname{Var} Y > \operatorname{Var} X$. Even though the drugs produce the same mean number of heartbeats per minute, they do not behave in the same way. The new drug is not as consistent in its effect as the old.

Note that the variance of a random variable reported alone is not very informative. Is a variance of 26.4 large or small? Only when this value is compared to the variance of a similar variable does it take on meaning. Hence variances are used often for comparative purposes to choose between two variables which otherwise appear to be identical. Also, note that the variance of a random variable is essentially a pure number whose associated units are often physically meaningless. When this occurs, the unit can be omitted. For example, the unit associated with the variance of Example 3.3.4 is a "squared heartbeat." This makes little sense, so in this case variance can be reported with no unit attached. To overcome this problem, a second measure of variability is employed. This measure is the nonnegative square root of the variance, and it is called the *standard deviation*. It has the advantage of having associated with it the same units as the original data.

Definition 3.3.3 (Standard deviation). Let X be a random variable with variance σ^2. The standard deviation of X, denoted by σ, is given by

$$\sigma = \sqrt{\text{Var } X} = \sqrt{\sigma^2}$$

Example 3.3.5. The standard deviations of variables X and Y of Example 3.3.4 are, respectively,

$$\sigma_X = \sqrt{\text{Var } X} = \sqrt{26.4} = 5.14 \text{ heartbeats per minute}$$

$$\sigma_Y = \sqrt{\text{Var } Y} = \sqrt{730.32} = 27.02 \text{ heartbeats per minute}$$

Just as there are three rules for expectation that help in simplifying complex expressions, there are three rules for variance. These rules parallel those for expectation. Rules 1 and 2 can be proved by using the rules for expectation (see Exercise 20). The proof of rule 3 must be deferred until the notion of "independent random variables" has been formalized.

Theorem 3.3.3 (Rules for variance). Let X and Y be random variables and c any real number. Then

1. $\text{Var } c = 0$
2. $\text{Var } cX = c^2 \text{ Var } X$
3. If X and Y are independent, then $\text{Var}(X + Y) = \text{Var } X + \text{Var } Y$

(Two variables are independent if knowledge of the value assumed by one gives no clue to the value assumed by the other.)

Example 3.3.6. Let X and Y be independent with $\sigma_X^2 = 9$ and $\sigma_Y^2 = 3$. Then

$$\begin{aligned}
\text{Var}[4X - 2Y + 6] &= \text{Var}[4X] + \text{Var}[-2Y] + \text{Var } 6 && \text{Rule 3} \\
&= 16 \text{ Var } X + 4 \text{ Var } Y + \text{Var } 6 && \text{Rule 2} \\
&= 16 \text{ Var } X + 4 \text{ Var } Y + 0 && \text{Rule 1} \\
&= 16(9) + 4(3) = 156
\end{aligned}$$

In this section we discussed three *theoretical* parameters associated with a random variable X. We showed not only how to determine their numerical values from knowledge of the density, but also how to interpret them physically. Keep these things in mind, for they play a major role in the study of statistical methods for analyzing experimental data.

3.4 MOMENT GENERATING FUNCTION AND THE GEOMETRIC DISTRIBUTION

Thus far we have considered properties common to all discrete random variables. We now turn our attention to the discussion of some particular types of discrete random variables that arise in the physical world. The variables form "families" in the sense that each member of the family is characterized by a density function of the same mathematical form, differing only with respect to the numerical value of some pertinent parameter(s).

Geometric Distribution

We begin by considering the family of *geometric* random variables. As you shall see, you have already encountered some random variables of this type even though the name "geometric random variable" was not mentioned at the time.

Geometric random variables arise in practice in experiments characterized by these properties:

Geometric properties

1. The experiment consists of a series of trials. The outcome of each trial can be classed as being either a "success" (s) or a "failure" (f). A trial with this property is called a *Bernoulli* trial.
2. The trials are identical and independent in the sense that the outcome of one trial has no effect on the outcome of any other. The probability of success, p, remains the same from trial to trial.
3. The random variable X denotes the number of trials needed to obtain the first success.

The sample space for an experiment such as that just described is

$$S = \{s, fs, ffs, fffs, \dots\}$$

Since the random variable X denotes the number of trials needed to obtain the first success, X assumes the values $1, 2, 3, 4, \dots$. To find the density for X, we look for a pattern. Note that

$$P[X = 1] = P[\text{success on first trial}] = p$$

$$P[X = 2] = P[\text{fail on first trial and succeed on second trial}]$$

Since the trials are independent, the latter probability can be found by multiplying. That is,

$$P[X = 2] = P[\text{fail on first trial and succeed on second trial}]$$

TABLE 3.7

x	1	2	3	4	5	$\cdots$
$f(x)$	p	$(1-p)p$	$(1-p)^2p$	$(1-p)^3p$	$(1-p)^4p$	$\cdots$

$$= P[\text{fail on first trial}]P[\text{succeed on second trial}]$$

$$= (1-p)(p)$$

Similarly,

$$P[X=3] = P[\text{fail on first trial and fail on second trial and succeed on third trial}]$$

$$= (1-p)(1-p)(p) = (1-p)^2 p$$

You should be able to see that the density for X is given by Table 3.7. As you can see, the probabilities given in row 2 of the table exhibit a definite pattern. This pattern can be expressed in closed form as

$$f(x) = (1-p)^{x-1}p \qquad x = 1, 2, 3, \ldots$$

We now define a geometric random variable as being any random variable with a density of this form.

Definition 3.4.1 (Geometric distribution). A random variable X is said to have a geometric distribution with parameter p if its density f is given by

$$f(x) = (1-p)^{x-1}p \qquad 0 < p < 1$$
$$x = 1, 2, 3, \ldots$$

The function f given in this definition is a density. It is obviously nonnegative. Furthermore,

$$\sum_{x=1}^{\infty} (1-p)^{x-1}p$$

is a geometric series with first term $a = p$ and common ratio $r = (1-p)$. Thus the series sums to

$$\frac{a}{1-r} = \frac{p}{1-(1-p)} = 1$$

as desired. From this argument the reason for the name "geometric" distribution should be apparent.

In Exercise 26 you are asked to verify that the general expression for the cumulative distributions function for a geometric random variable is

$$F(x) = 1 - q^{[x]}$$

where q is the probability of failure and $[x]$ is the greatest integer less than or equal to x.

Example 3.4.1. Random digits are integers selected from among $\{0, 1, 2, 3, 4, 5, 6, 7, 8, 9\}$ one at a time in such a way that at each stage in the selection process the integer chosen is just as likely to be one digit as any other. In simulation experiments it is often necessary to generate a series of random digits. This can be done in a number of ways, the most common being by means of a computerized random number generator. In generating such a series, let X denote the number of trials needed to obtain the first zero. This experiment consists of a series of independent, identical trials with "success" being the generation of a zero. The probability of success is $p = 1/10$. Since X denotes the number of trials needed to obtain the first success, X is a geometric random variable. Its density is found by substituting the value $1/10$ for p in the expression for f given in Definition 3.4.1. That is,

$$f(x) = (1-p)^{x-1}p \qquad x = 1, 2, 3, \ldots$$

or

$$f(x) = (9/10)^{x-1}1/10 \qquad x = 1, 2, 3, \ldots$$

The cumulative distribution function for X is given by

$$F(x) = 1 - (.9)^{[x]}$$

Finding the mean of a geometric random variable from the definition is tricky! Consider the next example.

Example 3.4.2. Let us find the mean of the random variable X, the number of trials needed to obtain a zero when generating a series of random digits. By Definition 3.3.1,

$$\mu = E[X] = \sum_{x=1}^{\infty} xf(x)$$

$$= \sum_{x=1}^{\infty} x(9/10)^{x-1}1/10$$

That is,

$$E[X] = 1/10 + 18/100 + 243/1000 + 2916/10{,}000 + \cdots$$

This series is not geometric. Consider the series $(9/10)E[X]$.

$$(9/10)E[X] = 9/100 + 162/1000 + 2187/10{,}000 + 26{,}244/100{,}000 + \cdots$$

Subtracting the latter from the former, we obtain

$$(1/10)E[X] = 1/10 + 9/100 + 81/1000 + 729/10{,}000 + \cdots$$

This series is geometric with first term $1/10$ and common ratio $9/10$. Thus

$$(1/10)E[X] = \frac{1/10}{1 - 9/10} = 1$$

or

$$E[X] = \frac{1}{1/10} = 10$$

Moment Generating Function

As we have seen, the two expectations $E[X]$ and $E[X^2]$ are very useful as they allow us to find the mean and variance of the random variable. These, and other expectations of the form $E[X^k]$ for k a positive integer, are called *ordinary moments*. Thus $E[X] = \mu$ is the first ordinary moment for X; $E[X^2]$ is its second ordinary moment. The preceding example shows that finding ordinary moments, even the first moment, from the definition of expectation is not always easy. Fortunately, it is often possible to obtain a function, called the *moment generating function*, which will enable us to find these moments with less effort.

Definition 3.4.2 (Moment generating function). Let X be a random variable with density f. The moment generating function for X (m.g.f.) is denoted by $m_X(t)$ and is given by

$$m_X(t) = E[e^{tX}]$$

provided this expectation is finite for all real numbers t in some open interval $(-h, h)$.

Since each geometric random variable has a density of the same general form, it is possible to find a general expression for the moment generating function for such a variable. This expression is given in Theorem 3.4.1.

Theorem 3.4.1 (Geometric moment generating function). Let X be a geometric random variable with parameter p. The moment generating function for X is given by

$$m_X(t) = \frac{pe^t}{1 - qe^t} \qquad t < -\ln q$$

where $q = 1 - p$.

Proof. The density for X is given by

$$f(x) = q^{x-1}p \qquad x = 1, 2, 3, \ldots$$

By definition

$$\begin{aligned} m_X &= E[e^{tX}] \\ &= \sum_{\text{all } x} e^{tx} f(x) \\ &= \sum_{x=1}^{\infty} e^{tx} q^{x-1} p \\ &= pq^{-1} \sum_{x=1}^{\infty} (qe^t)^x \end{aligned}$$

The series on the right is a geometric series with first term qe^t and common ratio qe^t. Thus

$$m_X(t) = pq^{-1}\left(\frac{qe^t}{1 - qe^t}\right)$$

$$= \frac{pe^t}{1 - qe^t}$$

provided $|r| = |qe^t| < 1$. Since the exponential function is nonnegative and $0 < q < 1$, this restriction implies that $qe^t < 1$. The inequality is solved for t as follows:

$$qe^t < 1$$

$$e^t < 1/q$$

$$\ln e^t < \ln 1/q$$

$$t < \ln 1 - \ln q$$

$$t < -\ln q$$

The following theorem shows how the moment generating function can be used to generate ordinary moments for a random variable X.

Theorem 3.4.2. Let $m_X(t)$ be the moment generating function for a random variable X. Then

$$\left.\frac{d^k m_X(t)}{dt^k}\right|_{t=0} = E[X^k]$$

Proof. Recall from elementary calculus that the Maclaurin series expansion for e^z is

$$e^z = 1 + z + z^2/2! + z^3/3! + z^4/4! + \cdots$$

Letting $z = tX$, the Maclaurin series expansion for e^{tX} is

$$e^{tX} = 1 + tX + (tX)^2/2! + (tX)^3/3! + (tX)^4/4! + \cdots$$

By taking the expected value of each side of this equation, we obtain

$$m_X(t) = E[e^{tX}] = E[1 + tX + t^2X^2/2! + t^3X^3/3! + t^4X^4/4! + \cdots]$$

$$= 1 + tE[X] + t^2/2!E[X^2] + t^3/3!E[X^3]$$

$$+ t^4/4!E[X^4] + \cdots$$

Differentiating this series term by term with respect to t, we see that

$$\frac{dm_X(t)}{dt} = E[X] + tE[X^2] + t^2/2!E[X^3] + t^3/3!E[X^4] + \cdots$$

When this derivative is evaluated at $t = 0$, every term except the first becomes 0.

Hence

$$\left.\frac{dm_X(t)}{dt}\right|_{t=0} = E[X]$$

Taking the second derivative of $m_X(t)$, we obtain

$$\frac{d^2m_X(t)}{dt^2} = E[X^2] + tE[X^3] + t^2/2!E[X^4] + \cdots$$

Evaluating this derivative at $t = 0$ yields

$$\left.\frac{d^2m_X(t)}{dt^2}\right|_{t=0} = E[X^2]$$

This procedure can be continued to show that

$$\left.\frac{d^km_X(t)}{dt^k}\right|_{t=0} = E[X^k]$$

for any positive integer k as desired.

Let us use the moment generating function to find a general expression for the mean and variance of a geometric distribution with parameter p.

Theorem 3.4.3. Let X be a geometric random variable with parameter p. Then

$$E[X] = 1/p \qquad \text{and} \qquad \operatorname{Var} X = q/p^2$$

Proof. For a geometric random variable with parameter p

$$m_X(t) = \frac{pe^t}{1 - qe^t}$$

$$\frac{dm_X(t)}{dt} = \frac{(1 - qe^t)pe^t + pe^tqe^t}{(1 - qe^t)^2}$$

$$= \frac{pe^t}{(1 - qe^t)^2}$$

Evaluating this derivative at $t = 0$, we obtain

$$E[X] = \left.\frac{dm_X(t)}{dt}\right|_{t=0} = \frac{p}{(1 - q)^2}$$

$$= p/p^2$$

$$= 1/p$$

Taking the second derivative of $m_X(t)$, we obtain

$$\frac{d^2 m_X(t)}{dt^2} = \frac{(1-qe^t)^2 pe^t + 2pe^t(1-qe^t)qe^t}{(1-qe^t)^4}$$

$$= \frac{pe^t(1-qe^t)[(1-qe^t)+2qe^t]}{(1-qe^t)^4}$$

$$= \frac{pe^t(1+qe^t)}{(1-qe^t)^3}$$

Evaluating this derivative at $t = 0$, we see that

$$E[X^2] = \left.\frac{d^2 m_X(t)}{dt^2}\right|_{t=0} = \frac{p(1+q)}{(1-q)^3} = \frac{(1+q)}{p^2}$$

Now

$$\text{Var } X = E[X^2] - (E[X])^2$$

$$= \frac{1+q}{p^2} - \frac{1}{p^2}$$

$$= \frac{q}{p^2}$$

We illustrate the use of these theorems by finding the moment generating function, mean, and variance for the random variable of Example 3.4.1.

Example 3.4.3. Consider the random variable X, the number of trials needed to obtain the first zero when generating a series of random digits. Since this random variable is geometric with parameter $p = 1/10$,

$$m_X(t) = \frac{pe^t}{1-qe^t} = \frac{(1/10)e^t}{1-(9/10)e^t}$$

$$\mu = E[X] = 1/p = 10$$

$$\sigma^2 = \text{Var } X = q/p^2 = \frac{9/10}{(1/10)^2} = 90$$

Note that this value for μ agrees with that obtained in Example 3.4.2.

The importance of the moment generating function for a random variable is not completely evident at this time. It does give us a way to find general expressions for the mean and variance as well as for the ordinary moments of an entire family of random variables. As we shall see later, the moment generating function, when it exists, serves as a fingerprint that completely identifies the random variable under study. This function will be used extensively in the remainder of this text.

Forgetfulness Property: Geometric

Example 3.4.4 (Forgetfulness Property of the Geometric Distribution). The geometric distribution has an interesting property, known as the "forgetfulness" property. To understand what this means, consider the following setting. Suppose that microprocessor chips are manufactured in such a way that 1 in every 20 is defective in some way. Thus for each chip the probability of a defect is .05. Chips are inspected one by one and X denotes the number of chips examined in order to find the first defective one. As described, X follows a geometric distribution with $p = .05$ and $E[X] = 20$. Suppose that 15 chips have already been inspected and no defects have been found. Thus we are given that $X \geq 16$. We now ask, "What is the probability that at least 5 more chips will be inspected before observing the first defect?" If this occurs, $X \geq 20$. We are being asked a conditional question. Namely, we want to find $P[X \geq 20 | X \geq 16]$. This probability is found as follows:

$$P[X \geq 20 | X \geq 16] = \frac{P[X \geq 20 \text{ and } X \geq 16]}{P[X \geq 16]}$$

$$= \frac{P[X \geq 20]}{P[X \geq 16]}$$

Now

$$P[X \geq 20] = 1 - P[X \leq 19] = 1 - F(19)$$

and

$$P[X \geq 16] = 1 - P[X \leq 15] = 1 - F(15)$$

Since $F(x) = 1 - q^x = 1 - (.95)^x$,

$$P[X \geq 20] = 1 - (1 - .95^{19})$$

and

$$P[X \geq 16] = 1 - (1 - .95^{15})$$

Division yields

$$P[X \geq 20 | X \geq 16] = \frac{.95^{19}}{.95^{15}} = .95^4$$

Where does the "forgetfulness" come into play? Notice that if we simply ask the question, "What is the probability that at least 5 chips must be inspected to find a defect?" the solution is as follows:

$$P[X \geq 5] = 1 - P[X \leq 4] = 1 - F(4)$$

Since $F[4] = 1 - .95^4$,

$$P[X \geq 5] = 1 - (1 - .95^4) = .95^4$$

This is the same result as that obtained earlier. The probability of having to inspect at least 5 chips *given that* 15 have already been tested is the same as the probability of having to inspect at least 5 chips at the start of the inspection process. The process has "forgotten" the fact that 15 chips have already been examined. At each stage of the game the process is essentially starting fresh. This result is somewhat counterintuitive. Since we know that $E[X] = 20$ and we have already seen 15 nondefective chips, it seems as though a defect is "due." However, the above computation shows that this is not the case.

Some distributions exhibit the forgetfulness property; others do not. To test for the property, we ask, "Is $P[X \geq x_1 + x_2 | X \geq x_1] = P[X \geq x_2]$? If so, the forgetfulness property holds; otherwise it does not.

3.5 BINOMIAL DISTRIBUTION

The next distribution to be studied is the *binomial* distribution. Once again, you have already seen some binomial random variables even though they were not labeled as such at the time. The theoretical basis for working with this distribution is the binomial theorem presented in most beginning algebra courses. The statement of this theorem is as follows:

Binomial theorem

For any two real numbers a and b and any positive integer n,

$$(a + b)^n = \sum_{k=0}^{n} \binom{n}{k} a^k b^{n-k}$$

where $\binom{n}{k}$ is given by $\dfrac{n!}{k!(n-k)!}$.

To recognize a situation that involves a binomial random variable, you must be familiar with the assumptions that underlie this distribution. These are given below:

Binomial properties

1. The experiment consists of a *fixed* number, n, of Bernoulli trials, trials that result in either a "success" (s) or a "failure" (f).
2. The trials are identical and independent and therefore the probability of success, p, remains the same from trial to trial.
3. The random variable X denotes the number of successes obtained in the n trials.

Once we realize that the binomial model is appropriate from the physical description of the experiment, we shall want to describe the behavior of the binomial random variable involved. To do so, we need to consider the density for the random variable. To get an idea of the general form for the binomial density, let us consider the case in which $n = 3$. The sample space for such an experiment is

$$S = \{fff, sff, fsf, ffs, ssf, sfs, fss, sss\}$$

Since the trials are independent, the probability assigned to each sample point is found by multiplying. For example, the probabilities assigned to the sample points *fff* and *sff* are $(1-p)(1-p)(1-p)=(1-p)^3$ and $p(1-p)(1-p)=p(1-p)^2$, respectively. The random variable X assumes the value 0 only if the experiment results in the outcome *fff*. That is,

$$P[X=0]=(1-p)^3$$

However, X assumes the value 1 if the experiment results in any one of the outcomes *sff*, *fsf*, or *ffs*. Thus

$$P[X=1]=3\cdot p(1-p)^2$$

Similarly,

$$P[X=2]=3\cdot p^2(1-p)$$

and

$$P[X=3]=p^3$$

It is evident that for $x=0,1,2,3$

$$P[X=x]=c(x)p^x(1-p)^{3-x}$$

where $c(x)$ denotes the number of sample points that correspond to x successes. Such a sample point is expressed as a permutation of three letters with x of these being s's and the rest, $3-x$, of these being f's. Using the formula for the number of permutations of indistinguishable objects studied in Chap. 1, we see that

$$c(x)=\frac{3!}{x!(3-x)!}=\binom{3}{x}$$

Thus the density for this binomial random variable is given by

$$f(x)=\binom{3}{x}p^x(1-p)^{3-x}\qquad x=0,1,2,3$$

To generalize this idea to n trials, we replace 3 by n to obtain the expression

$$f(x)=\binom{n}{x}p^x(1-p)^{n-x}\qquad x=0,1,2,\ldots,n$$

This suggests the formal definition of the binomial distribution.

Definition 3.5.1 (Binomial distribution). A random variable X has a binomial distribution with parameters n and p if its density is given by

$$f(x)=\binom{n}{x}p^x(1-p)^{n-x}\qquad x=0,1,2,\ldots,n$$
$$0<p<1$$

where n is a positive integer.

To see that the function given in this definition is a density, note that it is nonnegative. Furthermore, by applying the binomial theorem with $k=x$, $a=p$, and $b = 1 - p$ it can be seen that

$$\sum_{x=0}^{n} \binom{n}{x} p^x (1-p)^{n-x} = [p + (1-p)]^n = 1$$

as desired.

Example 3.5.1. Recent studies of German air traffic controllers have shown that it is difficult to maintain accuracy when working for long periods of time on data display screens. A surprising aspect of the study is that the ability to detect spots on a radar screen decreases as their appearance becomes too rare. The probability of correctly identifying a signal is approximately .9 when 100 signals arrive per 30-minute period. This probability drops to .5 when only 10 signals arrive at random over a 30-minute period. The hypothesis is that unstimulated minds tend to wander. Let X denote the number of signals correctly identified in a 30-minute time span in which 10 signals arrive. This experiment consists of a series of $n = 10$ independent and identical Bernoulli trials with "success" being the correct identification of a signal. The probability of success is $p = 1/2$. Since X denotes the number of successes in a fixed number of trials, X is binomial. Its density is found by letting $n = 10$ and $p = 1/2$ in the expression for f given in Definition 3.5.1. That is,

$$f(x) = \binom{n}{x} p^x (1-p)^{n-x} \qquad x = 0, 1, 2, \ldots, n$$

or

$$f(x) = \binom{10}{x} (1/2)^x (1/2)^{10-x} \qquad x = 0, 1, 2, \ldots, 10$$

(Based on a study reported in "Human Aspects of Quality Assurances" by W. E. Masing, *Quality Assurance*, vol. 8, no. 2, June 1982, p. 35.)

The next theorem summarizes other theoretical properties of the binomial distribution. Its proof is left as an exercise (Exercise 43).

Theorem 3.5.1. Let X be a binomial random variable with parameters n and p.

1. The moment generating function for X is given by

$$m_X(t) = (q + pe^t)^n \qquad q = 1 - p$$

2. $E[X] = \mu = np$
3. $\operatorname{Var} X = \sigma^2 = npq$

Example 3.5.2. The random variable X, the number of radar signals properly identified in a 30-minute period, is a binomial random variable with parameters $n = 10$ and $p = 1/2$. The moment generating function for this random variable is

$$m_X(t) = (1/2 + 1/2e^t)^{10}$$

Its mean is $\mu = np = 10(1/2) = 5$ and its variance is $\sigma^2 = npq = 10(1/2)(1/2) = 10/4$.

In statistical studies we shall usually be interested in computing the probability that the random variable assumes certain values. This probability can be computed from the density function, f, or from the cumulative distribution function, F. Since the binomial distribution comes into play in such a wide variety of physical applications, tables of the cumulative distribution function for selected values of n and p have been compiled. Table I of App. A is one such table. That is, Table I gives the values of

$$F(t) = \sum_{x=0}^{[t]} \binom{n}{x} p^x (1-p)^{n-x}$$

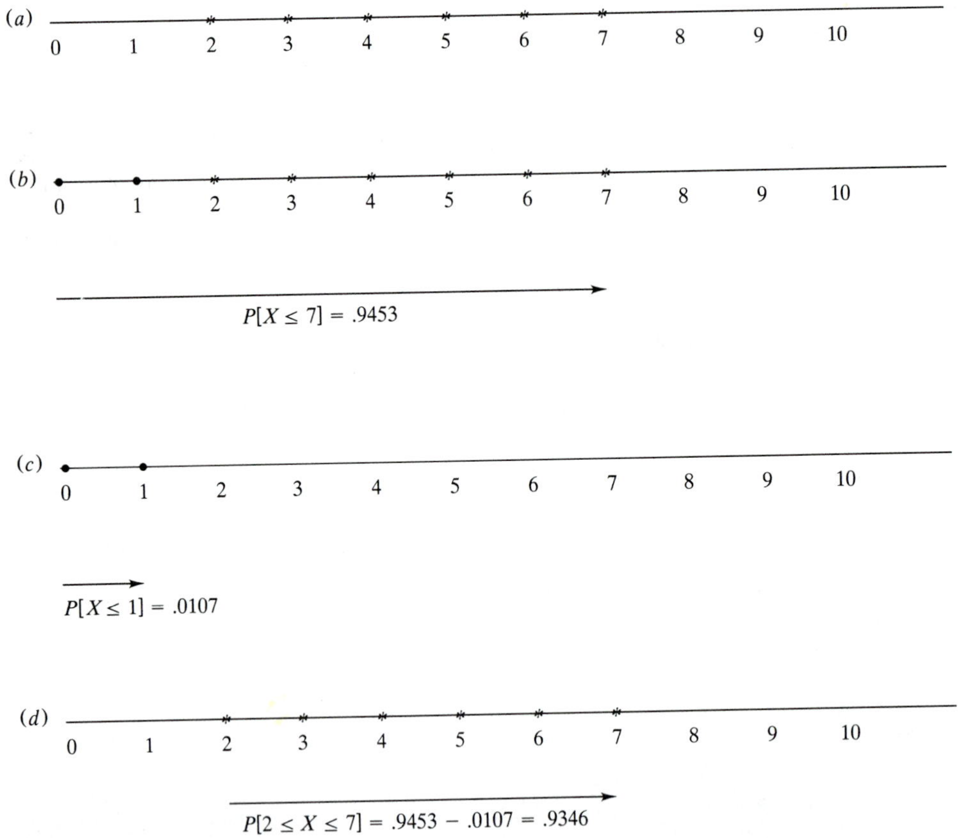

FIGURE 3.2
(a) The probability that X lies between 2 and 7 inclusive is the probability associated with the starred points; (b) $P[X \le 7] = .9453$ includes the probability associated with 0 and 1; (c) the probability associated with the unwanted points 0 and 1 is .0107; (d) the desired probability is found by subtraction.

for selected values of n and p, where $[t]$ represents the greatest integer less than or equal to t. Its use is illustrated in the following example.

Example 3.5.3. Let X denote the number of radar signals properly identified in a 30-minute time period in which 10 signals are received. Assuming that X is binomial with $n = 10$ and $p = 1/2$, find the probability that at most seven signals will be identified correctly. This probability can be found by summing the density from $x = 0$ to $x = 7$. That is,

$$P[X \leq 7] = \sum_{x=0}^{7} \binom{10}{x} (1/2)^x (1/2)^{10-x}$$

Evaluating this probability directly entails a large amount of arithmetic. However, its value can be read from Table I of App. A. We first look at the group of values labeled $n = 10$. The desired probability of .9453 is found in the column labeled .5 and the row labeled 7. That is,

$$P[X \leq 7] = F(7) = .9453$$

Other probabilities can be found. For example, find $P[2 \leq X \leq 7]$. Figure 3.2 suggests how this is done. Notice that in Fig. 3.2 we want the probability associated with points that are starred. To determine the desired probability, we first find the number 7 in Table I of App. A. Since the table is cumulative, the probability given, .9453, is the probability that X is at most 7. This probability includes the probability that $X = 0$ or $X = 1$. Since we did not want to include those values, $P[X \leq 1] = F(1) = .0107$ must be subtracted from .9453. Thus

$$\begin{aligned} P[2 \leq X \leq 7] &= P[X \leq 7] - P[X < 2] \\ &= P[X \leq 7] - P[X \leq 1] \\ &= F(7) - F(1) \\ &= .9453 - .0107 \\ &= .9346 \end{aligned}$$

Later in the text we shall show ways of approximating binomial probabilities when the values of n and p are such that no appropriate binomial table is available.

3.6 NEGATIVE BINOMIAL DISTRIBUTION (OPTIONAL)

The negative binomial distribution is a distribution that can be thought of as a "reversal" of the binomial distribution. In the binomial setting the random variable X represents the number of successes obtained in a series of n independent and identical Bernoulli trials; the number of trials is *fixed* and the number of successes will *vary* from experiment to experiment. The negative binomial random variable represents the number of trials needed to obtain exactly r successes; here, the number of successes is *fixed* and the number of trials will *vary* from experiment to experiment. In particular, the negative binomial random variable arises in situations characterized by these properties:

Negative binomial properties

1. The experiment consists of a series of independent and identical Bernoulli trials, each with probability p of success.
2. The trials are observed until exactly r successes are obtained, where r is fixed by the experimenter.
3. The random variable X is the number of trials needed to obtain the r successes.

It is not hard to derive the density function for X. To do so, let us consider a setting in which $r = 3$. Typical outcomes for such an experiment are

$$ssffffs \quad sffffss \quad ffffsss \quad sss \quad ssfs$$

Here X assumes the values 7, 7, 7, 3, and 4, respectively. There are several things to notice immediately. First, each outcome must end with a successful trial. Second, the remaining $x - 1$ trials must result in exactly two successes and $x - 3$ failures *in some order*. Third, different outcomes can yield identical values for X. To determine the number of outcomes that result in a given value of X, we ask, "How many permutations can be formed consisting of $x - 1$ objects of which exactly two represent success and the rest, $x - 3$, represent failure?" Exercise 23 of Chap. 1 can be applied to see that the answer to this question is $\binom{x-1}{2}$. For example, there are $\binom{6}{2} = 15$ ways in which X can assume the value 7. Three of these outcomes are given above. Since trials are independent with probability p of success and probability $1 - p$ of failure, the probability of an outcome for which $X = x$ is given by

$$P[X = x] = \binom{x-1}{2}(1-p)^{x-3}p^3 \qquad x = 3, 4, 5, \ldots$$

You can use this expression to verify that the probability that $X = 7$ is

$$\binom{6}{2}(1-p)^4 p^3$$

The argument given for $r = 3$ can be generalized easily. We simply replace 3 by r and 2 by $r - 1$ in the argument given to obtain the following definition for the negative binomial random variable.

Definition 3.6.1 (Negative binomial distribution). A random variable X is said to have a negative binomial distribution with parameters p and r if its density f is given by

$$f(x) = \binom{x-1}{r-1}(1-p)^{x-r}p^r \qquad \begin{array}{l} r = 1, 2, 3, \ldots \\ x = r, r+1, r+2, \ldots \end{array}$$

A Taylor series expansion in q where $q = 1 - p$ is used to show that this density sums to 1 as required. The proof is a bit tricky and is outlined in Exercise 45.

Theorem 3.6.1 gives the moment generating function for the negative binomial distribution. Its derivation is also based on a Taylor series expansion. (See Exercises 47 and 48.) The expectations stated in the theorem are obtained from the moment generating function.

Theorem 3.6.1. Let X be a negative binomial random variable with parameters r and p. Then

1. the moment generating function for X is given by

$$m_X(t) = \frac{(pe^t)^r}{(1 - qe^t)^r} \qquad q = 1 - p$$

2. $E[X] = r/p$

3. $\mathrm{Var}(X) = rq/p^2$

An example will illustrate the use of this distribution in a practical setting.

Example 3.6.1. Cotton linters used in the production of rocket propellant are subjected to a nitration process that enables the cotton fibers to go into solution. The process is 90% effective in that the material produced can be shaped as desired in a later processing stage with probability .9. What is the probability that exactly 20 lots will be produced in order to obtain the third defective lot? Here "success" is obtaining a defective lot, and hence $p = .1$ and $r = 3$. The probability that $X = 20$ is given by

$$f(20) = \binom{19}{2}(.9)^{17}(.1)^3$$

The expected value of X is $r/p = 3/.1$ or 30 and the variance of X is $rq/p^2 = 3(.9)/(.1)^2 = 270$. (Based on a study to compare different sources of cotton linters conducted by the Radford University Statistical Consulting Service for the Radford Army Ammunition Plant.)

One other point should be made. When $r = 1$, the negative binomial distribution reduces to the geometric distribution studied earlier. (See Exercise 51.)

3.7 HYPERGEOMETRIC DISTRIBUTION

Sampling from a finite population can be done in one of two ways. An item can be selected, examined, and returned to the population for possible reselection; or it can be selected, examined, and kept, thus preventing its reselection in subsequent draws. The former is called *sampling with replacement*, whereas the

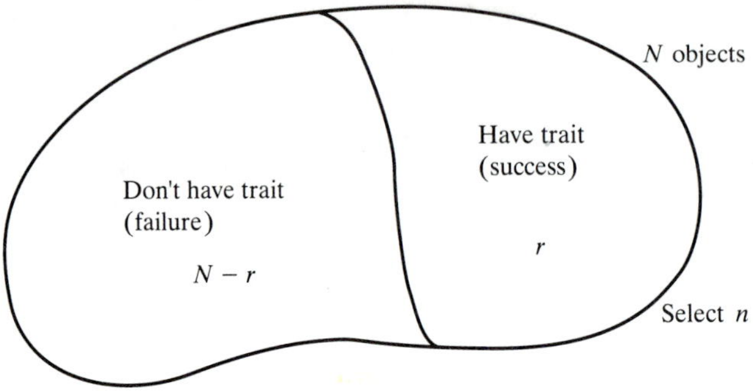

FIGURE 3.3

latter is called *sampling without replacement*. Sampling with replacement guarantees that the draws are independent. In sampling without replacement the draws are *not* independent. Thus if we sample without replacement, the random variable X, the number of successes in n draws, is no longer binomial. Rather, it follows a distribution known as the *hypergeometric distribution*.

To derive the density for this distribution, suppose that we have a group of N objects and that r of these objects have a trait of interest to us. We are to select n objects from the group randomly without replacement. Let X denote the number of objects chosen that have the trait. The idea is depicted in Fig. 3.3. Since we are not interested in the order in which the items are selected, we can use combinatorial techniques to conclude that there are $\binom{N}{n}$ ways to choose the n objects. In a random selection we are just as likely to obtain one set of n objects as any other. That is, there are $\binom{N}{n}$ equally likely ways in which this experiment can proceed. In order to have x successes, we must select exactly x objects from the r objects with the trait of interest. This can be done in $\binom{r}{x}$ ways. We must select the remaining $n - x$ objects from the $N - r$ objects that do not have the trait; this can be done in $\binom{N-r}{n-x}$ ways. Using classical probability and the multiplication rule for counting, we obtain

$$P[X = x] = \frac{\begin{array}{c}\text{number of ways to select } x \text{ objects with the trait}\\ \text{and } n - x \text{ objects without the trait}\end{array}}{\text{number of ways the experiment can proceed}}$$

$$= \frac{\binom{r}{x}\binom{N-r}{n-x}}{\binom{N}{n}}$$

This argument suggests the definition of the hypergeometric distribution.

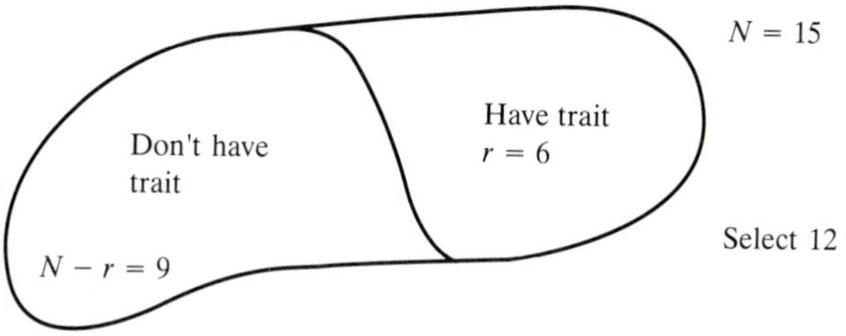

FIGURE 3.4

Definition 3.7.1 (Hypergeometric distribution). A random variable X has a hypergeometric distribution with parameters N, n, and r if its density is given by

$$f(x) = \frac{\binom{r}{x}\binom{N-r}{n-x}}{\binom{N}{n}} \qquad \max[0, n-(N-r)] \le x \le \min(n, r)$$

where N, r, and n are positive integers.

Notice the unusual bounds for X. A simple numerical example should show you why these bounds are as stated.

Example 3.7.1. Suppose that X is hypergeometric with $N = 15$, $r = 6$, and $n = 12$. This situation is depicted in Fig. 3.4. Since only six items have the desired trait, X cannot exceed 6. Note that $6 = \min(n, r) = \min(12, 6)$. Since we can select at most nine items from among those without the trait, we must select at least three items from among those with the trait. Note that

$$3 = \max[0, n-(N-r)] = \max[0, 12-(15-6)] = \max[0, 3]$$

Just be careful when stating the bounds for a hypergeometric random variable. They are tricky! Since the bounds for X are unusual, the theoretical development of the hypergeometric distribution is not easy. However, it can be shown that

$$E[X] = n\left(\frac{r}{N}\right) \qquad \text{and} \qquad \text{Var}\, X = n\left(\frac{r}{N}\right)\left(\frac{N-r}{N}\right)\left(\frac{N-n}{N-1}\right)$$

Example 3.7.2. A foundry ships engine blocks in lots of size 20. Since no manufacturing process is perfect, defective blocks are inevitable. However, to detect the defect, the block must be destroyed. Thus we cannot test each block. Before accepting a lot, three items are selected and tested. Suppose that a given lot actually contains five defective items. Let X denote the number of defective items sampled. The density for X is

$$f(x) = \frac{\binom{5}{x}\binom{15}{3-x}}{\binom{20}{3}} \qquad x = 0, 1, 2, 3$$

The expected number of defective blocks in a sample of size 3 is

$$E[X] = n\left(\frac{r}{N}\right) = 3\left(\frac{5}{20}\right) = \frac{3}{4}$$

The variance for X is

$$\begin{aligned}\operatorname{Var} X &= n\left(\frac{r}{N}\right)\left(\frac{N-r}{N}\right)\left(\frac{N-n}{N-1}\right)\\ &= 3\left(\frac{5}{20}\right)\left(\frac{15}{20}\right)\left(\frac{17}{19}\right)\\ &= \frac{153}{304}\end{aligned}$$

If the number of items sampled (n) is small relative to the number of objects from which the sample is drawn (N), then the binomial distribution can be used to approximate hypergeometric probabilities. A rule of thumb is that the approximation is usually satisfactory if $n/N \le .05$. The proof of this result depends upon Stirling's formula, which is studied in courses in advanced calculus. We shall not attempt the proof here. However, the result should not be surprising. If n is small relative to N, then the composition of the sampled group does not change much from trial to trial even though we are keeping the sampled items. Thus the probability of success is not changing much from trial to trial, and for all practical purposes it can be viewed as being constant. Thus the distribution of X, the number of successes obtained in n draws, can be approximated by the binomial distribution with parameters n and $p = r/N$.

Example 3.7.3. During the course of an hour 1000 bottles of beer are filled by a particular machine. Each hour a sample of 20 bottles is randomly selected and the number of ounces of beer per bottle is checked. Let X denote the number of bottles selected that are underfilled. Suppose that during a particular hour 100 underfilled bottles are produced. Find the probability that at least 3 underfilled bottles will be among those sampled. The exact value of this probability is given by

$$\begin{aligned}P[X \ge 3] &= 1 - P[X < 3]\\ &= 1 - P[X \le 2]\\ &= 1 - P[X = 0] - P[X = 1] - P[X = 2]\\ &= 1 - \frac{\binom{100}{0}\binom{900}{20}}{\binom{1000}{20}} - \frac{\binom{100}{1}\binom{900}{19}}{\binom{1000}{20}} - \frac{\binom{100}{2}\binom{900}{18}}{\binom{1000}{20}} = .3224\end{aligned}$$

As you can see, calculating this probability directly, even with the aid of a calculator, is time-consuming. However, since $n/N = 20/1000 \le .05$, our rule of thumb indicates that this probability can be approximated by using the binomial distribution with parameters $n = 20$ and $p = r/N = 100/1000 = .1$. From Table I of App. A, the cumulative binomial table, we find that

$$\begin{aligned} P[X \geq 3] &= 1 - P[X < 3] \\ &= 1 - P[X \leq 2] \\ &= 1 - .6769 \\ &= .3231 \end{aligned}$$

3.8 POISSON DISTRIBUTION

The last discrete family to be considered is the family of *Poisson* random variables, named for the French mathematician Simeon Denis Poisson (1781–1840). The Maclaurin series expansion for the function e^z studied in beginning calculus courses provides the theoretical basis for this distribution. This series is given by

Maclaurin series

For z a real number,

$$e^z = 1 + z + z^2/2! + z^3/3! + z^4/4! + \cdots$$

We begin by considering the mathematical properties of this important family of random variables.

Definition 3.8.1 (Poisson distribution). A random variable X is said to have a Poisson distribution with parameter k if its density f is given by

$$f(x) = \frac{e^{-k}k^x}{x!} \qquad \begin{array}{l} x = 0, 1, 2, \ldots \\ k > 0 \end{array}$$

The function f given in this definition is nonnegative. To see that it sums to 1, note that

$$\sum_{x=0}^{\infty} \frac{e^{-k}k^x}{x!} = e^{-k}(1 + k + k^2/2! + k^3/3! + \cdots)$$

The series on the right is the Maclaurin series for e^k. Thus

$$\sum_{x=0}^{\infty} \frac{e^{-k}k^x}{x!} = e^{-k}e^k = e^0 = 1$$

as desired.

The moment generating function for this distribution is easy to obtain as is its mean and variance. The following theorem gives these results. Its proof is outlined as an exercise. (Exercise 69.)

Theorem 3.8.1. Let X be a Poisson random variable with parameter k.

1. The moment generating function for X is given by

$$m_X(t) = e^{k(e^t - 1)}$$

2. $E[X] = k$
3. $\operatorname{Var} X = k$

Poisson random variables usually arise in connection with what are called *Poisson processes*. Poisson processes involve observing discrete events in a continuous "interval" of time, length, or space. We use the word "interval" in describing the general Poisson process with the understanding that we may not be dealing with an interval in the usual mathematical sense. For example, we might observe the number of white blood cells in a drop of blood. The discrete event of interest is the observation of a white cell, whereas the continuous "interval" involved is a drop of blood. We might observe the number of times radioactive gases are emitted from a nuclear power plant during a 3-month period. The discrete event of concern is the emission of radioactive gases. The continuous interval consists of a period of 3 months. The variable of interest in a Poisson process is X, the number of occurrences of the event in an interval of length s units. Although the derivation is a bit tricky, it can be shown using differential equations that X is a Poisson random variable with parameter $k = \lambda s$, where λ is a positive number that characterizes the underlying Poisson process. To understand the physical significance of the constant λ, note that by Definition 3.8.1 the density for X is given by

$$f(x) = \frac{e^{-\lambda s}(\lambda s)^x}{x!} \qquad x = 0, 1, 2, 3, \ldots$$

By Theorem 3.8.1 the expected value of X is λs. That is, the average number of occurrences of the event of interest in an interval of s units is λs. Thus the average number of occurrences of the event in 1 unit of time, length, area, or space is $\lambda s/s = \lambda$. That is, physically, the *parameter λ of a Poisson process represents the average number of occurrences of the event in question per measurement unit*.

These concepts are illustrated in Example 3.8.1.

Example 3.8.1. The white blood cell count of a healthy individual can average as low as 6000 per cubic millimeter of blood. To detect a white-cell deficiency, a .001 cubic millimeter drop of blood is taken and the number of white cells X is found. How many white cells are expected in a healthy individual? If at most two are found, is there evidence of a white cell deficiency?

This experiment can be viewed as involving a Poisson process. The discrete event of interest is the occurrence of a white cell; the continuous interval is a drop of blood.

Let the measurement unit be a cubic millimeter; then $s = .001$ and λ, the average number of occurrences of the event per unit, is 6000. Thus X is a Poisson random variable with parameter $\lambda s = 6000(.001) = 6$. By Theorem 3.8.1, $E[X] = \lambda s = 6$. In a healthy individual we would expect, on the average, to see six white cells. How rare is it to see at most two? That is, What is $P[X \leq 2]$? From Definition 3.8.1,

$$P[X \leq 2] = \sum_{x=0}^{2} f(x) = \sum_{x=0}^{2} \frac{e^{-6}6^x}{x!}$$

$$= \frac{e^{-6}6^0}{0!} + \frac{e^{-6}6^1}{1!} + \frac{e^{-6}6^2}{2!}$$

Evaluating this type of expression directly does entail some arithmetic.

Once again, because of the wide appeal of the Poisson model, the values of the cumulative distribution function for selected values of the parameter $k = \lambda s$ are tabulated. Table II of App. A is one such table. The desired probability of .062 is found by looking under the column labeled $k = 6$ in the row labeled 2. Is there evidence of a white-cell deficiency? There are no rules that say at what point probabilities are considered to be small. To answer this question, a value judgment must be made. If you consider .062 to be small, then the natural conclusion is that the individual does have a white-cell deficiency.

Poisson Approximation to the Binomial Distribution

One other important application of the Poisson density should be mentioned. Occasionally one encounters a binomial random variable for which n is very large. If binomial tables do not exist for this value of n, then calculating probabilities associated with the variable becomes cumbersome. It is often possible to approximate these binomial probabilities via an appropriately chosen Poisson distribution. The following theorem, first presented in 1837 by Poisson, shows that as n becomes large and p becomes small, the binomial density approaches that of the Poisson. Its proof is based on the following result usually proved in courses in advanced calculus [30]:

$$\lim_{n \to \infty} \left(1 + \frac{z}{n}\right)^n = e^z \qquad \text{for } z \text{ real}$$

Theorem 3.8.2 (Poisson approximation to the binomial distribution). Let X be binomial with parameters n and p. If the parameter n approaches infinity and p approaches 0 in such a way that np remains constant at some value $k > 0$, then

$$P[X = x] = \binom{n}{x} p^x (1-p)^{n-x} \doteq \frac{e^{-k}k^x}{x!}$$

Proof. Let X be binomial with parameters n and p and assume that $np = k > 0$. By Definition 3.5.1,

$$P[X = x] = \binom{n}{x} p^x (1-p)^{n-x}$$

Let $k = np$ or $p = k/n$, then

$$P[X=x] = \binom{n}{x}\left(\frac{k}{n}\right)^x\left(\frac{1-k}{n}\right)^{n-x}$$

$$= \frac{n!}{x!(n-x)!}\frac{k^x}{n^x}\left(\frac{1-k}{n}\right)^n\frac{1}{(1-k/n)^x}$$

$$= \frac{k^x}{x!}\frac{n(n-1)(n-2)\cdots(n-x+1)(n-x)!}{n\cdot n\cdot n\cdots n(n-x)!}\frac{(1-k/n)^n}{(1-k/n)^x}$$

$$= \frac{k^x}{x!}(1)\left(\frac{1-1}{n}\right)\left(\frac{1-2}{n}\right)\cdots\left(1-\frac{x-1}{n}\right)\frac{(1-k/n)^n}{(1-k/n)^x}$$

Now let n approach infinity to obtain

$$P[X=x] \doteq \frac{\lim_{n\to\infty} k^x}{x!}\frac{(1)(1-1/n)(1-2/n)\cdots\left(1-\dfrac{x-1}{n}\right)}{(1-k/n)^x}\left(\frac{1-k}{n}\right)^n$$

$$= \frac{k^x}{x!\lim_{n\to\infty}}\frac{(1)(1-1/n)(1-2/n)\cdots\left(1-\dfrac{x-1}{n}\right)}{(1-k/n)^x}$$

$$\times \lim_{n\to\infty}\left[1+\left(-\frac{k}{n}\right)\right]^n$$

$$= \frac{e^{-k}k^x}{x!}$$

Note that this theorem states that the desired binomial probability is approximately equal to a Poisson probability with the Poisson parameter $k = n\cdot p$. This approximation is usually good if $n \geq 20$ and $p \leq .05$ and very good if $n \geq 100$ and $np \leq 10$. Since the approximation is used when n is large and p is small, the Poisson distribution is often called the distribution of "rare" events. Its use is illustrated in the next example.

Example 3.8.2. In the manufacture of electronic circuits, ceramic plates are drilled to provide pathways called "vias" from one surface to another. A typical plate is about the size of a playing card and may require 10,000 vias each as small as a pinpoint. Once, these vias were drilled using diamond drills. New technology uses lasers to produce these precisely positioned pathways. Suppose that the probability of incorrectly positioning a via is only 1/20,000. What is the probability that a randomly selected plate will have no improperly positioned vias?

Let X denote the number of vias that are positioned incorrectly. X is binomial with n = 10,000 and p = 1/20,000. We want to approximate

$$P[X=0] = \binom{10{,}000}{0}(1/20{,}000)^0(19{,}999/20{,}000)^{10{,}000} = .6065$$

Since n is large and p is small, this binomial probability can be approximated by a Poisson probability with the Poisson parameter $k = np$ = 10,000(1/20,000) = .5.

TABLE 3.8

Discrete distributions: A summary

Name	Density		Moment generating function	Mean	Variance
Geometric	$(1-p)^{x-1}p$	$x = 1, 2, 3, \ldots$ $0 < p < 1$	$\dfrac{pe^t}{1-qe^t}$	$\dfrac{1}{p}$	$\dfrac{q}{p^2}$
Uniform	$\dfrac{1}{n}$	$x = x_1, x_2, \ldots, x_n$ n a positive integer	$\dfrac{\sum_{i=1}^n e^{tx_i}}{n}$	$\dfrac{\sum_{i=1}^n x_i}{n}$	$\dfrac{\sum_{i=1}^n x_i^2}{n} - \left(\dfrac{\sum_{i=1}^n x_i}{n}\right)^2$
Binomial	$\binom{n}{x} p^x(1-p)^{n-x}$	$x = 0, 1, 2, \ldots, n$ $0 < p < 1$ n a positive integer	$(q + pe^t)^n$	np	$np(1-p)$
Bernoulli or point binomial	$p^x(1-p)^{1-x}$	$x = 0, 1$ $0 < p < 1$	$q + pe^t$	p	$p(1-p)$
Hypergeometric	$\dfrac{\binom{r}{x}\binom{N-r}{n-x}}{\binom{N}{n}}$	$\max[0, n-(N-r)]$ $\leq x \leq \min(n, r)$		$n\dfrac{r}{N}$	$n\dfrac{r}{N}\left(\dfrac{N-r}{N}\right)\left(\dfrac{N-n}{N-1}\right)$
Negative binomial	$\binom{x-1}{r-1}(1-p)^{x-r}p^r$	$x = r, r+1, r+2, \ldots,$ $0 < p < 1$	$\dfrac{(pe^t)^r}{(1-qe^t)^r}$	$\dfrac{r}{p}$	$\dfrac{r(1-p)}{p^2}$
Poisson	$\dfrac{e^{-k}k^x}{x!}$	$x = 0, 1, 2, \ldots$ $k > 0$	$e^{k(e^t-1)}$	k	k

That is,

$$P[X = 0] \doteq \frac{e^{-.5}(.5)^0}{0!}$$

From Table II of App. A this probability is approximately .607. (Based on a report in *Laser Focus*, December 1982, p. 26.)

Table 3.8 summarizes the discrete distributions that have been presented.

3.9 SIMULATING A DISCRETE DISTRIBUTION (OPTIONAL)

In designing operating systems of various types, one often needs to simulate the system before it is built. Simulation is usually done with the aid of a computer. However, the idea behind simulation can be illustrated by using a random digit table. A portion of such a table is given in Table III of App. A. Its use is illustrated in the following example.

Example 3.9.1. Table 3.9 presents a portion of the random digit table in the appendix. Let us read a sequence of random two-digit numbers from this table. To do so, we must get a random start. This can be done by writing the integers 1 through 14 on slips of paper, placing the slips in a bowl, stirring, and drawing one slip at random from the bowl. The number selected identifies the column in which our starting number is located. In a similar way, we can select the row in which the starting number is located. Suppose that this process results in the selection of column 2 and row 5. This identifies the random starting point as 39975.

Since we want two-digit numbers, we need only read the first two digits of this number. Thus our first random number is 39. Since a random digit table is constructed in such a way that the digit appearing at each position in the table is just as likely to be one digit as any other, the table can be read in any way. Let us agree to read down the second column so that the next four two-digit numbers are 06, 72, 91, and 14.

TABLE 3.9

Row	Column / Random digits (1)	(2)	(3)
1	10480	15011	01536
2	22368	46573	25595
3	24130	48360	22527
4	42167	93093	06243
5	37570	39975	81837
6	77921	06907	11008
7	99562	72905	56420
8	96301	91977	05463
9	89579	14342	63661
10	85485	36857	43342

The next example illustrates the use of a random digit table in a simple simulation experiment.

Example 3.9.2. Suppose that at a particular airport planes arrive at an average rate of one per minute and depart at the same average rate. We are interested in simulating the behavior of the random variable Z, the number of planes on the ground at a given time. We will simulate Z for five consecutive one-minute periods. Note that for each of these periods the random variables X, the number of arrivals, and Y, the number of departures, are each Poisson variables with parameter $k = 1$. The density for X and Y is obtained from Table II of App. A and is shown below:

$$P[X = 0] = P[Y = 0] = .368$$

$$P[X = 1] = P[Y = 1] = .368$$

$$P[X = 2] = P[Y = 2] = .184$$

$$P[X = 3] = P[Y = 3] = .061$$

$$P[X = 4] = P[Y = 4] = .015$$

$$P[X = 5] = P[Y = 5] = .003$$

$$P[X = 6] = P[Y = 6] = .001$$

$$P[X > 6] = P[Y > 6] \doteq 0$$

There are 1000 possible three-digit numbers. We divide them into seven categories to reflect the above probabilities. This division is shown in Table 3.10. To perform the simulation, we read a total of 10 random three-digit numbers using the procedure demonstrated in Example 3.9.1. Assume that at the beginning of the simulation there are 100 planes on the ground and that our random starting point is the number 01536 found in line 1 and column 3 of Table 3.9. The first number read corresponds to the arrivals during the first minute of observation, the second to the departures during this time span, and so forth. The results of the simulation are shown in Table 3.11. If this simulation were continued over a long period of

TABLE 3.10

Random number	Number of arrivals (x)	Number of departures (y)	$P[X = x] = P[Y = y]$
000–367	0	0	.368
368–735	1	1	.368
736–919	2	2	.184
920–980	3	3	.061
981–995	4	4	.015
996–998	5	5	.003
999	6	6	.001

TABLE 3.11

Time span, min	Random 3-digit number	Number of arrivals (x)	Number of departures (y)	Number on ground at end of time period (z)
1	015	0		100
	255		0	100
2	225	0		
	062		0	100
3	818	2		
	110		0	102
4	564	1		
	054		0	103
5	636	1		
	433		1	103

time, we could begin to answer such questions as: "On the average, how many planes are on the ground at a given time?" and "How much variability is there in the number of planes on the ground?"

CHAPTER SUMMARY

In this chapter we introduced the concept of a random variable and showed you how to distinguish a discrete random variable from one that is not discrete. We studied two functions, the density function and the cumulative distribution function, that are used to compute probabilities. The density gives the probability that X assumes a specific value x; the cumulative distribution gives the probability that X assumes a value less than or equal to x. The concept of expected value was introduced and used to define three important parameters, the mean (μ), the variance (σ^2), and the standard deviation (σ). The mean is a measure of the center of location of the distribution; the variance and standard deviation measure the variability of the random variable about its mean. The moment generating function was introduced as a means of finding the mean and variance of X. Special discrete distributions that find extensive use in all areas of application were presented. These are the geometric, hypergeometric, negative binomial, binomial, Bernoulli, uniform, and Poisson distributions. We also discussed briefly how to simulate a discrete distribution. We introduced and defined terms that you should know. These are:

Random variable
Discrete random variable
Discrete density
Cumulative distribution
Expected value
Mean
Variance
Standard deviation
Bernoulli trial
Moment generating function
Sampling with replacement
Sampling without replacement

EXERCISES

Section 3.1

In each of the following, identify the variable as discrete or not discrete.

1. T: the turnaround time for a computer job (the time it takes to run the program and receive the results).
2. M: the number of meteorites hitting a satellite per day.
3. N: the number of neutrons expelled per thermal neutron absorbed in fission of uranium-235.
4. Neutrons emitted as a result of fission are either prompt neutrons or delayed neutrons. Prompt neutrons account for about 99% of all neutrons emitted and are released within 10^{-14} s of the instant of fission. Delayed neutrons are emitted over a period of several hours. Let D denote the time at which a delayed neutron is emitted in a fission reaction.
5. Electrical resistance is the opposition which is offered by electrical conductors to the flow of current. The unit of resistance is the ohm. For example, a $2\frac{1}{2}$-inch electric bell will usually have a resistance somewhere between 1.5 and 3 ohms. Let O denote the actual resistance of a randomly selected bell of this type.
6. The number of power failures per month in the Tennessee Valley power network.

Section 3.2

7. Grafting, the uniting of the stem of one plant with the stem or root of another, is widely used commercially to grow the stem of one variety that produces fine fruit on the root system of another variety with a hardy root system. Most Florida sweet oranges grow on trees grafted to the root of a sour orange variety. The density for X, the number of grafts that fail in a series of five trials, is given by Table 3.12.
   (*a*) Find $f(5)$.
   (*b*) Find the table for F.
   (*c*) Use F to find the probability that at most three grafts fail; that at least two grafts fail.
   (*d*) Use F to verify that the probability of exactly three failures is .03.

TABLE 3.12

x	0	1	2	3	4	5
$f(x)$	.7	.2	.05	.03	.01	?

8. In blasting soft rock such as limestone, the holes bored to hold the explosives are drilled with a Kelly bar. This drill is designed so that the explosives can be packed into the hole before the drill is removed. This is necessary since in soft rock the hole often collapses as the drill is removed. The bits for these drills must be changed fairly often. Let X denote the number of holes that can be drilled per bit. The

density for X is given in Table 3.13. (Based on data reported in *The Explosives Engineer*, vol. 1, 1976, p. 12.)

(*a*) Find $f(8)$.

(*b*) Find the table for F.

(*c*) Use F to find the probability that a randomly selected bit can be used to drill between three and five holes inclusive.

(*d*) Find $P[X \leq 4]$ and $P[X < 4]$. Are these probabilities the same?

(*e*) Find $F(-3)$ and $F(10)$. *Hint:* Express these in terms of the probabilities that they represent and their values will become obvious.

TABLE 3.13

x	1	2	3	4	5	6	7	8
$f(x)$	.02	.03	.05	.2	.4	.2	.07	?

9. Consider Example 1.2.1. Let X denote the number of computer systems operable at the time of the launch. Assume that the probability that each system is operable is .9.

(*a*) Use the tree of Fig. 1.2 to find the density table.

(*b*) There is a pattern to the probabilities in the density table. In particular,

$$f(x) = k(x)(.9)^x(.1)^{3-x}$$

where $k(x)$ gives the number of paths through the tree yielding a particular value for X. Use Exercise 23 of Chap. 1 to express $k(x)$ in terms of the number of computers available and the number operable.

(*c*) Find the table for F.

(*d*) Use F to find the probability that at least one system is operable at launch time.

(*e*) Use F to find the probability that at most one system is operable at the time of the launch.

10. It is known that the probability of being able to log on to a computer from a remote terminal at any given time is .7. Let X denote the number of attempts that must be made to gain access to the computer.

(*a*) Find the first four terms of the density table.

(*b*) Find a closed-form expression for $f(x)$.

(*c*) Find $P[X = 6]$.

(*d*) Find a closed-form expression for $F(x)$.

(*e*) Use F to find the probability that at most four attempts must be made to gain access to the computer.

(*f*) Use F to find the probability that at least five attempts must be made to gain access to the computer.

In parts (*c*), (*d*), and (*e*) of each of the next two exercises we point out the necessary and sufficient conditions for a function F to be a cumulative distribution function for a discrete random variable.

*11. Even though there is no closed-form expression for the cumulative distribution function of Exercise 7, we can rewrite it as follows:

$$F(x) = \begin{cases} 0 & x < 0 \\ .70 & 0 \le x < 1 \\ .90 & 1 \le x < 2 \\ .95 & 2 \le x < 3 \\ .98 & 3 \le x < 4 \\ .99 & 4 \le x < 5 \\ 1.00 & x \ge 5 \end{cases}$$

(*a*) Draw the graph of this function. Recall from elementary calculus that graphs of this form are called step functions.
(*b*) Is F a continuous function?
(*c*) Is F a right continuous function?
(*d*) What is $\lim_{x \to \infty} F(x)$? What is $\lim_{x \to -\infty} F(x)$?
(*e*) Is F nondecreasing?

*12. (*a*) Express the cumulative distribution function F of Exercise 8 in the manner shown in Exercise 11 and draw the graph of F.
(*b*) Is F a continuous function?
(*c*) Is F right continuous?
(*d*) What is $\lim_{x \to \infty} F(x)$? What is $\lim_{x \to -\infty} F(x)$?
(*e*) Is F nondecreasing?

*13. State the conditions that are necessary and sufficient for a function F to be a cumulative distribution function for a discrete random variable.

Section 3.3

14. In an experiment to graft Florida sweet orange trees to the root of a sour orange variety, a series of five trials is conducted. Let X denote the number of grafts that fail. The density for X is given in Table 3.12.
(*a*) Find $E[X]$.
(*b*) Find μ_X.
(*c*) Find $E[X^2]$.
(*d*) Find Var X.
(*e*) Find σ_X^2.
(*f*) Find the standard deviation for X.
(*g*) What physical unit is associated with σ_X?

15. The density for X, the number of holes that can be drilled per bit while drilling into limestone is given in Table 3.13.
(*a*) Find $E[X]$ and $E[X^2]$.
(*b*) Find Var X and σ_X.
(*c*) What physical unit is associated with σ_X?

16. Use the density derived in Exercise 9 to find the expected value and variance for X, the number of computer systems operable at the time of the launch. Can you express $E[X]$ and Var X in terms of n, the number of systems available, and p, the probability that a given system will be operable?

*17. The probability p of being able to log on to a computer from a remote terminal at any given time is .7. Let X denote the number of attempts that must be made to gain access to the computer. Find $E[X]$. Can you express $E[X]$ in terms of p?

Hint: The series $\Sigma_{x=1}^{\infty} x(.7)(.3)^{x-1} = E[X]$ is not geometric. To find $E[X]$, expand this series and the series $.3E[X]$. Subtract the two to form the series $.7E[X]$. Evaluate this *geometric* series and solve for $E[X]$.

***18.** The probability that a cell will fuse in the presence of polyethylene glycol is $1/2$. Let Y denote the number of cells exposed to antigen-carrying lymphocytes to obtain the first fusion. Use the method of Exercise 17 to find $E[Y]$.

***19.** Let X be a discrete random variable with density f. Let c be any real number. Show that

(*a*) $E[c] = c$. *Hint:* Remember that constants can be factored from summations and that $\Sigma_{\text{all } x} f(x) = 1$.

(*b*) $E[cX] = cE[X]$.

***20.** Use the rules for expectation to verify that $\text{Var}\, c = 0$ and $\text{Var}\, cX = c^2 \text{Var}\, X$ for any real number c. *Hint:* $\text{Var}\, c = E[c^2] - (E[c])^2$.

21. Let X and Y be independent random variables with $E[X] = 3$, $E[X^2] = 25$, $E[Y] = 10$ and $E[Y^2] = 164$.

(*a*) Find $E[3X + Y - 8]$.

(*b*) Find $E[2X - 3Y + 7]$.

(*c*) Find $\text{Var}\, X$.

(*d*) Find σ_X.

(*e*) Find $\text{Var}\, Y$.

(*f*) Find σ_Y.

(*g*) Find $\text{Var}[3X + Y - 8]$.

(*h*) Find $\text{Var}[2X - 3Y + 7]$.

(*i*) Find $E[(X - 3)/4]$ and $\text{Var}[(X - 3)/4]$.

(*j*) Find $E[(Y - 10)/8]$ and $\text{Var}[(Y - 10)/8]$.

(*k*) The results of parts (*i*) and (*j*) are not coincidental. Can you generalize and verify the conjecture suggested by these two exercises?

***22.** Consider the function f defined by

$$f(x) = (1/2)2^{-|x|} \qquad x = \pm 1, \pm 2, \pm 3, \pm 4, \ldots$$

(*a*) Verify that this is the density for a discrete random variable X. *Hint:* Expand the series $\Sigma_{\text{all } x} f(x)$ for a few terms. A recognizable series will develop!

(*b*) Let $g(X) = (-1)^{|X|-1}[2^{|X|}/(2|X| - 1)]$. Show that $\Sigma_{\text{all } x} g(x)f(x) < \infty$. *Hint:* Expand the series for a few terms. You will obtain an alternating series that can be shown to converge.

(*c*) Show that $\Sigma_{\text{all } x} |g(x)| f(x)$ does not converge. This will show that $E[g(X)]$ does not exist. *Hint:* Expand the series for a few terms. You will obtain a series that is term by term larger than the diverging harmonic type series $(1/3)\Sigma_{x=1}^{\infty} 1/x$.

***23.** (*An application to sort algorithms.*) In studying various sort algorithms in computer science, it is of interest to compare their efficiency by estimating the average number of interchanges needed to sort random arrays of various sizes. It is also of interest to compare these estimated averages to the "ideal" average, where by "ideal" we mean the expected minimum number of interchanges needed to sort the array. In this exercise you will derive this ideal average. (American Mathematical Association of Two-Year Colleges, "A Note on the Minimum Number of Interchanges Needed to Sort a Random Array," with T. McMillan, I. Liss, and J. Milton, Fall 1990.)

(*a*) Consider a random array of length n. When the positions of exactly two elements of the array are exchanged, we say that an "interchange" has taken place. Let X_n denote the minimum number of interchanges necessary to sort an array of size n. Note that

$$X_n = X_{n-1} + I$$

where $I = 0$ if the last element of the array is in the correct position and $I = 1$ otherwise. Argue that $P[I = 0] = 1/n$ and $P[I = 1] = 1 - (1/n)$.

(*b*) Show that

$$E[I] = 1 - \frac{1}{n}$$

(*c*) Argue that

$$E[X_n] = E[X_{n-1}] + 1 - \frac{1}{n}$$

$$E[X_{n-1}] = E[X_{n-2}] + 1 - \frac{1}{n-1}$$

$$E[X_{n-2}] = E[X_{n-3}] + 1 - \frac{1}{n-2}$$

$$\vdots$$

$$E[X_3] = E[X_2] + 1 - \frac{1}{3}$$

$$E[X_2] = E[X_1] + 1 - \frac{1}{2}$$

$$E[X_1] = 0$$

(*d*) Use a recursive argument to show that

$$E[X_n] = (n-1) - \sum_{i=2}^{n} \frac{1}{i}$$

(*e*) Illustrate the expression given in part (*d*) by finding $E[X_5]$.

(*f*) Elementary calculus can be used to approximate $E[X_n]$ by noting that

$$\sum_{i=2}^{n} \frac{1}{i} \doteq \int_{1.5}^{n+.5} \frac{1}{t}\, dt$$

Use this idea to approximate $E[X_5]$ and to compare the result to the exact solution found in part (*e*).

(*g*) A random digit generator is used to generate sets of 100 different three-digit numbers lying between 0 and 1. What is the ideal average number of interchanges needed to sort such an array?

Section 3.4

24. The probability that a wildcat well will be productive is 1/13. Assume that a group is drilling wells in various parts of the country so that the status of one well has no bearing on that of any other. Let X denote the number of wells drilled to obtain the first strike.

(*a*) Verify that X is geometric and identify the value of the parameter p.

(*b*) What is the exact expression for the density for X?

(c) What is the exact expression for the moment generating function for X?
(d) What are the numerical values of $E[X]$, $E[X^2]$, σ^2, and σ?
(e) Find $P[X \geq 2]$.
(f) Suppose that 10 wells have been drilled and no strikes have occurred. What is the probability that at least 2 more wells must be drilled in order to obtain the first strike? What property can be used to answer this question?

25. The zinc-phosphate coating on the threads of steel tubes used in oil and gas wells is critical to their performance. To monitor the coating process, an uncoated metal sample with known outside area is weighed and treated along with the lot of tubing. This sample is then stripped and reweighed. From this it is possible to determine whether or not the proper amount of coating was applied to the tubing. Assume that the probability that a given lot is unacceptable is .05. Let X denote the number of runs conducted to produce an unacceptable lot. Assume that the runs are independent in the sense that the outcome of one run has no effect on that of any other. (Based on a report in *American Machinist*, November 1982, p. 81.)
(a) Verify that X is geometric. What is "success" in this experiment? What is the numerical value of p?
(b) What is the exact expression for the density for X?
(c) What is the exact expression for the moment generating function for X?
(d) What are the numerical values of $E[X]$, $E[X^2]$, σ^2, and σ?
(e) Find the probability that the number of runs required to produce an unacceptable lot is at least 3.
(f) Suppose that 19 lots have been inspected and that all 19 are deemed to be acceptable. What is the probability that at least 21 acceptable lots are produced prior to obtaining the first defective lot?

26. Let X be geometric with probability of success p. Prove that when x is a positive integer, $F(x) = 1 - q^x$. Verify that this result holds true for the density given in Example 3.2.4. Argue that, in general, $F(x) = 1 - q^{[x]}$.

27. Find the expression for the cumulative distribution function for the random variable of Exercise 25. Use this function to find the probability that at most three runs are required to produce an unacceptable lot.

28. A system used to read electric meters automatically requires the use of a 128-bit computer message. Occasionally random interference causes a digit reversal resulting in a transmission error. Assume that the probability of a digit reversal for each bit is $1/1000$. Let X denote the number of transmission errors per 128-bit message sent. Is X geometric? If not, what geometric property fails?

29. Verify that the random variable X of Exercise 17 is geometric. Use Theorem 3.4.3 to find $E[X]$ and compare your answer to that obtained in Exercise 17.

30. Verify that the random variable Y of Exercise 18 is geometric. Use Theorem 3.4.3 to find $E[Y]$ and compare your answer to that obtained in Exercise 18.

31. Consider the random variable X whose density is given by

$$f(x) = \frac{(x-3)^2}{5} \qquad x = 3, 4, 5$$

(a) Verify that this function is a density for a discrete random variable.
(b) Find $E[X]$ directly. That is, evaluate $\sum_{\text{all } x} xf(x)$.
(c) Find the moment generating function for X.

(*d*) Use the moment generating function to find $E[X]$, thus verifying your answer to part (*b*) of this exercise.
(*e*) Find $E[X^2]$ directly. That is, evaluate $\sum_{\text{all } x} x^2 f(x)$.
(*f*) Use the moment generating function to find $E[X^2]$, thus verifying your answer to part (*e*) of this exercise.
(*g*) Find σ^2 and σ.

32. A discrete random variable has moment generating function

$$m_X(t) = e^{2(e^t - 1)}$$

(*a*) Find $E[X]$.
(*b*) Find $E[X^2]$.
(*c*) Find σ^2 and σ.

***33.** Let X have a geometric distribution with parameter p.
(*a*) Show that the probability that X is odd is $p/(1 - q^2)$, where $q = 1 - p$. *Hint:* If x is odd, then x can be expressed in the form $x = 2m - 1$ for $m = 1, 2, 3, \ldots$.
(*b*) Show that the probability that X is odd is never $1/2$ regardless of the value chosen for p.

34. (*Discrete uniform distribution.*) A discrete random variable is said to be *uniformly* distributed if it assumes a finite number of values with each value occurring with the same probability. If we consider the generation of a single random digit, then Y, the number generated, is uniformly distributed with each possible digit occurring with probability $1/10$. In general, the density for a uniformly distributed random variable is given by

$$f(x) = 1/n \qquad \begin{array}{l} n \text{ a positive integer} \\ x = x_1, x_2, x_3, \ldots, x_n \end{array}$$

(*a*) Find the moment generating function for a discrete uniform random variable.
(*b*) Use the moment generating function to find $E[X]$, $E[X^2]$, and σ^2.
(*c*) Find the mean and variance for the random variable Y, the number obtained when a random digit generator is activated once. *Hint:* The sum of the first n positive integers is $n(n + 1)/2$; the sum of the squares of the first n positive integers is $n(n + 1)(2n + 1)/6$.

35. Let the density for X be given by

$$f(x) = ce^{-x} \qquad x = 1, 2, 3, \ldots$$

(*a*) Find the value of c that makes this a density.
(*b*) Find the moment generating function for X.
(*c*) Use $m_X(t)$ to find $E[X]$.

Section 3.5

36. Let X be binomial with parameters $n = 15$ and $p = .2$.
(*a*) Find the expression for the density for X.
(*b*) Find the expression for the moment generating function for X.
(*c*) Find $E[X]$ and Var X.
(*d*) Find $E[X]$, $E[X^2]$, and Var X using the moment generating function, thus verifying your answer to part (*c*) of this exercise.

(*e*) Find $P[X \leq 1]$ by evaluating the density directly. Compare your answer to that given in Table I of App. A.

(*f*) Draw dot diagrams similar to that of Fig. 3.2 to illustrate each of these probabilities, and find the probabilities using Table I of App. A.

$$P[X \leq 5] \qquad P[X \geq 3]$$
$$P[X < 5] \qquad F(9)$$
$$P[2 \leq X \leq 7] \qquad F(20)$$
$$P[2 \leq X < 7] \qquad P[X = 10]$$

37. Albino rats used to study the hormonal regulation of a metabolic pathway are injected with a drug that inhibits body synthesis of protein. The probability that a rat will die from the drug before the experiment is over is .2. If 10 animals are treated with the drug, how many are expected to die before the experiment ends? What is the probability that at least eight will survive? Would you be surprised if at least five died during the course of the experiment? Explain, based on the probability of this occurring.

38. Consider Example 1.2.1. The random variable X is the number of computer systems operable at the time of a space launch. The systems are assumed to operate independently. Each is operable with probability .9.

(*a*) Argue that X is binomial and find its density. Compare your answer to that obtained in Exercise 9(*b*).

(*b*) Find $E[X]$ and Var X.

39. In humans, geneticists have identified two sex chromosomes, R and Y. Every individual has an R chromosome, and the presence of a Y chromosome distinguishes the individual as male. Thus the two sexes are characterized as RR (female) and RY (male). Color blindness is caused by a recessive allele on the R chromosome, which we denote by r. The Y chromosome has no bearing on color blindness. Thus relative to color blindness, there are three genotypes for females and two for males:

Female	**Male**
RR (normal)	RY (normal)
Rr (carrier)	rY (color-blind)
rr (color-blind)	

A child inherits one sex chromosome randomly from each parent.

(*a*) A carrier of color blindness parents a child with a normal male. Construct a tree to represent the possible genotypes for the child. Use the tree to find the probability that a given child will be a color-blind male.

(*b*) If the couple has five children, what is the expected number of color-blind males? What is the probability that three or more will be color-blind males?

40. In scanning electron microscopy photography, a specimen is placed in a vacuum chamber and scanned by an electron beam. Secondary electrons emitted from the specimen are collected by a detector and an image is displayed on a cathode-ray tube. This image is photographed. In the past a 4-× 5-inch camera has been used. It is thought that a 35-millimeter (mm) camera can obtain the same clarity. This type of camera is faster and more economical than the 4-× 5-inch variety. (Based on a report entitled "Adaptation of a Thirty-Five Millimeter Photographic System for a Scanning Electron Microscope," E. A. Lawton, *Biological Photography*, vol. 50, no. 3, July 1982, p. 65.)

(*a*) Photographs of 15 specimens are made using each camera system. These unmarked photographs are judged for clarity by an impartial judge. The judge is asked to select the better of the two photographs from each pair. Let X denote the number selected taken by a 35-mm camera. If there is really no difference in clarity and the judge is randomly selecting photographs, what is the expected value of X?

(*b*) Would you be surprised if the judge selected 12 or more photographs taken by the 35-mm camera? Explain, based on the probability involved.

(*c*) If $X \geq 12$, do you think that there is reason to suspect that the judge is not selecting the photographs at random?

41. It has been found that 80% of all printers used on home computers operate correctly at the time of installation. The rest require some adjustment. A particular dealer sells 10 units during a given month.

(*a*) Find the probability that at least nine of the printers operate correctly upon installation.

(*b*) Consider 5 months in which 10 units are sold per month. What is the probability that at least 9 units operate correctly in each of the 5 months?

42. It is possible for a computer to pick up an erroneous signal that does not show up as an error on the screen. The error is called a silent error. A particular terminal is defective, and when using the system word processor, it introduces a silent paging error with probability .1. The word processor is used 20 times during a given week.

(*a*) Find the probability that no silent paging errors occur.

(*b*) Find the probability that at least one such error occurs.

(*c*) Would it be unusual for more than four such errors to occur? Explain, based on the probability involved. P(X ≥ 4)

43. (*a*) Find the moment generating function for a binomial random variable with parameters n and p. *Hint:* Let

$$\binom{n}{x} e^{tx} p^x (1-p)^{n-x} = \binom{n}{x} (pe^t)^x (1-p)^{n-x}$$

and apply the binomial theorem.

(*b*) Use $m_X(t)$ to show that $E[X] = np$.

(*c*) Use $m_X(t)$ to show that $E[X^2] = n^2p^2 - np^2 + np$.

(*d*) Show that Var $X = npq$, where $q = 1 - p$.

***44.** Find the mean value for a binomial random variable with parameters n and p from the definition. That is, evaluate

$$\sum_{x=0}^{n} x \binom{n}{x} p^x (1-p)^{n-x}$$

Hint:

$$\sum_{x=0}^{n} x \binom{n}{x} p^x (1-p)^{n-x} = \sum_{x=1}^{n} x \binom{n}{x} p^x (1-p)^{n-x}$$

Now let $z = x - 1$ and evaluate

$$\sum_{z=0}^{n-1} (z+1)\binom{n}{z+1} p^{z+1}(1-p)^{n-(z+1)}$$

45. (*Point binomial or Bernoulli distribution.*) Assume that an experiment is conducted and that the outcome is considered to be either a success or a failure. Let p denote the probability of success. Define X to be 1 if the experiment is a success and 0 if it is a failure. X is said to have a *point binomial* or a *Bernoulli* distribution with parameter p.

(*a*) Argue that X is a binomial random variable with $n = 1$.

(*b*) Find the density for X.

(*c*) Find the moment generating function for X.

(*d*) Find the mean and variance for X.

(*e*) In DNA replication errors can occur that are chemically induced. Some of these errors are "silent" in that they do not lead to an observable mutation. Growing bacteria are exposed to a chemical that has probability .14 of inducing an observable error. Let X be 1 if an observable mutation results and let X be 0 otherwise. Find $E[X]$.

46. A binomial random variable has mean 5 and variance 4. Find the values of n and p that characterize the distribution of this random variable.

Section 3.6

***47.** In this exercise you will show that the density for the negative binomial distribution sums to 1.

(*a*) Show that

$$\sum_{x=r}^{\infty}\binom{x-1}{r-1}q^{x-r}p^r = \sum_{z=0}^{\infty}\binom{z+r-1}{r-1}q^z p^r$$

where $q = 1 - p$. *Hint:* Let $z = x - r$.

(*b*) Show that the Taylor series expansion for $h(q) = 1/(1-q)^r$ about 0 is given by

$$1 + rq + \frac{r(r+1)q^2}{2} + \frac{r(r+1)(r+2)q^3}{3!} + \cdots = 1/(1-q)^r$$

(*c*) Show that

$$\sum_{z=0}^{\infty}\binom{z+r-1}{r-1}q^z = 1/(1-q)^r$$

(*d*) Show that

$$\sum_{z=0}^{\infty}\binom{z+r-1}{r-1}q^z p^r = 1$$

Hint: Factor the term p^r from the expression and use part (*c*) with $q = 1 - p$.

***48.** In this exercise you will derive the moment generating function for the negative binomial distribution with parameters r and p.

(*a*) Show that

$$m_X(t) = \sum_{x=r}^{\infty} e^{tx}\binom{x-1}{r-1}q^{x-r}p^r = (pe^t)^r \sum_{z=0}^{\infty}\binom{z+r-1}{r-1}(qe^t)^z$$

Hint: Let $z = x - r$.

(*b*) Use the idea given in Exercise 47(*c*) to show that

$$\sum_{z=0}^{\infty}\binom{z+r+1}{r-1}(qe^t)^z = 1/(1-qe^t)^r$$

(*c*) Show that

$$m_X(t) = \frac{(pe^t)^r}{(1-qe^t)^r}$$

***49.** Use the moment generating function to show that the mean of a negative binomial distribution with parameters r and p is r/p.

***50.** Use the moment generating function to show that $E[X^2] = (r^2 + rq)/p^2$ and that $\text{Var } X = rq/p^2$ for the negative binomial distribution with parameters r and p.

***51.** Show that the geometric distribution is a special case of the negative binomial distribution with $r = 1$. Find the mean and variance of a geometric random variable with parameter p using Exercises 49 and 50. Compare your answer with the results of Theorem 3.4.3.

***52.** A vaccine for desensitizing patients to bee stings is to be packed with three vials in each box. Each vial is checked for strength before packing. The probability that a vial meets specifications is .9. Let X denote the number of vials that must be checked to fill a box. Find the density for X and its mean and variance. Would you be surprised if seven or more vials have to be tested to find three that meet specifications? Explain, based on the probability of this occurrence.

***53.** Some characteristics in animals are said to be sex-influenced. For example, the production of horns in sheep is governed by a pair of alleles, H and h. The allele H for the production of horns is dominant in males but recessive in females. The allele h for hornlessness is dominant in females and recessive in males. Thus, given a heterozygous male (Hh) and a heterozygous female (Hh), the male will have horns but the female will be hornless. Assume that two such animals mate and the offspring is just as likely to be male as female. The lamb inherits one gene for horns randomly from each parent. Use a tree diagram to show that the probability that a lamb will be a hornless female is $3/8$. Find the average number of lambs born to obtain the second hornless female. Would you be surprised if at most five lambs were born to obtain the second hornless female? Explain.

Section 3.7

54. Suppose that X is hypergeometric with $N = 20$, $r = 17$, and $n = 5$. What are the possible values for X? What is $E[X]$ and $\text{Var } X$?

55. Suppose that X is hypergeometric with $N = 20$, $r = 3$, and $n = 5$. What are the possible values for X? What is $E[X]$ and $\text{Var } X$?

56. Suppose that X is hypergeometric with $N = 20$, $r = 10$, and $n = 5$. What are the possible values for X? What is $E[X]$ and $\text{Var } X$?

57. Twenty microprocessor chips are in stock. Three have etching errors that cannot be detected by the naked eye. Five chips are selected and installed in field equipment.
(*a*) Find the density for X, the number of chips selected that have etching errors.
(*b*) Find $E[X]$ and $\text{Var } X$.
(*c*) Find the probability that no chips with etching errors will be selected.
(*d*) Find the probability that at least one chip with an etching error will be chosen.

58. Production line workers assemble 15 automobiles per hour. During a given hour, four are produced with improperly fitted doors. Three automobiles are selected at random and inspected. Let X denote the number inspected that have improperly fitted doors.
(a) Find the density for X.
(b) Find $E[X]$ and Var X.
(c) Find the probability that at most one will be found with improperly fitted doors.

59. A distributor of computer software wants to obtain some customer feedback concerning its newest package. Three thousand customers have purchased the package. Assume that 600 of these customers are dissatisfied with the product. Twenty customers are randomly sampled and questioned about the package. Let X denote the number of dissatisfied customers sampled.
(a) Find the density for X.
(b) Find $E[X]$ and Var X.
(c) Set up the calculations needed to find $P[X \leq 3]$.
(d) Use the binomial tables to approximate $P[X \leq 3]$.

60. A random telephone poll is conducted to ascertain public opinion concerning the construction of a nuclear power plant in a particular community. Assume that there are 150,000 numbers listed for private individuals and that 90,000 of these would elicit a negative response if contacted. Let X denote the number of negative responses obtained in 15 calls.
(a) Find the density for X.
(b) Find $E[X]$ and Var X.
(c) Set up the calculations needed to find $P[X \geq 6]$.
(d) Use the binomial tables to approximate $P[X \geq 6]$.

Section 3.8

61. Let X be a Poisson random variable with parameter $k = 10$.
(a) Find $E[X]$.
(b) Find Var X.
(c) Find σ_X.
(d) Find the expression for the density for X.
(e) Find $P[X \leq 4]$.
(f) Find $P[X < 4]$.
(g) Find $P[X = 4]$.
(h) Find $P[X \geq 4]$.
(i) Find $P[4 \leq X \leq 9]$.

62. A particular nuclear plant releases a detectable amount of radioactive gases twice a month on the average. Find the probability that there will be at most four such emissions during a month. What is the expected number of emissions during a 3-month period? If, in fact, 12 or more emissions are detected during a 3-month period, do you think that there is a reason to suspect the reported average figure of twice a month? Explain, on the basis of the probability involved.

63. Geophysicists determine the age of a zircon by counting the number of uranium fission tracks on a polished surface. A particular zircon is of such an age that the average number of tracks per square centimeter is five. What is the probability that a 2-centimeter-square sample of this zircon will reveal at most three tracks, thus leading to an underestimation of the age of the material?

64. California is hit by approximately 500 earthquakes that are large enough to be felt every year. However, those of destructive magnitude occur on the average once every year. Find the probability that California will experience at least one earthquake of this magnitude during a 6-month period. Would it be unusual to have 3 or more earthquakes of destructive magnitude in a 6-month period? Explain, based on the probability of this occurring. (Based on data presented in Robert Iacopi, *Earthquake Country*, Lana Books, Menlo Park, Calif., 1971.)

65. Load-bearing structures in underground mines are often required to carry additional loads while mining operations are in progress. As the structures adjust to this new weight small-scale displacements take place that result in the release of seismic and acoustic energy, called *rock noise*. This energy can be detected using special geophysical equipment. Assume that in a particular mine the average number of rock noises recorded during normal activity is 3 per hour. Would you consider it unusual if more than 10 were detected in a 2-hour period? Explain, based on the probability involved. (Based on "A multichannel rock noise monitoring system," T. Gowd and M. S. Rao, *Journal of Mines, Metals and Fuels*, September 1981, pp. 288–290.)

66. A burr is a thin ridge or rough area that occurs when shaping a metal part. These must be removed by hand or by means of some newer method such as water jets, thermal energy, or electrochemical processing before the part can be used. Assume that a part used in automatic transmissions typically averages two burrs each. What is the probability that the total number of burrs found on seven randomly selected parts will be at most four? (Based on "Advances in Deburring," B. Hignett, *Production Engineering*, December 1982, pp. 44–47.)

67. Cast iron is an alloy composed primarily of iron together with smaller amounts of other elements, including carbon, silicon, sulfur, and phosphorus. The carbon occurs as graphite, which is soft, or iron carbide, which is very hard and brittle. The type of cast iron produced is determined by the amount and distribution of carbon in the iron. Five types of cast iron are identifiable. These are gray, compacted graphite, ductile, malleable, and white. In malleable cast iron the carbon is present as discrete graphite particles. Assume that in a particular casting these particles average 20 per square inch. Would it be unusual to see a 1/4-inch-square area of this casting with fewer than two graphite particles? Explain, based on the probability involved. (Based on "Space Age Metal: Cast Iron," J. Lalich, *Mines Magazine*, February 1982, pp. 2–6.)

68. A Poisson random variable is such that it assumes the values 0 and 1 with equal probability. Find the value of the Poisson parameter k for this variable.

69. Prove Theorem 3.8.1. *Hint:* Note that

$$m_X(t) = E[e^{tX}] = \sum_{x=0}^{\infty} \frac{e^{tx}e^{-k}k^x}{x!} = \sum_{x=0}^{\infty} \frac{e^{-k}(ke^t)^x}{x!}$$

and use the Maclaurin series.

70. Let X be binomial with $n = 20$ and $p = .05$. Find $P[X = 0]$ using the binomial density and compare your answer to that obtained by using the Poisson approximation to this probability. Do you think that the error in the approximation is large?

71. In *Escherichia coli*, a bacterium often found in the human digestive tract, 1 cell in every 10^9 will mutate from streptomycin sensitivity to streptomycin resistance. This mutation can cause the individual involved to become resistant to the antibiotic streptomycin. In observing 2 billion (2×10^9) such cells, what is the probability that none will mutate? What is the probability that at least one will mutate?

72. The spontaneous flipping of a bit stored in a computer memory is called a "soft fail." Soft fails are rare, averaging only one per million hours per chip. However, the probability of a soft fail is increased when the chip is exposed to α particles (helium nuclei), which occur naturally in the environment. Assume that the probability of a soft fail under these conditions is $1/1000$. If a chip containing 6000 bits is exposed to α particles, what is the probability that there will be at least one soft fail? Would you be surprised if there were more than five soft fails? Explain, based on the probability of this occurring. (*McGraw-Hill Yearbook of Science and Technology*, 1981, p. 142.)

Section 3.9

73. Use Table II of App. A to simulate the arrival and departure of planes to the airport described in Example 3.9.2 for 10 more 1-minute periods. Based on these data, approximate the average number of planes on the ground at a given time by finding the arithmetic average of the values of Z simulated in the experiment.

74. Consider the random variable X, the number of runs conducted to produce an unacceptable lot when coating steel tubes (see Exercise 25.) X is geometric with $p = .05$. Divide the 100 possible two-digit numbers into two categories, with numbers 00–04 denoting the production of an unacceptable lot and the remaining numbers denoting the production of an acceptable lot. Simulate the experiment of producing lots until an unacceptable one is obtained 10 times. Record the value obtained for X in each simulation. Based on these data, approximate the average value of X. Does your approximate value lie close to the theoretical mean value of 20? If not, run the simulation 10 more times. Is the arithmetic average of your observed values for X closer to 20 this time?

REVIEW EXERCISES

75. A large microprocessor chip contains multiple copies of circuits. If a circuit fails, the chip knows it and knows how to select the proper logic to repair itself. The average number of defects per chip is 300. What is the probability that 10 or fewer defects will be found in a randomly selected region that comprises 5% of the total surface area? What is the probability that more than 10 defects are found? ("Self-Repairing Chips," *Datamation*, May 1983, p. 68.)

76. When a program is submitted to the computer in a time-sharing system, it is processed on a space-available basis. Past experience shows that a program submitted to one such system is accepted for processing within 1 minute with probability .25. Assume that during the course of a day five programs are submitted with enough time between submissions to ensure independence. Let X denote the number of programs accepted for processing within 1 minute.

(*a*) Find $E[X]$ and Var X.

(*b*) Find the probability that none of these programs will be accepted for processing within 1 minute.

(*c*) Five programs are submitted on each of two consecutive days. What is the probability that no programs will be accepted for processing within 1 minute during this two-day period?

77. A new type of brake lining is being studied. It is thought that the lining will last for at least 70,000 miles on 90% of the cars in which it is used. Laboratory trials are conducted to simulate the driving experience of 100 cars in which this lining is used. Let X denote the number of cars whose brakes must be relined before the 70,000-mile mark.

(*a*) What is the distribution of X? What is $E[X]$?

(*b*) What distribution can be used to approximate probabilities for X?

(*c*) Suppose that we agree that the 90% figure is too high if 17 or more of the 100 cars require a relinement prior to the 70,000-mile mark. What is the probability that we will come to this conclusion by chance even though the 90% figure is correct?

79. A bank of guns fires on a target one after the other. Each has probability $1/4$ of hitting the target on a given shot. Find the probability that the second hit comes before the seventh gun fires.

77. In a video game the player attempts to capture a treasure lying behind one of five doors. The location of the treasure varies randomly in such a way that at any given time it is just as likely to be behind one door as any other. When the player knocks on a given door, the treasure is his if it lies behind that door. Otherwise he must return to his original starting point and approach the doors through a dangerous maze again. Once the treasure is captured, the game ends. Let X denote the number of trials needed to capture the treasure. Find the average number of trials needed to capture the treasure. Find $P[X \leq 3]$. Find $P[X > 3]$.

80. An automobile repair shop has 10 rebuilt transmissions in stock. Three are not in correct working order and have an internal defect that will cause trouble within the first 1000 miles of operation. Four of these transmissions are randomly selected and installed in customers' cars. Find the probability that no defective transmissions are installed. Find the probability that exactly one defective transmission is installed.

81. A computer terminal can pick up an erroneous signal from the keyboard that does not show up on the screen. This creates a silent error that is difficult to detect. Assume that for a particular keyboard the probability that this will occur per entry is $1/1000$. In 12,000 entries find the probability that no silent errors occur. Find the probability of at least one silent error.

82. It is thought that 1 of every 10 cars on the road has a speedometer that is miscalibrated to the extent that it reads at least 5 miles per hour low. During the course of a day 15 drivers are stopped and charged with exceeding the speed limit by at least 5 miles per hour. Would you be surprised to find that at least 5 of the cars involved have miscalibrated speedometers? Explain, based on the probability of observing a result this unusual by chance.

83. Let

$$f(x) = \frac{x^2}{14} \qquad x = 1, 2, 3$$

(*a*) Show that f is the density for a discrete random variable.

(*b*) Find $E[X]$ and $E[X^2]$ from the definition of these terms.

(*c*) Find $m_X(t)$.

(*d*) Use $m_X(t)$ to verify your answers to part (*b*).

(*e*) Find Var X and σ.

84. Find the expression for the cumulative distribution function for the random variable of Exercise 24. Use this function to find the probability that at least three wells must be drilled to obtain the first strike.

85. Consider the moment generating function given below. In each case, state the name of the distribution involved and the numerical value of the parameters that identify the distribution. For example, if the distribution is binomial, state the value of n and p; if geometric, give the value of p.

(a) $(.2 + .8e^t)^{10}$

(b) $e^{5(e^t 1)}$

(c) $(.7 + .3e^t)$

(d) $\dfrac{.6e^t}{1 - .4e^t}$

(e) $\dfrac{(.3e^t)^5}{(1 - .7e^t)^5}$

(f) $e^{e^t 1}$

86. For each of the distribution in Exercise 85, give the numerical values of the mean, variance, and standard deviation.

87. Consider the problem of Example 1.2.3. Assume that sampling is independent and that at each stage the probability of obtaining a defective part when the process is working correctly is .01. Let X denote the number of samples taken to obtain the first defective part.

(a) Find the density for X.

(b What is the average value of X?

(c) What is the equation for the cumulative distribution function for X? Use F to find the probability that the first defective part will be found on or before the 90th sample.

CHAPTER 4

CONTINUOUS DISTRIBUTIONS

In Chap. 3 we learned to distinguish a discrete random variable from one that is not discrete. In this chapter we consider a large class of nondiscrete random variables. In particular, we consider random variables that are called *continuous*. We first study the general properties of variables of the continuous type and then present some important families of continuous random variables.

4.1 CONTINUOUS DENSITIES

In Chap. 3 we considered the random variable T, the time of the peak demand for electricity at a particular power plant. We agreed that this random variable is not discrete since, "a priori"—before the fact—we cannot limit the set of possible values for T to some finite or countably infinite collection of times. Time is measured continuously and T can conceivably assume any value in the time interval $[0, 24)$, where 0 denotes 12 midnight one day and 24 denotes 12 midnight the next day. Furthermore, if we ask *before* the day begins, What is the probability that the peak demand will occur exactly 12.013 278 650 931 271? The answer is 0. It is virtually impossible for the peak load to occur at this split second in time, not the slightest bit earlier or later. These two properties, possible values occurring as intervals and the a priori probability of assuming any specific value being 0, are the characteristics that identify a random variable as being continuous. This leads us to our next definition.

Definition 4.1.1 (Continuous random variable). A random variable is continuous if it can assume any value in some interval or intervals of real numbers and the probability that it assumes any specific value is 0.

Note that the statement that the probability that a continuous random variable assumes any specific value is 0 is essential to the definition. Discrete variables have no such restriction. For this reason, we calculate probabilities in the continuous case differently than we do in the discrete case. In the discrete case we defined a function f, called the density, which enabled us to compute probabilities associated with the random variable X. This function is given by

$$f(x) = P[X = x] \qquad x \text{ real}$$

This definition cannot be used in the continuous case because $P[X = x]$ is always 0. However, we do need a function that will enable us to compute probabilities associated with a continuous random variable. Such a function is also called a density.

Definition 4.1.2 (Continuous density). Let X be a continuous random variable. A function f such that

1. $f(x) \geq 0 \qquad \text{for } x \text{ real}$
2. $\int_{-\infty}^{\infty} f(x)\,dx = 1$
3. $P[a \leq X \leq b] = \int_a^b f(x)\,dx \qquad \text{for } a \text{ and } b \text{ real}$

is called a density for X.

Although this definition may look arbitrary at first glance, it is not. Note that, as in the discrete case, f is defined over the entire real line and is nonnegative. Recall from elementary calculus that integration is the natural extension of summation in the sense that the integral is the limit of a sequence of Riemann sums. In the discrete case we require that $\Sigma_{\text{all } x} f(x) = 1$. The natural extension of this requirement to the continuous case is property 2, namely, that

$$\int_{-\infty}^{\infty} f(x)\,dx = 1$$

In the discrete case we find the probability that X assumes a value in some set A by summing $f(x)$ over all values of x in A. That is,

$$P[X \in A] = \sum_{x \in A} f(x).$$

In the continuous case we shall be interested in finding the probability that X assumes values in some interval $[a, b]$. Replacing A by $[a, b]$ and substituting integration for summation in the previous expression suggest property 3 of Definition 4.1.2. That is,

$$P[a \le X \le b] = \int_a^b f(x)\,dx$$

It is evident that the term "density" in the continuous case is just an extension of the ideas presented in the discrete case, with summation being replaced by integration. This is an important notion as it will allow us to define the concept of expected value in the continuous case quite naturally.

Example 4.1.1. The lead concentration in gasoline currently ranges from .1 to .5 grams per liter. What is the probability that the lead concentration in a randomly selected liter of gasoline will lie between .2 and .3 grams inclusive? To answer this question, we need a density, f, for the random variable X, the number of grams of lead per liter of gasoline. Consider the function

$$f(x) = \begin{cases} 12.5x - 1.25 & .1 \le x \le .5 \\ 0 & \text{elsewhere} \end{cases}$$

The graph of f is shown in Fig. 4.1. The function is nonnegative. Furthermore,

$$\begin{aligned}\int_{-\infty}^{\infty} f(x)\,dx &= \int_{.1}^{.5}(12.5x - 1.25)\,dx \\ &= \left[\frac{12.5x^2}{2} - 1.25x\right]_{.1}^{.5} \\ &= \left[\frac{12.5(.5)^2}{2} - 1.25(.5)\right] - \left[\frac{12.5(.1)^2}{2} - 1.25(.1)\right] \\ &= .9375 - (-.0625) = 1\end{aligned}$$

Thus f satisfies properties 1 and 2 of Definition 4.1.2. Property 3 allows us to use f

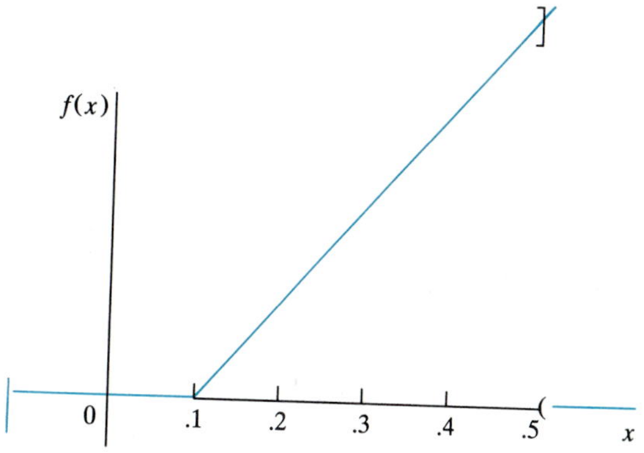

FIGURE 4.1
Graph of

$$(x) = \begin{cases} 12.5x - 1.25 & .1 \le x \le .5 \\ 0 & \text{elsewhere} \end{cases}$$

to find the desired probability. In particular,

$$P[.2 \le X \le .3] = \int_{.2}^{.3} f(x)\,dx$$

$$= \int_{.2}^{.3} (12.5x - 1.25)\,dx$$

$$= \left[\frac{12.5x^2}{2} - 1.25x\right]_{.2}^{.3}$$

$$= \left[\frac{12.5(.3)^2}{2} - 1.25(.3)\right] - \left[\frac{12.5(.2)^2}{2} - 1.25(.2)\right]$$

$$= .1875$$

(Based on data reported in *Petroleum Review*, August 1982, p. 45.)

There are several important points to be made concerning the density in the continuous case. First, we shall follow the convention of defining f only over intervals for which $f(x)$ may be nonzero. For values of x not explicitly mentioned $f(x)$ is assumed to be 0. In Example 4.1.1 we could have written f as

$$f(x) = 12.5x - 1.25 \qquad .1 \le x \le .5$$

with the understanding that $f(x) = 0$ elsewhere. Second, since the integral of a nonnegative function can be thought of as an area, properties 2 and 3 of Definition 4.1.2 can be expressed in terms of areas. In particular, property 2 requires that *the total area under the graph of f be 1*. Property 3 implies that the probability that the variable assumes a value between two points a and b is *the area under the graph of f between $x = a$ and $x = b$*. These ideas as they apply to Example 4.1.1 are demonstrated in Figs. 4.2(a) and (b), respectively. Third, since $P[X = a] = P[X = b] = 0$ in the continuous case,

$$P[a \le X \le b] = P[a \le X < b] = P[a < X \le b] = P[a < X < b].$$

In Example 4.1.1 the probability that the lead concentration in a liter of gasoline lies between .2 and .3 gram inclusive, $P[.2 \le X \le .3]$, is the same as $P[.2 < X < .3]$, the probability that it lies strictly between .2 and .3 gram. See Fig. 4.2(c). Fourth, properties 1 and 2 of Definition 4.1.2 are necessary and sufficient conditions for a function to be a density for a continuous random variable X. However, the density chosen for X cannot be just any function satisfying these conditions. It should be a function that assigns reasonable probabilities to events via property 3 of Definition 4.1.2. Whether or not the function f given in Example 4.1.1 satisfies this criteria is debatable. It was chosen for illustrative purposes only. Finding an appropriate density is not always easy. Some methods for helping in the selection of a density are discussed in Chap. 6.

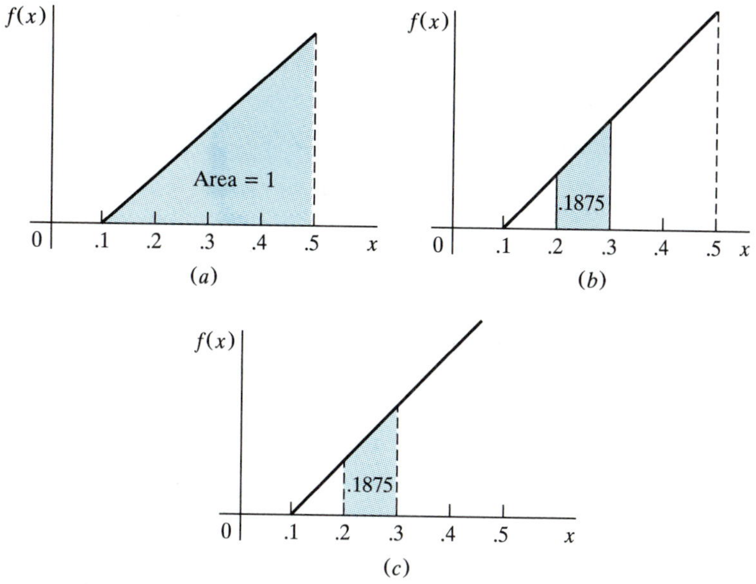

FIGURE 4.2
(*a*) $\int_{-\infty}^{\infty} f(x)\,dx = 1$ implies that the total area under the graph of f is 1; (*b*) $P[.2 \le X \le .3] = \int_{.2}^{.3}(12.5x - 1.25)\,dx = .1875$ implies that the area under the graph of f between $x = .2$ and $x = .3$ is .1875; (*c*) $P[.2 < X < .3] = P[.2 \le X \le .3] = .1875$.

Cumulative Distribution

The idea of a cumulative distribution function in the continuous case is useful. It is defined exactly as in the discrete case although found by using integration rather than summation.

Definition 4.1.3 (Cumulative distribution—continuous). Let X be continuous with density f. The cumulative distribution function for X, denoted by F, is defined by

$$F(x) = P[X \le x] \qquad x \text{ real}$$

To find $F(x)$ for a specific real number x, we integrate the density over all real numbers that are less than or equal to x. Computationally,

$$P[X \le x] = F(x) = \int_{-\infty}^{x} f(t)\,dt \qquad x \text{ real}$$

Graphically, this probability corresponds to the area under the graph of the density to the left of and including the point x. Exercise 14 points out the mathematical properties of F.

Example 4.1.2. The density for the random variable X, the lead content in a liter of gasoline, is

$$f(x) = 12.5x - 1.25 \qquad .1 \le x \le .5$$

The cumulative distribution function for X is

$$P[X \le x] = F(x) = \int_{-\infty}^{x} f(t)\,dt$$

For $x < .1$ this integral has value 0 since for these values of x, $f(t)$ is itself 0. For $.1 \le x \le .5$,

$$\begin{aligned} F(x) = \int_{-\infty}^{x} f(t)\,dt &= \int_{.1}^{x} (12.5t - 1.25)\,dt \\ &= \left[\frac{12.5t^2}{2} - 1.25t\right]_{.1}^{x} \\ &= 6.25x^2 - 1.25x + .0625 \end{aligned}$$

For $x > .5$ the integral has value 1 since for these values of x we have integrated the density over its entire set of possible values. Summarizing, F is given by

$$F(x) = \begin{cases} 0 & x < .1 \\ 6.25x^2 - 1.25x + .0625 & .1 \le x \le .5 \\ 1 & x > .5 \end{cases}$$

What is the probability that the lead concentration in a randomly selected liter of gasoline will lie between .2 and .3 gram per liter? To answer this question, we rewrite it in terms of the cumulative distribution

$$\begin{aligned} P[.2 \le X \le .3] &= P[X \le .3] - P[X < .2] \\ &= P[X \le .3] - P[X \le .2] \qquad (X \text{ is continuous}) \\ &= F(.3) - F(.2) \end{aligned}$$

By substitution,

$$F(.3) = 6.25(.3)^2 - 1.25(.3) + .0625 = .2500$$

$$F(.2) = 6.25(.2)^2 - 1.25(.2) + .0625 = .0625$$

Thus

$$\begin{aligned} P[.2 \le X \le .3] &= F(.3) - F(.2) \\ &= .2500 - .0625 = .1875 \end{aligned}$$

Note that this agrees with the result obtained in Example 4.1.1 using direct integration. Note also that $F(.3)$ gives the area to the left of .3 shown in Fig. 4.3(a); $F(.2)$ gives the area to the left of .2 shown in Fig. 4.3(b). When we

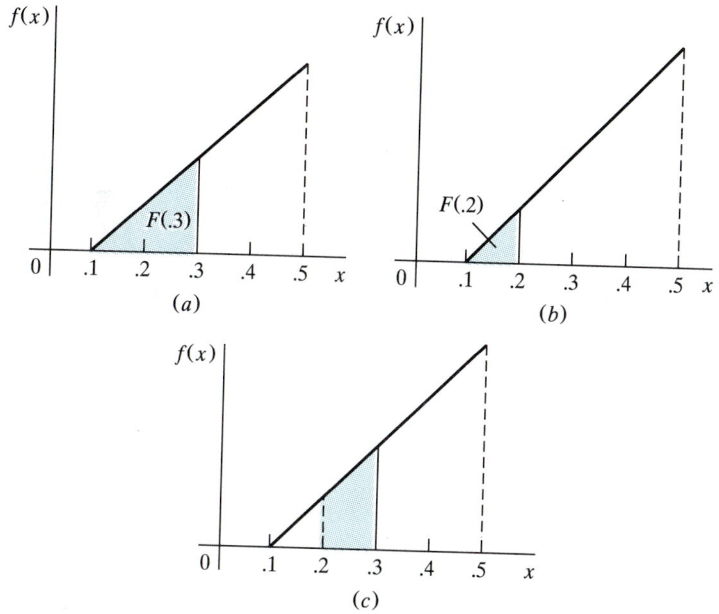

FIGURE 4.3
(a) $F(.3) = P[X \leq .3]$; (b) $F(.2) = P[X \leq .2]$; (c) $F(.3) - F(.2) = P[.2 \leq X \leq .3]$.

form the difference $F(.3) - F(.2)$, we naturally obtain the area between .2 and .3 given in Fig. 4.3(c).

Uniform Distribution

Perhaps the simplest continuous distribution with which to work is the *uniform* distribution. This distribution parallels the discrete uniform distribution presented in Exercise 34 of Chap. 3 in that, in a sense, events occur with equal or uniform probability. Since it is easy and instructive to develop the properties of this family of random variables directly from the definition, we leave the derivations to you. Important properties and applications are given in Exercises 5, 6, 10, 11, 18, 19, and 23.

4.2 EXPECTATION AND DISTRIBUTION PARAMETERS

In this section we define the term *expected value for continuous random variables*. We also discuss how to use the definition to find the moment generating function, the mean and the variance of a variable of the continuous type. As you will see, the definition parallels that given in the discrete case, with the summation operation being replaced by integration.

Definition 4.2.1 (Expected value). Let X be a continuous random variable with density f. Let $H(X)$ be a random variable. The expected value of $H(X)$, denoted by $E[H(X)]$, is given by

$$E[H(X)] = \int_{-\infty}^{\infty} H(x) f(x)\, dx$$

provided

$$\int_{-\infty}^{\infty} |H(x)| f(x)\, dx$$

is finite.

We illustrate the use of this definition by finding the mean and variance of the random variable X of Example 4.1.1.

Example 4.2.1. The density for X, the lead concentration in gasoline in grams per liter, is given by

$$f(x) = 12.5x - 1.25 \qquad .1 \le x \le .5$$

The mean or expected value of X is

$$\begin{aligned}
\mu = E[X] &= \int_{-\infty}^{\infty} x f(x)\, dx \\
&= \int_{.1}^{.5} x(12.5x - 1.25)\, dx \\
&= \left[\frac{12.5x^3}{3} - \frac{1.25x^2}{2} \right]_{.1}^{.5} \\
&= \left[\frac{(12.5)(.5)^3}{3} - \frac{1.25(.5)^2}{2} \right] - \left[\frac{12.5(.1)^3}{3} - \frac{1.25(.1)^2}{2} \right] \\
&\doteq .3667 \text{ g/liter}
\end{aligned}$$

Since integration is over an interval of finite length

$$\int_{-\infty}^{\infty} |x| f(x)\, dx$$

exists. We can conclude that, on the average, a liter of gasoline contains approximately .3667 g of lead. How much variability is there from liter to liter? To answer this question, we find $E[X^2]$ and apply Theorem 3.3.2 to find the variance of X:

$$\begin{aligned}
E[X^2] &= \int_{-\infty}^{\infty} x^2 f(x)\, dx \\
&= \int_{.1}^{.5} x^2(12.5x - 1.25)\, dx \\
&= \left[\frac{12.5x^4}{4} - \frac{1.25x^3}{3} \right]_{.1}^{.5} \doteq .1433
\end{aligned}$$

By Theorem 3.3.2,

$$\operatorname{Var} X = E[X^2] - (E[X])^2 \doteq .1433 - (.3667)^2 \doteq .00883$$

The standard deviation of X is

$$\sigma = \sqrt{\operatorname{Var} X} = \sqrt{.00883} = .09396 \text{ g/liter}$$

As in the discrete case, the moment generating function for a continuous random variable X is defined as $E[e^{tX}]$ provided this expectation exists for t in some open interval about 0. Its use is illustrated in the following example.

Example 4.2.2. The spontaneous flipping of a bit stored in a computer memory is called a "soft fail." Let X denote the time in millions of hours before the first soft fail is observed. Suppose that the density for X is given by

$$f(x) = e^{-x} \qquad x > 0$$

The mean and variance for X can be found directly using the method of Example 4.2.1. However, to find $E[X]$ and $E[X^2]$, integration by parts is required. This method of integration, although not difficult, is time-consuming. Let us find the moment generating function for X and use it to compute the mean and variance. By definition,

$$m_X(t) = E[e^{tX}] = \int_{-\infty}^{\infty} e^{tx} f(x)\,dx$$

In this case,

$$\begin{aligned} m_X(t) &= \int_0^{\infty} e^{tx} e^{-x}\,dx \\ &= \int_0^{\infty} e^{(t-1)x}\,dx \\ &= \frac{1}{t-1} e^{(t-1)x} \Big|_0^{\infty} \end{aligned}$$

Assume that $|t| < 1$. This guarantees that the exponent $(t-1)x < 0$, allowing us to evaluate the above integral. In particular,

$$m_X(t) = \frac{1}{1-t} \qquad |t| < 1$$

Since $e^{tx} > 0$, $|e^{tx}| = e^{tx}$. Thus the above argument has shown that

$$\int_{-\infty}^{\infty} |e^{tx}| f(x)\,dx$$

exists, as required in Definition 4.2.1. To use $m_X(t)$ to find $E[X]$ and $E[X^2]$, we apply Theorem 3.4.2. Note that

$$\frac{dm_X(t)}{dt} = \frac{d(1-t)^{-1}}{dt} = (1-t)^{-2}$$

$$\frac{d^2m_X(t)}{dt^2} = 2(1-t)^{-3}$$

$$E[X] = \left.\frac{dm_X(t)}{dt}\right|_{t=0} = 1$$

$$E[X^2] = \left.\frac{d^2m_X(t)}{dt^2}\right|_{t=0} = 2$$

$$\operatorname{Var} X = E[X^2] - (E[X])^2 = 2 - 1^2 = 1$$

The average or mean time that one must wait to observe the first soft fail is 1 million hours. The variance in waiting time is 1 and the standard deviation is 1 million hours.

To find the distribution parameters μ, σ^2, and σ, we can use either Definition 4.2.1 or the moment generating function technique. In practice, use whichever method is easier. Exercises 20 and 21 point out some interesting aspects of the mean and variance in the case of a continuous random variable.

4.3 GAMMA DISTRIBUTION

In this section we consider the gamma distribution. This distribution is especially important in that it allows us to define two families of random variables, the exponential and chi-squared, that are used extensively in applied statistics. The theoretical basis for the gamma distribution is the gamma function, a mathematical function defined in terms of an integral.

Definition 4.3.1 (Gamma function). The function Γ defined by

$$\Gamma(\alpha) = \int_0^\infty z^{\alpha-1}e^{-z}\,dz \qquad \alpha > 0$$

is called the gamma function.

Theorem 4.3.1 presents two numerical properties of the gamma function that are useful in evaluating the function for various values of α. Its proof is outlined in Exercise 26.

Theorem 4.3.1 (Properties of the gamma function)

1. $\Gamma(1) = 1$.
2. For $\alpha > 1$, $\Gamma(\alpha) = (\alpha - 1)\Gamma(\alpha - 1)$.

The use of Theorem 4.3.1 is illustrated in the next example.

Example 4.3.1

(*a*) Evaluate $\int_0^\infty z^3 e^{-z}\,dz$. To evaluate this integral using the methods of elementary calculus requires repeated applications of integration by parts. To evaluate the integral quickly, rewrite it as

$$\int_0^\infty z^3 e^{-z}\,dz = \int_0^\infty z^{4-1} e^{-z}\,dz$$

The integral on the right is Γ(4). By applying Theorem 4.3.1 repeatedly, it can be seen that

$$\begin{aligned}\int_0^\infty z^3 e^{-z}\,dz = \Gamma(4) &= 3 \cdot \Gamma(3)\\ &= 3 \cdot 2 \cdot \Gamma(2)\\ &= 3 \cdot 2 \cdot 1 \cdot \Gamma(1)\\ &= 3 \cdot 2 \cdot 1 = 6\end{aligned}$$

(*b*) Evaluate $\int_0^\infty (1/54)x^2 e^{-x/3}\,dx$. To evaluate this integral, we make a change of variable, a technique that is used extensively in deriving the properties of the gamma distribution. In particular, let $z = x/3$ or $3z = x$. Then $3\,dz = dx$ and the problem becomes

$$\begin{aligned}\int_0^\infty (1/54)x^2 e^{-x/3}\,dx &= \int_0^\infty 1/54(3z)^2 e^{-z} 3\,dz\\ &= 27/54 \int_0^\infty z^2 e^{-z}\,dz\end{aligned}$$

However,

$$\begin{aligned}\int_0^\infty z^2 e^{-z}\,dz = \int_0^\infty z^{3-1} e^{-z}\,dz &= \Gamma(3)\\ &= 2 \cdot \Gamma(2)\\ &= 2 \cdot 1 \cdot \Gamma(1)\\ &= 2 \cdot 1 = 2\end{aligned}$$

Thus

$$\int_0^\infty (1/54)x^2 e^{-x/3}\,dx = 27/54 \cdot 2 = 1$$

Note that since the nonnegative function

$$f(x) = (1/54)x^2 e^{-x/3}$$

has been shown to integrate to 1, it can be thought of as being a density for a continuous random variable X.

It is now possible to define the gamma distribution.

Gamma Distribution

Definition 4.3.2 (Gamma distribution). A random variable X with density

$$u > f(x) = \frac{1}{\Gamma(\alpha)\beta^{\alpha}} x^{\alpha-1} e^{-x/\beta} \qquad \begin{matrix} x > 0 \\ \alpha > 0 \\ \beta > 0 \end{matrix}$$

is said to have a gamma distribution with parameters α and β.

Although the mean and variance of a gamma random variable can be found easily from the definitions of these parameters (see Exercise 31), we shall use the moment generating function technique. As you will see later, it is very helpful to know the form of the moment generating function for a random variable.

Theorem 4.3.2. Let X be a gamma random variable with parameters α and β. Then

1. The moment generating function for X is given by

$$m_X(t) = (1 - \beta t)^{-\alpha} \qquad t < 1/\beta$$

2. $E[X] = \alpha\beta$

3. $\text{Var } X = \alpha\beta^2$

Proof. By definition,

$$m_X(t) = E[e^{tX}] = \int_0^{\infty} e^{tx} \frac{1}{\Gamma(\alpha)\beta^{\alpha}} x^{\alpha-1} e^{-x/\beta}\, dx$$

$$= \frac{1}{\Gamma(\alpha)\beta^{\alpha}} \int_0^{\infty} x^{\alpha-1} e^{-(1/\beta - t)x}\, dx$$

Let $z = (1 - \beta t)x/\beta$ so that $x = \beta z/(1 - \beta t)$. Then $\beta\, dz = (1 - \beta t)\, dx$ and $dx = \beta\, dz/(1 - \beta t)$. By making these substitutions, we obtain

$$m_X(t) = \frac{1}{\Gamma(\alpha)\beta^{\alpha}} \int_0^{\infty} \left(\frac{\beta z}{1 - \beta t}\right)^{\alpha-1} e^{-z} \frac{\beta\, dz}{(1 - \beta t)}$$

$$= \frac{1}{\Gamma(\alpha)\beta^{\alpha}} \frac{\beta^{\alpha}}{(1 - \beta t)^{\alpha}} \int_0^{\infty} z^{\alpha-1} e^{-z}\, dz$$

The integral on the right is $\Gamma(\alpha)$. Therefore

$$m_X(t) = \frac{1}{\Gamma(\alpha)\beta^{\alpha}} \frac{\beta^{\alpha}}{(1 - \beta t)^{\alpha}} \Gamma(\alpha) = (1 - \beta t)^{-\alpha}$$

We restrict t to be less than $1/\beta$ to avoid possible division by 0.

The proofs of parts 2 and 3 are straightforward applications of Theorem 3.4.2 and are left as exercises. (See Exercise 30.)

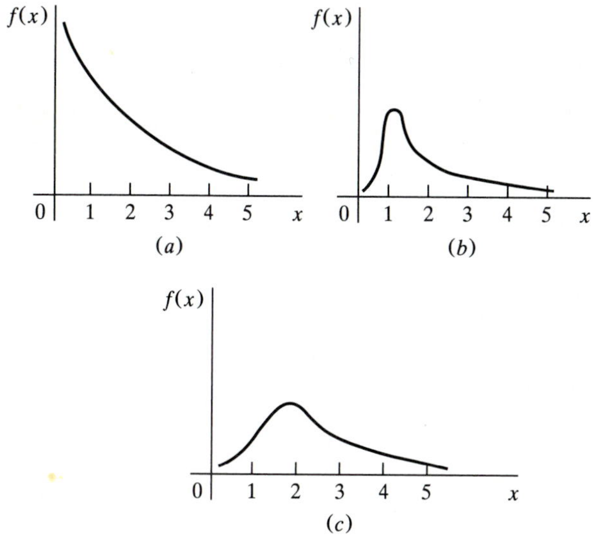

FIGURE 4.4
(a) $\alpha = 1$, $\beta = 1$, $\mu_X = 1$, $\sigma_X^2 = 1$; (b) $\alpha = 2$, $\beta = 1$, $\mu_X = 2$, $\sigma_X^2 = 2$; (c) $\alpha = 2$, $\beta = 2$, $\mu_X = 4$, $\sigma_X^2 = 8$.

Figure 4.4 shows the graphs of some gamma densities for a few values of α and β. Note that α and β both play a role in determining the mean and the variance of the random variable. Note also that the curves are not symmetric and are located entirely to the right of the vertical axis. It can be shown that for $\alpha > 1$, the maximum value of the density occurs at the point $x = (\alpha - 1)\beta$. (See Exercise 32.)

Exponential Distribution

As mentioned earlier, the gamma distribution gives rise to a family of random variables known as the *exponential* family. These variables are each gamma random variables with $\alpha = 1$. The density for an exponential random variable therefore assumes the form

Exponential density

$$f(x) = \frac{1}{\beta}e^{-x/\beta} \qquad \begin{array}{l} x > 0 \\ \beta > 0 \end{array}$$

The graph of a typical exponential density is shown in Fig. 4.4(a). This distribution arises often in practice in conjunction with the study of Poisson processes, which were discussed in Sec. 3.8. Recall that in a Poisson process discrete events are being observed over a continuous time interval. If we let W

denote the time of the occurrence of the first event, then W is a continuous random variable. Theorem 4.3.3 shows that W has an exponential distribution.

Theorem 4.3.3. Consider a Poisson process with parameter λ. Let W denote the time of the occurrence of the first event. W has an exponential distribution with $\beta = 1/\lambda$.

Proof. The distribution function F for W is given by

$$F(w) = P[W \le w] = 1 - P[W > w]$$

The first occurrence of the event will take place after time w only if no occurrences of the event are recorded in the time interval $[0, w]$. Let X denote the number of occurrences of the event in this time interval. X is a Poisson random variable with parameter λw. Thus

$$P[W > w] = P[X = 0] = \frac{e^{-\lambda w}(\lambda w)^0}{0!} = e^{-\lambda w}$$

By substitution we obtain

$$F(w) = 1 - P[W > w] = 1 - e^{-\lambda w}$$

Since in the continuous case the derivative of the cumulative distribution function is the density (see Exercise 14),

$$F'(w) = f(w) = \lambda e^{-\lambda w}$$

This is the density for an exponential random variable with $\beta = 1/\lambda$.

The next example illustrates the use of this theorem.

Example 4.3.2. Some strains of paramecia produce and secrete "killer" particles that will cause the death of a sensitive individual if contact is made. All paramecia unable to produce killer particles are sensitive. The mean number of killer particles emitted by a killer paramecium is 1 every 5 hours. In observing such a paramecium, what is the probability that we must wait at most 4 hours before the first particle is emitted? Considering the measurement unit to be one hour, we are observing a Poisson process with $\lambda = 1/5$. By Theorem 4.3.3, W, the time at which the first killer particle is emitted, has an exponential distribution with $\beta = 1/\lambda = 5$. The density for W is

$$f(w) = (1/5)e^{-w/5} \qquad w > 0$$

The desired probability is given by

$$\begin{aligned} P[W \le 4] &= \int_0^4 (1/5)e^{-w/5}\,dw \\ &= -e^{-w/5}\Big|_0^4 \\ &= 1 - e^{-4/5} \doteq .5507 \end{aligned}$$

Since an exponential random variable is also a gamma random variable, the average time that we must wait until the first killer particle is emitted is

$$E[W] = \alpha\beta = 1 \cdot 5 = 5 \text{ hours}$$

Forgetfulness Property: Exponential

The forgetfulness property of the geometric distribution was discussed in Sec. 3.4. It can be shown that the exponential distribution also exhibits this characteristic. For example, suppose that W denotes the time of the emission of the first killer particle as described in Example 4.3.2. Suppose that the process has already been observed for 3 hours and no killer particles have been seen so that $W > 3$. We now ask, "What is the probability that we must wait at least 4 more hours before observing such a particle?" Probability sought is $P[W \geq 7 | W > 3]$. The forgetfulness property implies that this probability is the same as the probability of having to wait at least 4 hours at the beginning of the observation period. That is,

$$P[W \geq 7 | W > 3] = P[W \geq 4].$$

In Example 4.3.2 it was shown that $P[W < 4] \doteq .5507$. Hence $P[W \geq 4] = 1 - .5507 = .4493$. Once again, the fact that nothing has happened for a while is "forgotten." The probability of no action is an interval of length 4 is the same at time 3 hours as it was at time 0. Our feeling that somehow a killer particle is "due" more at time 3 hours than at time 0 is not justifiable. You are asked to verify the forgetfulness property in general in Exercise 35.

Chi-Squared Distribution

The gamma distribution gives rise to another important family of random variables, namely, the *chi-squared* family. This distribution is used extensively in applied statistics. Among other things, it provides the basis for making inferences about the variance of a population based on a sample. At this time we consider only the theoretical properties of the chi-squared distribution. You will see many examples of its use in later chapters.

Definition 4.3.3 (Chi-squared distribution). Let X be a gamma random variable with $\beta = 2$ and $\alpha = \gamma/2$ for γ a positive integer. X is said to have a chi-squared distribution with γ degrees of freedom. We denote this variable by X^2_γ.

Note that a chi-squared random variable is completely specified by stating its degrees of freedom. By applying Theorem 4.3.2, we see that the mean of a chi-squared random variable is γ, its degrees of freedom; its variance is 2γ, twice its degrees of freedom. Figure 4.4(c) gives the graph of the density of a chi-squared random variable with 4 degrees of freedom.

Since the chi-squared distribution arises so often in practice, extensive tables of its cumulative distribution function have been derived. One such table is Table IV of App. A. In the table degrees of freedom appear as row headings, probabilities appear as column headings, and points associated with those probabilities are listed in the body of the table. Notationally, we shall use χ^2_r to

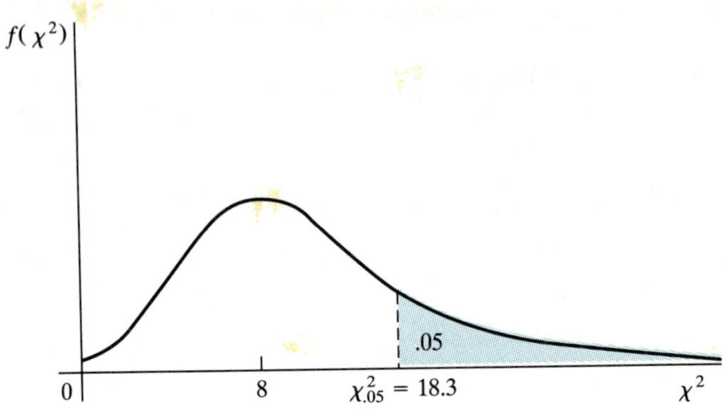

FIGURE 4.5
$P[X^2_{10} \geq \chi^2_{.05}] = .05$ and $P[X^2_{10} < \chi^2_{.05}] = .95$.

denote that point associated with a chi-squared random variable such that

$$P\left[X^2_\gamma \geq \chi^2_r\right] = r$$

That is, χ^2_r is the point such that the area to its *right* is r. Technically speaking, we should write $\chi^2_{r,\gamma}$ since the value of the point does depend on both the probability desired and the number of degrees of freedom associated with the random variable. However, in applications the value of γ will be obvious. Therefore to simplify notation, we use only a single subscript. The use of this notation is illustrated in the following example.

Example 4.3.3. Consider a chi-squared random variable with 10 degrees of freedom. Find the value of $\chi^2_{.05}$. This point is shown in Fig. 4.5. By definition the area to the right of this point is .05; the area to its left is .95. The column probabilities in Table IV give the area to the *left* of the point listed. Thus to find $\chi^2_{.05}$, we look in row 10 and column .95 and see that $\chi^2_{.05} = 18.3$.

4.4 NORMAL DISTRIBUTION

The normal distribution is a distribution that underlies many of the statistical methods used in data analysis. It was first described in 1733 by De Moivre as being the limiting form of the binomial density as the number of trials becomes infinite. This discovery did not get much attention, and the distribution was "discovered" again by both Laplace and Gauss a half-century later. Both men dealt with problems of astronomy, and each derived the normal distribution as a distribution that seemingly described the behavior of errors in astronomical measurements. The distribution is often referred to as the "gaussian" distribution.

Definition 4.4.1 (Normal distribution). A random variable X with density

$$f(x) = \frac{1}{\sqrt{2\pi}\,\sigma} e^{-(1/2)[(x-\mu)/\sigma]^2} \qquad \begin{array}{l} -\infty < x < \infty \\ -\infty < \mu < \infty \\ \sigma > 0 \end{array}$$

is said to have a normal distribution with parameters μ and σ.

One implication of this definition is that

$$\int_{-\infty}^{\infty} \frac{1}{\sqrt{2\pi}\,\sigma} e^{-(1/2)[(x-\mu)/\sigma]^2}\,dx = 1$$

To verify this requires a transformation to polar coordinates. This technique is beyond the mathematical level assumed here. A detailed proof can be found in [49]. Note that Definition 4.4.1 states only that μ is a real number and that σ is positive. As you might suspect from the notation used, the parameters that appear in the equation for the density for a normal random variable are, in fact, its mean and its standard deviation. This can be verified once we know the moment generating function for X. Our next theorem gives us the form for this important function.

Theorem 4.4.1. Let X be normally distributed with parameters μ and σ. The moment generating function for X is given by

$$m_X(t) = e^{\mu t + \sigma^2 t^2/2}$$

Proof. By definition,

$$\begin{aligned} m_X(t) = E[e^{tX}] &= \int_{-\infty}^{\infty} e^{tx} \frac{1}{\sqrt{2\pi}\,\sigma} e^{-(1/2)[(x-\mu)/\sigma]^2}\,dx \\ &= \frac{1}{\sqrt{2\pi}\,\sigma} \int_{-\infty}^{\infty} e^{tx-(1/2)[(x-\mu)/\sigma]^2}\,dx \end{aligned}$$

We complete the square in the exponent as follows:

$$\begin{aligned} &tx - (1/2)[(x-\mu)/\sigma]^2 \\ &\quad = tx - [1/2\sigma^2](x^2 - 2\mu x + \mu^2) \\ &\quad = -[1/2\sigma^2](x^2 - 2\mu x - 2\sigma^2 tx + \mu^2) \\ &\quad = -[1/2\sigma^2]\left[x^2 - 2(\mu + \sigma^2 t)x + (\mu + \sigma^2 t)^2 - (\mu + \sigma^2 t)^2 + \mu^2\right] \\ &\quad = -[1/2\sigma^2][x - (\mu + \sigma^2 t)]^2 \\ &\qquad + [1/2\sigma^2](\mu^2 + 2\mu\sigma^2 t + \sigma^4 t^2) - [1/2\sigma^2]\mu^2 \\ &\quad = -[1/2\sigma^2][x - (\mu + \sigma^2 t)]^2 + \mu t + \sigma^2 t^2/2 \end{aligned}$$

By substituting this expression into that for $m_X(t)$, we obtain

$$m_X(t) = \frac{1}{\sqrt{2\pi}\,\sigma}\int_{-\infty}^{\infty} e^{-[1/2\sigma^2][x-(\mu+\sigma^2 t)]^2+\mu t+\sigma^2 t^2/2}\,dx$$

$$= \frac{1}{\sqrt{2\pi}\,\sigma}\int_{-\infty}^{\infty} e^{-(1/2)[x-(\mu+\sigma^2 t)/\sigma]^2} e^{\mu t+\sigma^2 t^2/2}\,dx$$

$$= e^{\mu t+\sigma^2 t^2/2}\int_{-\infty}^{\infty} \frac{1}{\sqrt{2\pi}\,\sigma} e^{-(1/2)[x-(\mu+\sigma^2 t)/\sigma]^2}\,dx$$

The function

$$f(x) = \frac{1}{\sqrt{2\pi}\,\sigma} e^{-(1/2)[x-(\mu+\sigma^2 t)/\sigma]^2}$$

is the density for a normal random variable with parameters σ and $\mu + \sigma^2 t$ and thus integrates to 1. This implies that

$$m_X(t) = e^{\mu t+\sigma^2 t^2/2}$$

as claimed.

It is now easy to show that the parameters that appear in the definition of the normal density are actually the mean and the standard deviation of the variable.

Theorem 4.4.2. Let X be a normal random variable with parameters μ and σ. Then μ is the mean of X and σ is its standard deviation.

Proof. The moment generating function for X is

$$m_X(t) = e^{\mu t+\sigma^2 t^2/2}$$

and

$$\frac{dm_X(t)}{dt} = e^{\mu t+\sigma^2 t^2/2}(\mu + \sigma^2 t)$$

By Theorem 3.4.2 the mean of X is given by

$$E[X] = \left.\frac{dm_X(t)}{dt}\right|_{t=0} = e^{\mu\cdot 0+\sigma^2 0^2/2}(\mu + \sigma^2\cdot 0) = \mu$$

as claimed. The proof of the remainder of the theorem is left as an exercise (Exercise 44).

The graph of the density of a normal random variable is a symmetric, bell-shaped curve centered at its mean. The points of inflection occur at $\mu \pm \sigma$. These facts can be verified easily (see Exercise 43).

Example 4.4.1. One of the major contributors to air pollution is hydrocarbons emitted from the exhaust system of automobiles. Let X denote the number of grams of hydrocarbons emitted by an automobile per mile. Assume that X is normally distributed with a mean of 1 gram and a standard deviation of .25 gram.

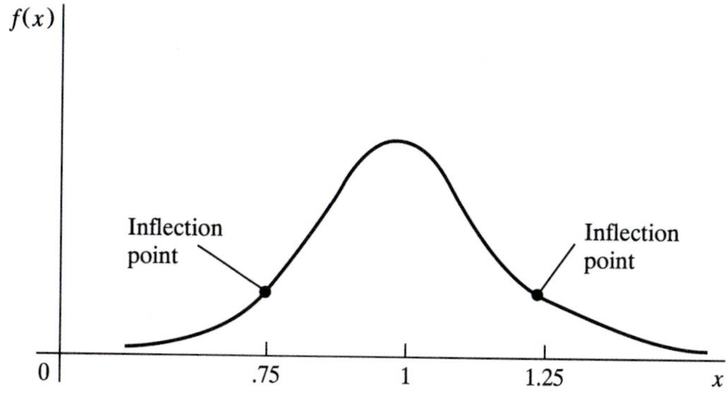

FIGURE 4.6
Graph of the density for a normal random variable with mean 1 and standard deviation .25.

The density for X is given by

$$f(x) = \frac{1}{\sqrt{2\pi}(.25)} e^{-(1/2)[(x-1)/.25]^2}$$

The graph of this density is a symmetric, bell-shaped curve centered at $\mu = 1$ with inflection points at $\mu \pm \sigma$, or $1 \pm .25$. A sketch of the density is given in Fig. 4.6.

One point must be made. Theoretically speaking, a normal random variable must be able to assume any value whatsoever. This is clearly unrealistic here. It is impossible for an automobile to emit a negative amount of hydrocarbons. When we say that X is normally distributed, we mean that over the range of physically reasonable values of X, the given normal curve yields acceptable probabilities. With this understanding, we can at least approximate, for example, the probability that a randomly selected automobile will emit between .9 and 1.54 grams of hydrocarbons by finding the area under the graph of f between these two points.

Standard Normal Distribution

There are infinitely many normal random variables each of which is uniquely characterized by the two parameters μ and σ. To calculate probabilities associated with a specific normal curve requires that one integrate the normal density over a particular interval. However, the normal density is not integrable in closed form. To find areas under the normal curve requires the use of numerical integration techniques. A simple algebraic transformation is employed to overcome this problem. By means of this transformation, called the *standardization procedure*, any question about any normal random variable can be transformed to an equivalent question concerning a normal random variable with mean 0 and standard deviation 1. This particular normal random variable is denoted by Z and is called the *standard normal* variable.

Theorem 4.4.3 (Standardization theorem). Let X be normal with mean μ and standard deviation σ. The variable $(X - \mu)/\sigma$ is standard normal.

You have already verified that the transformation yields a random variable with mean 0 and standard deviation 1 (see Chap. 3, Exercise 21). To prove that the transformed variable is normal requires the use of moment generating function techniques to be introduced in Chap. 7.

The cumulative distribution function for the standard normal random variable is given in Table V of App. A. The use of the standardization theorem and this table is illustrated in the following example.

Example 4.4.2. Let X denote the number of grams of hydrocarbons emitted by an automobile per mile. Assuming that X is normal with $\mu = 1$ gram and $\sigma = .25$ gram, find the probability that a randomly selected automobile will emit between .9 and 1.54 grams of hydrocarbons per mile. The desired probability is shown in Fig. 4.7. To find $P[.9 \le X \le 1.54]$, we first standardize by subtracting the mean of 1 and dividing by the standard deviation of .25 across the inequality. That is,

$$P[.9 \le X \le 1.54] = P[(.9 - 1)/.25 \le (X - 1)/.25 \le (1.54 - 1)/.25]$$

The random variable $(X - 1)/.25$ is now Z. Therefore the problem is to find $P[-.4 \le Z \le 2.16]$ from Table V. We first express the desired probability in terms of the cumulative distribution as follows:

$$\begin{aligned} P[-.4 \le Z \le 2.16] &= P[Z \le 2.16] - P[Z < -.4] \\ &= P[Z \le 2.16] - P[Z \le -.4] \qquad (Z \text{ is continuous}) \\ &= F(2.16) - F(-.4) \end{aligned}$$

$F(2.16)$ is found by locating the first two digits (2.1) in the column headed z; since the third digit is 6, the desired probability of .9846 is found in the row labeled 2.1 and the column labeled .06. Similarly, $F(-.4)$ or .3446 is found in the row labeled -0.4 and the column labeled .00. We now see that the probability that a randomly

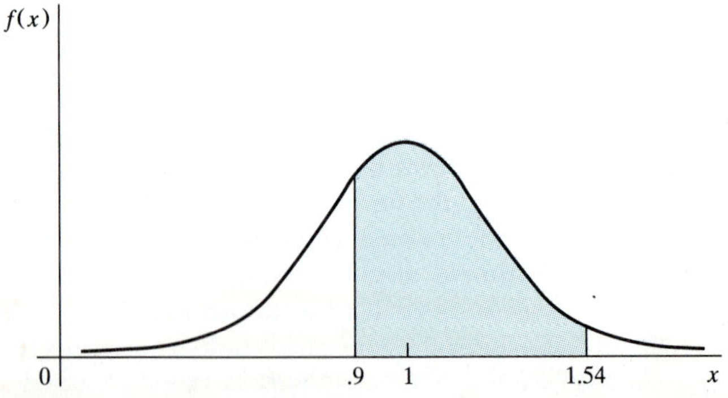

FIGURE 4.7
Shaded area = $P[.9 \le X \le 1.54]$.

selected automobile will emit between .9 and 1.54 grams of hydrocarbons per mile is

$$\begin{aligned} P[.9 \leq X \leq 1.54] &= P[-.4 \leq Z \leq 2.16] \\ &= F(2.16) - F(-.4) \\ &= .9846 - .3446 = .64 \end{aligned}$$

Interpreting this probability as a percentage, we can say that 64% of the automobiles in operation emit between .9 and 1.54 grams of hydrocarbons per mile driven.

We shall have occasion to read Table V in reverse. That is, given a particular probability r we shall need to find the point with r of the area to its right. This point is denoted by z_r. Thus, notationally, z_r denotes that point associated with a standard normal random variable such that

$$P[Z \geq z_r] = r$$

To see how this need arises, consider Example 4.4.3.

Example 4.4.3. Let X denote the amount of radiation that can be absorbed by an individual before death ensues. Assume that X is normal with a mean of 500 roentgens and a standard deviation of 150 roentgens. Above what dosage level will only 5% of those exposed survive? Here we are asked to find the point x_0 shown in Fig. 4.8. In terms of probabilities, we want to find the point x_0 such that

$$P[X \geq x_0] = .05$$

Standardizing gives

$$\begin{aligned} P[X \geq x_0] &= P\left[\frac{X - 500}{150} \geq \frac{x_0 - 500}{150}\right] \\ &= P\left[Z \geq \frac{x_0 - 500}{150}\right] = .05 \end{aligned}$$

Thus $(x_0 - 500)/150$ is the point on the standard normal curve with 5% of the area under the curve to its right and 95% to its left. That is, $(x_0 - 500)/150$ is the point $z_{.05}$. From Table V the numerical value of this point is approximately 1.645

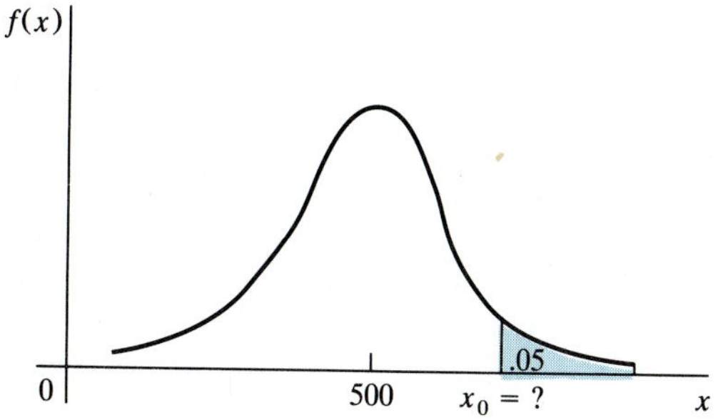

FIGURE 4.8
$P[X \geq x_0] = .05.$

(we have interpolated). Equating these, we get

$$\frac{x_0 - 500}{150} = 1.645$$

Solving this equation for x_0 gives the desired dosage level:

$$x_0 = 150(1.645) + 500 = 746.75 \text{ roentgens}$$

4.5 NORMAL PROBABILITY RULE AND CHEBYSHEV'S INEQUALITY (OPTIONAL)

It is sometimes useful to have a quick way of determining which values of a random variable are common and which are considered to be rare. In the case of a normally distributed random variable, a rule of thumb, called the *normal probability rule*, can be developed easily. This rule is given in Theorem 4.5.1.

Theorem 4.5.1 (Normal probability rule). Let X be normally distributed with parameters μ and σ. Then

$$P[-\sigma < X - \mu < \sigma] \doteq .68$$

$$P[-2\sigma < X - \mu < 2\sigma] \doteq .95$$

$$P[-3\sigma < X - \mu < 3\sigma] \doteq .99$$

Proof. Note that division by σ yields

$$P[-\sigma < X - \mu < \sigma] = P\left[-1 < \frac{X - \mu}{\sigma} < 1\right]$$

By Theorem 4.4.3, $(X - \mu)/\sigma$ follows the standard normal distribution. From Table V of App. A,

$$P[-1 < Z < 1] = .8413 - .1587 = .6826$$

This probability can be rounded to .68. The other results given in the theorem are proved similarly.

The normal probability rule can be expressed in terms of percentages. In particular, it implies that in repeated sampling from a normal distribution approximately 68% of the observed values of X should lie within 1 standard deviation of its mean; 95% should lie within two standard deviations, and 99% within 3 standard deviations of the mean. Thus an observed value that falls farther than 3 standard deviations from μ is indeed rare since such values occur with probability .01. This rule will be used later to obtain a quick estimate of the standard deviation of a normally distributed random variable.

Chebyshev's Inequality

A second rule of thumb that can be used to gauge the rarity of observed values of a random variable is *Chebyshev's inequality*. This inequality was derived by the Russian probabilist P. L. Chebyshev (Tchebysheff, 1821–1894). The inequality differs from the normal probability rule in that it does *not* require that the random variable involved be normally distributed. Although we shall prove the theorem in the continuous setting, continuity is *not* required. The inequality holds for any random variable.

Theorem 4.5.2 (Chebyshev's inequality). Let X be a random variable with mean μ and standard deviation σ. Then for any positive number k,

$$P[|X-\mu| < k\sigma] \geq 1 - \frac{1}{k^2}$$

Proof. Assume that X is continuous with mean μ, standard deviation σ, and density f. By definition,

$$\sigma^2 = \mathrm{Var}(X) = \int_{-\infty}^{\infty} (x-\mu)^2 f(x)\,dx$$

Let $k > 0$, $c = k^2\sigma^2$, and note that

$$\sigma^2 = \int_{-\infty}^{\mu-\sqrt{c}} (x-\mu)^2 f(x)\,dx + \int_{\mu-\sqrt{c}}^{\mu+\sqrt{c}} (x-\mu)^2 f(x)\,dx + \int_{\mu+\sqrt{c}}^{\infty} (x-\mu)^2 f(x)\,dx$$

Since $(x-\mu)^2 f(x) \geq 0$,

$$\int_{\mu-\sqrt{c}}^{\mu+\sqrt{c}} (x-\mu)^2 f(x)\,dx \geq 0$$

and thus

$$\sigma^2 \geq \int_{-\infty}^{\mu-\sqrt{c}} (x-\mu)^2 f(x)\,dx + \int_{\mu+\sqrt{c}}^{\infty} (x-\mu)^2 f(x)\,dx$$

Note that over both regions of integration $(x-\mu)^2 \geq c$, and so it can be concluded that

$$\sigma^2 \geq \int_{-\infty}^{\mu-\sqrt{c}} cf(x)\,dx + \int_{\mu+\sqrt{c}}^{\infty} cf(x)\,dx$$

In terms of probabilities,

$$\sigma^2 \geq cP\left[X \leq \mu - \sqrt{c}\right] + cP\left[X \geq \mu + \sqrt{c}\right]$$

or $$\sigma^2 \geq c\left\{P\left[X-\mu \leq -\sqrt{c}\right] + P\left[X-\mu \geq \sqrt{c}\right]\right\}$$

This inequality can be rewritten to conclude that

$$P\left[X-\mu \leq -\sqrt{c}\right] + P\left[X-\mu \geq \sqrt{c}\right] \leq \frac{\sigma^2}{c}$$

or that $$P\left[-\sqrt{c} \leq X-\mu \leq \sqrt{c}\right] \geq 1 - \frac{\sigma^2}{c}$$

Since $c = k^2\sigma^2$ where k and c are each nonnegative, $\sqrt{c} = k\sigma$. Substitution yields

$$P[-k\sigma \le X - \mu \le k\sigma] \ge 1 - \frac{1}{k^2}$$

or

$$P[|X - \mu| \le k\sigma] \ge 1 - \frac{1}{k^2}$$

Since X is continuous, we can conclude that

$$P[|X - \mu| < k\sigma] \ge 1 - \frac{1}{k^2}$$

as claimed. The proof in the discrete case is similar with summation replacing integration.

Some examples will clarify the difference between Theorems 4.5.1 and 4.5.2.

Example 4.5.1. The viscosity of a fluid can be measured roughly by dropping a small ball into a calibrated tube containing the fluid and observing X, the time that it takes for the ball to drop a measured distance. Assume that this random variable is normally distributed with a mean of 20 s and a standard deviation of .5 s. By the normal probability rule, approximately 95% of the observed values of X will lie within 1 s (2 standard deviations) of the mean. That is, X will fall between 19 and 21 s with probability .95. Since Chebyshev's inequality applies to any random variable, it is appropriate here. This inequality guarantees that X will fall between 19 and 21 s (within $k = 2$ standard deviations of its mean) with probability *at least* $1 - 1/k^2 = .75$. Note that when the random variable in question is normally distributed, the normal probability rule yields a stronger statement than does Chebyshev's inequality.

Example 4.5.2. The safety record of an industrial plant is measured in terms of M, the total man-hours worked without a serious accident. Past experience indicates that M has a mean of 2 million with a standard deviation of .1 million. A serious accident has just occurred. Would it be unusual for the next serious accident to occur within the next 1.6 million man-hours? To answer this question, we must assess $P[M \le 1.6]$. Since we have no reason to assume that M is normally distributed, the normal probability rule is inappropriate here. However, we know from Chebyshev's inequality with $k = 4$ that

$$P[1.6 < M < 2.4] \ge 1 - (1/16) = .9375$$

This implies that

$$P[M \le 1.6] + P[M \ge 2.4] \le .0625$$

Since it is possible for M to exceed 2.4, we can safely say that

$$P[M \le 1.6] < .0625$$

No stronger statement can be made without some knowledge of the shape of the density of M. However, if it is known that the density is symmetric, then we can go one step further and state that

$$P[M \le 1.6] \le .0625/2 = .03125$$

4.6 NORMAL APPROXIMATIONS

We saw in Sec. 3.8 that the Poisson density can be used to approximate binomial probabilities when n is large. The normal curve can be used for the same purpose. To see how this can be done, we consider four binomial random variables each with probability of success .4 but with differing values for n. The densities for these variables, obtained from Table I of App. A, together with a sketch for each, are given in Fig. 4.9(a) to (d).

The point to note from these diagrams is made in Fig. 4.9(d). Namely, it is not hard to imagine a smooth bell curve that closely fits the block diagram shown. This suggests that binomial probabilities represented by one or more blocks in the diagram can be approximated reasonably well by a carefully selected area under an appropriately chosen normal curve. Which of the infinitely many normal curves is appropriate? Common sense indicates that the normal variable selected should have the same mean and variance as the binomial variable that it approximates. Theorem 4.6.1 summarizes these ideas.

Theorem 4.6.1 (Normal approximation to the binomial distribution). Let X be binomial with parameters n and p. For large n, X is approximately normal with mean np and variance $np(1-p)$.

The proof of this theorem is based on the Central Limit Theorem, which will be considered in Chap. 7. Admittedly, Theorem 4.6.1 is a bit vague in the sense that the word "large" is not well defined. In the strictest mathematical sense, "large" means as n approaches infinity. For most practical purposes the approximation is acceptable for values of n and p such that either $p \leq .5$ and $np > 5$ or $p > .5$ and $n(1-p) > 5$.

Example 4.6.1. A study is performed to investigate the connection between maternal smoking during pregnancy and birth defects in children. Of the mothers studied, 40% smoke and 60% do not. When the babies were born, 20 were found to have some sort of birth defect. Let X denote the number of children whose mother smoked while pregnant. If there is no relationship between maternal smoking and birth defects, then X is binomial with $n = 20$ and $p = .4$. What is the probability that 12 or more of the affected children had mothers who smoked?

To answer this question, we need to find $P[X \geq 12]$ under the assumption that X is binomial with $n = 20$ and $p = .4$. This probability, .0565, can be found from Table I of App. A. Note that since $p = .4 \leq .5$ and $np = 20(.4) = 8 > 5$, the normal approximation should give a result quite close to .0565. We shall approximate probabilities associated with X using a normal random variable Y with mean $np = 20(.4) = 8$ and standard deviation $\sqrt{np(1-p)} = \sqrt{20(.4)(.6)} = \sqrt{4.8}$.

The exact probability of .0565 is given by the sum of the areas of the blocks centered at 12, 13, 14, 15, 16, 17, 18, 19, and 20, as shown in Fig. 4.10. The approximate probability is given by the area under the normal curve shown above 11.5. That is,

$$P[X \geq 12] \doteq P[Y \geq 11.5]$$

$n = 5$, $p = .4$

x	$f(x)$
0	.0778
1	.2592
2	.3456
3	.2304
4	.0768
5	.0102

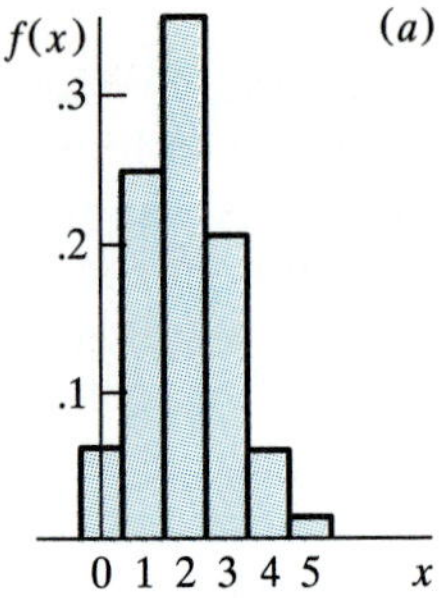

$n = 10$, $p = .4$

x	$f(x)$
0	.0060
1	.0404
2	.1209
3	.2150
4	.2508
5	.2007
6	.1114
7	.0425
8	.0106
9	.0016
10	.0001

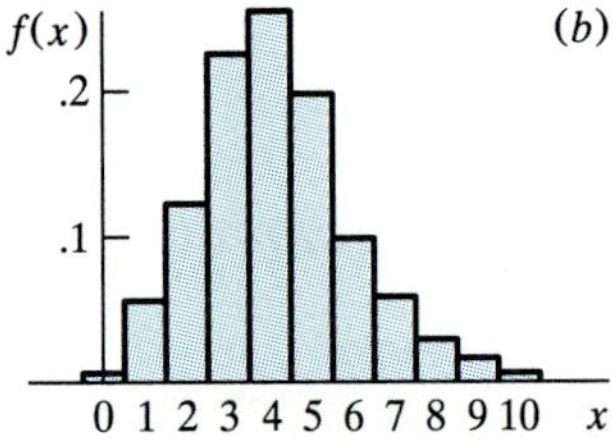

$n = 15$, $p = .4$

x	$f(x)$	x	$f(x)$
0	.0005	8	.1181
1	.0047	9	.0612
2	.0219	10	.0245
3	.0634	11	.0074
4	.1268	12	.0016
5	.1859	13	.0003
6	.2066	14	~ 0
7	.1771	15	~ 0

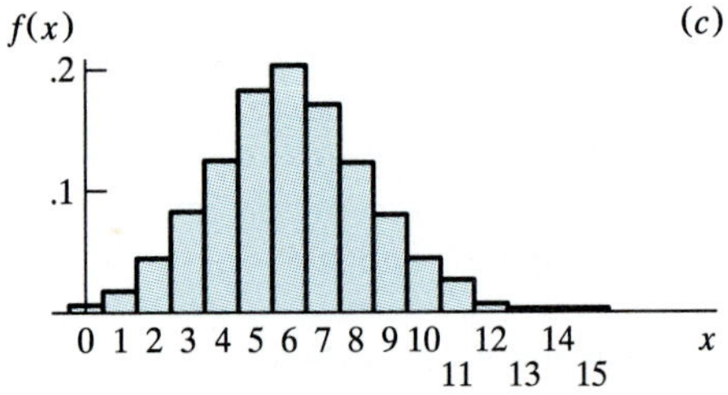

$n = 20$, $p = .4$

x	$f(x)$	x	$f(x)$
0	~ 0	11	.0710
1	.0005	12	.0355
2	.0031	13	.0145
3	.0124	14	.0049
4	.0350	15	.0013
5	.0746	16	.0003
6	.1244	17	~ 0
7	.1659	18	~ 0
8	.1797	19	~ 0
9	.1597	20	~ 0
10	.1172		

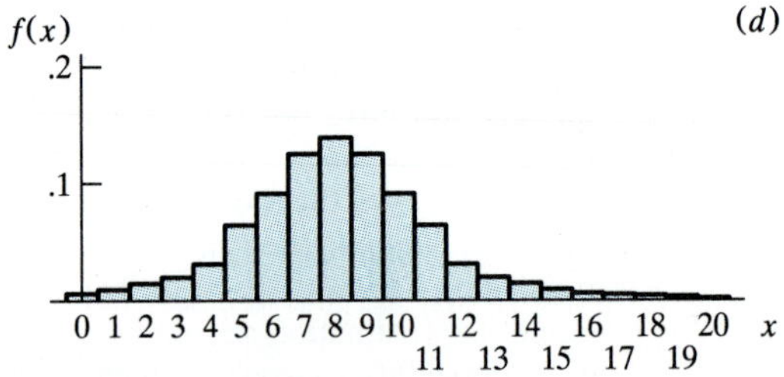

FIGURE 4.9
Density for X binomial: (*a*) $n = 5$, $p = .4$; (*b*) $n = 10$, $p = .4$; (*c*) $n = 15$, $p = .4$; (*d*) $n = 20$, $p = .4$.

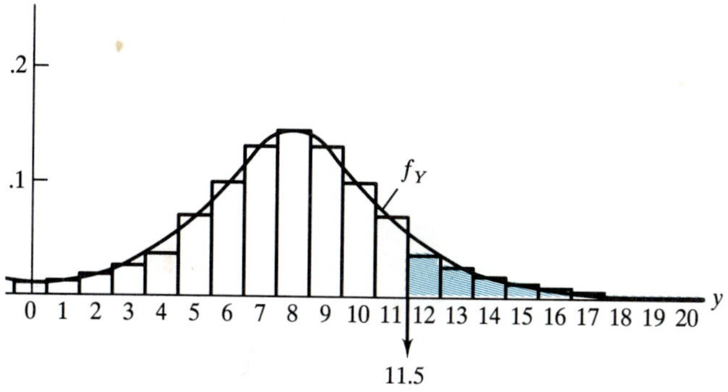

FIGURE 4.10
$P[X \geq 12]$ = area of shaded blocks $\doteq$ area under curve beyond 11.5.

The number .5 is called the *half-unit correction* for continuity. It is subtracted from 12 in the approximation because otherwise half the area of the block centered at 12 will be inadvertently ignored, leading to an unnecessary error in the calculation. From this point on the calculation is routine:

$$
\begin{aligned}
P[X \geq 12] &\doteq P[Y \geq 11.5] \\
&= P\left[\frac{Y-8}{\sqrt{4.8}} \geq \frac{11.5-8}{\sqrt{4.8}}\right] \\
&= P[Z \geq 1.59] \\
&= 1 - .9441 = .0559
\end{aligned}
$$

Note that even with n as small as 20, the approximated value of .0559 compares quite favorably with the exact value of .0565. In practice, of course, one would not approximate a probability that could be found directly from a binomial table. This was done here only for comparative purposes.

Poisson probabilities can also be approximated with the help of a normal curve. The method for doing so is outlined in Exercise 56.

4.7 WEIBULL DISTRIBUTION AND RELIABILITY

In 1951 W. Weibull introduced a distribution that has been found to be useful in a variety of physical applications. It arises quite naturally in the study of reliability as we shall show. The most general form for the Weibull density is given by

$$
f(x) = \alpha\beta(x-\gamma)^{\beta-1}e^{-\alpha(x-\gamma)^{\beta}} \qquad \begin{array}{l} x > \gamma \\ \alpha > 0 \\ \beta > 0 \end{array}
$$

The implication of this definition of the density is that there is some minimum or "threshold" value γ below which the random variable X cannot fall. In most physical applications this value is 0. For this reason, we shall define the Weibull density with this fact in mind. Be careful when reading scientific literature to note the form of the Weibull density being used.

Definition 4.7.1 (Weibull distribution). A random variable X is said to have a Weibull distribution with parameters α and β if its density is given by

$$f(x) = \alpha\beta x^{\beta-1}e^{-\alpha x^{\beta}} \qquad \begin{array}{l} x > 0 \\ \alpha > 0 \\ \beta > 0 \end{array}$$

It is easy to verify that the function given in Definition 4.7.1 is a density. (See Exercise 61.) We shall find the mean of this distribution directly rather than by means of the moment generating function.

Theorem 4.7.1. Let X be a Weibull random variable with parameters α and β. The mean and variance of X are given by

$$\mu = \alpha^{-1/\beta}\Gamma(1 + 1/\beta)$$

and

$$\sigma^2 = \alpha^{-2/\beta}\Gamma(1 + 2/\beta) - \mu^2$$

Proof. By Definition 4.2.1,

$$E[X] = \int_0^\infty x\alpha\beta x^{\beta-1}e^{-\alpha x^{\beta}}\,dx$$

$$= \int_0^\infty \alpha\beta x^{\beta}e^{-\alpha x^{\beta}}\,dx$$

Let $z = \alpha x^{\beta}$. This implies that

$$x = (z/\alpha)^{1/\beta} \qquad \text{and} \qquad dx = (1/\alpha\beta)(z/\alpha)^{1/\beta - 1}\,dz$$

By substitution, it is seen that

$$E[X] = \int_0^\infty \alpha\beta(z/\alpha)e^{-z}(1/\alpha\beta)(z/\alpha)^{1/\beta-1}\,dz$$

$$= \int_0^\infty (z/\alpha)^{1/\beta}e^{-z}\,dz$$

$$= \alpha^{-1/\beta}\int_0^\infty z^{1/\beta}e^{-z}\,dz$$

The integral on the right is, by definition, $\Gamma(1 + 1/\beta)$. (See Definition 4.3.1.) Thus we have shown that the mean of the Weibull distribution is

$$\mu = E[X] = \alpha^{-1/\beta}\Gamma(1 + 1/\beta)$$

as claimed. The remainder of the proof is outlined as an exercise. (See Exercises 62 and 63.)

The graph of the Weibull density varies depending on the values of α and β. The general shape resembles that of the gamma density with the curve becoming more symmetric as the value of β increases.

Example 4.7.1. Let X be a Weibull random variable with $\beta = 1$. The density for X is

$$f(x) = \alpha e^{-\alpha x} \qquad x > 0, \quad \alpha > 0$$

Note that this is the density for an *exponential* random variable. That is, the exponential distribution is a special case of the Weibull distribution with $\beta = 1$. By Theorem 4.7.1

$$\mu = \alpha^{-1/\beta}\Gamma(1 + 1/\beta) = (1/\alpha)\Gamma(2) = 1/\alpha \cdot 1! = 1/\alpha$$

$$\begin{aligned}\sigma^2 &= \alpha^{-2/\beta}\Gamma(1 + 2/\beta) - \mu^2 \\ &= 1/\alpha^2\Gamma(3) - (1/\alpha)^2 \\ &= 2/\alpha^2 - 1/\alpha^2 = 1/\alpha^2\end{aligned}$$

Note that these results are consistent with those obtained by viewing this random variable as being exponential. (See Exercise 33.)

Reliability

As we have said, the Weibull distribution frequently arises in the study of reliability. Reliability studies are concerned with assessing whether or not a system functions adequately under the conditions for which it was designed. Interest centers on describing the behavior of the random variable X, the time to failure of a system that cannot be repaired once it fails to operate. Three functions come into play when assessing reliability. These are the failure density f, the reliability function R, and ρ, the failure or hazard rate of the distribution. To understand how these functions are defined, consider some system being put into operation at time $t = 0$. We observe the system until it eventually fails. Let X denote the time of the failure. This random variable is continuous and "a priori" can assume any value in the interval $(0, \infty)$. The density f, for X, is called the *failure density* for the component. The *reliability function*, R, is defined to be the probability that the component will not fail before time t. Thus

$$\begin{aligned}R(t) &= 1 - P[\text{component will fail before time } t] \\ &= 1 - \int_0^t f(x)\,dx \\ &= 1 - F(t)\end{aligned}$$

where F is the cumulative distribution function for X. To define ρ, the hazard rate function, consider a time interval $[t, t + \Delta t]$ of length Δt. We define the force of mortality or hazard rate function over this interval by

$$\rho(t) = \lim_{\Delta t \to 0} P(t \le X \le t + \Delta t | t \le X) \frac{1}{\Delta t}$$

$$= \lim_{\Delta t \to 0} \frac{\text{probability of failure in } [t, t + \Delta t]}{\text{probability of failure in } [t, \infty)} \cdot \frac{1}{\Delta t}$$

Notice that

$$\lim_{\Delta t \to 0} \frac{\text{probability of failure in } [t, t + \Delta t]}{\Delta t} = \lim_{\Delta t \to 0} \frac{F(t + \Delta t) - F(t)}{\Delta t}$$

This, by definition, is the derivative of the cumulative distribution function for X. Since the derivative of a function in general can be interpreted as giving the "instantaneous rate of change" of the function, this portion of the definition of $\rho(t)$ gives the instantaneous rate of change of F at time t. Since a cumulative distribution function cannot decrease, the derivative of F will always be nonnegative. Its magnitude tells us how fast failures are occurring at any given time. A large value of $F'(t)$ implies a steep curve at t, which in turn implies that failures are coming rapidly in an interval near t; a small value of $F'(t)$ implies that failures are occurring at a slower pace. (See Fig. 4.11.) Thus we can say that $\rho(t)$ gives us a picture of the instantaneous rate of failure at times t given that the system was operable prior to this time.

Theorem 4.7.2 relates the three functions f, R, and ρ.

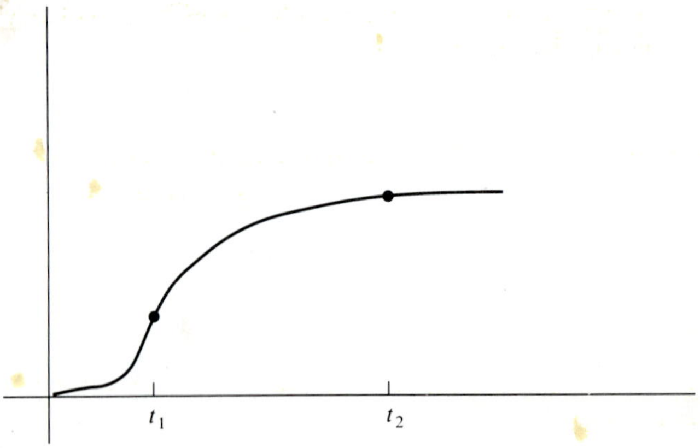

FIGURE 4.11
At time t_1, the graph of F is steep. Failures are likely to occur in the time interval near t_1. At time t_2, the graph of F is rather flat. Failures are not highly likely in the time interval near t_2.

Theorem 4.7.2. Let X be a random variable with failure density f, reliability function R, and hazard rate function ρ. Then

$$\rho(t) = \frac{f(t)}{R(t)}$$

Proof. By definition,

$$\begin{aligned}
\rho(t) &= \lim_{\Delta t \to 0} \frac{\text{probability of failure in } [t, t + \Delta t]}{\text{probability of failure in } [t, \infty]} \cdot \frac{1}{\Delta t} \\
&= \lim_{\Delta t \to 0} \frac{\int_t^{t+\Delta t} f(x)\, dx}{\int_t^{\infty} f(x)\, dx} \cdot \frac{1}{\Delta t} \\
&= \lim_{\Delta t \to 0} \frac{F(t + \Delta t) - F(t)}{1 - F(t)} \cdot \frac{1}{\Delta t} \\
&= \lim_{\Delta t \to 0} \frac{F(t + \Delta t) - F(t)}{\Delta t} \cdot \frac{1}{R(t)} \\
&= \frac{F'(t)}{R(t)} = \frac{f(t)}{R(t)}
\end{aligned}$$

The job of the scientist is to find the form of these functions for the problem at hand. In practice, one often begins by assuming a particular form for the hazard rate function based on empirical evidence. To do so, one must have some practical way to interpret ρ. A rough interpretation is as follows:

1. If ρ is increasing over an interval, then as time goes by a failure is more likely to occur. This normally happens for systems which begin to fail primarily due to wear.
2. If ρ is decreasing over an interval, then as time goes by a failure is less likely to occur than it was earlier in the time interval. This happens in situations in which defective systems tend to fail early. As time goes by, the hazard rate for a well-made system decreases.
3. A steady hazard rate is expected over the useful life span of a component. A failure tends to occur during this period due mainly to random factors.

Since one often has an idea of the form only of ρ, the natural question to ask is: "Can we derive the failure density and the reliability function from knowledge of ρ?" Theorem 4.7.3 shows how this can be done.

Theorem 4.7.3. Let X be a random variable with failure density f, reliability function R, and hazard rate ρ. Then

$$R(t) = \exp\left[-\int_0^t \rho(x)\, dx\right]$$

and $f(t) = \rho(t)R(t)$.

Proof. Note that since $R(x) = 1 - F(x)$, $R'(x) = -F'(x)$. Therefore

$$\rho(x) = \frac{f(x)}{R(x)} = \frac{F'(x)}{R(x)} = \frac{-R'(x)}{R(x)}$$

We integrate each side of this equation to obtain

$$\int_0^t \rho(x)\,dx = -\int_0^t \frac{R'(x)}{R(x)}\,dx = -[\ln R(t) - \ln R(0)]$$

Note that $R(0) = 1$ since the component will not fail before time $t = 0$, the moment that it is put into operation. Since $\ln R(0) = \ln 1 = 0$, we see that

$$-\int_0^t \rho(x)\,dx = \ln R(t)$$

or that

$$\exp\left[-\int_0^t \rho(x)\,dx\right] = e^{\ln R(t)} = R(t)$$

as claimed.

Example 4.7.2 illustrates the use of Theorem 4.7.3 and shows how the Weibull distribution arises in reliability studies.

Example 4.7.2. One hazard rate function in widespread use is the function

$$\rho(t) = \alpha\beta t^{\beta-1} \qquad t > 0 \quad \alpha > 0 \quad \beta > 0$$

This function has the property that if $\beta = 1$, the hazard rate is constant indicating that the occurrence of a failure is due primarily to random factors; if $\beta > 1$, the hazard rate is increasing, indicating that a failure is due primarily to a system wearing out over time; if $\beta < 1$, the hazard rate is decreasing, indicating that an early failure is likely due to a malfunctioning system. (See Exercise 64.) The reliability function is given by

$$R(t) = \exp\left[-\int_0^t \alpha\beta x^{\beta-1}\,dx\right]$$

$$= \exp\left[-\alpha x^{\beta}\Big|_0^t\right] = e^{-\alpha[t^{\beta} - 0^{\beta}]} = e^{-\alpha t^{\beta}}$$

The failure density is given by

$$f(t) = \rho(t)R(t) = \alpha\beta t^{\beta-1}e^{-\alpha t^{\beta}}$$

This is the density for a Weibull random variable with parameters α and β.

Reliability of Series and Parallel Systems

Components in multiple component systems can be installed within the system in various ways. Many systems are arranged in a "series" configuration, some are in "parallel," and others are combinations of the two designs. A system whose components are arranged in such a way that the system fails whenever *any* of its components fail is called a *series* system; if the system fails only if *all* of its components fail, then it is said to be arranged in *parallel*.

Recall that the reliability function for a component is the probability that it will not fail before time t. Consider a system consisting of k components connected in series. Let $R_i(t)$ denote the reliability of component i and assume that the components are independent in the sense that the reliability of one is unaffected by the reliability of the others. The reliability of the entire system is the probability that the system will not fail before time t. The system will not fail if and only if no component fails before time t. Thus the reliability of the system, $R_s(t)$, is given by

$$R_s(t) = \prod_{i=1}^{k} R_i(t)$$

The next two examples illustrate the use of this equation.

Example 4.7.3. Consider a system with five components connected in series. If each component has reliability .95 at time t, then the system reliability at that time is $R_s(t) = (.95)^5 = .774$.

Example 4.7.4. Suppose that we are designing a system of five independent components and we want the system reliability at time t to be at least .95. If the reliability of each component at t is to be the same, what is the minimum reliability required per component? Here we want $x^5 \geq .95$, where x is the reliability of each component. The solution is $x = (.95)^{1/5} = .9898$.

A more practical design for most equipment is the parallel system. Consider k independent components arranged in parallel. When the first fails, the second is used; when the second fails, the third comes on line. This continues until the last component fails, at which time the system fails. The system reliability at time t in this case is the probability that at least one of the k components does not fail before time t. This probability is given by

$$R_s(t) = 1 - P\,[\text{all components fail}]$$

$$= 1 - \prod_{i=1}^{k} [1 - R_i(t)]$$

It should be noted that in both series and parallel systems the reliability of individual components can differ. Example 4.7.5 illustrates a system that makes use of both types of configurations.

Example 4.7.5. Consider a system consisting of eight independent components connected as shown in Fig. 4.12. Note that the system basically consists of five assemblies in series, where assembly I consists of component 1; assembly II consists of components 2 and 3 in parallel; assembly III consists of components 4, 5, and 6 in parallel; assemblies IV and V consist of components 7 and 8,

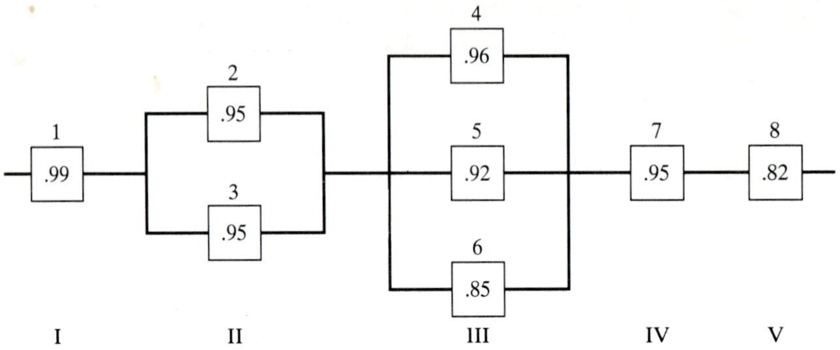

FIGURE 4.12
System with five assemblies, with assemblies II and III in parallel.

respectively. To calculate the system reliability, we first calculate the reliability of the two parallel assemblies. The reliability of assembly II is

$$\left[1 - (1 - .95)^2\right] = .9975$$

that of assembly III is

$$[1 - (1 - .96)(1 - .92)(1 - .85)] = .99952$$

The system reliability is the product of the reliabilities of the five assemblies and is given by

$$R_s(t) = (.99)(.9975)(.99952)(.95)(.82) = .7689$$

It is evident that a system with many independent components connected in series may have a very low system reliability even if each component alone is highly reliable. For example, a system of 20 components, each with a reliability of .95 connected in series, has a system reliability of $(.95)^{20} = .358$. One way to increase system reliability is to replace single components with several similar components arranged in parallel. Of course, the cost of providing this sort of redundancy is usually high.

4.8 TRANSFORMATION OF VARIABLES (OPTIONAL)

Consider a continuous random variable X with density f_X. Suppose that interest centers on some random variable Y, where Y is a function of X. Can we determine the density for Y based on knowledge of the distribution of X? The next theorem allows us to answer this question whenever Y is a strictly monotonic function of X.

Theorem 4.8.1. Let X be a continuous random variable with density f_X. Let $Y = g(X)$, where g is strictly monotonic and differentiable. The density for Y is denoted by f_Y and is given by

$$f_Y(y) = f_X(g^{-1}(y))\left|\frac{dg^{-1}(y)}{dy}\right|$$

Proof. Assume that $Y = g(X)$ is a strictly decreasing function of X. By definition the cumulative distribution function for Y is

$$F_Y(y) = P[Y \le y] = P[g(X) \le y]$$

Since g is strictly decreasing, g^{-1} exists and is also decreasing. Therefore

$$\begin{aligned} P[g(X) \le y] &= P\left[g^{-1}(g(X)) \ge g^{-1}(y)\right] \\ &= P\left[X \ge g^{-1}(y)\right] \\ &= 1 - P\left[X \le g^{-1}(y)\right] \end{aligned}$$

By definition $P[X \le g^{-1}(y)] = F_X(g^{-1}(y))$, and thus substitution yields

$$F_Y(y) = 1 - F_X(g^{-1}(y))$$

Since the derivative of the cumulative distribution function yields the density,

$$f_Y(y) = (-1)f_X(g^{-1}(y))\frac{dg^{-1}(y)}{dy}$$

Note that since g^{-1} is decreasing, $dg^{-1}(y)/dy < 0$ and

$$-\frac{dg^{-1}(y)}{dy} = \left|\frac{dg^{-1}(y)}{dy}\right|$$

By substitution $f_Y(y)$ can be written as

$$f_Y(y) = f_X(g^{-1}(y))\left|\frac{dg^{-1}(y)}{dy}\right|$$

as claimed. The proof in the case in which g is increasing is similar and is left as an exercise. (See Exercise 69.)

An example will illustrate the idea.

Example 4.8.1. Let X be a random variable with density

$$f_X(x) = 2x \qquad 0 < x < 1$$

and let $g(X) = Y = 3X + 6$. Since $g(x) = 3x + 6$ is strictly increasing and differentiable, Theorem 4.8.1 is applicable. To obtain the expression for g^{-1}, we solve the equation $y = 3x + 6$ for x and see that

$$x = g^{-1}(y) = \frac{y - 6}{3}$$

and $$\frac{dg^{-1}(y)}{dy} = \frac{1}{3}$$

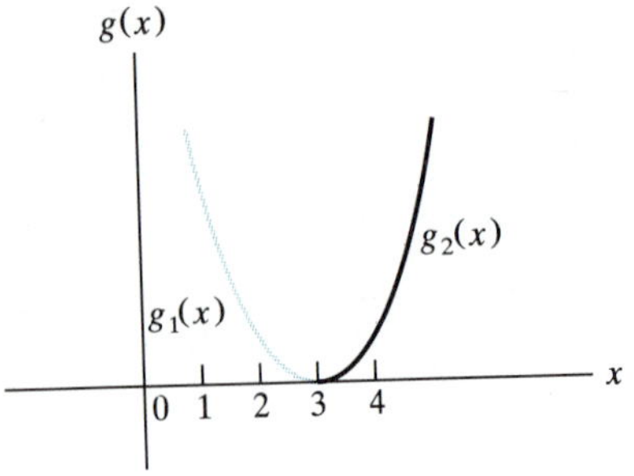

FIGURE 4.13
Graph of $g(x) = (x-3)^2, 0 < x < 4$ partitioned into two monotonic functions $g_1(x) = (x-3)^2, 0 < x \le 3$ and $g_2(x) = (x-3)^2, 3 < x < 4$.

An application of Theorem 4.8.1 yields

$$f_Y(y) = f_X(g^{-1}(y))\frac{dg^{-1}(y)}{dy}$$

or
$$f_Y(y) = 2\frac{(y-6)}{3}\cdot\frac{1}{3} = \frac{2}{9}(y-6) \qquad 6 < y < 9$$

It should be pointed out that the results given in Theorem 4.8.1 can be applied to piecewise monotonic functions as well as to those that are strictly monotonic. In this case several different equations might be required to define the density for Y. The idea is illustrated in the next example.

Example 4.8.2. Let X be uniformly distributed over $(0,4)$ and let $g(X) = Y = (X-3)^2$. The graph of this function is shown in Fig. 4.13. Note that since g is strictly decreasing on $(0,3)$ and strictly increasing on $(3,4)$, it is piecewise monotonic. It can be defined in terms of the two one-to-one functions g_1 and g_2 given by

$$g(x) = \begin{cases} g_1(x) = (x-3)^2 & 0 < x \le 3 \\ g_2(x) = (x-3)^2 & 3 < x < 4 \end{cases}$$

Each of the functions g_1 and g_2 is invertible and their inverses are given by

$$g_1^{-1}(y) = 3 - \sqrt{y} \qquad 0 \le y < 9$$

$$g_2^{-1}(y) = 3 + \sqrt{y} \qquad 0 < y < 1$$

Functions h_1 and h_2 used to determine the density for Y are found by applying Theorem 4.8.1 to each of the above. With this done, we then add those functions having common domains to obtain the final expression for f_Y. In this case the density is formed from the functions

$$h_1(y) = f_X(g_1^{-1}(y))\left|\frac{dg_1^{-1}(y)}{dy}\right| = \frac{1}{4}\left|-\frac{1}{2\sqrt{y}}\right| = \frac{1}{8\sqrt{y}}$$

$$h_2(y) = f_X(g_2^{-1}(y))\left|\frac{dg_2^{-1}(y)}{dy}\right| = \frac{1}{4}\left|\frac{1}{2\sqrt{y}}\right| = \frac{1}{8\sqrt{y}}$$

To obtain the density for Y, we note that the interval $[0,1)$ is common to the domain of both h_1 and h_2. Thus

$$f_Y(y) = h_1(y) + h_2(y) = \frac{2}{8\sqrt{y}} \qquad 0 \le y < 1$$

The interval [1, 9] is contained only in the domain of h_2. Hence

$$f_Y(y) = h_2(y) = \frac{1}{8\sqrt{y}} \qquad 1 \le y < 9$$

You can verify for yourself that f_Y is a valid density.

4.9 SIMULATING A CONTINUOUS DISTRIBUTION (OPTIONAL)

In Sec. 3.9 we showed how to simulate a discrete distribution using a random digit table. The table also can be used to simulate a continuous distribution. The idea is as follows:

1. We find the cumulative distribution function F for the random variable and its inverse.
2. We select a random two- (or three-) digit number from Table III of App. A and interpret this number as a probability, that is, as a number between 0 and 1.
3. We evaluate F^{-1} at this randomly selected point to obtain a randomly generated value for the random variable X.

This procedure is illustrated in Example 4.9.1.

Example 4.9.1. Consider the random variable X, the time to failure of a computer chip. Assume that X has a Weibull distribution with parameters $\alpha = .02$ and $\beta = 1$. The density for X is

$$f(x) = .02e^{-.02x} \qquad x > 0$$

and its cumulative distribution is

$$y = F(x) = 1 - e^{-.02x}$$

The inverse of F is found by solving this equation for x as follows:

$$\begin{aligned} y &= 1 - e^{-.02x} \\ e^{-.02x} &= 1 - y \\ -.02x &= \ln(1 - y) \\ x &= \frac{-\ln(1 - y)}{.02} \end{aligned}$$

To simulate an observation on X, we select a random two-digit number from Table III of App. A. Suppose the number selected is 77, which is interpreted as the probability $y = .77$. For this value of y our simulated observation on X is

$$x = \frac{-\ln(1 - .77)}{.02} = 73.48 \text{ years}$$

This procedure can be repeated to generate as many random values for X as desired. Figure 4.14 illustrates this procedure graphically.

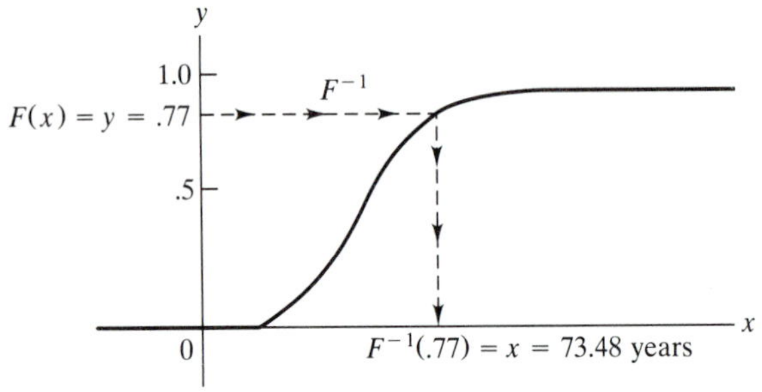

FIGURE 4.14
$F(x) = y = .77$ if and only if $F^{-1}(.77) = x = 73.48$ years.

CHAPTER SUMMARY

In this chapter we considered the general properties underlying random variables of the continuous type. These are random variables that assume their values in intervals of real numbers rather than at isolated points. The density function was introduced as a means of computing probabilities. These densities are defined in such a way that probabilities correspond to areas. The ideas of expected value and moment generating function were defined by replacing the summation operation, used in the discrete case, with integration. A number of continuous distributions were studied. See Table 4.1. The gamma distribution was presented. We noted that the exponential distribution and the chi-squared distribution are special cases of the gamma distribution. We studied the normal distribution and showed how to use this distribution to approximate binomial and Poisson probabilities. The Weibull distribution was introduced and its use in reliability studies was examined. Other distributions that were considered briefly are the log-normal, uniform, and Cauchy distributions. We saw how to simulate continuous distributions. We introduced and defined important terms that you should know. These are:

Continuous random variable	Continuous density
Continuous distribution function	Gamma function
Half-unit correction	Failure density
Reliability function	Hazard rate function
Standard normal	

In the last two chapters we have presented some commonly encountered discrete and continuous distributions and have looked at some of the relationships that exist among them. The chart given in Fig. 4.15 summarizes the results that have been obtained. It is an adaptation of the more complete chart developed by Lawrence Leemis in "Relationships Among Common Univariate Distributions," *The American Statistician*, May 1986, vol. 40, no. 2. (Used with permission of the author.) In the chart two types of relationships, namely, special cases and approximations, are depicted. Special cases are indicated by a solid arrow and approximations are shown by a dashed arrow. In each case the name of the distribution along with its associated parameters are given.

TABLE 4.1

Continuous distributions

Name	Density		Moment generating function	Mean	Variance
Gamma	$\frac{1}{\Gamma(\alpha)\beta^{\alpha}} x^{\alpha-1} e^{-x/\beta}$	$\alpha > 0$ $\beta > 0$ $x > 0$	$(1-\beta t)^{-\alpha}$	$\alpha\beta$	$\alpha\beta^2$
Exponential	$\frac{1}{\beta} e^{-x/\beta}$	$x > 0$ $\beta > 0$	$(1-\beta t)^{-1}$	β	β^2
Chi-squared	$\frac{1}{\Gamma(\gamma/2)2^{\gamma/2}} x^{\gamma/2-1} e^{-x/2}$	$x > 0$ γ a positive integer	$(1-2t)^{-\gamma/2}$	γ	2γ
Uniform	$\frac{1}{b-a}$	$a < x < b$	$\frac{e^{tb}-e^{ta}}{t(b-a)}$ $t \neq 0$ 1 $t = 0$	$\frac{a+b}{2}$	$\frac{(b-a)^2}{12}$
Cauchy	$\frac{1}{\pi} \frac{a}{a^2+(x-b)^2}$	$-\infty < x < \infty$ $-\infty < b < \infty$ $a > 0$	Does not exist	Does not exist	Does not exist
Normal	$\frac{1}{\sqrt{2\pi}\sigma} \exp\left[-1/2\left(\frac{x-\mu}{\sigma}\right)^2\right]$	$-\infty < x < \infty$ $\sigma > 0$ $-\infty < \mu < \infty$	$e^{\mu t+\sigma^2 t^2/2}$	μ	σ^2
Weibull	$\alpha\beta x^{\beta-1} e^{-\alpha x^{\beta}}$	$x > 0$ $\alpha > 0$ $\beta > 0$		$\alpha^{-1/\beta}\Gamma\left(1+\frac{1}{\beta}\right)$	$\alpha^{-2/\beta}\Gamma\left(1+\frac{2}{\beta}\right) - \mu^2$

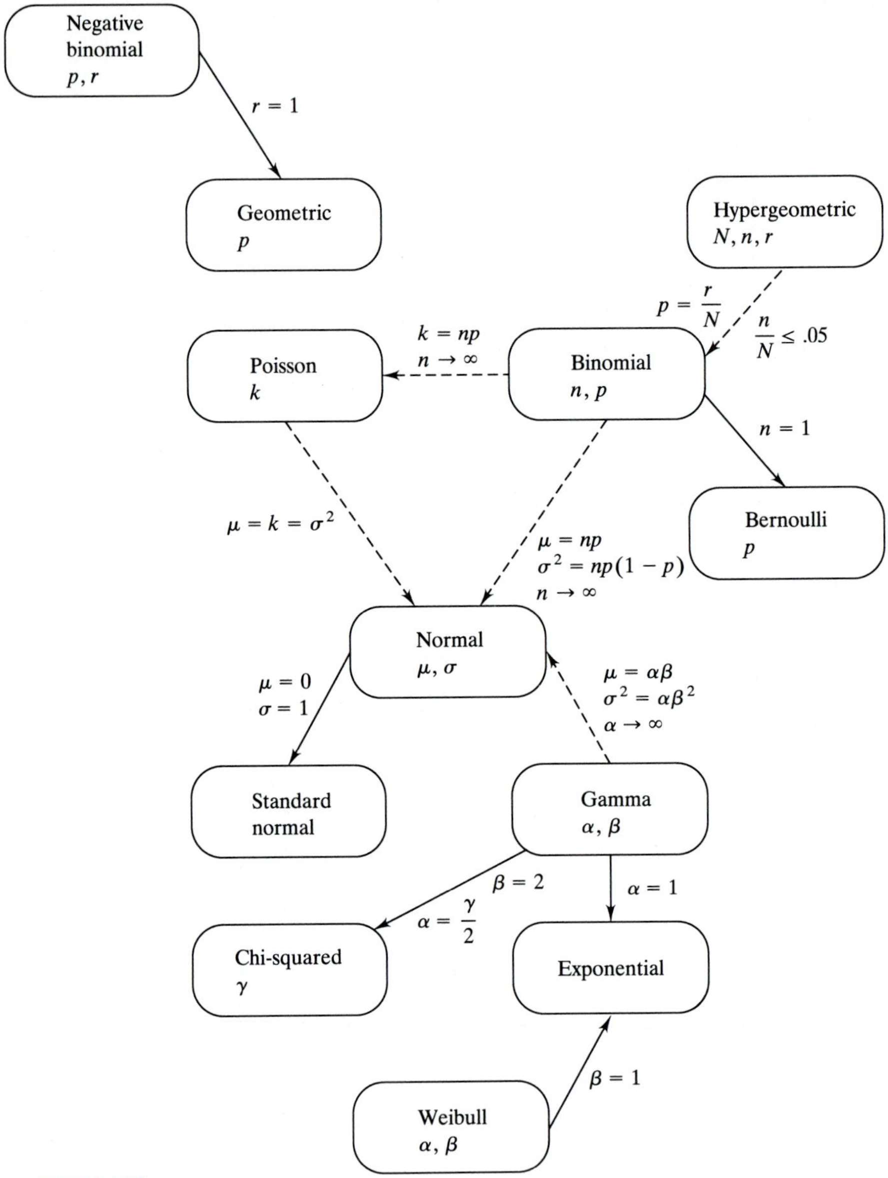

FIGURE 4.15
Some interrelationships among common distributions.

EXERCISES

Section 4.1

1. Consider the function

$$f(x) = kx \qquad 2 \le x \le 4$$

(*a*) Find the value of k that makes this a density for a continuous random variable.
(*b*) Find $P[2.5 \le X \le 3]$.
(*c*) Find $P[X = 2.5]$.
(*d*) Find $P[2.5 < X \le 3]$.

2. Consider the areas shown in Fig. 4.16. In each case, state what probability is being depicted. What is the relationship between the areas depicted in Figs. 4.16(*a*) and (*b*)? Between those in Figs. 4.16(*d*) and (*e*)

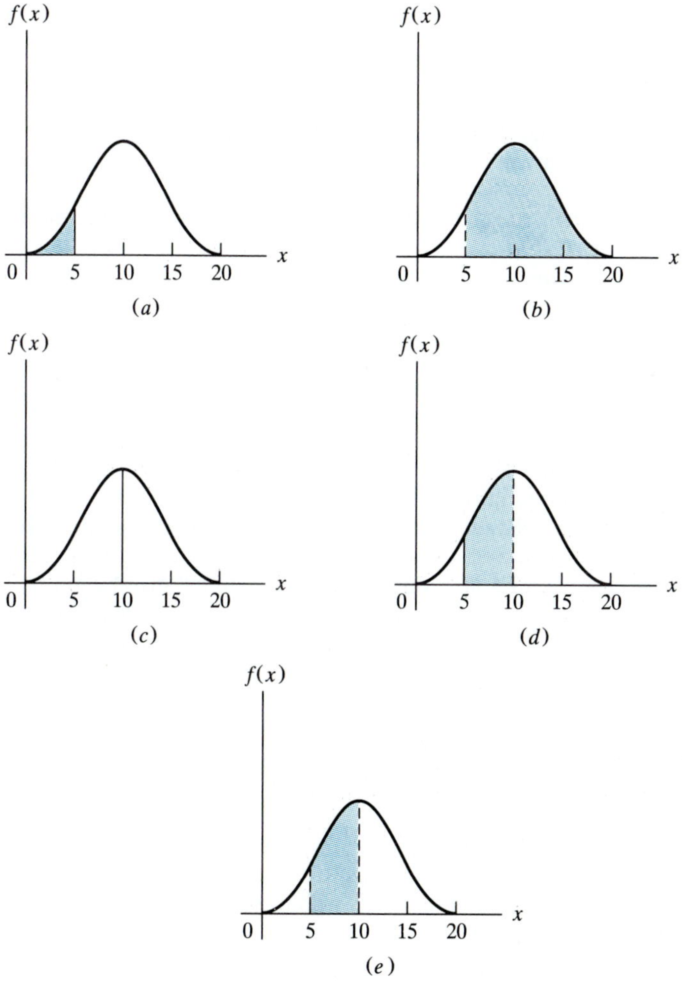

FIGURE 4.16

3. Let X denote the length in minutes of a long-distance telephone conversation. Assume that the density for X is given by

$$f(x) = (1/10)e^{-x/10} \qquad x > 0$$

(a) Verify that f is a density for a continuous random variable.
(b) Assuming that f adequately describes the behavior of the random variable X, find the probability that a randomly selected call will last at most 7 minutes; at least 7 minutes; exactly 7 minutes.
(c) Would it be unusual for a call to last between 1 and 2 minutes? Explain, based on the probability of this occurring.
(d) Sketch the graph of f and indicate in the sketch the area corresponding to each of the probabilities found in part (b).

4. Some plastics in scrapped cars can be stripped out and broken down to recover the chemical components. The greatest success has been in processing the flexible polyurethane cushioning found in these cars. Let X denote the amount of this material, in pounds, found per car. Assume that the density for X is given by

$$f(x) = \frac{1}{\ln 2}\frac{1}{x} \qquad 25 \leq x \leq 50$$

(Based on a report in *Design Engineering*, February 1982, p. 7.)
(a) Verify that f is a density for a continuous random variable.
(b) Use f to find the probability that a randomly selected auto will contain between 30 and 40 pounds of polyurethane cushioning.
(c) Sketch the graph of f and indicate in the sketch the area corresponding to the probability found in part (b).

5. (*Continuous uniform distribution.*) A random variable X is said to be uniformly distributed over an interval (a, b) if its density is given by

$$f(x) = \frac{1}{b - a} \qquad a < x < b$$

(a) Show that this is a density for a continuous random variable.
(b) Sketch the graph of the uniform density.
(c) Shade the area in the graph of part (b) that represents $P[X \leq (a + b)/2]$.
(d) Find the probability pictured in part (c).
(e) Let (c, d) and (e, f) be subintervals of (a, b) of equal length. What is the relationship between $P[c \leq X \leq d]$ and $P[e \leq X \leq f]$? Generalize the idea suggested by this example, thus justifying the name "uniform" distribution.

6. When a pair of coils is placed around a homing pigeon and a magnetic field applied that reverses the earth's field, it is thought that the bird will become disoriented. Under these circumstances it is just as likely to fly in one direction as in any other. Let θ denote the direction in radians of the bird's initial flight. See Fig. 4.17. θ is uniformly distributed over the interval $[0, 2\pi]$.
(a) Find the density for θ.
(b) Sketch the graph of the density. The uniform distribution is sometimes called the "rectangular" distribution. Do you see why?
(c) Shade the area corresponding to the probability that a bird will orient within $\pi/4$ radians of home, and find this area using plane geometry.
(d) Find the probability that a bird will orient within $\pi/4$ radians of home by integrating the density over the appropriate region(s), and compare your answer to that obtained in part (c).

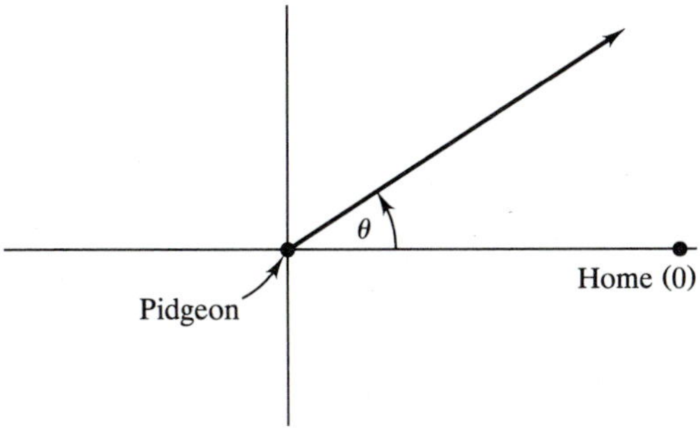

FIGURE 4.17
θ = direction of the initial flight of a homing pigeon measured in radians.

(*e*) If 10 birds are released independently and at least seven orient within $\pi/4$ radians of home, would you suspect that perhaps the coils are not disorienting the birds to the extent expected? Explain, based on the probability of this occurring.

7. Use Definition 4.1.2 to show that for a continuous random variable X, $P[X = a] = 0$ for every real number a. *Hint:* Write $P[X = a]$ as $P[a \leq X \leq a]$.

8. Express each of the probabilities depicted in Fig. 4.16 in terms of the cumulative distribution function F.

9. Consider the random variable of Exercise 1.
(*a*) Find the cumulative distribution function F.
(*b*) Use F to find $P[2.5 \leq X \leq 3]$, and compare your answer to that obtained previously.
*(*c*) Sketch the graph of F. Is F right continuous? Is F continuous? What is $\lim_{x \to -\infty} F(x)$? What is $\lim_{x \to \infty} F(x)$? Is F nondecreasing?
*(*d*) Find $dF(x)/dx$ for $x \in (2, 4)$. Does your answer look familiar?

10. (*Uniform distribution.*) Find the general expression for the cumulative distribution function for a random variable X that is uniformly distributed over the interval (a, b). See Exercise 5.

11. (*Uniform distribution.*) Consider the random variable of Exercise 6.
(*a*) Use Exercise 10 to find the cumulative distribution function F.
*(*b*) Sketch the graph of F. Is F right continuous? Is F continuous? What is $\lim_{x \to -\infty} F(x)$? What is $\lim_{x \to \infty} F(x)$? Is F nondecreasing?
*(*c*) Find $dF(x)/dx$ for $x \in (0, 2\pi)$. Does your answer look familiar?

12. Find the cumulative distribution function for the random variable of Exercise 3. Use F to find $P[1 \leq X \leq 2]$, and compare your answer to that obtained previously.

13. Find the cumulative distribution function for the random variable of Exercise 4. Use F to find $P[30 \leq X \leq 40]$, and compare your answer to that obtained previously.

***14.** In Exercise 13 of Chap. 3 the mathematical properties of the cumulative distribution function for discrete random variables were pointed out. In the continuous case similar properties hold. The results of Exercises 9 and 11 are not coincidental! It can be shown that the cumulative distribution function F for any continuous random variable has these characteristics:
(i) F is continuous.
(ii) $\lim_{x \to -\infty} F(x) = 0$ and $\lim_{x \to \infty} F(x) = 1$.
(iii) F is nondecreasing.
(iv) $dF(x)/dx = f(x)$ for all values of x for which this derivative exists.

(*a*) Consider the function F defined by

$$F(x) = \begin{cases} 0 & x < -1 \\ x + 1 & -1 \le x \le 0 \\ 1 & x > 0 \end{cases}$$

Does F satisfy properties (i) to (iii) given above? If so, what is f? If not, what property fails?

(*b*) Consider the function defined by

$$F(x) = \begin{cases} 0 & x \le 0 \\ x^2 & 0 < x \le 1/2 \\ (1/2)x & 1/2 < x \le 1 \\ 1 & x > 1 \end{cases}$$

Does F satisfy properties (i) to (iii) given above? If so, what is f? If not, what property fails?

Section 4.2

15. Consider the random variable X with density

$$f(x) = (1/6)x \qquad 2 \le x \le 4$$

(*a*) Find $E[X]$.
(*b*) Find $E[X^2]$.
(*c*) Find σ^2 and σ.

16. Let X denote the amount in pounds of polyurethane cushioning found in a car. (See Exercise 4.) The density for X is given by

$$f(x) = \frac{1}{\ln 2}\frac{1}{x} \qquad 25 \le x \le 50$$

Find the mean, variance, and standard deviation for X.

17. Let X denote the length in minutes of a long-distance telephone conversation. The density for X is given by

$$f(x) = (1/10)e^{-x/10} \qquad x > 0$$

(*a*) Find the moment generating function, $m_X(t)$.
(*b*) Use $m_X(t)$ to find the average length of such a call.
(*c*) Find the variance and standard deviation for X.

18. (*Uniform distribution.*) The density for a random variable X distributed uniformly over (a, b) is

$$f(x) = \frac{1}{b - a} \qquad a < x < b$$

Use Definition 4.2.1 to show that

$$E[X] = \frac{a + b}{2} \qquad \text{and} \qquad \operatorname{Var} X = \frac{(b - a)^2}{12}$$

19. (*Uniform distribution.*) Let θ denote the direction in radians of the flight of a bird whose sense of direction has been disoriented as described in Exercise 6. Assume that θ is uniformly distributed over the interval $[0, 2\pi]$. Use the results of Exercise 18 to find the mean, variance, and standard deviation of θ.

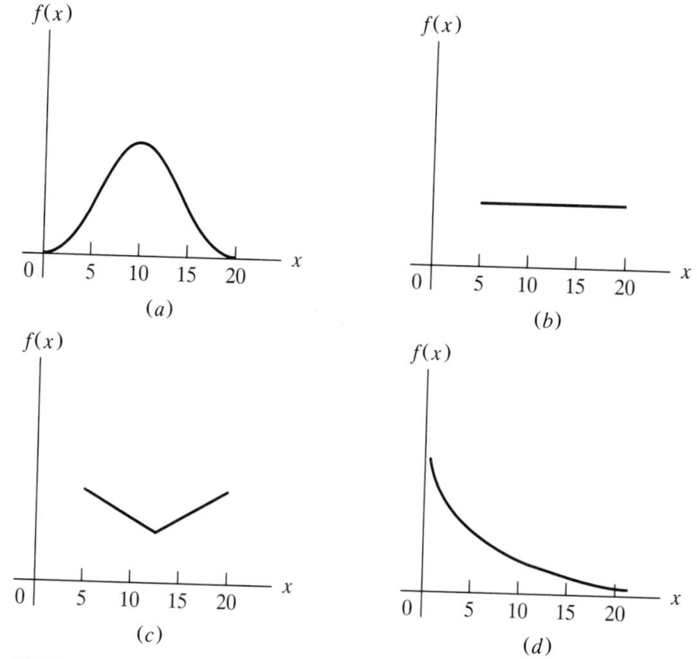

FIGURE 4.18

20. Let X be continuous with density f. Imagine cutting out of a piece of thin rigid metal the region bounded by the graph of f and the x axis, and attempting to balance this region on a knife-edge held parallel to the vertical axis. The point at which the region would balance, if such a point exists, is the mean of X. Thus μ_X is a "location" parameter in that it indicates the position of the center of the density along the x axis. Figure 4.18 gives the graphs of the densities of four continuous random variables whose means do exist. In each case, approximate the value of μ_X from the graph.

21. In the continuous case variance is a "shape" parameter in the sense that a random variable with small variance will have a compact density; one with a large variance will have a density that is rather spread out or flat. Consider the two densities given in Fig. 4.19. What is μ_X? What is μ_Y? Which random variable has the larger variance?

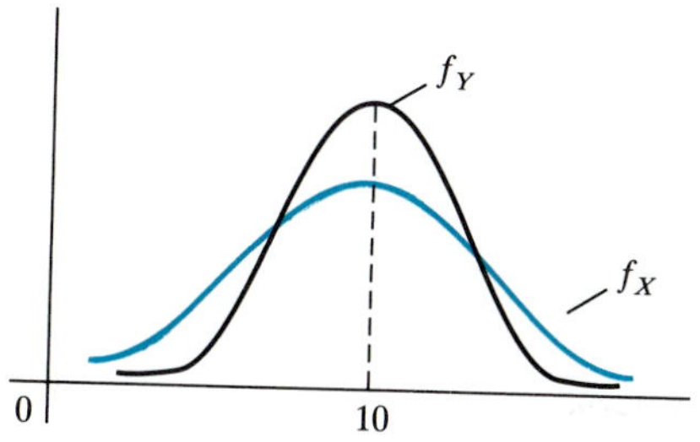

FIGURE 4.19

★22. (*Cauchy distribution.*) A random variable X with density

$$f(x) = \frac{1}{\pi}\frac{a}{a^2 + (x-b)^2} \qquad \begin{array}{l} -\infty < x < \infty \\ -\infty < b < \infty \\ a > 0 \end{array}$$

is said to have a Cauchy distribution with parameters a and b. This distribution is interesting in that it provides an example of a continuous random variable whose mean does not exist. Let $a = 1$ and $b = 0$ to obtain a special case of the Cauchy distribution with density

$$f(x) = \frac{1}{\pi}\frac{1}{1+x^2} \qquad -\infty < x < \infty$$

Show that $\int_{-\infty}^{\infty}|x|f(x)\,dx$ does not exist, thus showing that $E[X]$ does not exist, *Hint:* Write

$$\int_{-\infty}^{\infty}|x|\frac{1}{\pi}\frac{1}{1+x^2}\,dx = \frac{1}{\pi}\int_{-\infty}^{0}\frac{-x}{1+x^2}\,dx + \frac{1}{\pi}\int_{0}^{\infty}\frac{x}{1+x^2}\,dx$$

and recall that $\int(du/u) = \ln|u|$.

★23. (*Uniform distribution.*) Let X be uniformly distributed over (a, b). (See Exercise 18.)

(a) Show that the moment generating function for X is given by

$$m_X(t) = \begin{cases} \dfrac{e^{tb} - e^{ta}}{t(b-a)} & t \neq 0 \\ 1 & t = 0 \end{cases}$$

Hint: When $t = 0$, $m_X(t) = E[e^{0 \cdot X}]$.

(b) Use $m_X(t)$ to find $E[X]$. *Hint:* Find

$$\frac{d}{dt}\left[\frac{e^{tb} - e^{ta}}{t(b-a)}\right]$$

and take the limit of this derivative as $t \to 0$ using L'Hospital's rule.

★24. Let the density for X be given by

$$f(x) = ce^{-|x|} \qquad -\infty < x < \infty$$

(a) Find the value of c that makes this a density.

(b) Show that

$$\frac{1}{2}\int_{-\infty}^{\infty}|x|e^{-|x|}\,dx$$

exists.

(c) Find $E[X]$.

(d) Show that

$$m_X(t) = \frac{-1}{t^2 - 1} \qquad -1 < t < 1$$

(e) Use $m_X(t)$ to find $E[X]$ and $E[X^2]$.

(f) Find Var X.

Section 4.3

25. Evaluate each of these integrals:
(*a*) $\int_0^\infty z^2 e^{-z}\, dz$
(*b*) $\int_0^\infty z^7 e^{-z}\, dz$
(*c*) $\int_0^\infty x^3 e^{-x/2}\, dx$
(*d*) $\int_0^\infty (1/16) x e^{-x/4}\, dx$

26. Prove Theorem 4.3.1. *Hint:* To prove part 1, evaluate $\Gamma(1)$ directly from the definition of the gamma function. To prove part 2, use integration by parts with

$$u = z^{\alpha-1} \qquad dv = \int e^{-z}\, dz$$

$$du = (\alpha - 1) z^{\alpha-2}\, dz \qquad v = -e^{-z}$$

Use L'Hospital's rule repeatedly to show that

$$-z^{\alpha-1} e^{-z}\Big|_0^\infty = 0$$

27. (*a*) Use Theorem 4.3.1 to evaluate $\Gamma(2)$, $\Gamma(3)$, $\Gamma(4)$, $\Gamma(5)$, and $\Gamma(6)$.
(*b*) Can you generalize the pattern suggested in part (*a*)?
(*c*) Does the result of part (*b*) hold even if $n = 1$?
(*d*) Evaluate $\Gamma(15)$ using the result of part (*b*).

28. Show that for $\alpha > 0$ and $\beta > 0$,

$$\int_0^\infty \frac{1}{\Gamma(\alpha)\beta^\alpha} x^{\alpha-1} e^{-x/\beta}\, dx = 1$$

thereby showing that the function given in Definition 4.3.2 is a density for a continuous random variable. *Hint:* Change the variable by letting $z = x/\beta$.

29. Let X be a gamma random variable with $\alpha = 3$ and $\beta = 4$.
(*a*) What is the expression for the density for X?
(*b*) What is the moment generating function for X?
(*c*) Find μ, σ^2, and σ.

30. Let X be a gamma random variable with parameters α and β. Use the moment generating function to find $E[X]$ and $E[X^2]$. Use these expectations to show that $\text{Var } X = \alpha\beta^2$.

31. Let X be a gamma random variable with parameters α and β.
(*a*) Use Definition 4.2.1, the definition of expected value, to find $E[X]$ and $E[X^2]$ directly. *Hint:* $z^\alpha = z^{(\alpha+1)-1}$ and $z^{\alpha+1} = z^{(\alpha+2)-1}$.
(*b*) Use the results of (*a*) to verify that $\text{Var } X = \alpha\beta^2$.

32. Show that the graph of the density for a gamma random variable with parameters α and β assumes its maximum value at $x = \beta(\alpha - 1)$ for $\alpha > 1$. Sketch a rough graph of the density for a gamma random variable with $\alpha = 3$ and $\beta = 4$. *Hint:* Find the first derivative of the density, set this derivative equal to 0, and solve for x.

33. Let X be an exponential random variable with parameter β. Find general expressions for the moment generating function, mean, and variance for X.

34. A particular nuclear plant releases a detectable amount of radioactive gases twice a month on the average. Find the probability that at least 3 months will elapse before the release of the first detectable emission. What is the average time that one must wait to observe the first emission?

35. The "forgetfulness" property of the exponential distribution says that the probability that we must wait a total of $w_1 + w_2$ units before the occurrence of an event given that we have already waited w_1 units is the same as the probability that we must wait w_2 units at the outset. That is,

$$P[W > w_1 + w_2 | W > w_1] = P[W > w_2]$$

Verify this statement for any exponential random variable W with parameter β.

36. Rock noise in an underground mine occurs at an average rate of three per hour. (See Exercise 65, Chap. 3.)

(*a*) Find the probability that no rock noise will be recorded for at least 30 minutes.

(*b*) Suppose that no rock noise has been heard for 15 minutes. Find the probability that another 30 minutes will elapse before the first rock noise is detected.

37. California is hit every year by approximately 500 earthquakes that are large enough to be felt. However, those of destructive magnitude occur, on the average, once a year.

(*a*) Find the probability that at least 3 months elapse before the first earthquake of destructive magnitude occurs. (See Exercise 64, Chap. 3.)

(*b*) Suppose that no destructive quake has occurred for 4 months. Find the probability that an additional 3 months will elapse before a destructive quake occurs.

38. Consider a chi-squared random variable with 15 degrees of freedom.

(*a*) What is the mean of X^2_{15}? What is its variance?

(*b*) What is the expression for the density for X^2_{15}?

(*c*) What is the expression for the moment generating function for X^2_{15}?

(*d*) Use Table IV of App. A to find each of the following:

$$P[X^2_{15} \le 5.23] \qquad P[6.26 \le X^2_{15} \le 27.5] \qquad \chi^2_{.05}$$

$$P[X^2_{15} \ge 22.3] \qquad \chi^2_{.01} \qquad \chi^2_{.95}$$

Section 4.4

39. Use Table V of App. A to find each of the following:

(*a*) $P[Z \le 1.57]$. (*b*) $P[Z < 1.57]$.

(*c*) $P[Z = 1.57]$. (*d*) $P[Z > 1.57]$.

(*e*) $P[-1.25 \le Z \le 1.75]$. (*f*) $z_{.10}$.

(*g*) $z_{.90}$.

(*h*) The point z such that $P[-z \le Z \le z] = .95$.

(*i*) The point z such that $P[-z \le Z \le z] = .90$.

40. The bulk density of soil is defined as the mass of dry solids per unit bulk volume. A high bulk density implies a compact soil with few pores. Bulk density is an important factor in influencing root development, seedling emergence, and aeration. Let X denote the bulk density of Pima clay loam. Studies show that X is normally distributed with $\mu = 1.5$ and $\sigma = .2$ g/cm^3. (*McGraw-Hill Yearbook of Science and Technology*, 1981, p. 361.)

(*a*) What is the density for X? Sketch a graph of the density function. Indicate on this graph the probability that X lies between 1.1 and 1.9. Find this probability.

(*b*) Find the probability that a randomly selected sample of Pima clay loam will have bulk density less than .9 g/cm^3.

(*c*) Would you be surprised if a randomly selected sample of this type of soil has a bulk density in excess of 2.0 g/cm^3? Explain, based on the probability of this occurring.

(*d*) What point has the property that only 10% of the soil samples have bulk density this high or higher?

(*e*) What is the moment generating function for X?

41. Most galaxies take the form of a flattened disc with the major part of the light coming from this very thin fundamental plane. The degree of flattening differs from galaxy to galaxy. In the Milky Way Galaxy most gases are concentrated near the center of the fundamental plane. Let X denote the perpendicular distance from this center to a gaseous mass. X is normally distributed with mean 0 and standard deviation 100 parsecs. (A parsec is equal to approximately 19.2 trillion miles.) (*McGraw-Hill Encyclopedia of Science and Technology*, vol. 6, 1971, p. 10.)

(*a*) Sketch a graph of the density for X. Indicate on this graph the probability that a gaseous mass is located within 200 parsecs of the center of the fundamental plane. Find this probability.

(*b*) Approximately what percentage of the gaseous masses are located more than 250 parsecs from the center of the plane?

(*c*) What distance has the property that 20% of the gaseous masses are at least this far from the fundamental plane?

(*d*) What is the moment generating function for X?

42. Among diabetics, the fasting blood glucose level X may be assumed to be approximately normally distributed with mean 106 milligrams per 100 milliliters and standard deviation 8 milligrams per 100 milliliters.

(*a*) Sketch a graph of the density for X. Indicate on this graph the probability that a randomly selected diabetic will have a blood glucose level between 90 and 122 mg/100 ml. Find this probability.

(*b*) Find $P[X \leq 120 \text{ mg}/100 \text{ ml}]$.

(*c*) Find the point that has the property that 25% of all diabetics have a fasting glucose level of this value or lower.

(*d*) If a randomly selected diabetic is found to have fasting blood glucose level in excess of 130, do you think there is cause for concern? Explain, based on the probability of this occurring naturally.

43. (*a*) Find the density for the standard normal random variable Z.

(*b*) Find $f'(z)$. Show that the only critical point for f occurs at $z = 0$. Use the first derivative test to show that f assumes its maximum value at $z = 0$.

(*c*) Find $f''(z)$. Show that the possible inflection points occur at $z = \pm 1$. Use the second derivative to show that f changes concavity at $z = \pm 1$, implying that the inflection points do occur when $z = \pm 1$.

(*d*) Let X be normal with parameters μ and σ. Let $(X - \mu)/\sigma = Z$. Use the results of parts (*b*) and (*c*) to verify that, in general, a normal curve assumes its maximum value at $x = \mu$ and has points of inflection at $x = \mu \pm \sigma$.

44. Let X be normal with parameters μ and σ. Use the moment generating function to find $E[X^2]$. Find Var X, thus completing the proof of Theorem 4.4.2.

***45.** (*Log-normal distribution.*) The log-normal distribution is the distribution of a random variable whose natural logarithm follows a normal distribution. Thus if X is a normal random variable then $Y = e^X$ follows a log-normal distribution. Complete the argument below, thus deriving the density for a log-normal random variable.

Let X be normal with mean μ and variance σ^2. Let G denote the cumulative distribution function for $Y = e^X$, and let F denote the cumulative distribution function for X.

(a) Show that $G(y) = F(\ln y)$.

(b) Show that $G'(y) = F'(\ln y)/y$.

(c) Use Exercise 14 part (iv) to show that the density for Y is given by

$$g(y) = \frac{1}{\sqrt{2\pi}\,\sigma y} \exp\left[-\frac{1}{2}\frac{(\ln y - \mu)^2}{\sigma^2}\right] \qquad \begin{array}{l} -\infty < \mu < \infty \\ \sigma > 0 \\ y > 0 \end{array}$$

Note that μ and σ are the mean and standard deviation of the underlying normal distribution; they are not the mean and standard deviation of Y itself.

*46. Let Y denote the diameter in millimeters of Styrofoam pellets used in packing. Assume that Y has a log-normal distribution with parameters $\mu = .8$ and $\sigma = .1$.

(a) Find the probability that a randomly selected pellet has a diameter that exceeds 2.7 millimeters.

(b) Between what two values will Y fall with probability approximately .95?

Section 4.5

47. Verify the normal probability rule.

48. The number of Btu's of petroleum and petroleum products used per person in the United States in 1975 was normally distributed with mean 153 million Btu and standard deviation 25 million Btu. Approximately what percentage of the population used between 128 and 178 million Btu during that year? Approximately what percentage of the population used in excess of 228 million Btu?

49. Reconsider Exercises 40(a), 41(a), and 42(a) in light of the normal probability rule.

50. For a normal random variable, $P[|X - \mu| < 3\sigma] \doteq .99$. What value is assigned to this probability via Chebyshev's inequality? Are the results consistent? Which rule gives a stronger statement in the case of a normal variable?

51. Animals have an excellent spatial memory. In an experiment to confirm this statement an eight-armed maze such as that shown in Fig. 4.20 is used. At the

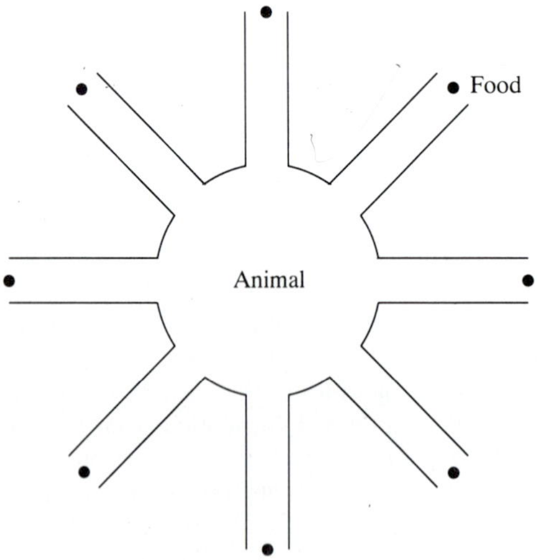

FIGURE 4.20
An eight-armed maze.

beginning of a test, one pellet of food is placed at the end of each arm. A hungry animal is placed at the center of the maze and is allowed to choose freely from among the arms. The optimal strategy is to run to the end of each arm exactly once. This requires that the animal remember where it has been. Let X denote the number of correct arms (arms still containing food) selected among its first eight choices. Studies indicate that $\mu = 7.9$. (*McGraw-Hill Encyclopedia of Science and Technology*, 1980.)

(*a*) Is X normally distributed?

(*b*) State and interpret Chebyshev's inequality in the context of this problem for $k = .5$, 1, 2, and 3. At what point does the inequality begin to give us some practical information?

Section 4.6

52. Let X be binomial with $n = 20$ and $p = .3$. Use the normal approximation to approximate each of the following. Compare your results with the values obtained from Table I of App. A.

(*a*) $P[X \leq 3]$.

(*b*) $P[3 \leq X \leq 6]$.

(*c*) $P[X \geq 4]$.

(*d*) $P[X = 4]$.

53. Although errors are likely when taking measurements from photographic images, these errors are often very small. For sharp images with negligible distortion, errors in measuring distances are often no larger than .0004 inch. Assume that the probability of a serious measurement error is .05. A series of 150 independent measurements are made. Let X denote the number of serious errors made.

(*a*) In finding the probability of making at least one serious error, is the normal approximation appropriate? If so, approximate the probability using this method.

(*b*) Approximate the probability that at most three serious errors will be made.

54. A chemical reaction is run in which the usual yield is 70%. A new process has been devised that should improve the yield. Proponents of the new process claim that it produces better yields than the old process more than 90% of the time. The new process is tested 60 times. Let X denote the number of trials in which the yield exceeds 70%.

(*a*) If the probability of an increased yield is .9, is the normal approximation appropriate?

(*b*) If $p = .9$, what is $E[X]$?

(*c*) If $p > .9$ as claimed, then, on the average, more than 54 of every 60 trials will result in an increased yield. Let us agree to accept the claim if X is at least 59. What is the probability that we will accept the claim if p is really only .9?

(*d*) What is the probability that we shall not accept the claim ($X \leq 58$) if it is true, and p is really .95?

55. Opponents of a nuclear power project claim that the majority of those living near a proposed site are opposed to the project. To justify this statement, a random sample of 75 residents is selected and their opinions are sought. Let X denote the number opposed to the project.

(*a*) If the probability that an individual is opposed to the project is .5, is the normal approximation appropriate?

(*b*) If $p = .5$, what is $E[X]$?

(*c*) If $p > .5$ as claimed, then, on the average, more than 37.5 of every 75 individuals are opposed to the project. Let us agree to accept the claim if X is at least 46. What is the probability that we shall accept the claim if p is really only .5?

(*d*) What is the probability that we shall not accept the claim ($X \leq 45$) even though it is true and p is really .7?

56. (*Normal approximation to the Poisson distribution.*) Let X be Poisson with parameter λs. Then for large values of λs, X is approximately normal with mean λs and variance λs. (The proof of this theorem is also based on the Central Limit Theorem and will be considered in Chap. 7.) Let X be a Poisson random variable with parameter $\lambda s = 15$. Find $P[X \leq 12]$ from Table II of App. A. Approximate this probability using a normal curve. Be sure to employ the half-unit correction factor.

57. The average number of jets either arriving at or departing from O'Hare Airport is one every 40 seconds. What is the approximate probability that at least 75 such flights will occur during a randomly selected hour? What is the probability that fewer than 100 such flights will take place in an hour?

Section 4.7

58. The length of time in hours that a rechargeable calculator battery will hold its charge is a random variable. Assume that this variable has a Weibull distribution with $\alpha = .01$ and $\beta = 2$.

(*a*) What is the density for X?

(*b*) What are the mean and variance for X? *Hint:* It can be shown that $\Gamma(\alpha) = (\alpha - 1)\Gamma(\alpha - 1)$ for any $\alpha > 1$. Furthermore, $\Gamma(1/2) = \sqrt{\pi}$.

(*c*) What is the reliability function for this random variable?

(*d*) What is the reliability of such a battery at $t = 3$ hours? At $t = 12$ hours? At $t = 20$ hours?

(*e*) What is the hazard rate function for these batteries?

(*f*) What is the failure rate at $t = 3$ hours? At $t = 12$ hours? At $t = 20$ hours?

(*g*) Is the hazard rate function an increasing or a decreasing function? Does this seem to be reasonable from a practical point of view? Explain.

59. Computer chips do not "wear out" in the ordinary sense. Assuming that defective chips have been removed from the market by factory inspection, it is reasonable to assume that these chips exhibit a constant hazard rate. Let the hazard rate be given by $\rho(t) = .02$. (Time is in years.)

(*a*) In a practical sense, what are the main causes of failure of these chips?

(*b*) What is the reliability function for chips of this type?

(*c*) What is the reliability of a chip 20 years after it has been put into use?

(*d*) What is the failure density for these chips?

(*e*) What type of random variable is X, the time to failure of a chip?

(*f*) What is the mean and variance for X?

(*g*) What is the probability that a chip will be operable for at least 30 years?

60. The random variable X, the time to failure (in thousands of miles driven) of the signal lights on an automobile has a Weibull distribution with $\alpha = .04$ and $\beta = 2$.

(*a*) Find the density, mean, and variance for X.

(*b*) Find the reliability function for X.

(*c*) What is the reliability of these lights at 5000 miles? At 10,000 miles?

(*d*) What is the hazard rate function?

(*e*) What is the hazard rate at 5000 miles? At 10,000 miles?

(*f*) What is the probability that the lights will fail during the first 3000 miles driven?

61. Show that for $\alpha > 0$ and $\beta > 0$,

$$\int_0^\infty \alpha\beta x^{\beta-1}e^{-\alpha x^\beta}\,dx = 1$$

thereby showing that the nonnegative function given in Definition 4.7.1 is a density for a continuous random variable. *Hint:* Let $z = \alpha x^\beta$.

62. Let X be a Weibull random variable with parameters α and β. Show that $E[X^2] = \alpha^{-2/\beta}\Gamma(1 + 2/\beta)$. *Hint:* In evaluating

$$\int_0^\infty x^2\alpha\beta x^{\beta-1}e^{-\alpha x^\beta}\,dx$$

let $z = \alpha x^\beta$. Evaluate the integral in a manner similar to that used in the proof of Theorem 4.7.1.

63. Use the result of Exercise 62 to find Var X for a Weibull random variable with parameters α and β, thus completing the proof of Theorem 4.7.1.

64. Consider the hazard rate function

$$\rho(t) = \alpha\beta t^{\beta-1} \qquad t > 0,\ \alpha > 0,\ \beta > 0$$

(*a*) Show that $\rho(t)$ is constant if $\beta = 1$.

(*b*) Find $\rho'(t)$. Argue that $\rho'(t) > 0$ if $\beta > 1$, thus producing an increasing hazard rate. Argue that $\rho'(t) < 0$ if $\beta < 1$, thus producing a decreasing hazard rate.

65. A system has eight components connected as shown in Fig. 4.21.

(*a*) Find the reliability of each of the parallel assemblies.

(*b*) Find the system reliability.

(*c*) Suppose that assembly II is replaced by two identical components in parallel, each with reliability .98. What is the reliability of the new assembly?

(*d*) What is the new system reliability after making the change suggested in part (*c*)?

(*e*) Make changes analogous to that of part (*c*) in each of the remaining single component assemblies. Compute the new system reliability.

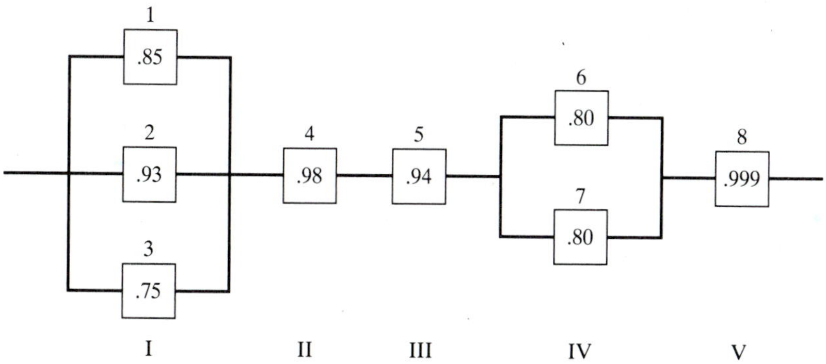

FIGURE 4.21

66. A system consists of two independent components connected in series. The life span of the first component follows a Weibull distribution with $\alpha = .006$ and $\beta = .5$; the second has a life span that follows the exponential distribution with $\beta = .00004$.
(*a*) Find the reliability of the system at 2500 hours.
(*b*) Find the probability that the system will fail before 2000 hours.
(*c*) If the two components are connected in parallel, what is the system reliability at 2500 hours?

67. Suppose that a missile can have several independent and identical computers, each with reliability .9 connected in parallel so that the system will continue to function as long as at least one computer is operating. If it is desired to have a system reliability of at least .999, how many computers should be connected in parallel?

68. Three independent and identical components, each with a reliability of .9, are to be used in an assembly.
(*a*) The assembly will function if at least one of the components is operable. Find the system reliability.
(*b*) The assembly will function if at least two of the components are operable. Find the reliability of the system.
(*c*) The assembly will function only if all three of the components are operable. Find the reliability of the system.

Section 4.8

69. Prove Theorem 4.8.1 in the case in which g is strictly increasing.

70. Let X be a random variable with density

$$f_X(x) = (1/4)x \qquad 0 \le x \le \sqrt{8}$$

and let $Y = X + 3$.
(*a*) Find $E[X]$, and then use the rules for expectation to find $E[Y]$.
(*b*) Find the density for Y.
(*c*) Use the density for Y to find $E[Y]$, and compare your answer to that found in part (*a*).

71. Let X be a random variable with density

$$f_X(x) = (1/4)xe^{-x/2} \qquad x \ge 0$$

and let $Y = (-1/2)X + 2$. Find the density for Y.

72. Let X be a random variable with density

$$f_X(x) = e^{-x} \qquad x > 0$$

and let $Y = e^X$. Find the density for Y.

73. Let C denote the temperature in degrees Celsius to which a computer will be subjected in the field. Assume that C is uniformly distributed over the interval $(15, 21)$. Let F denote the field temperature in degrees Fahrenheit so that $F = (9/5)C + 32$. Find the density for F.

74. Let X denote the velocity of a random gas molecule. According to the Maxwell-Boltzmann law, the density for X is given by

$$f_X(x) = cx^2e^{-\beta x^2} \qquad x > 0$$

Here c is a constant that depends on the gas involved and β is a constant whose value depends on the mass of the molecule and its absolute temperature. The kinetic energy of the molecule, Y, is given by $Y = (1/2)mX^2$ where $m > 0$. Find the density for Y.

75. Let X be a continuous random variable with density f_X, and let $Y = X^2$.

(*a*) Show that for $y \geq 0$,

$$F_Y(y) = P\left[-\sqrt{y} \leq X \leq \sqrt{y}\right]$$

(*b*) Show that for $y \geq 0$,

$$F_Y(y) = F_X\left(\sqrt{y}\right) - F_X\left(-\sqrt{y}\right)$$

(*c*) Use the technique given in the proof of Theorem 4.8.1 to show that

$$f_Y(y) = 1/\left(2\sqrt{y}\right)\left[f_X\left(\sqrt{y}\right) + f_X\left(-\sqrt{y}\right)\right]$$

(*d*) Use the technique illustrated in Example 4.8.2 to show that

$$f_Y(y) = 1/\left(2\sqrt{y}\right)\left[f_X\left(\sqrt{y}\right) + f_X\left(-\sqrt{y}\right)\right]$$

76. Let Z be a standard normal random variable and let $Y = Z^2$.

(*a*) Show that $\Gamma(1/2) = \int_0^\infty x^{-1/2}e^{-x}\,dx$.

(*b*) Show that $\Gamma(1/2) = \sqrt{\pi}$. *Hint:* Use the results of part (*a*) with $x = t^2/2$ and make use of the fact that the standard normal density integrates to 1 when integrated over the set of real numbers.

(*c*) Use the results of Exercise 75 to find f_Y.

(*d*) Argue that Y follows a chi-squared distribution with 1 degree of freedom.

77. Let X be normally distributed with mean μ and variance σ^2. Let $Y = e^X$. Show that Y follows the log-normal distribution. (See Exercise 45.)

78. Let Z be a standard normal random variable and let $Y = 2Z^2 - 1$. Find the density for Y.

Section 4.9

79. Use Table III of App. A to generate nine more observations on the random variable X, the time to failure of a computer chip. (See Example 4.9.1.) Based on these data, approximate the average time to failure by finding the arithmetic average of the values of X simulated in the experiment. Does this value agree well with the theoretical mean value of 50 years?

80. Simulate 20 observations on the random variable X, the time to failure of the signal lights on an automobile. (See Exercise 60.) Approximate the average time to failure for these lights based on the simulated data. Does this value agree well with the theoretical mean value for X?

81. A satellite has malfunctioned and is expected to reenter the earth's atmosphere sometime during a 4-hour period. Let X denote the time of reentry. Assume that X is uniformly distributed over the interval $[0, 4]$. Simulate 20 observations on X. (See Exercise 18.)

REVIEW EXERCISES

82. Let X be a continuous random variable with density

$$f(x) = cx^2 \qquad -3 \leq x \leq 3$$

(*a*) Assuming that $f(x) = 0$ elsewhere, find the value of c that makes this a density.

(*b*) Find $E[X]$ and $E[X^2]$ from the definitions of these terms.

(*c*) Find Var X and σ.
(*d*) Find $P[X \leq 2]$; $P[-1 \leq X \leq 2]$; $P[X > 1]$ by direct integration.
(*e*) Find the closed-form expression for the cumulative distribution function F.
(*f*) Use F to find each of the probabilities of part (*d*), and compare your answers to those obtained earlier.

83. Find $\int_0^\infty z^{10} e^{-z}\, dz$.

84. A computer firm introduces a new home computer. Past experience shows that the random variable X, the time of peak demand measured in months after its introduction, follows a gamma distribution with variance 36.
(*a*) If the expected value of X is 18 months, find α and β.
(*b*) Find $P[X \leq 7.01]$; $P[X \geq 26]$; $P[13.7 \leq X \leq 31.5]$.

85. Let X denote the lag time in a printing queue at a particular computer center. That is, X denotes the difference between the time that a program is placed in the queue and the time at which printing begins. Assume that X is normally distributed with mean 15 minutes and variance 25.
(*a*) Find the expression for the density for X.
(*b*) Find the probability that a program will reach the printer within 3 minutes of arriving in the queue.
(*c*) Would it be unusual for a program to stay in the queue between 10 and 20 minutes? Explain, based on the approximate probability of this occurring. You do not have to use the Z table to answer this question!
(*d*) Would you be surprised if it took longer than 30 minutes for the program to reach the printer? Explain, based on the probability of this occurring.

86. A computer center maintains a telephone consulting service to troubleshoot for its users. The service is available from 9 a.m. to 5 p.m. each working day. Past experience shows that the random variable X, the number of calls received per day, follows a Poisson distribution with $\lambda = 50$. For a given day, find the probability that the first call of the day will be received by 9:15 a.m.; after 3 p.m.; between 9:30 a.m. and 10 a.m.

87. Let $H(X) = X^2 + 3X + 2$. Find $E[H(X)]$ if
(*a*) X is normally distributed with mean 3 and variance 4.
(*b*) X has a gamma distribution with $\alpha = 2$ and $\beta = 4$.
(*c*) X has a chi-squared distribution with 10 degrees of freedom.
(*d*) X has an exponential distribution with $\beta = 5$.
(*e*) X has a Weibull distribution with $\alpha = 2$ and $\beta = 1$.

88. Let the density for the continuous random variable X be given by

$$f(x) = 1/2e^{-|x|} \qquad -\infty < x < \infty$$

(*a*) Show that $\int_{-\infty}^{\infty} f(x)\, dx = 1$.
(*b*) Show that

$$m_X(t) = (1/2)[1/(t+1) - 1/(t-1)] \qquad -1 < t < 1$$

(*c*) Use $m_X(t)$ to show that $E[X] = 0$.

89. Let X denote the time to failure in years of a telephone modem used to access a mainframe computer from a remote terminal. Assume that the hazard rate function for X is given by

$$\rho(t) = \alpha\beta t^{\beta - 1}$$

where $\alpha = 2$ and $\beta = 1/5$.

(*a*) Find the failure density for X.
(*b*) Find the expected value of X.
(*c*) Find the reliability function for X.
(*d*) Find the probability that the modem will last for at least 2 years.
(*e*) What is the hazard rate at $t = 1$ year?
(*f*) Describe roughly the theoretical pattern in the causes of failure in these modems.

90. Past evidence shows that when a customer complains of an out-of-order phone there is an 8% chance that the problem is with the inside wiring. During a 1-month period, 100 complaints are lodged. Assume that there have been no wide-scale problems that could be expected to affect many phones at once, and that, for this reason, these failures are considered to be independent. Find the expected number of failures due to a problem with the inside wiring. Find the probability that at least 10 failures are due to a problem with the inside wiring. Would it be unusual if at most 5 were due to problems with the inside wiring? Explain, based on the probability of this occurring.

91. The cumulative distribution function for a continuous random variable X is defined by

$$F(x) = \begin{cases} 0 & x < 0 \\ \dfrac{x^3 + x^2}{2} & 0 \le x \le 1 \\ 1 & x > 1 \end{cases}$$

Find the density for X.

92. The density for a continuous random variable is given by

$$f(x) = xe^{-x} \qquad 0 < x < \infty$$

(*a*) Show that $\int_0^\infty xe^{-x}\,dx = 1$. *Hint:* Use the gamma function.
(*b*) Find $E[X]$, $E[X^2]$, and Var X.
(*c*) Show that $m_X(t) = 1/(1-t)^2$, where $t < 1$.
(*d*) Use $m_X(t)$ to find $E[X]$.

93. An electronic counter records the number of vehicles exiting the interstate at a particular point. Assume that the average number of vehicles leaving in a 5-minute period is 10. Approximate the probability that between 100 and 120 vehicles inclusive will exit at this point in a 1-hour period.

94. Consider the following moment generating functions. In each case, identify the distribution involved completely. Be sure to specify the numerical value of all parameters that identify the distribution. For example, if X is normal, give the numerical value of μ and σ^2; if gamma, state α and β.
(*a*) $e^{3t+16t^2/2}$
(*b*) $(1-3t)^{-7}$
(*c*) $(1-2t)^{-12}$
(*d*) $\dfrac{e^{3t}-e^t}{2t}$
(*e*) $e^{t^2/2}$
(*f*) $(1-7t)^{-1}$
(*g*) $e^{3t+t^2/2}$

95. For each random variable in Exercise 94, state the numerical value of the average for X and its variance.

CHAPTER 5

JOINT DISTRIBUTIONS

Thus far interest has centered on a single random variable of either the discrete or the continuous type. Such random variables are called *univariate*. Problems do arise in which two random variables are to be studied simultaneously. For example, we might wish to study the yield of a chemical reaction in conjunction with the temperature at which the reaction is run. Typical questions to ask are: "Is the yield independent of the temperature?" or, "What is the average yield if the temperature is 40°C?" To answer questions of this type, we need to study what are called *two-dimensional* or *bivariate random variables* of both the discrete and continuous type. In this chapter we present a brief introduction to the basic theoretical concepts underlying these variables. These concepts form the basis for the study of regression analysis and correlation, topics of extreme importance in applied statistics. (See Chaps. 11 and 12.)

5.1 JOINT DENSITIES AND INDEPENDENCE

We begin by considering two-dimensional random variables and their density functions. The definitions presented here are natural extensions of those presented for a single random variable in Chaps. 3 and 4. (See Definition 3.2.1 and 4.1.2.)

Definition 5.1.1 (Discrete joint density). Let X and Y be discrete random variables. The ordered pair (X, Y) is called a two-dimensional discrete random variable. A function f_{XY} such that

$$f_{XY}(x, y) = P[X = x \text{ and } Y = y]$$

is called the joint density for (X, Y).

Again, let us point out that in the discrete case some statisticians prefer to use the term "probability function" or "probability mass function" rather than the term "density." We shall use the term "density" and the notation f_{XY} in both the discrete and the continuous cases for consistency of notation and terminology.

Note that the purpose of the density here is the same as in the past—to allow us to compute the probability that the random variable (X, Y) will assume specific values. As in the one-dimensional case, f_{XY} is nonnegative since it represents a probability. Furthermore, if the density is summed over all possible values of X and Y, it must sum to 1. That is, the necessary and sufficient conditions for a function to be a joint density for a two-dimensional discrete random variable are

1. $f_{XY}(x, y) \geq 0$
2. $\sum_{\text{all } x} \sum_{\text{all } y} f_{XY}(x, y) = 1$

The joint density in the discrete case is sometimes expressed in closed form. However, it is more common to present the density in table form.

Example 5.1.1. In an automobile plant two tasks are performed by robots. The first entails welding two joints; the second, tightening three bolts. Let X denote the number of defective welds and Y the number of improperly tightened bolts produced per car. Since X and Y are each discrete (X, Y) is a two-dimensional discrete random variable. Past data indicates that the joint density for (X, Y) is as shown in Table 5.1. Note that each entry in the table is a number between 0 and 1 and therefore can be interpreted as a probability. Furthermore,

$$\sum_{x=0}^{2} \sum_{y=0}^{3} f_{XY}(x, y) = .840 + .030 + .020 + \cdots + .001 = 1$$

as required. The probability that there will be no errors made by the robots is given by

$$P[X = 0 \text{ and } Y = 0] = f_{XY}(0, 0) = .840$$

The probability that there will be exactly one error made is

$$\begin{aligned} P[X = 1 \text{ and } Y = 0] + P[X = 0 \text{ and } Y = 1] &= f_{XY}(1, 0) + f_{XY}(0, 1) \\ &= .060 + .030 \\ &= .09 \end{aligned}$$

TABLE 5.1

x/y	0	1	2	3
0	.840	.030	.020	.010
1	.060	.010	.008	.002
2	.010	.005	.004	.001

The probability that there will be no improperly tightened bolts is $P[Y = 0]$. Note that this probability, which concerns only the random variable Y, can be obtained by summing $f_{XY}(x, 0)$ over all values of X. That is,

$$\begin{aligned} P[Y = 0] &= \sum_{x=0}^{2} f_{XY}(x, 0) \\ &= P[X = 0 \text{ and } Y = 0] + P[X = 1 \text{ and } Y = 0] \\ &\quad + P[X = 2 \text{ and } Y = 0] \\ &= .840 + .060 + .010 = .91 \end{aligned}$$

Marginal Distributions: Discrete

Given the joint density for a two-dimensional discrete random variable (X, Y), it is easy to derive the individual densities for X and Y. The manner in which this is done is suggested by the method used to answer the last question posed in Example 5.1.1. To find the density for Y alone, we sum the joint density over all values of X; to find the density for X alone, we sum over Y. When the joint density is given in table form, it is customary to report the individual densities for X and Y in the margins of the joint density table. For this reason, the densities for X and Y alone are called *marginal densities*. This idea is formalized in Definition 5.1.2.

Definition 5.1.2 (Discrete marginal densities). Let (X, Y) be a two-dimensional discrete random variable with joint density f_{XY}. The marginal density for X, denoted by f_X, is given by

$$f_X(x) = \sum_{\text{all } y} f_{XY}(x, y)$$

The marginal density for Y, denoted by f_Y, is given by

$$f_Y(y) = \sum_{\text{all } x} f_{XY}(x, y)$$

Example 5.1.2. Table 5.2 gives the joint density for the random variable (X, Y) of Example 5.1.1. It also displays the marginal densities for X, the number of defective welds, and Y, the number of improperly tightened bolts per car. Note that the marginal density for X is obtained by summing across the rows of the table; that for Y is obtained by summing down the columns.

TABLE 5.2

x/y	0	1	2	3	$f_X(x)$
0	.840	.030	.020	.010	.900
1	.060	.010	.008	.002	.080
2	.010	.005	.004	.001	.020
$f_Y(y)$	.910	.045	.032	.013	1.000

Joint and Marginal Distributions: Continuous

The idea of a two-dimensional continuous random variable and continuous joint density can be developed by extending Definition 4.1.1 to more than one variable.

Definition 5.1.3 (Continuous joint density). Let X and Y be continuous random variables. The ordered pair (X, Y) is called a two-dimensional continuous random variable. A function f_{XY} such that

1. $f_{XY}(x, y) \geq 0 \qquad -\infty < x < \infty$
   $\qquad\qquad\qquad\qquad -\infty < y < \infty$

2. $\int_{-\infty}^{\infty} \int_{-\infty}^{\infty} f_{XY}(x, y)\, dy\, dx = 1$

3. $P[a \leq X \leq b \text{ and } c \leq Y \leq d] = \int_a^b \int_c^d f_{XY}(x, y)\, dy\, dx$

for a, b, c, d real is called the joint density for (X, Y).

Even though the joint density is defined for all real values x and y, we shall follow the convention of specifying its equation only over those regions for which it may be nonzero. Recall that in the case of a single continuous random variable, probabilities correspond to areas. In the case of a two-dimensional continuous random variable, probabilities correspond to *volumes*. These ideas are illustrated in Example 5.1.3.

Example 5.1.3. In a healthy individual age 20 to 29 years, the calcium level in the blood, X, is usually between 8.5 and 10.5 milligrams per deciliter (mg/dl) and the cholesterol level, Y, is usually between 120 and 240 mg/dl. Assume that for a healthy individual in this age group the random variable (X, Y) is uniformly distributed over the rectangle whose corners are (8.5, 120), (8.5, 240), (10.5, 120), (10.5, 240). That is, assume that the joint density for (X, Y) is

$$f_{XY}(x, y) = c \qquad 8.5 \leq x \leq 10.5$$
$$120 \leq y \leq 240$$

To be a density, c must be chosen so that

$$\int_{8.5}^{10.5} \int_{120}^{240} c\, dy\, dx = 1$$

That is, c must be chosen so that the volume of the rectangular solid shown in Fig. 5.1(a) is 1. To find c, we can use geometry or complete the indicated integration as shown below.

$$\int_{8.5}^{10.5} \int_{120}^{240} c\, dy\, dx = 1$$

$$c\int_{8.5}^{10.5} (240 - 120)\, dx = 1$$

$$120c(10.5 - 8.5) = 1$$

$$240c = 1$$

$$c = 1/240$$

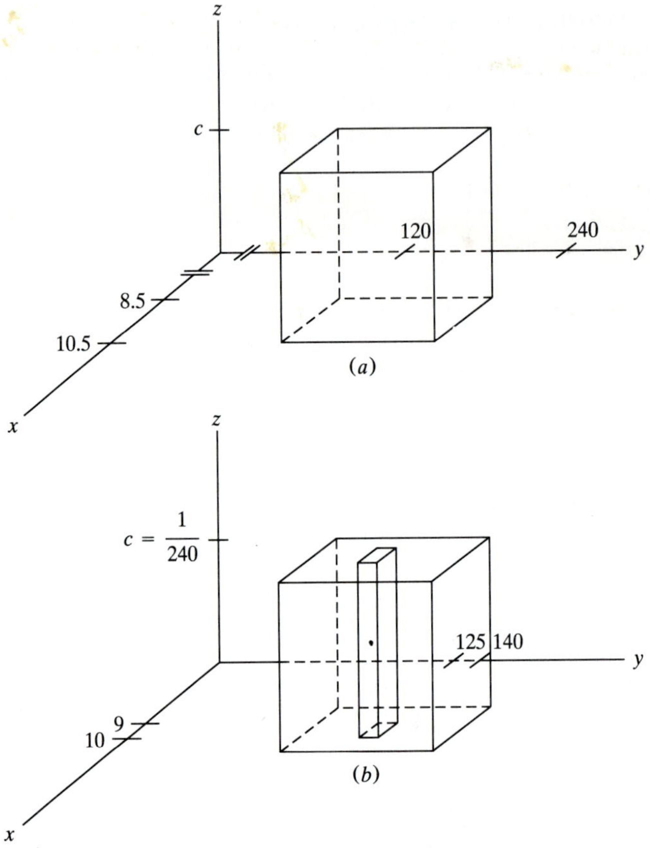

FIGURE 5.1
(*a*) Volume of the solid whose base is a rectangle with corners (8.5, 120), (8.5, 240), (10.5, 120), (10.5, 240) and height c is 1; (*b*) $P[9 \le X \le 10 \text{ and } 125 \le Y \le 140]$ = volume of solid whose base is a rectangle with corners (9, 125), (9, 140), (10, 125), (10, 140) and height $c = 1/240$.

Let us now use the joint density to find the probability that an individual's calcium level will lie between 9 and 10 mg/dl, whereas the cholesterol level is between 125 and 140 mg/dl. This probability corresponds to the volume of the solid shown in Fig. 5.1(*b*). This probability is

$$P[9 \le X \le 10 \text{ and } 125 \le Y \le 140] = \int_9^{10} \int_{125}^{140} 1/240 \, dy \, dx$$

$$= 1/240 \int_9^{10} (140 - 125) \, dx$$

$$= 15/240$$

To define "marginal" densities in the continuous case, we replace summation by integration. This yields the following definition.

Definition 5.1.4 (Continuous marginal densities). Let (X, Y) be a two-dimensional continuous random variable with joint density f_{XY}. The marginal density for X, denoted by f_X, is given by

$$f_X(x) = \int_{-\infty}^{\infty} f_{XY}(x, y)\, dy$$

The marginal density for Y, denoted by f_Y, is given by

$$f_Y(y) = \int_{-\infty}^{\infty} f_{XY}(x, y)\, dx$$

We illustrate the idea of marginal densities in Examples 5.1.4 and 5.1.5.

Example 5.1.4. Let X denote an individual's blood calcium level and Y his or her blood cholesterol level. The joint density for (X, Y) is

$$f_{XY}(x, y) = 1/240 \qquad \begin{array}{l} 8.5 \le x \le 10.5 \\ 120 \le y \le 240 \end{array}$$

The marginal densities for X and Y are

$$f_X(x) = \int_{120}^{240} 1/240\, dy = 1/2 \qquad 8.5 \le x \le 10.5$$

$$f_Y(y) = \int_{8.5}^{10.5} 1/240\, dx = 2/240 \qquad 120 \le y \le 240$$

To find the probability that a healthy individual has a cholesterol level between 150 and 200, we can use either the joint density or the marginal density for Y. That is,

$$P[150 \le Y \le 200] = \int_{8.5}^{10.5} \int_{150}^{200} 1/240\, dy\, dx = 100/240$$

or

$$P[150 \le Y \le 200] = \int_{150}^{200} 2/240\, dy = 100/240$$

Note that both X and Y are uniformly distributed.

Example 5.1.5. In studying the behavior of air support roofs, the random variables X, the inside barometric pressure (in inches of mercury), and Y, the outside pressure, are considered. Assume that the joint density for (X, Y) is given by

$$f_{XY}(x, y) = c/x \qquad \begin{array}{l} 27 \le y \le x \le 33 \\ c = 1/(6 - 27 \ln 33/27) \doteq 1.72 \end{array}$$

The region in the plane over which this joint density is defined is shown in Fig. 5.2(a). The marginal densities for X and Y are given by

$$f_X(x) = \int_{27}^{x} c/x\, dy = (c/x)y \Big|_{27}^{x} = c(1 - 27/x) \qquad 27 \le x \le 33$$

$$f_Y(y) = \int_{y}^{33} c/x\, dx = c(\ln 33 - \ln y) \qquad 27 \le y \le 33$$

Let us find the probability that the inside pressure is at most 30 and the outside pressure is at most 28. That is, let us find $P[X \le 30 \text{ and } Y \le 28]$. The region over which the joint density is to be integrated is shown in Fig. 5.2(b). Integration can be done with respect to y and then x or vice versa. In the former case the problem

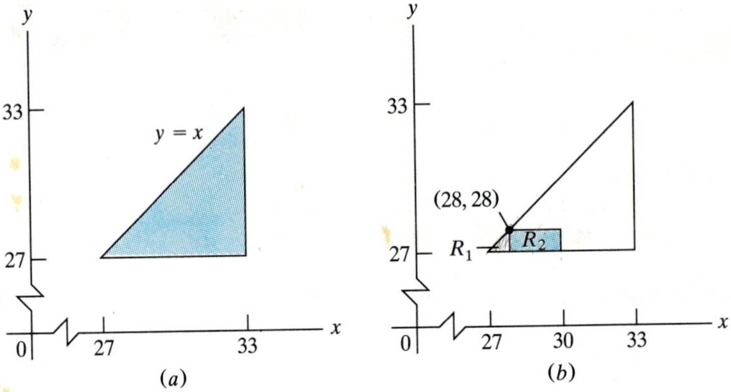

FIGURE 5.2
(a) The joint density $f(x, y) = c/x$ is defined over the triangular region bounded by $y = 27$, $y = x$, and $x = 33$.
(b)

$$P[X \le 30 \text{ and } Y \le 28] = \iint_{R_1} c/x \, dy \, dx + \iint_{R_2} c/x \, dy \, dx$$

$$= \int_{27}^{28} \int_{27}^{x} c/x \, dy \, dx + \int_{28}^{30} \int_{27}^{28} c/x \, dy \, dx$$

or

$$P[X \le 30 \text{ and } Y \le 28] = \int_{27}^{28} \int_{y}^{30} c/x \, dx \, dy.$$

must be split into two pieces since the boundaries for y change at the point (28, 28). In the latter case integration can be accomplished more easily. The integrals required in the two cases are

Case I:

$$P[X \le 30 \text{ and } Y \le 28] = \int_{27}^{28} \int_{27}^{x} c/x \, dy \, dx + \int_{28}^{30} \int_{27}^{28} c/x \, dy \, dx$$

Case II:

$$P[X \le 30 \text{ and } Y \le 28] = \int_{27}^{28} \int_{y}^{30} c/x \, dx \, dy$$

Since case II requires less effort, we find $P[X \le 30 \text{ and } Y \le 28]$ as follows:

$$P[X \le 30 \text{ and } Y \le 28] = \int_{27}^{28} \int_{y}^{30} c/x \, dx \, dy$$

$$= c\int_{27}^{28} [\ln 30 - \ln y] \, dy$$

$$= c\left[y \ln 30 \Big|_{27}^{28} - \int_{27}^{28} \ln y \, dy \right]$$

$$= c\left[\ln 30 - (y \ln y - y)\Big|_{27}^{28} \right]$$

$$= c[\ln 30 - 28 \ln 28 + 27 \ln 27 + 1]$$

$$\doteq c(.09) = 1.72(.09) = .15$$

It is left as an exercise to show that the same result is obtained via case I. (See Exercise 6.)

Independence

There is one other point to be made in this section. Recall that two events are independent if knowledge of the fact that one has occurred gives us no clue as to the likelihood that the other will occur. Suppose that X and Y are discrete random variables such that knowledge of the value assumed by one gives us no clue as to the value assumed by the other. We would like to think of these random variables as being "independent" and would like a mathematical characterization of this property. The characterization is suggested by the following argument. Let X and Y be discrete. Let A_1 denote the event that $X = x$ and let A_2 denote the event that $Y = y$. If X and Y are independent in the intuitive sense, then A_1 and A_2 are independent events. By Definition 2.3.1

$$P[A_1 \cap A_2] = P[A_1]P[A_2]$$

Substituting, we see that

$$P[X = x \text{ and } Y = y] = P[X = x]P[Y = y]$$

or

$$f_{XY}(x, y) = f_X(x)f_Y(y)$$

It seems that, at least in the discrete case, independence implies that the *joint density can be expressed as the product of the marginal densities*. This idea provides the basis for the definition of the term "independent random variables" in both the discrete and continuous cases.

Definition 5.1.5 (Independent random variables). Let X and Y be random variables with joint density f_{XY} and marginal densities f_X and f_Y, respectively. X and Y are independent if and only if

$$f_{XY}(x, y) = f_X(x)f_Y(y)$$

for all x and y.

Example 5.1.6

(*a*) The random variables X, the number of defective welds, and Y, the number of improperly tightened bolts per car of Examples 5.1.1 and 5.1.2, are not independent. To verify this, note that from Table 5.2

$$f_{XY}(0, 0) = .84 \neq .9(.91) = .819 = f_X(0)f_Y(0)$$

(*b*) The random variables X, an individual's blood calcium level, and Y, his or her blood cholesterol level as described in Examples 5.1.3 and 5.1.4, are independent. To verify this, note that

$$f_{XY}(x, y) = 1/240 = 1/2 \cdot 2/240 = f_X(x)f_Y(y)$$

An important point should be made here. The assumption that (X, Y) is uniformly distributed leads to the conclusion that X and Y are independent. If this conclusion is *medically unsound*, then another more realistic density should be sought to describe the behavior of the two-dimensional random variable (X, Y).

(*c*) The random variables X and Y, the inside and outside pressure, respectively, on an air support roof of Example 5.1.5 are not independent. This is seen by noting that

$$f_{XY}(x, y) = c/x \neq c(1 - 27/x)c(\ln 33 - \ln y) = f_X(x)f_Y(y)$$

The assumption of nonindependence here is realistic from a physical point of view.

The exercises for Sec. 5.1 provide some practice in dealing with these theoretical ideas. You will see their relationship to data analysis in chapters to come.

5.2 EXPECTATION AND COVARIANCE

In this section we introduce the idea of *expectation* in the case of a two-dimensional random variable. We also study a specific expectation, called the *covariance*, that is useful in describing the behavior of one variable relative to another.

We begin by extending Definitions 3.3.1 and 4.2.1 to the two-dimensional case.

Definition 5.2.1 (Expected value). Let (X, Y) be a two-dimensional random variable with joint density f_{XY}. Let $H(X, Y)$ be a random variable. The expected value of $H(X, Y)$, denoted by $E[H(X, Y)]$ is given by

1. $E[H(X,Y)] = \sum_{\text{all } x} \sum_{\text{all } y} H(x, y) f_{XY}(x, y)$

provided $\sum_{\text{all } x} \sum_{\text{all } y} |H(x, y)| f_{XY}(x, y)$ exists for (X, Y) discrete;

2. $E[H(X,Y)] = \int_{-\infty}^{\infty} \int_{-\infty}^{\infty} H(x, y) f_{XY}(x, y)\, dy\, dx$

provided $\int_{-\infty}^{\infty} \int_{-\infty}^{\infty} |H(x, y)| f_{XY}(x, y)\, dy\, dx$ exists for (X, Y) continuous.

Examples 5.2.1 and 5.2.2 illustrate the use of this definition.

Example 5.2.1. The joint density for the random variable (X, Y) of Example 5.1.1 is given in Table 5.3. X denotes the number of defective welds and Y, the number of improperly tightened bolts produced per car by assembly line robots. Let us use

TABLE 5.3

x/y	0	1	2	3	$f_X(x)$
0	.840	.030	.020	.010	.900
1	.060	.010	.008	.002	.080
2	.010	.005	.004	.001	.020
$f_Y(y)$	.910	.045	.032	.013	1.000

Definition 5.2.1 to find $E[X]$, $E[Y]$, $E[X+Y]$, and $E[XY]$.

$$E[X] = \sum_{x=0}^{2} \sum_{y=0}^{3} x f_{XY}(x, y)$$

$$= 0(.840) + 0(.030) + 0(.020) + 0(.010) + 1(.060) + \cdots + 2(.001)$$

$$= .12$$

$$E[Y] = \sum_{x=0}^{2} \sum_{y=0}^{3} y f_{XY}(x, y)$$

$$= 0(.840) + 1(.030) + 2(.020) + 3(.010) + 0(.060) + \cdots + 3(.001)$$

$$= .148$$

$$E[X+Y] = \sum_{x=0}^{2} \sum_{y=0}^{3} (x+y) f_{XY}(x, y)$$

$$= (0+0)(.840) + (0+1)(.030) + (0+2)(.020) + \cdots + (2+3)(.001)$$

$$= .268$$

$$E[XY] = \sum_{x=0}^{2} \sum_{y=0}^{3} xy f_{XY}(x, y)$$

$$= (0 \cdot 0)(.840) + (0 \cdot 1)(.030) + (0 \cdot 2)(.020) + \cdots + (2 \cdot 3)(.001)$$

$$= .064$$

There are two points to be made. First, both $E[X]$ and $E[Y]$ were found via the joint density and Definition 5.2.1. These expectations could have been found just as easily from the marginal densities and Definition 3.3.1. (See Exercise 18.) Second, note that $E[X+Y] = E[X] + E[Y]$. This result is consistent with the rules of expectation given in Theorem 3.3.1.

Example 5.2.2. The joint density for the random variable (X, Y), where X denotes the calcium level and Y denotes the cholesterol level in the blood of a healthy individual, is given by

$$f_{XY}(x, y) = 1/240 \qquad \begin{array}{l} 8.5 \le x \le 10.5 \\ 120 \le y \le 240 \end{array}$$

For these variables,

$$E[X] = \int_{-\infty}^{\infty}\int_{-\infty}^{\infty} x f_{XY}(x, y)\, dy\, dx$$

$$= \int_{8.5}^{10.5}\int_{120}^{240} x(1/240)\, dy\, dx$$

$$= \int_{8.5}^{10.5} (1/2)x\, dx = x^2/4 \Big|_{8.5}^{10.5} = 9.5 \text{ mg/dl}$$

$$E[Y] = \int_{-\infty}^{\infty}\int_{-\infty}^{\infty} y f_{XY}(x, y)\, dy\, dx$$

$$= \int_{8.5}^{10.5}\int_{120}^{240} y(1/240)\, dy\, dx$$

$$= 1/240 \int_{8.5}^{10.5} y^2/2 \Big|_{120}^{240} dx$$

$$= 1/240 \int_{8.5}^{10.5} 21{,}600\, dx = 180 \text{ mg/dl}$$

$$E[XY] = \int_{-\infty}^{\infty}\int_{-\infty}^{\infty} xy f_{XY}(x, y)\, dy\, dx$$

$$= \int_{8.5}^{10.5}\int_{120}^{240} xy(1/240)\, dy\, dx$$

$$= 1/240 \int_{8.5}^{10.5} xy^2/2 \Big|_{120}^{240} dx$$

$$= 1/240 \int_{8.5}^{10.5} 21{,}600x\, dx$$

$$= (21{,}600/240)(x^2/2)\Big|_{8.5}^{10.5} = 1710$$

Covariance

Occasionally the expected value of a function of X and Y is of interest in its own right. For instance, in Example 5.2.1, $E[X + Y]$ gives the theoretical average number of errors made by the robots overall. However, we shall be concerned primarily with those expectations that are needed to compute the covariance between X and Y. This term is defined as follows.

Definition 5.2.2 (Covariance). Let X and Y be random variables with means μ_X and μ_Y respectively. The covariance between X and Y, denoted by $\text{Cov}(X, Y)$ or σ_{XY} is given by

$$\text{Cov}(X, Y) = E[(X - \mu_X)(Y - \mu_Y)]$$

Note that if small values of X tend to be associated with small values of Y and large values of X with large values of Y, then $X - \mu_X$ and $Y - \mu_Y$ will usually have the same algebraic signs. This implies that $(X - \mu_X)(Y - \mu_Y)$ will be positive, yielding a positive covariance. If the reverse is true and small values of X tend to be associated with large values of Y and vice versa, then $X - \mu_X$ and $Y - \mu_Y$ will usually have opposite algebraic signs. This results in a negative value for $(X - \mu_X)(Y - \mu_Y)$, yielding a negative covariance. In this sense covariance is an indication of how X and Y vary relative to one another.

Covariance is seldom computed from Definition 5.2.2. Rather, we apply the following computational formula whose derivation is left as an exercise. (See Exercise 24.)

Theorem 5.2.1 (Computational formula for covariance)

$$\text{Cov}(X, Y) = E[XY] - E[X]E[Y]$$

We illustrate the use of Theorem 5.2.1 by finding the covariance for the random variables of Examples 5.2.1 and 5.2.2.

Example 5.2.3

(*a*) The covariance between X, the number of defective welds, and Y, the number of improperly tightened bolts of Example 5.2.1, is given by

$$\begin{aligned}\text{Cov}(X, Y) &= E[XY] - E[X]E[Y]\\ &= .064 - (.12)(.148) = .046\end{aligned}$$

Since $\text{Cov}(X, Y) > 0$, there is a tendency for large values of X to be associated with large values of Y and vice versa. That is, a car with an above average number of defective welds tends also to have an above average number of improperly tightened bolts and vice versa.

(*b*) The covariance between X, an individual's blood calcium level, and Y, his or her blood cholesterol level, has covariance given by

$$\begin{aligned}\text{Cov}(X, Y) &= E[XY] - E[X]E[Y]\\ &= 1710 - (9.5)(180) = 0\end{aligned}$$

A covariance of 0 implies that knowledge that X assumes a value above its mean gives us no indication as the value of Y relative to its mean.

The fact that the covariance between X and Y is 0 in Example 5.2.2 is not a coincidence. It is, of course, due to the fact that $E[XY] = E[X]E[Y]$. It can be shown that this property will hold whenever the random variables X and Y are independent, as they are in Example 5.2.2. This important result is formalized in the following theorem.

Theorem 5.2.2. Let (X, Y) be a two-dimensional random variable with joint density f_{XY}. If X and Y are independent then

$$E[XY] = E[X]E[Y]$$

Proof. We shall prove this theorem in the continuous case. The proof in the discrete case is similar. Assume that (X, Y) has joint density f_{XY} and that X and Y are independent. Let f_X and f_Y denote the marginal densities for X and Y,

TABLE 5.4

x/y	-2	-1	1	2	$f_X(x)$
1	0	1/4	1/4	0	1/2
4	1/4	0	0	1/4	1/2
$f_Y(y)$	1/4	1/4	1/4	1/4	1

respectively. By Definition 5.2.1,

$$\begin{aligned} E[XY] &= \int_{-\infty}^{\infty}\int_{-\infty}^{\infty} xyf_{XY}(x,y)\,dy\,dx \\ &= \int_{-\infty}^{\infty}\int_{-\infty}^{\infty} xyf_X(x)f_Y(y)\,dy\,dx \qquad (X \text{ and } Y \text{ are independent}) \\ &= \int_{-\infty}^{\infty} xf_X(x)\int_{-\infty}^{\infty} yf_Y(y)\,dy\,dx \\ &= \int_{-\infty}^{\infty} xf_X(x)E[Y]\,dx \\ &= E[Y]\int_{-\infty}^{\infty} xf_X(x)\,dx = E[Y]E[X] \end{aligned}$$

An immediate consequence of this theorem is the result that we have already noted and observed relative to Example 5.2.2. In particular, *if X and Y are independent, then* $\text{Cov}(X,Y) = 0$. Unfortunately, the converse of this statement is not true. That is, we *cannot* conclude that a zero covariance implies independence. The next example verifies this contention.

Example 5.2.4. The joint density for (X, Y) is given in Table 5.4, from which we see that $E[X] = 5/2$, $E[Y] = 0$, and $E[XY] = 0$, yielding a covariance of 0. It is also easy to see that X and Y are *not* independent. The value assumed by Y does have an effect on that assumed by X. In fact, $X = Y^2$. The value of Y completely determines the value of X!

Covariance gives us only a very rough idea of the relationship between X and Y. We are concerned only with its algebraic sign and not with its magnitude. However, covariance is used to define another measure of the relationship between X and Y which is easier to interpret. This measure, called the *correlation*, is discussed in the next section.

5.3 CORRELATION

Recall that the covariance between X and Y gives only a rough indication of any association that may exist between X and Y. No attempt is made to describe the type or strength of the association. Often it is of interest to know whether or not two random variables are *linearly* related. One measure used to determine this is the Pearson coefficient of correlation, ρ. In this section we define this theoretical measure of linearity; in Chap. 11 we shall discuss how to estimate its value from a data set.

Definition 5.3.1 (Pearson coefficient of correlation). Let X and Y be random variables with means μ_X and μ_Y and variances σ_X^2 and σ_Y^2, respectively. The correlation, ρ_{XY}, between X and Y is given by

$$\rho_{XY} = \frac{\operatorname{Cov}(X, Y)}{\sqrt{(\operatorname{Var} X)(\operatorname{Var} Y)}}$$

Since we already know how to calculate each of the terms appearing in the above definition, calculating ρ_{XY} (or ρ) from the joint density for (X, Y) is easy. The question is, "How do we interpret ρ once we know its numerical value?" To interpret ρ, we must know its range of possible values. The next theorem shows that, unlike the covariance which can assume any real value, the correlation coefficient is bounded.

Theorem 5.3.1. The correlation coefficient ρ_{XY} for any two random variables X and Y lies between -1 and 1 inclusive.

Proof. Let Z and W denote random variables such that $E[Z^2] \neq 0$ and $E[W^2] \neq 0$. Let a denote any real number. Note that since the random variable $(aW - Z)^2 \geq 0$, its mean is nonnegative. That is,

$$E\left[(aW - Z)^2\right] = a^2E[W^2] - 2aE[WZ] + E[Z^2] \geq 0$$

Let $a = E[WZ]/E[W^2]$. Substituting, we can conclude that

$$\frac{(E[WZ])^2}{(E[W^2])^2}E[W^2] - 2\frac{E[WZ]}{E[W^2]}E[WZ] + E[Z^2] \geq 0$$

or

$$-\frac{(E[WZ])^2}{E[W^2]} + E[Z^2] \geq 0$$

This implies that

$$\frac{(E[WZ])^2}{E[W^2]E[Z^2]} \leq 1$$

Now let $W = X - \mu_X$ and $Z = Y - \mu_Y$. Substituting into the above inequality, we can conclude that

$$\frac{\left(E[(X-\mu_X)(Y-\mu_Y)]\right)^2}{\left(E\left[(X-\mu_X)^2\right]\right)\left(E\left[(Y-\mu_Y)^2\right]\right)} = \rho_{XY}^2 \leq 1$$

Solving for ρ_{XY}, we see that $|\rho_{XY}| \leq 1$ or that $-1 \leq \rho_{XY} \leq 1$ as claimed.

The next theorem indicates how ρ measures linearity. The point of the theorem is twofold. First, if there is a linear relationship between X and Y, then this fact is reflected in a correlation coefficient of 1 or -1. Second, if $\rho = 1$ or -1, then a linear relationship exists between X and Y. The formal statement of this result is given in Theorem 5.3.2.

Theorem 5.3.2. Let X and Y be random variables with correlation coefficient ρ_{XY}. Then $|\rho_{XY}| = 1$ if and only if $Y = \beta_0 + \beta_1 X$ for some real numbers β_0 and $\beta_1 \neq 0$.

Proof. We shall show that if $|\rho_{XY}| = 1$, then X and Y are linearly related. The proof of the converse is straightforward and is outlined as an exercise (Exercise 35). Assume that $|\rho_{XY}| = 1$. We can reverse the steps given in the proof of Theorem 5.3.1, replacing inequality with equality at each step to conclude that $E[(aW - Z)^2] = 0$. For the mean of a nonnegative random variable to be 0, the variable must equal 0 with probability 1. That is, $P[(aW - Z)^2 = 0] = 1$. This in turn implies that $P[aW - Z = 0] = 1$ or that $P[aW = Z] = 1$. Let $W = X - \mu_X$ and $Z = Y - \mu_Y$ to conclude that $P[aX - a\mu_X = Y - \mu_X] = 1$. Rewriting this expression, we can conclude that $P[Y = \mu_Y - a\mu_X + aX] = 1$. That is, $P[Y = \beta_0 + \beta_1 X] = 1$, where $\beta_0 = \mu_Y - a\mu_X$ and $\beta_1 = a$. This means that points not on the line $Y = \beta_0 + \beta_1 X$ occur with 0 probability, and the proof is complete.

If $\rho = 1$, then we say that X and Y have *perfect positive* correlation. Perfect position correlation implies that $Y = \beta_0 + \beta_1 X$, where $\beta_1 > 0$. This in turn implies that small values of X are associated with small values of Y, and large values of X with large values of Y. Perfect negative correlation implies that $Y = \beta_0 + \beta_1 X$, where $\beta_1 < 0$. Practically speaking, this means that small values of X are associated with large values of Y and vice versa. Unfortunately, random variables seldom assume the easily interpretable values of 1 or -1. However, values of ρ near 1 or -1 do occur and indicate a linear trend. That is, they indicate that, even though no single straight line passes through the points of positive probability, there is a straight line passing through the graph with the property that most of the probability is associated with points lying on or near this straight line. It is equally important to realize what Theorem 5.3.2 is not saying. If $\rho = 0$, we say that X and Y are uncorrelated, but we are *not* saying that they are unrelated. We are saying that if a relationship exists, then it is *not linear*. These ideas are illustrated in Fig. 5.3.

Example 5.3.1. To find the correlation between X, the number of defective welds, and Y, the number of improperly tightened bolts produced per car by assembly line robots, we use Table 5.3 to compute $E[X^2]$ and $E[Y^2]$. For these variables

$$E[X^2] = 0^2(.90) + 1^2(.08) + 2^2(.02) = .16$$

$$E[Y^2] = 0^2(.910) + 1^2(.045) + 2^2(.032) + 3^2(.013) = .29$$

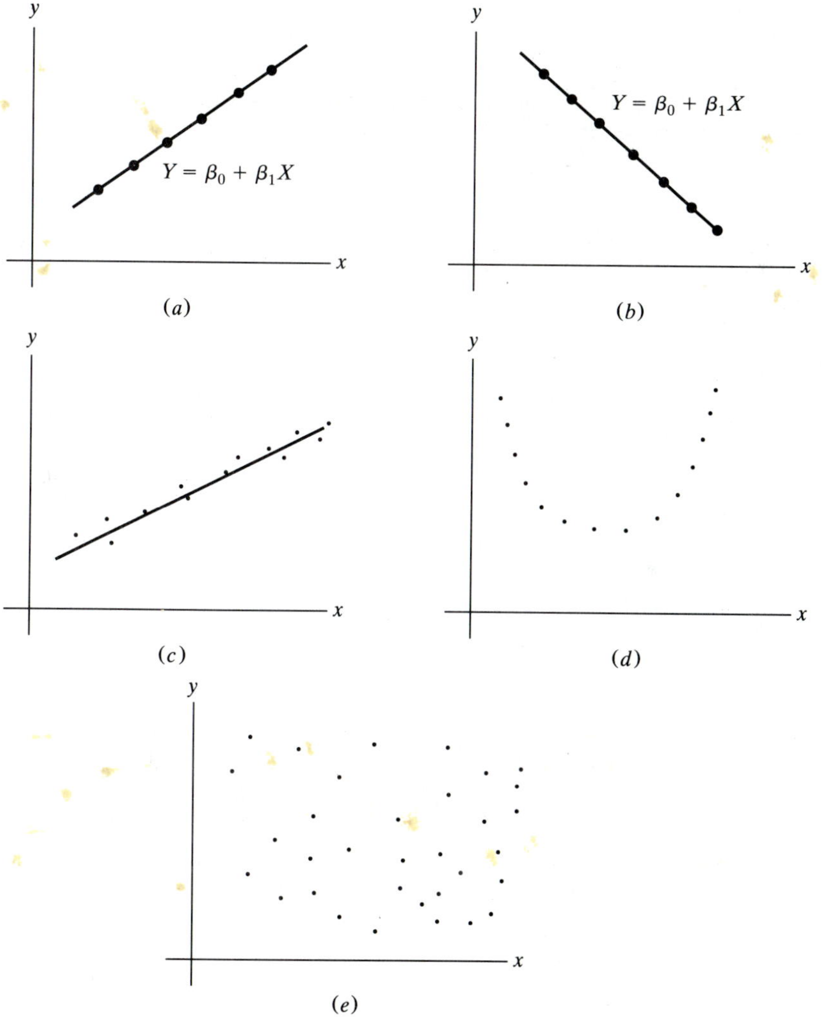

FIGURE 5.3
(a) Perfect positive correlation: $\rho = 1$, $\beta > 0$, all points lie on a straight line with positive slope; (b) perfect negative correlation: $\rho = -1$, $\beta < 0$, all points lie on a straight line with negative slope; (c) ρ near 1, points exhibit a linear trend; (d) uncorrelated: $\rho = 0$, points indicate a relationship between X and Y, but the relationship is not linear; (e) uncorrelated: $\rho = 0$, points are randomly scattered.

In Example 5.2.1, we found that $E[X] = .12$ and $E[Y] = .148$. Therefore

$$\text{Var } X = E[X^2] - (E[X])^2 = .16 - (.12)^2 \doteq .146$$

$$\text{Var } Y = E[Y^2] - (E[Y])^2 = .29 - (.148)^2 \doteq .268$$

In Example 5.2.3 we found that $\text{Cov}(X, Y) = .046$. By Definition 5.3.1,

$$\rho_{XY} = \frac{\text{Cov}(X,Y)}{\sqrt{\text{Var } X \text{ Var } Y}} = \frac{.046}{\sqrt{(.146)(.268)}} \doteq .23$$

Since this value does not appear to lie close to 1, we would not expect the observed values of X and Y to exhibit a strong linear trend.

Exercise 36 points out the relationship between correlation and independence.

5.4 CONDITIONAL DENSITIES AND REGRESSION

In this section we consider two topics that are closely related. These are *conditional densities* and *regression*. To see what is to be done, let us reconsider Example 5.1.5.

Example 5.4.1. In Example 5.1.5 we considered the random variable (X, Y) where X is the inside and Y the outside barometric pressure on an air support roof. Suppose we are interested in studying the inside pressure when the outside pressure is fixed at $y = 30$. There are three important points to understand:

1. The inside pressure will vary even though the outside pressure is constant. Therefore it makes sense to talk about "the random variable X given that $y = 30$." We shall denote this new random variable by $X|y = 30$.
2. Since $X|y = 30$ is a random variable in its own right, it has a probability distribution. Therefore it makes sense to ask, "What is the density for $X|y = 30$?" We shall call this density the "conditional density for X given that $y = 30$" and shall denote it by $f_{X|y=30}$.
3. Since the inside pressure varies even though the outside pressure is constant, it makes sense to ask, "What is the mean or average pressure on the inside of the roof when the outside pressure is 30?" That is, we can ask, "What is the mean value for the random variable $X|y = 30$?" This mean value is denoted by $E[X|y = 30]$ or $\mu_{X|y=30}$.

In general, the conditional density for X given $Y = y$, denoted by $f_{X|y}$, is a function that allows us to find the probability that X assumes specific values based on knowledge of the value assumed by the random variable Y. To see how to define $f_{X|y}$ let us assume that (X, Y) is discrete with joint density f_{XY} and marginal densities f_X and f_Y. Let A_1 denote the event that $X = x$ and A_2

denote the event that $Y = y$. From Definition 2.2.1,

$$P[A_1|A_2] = \frac{P[A_1 \cap A_2]}{P[A_2]}$$

Substituting, we see that

$$P[X = x|Y = y] = \frac{P[X = x \text{ and } Y = y]}{P[Y = y]} = \frac{f_{XY}(x, y)}{f_Y(y)}$$

In the discrete case the conditional density for X given $Y = y$, is the ratio of the joint density for (X, Y) to the marginal density for Y. This observation provides the motivation for the definition of the term "conditional density" in both the discrete and continuous cases. In the formal definition note that the roles of X and Y can be reversed.

Definition 5.4.1 (Conditional density). Let (X, Y) be a two-dimensional random variable with joint density f_{XY} and marginal densities f_X and f_Y. Then

1. The conditional density for X given $Y = y$, denoted by $f_{X|y}$ is given by

$$f_{X|y}(x) = \frac{f_{XY}(x, y)}{f_Y(y)} \qquad f_Y(y) > 0$$

2. The conditional density for Y given $X = x$, denoted by $f_{Y|x}$, is given by

$$f_{Y|x}(y) = \frac{f_{XY}(x, y)}{f_X(x)} \qquad f_X(x) > 0$$

The use of this definition is illustrated in Example 5.4.2.

Example 5.4.2. The joint density for the random variable (X, Y), where X is the inside and Y is the outside pressure on an air support roof, is given by

$$f_{XY}(x, y) = c/x \qquad 27 \le y \le x \le 33$$
$$c = 1/(6 - 27 \ln 33/27)$$

From Example 5.1.5 the marginal densities for X and Y are

$$f_X(x) = c(1 - 27/x) \qquad 27 \le x \le 33$$

and

$$f_Y(y) = c(\ln 33 - \ln y) \qquad 27 \le y \le 33$$

The conditional density for X given $Y = y$ is

$$f_{X|y}(x) = \frac{f_{XY}(x, y)}{f_Y(y)}$$

$$= \frac{c/x}{c(\ln 33 - \ln y)} = \frac{1}{x(\ln 33 - \ln y)} \qquad y \le x \le 33$$

To find the probability that the inside pressure exceeds 32 given that the outside pressure is 30, we let $y = 30$ in the above expression. We then integrate the conditional density over values of X that exceed 32. That is,

$$P[X > 32 | y = 30] = \int_{32}^{33} \frac{1}{x(\ln 33 - \ln 30)} \, dx$$

$$= \left. \frac{\ln x}{\ln 33 - \ln 30} \right|_{32}^{33}$$

$$= \frac{\ln 33 - \ln 32}{\ln 33 - \ln 30} \doteq .32$$

To find the expected or mean value of X given $y = 30$ we apply Definition 4.2.1 to the random variable $X|y = 30$. That is,

$$E[X|y = 30] = \mu_{X|y=30} = \int_{-\infty}^{\infty} x f_{X|y=30} \, dx$$

$$= \int_{30}^{33} x \frac{1}{x(\ln 33 - \ln 30)} \, dx$$

$$= \int_{30}^{33} \frac{1}{\ln 33 - \ln 30} \, dx$$

$$= \frac{3}{\ln 33 - \ln 30} \doteq 31.48$$

When the outside pressure on the roof is 30, the average value of the inside pressure is 31.48 inches of mercury.

Curves of Regression

In the previous example note that we did not find the mean for X. We found the mean for X when $y = 30$. The mean value obtained depended on the value chosen for Y. In general, the mean of X given $Y = y$ or $\mu_{X|y}$ is a *function of* y. When this function is graphed, we obtain what is called the *curve of regression of X on Y*. This term is defined formally in Definition 5.4.2. Note that, once again, the roles of X and Y can be reversed.

Definition 5.4.2 (Curve of regression). Let (X, Y) be a two-dimensional random variable.

1. The graph of the mean value of X given $Y = y$, denoted by $\mu_{X|y}$, is called the curve of regression of X on Y.
2. The graph of the mean value of Y given $X = x$, denoted by $\mu_{Y|x}$, is called the curve of regression of Y on X.

We illustrate the use of this definition by finding the curve of regression of X on Y and the curve of regression of Y on X for the random variable (X, Y) of Example 5.4.2.

Example 5.4.3. The conditional density for X given $Y = y$, where X is the inside and Y is the outside pressure on an air support roof, is given by

$$f_{X|y}(x) = \frac{1}{x(\ln 33 - \ln y)} \qquad y \le x \le 33$$

The equation for the curve of regression of X on Y is given by

$$\mu_{X|y} = \int_y^{33} x \frac{1}{x(\ln 33 - \ln y)}\, dx$$

$$= \int_y^{33} \frac{1}{\ln 33 - \ln y}\, dx$$

$$= \frac{33 - y}{\ln 33 - \ln y}$$

Note that this equation is *nonlinear*. Its graph is not a straight line. A sketch of the graph is found by plotting $\mu_{X|y}$ for selected values of y. The graph is shown in Fig. 5.4(a). The conditional density for Y given $X = x$ is

$$f_{Y|x}(y) = \frac{f_{XY}(x, y)}{f_X(x)}$$

$$= \frac{c/x}{c(1 - 27/x)}$$

$$= \frac{1}{x - 27} \qquad 27 \le y \le x$$

The equation for the curve of regression of Y on X is given by

$$\mu_{Y|x} = \int_{27}^{x} y \frac{1}{x - 27}\, dy$$

$$= \left. \frac{y^2}{2(x - 27)} \right|_{27}^{x}$$

$$= \frac{x^2 - 27^2}{2(x - 27)}$$

$$= (1/2)(x + 27)$$

Note that this equation is *linear*. Its graph is the straight line shown in Fig. 5.4(b). These curves can be used now to find the mean of X for any specified value of Y or vice versa. For example, the average value of Y, the outside pressure, given that the inside pressure is 29 is

$$\mu_{Y|x=29} = (1/2)(x + 27) = (1/2)(56) = 28 \text{ inches of mercury}$$

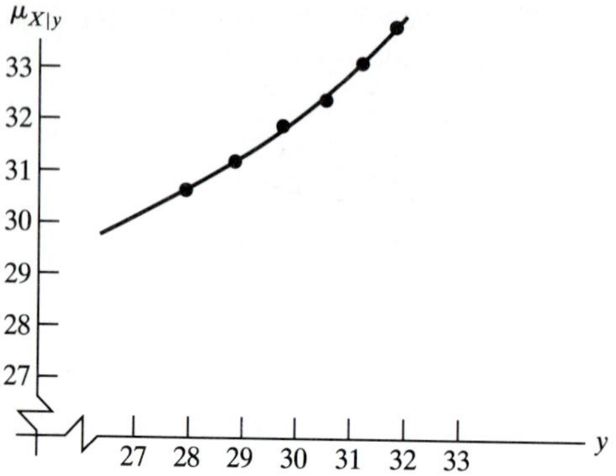

y	$\mu_{X\|y}$
27	29.90
28	30.43
29	30.95
30	31.48
31	31.98
32	32.50
32.5	32.74

(*a*)

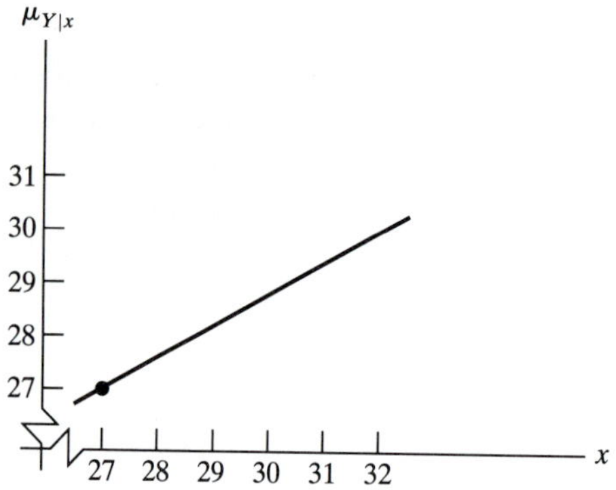

(*b*)

FIGURE 5.4
(*a*) A nonlinear curve of regression: $\mu_{X|y} = (33 - y)/(\ln 33 - \ln y)$; (*b*) a linear curve of regression: $\mu_{Y|x} = (1/2)(x + 27)$.

We have introduced only the basic ideas underlying the topic of regression. To find the theoretical regression curves, you must *know* the joint density for (X, Y). In practice, this density is seldom known with certainty. Thus, in practice, we are forced to approximate these theoretical curves from a data set —a set of observations on the random variable (X, Y). Methods for doing so are presented in Chaps. 11 and 12.

5.5 TRANSFORMATION OF VARIABLES (OPTIONAL)

In Sec. 4.8 we considered the problem of transforming continuous variables in the univariate case. That is, given a continuous random variable X whose density is known, we saw how to find the density for the random variable Y,

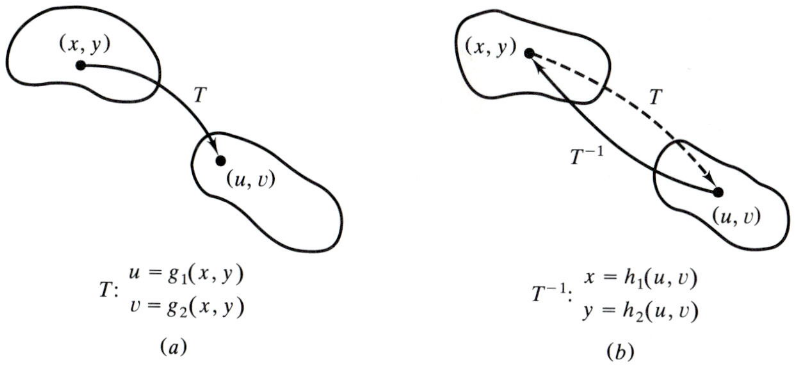

FIGURE 5.5
(a) T maps from the xy plane into the uv plane; (b) T^{-1} maps from the uv plane into the xy plane.

where Y is a function of X. Here we reconsider the problem in the bivariate case. To do so, we must first introduce the notation of Jacobians.

Suppose that we are working in the xy plane and that u and v are variables, each of which is a function of x and y. That is,

$$u = g_1(x, y) \qquad \text{and} \qquad v = g_2(x, y)$$

These two equations define a transformation T from some region in the xy plane into the uv plane as pictured in Fig. 5.5(a). Assume that g_1 and g_2 have continuous partial derivatives with respect to x and y. The Jacobian of T is denoted by J_T and is given by the following determinant:

$$J_T = \begin{vmatrix} \dfrac{\partial u}{\partial x} & \dfrac{\partial u}{\partial y} \\ \dfrac{\partial v}{\partial x} & \dfrac{\partial v}{\partial y} \end{vmatrix}$$

Example 5.5.1 illustrates the idea.

Example 5.5.1. Consider the transformation T from the xy plane into the uv plane defined by

$$u = g_1(x, y) = (3y - x)/6$$
$$v = g_2(x, y) = x/3$$

The Jacobian of T is

$$J_T = \begin{vmatrix} \dfrac{\partial u}{\partial x} & \dfrac{\partial u}{\partial y} \\ \dfrac{\partial v}{\partial x} & \dfrac{\partial v}{\partial y} \end{vmatrix} = \begin{vmatrix} -1/6 & 1/2 \\ 1/3 & 0 \end{vmatrix} = (-1/6)(0) - (1/2)(1/3) = -(1/6)$$

If a transformation T is one-to-one, then it is invertible. Assume that the inverse transformation, T^{-1}, is defined by the equations

$$x = h_1(u, v) \qquad \text{and} \qquad y = h_2(u, v)$$

and that h_1 and h_2 have continuous partial derivatives. [See Fig. 5.5(*b*).] The Jacobian of this inverse transformation is given by the determinant

$$\begin{vmatrix} \dfrac{\partial x}{\partial u} & \dfrac{\partial x}{\partial v} \\ \dfrac{\partial y}{\partial u} & \dfrac{\partial y}{\partial v} \end{vmatrix}$$

This is the sort of Jacobian that will be useful to us in the statistical setting.

Assume that we have two continuous random variables X and Y whose joint density f_{XY} is known. Let U and V be random variables, each of which is a function of X and Y. We want to determine the form of f_{UV}, the joint density for (U, V), based on knowledge of the form of f_{XY}. The method for doing so parallels Theorem 4.8.1 and is given in Theorem 5.5.1.

Theorem 5.5.1. Let (X, Y) be continuous with joint density f_{XY}. Let

$$U = g_1(X, Y) \qquad \text{and} \qquad V = g_2(X, Y)$$

where g_1 and g_2 define a one-to-one transformation. Let the inverse transformation be defined by

$$X = h_1(U, V) \qquad \text{and} \qquad Y = h_2(U, V)$$

where h_1 and h_2 have continuous first partial derivatives. Then the joint density for (U, V) is given by

$$f_{UV}(u, v) = f_{XY}(h_1(u, v), h_2(u, v))|J|$$

where $J \neq 0$ is the Jacobian of the inverse transformation. That is,

$$J = \begin{vmatrix} \dfrac{\partial x}{\partial u} & \dfrac{\partial x}{\partial v} \\ \dfrac{\partial y}{\partial u} & \dfrac{\partial y}{\partial v} \end{vmatrix}$$

It is easy to see that Theorem 4.8.1 is a special case of this theorem with f_X corresponding to f_{XY}, $g^{-1}(y)$ playing the role of the inverse transformation, and $|dg^{-1}(y)/dy|$ being equivalent to the absolute value of the Jacobian of the inverse transformation.

Example 5.5.2. Assume that X and Y are independent uniformly distributed random variables over $(0, 2)$ and $(0, 3)$, respectively. The joint density for (X, Y) is given by

$$f_{XY}(x, y) = 1/6 \qquad \begin{array}{l} 0 < x < 2 \\ 0 < y < 3 \end{array}$$

Let $U = X - Y$ and $V = X + Y$. What is the joint density for (U, V)? To apply Theorem 5.5.1, we first note that the transformation

$$T: \begin{cases} U = X - Y \\ V = X + Y \end{cases}$$

is a linear transformation from the xy plane into the uv plane. A result from advanced calculus states that a linear transformation from two-dimensional space into two-dimensional space is one-to-one whenever the determinant of its matrix of coefficients is not zero. Here the determinant is

$$\begin{vmatrix} 1 & -1 \\ 1 & 1 \end{vmatrix} = (1)(1) - (1)(-1) = 2$$

so T is invertible. The inverse transformation is found by solving the above system of equations for X and Y. Here T^{-1} is given by

$$T^{-1}: \begin{cases} X = (U + V)/2 \\ Y = (V - U)/2 \end{cases}$$

The Jacobian of T^{-1} is

$$\begin{vmatrix} \dfrac{\partial x}{\partial u} & \dfrac{\partial x}{\partial v} \\ \dfrac{\partial y}{\partial u} & \dfrac{\partial y}{\partial v} \end{vmatrix} = \begin{vmatrix} 1/2 & 1/2 \\ -1/2 & 1/2 \end{vmatrix} = (1/2)(1/2) - (1/2)(-1/2) = 1/2$$

By Theorem 5.5.1,

$$\begin{aligned} f_{UV}(u, v) &= f_{XY}((v + u)/2, (v - u)/2)|J| \\ &= (1/6)(1/2) = 1/12 \end{aligned}$$

To find the set of values for which $f_{UV} > 0$, we note that since $0 < x < 2$ and $0 < y < 3$, (X, Y) lies in the rectangle shown in Fig. 5.6(a). It is easy to see that

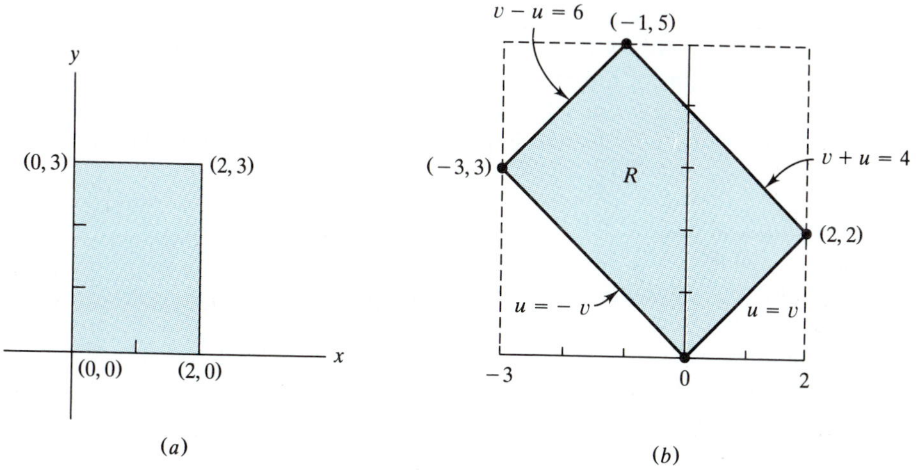

FIGURE 5.6
(a) (X, Y) lies in the rectangle with corners (0, 0), (2, 0), (0, 3), (2, 3); (b) (U, V) lies in the region R.

$U = X - Y$ must lie between -3 and 2 and that $V = X + Y$ must lie between 0 and 5. Furthermore, U and V must satisfy the inequalities

$$0 < (v + u)/2 < 2$$
$$0 < (v - u)/2 < 3$$

or

$$0 < v + u < 4$$
$$0 < v - u < 6$$

Solving these inequalities simultaneously yields the region R shown in Fig. 5.6(*b*). Thus the density for (U, V) is given by

$$f_{UV}(u, v) = 1/12 \qquad (u, v) \in R$$

We leave it to you to verify that f_{UV} is, in fact, a valid density.

Other transformation theorems can be derived from Theorem 5.5.1. Some of these are given in Exercises 48, 50, and 51. For a more detailed discussion of this topic, please see [49].

CHAPTER SUMMARY

In this chapter we considered random variables of more than one dimension. Emphasis was on random variables of two dimensions. The joint density was defined by extending the notion of a density for a single variable in a logical way. This function was used to calculate probabilities associated with two-dimensional random variables (X, Y). We saw how to obtain the marginal densities for both X and Y from the joint density. These marginal densities are the usual densities for X or Y when considered alone. The correlation coefficient ρ was introduced as a measure of linearity between X and Y. The notion of independence between X and Y was defined formally and its relationship to ρ was investigated. We saw how to define the conditional densities for X given Y and Y given X from knowledge of the joint density for (X, Y) and the marginal densities for X and Y. The conditional densities were used to find the equations for the curves of regression of Y on X and X on Y. These regression curves are the graphs of the mean value of Y as a function of X or vice versa. We saw that these curves may be linear or nonlinear.

We introduced and defined important terms that you should know. These are:

Two-dimensional discrete random variable
Two-dimensional continuous random variable
Discrete joint density
Discrete marginal density
Independent random variables
Covariance
Perfect positive correlation
Uncorrelated
Curve of regression
n-dimensional discrete random variable
n-dimensional continuous random variable
Bivariate normal distribution
Continuous joint density
Continuous marginal density
Expected value of $H(X, Y)$
Correlation coefficient
Perfect negative correlation
Conditional density

EXERCISES

Section 5.1

1. Use Table 5.2 to find each of these probabilities:

(*a*) The probability that exactly two defective welds and one improperly tightened bolt will be produced by the robots.

(*b*) The probability that at least one defective weld and at least one improperly tightened bolt will be produced.

(*c*) The probability that at most one defective weld will be produced.

(*d*) The probability that at least two improperly tightened bolts will be produced.

2. In conducting an experiment in the laboratory, temperature gauges are to be used at four junction points in the equipment setup. These four gauges are randomly selected from a bin containing seven such gauges. Unknown to the scientist, three of the seven gauges give improper temperature readings. Let X denote the number of defective gauges selected and Y the number of nondefective gauges selected. The joint density for (X, Y) is given in Table 5.5.

(*a*) The values given in Table 5.5 can be derived by realizing that the random variable X is hypergeometric. Use the results of Sec. 3.7 to verify the values given in Table 5.5.

(*b*) Find the marginal densities for both X and Y. What type of random variable is Y?

(*c*) Intuitively speaking, are X and Y independent? Justify your answer mathematically.

3. The joint density for (X, Y) is given by

$$f_{XY}(x, y) = 1/n^2 \qquad \begin{aligned} x &= 1, 2, 3, \ldots, n \\ y &= 1, 2, 3, \ldots, n \end{aligned}$$

(*a*) Verify that $f_{XY}(x, y)$ satisfies the conditions necessary to be a density.

(*b*) Find the marginal densities for X and Y.

(*c*) Are X and Y independent?

***4.** The joint density for (X, Y) is given by

$$f_{XY}(x, y) = 2/n(n+1) \qquad 1 \le y \le x \le n \qquad n \text{ a positive integer}$$

(*a*) Verify that $f_{XY}(x, y)$ satisfies the conditions necessary to be a density. *Hint:* The sum of the first n integers is given by $n(n + 1)/2$.

(*b*) Find the marginal densities for X and Y. *Hint:* Draw a picture of the region over which (X, Y) is defined.

(*c*) Are X and Y independent?

(*d*) Assume that $n = 5$. Use the joint density to find $P[X \le 3 \text{ and } Y \le 2]$. Find $P[X \le 3]$ and $P[Y \le 2]$. *Hint:* Draw a picture of the region over which (X, Y) is defined.

TABLE 5.5

x/y	0	1	2	3	4
0	0	0	0	0	1/35
1	0	0	0	12/35	0
2	0	0	18/35	0	0
3	0	4/35	0	0	0

TABLE 5.6

x/y	0	1	2	3
0	.400	.100	.020	.005
1	.300	.040	.010	.004
2	.040	.010	.009	.003
3	.009	.008	.007	.003
4	.008	.007	.005	.002
5	.005	.002	.002	.001

5. The two most common types of errors made by programmers are syntax errors and errors in logic. For a simple language such as BASIC the number of such errors is usually small. Let X denote the number of syntax errors and Y the number of errors in logic made on the first run of a BASIC program. Assume that the joint density for (X, Y) is as shown in Table 5.6.

(*a*) Find the probability that a randomly selected program will have neither of these types of errors.

(*b*) Find the probability that a randomly selected program will contain at least one syntax error and at most one error in logic.

(*c*) Find the marginal densities for X and Y.

(*d*) Find the probability that a randomly selected program contains at least two syntax errors.

(*e*) Find the probability that a randomly selected program contains one or two errors in logic.

(*f*) Are X and Y independent?

6. Consider Example 5.1.5. Verify that $P[X \leq 30 \text{ and } Y \leq 28] \doteq .15$ by integrating the joint density first with respect to y, then with respect to x.

7. (*a*) Use the joint density of Example 5.1.5 to find the probability that the inside pressure on the roof will be greater than 30, whereas the outside pressure is less than 32.

(*b*) Use the marginal density for X to find $P[X \leq 28]$.

(*c*) Use the marginal density for Y to find $P[Y > 30]$.

8. Let X denote the temperature (°C) and let Y denote the time in minutes that it takes for the diesel engine on an automobile to get ready to start. Assume that the joint density for (X, Y) is given by temp

$$f_{XY}(x, y) = c(4x + 2y + 1) \qquad \begin{array}{l} 0 \leq x \leq 40 \\ 0 \leq y \leq 2 \end{array}$$

time

(*a*) Find the value of c that makes this a density.

(*b*) Find the probability that on a randomly selected day the air temperature will exceed 20°C and it will take at least 1 minute for the car to be ready to start.

(*c*) Find the marginal densities for X and Y.

(*d*) Find the probability that on a randomly selected day it will take at least one minute for the car to be ready to start.

(*e*) Find the probability that on a randomly selected day the air temperature will exceed 20°C.

(*f*) Are X and Y independent? Explain on a mathematical basis.

9. An engineer is studying early morning traffic patterns at a particular intersection. The observation period begins at 5:30 a.m. Let X denote the time of arrival of the first vehicle from the north-south direction; let Y denote the first arrival time from the east-west direction. Time is measured in fractions of an hour after 5:30 a.m. Assume that the density for (X, Y) is given by

$$f_{XY}(x, y) = 1/x \qquad 0 < y < x < 1.$$

(*a*) Verify that this is a joint density for a two-dimensional random variable.
(*b*) Find $P[X \leq .5 \text{ and } Y \leq .25]$.
(*c*) Find $P[X > .5 \text{ or } Y > .25]$.
(*d*) Find $P[X \geq .5 \text{ and } Y \geq .5]$.
(*e*) Find the marginal densities for X and Y.
(*f*) Find $P[X \leq .5]$.
(*g*) Find $P[Y \leq .25]$.
(*h*) Are X and Y independent? Explain.

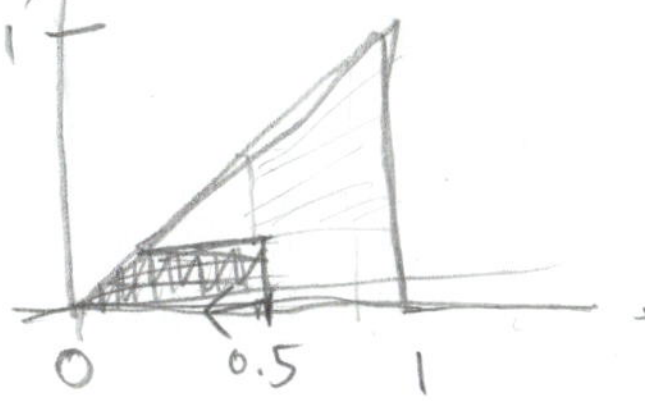

10. The joint density for (X, Y) is given by

$$f_{XY}(x, y) = x^3y^3/16 \qquad 0 \leq x \leq 2, 0 \leq y \leq 2$$

(*a*) Find the marginal densities for X and Y.
(*b*) Are X and Y independent?
(*c*) Find $P[X \leq 1]$.
(*d*) If it is known that $Y = 1$, what is $P[X \leq 1]$? (Do not use any computation to answer this question!)

11. Economic conditions cause fluctuations in the prices of raw commodities as well as in finished products. Let X denote the price paid for a barrel of crude oil by the initial carrier and let Y denote the price paid by the refinery purchasing the product from the carrier. Assume that the joint density for (X, Y) is given by

$$f_{XY}(x, y) = c \qquad 20 < x < y < 40$$

(*a*) Find the value of c that makes this a joint density for a two-dimensional random variable.
(*b*) Find the probability that the carrier will pay at least \$25 per barrel and the refinery will pay at most \$30 per barrel for the oil.
(*c*) Find the probability that the price paid by the refinery exceeds that of the carrier by at least \$10 per barrel.
(*d*) Find the marginal densities for X and Y.
(*e*) Find the probability that the price paid by the carrier is at least \$25.
(*f*) Find the probability that the price paid by the refinery is at most \$30.
(*g*) Are X and Y independent? Explain.

***12.** (*n-dimensional discrete random variables.*) Random variables of dimension $n > 2$ can be defined and studied by extending the definitions presented in the two-dimensional case in a logical way. For example, an n-tuple $(X_1, X_2, X_3, \ldots, X_n)$ in which each of the random variables $X_1, X_2, X_3, \ldots, X_n$ is a discrete random variable is called an n-dimensional discrete random variable. The density for such a random variable is given by

$$f(x_1, x_2, x_3, \ldots, x_n) = P[X_1 = x_1, X_2 = x_2, X_3 = x_3, \ldots, X_n = x_n]$$

This problem entails the use of a three-dimensional random variable.

Items coming off an assembly line are classed as being either nondefective, defective but salvageable, or defective and nonsalvageable. The probabilities of observing items in each of these categories are .9, .08, and .02, respectively. The probabilities do not change from trial to trial. Twenty items are randomly selected and classified. Let X_1 denote the number of nondefective items obtained, X_2 the number of defective but salvageable items obtained, and X_3 the number of defective and nonsalvageable items obtained.

(*a*) Find $P[X_1 = 15, X_2 = 3, X_3 = 2]$. *Hint:* Use the formula for the number of permutations of indistinguishable objects, Exercise 23, Chap. 1, to count the number of ways to get this sort of split in a sequence of 20 trials.

(*b*) Find the general formula for the density for (X_1, X_2, X_3).

*13. (*n-dimensional continuous random variables.*) An n-tuple $(X_1, X_2, X_3, \ldots, X_n)$, where each of the random variables $X_1, X_2, \ldots, X_n$ is continuous, is called an n-dimensional continuous random variable. The density for an n-dimensional continuous random variable is defined by extending Definition 5.1.3 in a natural way. State the three properties that identify a function as a density for $(X_1, X_2, X_3, \ldots, X_n)$.

*14. Let $f(x_1, x_2, x_3) = c(x_1 \cdot x_2 \cdot x_3)$ for $0 \le x_1 \le 1$, $0 \le x_2 \le 1$, $0 \le x_3 \le 1$. Find the value of c that makes this a density for the three-dimensional random variable (X_1, X_2, X_3).

Section 5.2

15. Four temperature gauges are randomly selected from a bin containing three defective and four nondefective gauges. Let X denote the number of defective gauges selected and Y the number of nondefective gauges selected. (See Exercise 2.) The joint density for (X, Y) is given in Table 5.5.

(*a*) From the physical description of the problem, should $\mathrm{Cov}(X, Y)$ be positive or negative?

(*b*) Find $E[X]$, $E[Y]$, $E[XY]$, and $\mathrm{Cov}(X, Y)$.

16. Let X denote the number of syntax errors and Y the number of errors in logic made on the first run of a BASIC program. (See Exercise 5.) The joint density for (X, Y) is given in Table 5.6.

(*a*) X and Y are not independent. Does this give any indication of the value of the covariance?

(*b*) Find $E[X]$, $E[Y]$, $E[XY]$, and $\mathrm{Cov}(X, Y)$. Give a rough physical interpretation of the covariance.

(*c*) Find $E[X + Y]$. What is the practical interpretation of this expectation?

17. Consider the random variable (X, Y) of Exercise 3. Without doing any additional computation, find $\mathrm{Cov}(X, Y)$.

18. Use the marginal densities given in Table 5.3 to compute $E[X]$ and $E[Y]$. Compare your results to those obtained in Example 5.2.1.

19. The joint density for (X, Y), where X is the inside and Y is the outside barometric pressure on an air support roof (see Example 5.1.5), is given by

$$f_{XY}(x, y) = c/x \qquad 27 \le y \le x \le 33$$
$$c = 1/(6 - 27 \ln 33/27) = 1.72$$

(*a*) Find $E[X]$, $E[Y]$, $E[XY]$, and $\mathrm{Cov}(X, Y)$.

(*b*) Find $E[X - Y]$. What is the practical physical interpretation of this expectation?

20. The joint density for (X, Y), where X is the temperature and Y is the time that it takes for a diesel engine on an automobile to get ready to start (see Exercise 8), is given by

$$f_{XY}(x, y) = (1/6640)(4x + 2y + 1) \qquad \begin{array}{l} 0 \le x \le 40 \\ 0 \le y \le 2 \end{array}$$

(*a*) From a physical standpoint, do you think Cov(X, Y) should be positive or negative?

(*b*) Find $E[X]$, $E[Y]$, $E[XY]$, and Cov(X, Y). 26.43, 1.008, 26.6, -0.151

21. The joint density for (X, Y), where X is the arrival time of the first vehicle from the north-south direction and Y is the arrival time of the first vehicle from the east-west direction at an intersection (see Exercise 9), is given by

$$f_{XY}(x, y) = 1/x \qquad 0 < y < x < 1$$

Find $E[X]$, $E[Y]$, $E[XY]$, and Cov(X, Y).

22. Find the covariance between the random variables X and Y of Exercise 10.

23. Let X denote the price paid for a barrel of crude oil by the initial carrier and let Y denote the price paid by the refinery purchasing the oil. (See Exercise 11.) The joint density for (X, Y) is given by

$$f_{XY}(x, y) = 1/200 \qquad 20 < x < y < 40$$

(*a*) From a physical standpoint, should Cov(X, Y) be positive or negative?

(*b*) Find $E[X]$, $E[Y]$, $E[XY]$, and Cov(X, Y).

(*c*) Find $E[Y - X]$. Interpret this expectation in a practical sense.

24. Show that Cov$(XY) = E[XY] - E[X]E[Y]$. *Hint:* By definition, Cov$(X, Y) = E[(X - \mu_X)(Y - \mu_Y)]$. Expand this product and apply the rules for expectation (Theorem 3.3.1). Remember that $\mu_X = E[X]$ and $\mu_Y = E[Y]$.

25. Prove that Var$(X + Y) = $ Var $X +$ Var $Y + 2\,$Cov(X, Y). *Hint:* Var$(X + Y) = E[(X + Y)^2] - (E[X + Y])^2$. Square these terms and apply the rules for expectation. (Theorem 3.3.1.)

26. Use the result of Exercise 25 to show that if X and Y are independent, then Var$(X + Y) = $ Var $X +$ Var Y. This proves the third rule for variance. (Theorem 3.3.3.)

27. Show that if $X = Y$, then Cov$(X, Y) = $ Var $X = $ Var Y.

***28.** Let the joint density for (X, Y) be given by

$$f(x, y) = \frac{1}{2(e - 1)}\left[\frac{1}{x} + \frac{1}{y}\right] \qquad 1 \le x \le e \qquad 1 \le y \le e$$

(*a*) Show that $\int_1^e \int_1^e f(x, y)\, dy\, dx = 1$.

(*b*) Find $E[X]$ and $E[Y]$.

(*c*) Find $E[XY]$.

(*d*) Are X and Y independent? Explain, based on your answers to parts (*b*) and (*c*) and Theorem 5.2.2.

Section 5.3

29. The joint density for (X, Y), where X denotes the number of defective and Y the number of nondefective temperature gauges selected from a bin containing three defective and four nondefective gauges, is given in Table 5.5. (See Exercise 2.)

(*a*) From the physical interpretation of the problem, should ρ_{XY} be positive or negative? Should ρ_{XY} be $+1$ or -1? Explain.

(*b*) Find $E[X^2]$ and $E[Y^2]$. Use the information from Exercise 15 to find ρ_{XY}.

In Exercises 30 to 34, find $E[X^2]$, $E[Y^2]$, Var X, Var Y, and ρ_{XY} for the random variables in the exercises referenced. In each case decide whether or not you would expect the graph of Y versus X to exhibit a strong linear trend.

30. Exercise 16.

31. Exercise 19.

32. Exercise 20.

33. Exercise 21.

34. Exercise 23.

35. Assume that $Y = \beta_0 + \beta_1 X$, $\beta_1 \neq 0$.
  (*a*) Show that $\mathrm{Cov}(X, Y) = \beta_1 \mathrm{Var}\, X$. *Hint:* $\mathrm{Cov}(X, Y) = E[X(\beta_0 + \beta_1 X)] - E[X] \times E[\beta_0 + \beta_1 X]$. Use the rules for expectation.
  (*b*) Show that $\mathrm{Var}\, Y = \beta_1^2 \mathrm{Var}\, X$. *Hint:* Use the rules for variance. (Theorem 3.3.3.)
  (*c*) Find ρ_{XY}.
  (*d*) Argue that $\rho_{XY} = 1$ if β_1, the slope of the line $Y = \beta_0 + \beta_1 X$, is positive and that $\rho_{XY} = -1$ if the slope of this line is negative.

36. Prove that if X and Y are independent, then $\rho_{XY} = 0$. Can we conclude that if X and Y are uncorrelated, then they are independent? Explain.

37. Without doing any additional computation, find ρ_{XY} for the random variables of Exercise 3.

38. What is the correlation between the random variables X and Y of Exercise 10?

Section 5.4

39. Consider Example 5.4.3.
  (*a*) What is the expected value of X when $y = 31$?
  (*b*) What is the expected value of Y when $x = 30$?

40. Consider Example 5.1.4.
  (*a*) Find $f_{X|y}$. Note that $f_{X|y} = f_X$. From a physical standpoint, can you explain why these densities are the same?
  (*b*) Find $f_{Y|x}$. Is $f_{Y|x} = f_Y$?
  (*c*) Find the curve of regression of X on Y and the curve of regression of Y on X. Are these curves linear?

***41.** Consider the random variable (X, Y) of Exercise 4.
  (*a*) Find the curve of regression of X on Y. Is the regression linear?
  (*b*) Assume that $n = 10$ and find the mean value of X when $y = 4$.
  (*c*) Find the curve of regression of Y on X. Is the regression linear?
  (*d*) Assume that $n = 10$ and find the mean value of Y when $x = 4$.

42. Consider the random variable (X, Y) of Exercise 9.
  (*a*) Find the curve of regression of X on Y. Is the regression linear?
  (*b*) Find the mean value of X when $y = .5$.
  (*c*) Find the curve of regression of Y on X. Is the regression linear?
  (*d*) Find the mean value of Y when $x = .75$.

43. Consider Exercise 11.
  (*a*) Find the curve of regression of X on Y. Is the regression linear?
  (*b*) Find the mean price paid by the carrier for a barrel of crude oil given that the refinery price is \$30 per barrel.

(*c*) Find the curve of regression of Y on X. Is the regression linear?
(*d*) Find the mean price paid by the refinery for a barrel of crude oil given that the carrier paid \$35 per barrel.

44. Note that if $|\rho| = 1$, then $Y = \beta_0 + \beta_1 X$. For fixed values of X, $Y|x = \beta_0 + \beta_1 x$. Argue that $\mu_{Y|x}$ is a linear function of x. That is, argue that if X and Y are perfectly correlated, then the curve of regression of Y on X is linear. Is the converse true? Explain.

Section 5.5

45. Consider the linear transformation T defined by

$$T: u = 2x + y$$
$$v = x + 3y$$

(*a*) Is this transformation invertible? If so, find the defining equations for T^{-1}.
(*b*) Find the Jacobian for T^{-1}.

46. Consider the linear transformation T defined by

$$T: u = 3x + 2y$$
$$v = x - y$$

(*a*) Is this transformation invertible? If so, find the defining equations for T^{-1}.
(*b*) Find the Jacobian for T^{-1}.

47. Assume that X and Y are independent and uniformly distributed over $(0, 1)$ and $(0, 2)$, respectively. Find the joint density for (U, V), where U and V are as defined in Exercise 45.

48. (*Distribution of one function of two continuous random variables.*) Let X and Y be continuous random variables with joint density f_{XY}. Let $U = X + Y$. Prove that f_U, the density for $X + Y$, is given by

$$f_U(u) = \int_{-\infty}^{\infty} f_{XY}(u - v, v)\, dv$$

Hint: Define a transformation T by

$$u = g_1(x, y) = x + y$$
$$v = g_2(x, y) = y$$

Follow the procedure given in Theorem 5.5.1 to obtain the joint density for (U, V). Integrate the joint density to obtain the marginal density for U.

49. Let X and Y be independent standard normal random variables. Let $U = X + Y$. Use Exercise 48 to prove that U follows a normal distribution with mean 0 and variance 2. *Hint:* In integrating over v, complete the square in the exponent and remember that a normal density integrated over the real line is equal to 1.

50. Let X and Y be continuous random variables with joint density f_{XY}. Let $U = XY$. Prove that f_U, the density for XY, is given by

$$f_U(u) = \int_{-\infty}^{\infty} f_{XY}(u/v, v)|1/v|\, dv$$

Hint: Let $u = g_1(x, y) = xy$ and $v = y$, and apply Theorem 5.5.1.

51. Let X and Y be continuous random variables with joint density f_{XY}. Let $U = X/Y$. Prove that f_U, the density for X/Y, is given by

$$f_U(u) = \int_{-\infty}^{\infty} f_{XY}(uv, v)|v|\, dv$$

Hint: Let $u = g_1(x, y) = x/y$ and $v = y$, and apply Theorem 5.5.1.

52. Let X and Y be independent exponentially distributed random variables with parameters β_1 and β_2, respectively.
(*a*) Find the joint density for (X, Y).
(*b*) Let $U = X + Y$, and verify that

$$f_U(u) = \int_0^u f_{XY}(u - v, v)\, dv$$

Hint: Remember that $0 < x < \infty$ and that $x = u - v$.
(*c*) Assume that $\beta_1 = 3$ and $\beta_2 = 1$. Show that

$$f_U(u) = e^{-u/3} - e^{-u/2} \qquad 0 < u < \infty$$

53. Let X and Y be independent uniformly distributed random variables over the intervals (0, 2) and (0, 3), respectively.
(*a*) Let $U = XY$ and find f_U.
(*b*) Let $U = X/Y$ and find f_U.

REVIEW EXERCISES

54. An electronic device is designed to switch house lights on and off at random times after it has been activated. Assume that the device is designed in such a way that it will be switched on and off exactly once in a 1-hour period. Let Y denote the time at which the lights are turned on and X the time at which they are turned off. Assume that the joint density for (X, Y) is given by

$$f_{XY}(x, y) = 8xy \qquad 0 < y < x < 1$$

(*a*) Verify that f_{XY} satisfies the conditions necessary to be a density.
(*b*) Find $E[XY]$.
(*c*) Find the probability that the lights will be switched on within 1/2 hour after being activated and then switched off again within 15 minutes.
(*d*) Find the marginal density for X. Find $E[X]$ and $E[X^2]$.
(*e*) Find the marginal density for Y. Find $E[Y]$ and $E[Y^2]$.
(*f*) Are X and Y independent?
(*g*) Find the conditional distribution of X given Y.
(*h*) Find the probability that the lights will be switched off within 45 minutes of the system being activated given that they were switched on 10 minutes after the system was activated.
(*i*) Find the curve of regression of X on Y. Is the regression linear?
(*j*) Find the expected time that the lights will be turned off given that they were turned on 10 minutes after the system was activated.
(*k*) Based on the physical description of the problem, would you expect ρ to be positive, negative, or 0? Explain. Verify by computing ρ.

TABLE 5.7

x/y	1	2	3	4
0	.059	.100	.050	.001
1	.093	.120	.082	.003
2	.065	.102	.100	.010
3	.050	.075	.070	.020

55. Verify that

$$f_{XY}(x, y) = xye^{-x}e^{-y} \qquad x > 0 \qquad y > 0$$

satisfies the conditions necessary to be a density for a continuous random variable (X, Y). Find the marginal densities for X and Y. Are X and Y independent? Find ρ_{XY}.

56. Let X denote the number of "do loops" in a Fortran program and Y the number of runs needed for a novice to debug the program. Assume that the joint density for (X, Y) is given in Table 5.7.

(a) Find the probability that a randomly selected program contains at most one "do loop" and requires at least two runs to debug the program.

(b) Find $E[XY]$.

(c) Find the marginal densities for X and Y. Use these to find the mean and variance for both X and Y.

(d) Find the probability that a randomly selected program requires at least two runs to debug given that it contains exactly one "do loop."

(e) Find Cov(X, Y). Find the correlation between X and Y. Based on the observed value of ρ, can you claim that X and Y are not independent? Explain.

57. Vehicles arrive at a highway toll booth at random instances from both the south and north. Assume that they arrive at average rates of five and three per 5-minute period, respectively. Let X denote the number arriving from the south during a 5-minute period and let Y denote the number arriving from the north during this same time. Assume that X and Y are independent.

(a) Find the joint density for (X, Y).

(b) Find the probability that a total of four vehicles arrives during a five-minute time period.

(c) Find the correlation between X and Y.

(d) Find the conditional density for X given $Y = y$.

***58.** (*Bivariate normal distribution.*) A random variable (X, Y) is said to have a bivariate normal distribution if its joint density is given by

$$f_{XY}(x, y) = \frac{\exp\left\{\dfrac{1}{-2(1-\rho^2)}\left[\left(\dfrac{x-\mu_X}{\sigma_X}\right)^2 - 2\rho\left(\dfrac{x-\mu_X}{\sigma_X}\right)\left(\dfrac{y-\mu_Y}{\sigma_Y}\right) + \left(\dfrac{y-\mu_Y}{\sigma_Y}\right)^2\right]\right\}}{2\pi\sigma_X\sigma_Y\sqrt{1-\rho^2}}$$

where x and y can assume any real value. The parameters $\mu_X, \mu_Y, \sigma_X, \sigma_Y$ denote the respective means and standard deviations for X and Y. The parameter ρ is the correlation coefficient. The name of this distribution comes from the fact that the marginal densities for X and Y are both normal. Show that in the case of a bivariate normal distribution, if $\rho = 0$, then X and Y are independent.

CHAPTER 6

DESCRIPTIVE STATISTICS

Thus far we have considered random variables from a theoretical point of view. We have studied two functions, the density and the cumulative distribution function, that enable us to predict the behavior of the variable in a probabilistic sense. We have also considered three parameters that characterize or describe a random variable, namely, μ, σ^2, and σ. In practice, the exact distribution of a random variable is seldom known. Rather, we must determine a reasonable form for the density and appropriate values for the distribution parameters from a data set. In this chapter we consider some simple graphical and analytic methods for doing so.

6.1 RANDOM SAMPLING

We begin by considering a typical problem that calls for a statistical solution. Suppose that we wish to study the performance of the lithium batteries used in a particular model of pocket calculator. The purpose of our study is to determine the mean effective life span of these batteries so that we can place a limited warrantee on them in the future. Since this type of battery has not been used in this model before, no one can tell us the distribution of the random variable, X, the life span of a battery. We must attempt to discover its distribution for ourselves. This is inherently a statistical problem. What characteristics identify it as such? Simply the following:

1. Associated with the problem is a large group of objects about which inferences are to be made. This group of objects is called the *population*.
2. There is at least one random variable whose behavior is to be studied relative to the population.
3. The population is too large to study in its entirety, or techniques used in the study are destructive in nature. In either case we must draw conclusions about the population based on observing only a portion or "sample" of objects drawn from the population.

In our example the population is large and hypothetical in the sense that it consists of all lithium batteries used in this model calculator in the past, present, and future. Since we cannot observe the life span of batteries not yet produced, the population obviously cannot be studied in its entirety! Furthermore, to determine the life span of a battery, it must be used until it fails. That is, the method of study destroys the object being studied. For these reasons, we must devise methods for approximating the characteristics of the life span of a lithium battery based on observing only a sample of these batteries.

To draw inferences about a population using statistical methods, the sample drawn should be "random." To understand what we mean by this term, let us return to our example. Here we have a large population that consists of all lithium batteries produced for a certain model of pocket calculator. Associated with the population is a random variable X. We do not know the form of its density, nor do we know its mean or variance. We want to select a subset of n batteries from the population "at random." That is, we want to select n batteries for study in such a way that the selection of one battery neither ensures nor precludes the selection of any other. In this way the selection of one battery is independent of the selection of any other. This collection of objects can be thought of as a "random sample."

Note that, prior to the actual selection of the batteries to be studied, X_i $(i = 1, 2, 3, \ldots, n)$, the life span of the ith battery selected is a random variable. It has the same distribution as X, the life span of batteries in the population. Furthermore, these random variables are independent in the sense that the value assumed by one has no effect on the value assumed by any of the others. The random variables $X_1, X_2, X_3, \ldots, X_n$ and can be thought of as a "random sample."

Once we have actually selected n batteries for study and have observed the life span of each battery, we shall have available n numbers, $x_1, x_2, x_3, \ldots, x_n$. These numbers are the observed values of the random variables $X_1, X_2, X_3, \ldots, X_n$ and can be thought of as a "random sample."

As you can see, the term "random sample" is used in three different but closely related ways in applied statistics. It may refer to the *objects* selected for

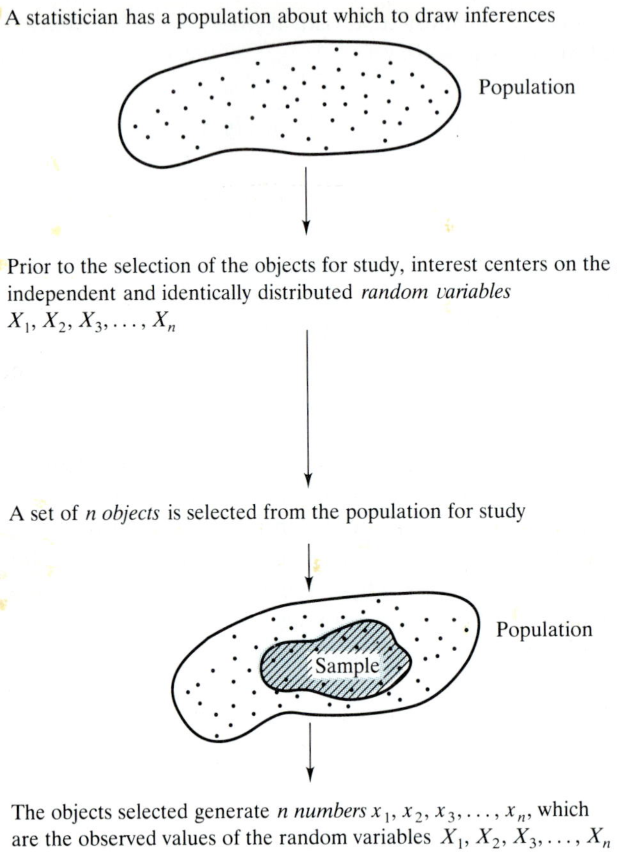

FIGURE 6.1
The objects selected generate *n numbers* $x_1, x_2, x_3, \ldots, x_n$, which are the observed values of the random variables $X_1, X_2, X_3, \ldots, X_n$.

study, to the *random variables* associated with the objects to be selected, or to the *numerical values* assumed by those variables. It is usually clear from the context of the discussion which is intended. These ideas are illustrated in Fig. 6.1.

Even though the term "random sample" is used in these three ways, the formal definition of the term is mathematical in nature. When we use the term in stating theoretical results, we mean the following.

> **Definition 6.1.1 (Random sample).** A random sample of size n from the distribution of X is a collection of n independent random variables, each with the same distribution as X.

The theorems and definitions presented later use the term "random sample" in the sense just described. When objects are selected from a finite population, this type of sample results only when sampling is done with replacement. That is, an object is drawn, observed, and placed back in the population for possible reselection. This ensures that $X_1, X_2, X_3, \ldots, X_n$ are indeed independent and identically distributed. Usually, sampling from a finite population is done without replacement. This means that the random variables $X_1, X_2, X_3, \ldots, X_n$ are not independent. However, if the sample is small relative to the population itself, then removal of a few items does not drastically alter the composition of the population. A generally accepted guideline is that for all practical purposes we may assume independence whenever the sample constitutes at most 5% of the population. If this is not true, then the techniques used to estimate parameters must be altered to take this into account. We shall be assuming that for all practical purposes $X_1, X_2, X_3, \ldots, X_n$ are independent in the discussions that follow.

Once a random sample has been drawn, we commonly use the data gathered to evaluate pertinent *statistics*. What is a statistic? Roughly speaking, a statistic is a random variable whose numerical value can be determined from a random sample. That is, a statistic is a random variable that is a function of the elements of a random sample $X_1, X_2, X_3, \ldots, X_n$. Typical statistics of interest to statisticians are $\sum_{i=1}^n X_i$, $\sum_{i=1}^n X_i^2$, $\sum_{i=1}^n X_i/n$, $\max_i\{X_i\}$, and $\min_i\{X_i\}$. These ideas are illustrated in Example 6.1.1.

Example 6.1.1. Consider the random variable X, the number of times per hour that a television signal is interrupted by random interference. Assume that this random variable has a Poisson distribution with unknown mean μ and unknown variance σ^2. To approximate the value of each of these parameters, we intend to observe the signal for ten randomly selected nonoverlapping one-hour periods over a week's time. Let X_i $(i = 1, 2, 3, \ldots, 10)$ denote the number of interruptions that occur during the ith observation period. The random variables $X_1, X_2, X_3, \ldots, X_{10}$ constitute a random sample size 10 from a Poisson distribution with unknown mean μ and unknown variance σ^2. When the experiment is conducted, these data result:

$$x_1 = 1 \quad x_3 = 0 \quad x_5 = 1 \quad x_7 = 0 \quad x_9 = 3$$

$$x_2 = 0 \quad x_4 = 2 \quad x_6 = 1 \quad x_8 = 0 \quad x_{10} = 0$$

The observed values of the statistics $\sum X_i$, $\sum X_i^2$, $\sum X_i/n$, $\max_i\{X_i\}$, and $\min_i\{X_i\}$ based on this sample are 8, 16, .8, 3, and 0, respectively. Note that the random variable $X_1 - \mu$ is *not* a statistic. Since μ is unknown, we cannot determine its numerical value from a random sample.

6.2 PICTURING THE DISTRIBUTION

When studying a random variable X, one important question to be answered is, "To which family of random variables does X belong?" That is, we need to determine whether X is binomial, Poisson, normal, exponential, or belongs to some other family of variables. In the discrete case it is often possible to determine the appropriate family from the physical description of the experiment. The only job left for the statistician is to approximate the values of the parameters that characterize the distribution. Continuous random variables are more difficult to handle. To determine the family to which such a variable belongs, we must get an idea of the *shape* of its density. For example, if the density appears to be flat, then it is reasonable to suspect that X is uniformly distributed; if it is bell-shaped, then X may be normally distributed.

Stem-and-Leaf Charts

Here we consider some graphical methods for studying the distribution of a continuous random variable. The first method entails constructing what is called a *stem-and-leaf* diagram. This method was first introduced by John Tukey in 1977 [50].

A stem-and-leaf diagram consists of a series of horizontal rows of numbers. Each row is labeled via a number called its stem; the other numbers in the rows are called *leaves*. There are no rigid rules as to how to construct such a diagram. Basically these steps are followed:

1. Choose some convenient numbers to serve as stems. The stems are usually the first one or two digits of the numbers in the data set.
2. Label the rows via the stems selected.
3. Reproduce the data set graphically by recording the digit following the stem as a leaf.
4. Turn the graph on its side to get an idea of the shape of the distribution.

These ideas are illustrated in Example 6.2.1.

Example 6.2.1. To study the random variable X, the life span in hours of the lithium battery in a particular model of pocket calculator, we obtain a random sample of 50 batteries and determine the life span of each we obtain. These data result:

4285	564	1278	205	3920
2066	604	209	602	1379
2584	14	349	3770	99
1009	4152	478	726	510
318	737	3032	3894	582
1429	852	1461	2662	308
981	1560	701	497	3367
1402	1786	1406	35	99
1137	520	261	2778	373
414	396	83	1379	454

To construct a "stem-and-leaf" diagram for these data, we first choose numbers to serve as "stems." It is often convenient to use the first digit of a number as its stem. If a three-digit number such as 318 is expressed as a four-digit number

(a) stem	(a) leaf	(b) stem	(b) leaf	(c) stem	(c) leaf
0		0		0	3945607853234720267400553034
1		1		1	04415724433
2		2		2	0567
3		3		3	07893
4		4	2	4	21

(*a*) (*b*) (*c*)

FIGURE 6.2

(*a*) The integers 0, 1, 2, 3, 4 form the stems for a stem-and-leaf diagram; (*b*) the number 4285 has a stem of 4 and a leaf of 2; (*c*) complete stem-and-leaf diagram for the sample of battery life spans of Example 6.2.1.

(0318) by including a leading zero, then this data set entails the use of the five stems 0, 1, 2, 3, 4. We shall use the second digit of a number as its "leaf." The diagram is constructed by listing the stems as a vertical column as shown in Fig. 6.2(*a*). The first observation, 4285, has a stem of 4 and a leaf of 2. It is represented in the diagram as shown in Fig. 6.2(*b*). The entire data set, recorded in the order in which the observations appear, is shown in Fig. 6.2(*c*).

Is it reasonable to assume that X is normally distributed? To answer this question, turn the stem-and-leaf diagram on its side and look for the bell-shape characteristic of a normal density. This bell shape is not present, leading us to suspect that X is *not* a member of the family of normal random variables.

Histograms and Ogives

The stem-and-leaf diagram provides a quick but rough look at the data set. A second and more thorough method for categorizing data is now considered. It entails breaking the data into "classes" or categories, determining the number of observations in each class, and constructing a graph to display these frequencies. This graph, called a *histogram*, is a vertical bar graph. It depicts the frequency distribution using bars constructed so that the area of each bar is proportional to the number of objects in the respective category. It also allows us to get an idea of the family of random variables to which the variable under study belongs by observing the shape of the histogram. These rules for creating categories, which will be illustrated using the data of Example 6.2.1, follow.

Rules for Breaking Data into Categories

1. Decide on the number of categories wanted. The number chosen depends on the number of observations available. Table 6.1 gives suggested guidelines for the number of categories to be used as a function of sample size. It is based on Sturges' rule, a formula developed by H. A. Sturges in 1926. (H. A. Sturges, "The Choice of a Class Interval," *Journal of the American Statistical Association*, vol. 21, 1926, pp. 65–66.)
2. Locate the largest observation and the smallest observation.
3. Find the difference between the largest and the smallest observations. Subtract in the order of the largest minus the smallest. This difference is called the *range* of the data.

4. Find the minimum length required to cover this range by dividing the range by the number of categories desired. This length is the minimum length required to cover the range if the lower boundary for the first category is taken to be the smallest data point. However, to ensure that no data point falls on a boundary, we shall define boundaries in such a way that they involve one more decimal place than the data. Hence we shall start the first category slightly *below* the first data point. By doing this, the minimum category length required to cover the range is not long enough to trap the largest data point in the last category. For this reason, the actual length used must be a little longer than minimum.
5. The actual category length to be used is found by rounding the minimum length *up* to the same number of decimal places as the data itself. If the minimum length by chance already has the same number of decimal places as the data, we shall round up 1 unit. For example, if we have data reported to one decimal place accuracy and the minimum length required to cover the range is found to be 1.7, we bump this up to 1.8 to obtain the actual category length to be used.
6. The lower boundary for the first category lies 1/2 unit below the smallest observation. Table 6.2 gives units and half units for various types of data sets.
7. The remaining category boundaries are found by adding the category length to the preceding boundary value.

Example 6.2.2. Consider the data of Example 6.2.1. The data set has 50 observations. From Table 6.1 we see that the suggested number of categories to be used is 6. Now we locate the largest data point (4285) and the smallest (14). These are used to find the range, the length of the interval containing all the data points. In this case the data are covered by an interval of length $4285 - 14 = 4271$ units. To find the minimum length required for each category, we divide this number by the number of categories desired. Here the minimum category length is $4271/6 \doteq$

TABLE 6.1
Suggested number of categories to be used in subdividing numeric data as a function of sample size

Sample size	Number of categories
Less than 16	Not enough data
16–31	5
32–63	6
64–127	7
128–255	8
256–511	9
512–1023	10
1024–2047	11
2048–4095	12
4096–8190	13

TABLE 6.2
Units and half units for data reported to the stated degree of accuracy

Data reported to nearest	Unit	1/2 unit
Whole number	1	.5
Tenth (1 decimal place)	.1	.05
Hundredth (2 decimal places)	.01	.005
Thousandth (3 decimal places)	.001	.000 5
Ten thousandth (4 decimal places)	.000 1	.000 05

711.83 units. To find the actual category length to be used in splitting the data, we round up the minimum length to the same number of decimal places as the data. Here the data are reported in whole numbers. Thus we round up the minimum length, 711.83, to the nearest whole number, 712. The categories actually used will be of length 712. The first category starts 1/2 unit below the smallest observation. From Table 6.2 we see that 1/2 unit is .5 in the case of integer data. That is, the lower boundary for the first category is 14 − .5 = 13.5. The remaining category boundaries are found by successively adding the category length (712) to the preceding boundary until all data points are covered. In this way we obtain the following six finite categories for the battery lives:

13.5 to 725.5	2149.5 to 2861.5
725.5 to 1437.5	2861.5 to 3573.5
1437.5 to 2149.5	3573.5 to 4285.5

Note that since the boundaries have one more decimal place than the data, no data point can fall on a boundary; each data point must fall into exactly one category. The data can be summarized now in table form by recording the number (frequency) and the percentage (relative frequency) of the observations in each category as shown in Table 6.3. From this table we can construct a histogram of the data. If the frequency per category is plotted along the vertical axis, the resulting bar graph is called a *frequency histogram*; if the vertical axis is used to plot the relative frequency per category, then the diagram is called a *relative frequency histogram*. Both plots provide a visual display of the data that conveys an idea of the shape of the density of the random variable X under study. The

TABLE 6.3

Category	Boundaries	Frequency	Relative frequency
1	13.5 to 725.5	24	24/50 = 48%
2	725.5 to 1437.5	12	12/50 = 24%
3	1437.5 to 2149.5	4	4/50 = 8%
4	2149.5 to 2861.5	3	3/50 = 6%
5	2861.5 to 3573.5	2	2/50 = 4%
6	3573.5 to 4285.5	5	5/50 = 10%

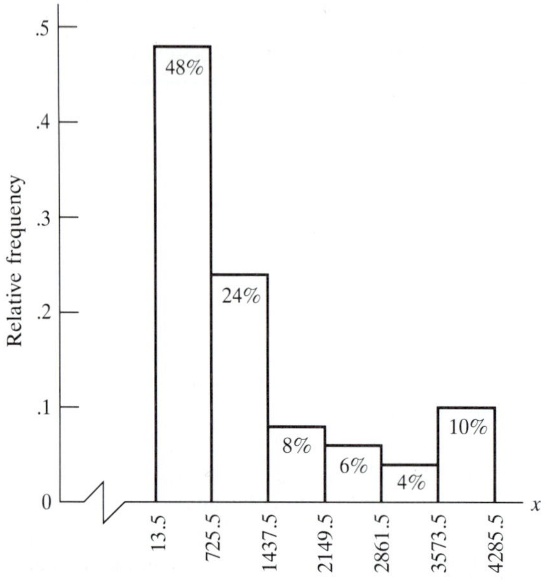

FIGURE 6.3
Relative frequency histogram for the sample of battery life spans of Example 6.2.1.

relative frequency histogram for the data of Example 6.2.1 is shown in Fig. 6.3. Since the histogram does not exhibit a bell shape, we see once again that these data do not support an assumption of normality. In fact, the distribution suggested by the data is the exponential distribution. In this case it is now the job of the researcher to estimate β, the parameter that describes this distribution. By so doing, we are able to estimate the density for X. This estimated density can then be used to approximate probabilities in the future.

In addition to the frequency distribution among categories, it is of interest to consider the cumulative frequency distribution of the observations. The cumulative frequency distribution is found by determining for each category the number and percentage of observations falling in or below that category. The cumulative distribution of the data of Example 6.2.1 is shown in Table 6.4.

When the random variable under study is continuous, the cumulative distribution can be used to construct a graph that approximates its cumulative

TABLE 6.4

Category	Boundaries	Frequency	Cumulative frequency	Relative cumulative frequency
1	13.5 to 725.5	24	24	24/50 = 48%
2	725.5 to 1437.5	12	36	36/50 = 72%
3	1437.5 to 2149.5	4	40	40/50 = 80%
4	2149.5 to 2861.5	3	43	43/50 = 86%
5	2861.5 to 3573.5	2	45	45/50 = 90%
6	3573.5 to 4285.5	5	50	50/50 = 100%

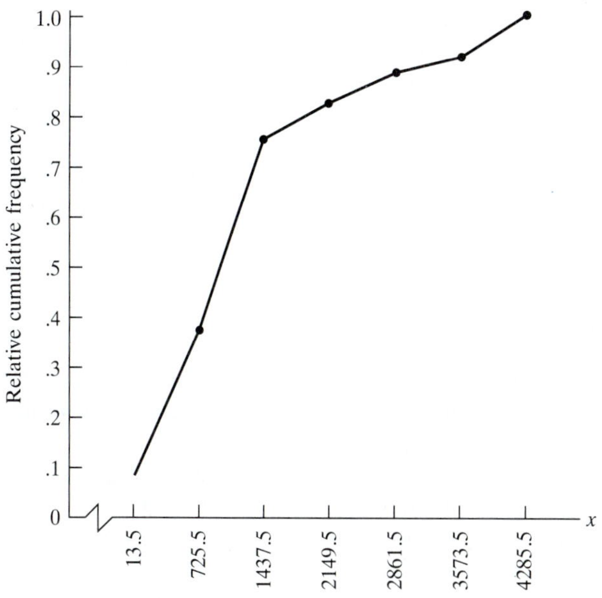

FIGURE 6.4
Relative cumulative frequency ogive for the sample of battery life spans of Example 6.2.1.

distribution function F. The graph is a line graph obtained by plotting the upper boundary of each category on the horizontal axis against the relative cumulative frequency. This type of graph is called a *relative cumulative frequency ogive*. The ogive for the data of Example 6.2.1 is shown in Fig. 6.4. From the ogive we can answer questions such as these, "Approximately what percentage of batteries

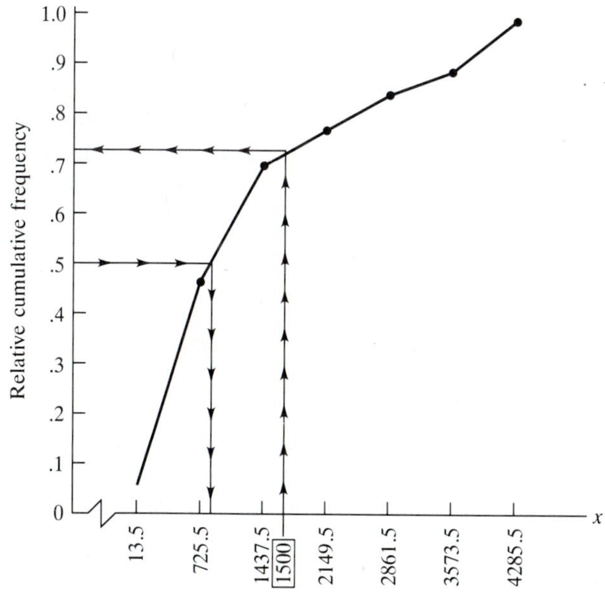

FIGURE 6.5
Projective method of approximating probabilities using a relative cumulative frequency ogive.

fail during the first 1500 hours of operation?", "What time represents the midway point in the sense that half the batteries fail on or before this time?"

The first question can be answered graphically by locating 1500 on the horizontal axis, projecting a vertical line up to the ogive, and then projecting a horizontal line over to the vertical axis as shown in Fig. 6.5. The desired percentage is seen to be approximately 72%. The second question is answered by locating .5 on the vertical axis and reversing the process. The answer is seen to be a little over 725 hours. (See Fig. 6.5.)

6.3 SAMPLE STATISTICS

We have seen that the behavior of a random variable X is determined by its density. We have also seen that the parameters μ, the theoretical average value of the random variable, and σ^2, its variability about the mean, are helpful in describing X. In the last section we considered some graphical methods for getting an idea of the shape of the density. In this section we consider some statistics that allow us to summarize a data set analytically. Since it is hoped that the data set reflects the population as a whole, these statistics also give us some idea of the values of the parameters that characterize X over the population under study. In particular, we consider two measures of location or central tendency in a data set, the *sample mean* and the *sample median*. We also consider three measures of variability within the data set, the *sample variance*, the *sample standard deviation*, and the *sample range*. The word "sample" is used to emphasize the fact that the data sets presented are based on experiments involving only a small portion of objects that constitute the population being studied. That is, they represent a random sample from the distribution of X.

Location Statistics

The mean or theoretical average value of X is our primary measure of the center of location of X. The primary measure of the center of location of a data set is its arithmetic average. Since we view a data set as a set of observations on X, the arithmetic average for a particular set of observations is just the observed value of the *statistic* $\sum_{i=1}^{n} X_i/n$. This statistic, called the *sample mean* is defined formally in the next definition.

Definition 6.3.1 (Sample mean). Let $X_1, X_2, X_3, \ldots, X_n$ be a random sample from the distribution of X. The statistic $\sum_{i=1}^{n} X_i/n$ is called the sample mean and is denoted by $\bar{X}$.

Note that μ_X and $\bar{X}$ are *not* the same. The parameter μ_X is the theoretical average value for X over the entire population; $\bar{X}$ is a statistic which, when evaluated over a particular random sample, gives the average value of X *for that sample*. It is hoped, of course, that the observed value of $\bar{X}$ is close to μ_X. In reporting sample means, we shall retain one more decimal place than that of the data. Rounding will be used rather than truncation.

Example 6.3.1. A random sample of size 9 yields the following observations on the random variable X, the coal consumption in millions of tons by electric utilities for a given year:

406	395	400	450	390	410	415	401	408

The observed value of the sample mean for these data is

$$\bar{x} = \sum_{i=1}^{n} x_i/n = (406 + 395 + 400 + \cdots + 408)/9$$

$$= 3675/9 \doteq 408.3 \text{ million tons}$$

The average value for X for this sample is 408.3 million tons. What is the average number of tons of coal used by electric utilities across the country in this particular year? That is, What is μ_X? Unfortunately, this question cannot be answered with certainty from this sample. However, the sample leads us to believe that μ_X lies close to 408.3 million tons. Admittedly, the word "close" is a bit vague. In Chap. 8 we shall consider a method for determining how close μ_X is likely to be to 408.3 million tons.

A second measure of the center of location of a random variable X is its *median*. The median of a random variable is its 50th percentile (see Exercise 12). That is, the median for X is that number M such that

$$P[X < M] \leq .50 \qquad \text{and} \qquad P[X \leq M] \geq .50$$

If X is continuous, then its median is the "halfway point" in the sense that an observation on X is just as likely to fall below M as it is to fall above it. We define the median for a sample with this in mind.

Definition 6.3.2. Let $x_1, x_2, \ldots, x_n$ be a sample of observations arranged in order from the smallest to the largest. The sample median is the middle observation if n is odd. It is the average of the two middle observations if n is even. We shall denote the median of a sample by $\tilde{x}$.

If n is small, it is easy to spot the middle of a data set. However, if n is large, it is useful to have a formula that pinpoints the location of the middle observation or observations. The formula is given below and its use is illustrated in Example 6.3.2.

$$\text{Median location} = \frac{n+1}{2}$$

Example 6.3.2. The nine observations on X, the coal consumption in millions of tons by electric utilities for a given year, arranged in order, are

390 395 400 401 406 408 410 415 450

The median location is $\frac{n+1}{2} = \frac{9+1}{2} = 5$. The median is the fifth data point in the ordered list. In this case, $\tilde{x} = 406$. This observation is the middle value in our ordered list. Note that this is the median for this data set. It gives us a *rough* idea of the median coal consumption across the country during the year.

Measures of Variability

Recall that we are usually concerned not only with the mean of a random variable, but also with its variance. The variance of a random variable, given by

$$\sigma^2 = E\left[(X - \mu)^2\right]$$

measures the variability of X about the population mean. We want to develop an analogous measure of variability within a sample. To do so, we parallel the logic used in defining σ^2. We do not know the value of the population mean, but we shall have available an observed value for the sample mean. We cannot observe the differences $(X - \mu)^2$ for all members of the population, but we can observe the difference $(X_i - \bar{X})^2$ for each element X_i of the random sample. Since σ^2 is an expectation, a theoretical average value, logic dictates that we replace this operation by an arithmetic average of sample values. That is, the natural measure of variability within a sample that parallels our definition of variability within the population is

$$\sum_{i=1}^{n} \frac{(X_i - \bar{X})^2}{n}$$

This method of measuring variability within a sample is acceptable. In fact, many electronic calculators with built-in statistical capability utilize this formula to compute the variance of a sample. In most cases we shall be using the variability in the sample to approximate σ^2. However, it can be shown that this statistic tends, on the average, to underestimate σ^2. To improve the situation, we divide $\sum_{i=1}^{n}(X_i - \bar{X})^2$ by $n - 1$ rather than by n. In this way we obtain a statistic that is unbiased for σ^2. The term "unbiased" is a technical term. It is defined formally in Sec. 7.1. Basically, it means "centered at the right spot." Since the sample variance is used to estimate σ^2, in this case "the right spot" is σ^2. Successive estimates for σ^2 based on the formula $[\sum_{i=1}^{n}(X_i - \bar{X})^2]/(n - 1)$ should be centered at σ^2. Figure 6.6 illustrates the expected behavior of the two statistics just discussed. So that the statistic used to estimate σ^2 will be unbiased for σ^2, we choose to define the variance of a sample as given in Definition 6.3.3. The definition of the term "sample standard deviation" follows logically.

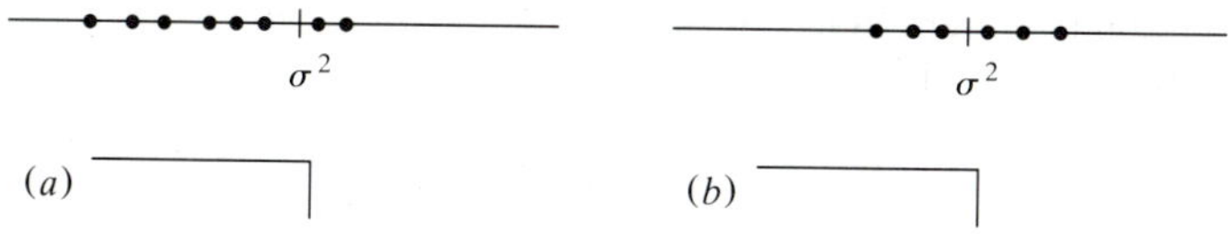

FIGURE 6.6
(a) The statistic $\sum_{i=1}^{n}(X_i - \bar{X})^2/n$ tends to underestimate σ^2. On the average, it will produce estimates that are a bit too small. It is not an unbiased estimator for σ^2; (b) the statistic $\sum_{i=1}^{n}(X_i - \bar{X})^2/(n-1)$ is unbiased for σ^2. On the average, it will produce estimates that are centered at σ^2.

Definition 6.3.3 (Sample variance and sample standard deviation). Let $X_1, X_2, X_3, \ldots, X_n$ be a random sample of size n from the distribution of X. Then the statistic

$$S^2 = \sum_{i=1}^{n} \frac{(X_i - \bar{X})^2}{n-1}$$

is called the *sample variance*. Furthermore, the statistic $S = \sqrt{S^2}$ is called the *sample standard deviation*.

Recall that when we computed the value of σ^2 in Chap. 3, the actual definition of the term "variance" was seldom used; a computational formula was developed that was arithmetically easier to handle than the definition. The same is true here. When S^2 is evaluated from a sample, Definition 6.3.3 is not commonly used. Rather, we use a computational formula.

Theorem 6.3.1 (A computational formula for S^2). Let $X_1, X_2, X_3, \ldots, X_n$ be a random sample of size n from the distribution of X. The sample variance is given by

$$S^2 = \frac{n \sum_{i=1}^{n} X_i^2 - \left(\sum_{i=1}^{n} X_i\right)^2}{n(n-1)}$$

The above formula was convenient before the advent of calculators with built-in statistical capabilities and statistical computer packages. Its derivation is based on the rules of summation and is left as an exercise (Exercise 23). You are encouraged to use whatever computing aids you have available to find $\bar{x}$, s^2, and s. However, the use of the formula is illustrated in Example 6.3.3. In reporting s^2, we shall retain two more decimal places than that of the data; s will be reported to one more decimal place. Rounding will be used.

Example 6.3.3. These data constitute a sample of observations on X, the coal consumption in millions of tons by electric utilities for a given year:

390 400 406 410 450 395 401 408 415

To compute the sample variance, we must evaluate the statistics $\sum_{i=1}^{n} X_i$ and $\sum_{i=1}^{n} X_i^2$ for this sample. The observed values are

$$\sum_{i=1}^{9} x_i = 3675 \qquad \sum_{i=1}^{9} x_i^2 = 1{,}503{,}051$$

The observed value of S^2 is

$$s^2 = \frac{9\sum_{i=1}^{9} x_i^2 - \left(\sum_{i=1}^{9} x_i\right)^2}{9(8)} = \frac{9(1{,}503{,}051) - (3675)^2}{9(8)} \doteq 303.25$$

Remember that variance is usually considered to be unitless because the physical unit attached to it is often meaningless. The observed value of S is

$$s = \sqrt{s^2} = \sqrt{303.25} \doteq 17.4 \text{ million tons}$$

Notice that the physical measurement unit associated with s matches that of the original data and that 17.4 million tons is the standard deviation for this sample. It is not the standard deviation in coal consumption for all electric utilities across the country for the given year. However, it does indicate that σ probably has a value close to 17.4 million tons.

The last sample statistic to be considered is the *sample range*. This statistic was used in categorizing data in Sec. 6.2. The sample range is defined to be the difference between the largest and the smallest observations. Thus the sample range for the data of Example 6.3.3 is $450 - 390 = 60$ million tons.

One word of caution is in order. We have assumed that the data set presented in this section represents a random sample drawn from a larger population because this is the situation most often encountered in practice. Occasionally you will encounter a data set that is *not* a sample. Rather, it represents an observation on X for *every* member of the population. If this is the case, then the population mean is just the arithmetic average of these observations; that is, $\mu = \bar{x}$. Furthermore, the population variance is given by

$$\sigma^2 = \sum_{i=1}^{n} \frac{(x_i - \bar{x})^2}{n}$$

Be careful! Be sure that you understand the nature of your data set before you begin to summarize its properties.

6.4 BOXPLOTS

In summarizing data, it is useful to report all the statistics considered in Sec. 6.3. This is especially true if the data set contains a value that is unusually large or unusually small. A value that appears to be atypical in that it seems to be far removed from the bulk of the data is called an *outlier* or a "wild" number. It is important to be able to detect such numbers and to understand the effect that they have on the usual sample statistics.

Outliers arise for two reasons: (1) They are legitimate observations whose values are simply unusually large or unusually small, or (2) they are the result of an error in measurement, poor experimental technique, or a mistake in recording or entering the data. In the first case it is suggested that the presence of the outlier be reported and that sample statistics be reported both with and without the outlier. In the second case the data point can be corrected if possible or else dropped from the data set.

Of the statistics presented thus far the sample mean, the variance, the standard deviation, and the range are adversely affected by the presence of an outlier; however, the sample median is not so affected. Thus in the presence of an outlier the sample median may be preferable to the sample mean measure of location. We say that the median is resistant to outliers.

Sometimes outliers are so obvious that their presence can be detected by inspection. However, it is useful to have an analytical and graphical technique for identifying values that are truly unusual. One such technique is the *boxplot*. Its construction is based on the interquartile range, a measure of variability that is resistant to outliers. The sample interquartile range, iqr, represents the length of the interval that contains roughly the middle 50% of the data. If the iqr is small, then much of the data lies close to the center of the distribution; if it is large, the data tend to be widely dispersed. These steps are used to calculate the iqr.

Finding the Sample Interquartile Range

1. Find the median location $(n + 1)/2$, where n is the sample size.
2. Truncate the median location by rounding it *down* to the nearest whole number.
3. Find the quartile location q by

$$q = \frac{\text{truncated median location} + 1}{2}$$

4. Find q_1 by counting up from the smallest data point to location q. If q is an integer, then q_1 is the data point in position q. If q is not an integer, then q_1 is the average of the data points in positions $q - .5$ and $q + .5$. Approximately 25% of the data will fall on or below q_1.
5. Find q_3 by counting down from the largest data point to position q as in part 4. Approximately 75% of the data will fall on or below q_3.
6. Define iqr by iqr $= q_3 - q_1$.

Example 6.4.1. A study of the type of sediment found at two different deep-sea drilling sites is conducted. The random variable of interest is the percentage by volume of cement found in core samples. By cement we mean dissolved and reprecipitated carbonate material. These data are obtained (based on information

1	0433223
1	86769
2	014
2	57
3	122
3	76
4	
4	9

FIGURE 6.7
Double stem-and-leaf diagram for the percentage by volume of cement in core samples taken at deep-sea drilling site I.

found in Andreas Wetzel, "Influence of Heat Flow on Ooze/Chalk Cementation: Quantification from Consolidation Parameters in DSDP Site 504 and 505 Sediments," *Journal of Sedimentary Petrology*, July 1989, pp. 539–547):

Site I, % cement			
10	21	12	12
20	13	24	36
31	18	17	16
37	16	32	13
14	49	25	19
13	32	27	

Site II, % cement		
1	10	14
9	21	19
15	17	13
25	22	20
24	12	23
15	20	18

The double stem-and-leaf diagram for the data of site I is shown in Fig. 6.7. The sample is size $n = 23$. The median location is $(n + 1)/2 = 12$. The quartile location is $q = (12 + 1)/2 = 6.5$. To find q_1, we use the stem-and-leaf diagram to locate the sixth and seventh data points counting from the smaller numbers up. These values are 13 and 14, respectively. Hence $q_1 = (13 + 14)/2 = 13.5$. To find q_3, we find the sixth and seventh data points counting from the higher numbers down. These points are 31 and 27, respectively, yielding $q_3 = (31 + 27)/2 = 29$. The sample interquartile range is $q_3 - q_1 = 29 - 13.5 = 15.5$.

Once the interquartile range has been found, it can be used to construct a boxplot. The *boxplot* is a graphical representation of a data set that gives a visual impression of location, spread, and the degree and direction of skewness. For an approximately bell-shaped distributions the *boxplot* also allows us to identify outliers. It is especially useful when we want to compare two or more data sets.

Constructing a Boxplot

1. A horizontal or vertical reference scale is constructed.
2. Find the sample median, q_1, q_3, and iqr.
3. Find two points f_1 and f_3, called *inner fences*, by

$$f_1 = q_1 - 1.5(\text{iqr})$$
$$f_3 = q_3 + 1.5(\text{iqr})$$

   These points will be used to identify outliers.

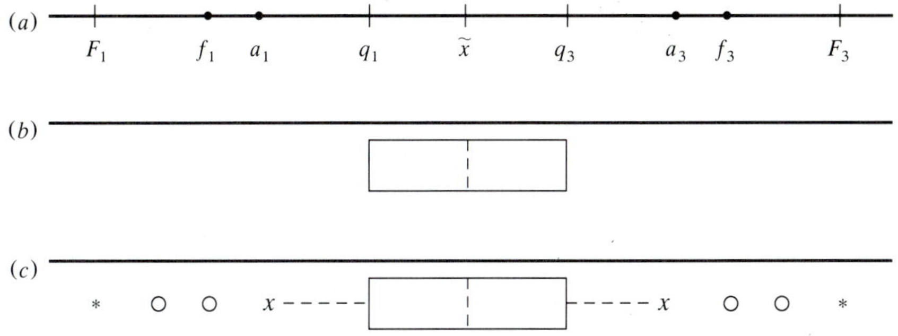

FIGURE 6.8
(a) Relative positions of median ($\tilde{x}$), quartiles (q_1 and q_3), adjacent values (a_1 and a_3), inner fences (f_1 and f_3), and outer fences (F_1 and F_3); (b) a box is drawn with ends at q_1 and q_3 and interior line at $\tilde{x}$; (c) adjacent values are indicated by x. Mild outliers are indicated by open circles; extreme outliers are given by asterisks.

4. Find two points a_1 and a_3, called *adjacent values*. The point a_1 is the data point that is closest to f_1 without lying below f_1 in value. The point a_3 is the data point that is closest to f_3 without lying above f_3 in value.
5. Find two points F_1 and F_3, called outer fences, by

$$F_1 = q_1 - 2(1.5)(\text{iqr})$$

$$F_3 = q_3 + 2(1.5)(\text{iqr})$$

6. Locate the points found thus far on the horizontal or vertical scale. Their relative positions are shown in Fig. 6.8(a).
7. Construct a box with ends at q_1 and q_3 with an interior line drawn at the median as shown in Fig. 6.8(b).
8. Indicate adjacent values by x and connect them to the box with dashed lines. Locate any data points falling between the inner and outer fences and denote these by open circles. These points are considered to be mild outliers. Indicate data points that fall beyond the outer fences with asterisks. These points are considered to be extreme outliers [see Fig. 6.8(c)].

The location of the midline of the box is an indication of the shape of the distribution. If the line is badly off center, then we know that the distribution is skewed in the direction of the longer end of the box.

Before we illustrate this technique, the notion of fences needs to be clarified. It can be shown that when sampling from a normal distribution, only about 7 values in every 1000 fall beyond the inner fences. You are asked to verify this result in Exercises 26 and 27. Since these values are very unusual, they are deemed to be outliers. Outliers must be treated with care since, as you have already seen, their presence can have a dramatic impact on $\bar{x}$, s^2, and s, the usual measures of location and variation. When an outlier is found, we should consider its source. Is it a legitimate data point whose value is simply unusually large or small? Is it a misrecorded value? Is it the result of some error or accident in experimentation? In the last two instances the point can be deleted from the data set and the analysis completed on the remaining data. In the first case we suggest that the presence of the outlier be made known and that statistics be reported both with and without the outlier. In this way the decision of whether or not to include the outlier in future analyses can be made by the researcher who is the subject matter expert.

Example 6.4.2. A study of posttraumatic amnesia after a closed head injury is conducted. One variable studied is the length of hospitalization in days. The stem-and-leaf diagram for the data is shown in Fig. 6.9. (Based on information found in Jerry Mysia et al., "Prospective Assessment of Posttraumatic Amnesia: A Comparison of GOAT and the OGMS," *Journal of Head Trauma Rehabilitation*, March 1990, pp. 65–77.) For these data the median location is $(n + 1)/2 = 11$ and the median is 40 days. Quartile location is $q =$ (truncated median location + 1) $/2 = 6$. The points q_1 and q_3 are 32 and 47, respectively. The interquartile range is iqr $= q_3 - q_1 = 15$. The inner fences are

$$f_1 = q_1 - 1.5(\text{iqr}) = 32 - 22.5 = 9.5 \qquad f_3 = q_3 + 1.5(\text{iqr}) = 47 + 22.5 = 69.5$$

The adjacent values are $a_1 = 12$ and $a_3 = 61$. The outer fences are

$$F_1 = q_1 - 2(1.5)(\text{iqr}) = 32 - 45 = -13 \qquad F_3 = q_3 + 2(1.5)(\text{iqr}) = 47 + 45 = 92$$

The data set contains two points, 8 and 89, that qualify as mild outliers. The point 108 qualifies as an extreme outlier. Notice that since F_1 is negative, it is physically impossible to see an extreme outlier on the lower end of the scale. The boxplot is shown in Fig. 6.10. Notice that the midline of the box is near its center, indicating a nearly symmetric distribution. Are the outliers real observations that must be taken into account or are they the result of errors in data collection? In this case it would be easy to check patient records to find the answer, and this should be done before proceeding with any further analysis of the data.

0	8
1	2
2	07
3	0256
4	00001257
5	02
6	1
7	
8	9
9	
10	8

FIGURE 6.9
Stem-and-leaf diagram for the data of Example 6.4.2. Data represent length of hospitalization in days of posttraumatic amnesia patients ($n = 21$).

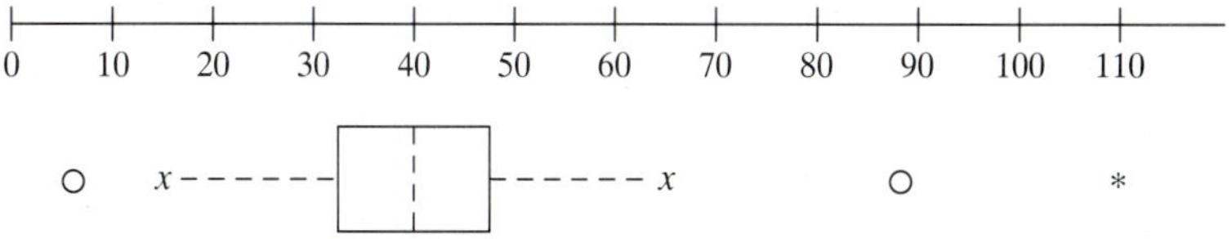

FIGURE 6.10
Boxplot for the data of Example 6.4.2.

As with any other statistical technique, the method given here for detecting outliers must be used with care. Since the location of the fences is chosen to detect unusual values when sampling from a normal distribution, this fact must be kept in mind when interpreting the boxplot. If the data set is large enough so that a histogram or a stem-and-leaf plot exhibits the bell characteristic of a normal curve, then legitimate data points that are flagged as outliers are unusual enough to warrant investigation. If the data set is small or appears to be drawn from a distribution that is not normal, then no real conclusions concerning outliers can be drawn. For example, the exponential distribution is far from symmetric and by nature has a long tail. In this case it is quite likely that the technique demonstrated in this section would flag the largest data point as an outlier. In fact, the point might not be unusual at all. David Hoaglin and John Tukey [50] have a nice discussion of the use of boxplots and outliers for distributions that are not normal.

CHAPTER SUMMARY

This chapter is a link between the study of probability in its own right and the use of probability in the study of applied statistics. We began by defining exactly what we mean by the term "random sample." In particular, we noted that the term is used in three ways. It can denote the objects sampled, the random variables associated with those objects, or the numerical values assumed by these random variables. We noted also that in this text we are assuming that either sampling is from an infinite population, sampling is done with replacement from a finite population, or sampling without replacement from a finite population is done in such a way that the sample constitutes at most 5% of the population. This ensures that it is reasonable to assume that the random variables $X_1, X_2, \ldots, X_n$ are, for all practical purposes, independent. We introduced three graphical methods for picturing the distribution of a data set. These methods, the stem-and-leaf chart, histograms, and boxplots, help to determine the type of random variable with which we are dealing. That is, they help us get an idea of the shape of the density f associated with the random variable. The relative cumulative frequency ogive was introduced as a means of approximating the cumulative distribution function, F, of a continuous random variable. We introduced some summary statistics that serve two purposes. They describe the data set at hand and they help approximate the value of corresponding parameters associated with the population from which the sample was drawn. We introduced and defined important terms that you should know. These are:

Population
Percentile
Median
Statistic
Decile
Sample variance
Frequency histogram
Relative frequency histogram
Relative cumulative frequency ogive
Inner fences
Outer fences
Adjacent values
Resistant statistic
Sample mean
Random sample
Quartile
Sample median
Stem and leaf
Interquartile range
Sample standard deviation
Sample range
Outlier
Mild outliers
Extreme outliers
Boxplots

EXERCISES

Section 6.1

In Exercises 1 through 5 a problem is described. In each case, decide whether a statistical study is appropriate. If so, explain why you think this is the case and identify the population(s) of interest.

1. A bridge is to be built across a deep canyon. An engineer is interested in determining the distribution of the random variable X, the maximum wind speed per day at the site so that the bridge can be designed to withstand potential stresses that will be placed upon it from this source.
2. A botanist thinks that indoleacetic acid is effective in stimulating the formation of roots in cuttings from lemon trees. In an experiment to verify this contention two groups of cuttings are to be used. One group is to be treated with a dilute solution of indoleacetic acid; the other is given only water. Later a comparison of the root systems of the two groups will be made.
3. An architectural firm is to sublet a contract for a wiring project. Seven electrical contractors are available for the job. We want to determine the average estimated cost of the job and the average projected time required to complete the job for these seven contractors.
4. A computer system has a number of remote terminals attached to it. To decide whether or not to increase this number, it is necessary to study the random variable X, the length of time expended per session by users of the terminals currently in place.
5. Prior to changing from the traditional 8-hour-a-day, 5-day-a-week work schedule to a 10-hour-a-day, 4-day-a-week schedule, the opinion of the 50,000 workers who would be affected is to be sought.
6. Air quality is of concern to everyone. It is judged by the number of micrograms of particulate present per cubic meter of air. Assume that this variable is normally distributed with unknown mean and unknown variance. Monitoring stations sample

air by sucking it through a thin fiberglass sheet that collects the fine particles suspended in the air. In a particular locality this is done for five randomly selected 24-hour periods each month. Thus each month a random sample of size $n = 5$ from a normal distribution is available.

(*a*) Consider the random variable X_1, the particulate level for the first 24-hour period studied during a given month. What is the distribution of this random variable?

(*b*) For a given month, these readings result:

$$x_1 = 45 \qquad x_2 = 50 \qquad x_3 = 62 \qquad x_4 = 57 \qquad x_5 = 70$$

For these data, evaluate the statistics ΣX_i, ΣX_i^2, $\Sigma X_i/n$, $\max_i\{X_i\}$, $\min_i\{X_i\}$.

(*c*) Is the random variable $X_5 - \mu$ a statistic? Is the random variable $(X_5 - \mu)/\sigma$ a statistic? Explain.

Section 6.2

7. A data set containing 70 observations, each reported to one decimal place, is to be split into seven categories. The largest observation is 75.1 and the smallest is 16.3.

(*a*) These data are covered by an interval of what length?

(*b*) Using the method outlined in this section, each category will be of what length?

(*c*) What is the lower boundary for the first category?

(*d*) What are the boundaries for each of the seven categories?

8. Acute exposure to cadmium produces respiratory distress and kidney and liver damage, and may even result in death. For this reason, the level of airborne cadmium dust and cadmium oxide fume in the air is monitored. This level is measured in milligrams cadmium per cubic meter of air. A sample of 35 readings yields these data (based on a report in *Environmental Management*, September 1981, p. 414):

.044	.030	.052	.044	.046
.020	.066	.052	.049	.030
.040	.045	.039	.039	.039
.057	.050	.056	.061	.042
.055	.037	.062	.062	.070
.061	.061	.058	.053	.060
.047	.051	.054	.042	.051

(*a*) Construct a stem-and-leaf diagram for these data. Use the numbers 02, 03, 04, 05, 06, and 07 as stems.

(*b*) Would you be surprised to hear someone claim that the random variable X, the cadmium level in the air, is normally distributed? Explain.

(*c*) Use the method outlined in this section to break these data into six categories. (Here a unit is .001 and a half unit is .0005.)

(*d*) Construct a frequency table and a relative frequency histogram for these data. Does the histogram exhibit the bell-shape characteristic of a normal density?

(*e*) Construct a cumulative frequency table and a relative cumulative frequency ogive for these data. Use the ogive to approximate that point above which 50% of the readings should fall.

9. Consider the stem-and-leaf diagram of Fig. 6.2. What family of distributions is suggested by the diagram?

10. Liquid products were first obtained from coal in England during the 1700s. Lamp oil was produced from coal in the United States as early as 1850, but the domestic coal chemicals industry did not develop until World War I. A modern coal-for-recovery system uses a battery of coke ovens to produce liquid products from the coal feed. These observations are obtained on the random variable X, the number of gallons of liquid product obtained per ton of coal feed (based on a report in *McGraw-Hill Yearbook of Science and Technology*, 1983, p. 37):

7.6	8.2	7.1	10.0	6.5	9.6
6.1	6.2	7.6	6.2	9.5	6.7
7.4	9.5	9.2	8.0	8.5	9.3
8.8	9.6	9.7	6.8	7.1	7.7
8.7	7.8	8.7	8.2	8.2	7.4
9.0	8.8	7.3	7.9	7.1	7.9
7.6	6.7	8.1	6.2	5.3	7.4
7.7	9.1	7.9	8.7	8.4	8.1

(*a*) Construct a stem-and-leaf diagram for these data. Use the numbers 5, 6, 7, 8, 9, 10 as stems.
(*b*) Is the assumption that X is normally distributed justifiable? Explain.
(*c*) Use the method outlined in this section to break these data into six categories.
(*d*) Construct a frequency table and a relative frequency histogram for these data. Does the histogram exhibit the bell-shape characteristic of a normal density?
(*e*) Construct a cumulative frequency table and a relative cumulative frequency ogive for these data. Use the ogive to approximate the probability that a randomly selected ton of coal will yield less than 7 gallons of liquid product.

11. Some efforts are currently being made to make textile fibers out of peat fibers. This would provide a source of cheap feedstock for the textile and paper industries. One variable being studied is X, the percentage ash content of a particular variety of peat moss. Assume that a random sample of 50 mosses yields these observations (based on data found in "Peat Fibre: Is There Scope in Textiles?" *Textile Horizons*, vol. 2, no. 10, October 1982, p. 24):

.5	1.8	4.0	1.0	2.0
1.1	1.6	2.3	3.5	2.2
2.0	3.8	3.0	2.3	1.8
3.6	2.4	.8	3.4	1.4
1.9	2.3	1.2	1.9	2.3
2.6	3.1	2.5	1.7	5.0
1.3	3.0	2.7	1.2	1.5
3.2	2.4	2.5	1.9	3.1
2.4	2.8	2.7	4.5	2.1
1.5	.7	3.7	1.8	1.7

(*a*) Construct a stem-and-leaf diagram for these data. Use the numbers 0, 1, 2, 3, 4, 5 as stems.
(*b*) Is there any reason to suspect that X is not normally distributed? Explain.
(*c*) Use the method outlined in this section to break these data into six categories.

(*d*) Construct a frequency table and a relative frequency histogram for these data. Does the histogram suggest that X might not be normally distributed? If so, what distribution might be appropriate?

(*e*) Construct a cumulative frequency table and a relative cumulative frequency ogive for these data. Use the ogive to approximate the probability that a randomly selected specimen of this variety of moss will have an ash content that exceeds 2%.

12. (*Percentiles.*) Let X be a random variable. The point $p_{k/100}$ $(k = 1, 2, 3, \ldots, 100)$ such that

$$P[X < p_{k/100}] \leq k/100 \qquad \text{and} \qquad P[X \leq p_{k/100}] \geq k/100$$

is called the *kth percentile* for X. For example, let X be binomial with $n = 20$ and $p = .5$. The 25th percentile for X is the point $p_{25/100} = 8$ since, from Table I of App. A, we see that

$$P[X < 8] = .1316 \leq .25 \qquad \text{and} \qquad P[X \leq 8] = .2517 \geq .25$$

(*a*) Let X be binomial with $n = 20$ and $p = .5$. Find the 60th percentile for X.

(*b*) Let X be Poisson with $\lambda s = 10$. Find the 30th percentile for X.

(*c*) Argue that in the case of a continuous random variable the kth percentile is that point such that $P[X \leq Pk/100] = k/100$.

(*d*) Let X be exponentially distributed with $\beta = 1$. Show that the 20th percentile for X is $-\ln .80$. *Hint:* Find the point p such that

$$\int_0^p e^{-x}\,dx = .20$$

13. (*Quartiles.*) The 25th, 50th, 75th, and 100th percentiles for X are called its first, second, third, and fourth *quartiles*, respectively.

(*a*) State the definition of the first quartile in terms of probabilities.

(*b*) Let X be binomial with $n = 20$ and $p = .5$. Find the first quartile for X.

(*c*) Let X be exponentially distributed with $\beta = 1$. Find the first quartile for X.

14. (*Deciles.*) The 10th, 20th, 30th, 40th, 50th, 60th, 70th, 80th, 90th, and 100th percentiles for X are called its *deciles*.

(*a*) State the definition of the 40th decile for X in terms of probabilities.

(*b*) Let X be Poisson with $\lambda s = 10$. Find the 6th decile for X.

(*c*) Let X be exponentially distributed with $\beta = 1$. Find the third decile for X.

15. The percentiles, quartiles, and deciles for a continuous random variable can be approximated from a relative cumulative frequency ogive using the projective method. For instance, in Fig. 6.5 we approximated the 50th percentile for X, the life span of a lithium battery, to be a little over 725 hours.

(*a*) Approximate the first quartile for X, the cadmium level in the air, using the data of Exercise 8.

(*b*) Approximate the fourth decile for X, the number of gallons of liquid product obtained per ton of coal fuel, using the data of Exercise 10.

(*c*) Approximate the 50th percentile for X, the percentage ash content for a particular variety of moss, using the data of Exercise 11.

16. In running computer programs on a time-sharing basis, the costs vary from session to session. These observations are obtained on the random variable X, the cost per session to the user:

\$1.08	.84	1.41	.99	.82
.89	.38	1.05	1.19	.65
1.09	1.03	.81	.55	.71
1.89	.47	.59	1.22	1.27
1.02	1.09	1.02	.86	1.23
1.23	.85	1.02	1.25	.80

Construct a relative cumulative frequency ogive for these data. Use the ogive to approximate the 50th percentile; the first quartile; the third quartile.

Section 6.3

17. Consider these data sets:

I				II			
1	3	2		1	2	4	1
2	5	4		2	5	2	5
4	3	3		1	5	5	3

(*a*) Find the sample mean and sample median for each data set.
(*b*) Find the sample range for each data set.
(*c*) Find the sample variance and sample standard deviation for each data set.
(*d*) Would you be surprised to hear someone claim that these data were drawn from the same population? Explain. *Hint:* Consider the shape of the distribution as well as the observed values of the sample statistics.

18. The observed values of the statistics $\sum_{i=1}^{50} X_i$ and $\sum_{i=1}^{50} X_i^2$ for the data of Example 6.2.1 are $\sum_{i=1}^{50} x_i = 63{,}707$ and $\sum_{i=1}^{50} x_i^2 = 154{,}924{,}261$.
(*a*) Would you be surprised to hear someone claim that the mean lifespan of the lithium batteries used in this model calculator is 1270 hours? Explain.
(*b*) Find the sample variance and sample standard deviation for these data.

19. Use the data of Example 6.1.1 to approximate the mean and variance of the random variable X, the number of times per hour that a television signal is interrupted by random interference.

20. Use the data of Exercise 8 to approximate the mean, variance, and standard deviation of the random variable X, the level of airborne cadmium dust and cadmium oxide fume. Assume that these approximations are fairly accurate. Between what two values would you expect approximately 95% of the readings to fall? Explain.

21. Use the data of Exercise 10 to approximate the mean, variance, and standard deviation of the random variable X, the number of gallons of liquid product obtained per ton of coal feed.

22. Use the data of Exercise 11 to approximate the mean, variance, and standard deviation of the random variable X, the percentage ash content of a particular variety of peat moss.

23. Prove Theorem 6.3.1. *Hint:* Begin by expanding the term $(X_i - \bar{X})^2$. Apply the rules of summation and remember that relative to summation on i, $\bar{X}$ is constant.

24. There have been many improvements made in lighting in the last 10 years. One new bulb, the Philips' Earth Light uses a compact screw-in fluorescent bulb with an electronic ballast incorporated in its base. It is thought to last 10 to 13 times longer than household bulbs used in the past. These data are obtained on the lifespan of a sample of these new bulbs (time is in thousands of hours):

9.1	10.1	9.0	11.4
10.5	9.5	12.0	9.1
12.2	13.1	10.0	9.3
9.0	9.6	11.1	9.1
13.3	10.7	9.1	9.0
9.0	11.0	9.2	11.6

(Based on information found in "Lighting Comes of Age with New Technology," *Research and Development*, November 1992, pp. 30–31.)

(*a*) Construct a stem-and-leaf diagram for these data and suggest a distribution from which these data might have been drawn.

(*b*) Based on these data, approximate the value of μ, the average life span of these bulbs.

(*c*) Approximate the median life span of these bulbs and explain exactly what this value means.

(*d*) Find the sample variance and sample standard deviation for these data.

(*e*) Criticize the following statement:
"Based on the normal probability rule, it is estimated that approximately 95% of all bulbs have a life span between 7,530 and 13,050 hours."

(*f*) Based on Chebyshev's inequality, what can be said about the proportion of bulbs whose life span is expected to fall between 7,530 and 13,050 hours?

***25.** (*Approximating σ via the range.*) The range can play an important role in the design of statistical studies. To obtain a prespecified degree of accuracy when estimating population parameters, an adequate sized sample must be drawn. Most formulas used to determine sample size require knowledge of σ, the population standard deviation. Often the researcher will not have an estimate of σ available but will have an idea of the expected range of his or her data. In Sec. 4.5 we saw that when sampling from a normal distribution,

$$P[-2\sigma < X - \mu < 2\sigma] \doteq .95$$

If X is not normally distributed, then Chebyshev's inequality can be applied to conclude that

$$P[-3\sigma < X - \mu < 3\sigma] \geq .89$$

That is, X always lies within at most 3 standard deviations of its mean with high probability. From this it can be concluded that the estimated range covers an interval of roughly 4σ for normally distributed random variables and 6σ otherwise. In the normal case an estimate of σ can be obtained by solving the equation

$$4\sigma \doteq \text{estimated range}$$

for σ. Thus we see that

$$\sigma \doteq (\text{estimated range})/4$$

when X is normally distributed. If X is not normally distributed, then

$$\sigma \doteq (\text{estimated range})/6$$

These data are obtained on the random variable X, the cpu time in seconds required to run a program using a statistical package:

6.2	5.8	4.6	4.9	7.1	5.2
8.1	.2	3.4	4.5	8.0	7.9
6.1	5.6	5.5	3.1	6.8	4.6
3.8	2.6	4.5	4.6	7.7	3.8
4.1	6.1	4.1	4.4	5.2	1.5

(*a*) Construct a stem-and-leaf diagram for these data. Is the assumption justified that X is normally distributed?
(*b*) Approximate σ via the sample standard deviation s.
(*c*) Find the sample range for these data and use it to approximate σ. Compare your result to that obtained in part (*b*).

Section 6.4

26. Consider the standard normal distribution.
(*a*) Use the Z table to verify that q_1 is approximately $-.67$ and q_3 is approximately .67.
(*b*) Find the interquartile range for Z and explain what this means.
(*c*) Verify that the inner fences for Z are $f_1 = -2.68$ and $f_3 = 2.68$.
(*d*) Verify that the probability that a standard normal random variable will fall beyond the inner fences is approximately .007.
(*e*) Find the outer fences for Z.
(*f*) Find the probability that a standard normal random variable will fall beyond the outer fences.

27. Let X be normally distributed with mean μ and variance σ^2.
(*a*) Verify that $q_3 = \mu + .67\sigma$ and that $q_1 = \mu - .67\sigma$.
(*b*) Find the interquartile range for X.
(*c*) Verify that the inner fences for X are $f_1 = \mu - 2.68\sigma$ and $f_3 = \mu + 2.68\sigma$.
(*d*) Verify that the probability that X will fall beyond the inner fences is approximately .007.

28. Temperature differences between the warm upper surface of the ocean and the colder deeper levels can be utilized to convert thermal energy to mechanical energy. This mechanical energy can in turn be used to produce electrical power using a vapor turbine. Let X denote the difference in temperature between the surface of the water and the water at a depth of 1 kilometer. Measurements are taken at 15 randomly selected sites in the Gulf of Mexico. These data result (based on a report in *McGraw-Hill Yearbook of Science and Technology*, 1983, p. 410):

22.5	23.8	23.2	22.8	10.1*
23.5	24.0	23.2	24.2	24.3
23.3	23.4	23.0	23.5	22.8

(*a*) Construct a double stem-and-leaf diagram for these data.
(*b*) Find the sample mean, sample median, and sample standard deviation for these data.

(*c*) Note that the starred observation in the data set is very different from the others. It is a potential outlier. Construct a boxplot for these data to verify that the value 10.1 does, in fact, qualify as an outlier.

(*d*) To see the effect of this outlier, drop it from the data set and calculate the sample mean, median, and standard deviation for the remaining 14 observations. Which measure is least affected by the presence of the outlier? Do you see why it is desirable to report both the mean and median of a data set?

29. Most homes utilize a variety of electronic equipment and appliances. For this reason, both suppliers and consumers of these products have become interested in product reliability. One aspect of reliability is the ability of the appliance to withstand power surges. In a study of this phenomena the following data are obtained on the strength of a surge in kilovolts required to damage or upset the appliance (based on figures found in "The Effects of Surges on Electronic Appliances," Stephen B. Smith and Ronald B. Standler, *IEEE Power Engineering Review*, July 1992, p. 50):

Clocks

1.1	3.5	3.2	4.0
3.6	1.5	2.3	3.0
4.7	4.0	4.9	2.9
2.6	2.5	2.7	4.2
2.4	3.7	3.8	6.0
5.0	1.8	5.6	5.1
3.9	3.8	3.7	3.5

Television receivers

2.0	5.0	4.5	5.4
4.2	4.3	5.1	5.6
5.2	4.7	4.4	5.8
5.2	7.8	5.4	4.9
4.6	4.6	5.0	4.8
5.3	5.2	5.3	5.9

dc power supplies

4.2	4.4	5.1	4.1	6.1
4.5	5.9	3.9	5.0	58
5.0	4.8	4.3	5.4	
4.7	5.1	5.2	5.6	
4.9	4.7	4.6	4.8	

(*a*) Sketch a double stem-and-leaf diagram for the clock data. Based on this diagram, would you be surprised to hear a claim that these data are drawn from an exponential distribution? Explain.

(*b*) Use the boxplot technique to check for outliers in the clock data. Based on your results, which measure of location, the sample mean or the sample median, is probably the better measure of the location of the bulk of the data for these data?

(*c*) Sketch a stem-and-leaf diagram for the television data. Use the stem 4 five times and the stem 5 five times. Based on this diagram, does there appear to be at least one outlier in the data set?

(*d*) Use the boxplot technique on the television data to test the suspicious points. Do you think that they are truly outliers? If so, are they mild outliers or extreme outliers. Which measure of variability, the sample variance or the iqr, is probably a better measure of the variability of the bulk of the data?

(*e*) Sketch a double stem-and-leaf diagram for the dc power supply data. These data contain an outlier due to a misplaced decimal point. Do you see it? Calculate the mean for the data using the bad data point as written. Now correct the data point and recalculate the sample mean. In light of this, explain what it means to say that $\bar{x}$ is not resistant to outliers.

REVIEW EXERCISES

30. Bricks are produced in lots of size 1000. Before shipping a lot, a sample of 25 bricks is selected and inspected for quality. Two random variables are of interest. These are X, the number of chips per brick, and Y, the hardness of the brick. Assume that hardness is measured on a continuous scale from 1 to 10 with larger numbers indicating a harder brick:

		x					*y*		
2	5	0	1	2	3.2	6.3	6.4	6.7	7.3
0	3	0	0	2	7.1	5.4	4.6	5.8	9.1
1	1	0	1	3	7.7	6.1	8.1	5.9	6.2
2	1	1	7	4	6.0	6.8	7.2	6.3	8.2
0	2	3	5	1	5.1	4.2	6.9	4.5	5.0

(*a*) What is the name of the family of random variables to which X belongs?

(*b*) Approximate the mean, variance, standard deviation, and median of X based on these data.

(*c*) Construct a stem-and-leaf diagram for the hardness measurements. Based on this diagram, would it be unrealistic to assume that Y is approximately normally distributed?

(*d*) Approximate the mean, variance, standard deviation, and median of Y.

31. In an attempt to study the problem of failure in field installed computer equipment, data is collected on fifty field trips made to repair equipment. The random variables studied are X, the time in hours required to locate and rectify the problem, and Y, the cause of the failure. We define Y by

$$Y = \begin{cases} 1 & \text{if the failure is due to a faulty microprocessor chip} \\ 0 & \text{otherwise} \end{cases}$$

These data are obtained:

			x								*y*		
1.52	1.83	2.25	4.73	2.89	1.49	1.34	0	0	0	0	0	0	1
2.15	2.66	2.79	1.35	1.54	4.59	4.27	0	0	0	0	0	0	0
3.91	2.76	3.03	3.52	5.97	1.45		0	0	0	0	0	0	
3.07	2.18	1.38	2.04	1.49	1.11		1	0	0	0	0	0	
1.24	4.84	2.82	3.16	4.58	3.28		0	0	0	0	0	0	
1.30	3.01	1.20	3.42	1.86	3.49		0	0	0	0	0	0	
3.93	2.56	2.63	5.60	4.60	5.34		0	0	0	0	0	0	
1.62	2.82	4.88	2.04	1.62	.24		0	0	1	0	0	0	

(*a*) Construct a relative frequency histogram for the data on the time required to locate and rectify the problem. Use six categories. Based on this histogram, would you be surprised to hear someone claim that X is approximately normally distributed? Explain.

(*b*) Approximate the mean, variance, and standard deviation for X.

(*c*) Construct a relative cumulative frequency ogive. Use this ogive to approximate the median for X. Approximately what percentage of problems can be located and rectified in 1.5 hours or less?

(*d*) Let p denote the probability that the failure is due to a faulty microprocessor chip. Assume that even though p is unknown its value is the same for each chip. Theoretically, Y follows a point binomial distribution with parameter p. What is the theoretical mean for Y? Approximate this mean based on these data. If asked to approximate the probability that a future failure is due to the failure of a microprocessor chip, what would you say?

(*e*) What is the theoretical variance for Y? Use your answer to part (*d*) to approximate the variance of Y. Use the sample variance to approximate σ_Y^2. Did you get the same result? Which answer is unbiased for σ_Y^2?

*(*f*) Use the technique of Exercise 25 to estimate σ_X. Compare your answer to that of part (*b*).

(*g*) Construct a boxplot for the data on x.

32. Most people are familiar with sparklers burned to celebrate New Year's Day and the Fourth of July. Two random variables are of interest. These are X, the length of the chemical coating that covers the tip of the sparkler, and Y the burn time of the sparkler in seconds. These data are obtained on these random variables (based on data gathered in 1993–1994 by students at Radford University and Virginia Polytechnic Institute and State University):

x (in)	y (s)	x (in)	y (s)
4.5	29	4.3	22
3.6	26	3.5	21
4.0	25	4.5	30
3.7	25	4.6	22
4.0	27	4.6	25
3.7	27	3.9	20
4.0	28	3.5	13
4.0	25	3.8	19
3.8	25	3.9	23
4.0	28	3.6	25
3.8	24	3.6	27
4.1	15	3.6	18
3.9	22	3.3	11
4.1	25	3.7	24
3.9	24	3.7	23
4.2	26	4.3	26
3.8	24	3.9	27

(*a*) Construct a stem-and-leaf diagram for the burn time data. Use each stem 5 times so that each stem will involve two leaves.

(*b*) Construct a boxplot for the burn time data. Are any data points flagged as outliers?

(*c*) Take a good look at the shape of the distribution as indicated by the stem-and-leaf diagram. Does the distribution appear to be skewed? If so, what is the direction of the skew? If we assume in this case that all data points are legitimate and not due to poor technique or recording errors, then from what family of random variables might these data have been drawn? Do you think that the "outliers" should be treated as such? Explain.

(*d*) Construct a stem-and-leaf diagram for the length data. Again, use each stem 5 times. Does the normality assumption appear fairly reasonable here?

(*e*) Construct a boxplot for the length data and comment on any outliers that might be identified.

COMPUTING SUPPLEMENT: SAS

I. Summary Statistics and Histograms

As can be seen, the calculations necessary to summarize even a relatively small data set can become tedious and time-consuming. To alleviate this problem, several computer systems for data analysis have been developed in recent years. Among the systems in widespread use are SPSS (Statistical Package for the Social Sciences, McGraw-Hill), BMD (Biomedical Computer Programs, University of California Press), MINITAB (Pennsylvania State University), and SAS (Statistical Analysis System, SAS Institute Inc.). To use any such system, one needs little background in computer science.

We present here a very brief introduction to SAS programming to give you some experience with computer packages. A MINITAB supplement is available in the Instructor's manual. Once this experience has been gained, it is not difficult to adjust to any of the other packages, for they are similar. We introduce SAS by presenting some sample programs that could be modified to analyze any of the data sets presented in this chapter. You should consult the appropriate expert at your own installation to determine the commands necessary to access SAS.

We begin by writing a program to obtain selected sample statistics and a histogram for the data of Example 6.2.1. The variable studied is X, the life span in hours of a lithium calculator battery.

The first step in analyzing data via the SAS package is to read the data and to get them into an SAS data set. This is done by means of a series of statements which name the data set (the DATA statement), describe the arrangement of the data (the INPUT statement), and signal the beginning of the data lines (the LINES statement). SAS statements may begin in any column and always end with a semicolon.

The name chosen for the data set should be a one-word name up to eight characters long. The first character must be a letter or an underscore. Later characters can be letters, characters, or underscores. It is usually helpful to choose a name that is in some way related to the data itself. To name the data set, the SAS key word DATA is used, followed by the name chosen. For instance, in our example we might choose to name the data set "battery." We would inform the computer of this choice by typing the statement

```
DATA BATTERY;
```

on a single line.

The next statement is the INPUT statement. This statement names the variables and describes the order in which they will appear on the data line. Variable names are chosen at the discretion of the programmer, but they must satisfy the same guidelines as those used in naming the data set. In our example there is only one variable, the life span of the battery. The input statement is simple in this case. We need only write

```
INPUT LIFESPAN;
```

This tells the computer that each data line will contain the value of only one variable, whose name is "life span."

The INPUT statement is followed by the LINES statement. The purpose of this statement is to signal SAS that the data follows immediately. This statement is written

```
LINES;
```

The data follow immediately with one observation per line. Since SAS recognizes the end of the data when it sees a semicolon, make sure that the first line after the data contains a semicolon. This can be done by entering the line containing only a semicolon. Thus far our program looks like this:

```
DATA BATTERY;          names the data set
INPUT LIFESPAN;        names the variable
LINES;                 signals beginning of data lines
    4285
    2066               data (one observation per line)
    2584
      :
    454
;                      signals the end of data
```

Now the data are in an SAS data set named "battery." We must tell the computer what to do with the data by means of one or more "procedure" statements. The procedure statement begins with the SAS key word PROC, followed by the name of the procedure desired. Most of the sample statistics introduced in this chapter can be obtained by using the "means" procedure. The key word for this procedure is MEANS, followed by the key words for those sample statistics desired. Some of these key words are as follows:

MAXDEC = *n*	*n* is an integer from 0 to 8 specifying the number of decimal places that will be used to print the results
MEAN	mean
STD	standard deviation
MIN	smallest value
MAX	largest value
RANGE	range
VAR	variance
STDERR	standard error of the mean
CV	coefficient of variation
N	number of observations
SUM	sum of observations
USS	sum of the squares of the observations

To find the sample statistics for our example and to have the results reported to two decimal places, we write

```
PROC MEANS MAXDEC = 2 MEAN VAR STD RANGE MAX MIN
                    SUM USS N;
```

A relative frequency histogram can be obtained by calling on the "chart" procedure. We can allow SAS to choose category boundaries for us, or we can reproduce the histogram found in the text. The statements required to do the former via vertical bars are

```
PROC CHART;
VBAR LIFESPAN/TYPE = PERCENT;
```

The first statement calls for a chart to be made; the second requests a vertical bar chart of the relative frequency distribution for the variable "life span." To obtain a horizontal histogram together with a distribution table such as that found in Example 6.2.2, we write

```
PROC CHART;
HBAR LIFESPAN/MIDPOINTS = 369.5  1081.5  1793.5  2505.5
                          3217.5  3929.5;
```

Notice that the midpoints listed are the midpoints of the categories found in Example 6.2.2. Each midpoint is the average of the respective category boundaries.

One further statement is desirable. This is a "title" statement that enables the programmer to print a title at the top of each page of output. The titles can contain up to 132 characters. The key word is TITLEn, where n gives the line number at which the title is to be printed. For instance, the statements

```
TITLE1  ANALYSIS;
TITLE2  OF;
TITLE3  BATTERY DATA;
```

would result in the word "analysis" being printed on line 1 of each page of output, the word "of" being printed on line 2 of each page, and the words "battery data" being printed on line 3. The entire program is as follows:

```
DATA BATTERY;
INPUT LIFESPAN;
LINES;
    4285
    2066
    2584
      .
      .
      .
     454
;
TITLE1  ANALYSIS;
TITLE2  OF;
TITLE3  BATTERY DATA;
PROC MEANS MAXDEC = 2 MEAN VAR STD RANGE MAX MIN
                      SUM USS N;
PROC CHART;
VBAR LIFESPAN/TYPE = PERCENT;

PROC CHART;
HBAR LIFESPAN/MIDPOINTS = 369.5  1081.5  1793.5  2505.5
                          3217.5  3929.5;
```

To adjust the program to handle another data set, one should change the name of the data set (BATTERY) and the input variable (LIFESPAN) to names appropriate to the new data. The title should also be changed.

The output of this program is as follows:

ANALYSIS OF
BATTERY DATA

Analysis Variable: LIFESPAN

N Obs	N	Minimum	Maximum	Sum	USS	Range	Mean	Variance	Std Dev.
50	50	14.00	4285.00	63707.00	154924261.00	4271.00	1274.14	1505155.59	1226.85

ANALYSIS OF
BATTERY DATA

PERCENTAGE OF LIFESPAN

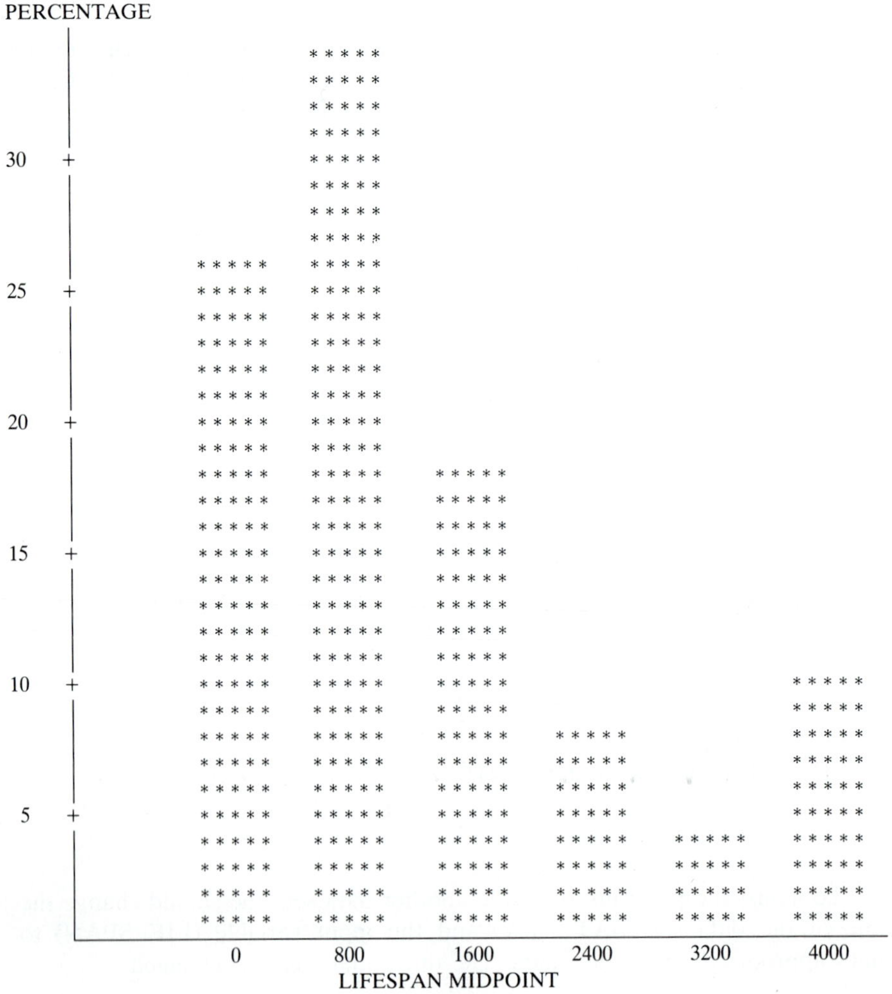

ANALYSIS
OF
BATTERY DATA

FREQUENCY OF LIFE SPAN

LIFE SPAN MIDPOINT		FREQ	CUM FREQ	PERCENT	CUM PERCENT
369.5	************************	24	24	48.00	48.00
1081.5	************	12	36	24.00	72.00
1793.5	****	4	40	8.00	80.00
2505.5	***	3	43	6.00	86.00
3217.5	**	2	45	4.00	90.00
3929.5	*****	5	50	10.00	100.00

5 10 15 20

FREQUENCY

CHAPTER 7

ESTIMATION

In Chap. 6, we found that once the family to which a random variable belongs is determined, the problem of approximating or *estimating* the numerical value of pertinent parameters remains. Even though we were able to define sample statistics that allow us to estimate the mean, variance, and standard deviation of a random variable in a logical manner, we were unable to assess their effectiveness. In this chapter we consider the mathematical properties of these statistics. We also present a brief introduction to the theory of estimation. The ideas developed here will be used extensively throughout the remainder of the text.

7.1 POINT ESTIMATION

In an estimation problem there is at least one parameter θ whose value is to be approximated on the basis of a sample. The approximation is done by using an appropriate statistic. A statistic used to approximate or estimate a population parameter θ is called a *point estimator* for θ and is denoted by $\hat{\theta}$ (the symbol ^ is called a "hat"); the numerical value assumed by this statistic when evaluated for a given sample is called a *point estimate* for θ. For example, in estimating the mean coal consumption by electric utilities for a given year (see Example 6.3.1), the statistic $\bar{X}$ was used. Thus $\bar{X}$ is a point estimator for μ and we write $\hat{\mu} = \bar{X}$. In Example 6.3.1 we evaluated this statistic for a particular sample and obtained the value 408.3 million tons. This number is called a *point estimate* for μ. Note that there is a difference in the terms "estimator" and "estimate." The estimator is the statistic used to generate the estimate; it is a random variable. An estimate is a number.

Once a logical point estimator for a parameter θ has been developed, the natural question to ask is, "How good is this estimator?" Obviously, we want the estimator to generate estimates that can be expected to be close in value to θ. This can be expected to occur if the estimator $\hat{\theta}$ possesses two properties. In particular, we would like:

1. $\hat{\theta}$ to be *unbiased* for θ.
2. $\hat{\theta}$ to have a small variance for large sample sizes.

The word "unbiased" was explained graphically in Chap. 6. Basically, it means "centered at the right spot" where the right spot is the parameter being estimated. The term "unbiased" is a technical term. To be able to prove analytically that an estimator $\hat{\theta}$ is an unbiased estimator for a parameter θ, we need a formal definition for the term. This definition is given here.

Definition 7.1.1 (Unbiased). An estimator $\hat{\theta}$ is an unbiased estimator for a parameter θ if and only if $E[\hat{\theta}] = \theta$.

Recall that $\hat{\theta}$ is a statistic; therefore it is also a random variable and, as such, has a mean, or expected, value. To say that $\hat{\theta}$ is unbiased for θ implies that the mean of the estimator $\hat{\theta}$ is equal to the parameter θ that it is estimating. Thus an estimator $\hat{\mu}$ is an unbiased estimator for μ if and only if $E[\hat{\mu}] = \mu$; an estimator $\hat{\sigma}^2$ is unbiased for σ^2 if and only if $E[\hat{\sigma}^2] = \sigma^2$; an estimator $\hat{\sigma}$ is unbiased for σ if and only if $E[\hat{\sigma}] = \sigma$. Let us reexamine the estimators $\bar{X}$, S^2, and S developed in Chap. 6 in light of this new definition.

Theorem 7.1.1. Let $X_1, X_2, X_3, \ldots, X_n$ be a random sample of size n from a distribution with mean μ. The sample mean, $\bar{X}$, is an unbiased estimator for μ.

Proof. By Definition 6.3.1,

$$E[\bar{X}] = E[1/n(X_1 + X_2 + X_3 + \cdots + X_n)]$$

By the Rules for Expectation (Theorem 3.3.1),

$$E[\bar{X}] = 1/n(E[X_1] + E[X_2] + E[X_3] + \cdots + E[X_n])$$

Since $X_1, X_2, X_3, \ldots, X_n$ constitutes a random sample from a distribution with mean μ, each of these random variables has mean μ. Therefore

$$E[\bar{X}] = 1/n\Big(\underbrace{\mu + \mu + \mu + \cdots + \mu}_{n \text{ terms}}\Big) = 1/n(n\mu) = \mu$$

and the proof is complete.

It is important to realize that since $\hat{\theta}$ is a statistic, in repeated sampling the estimates generated will vary from sample to sample. To say that $\hat{\theta}$ is unbiased for θ implies that these estimates vary about θ; it also implies that the *average* value of these estimates can be expected to lie reasonably close to θ. For example, since $\bar{X}$ is unbiased for μ, for k repetitions of an experiment the observed sample means $\bar{x}_1, \bar{x}_2, \bar{x}_3, \ldots, \bar{x}_k$ will vary about μ and the *average* value of these k estimates should lie reasonably close to μ.

It is equally important to understand what the term "unbiased" does *not* imply. It does not imply that any *one* estimate will be close in value to the parameter being estimated. In reference to Example 6.3.1, the estimated mean coal consumption by electric utilities was $\hat{\mu} = \bar{x} \doteq 408.3$ million tons. This estimate is unbiased in the sense that it was generated by means of the unbiased estimator $\bar{X}$. This *alone* does not guarantee that the actual mean coal consumption by electric utilities across the country is anywhere close to 408.3 million tons. This is unfortunate. Usually, statistical studies are not repeated over and over so that the estimates obtained can be averaged. In general, only one sample is drawn; one estimate is obtained. To have some assurance that this estimate is close in value to θ, the parameter being estimated, ideally the estimator used not only should be unbiased, but also it should have a small variance for large sample sizes. In this way even though the estimated values fluctuate about θ, the variability is small. Each estimate produced can be expected to be fairly close in value to θ. Theorem 7.1.2 shows that $\bar{X}$ has this property.

Theorem 7.1.2. Let $\bar{X}$ be the sample mean based on a random sample of size n from a distribution with mean μ and variance σ^2. Then

$$\operatorname{Var} \bar{X} = \frac{\sigma^2}{n}$$

The proof of this theorem is based on the Rules for Variance (Theorem 3.3.3) and is similar to that of Theorem 7.1.1. Note that since σ^2 is constant, as the sample size n increases, the variance of $\bar{X}$, σ^2/n, decreases and can be made as small as we wish by choosing n sufficiently large. This implies that a sample mean based on a large sample can be expected to lie reasonably close to μ; one based on a small sample may vary widely from the actual population mean. This points out the advantages of working with a large sample and the danger of placing too much emphasis on conclusions drawn from small samples. Keep in mind that many of the examples and exercises presented in this text are based on small samples. This is done for illustrative purposes only. We do *not* mean to imply that samples this small are common in research.

Since the standard deviation of any random variable is the square root of its variance, the standard deviation of the sample mean is the square root of the variance of $\bar{X}$. Thus the standard deviation of $\bar{X}$ is $\sqrt{\sigma^2/n} = \sigma/\sqrt{n}$. This standard deviation is often referred to as the *standard error of the mean*.

In Chap. 6 we defined the sample variance S^2 by dividing $\sum_{i=1}^{n}(X_i - \bar{X})^2$ by $n - 1$. This was done so that the resulting estimator would be unbiased for σ^2. We now prove that this is the case.

Theorem 7.1.3. Let S^2 be the sample variance based on a random sample of size n from a distribution with mean μ and variance σ^2. S^2 is an unbiased estimator for σ^2.

Proof. By definition,

$$E[S^2] = E\left[\sum_{i=1}^{n} \frac{(X_i - \bar{X})^2}{n-1}\right]$$

$$= \frac{1}{n-1}E\left[\sum_{i=1}^{n} (X_i - \mu + \mu - \bar{X})^2\right]$$

$$= \frac{1}{n-1}E\left[\sum_{i=1}^{n} (X_i - \mu)^2 - 2(\bar{X} - \mu)\sum_{i=1}^{n} (X_i - \mu) + \sum_{i=1}^{n} (\bar{X} - \mu)^2\right]$$

$$= \frac{1}{n-1}E\left[\sum_{i=1}^{n} (X_i - \mu)^2 - 2(\bar{X} - \mu)\frac{n(\sum X_i - n\mu)}{n} + n(\bar{X} - \mu)^2\right]$$

$$= \frac{1}{n-1}E\left[\sum_{i=1}^{n} (X_i - \mu)^2 - 2n(\bar{X} - \mu)^2 + n(\bar{X} - \mu)^2\right]$$

$$= \frac{1}{n-1}E\left[\sum_{i=1}^{n} (X_i - \mu)^2 - n(\bar{X} - \mu)^2\right]$$

$$= \frac{1}{n-1}\left[\sum_{i=1}^{n} E\left[(X_i - \mu)^2\right] - nE\left[(\bar{X} - \mu)^2\right]\right]$$

Note that since $X_1, X_2, X_3, \ldots, X_n$ is a random sample from a distribution with variance σ^2, $E[(X_i - \mu)^2] = \sigma^2$ for each $i = 1, 2, 3, \ldots, n$. Note that by Theorems 7.1.2 and 7.1.1, $\text{Var}\,\bar{X} = E[(\bar{X} - \mu)^2] = \sigma^2/n$. By substitution, we obtain

$$E[S^2] = \frac{1}{n-1}\left[\sum_{i=1}^{n} \sigma^2 - n\sigma^2/n\right]$$

$$= \frac{1}{n-1}(n\sigma^2 - \sigma^2) = \sigma^2$$

and the proof is complete.

It should be noted that even though S^2 is an unbiased estimator for σ^2, it can be shown that S is not unbiased for σ (see Exercise 8). This emphasizes the fact that unbiasedness is desirable in an estimator but not essential.

Notice that the logical estimator for $\sigma/\sqrt{n}$, the standard error of the mean, is $S/\sqrt{n}$. This statistic is called the *sample standard error*. It will be used extensively in Chap. 8.

7.2 THE METHOD OF MOMENTS AND MAXIMUM LIKELIHOOD

In this section we consider two methods for deriving point estimators for distribution parameters. The first, called the *method of moments*, is a simple method that was first proposed by Karl Pearson in 1894. The second, called the *method of maximum likelihood*, is more complex. It was used by C. F. Gauss to solve isolated problems over 170 years ago. In the early 1900s the method was formalized by R. A. Fisher and has been used extensively since that time.

To begin, we note that terms of the form $E[X^k]$ $(k = 1, 2, 3, \ldots)$ are called the *kth moments* for X. Since an expectation is a theoretical average, logic implies that the moments for X can be estimated via an arithmetic average. That is, an estimator M_k for $E[X^k]$ based on a random sample of size n is

$$M_k = \sum_{i=1}^{n} \frac{X_i^k}{n}$$

The method of moments exploits the fact that in many cases the moments for X can be expressed as a function of θ, the parameter to be estimated. We can often obtain a reasonable estimator for θ by replacing the theoretical moments by their estimators and solving the resulting equation for $\hat{\theta}$.

You have already used the technique quite naturally in solving some of the problems in the last section! We now formalize the idea. The technique is illustrated by finding the method of moments estimator for the parameter p of a binomial random variable.

Example 7.2.1. A forester plants five rows of 20 pine seedlings, each to serve as eventual windbreaks. The soil and wind conditions to which the seedlings are subjected are identical. The variable being studied is X, the number of seedlings per row that survive the first winter. We are dealing with a random sample of size $m = 5$ from a binomial distribution with parameters $n = 20$ and p unknown. We want to use the method of moments to derive an estimator for p. To do so, note that since X is binomial,

$$E[X] = np = 20p$$

We now replace the first moment of X, $E[X]$, by its estimator $M_1 = (\sum_{i=1}^{5} X_i)/5 = \bar{X}$ to obtain the equation

$$\bar{X} = 20\hat{p}$$

This equation is solved for $\hat{p}$ to obtain the estimator

$$\hat{p} = \bar{X}/20$$

When the experiment is conducted, these data result:

$$x_1 = 18 \qquad x_3 = 15 \qquad x_5 = 20$$
$$x_2 = 17 \qquad x_4 = 19$$

For these data $\bar{x} = (\sum_{i=1}^{5} x_i)/5 = 17.8$. The method of moments estimate for p, the probability that a seedling will survive the first winter is

$$\hat{p} = \bar{x}/20 = 17.8/20 = .89$$

Occasionally there are two parameters θ_1 and θ_2 to be estimated from a single sample. To use the method of moments in this case, we must obtain two equations relating the moments of the distribution to these parameters. We then replace the theoretical moments by their estimators and solve the resulting equations simultaneously for $\hat{\theta}_1$ and $\hat{\theta}_2$. This idea is illustrated by finding estimators for α and β, the parameters that identify the gamma distribution.

Example 7.2.2. Let $X_1, X_2, X_3, \ldots, X_n$ be a random sample from a gamma distribution with parameters α and β. From Theorem 4.3.2 we know that $E[X] = \alpha\beta$ and $\operatorname{Var} X = \alpha\beta^2$. Recall that since $\operatorname{Var} X = E[X^2] - (E[X])^2$, the first two moments of X are functions of α and β. The equations relating the moments to these unknown parameters are

$$E[X] = \alpha\beta$$
$$E[X^2] - (E[X])^2 = \alpha\beta^2$$

We now replace $E[X]$ and $E[X^2]$ by their estimators M_1 and M_2, respectively, to obtain

$$M_1 = \hat{\alpha}\hat{\beta}$$
$$M_2 - M_1^2 = \hat{\alpha}\hat{\beta}^2$$

Solving this set of equations simultaneously, we see that

$$M_2 - M_1^2 = M_1\hat{\beta}$$

This implies that

$$\hat{\beta} = (M_2 - M_1^2)/M_1$$

and

$$\hat{\alpha} = M_1/\hat{\beta} = M_1^2/(M_2 - M_1^2)$$

Maximum Likelihood Estimators

The maximum likelihood method for deriving estimators is more complex than the method of moments. However, it is based on an appealing notion. Recall that the density f for a random variable X usually has at least one parameter θ associated with it. Assume that we have a random sample $x_1, x_2, x_3, \ldots, x_n$ available. The method of maximum likelihood in a sense picks out of all the possible values of θ the one most likely to have produced these observations. Before formalizing the method, let us demonstrate the idea in a simple context.

Example 7.2.3. Water samples of a specified size are taken from a river suspected of having been polluted by improper treatment procedures at an upstream sewage disposal plant. Let X denote the number of coliform organisms found per sample and assume that X is a Poisson random variable with parameter k. Let $x_1, x_2, x_3, \ldots, x_n$ be a random sample from the distribution of X. We want to determine the value of k that gives the highest probability of observing this sample. Since random sampling implies independence,

$$
\begin{aligned}
&P[X_1 = x_1, X_2 = x_2, \ldots, X_n = x_n] \\
&\quad = P[X_1 = x_1]P[X_2 = x_2] \cdots P[X_n = x_n] \\
&\quad = \prod_{i=1}^{n} P[X_i = x_i]
\end{aligned}
$$

Recall that the density for X is given by

$$P[X = x] = f(x) = \frac{e^{-k}k^x}{x!} \qquad x = 0, 1, 2, \ldots$$

Therefore the probability of obtaining the given sample is

$$\prod_{i=1}^{n} P[X_i = x_i] = \prod_{i=1}^{n} f(x_i) = \prod_{i=1}^{n} \frac{e^{-k}k^{x_i}}{x_i!}$$

Note that this probability is a function of k, which we denote by $L(k)$. Using the laws of exponents,

$$L(k) = \frac{e^{-nk}k^{\sum_{i=1}^{n} x_i}}{\prod_{i=1}^{n} x_i!}$$

This function is called the "likelihood function." It gives us the probability of observing the values $x_1, x_2, \ldots, x_n$ as a function of the parameter k. We want to find the value of k that maximizes this probability. That is, of all the possible values for k, we want to find the one that gives us the highest probability of observing the values that we did observe. To find this value of k, we use elementary calculus to maximize the likelihood function. This can be done directly. However, to simplify the process, we first take the natural logarithm of $L(k)$ and use the laws of logarithms to simplify the resulting expression

$$\ln L(k) = -nk + \sum_{i=1}^{n} x_i \ln k - \ln \prod_{i=1}^{n} x_i!$$

The value of k that maximizes $\ln L(k)$ also maximizes $L(k)$. Therefore to complete the derivation, we differentiate $\ln L(k)$ with respect to k, set the derivative equal to 0, and solve for k:

$$\frac{d \ln L(k)}{dk} = -n + \left(\sum_{i=1}^{n} x_i \right) \Big/ k$$

$$0 = -n + \left(\sum_{i=1}^{n} x_i \right) \Big/ k$$

$$k = \left(\sum_{i=1}^{n} x_i \right) \Big/ n = \bar{x}$$

Since this procedure does not give us the exact value of k but rather provides a logical method for estimating k, we write $\hat{k} = \bar{X}$. That is, the sample mean is the "maximum likelihood estimator" for the parameter k of a Poisson random variable.

Suppose that a random sample of size 4 yields these data:

$$x_1 = 12 \qquad x_2 = 15 \qquad x_3 = 16 \qquad x_4 = 17$$

Since the value of k that is most likely to have produced this sample is $\bar{x} = 15$, it is natural to take this value as our estimate for k.

Although our example involves a discrete random variable, the same general method is used in the continuous case. This method is summarized as follows:

1. Obtain a random sample $x_1, x_2, x_3, \ldots, x_n$ from the distribution of a random variable X with density f and associated parameter θ.
2. Define a function $L(\theta)$ by

$$L(\theta) = \prod_{i=1}^{n} f(x_i)$$

This function is called the *likelihood function* for the sample.

3. Find the expression for θ that maximizes the likelihood function. This can be done directly or by maximizing $\ln L(\theta)$.
4. Replace θ by $\hat{\theta}$ to obtain an expression for the maximum likelihood estimator for θ.
5. Find the observed value of this estimator for a given sample.

As with the method of moments, the maximum likelihood procedure can be applied when the density for X is characterized by two parameters. We illustrate the technique by finding the maximum likelihood estimators for μ and σ^2, the mean and variance of a normal random variable.

Example 7.2.4. Let $x_1, x_2, x_3, \ldots, x_n$ be a random sample from a normal distribution with mean μ and variance σ^2. The density for X is

$$f(x) = \frac{1}{\sqrt{2\pi}\,\sigma} e^{-(1/2)[(x-\mu)/\sigma]^2}$$

The likelihood function for the sample is a function of both μ and σ. In particular,

$$L(\mu, \sigma) = \prod_{i=1}^{n} \frac{1}{\sqrt{2\pi}\,\sigma} e^{-(1/2)[(x_i-\mu)/\sigma]^2}$$

$$= \left(\frac{1}{\sqrt{2\pi}}\right)^n \left(\frac{1}{\sigma}\right)^n e^{-(1/2\sigma^2)\sum_{i=1}^{n}(x_i-\mu)^2}$$

The logarithm of the likelihood function is

$$\ln L(\mu, \sigma) = -n \ln \sqrt{2\pi} - n \ln \sigma - (1/2\sigma^2) \sum_{i=1}^{n} (x_i - \mu)^2$$

To maximize this function, we take the partial derivatives with respect to μ and σ, set these derivatives equal to 0, and solve the equations simultaneously for μ and σ:

$$\begin{cases} \dfrac{\partial \ln L(\mu,\sigma)}{\partial \mu} = \dfrac{1}{\sigma^2}\sum_{i=1}^{n}(x_i-\mu) \\ \dfrac{\partial \ln L(\mu,\sigma)}{\partial \sigma} = \dfrac{-n}{\sigma} + \sum_{i=1}^{n}\dfrac{(x_i-\mu)^2}{\sigma^3} = \dfrac{-n\sigma^2+\sum_{i=1}^{n}(x_i-\mu)^2}{\sigma^3} \end{cases}$$

$$\begin{cases} \dfrac{1}{\sigma^2}\sum_{i=1}^{n}(x_i-\mu) = 0 \\ \dfrac{-n\sigma^2+\sum_{i=1}^{n}(x_i-\mu)^2}{\sigma^3} = 0 \end{cases}$$

$$\begin{cases} \sum_{i=1}^{n} x_i - n\mu = 0 \quad \text{or} \quad \mu = \left(\sum_{i=1}^{n} x_i\right)\Big/ n = \bar{x} \\ -n\sigma^2 + \sum_{i=1}^{n}(x_i-\mu)^2 = 0 \quad \text{or} \quad \sigma^2 = \left[\sum_{i=1}^{n}(x_i-\mu)^2\right]\Big/ n \end{cases}$$

Realizing that these are not the true values of μ and σ^2 but are only estimates, we see that the maximum likelihood estimators for these parameters are

$$\hat{\mu} = \bar{X}$$

$$\hat{\sigma}^2 = \left[\sum_{i=1}^{n}(X_i-\bar{X})^2\right]\Big/ n$$

The method of moments estimator for a parameter and the maximum likelihood estimator often agree. However, if they do not, the maximum likelihood estimator is usually preferred.

7.3 FUNCTIONS OF RANDOM VARIABLES—DISTRIBUTION OF $\bar{X}$

There is one drawback to point estimation. It yields a single value for the unknown parameter θ. Is there any assurance that this estimate is even close in value to θ? The best answer is that in most cases the point estimators used are logical. To get an idea not only of the value of the parameter being estimated, but also of the accuracy of the estimate, researchers turn to the method of *interval estimation* or *confidence intervals*. An interval estimator is what the name implies. It is a random interval, an interval whose endpoints L_1 and L_2 are each statistics. It is used to determine a numerical interval based on a sample. It is hoped that the numerical interval obtained will contain the population parameter being estimated. By expanding from a point to an interval, we create a little room for error and in so doing gain the ability, based on probability theory, to report the confidence that we have in the estimate.

In later chapters we shall derive confidence intervals for many important parameters. To do so, we must know the distribution of some key random variables. In this section we consider a technique for identifying the distribution of a random variable from its moment generating function. This technique depends on the result given in Theorem 7.3.1.

Theorem 7.3.1. Let X and Y be random variables with moment generating functions $m_X(t)$ and $m_Y(t)$, respectively. If $m_X(t) = m_Y(t)$ for all t in some open interval about 0, then X and Y have the same distribution.

The proof of this theorem is based on transform theory and is beyond the scope of this text. The theorem implies that the moment generating function, when it exists, serves as a "fingerprint" for the random variable. We illustrate this idea by proving Theorem 4.4.3, the "standardization" theorem for normal random variables.

Example 7.3.1. Let X be a normal random variable with mean μ and variance σ^2. Recall from Theorem 4.4.1 that the moment generating function for X is

$$m_X(t) = E[e^{tX}] = e^{\mu t + \sigma^2 t^2/2}$$

normal random variable

The moment generating function for a standard normal random variable Z is

$$m_Z(t) = e^{0t + (1)^2 t^2/2} = e^{t^2/2}$$

Let $Y = (X - \mu)/\sigma = (1/\sigma)X - \mu/\sigma$. The moment generating function for Y is given by

$$\begin{aligned} m_Y(t) = E[e^{tY}] &= E[e^{(t/\sigma)X - (\mu/\sigma)t}] \\ &= E[e^{(t/\sigma)X} e^{(-\mu/\sigma)t}] \\ &= e^{(-\mu/\sigma)t} E[e^{(t/\sigma)X}] \end{aligned}$$

Note that $E[e^{(t/\sigma)X}] = m_X(t/\sigma) = e^{\mu(t/\sigma) + \sigma^2 t^2/2\sigma^2}$. Substituting, we obtain

$$m_Y(t) = e^{(-\mu/\sigma)t} e^{(\mu/\sigma)t + t^2/2} = e^{t^2/2} = m_Z(t)$$

We have shown that Y and Z have the same moment generating function. By Theorem 7.3.1 these variables have the same distribution. In particular, they are both *standard normal* random variables.

Many of the statistics used in data analysis entail summing a collection of random variables. The following theorem together with Theorem 7.3.1 will help to determine the distribution of such statistics.

Theorem 7.3.2. Let X_1 and X_2 be independent random variables with moment generating functions $m_{X_1}(t)$ and $m_{X_2}(t)$, respectively. Let $Y = X_1 + X_2$. The moment generating function for Y is given by

$$m_Y(t) = m_{X_1}(t) m_{X_2}(t)$$

Proof. By definition

$$m_Y(t) = E[e^{tY}] = E[e^{tX_1 + tX_2}] = E[e^{tX_1}e^{tX_2}]$$

Since X_1 and X_2 are independent, e^{tX_1} and e^{tX_2} are also independent. By Theorem 5.2.2

$$m_Y(t) = E[e^{tX_1}e^{tX_2}] = E[e^{tX_1}]E[e^{tX_2}] = m_{X_1}(t)m_{X_2}(t)$$

This theorem can be extended easily to include a sum of more than two random variables. That is, we can say that the moment generating function for the sum of a finite number of *independent* random variables is the product of the moment generating functions of the individual variables. The requirement that the random variables be independent is not restrictive since in most cases the sum of interest is a function of the elements of a random sample. The term "random sample" implies independence. (See Definition 6.1.1.) Theorem 7.3.2 is illustrated by showing that the sum of a collection of independent normal random variables is normal.

Example 7.3.2. Let $X_1, X_2, X_3, \ldots, X_n$ be independent normal random variables with means $\mu_1, \mu_2, \mu_3, \ldots, \mu_n$ and variances $\sigma_1^2, \sigma_2^2, \sigma_3^2, \ldots, \sigma_n^2$, respectively. Let $Y = X_1 + X_2 + X_3 + \cdots + X_n$. Note that the moment generating function for X_i is given by

$$m_{X_i}(t) = e^{\mu_i t + \sigma_i^2 t^2/2} \qquad i = 1, 2, 3, \ldots, n$$

and the moment generating function for Y is

$$m_Y(t) = \prod_{i=1}^{n} m_{X_i}(t) = \exp\left[\left(\sum_{i=1}^{n} \mu_i\right)t + \left(\sum_{i=1}^{n} \sigma_i^2\right)t^2\Big/2\right]$$

The function on the right is the moment generating function for a normal random variable with mean $\mu = \sum_{i=1}^{n}\mu_i$ and variance $\sigma^2 = \sum_{i=1}^{n}\sigma_i^2$.

Distribution of $\bar{X}$

One of the more useful statistics that we have studied is $\bar{X}$, the sample mean. Since $\bar{X}$ is a statistic, it is also a random variable. It makes sense to ask, "What is the distribution of $\bar{X}$?" We have already seen that the center of location for $\bar{X}$ is μ, the mean of the population from which the sample is drawn. We have also seen that its variance is σ^2/n, the original population variance divided by the sample size. We have not yet mentioned the type of distribution possessed by this statistic. Does $\bar{X}$ follow some distribution such as the gamma, uniform, or normal that we have already studied, or must we introduce a new distribution now? The next theorem, whose derivation is outlined in Exercise 38, will help us to answer this question.

Theorem 7.3.3. Let X be a random variable with moment generating function $m_X(t)$. Let $Y = \alpha + \beta X$. The moment generating function for Y is

$$m_Y(t) = e^{\alpha t} m_X(\beta t)$$

We illustrate the use of this theorem in a numerical context.

Example 7.3.3. Let X denote the maximum wind speed per day recorded at the weather station of a particular locality. Assume that X is normally distributed with mean 10 miles per hour (mph) and standard deviation 4 mph. Engineers are constructing a bridge over a deep canyon in the area. They suspect that the maximum wind speed at the bridge site is given by $Y = 2X - 5$. What is the distribution of Y? To answer this question we first note that the moment generating function for X is

$$m_X(t) = e^{\mu t + \sigma^2 t^2/2}$$
$$= e^{10t + 16t^2/2}$$

We next apply Theorem 7.3.3 with $\alpha = -5$ and $\beta = 2$ to see that the moment generating function for Y is

$$m_Y(t) = e^{-5t} e^{10(2t) + 16(2t)^2/2}$$
$$= e^{15t + 64t^2/2}$$

This is the moment generating function for a normal random variable with mean 15 mph and variance 64. Since the moment generating function for a random variable is its fingerprint, we know that the maximum speed at the bridge site is normally distributed with an average speed of 15 mph and a standard deviation of 8 mph.

Theorem 7.3.3 is interesting in its own right, but its primary purpose at this time is to help us derive the next very important theorem. This theorem answers the question posed earlier concerning the distribution of $\bar{X}$. In particular, it assures us that when sampling from a *normal* distribution the random variable $\bar{X}$ will itself be *normally* distributed.

Theorem 7.3.4 (Distribution of $\bar{X}$—normal population). Let $X_1, X_2, \ldots, X_n$ be a random sample of size n from a normal distribution with mean μ and variance σ^2. Then $\bar{X}$ is normally distributed with mean μ and variance σ^2/n.

The derivation of this theorem is not hard. It is outlined in Exercises 38 to 42. We feel that by working through the derivation for yourself you will have a better understanding of the point being made. The other exercises presented are also important. They contain some results that will have major practical consequences later. Be sure to give them all a try!

7.4 INTERVAL ESTIMATION AND THE CENTRAL LIMIT THEOREM

As mentioned previously, point estimation does not give us the ability to report the accuracy of our estimate. To do this, we must turn to the method of interval estimation. The statistics used to extend a point estimate for a parameter θ to an interval of values that should contain the true value of θ vary from parameter to parameter. However, the method for deriving these statistics is basically the same in each case. In this section we illustrate the method by deriving a "confidence interval" for the mean of a normal random variable when its variance is assumed to be known. In later chapters we apply the general technique illustrated here to find confidence intervals for other important parameters.

The term "confidence interval" is a technical term which we now define.

Definition 7.4.1 (Confidence interval). A $100(1-\alpha)\%$ confidence interval for a parameter θ is a random interval $[L_1, L_2]$ such that

$$P[L_1 \le \theta \le L_2] \doteq 1 - \alpha$$

regardless of the value of θ.

One general statement will guide in the construction of most of the confidence intervals presented in this text:

To construct a $100(1-\alpha)\%$ confidence interval for a parameter θ, we shall find a random variable whose expression involves θ and whose probability distribution is known at least approximately.

Confidence Interval on the Mean: Variance Known

To use this guideline to find a $100(1-\alpha)\%$ confidence interval for the mean of a normal random variable whose variance is known, we must find a random variable whose expression involves μ and whose distribution is known. This is easy to do. Note that in Theorem 7.3.4, we showed that under the given conditions the sample mean, $\bar{X}$, is normally distributed with mean μ and variance σ^2/n. This implies that the random variable

$$\frac{\bar{X} - \mu}{\sigma/\sqrt{n}}$$

is *standard normal*. Note that this random variable involves the parameter μ and its distribution is known. We illustrate how this random variable can be used to generate a 95% confidence interval for μ. The technique used can be generalized easily to obtain any desired degree of confidence.

Example 7.4.1. Acute myeloblastic leukemia is among the most deadly of cancers. Past experience indicates that the time in months that a patient survives after initial diagnosis of the disease is normally distributed with a mean of 13 months and a standard deviation of 3 months. A new treatment is being investigated which should prolong the average survival time without affecting variability. Let $X_1, X_2, X_3, \ldots, X_n$ denote a random sample from the distribution of X, the survival time under the new treatment. We are assuming that X is normally distributed with $\sigma^2 = 9$ and μ unknown. We want to find statistics L_1 and L_2 so that $P[L_1 \le \mu \le L_2] \doteq .95$. To do so, consider the partition of the standard normal

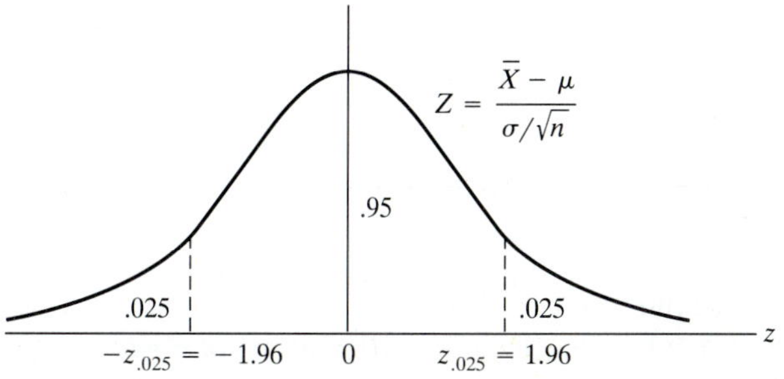

FIGURE 7.1
Partition of Z to obtain a 95% confidence interval for μ.

curve shown in Fig. 7.1. It can be seen that

$$P[-1.96 \le Z \le 1.96] = .95$$

In this case $Z = (\overline{X} - \mu)/(\sigma/\sqrt{n})$, and hence we may conclude that

$$P\left[-1.96 \le \frac{\overline{X} - \mu}{\sigma/\sqrt{n}} \le 1.96\right] = .95$$

To find L_1 and L_2, we algebraically isolate μ in the center of the preceding inequality as follows:

$$P\left[-1.96\sigma/\sqrt{n} \le \overline{X} - \mu \le 1.96\sigma/\sqrt{n}\right] = .95$$

$$P\left[-\overline{X} - 1.96\sigma/\sqrt{n} \le -\mu \le -\overline{X} + 1.96\sigma/\sqrt{n}\right] = .95$$

$$P\left[\overline{X} - 1.96\sigma/\sqrt{n} \le \mu \le \overline{X} + 1.96\sigma/\sqrt{n}\right] = .95$$

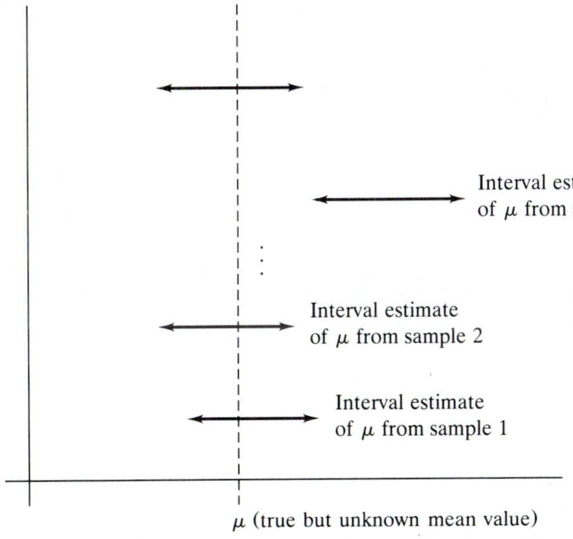

FIGURE 7.2
Of the intervals constructed by using $[L_1, L_2]$, 95% are expected to contain μ, the true but unknown population mean.

From this we see that the lower and upper bounds for a 95% confidence interval are

$$L_1 = \bar{X} - 1.96\sigma/\sqrt{n} \qquad L_2 = \bar{X} + 1.96\sigma/\sqrt{n}$$

These statistics have the property that in repeated sampling from the population, 95% of the numerical intervals generated are expected to contain μ; by chance, 5% will not. This idea is illustrated in Fig. 7.2.

Note that since we are assuming that σ^2 is known, the confidence bounds, $\bar{X} \pm 1.96\sigma/\sqrt{n}$, just derived are *statistics*. Given a particular set of observations on X, their numerical values can be determined easily as demonstrated in Example 7.4.2.

Example 7.4.2. When the experiment of Example 7.4.1 is conducted, these observations on X, the survival time under the new treatment, result:

8.0	13.6	13.2	13.6
12.5	14.2	14.9	14.5
13.4	8.6	11.5	16.0
14.2	19.0	17.9	17.0

Based on these data, $\hat{\mu} = \bar{x} = 13.88$ months. This point estimate is extended to a 95% confidence interval by evaluating the statistics L_1 and L_2. In particular,

$$\begin{aligned} L_1 = \bar{x} - 1.96\sigma/\sqrt{n} &= 13.88 - 1.96(3/\sqrt{16}) \\ &= 13.88 - 1.47 \\ &= 12.41 \text{ months} \\ L_2 = \bar{x} + 1.96\sigma/\sqrt{n} &= 13.88 + 1.47 \\ &= 15.35 \text{ months} \end{aligned}$$

Based on these data, the interval estimate for μ is [12.41, 15.35]. Does the true mean survival time for patients receiving the new treatment really lie between 12.41 and 15.35 months? Unfortunately, there is no way of knowing. The interval [12.41, 15.35] is a 95% confidence interval. This means that the procedure used is expected to trap μ 95% of the time. We hope that the interval obtained from our particular sample does so.

To obtain the general formula for a $100(1 - \alpha)\%$ confidence interval on the mean of a normal random variable whose variance is known, we need only to partition the standard normal curve as shown in Fig. 7.3. The algebraic

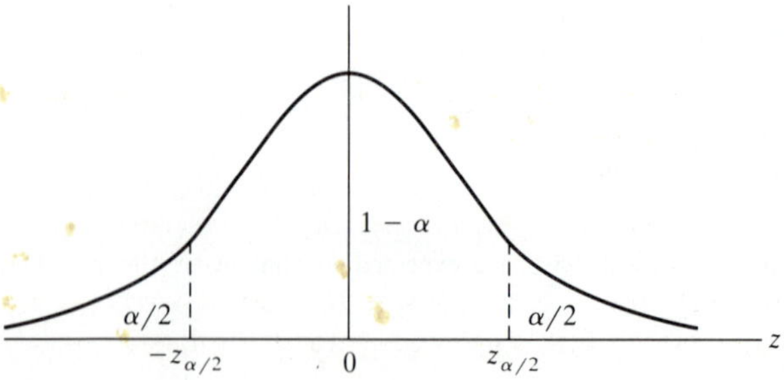

FIGURE 7.3
Partition of Z to obtain a $100(1 - \alpha)\%$ confidence interval for μ.

argument of Example 7.4.1 goes through exactly as presented with the point $z_{.025} = 1.96$ being replaced by $z_{\alpha/2}$. This change results in the general formula given in Theorem 7.4.1.

Theorem 7.4.1 [100(1 − α)% Confidence interval on μ when σ^2 is known]. Let $X_1, X_2, X_3, \ldots, X_n$ be a random sample of size n from a normal distribution with mean μ and variance σ^2. A $100(1-\alpha)\%$ confidence interval on μ is given by

$$\bar{X} \pm z_{\alpha/2}\sigma/\sqrt{n}$$

Let us point out that the preceding confidence interval is very idealistic. It is usable only in settings in which the population standard deviation, σ, is known. In practice, this is seldom the case. In most real life problems both μ and σ^2 must be estimated from available data. When this occurs, the previous confidence interval is *not* appropriate. In Sec. 8.2 we shall show how to overcome this problem. Meanwhile, view this interval as a prototype for confidence intervals in general. It is useful as an aid for understanding how confidence intervals are derived and interpreted.

There are several things to notice concerning the preceding formula. First, every confidence interval on μ is centered at $\bar{x}$, the unbiased point estimate for μ. Second, the length of the confidence interval is dependent on three factors. These are the desired confidence, the amount of variability in X, and the sample size (n). The desired confidence determines the value of the z point used. The higher the confidence desired, the larger this value becomes. When a random variable displays a high degree of variability, it is hard to predict its behavior. Thus the larger σ becomes, the longer the confidence interval must become. Sample size works in reverse. With all other factors held constant, as n increases, the length of the confidence interval decreases. We can say that the length of a confidence interval on μ is directly proportional to σ and to the confidence desired and inversely proportional to the sample size.

Central Limit Theorem

There is one further point to be made. Theorem 7.4.1 does require that the base variable X be normal. If this condition is not satisfied, then the confidence bounds given can be used as long as the sample is not too small. Empirical studies have shown that for samples as small as 25, the above bounds are usually satisfactory even though approximate. This is due to a remarkable theorem, first formulated in the early nineteenth century by Laplace and Gauss. This theorem, known as the *Central Limit Theorem*, gives the distribution of $\bar{X}$ when sampling from a distribution that is not necessarily normal.

Theorem 7.4.2 (Central Limit Theorem). Let $X_1, X_2, \ldots, X_n$ be a random sample of size n from a distribution with mean μ and variance σ^2. Then for large n, $\bar{X}$ is approximately normal with mean μ and variance σ^2/n. Furthermore, for large n, the random variable $(\bar{X} - \mu)/(\sigma/\sqrt{n})$ is approximately standard normal.

Please note the differences between the Central Limit Theorem and Theorem 7.3.4. The former does not require that sampling be from a normal distribution, whereas normality is assumed in the latter; the former claims that $\bar{X}$ will be approximately normally distributed for large sample sizes, whereas the latter claims that $\bar{X}$ will be exactly normally distributed regardless of the sample size involved.

The Central Limit Theorem is important to us for two reasons. First it allows us to make inferences on the mean of a distribution based on relatively large samples without having to be overly concerned as to whether or not we are sampling from a normal distribution. Second, it allows us to justify analytically the normal approximations to discrete distributions presented in Chap. 4. Recall that in that chapter we considered the normal approximation to the binomial distribution (Theorem 4.6.1). At that time the approximation was justified graphically. We can now give a more rigorous argument to justify this approximation procedure.

Example 7.4.3. Let $X_1, X_2, \ldots, X_n$ be a random sample drawn from a point binomial distribution (see Exercise 45, Chap. 3). Recall that each of these random variables is binomial with parameters 1 and p. Each has mean p, variance $p(1-p)$, and moment generating function of the form $q + pe^t$. Let $X = \sum_{i=1}^{n} X_i$. Since $X_1, X_2, \ldots, X_n$ are independent, the moment generating function for X is given by

$$m_X(t) = \prod_{i=1}^{n} (q + pe^t) = (q + pe^t)^n$$

This is the moment generating function for a binomial random variable with parameters n and p. By the Central Limit Theorem $\bar{X} = (\sum_{i=1}^{n} X_i)/n = X/n$ is approximately normal with mean p and variance $p(1-p)/n$. Now consider the binomial random variable $n(X/n) = X$. Since X is a linear function of the approximately normal random variable X/n, we can apply Exercise 41 with $a_1 = n$ and $a_i = 0$, $i \neq 1$, to conclude that X is approximately normal with mean np and variance $[n^2 p(1-p)]/n = np(1-p)$.

Exercises 49, 50, 55, 56, 58, 61, and 62 will give you practice in the application of the Central Limit Theorem.

CHAPTER SUMMARY

In this chapter we considered the ideas of point and interval estimation. We introduced three types of point estimators. These are unbiased estimators, method of moments estimators, and maximum likelihood estimators. Unbiased estimators are estimators whose mean value is equal to the parameter being estimated. We showed that $\bar{X}$ is unbiased for μ, that S^2 is unbiased for σ^2, but that S is not unbiased for σ. Method of moments estimators are derived by noting that the parameters that characterize a distribution are often functions of the kth moments of the distribution. Maximum likelihood estimators are found by choosing the value of the parameter θ that maximizes the likelihood function. In this way in some sense we pick out of all possible values of θ the one that is most likely to have produced the observed data.

In order to develop the idea of interval estimation, we introduced some theorems that help us to determine the distribution of a random variable. In particular, we noted that the moment generating function for a random variable is its "fingerprint." To determine its distribution we look at its moment generating function. This technique was used to verify the standardization theorem used in earlier chapters. It was also used to show that a linear function of independent normal random variables is normal, that a sum of independent chi-squared random variables is chi-squared and that $\bar{X}$ is normally distributed when sampling from a normal distribution. (See Exercise 42.)

We introduced the general concept of a $100(1 - \alpha)\%$ confidence interval on a parameter θ. This is a random interval, an interval of the form $[L_1, L_2]$, where L_1 and L_2 are statistics with the property that "a priori" θ will be trapped between L_1 and L_2 with probability $1 - \alpha$. We used information just developed on the distribution of $\bar{X}$ to develop specific formulas for constructing a $100(1 - \alpha)\%$ confidence interval on the mean of a normal distribution. Finally, we considered the Central Limit Theorem. This theorem concerns the approximate distribution of $\bar{X}$ when sampling from a nonnormal distribution. It allows us to make inferences on the mean of any distribution when relatively large samples are available. It also allows us to justify some of the approximation techniques presented earlier in the text.

We introduced and defined important terms that you should know. These are:

Point estimator	Point estimate
Unbiased	Weighted mean
kth moments	Likelihood function
Confidence interval or interval estimator	Interval estimate
Methods of moments estimator	Sample standard error
Standard error of the mean	Maximum likelihood estimator
Central Limit Theorem	

EXERCISES

Section 7.1

1. Let $X_1, X_2, X_3, \ldots, X_{20}$ be a random sample from a distribution with mean 8 and variance 5. Find the mean and variance of $\bar{X}$.

2. Let $X_1, X_2, X_3, \ldots, X_{15}$ be a random sample from a Poisson distribution with parameter λs. Give an unbiased estimator for this parameter.

3. Let X denote the number of paint defects found in a square yard section of a car body painted by a robot. These data are obtained:

8	5	0	10
0	3	1	12
2	7	9	6

Assume that X has a Poisson distribution with parameter λs.

(*a*) Find an unbiased estimate for λs.

(*b*) Find an unbiased estimate for the average number of flaws per square yard.

(*c*) Find an unbiased estimate for the average number of flaws per square foot.

4. An interactive computer system is available at a large installation. Let X denote the number of requests for this system received per hour. Assume that X has a Poisson distribution with parameter λs. These data are obtained:

25	20	20
30	24	15
10	23	4

(*a*) Find an unbiased estimate for λs.

(*b*) Find an unbiased estimate for the average number of requests received per hour.

(*c*) Find an unbiased estimate for the average number of requests received per quarter hour.

5. Let X_1, X_2, X_3, X_4, X_5 be a random sample from a binomial distribution with $n = 10$ and p unknown.

(*a*) Show that $\bar{X}/10$ is an unbiased estimator for p.

(*b*) Estimate p based on these data: 3, 4, 4, 5, 6.

6. An experiment is conducted to study the effect of a power surge on data stored in a digital computer. A "word" is a sequence of 8 bits. Each bit is either "on" (activated) or "off" (not activated) at any given time. Twenty 8-bit words are stored and a power surge is induced. Let X denote the number of bit reversals that result per word. Assume that X is binomially distributed with $n = 8$ and p, the probability of a bit reversal, unknown. These data result:

1	0	0	0
0	0	1	1
0	1	2	1
1	0	1	0
2	2	3	0

(*a*) Find an unbiased estimate for p.

(*b*) Based on the estimate for p just found, approximate the probability that in another 8-bit word a similar power surge will result in no bit reversals.

(*c*) A data line utilizes 64 bits. Based on the estimate for p just found, approximate the probability that at most one bit reversal will occur.

7. Stress tests are conducted on fiberglass rods used in communications networks. The random variable studied is X, the distance in inches from the anchored end of the rod to the crack location when the rod is subjected to extreme stress. Assume that

X is uniformly distributed over the interval $(0, b)$. These data are obtained on 10 test rods:

10	7	11	12	8
8	9	10	9	13

(*a*) Find an unbiased estimate for the average distance from the anchored end of the rod to the crack.
(*b*) Find an unbiased estimate for the variance of X.
(*c*) Find an unbiased estimate for b.
(*d*) Find an estimate for σ, the standard deviation of X. Is this estimate unbiased?

8. Note that S is a statistic, and unless X is constant, its value will vary from sample to sample. Therefore $\operatorname{Var} S > 0$. To show that S is not unbiased for σ, use proof by contradiction. That is, assume that $E[S] = \sigma$ and obtain a contradiction. *Hint:* Use Theorem 3.3.2.

9. (*Weighted means.*) Assume that one has k independent random samples of sizes $n_1, n_2, n_3, \ldots, n_k$ from the same distribution. These samples generate k unbiased estimators for the mean, namely, $\bar{X}_1, \bar{X}_2, \bar{X}_3, \ldots, \bar{X}_k$.
(*a*) Show that the arithmetic average of these estimators, $(\bar{X}_1 + \bar{X}_2 + \bar{X}_3 + \cdots + \bar{X}_k)/k$, is also unbiased for μ.
(*b*) Certain mineral elements required by plants are classed as macronutrients. Macronutrients are measured in terms of their percentage of the dry weight of the plant. Proportions of each element vary in different species and in the same species grown under differing conditions. One macronutrient is sulfur. In a study of winter cress, a member of the mustard family, these data, based on three independent random samples, are obtained:

$$\bar{x}_1 = .8 \qquad \bar{x}_2 = .95 \qquad \bar{x}_3 = .7$$

$$n_1 = 9 \qquad n_2 = 3 \qquad n_3 = 200$$

Use the result of part (*a*) to obtain an unbiased estimate for μ, the mean proportion of sulfur by dry weight in winter cress. By averaging the three values .8, .95, and .7 to obtain the estimate for μ, each sample is being given the equal importance or "weight." Does this seem reasonable in this problem? Explain.
(*c*) To take sample sizes into account, a "weighted" mean is used. This estimator, $\hat{\mu}_W$, is given by

$$\hat{\mu}_W = \frac{n_1\bar{X}_1 + n_2\bar{X}_2 + \cdots + n_k\bar{X}_k}{n_1 + n_2 + \cdots + n_k}$$

Show that $\hat{\mu}_W$ is an unbiased estimator for μ.
(*d*) Use the data of part (*b*) to find the weighted estimate for the mean proportion of sulfur by dry weight in winter cress. Compare your answer to the estimate found in part (*b*).

10. Let X denote the number of heads obtained when a fair coin is tossed 4 times.
(*a*) What is $E[X]$ and $\operatorname{Var} X$?
(*b*) Perform the experiment of tossing a fair coin 4 times and recording the number of heads obtained 10 times. You thus obtain a random sample of size 10 from a binomial distribution with $n = 4$ and $p = 1/2$.

(*c*) Based on your 10 observations, estimate the mean and variance of X. Compare your answers to those of your classmates. Do the observed values of $\bar{X}$ fluctuate about the theoretical mean of 2? Do the observed values of S^2 fluctuate about the theoretical variance of 1?

(*d*) Average the values of $\bar{X}$ that you have available. Is the average value close to 2? Average the values of S^2 that you have available. Is the average value of S^2 close to 1?

11. Let X denote the number of heads obtained when a fair coin is tossed 4 times. Perform this experiment 3 times and record the value of X for each set of four tosses. In this way you obtain a single sample of size 3 from a binomial distribution with $n = 4$ and $p = 1/2$.

(*a*) Find the numerical value of $\bar{X}$ for your sample.

(*b*) Repeat the experiment 9 more times, recording the value of $\bar{X}$ each time.

(*c*) What is $E[\bar{X}]$? Average your 10 values of $\bar{X}$. Is the average value close to the theoretical mean of 2?

(*d*) What is Var $\bar{X}$? Find the value of S^2 for the 10 observations on $\bar{X}$. Does this value lie close to the theoretical value of $1/3$?

12. Consider the experiment of rolling a pair of fair dice until a sum of 7 is obtained. Let X denote the number of trials needed to obtain a sum of 7.

(*a*) Notice that X is discrete. What is the distribution of X?

(*b*) What is the theoretical average value of X? That is, what is μ?

(*c*) What is the theoretical variance of X? That is, what is σ^2?

(*d*) Perform the experiment described 25 times and thus obtain a sample of size $n = 25$ observations on X. Plot a stem-and-leaf diagram for your data. Does the distribution appear to be symmetric? Use your data to obtain unbiased estimates for μ and σ^2. Compare your answers to the true values of these parameters found in parts (*b*) and (*c*), respectively.

(*e*) Consider the random variable $\bar{X}$, the average number of trials needed to roll a sum of 7 based on 25 trials. What is $E[\bar{X}]$? What is Var $\bar{X}$?

(*f*) Pool the class observations on $\bar{X}$. Plot these values on a number line. Do they fluctuate about μ as expected? Find the average value of these observed $\bar{X}$ values. Is it close to μ as expected? Find the variance of the $\bar{X}$ values. Is this sample variance close in value to $\sigma^2/25$ as expected?

13. Ozone levels around Los Angeles have been measured as high as 220 parts per billion (ppb). Concentrations this high can cause the eyes to burn and are a hazard to both plant and animal life. These data are obtained on the ozone level in a forested area near Seattle, Washington (based on information found in "Twigs," *American Forests*, April 1990, p. 71):

160	176	160	180	167	164
165	163	162	168	173	179
170	196	185	163	162	163
172	162	167	161	169	178
161					

(*a*) Construct a double stem-and-leaf diagram for these data. Do these data appear to be skewed? If so, in which direction?

(*b*) Construct a boxplot for these data and identify the potential outlier that is flagged by this technique. Assume that the point in question is a legitimate data point. In this case, do you believe that it is truly an outlier or probably simply a natural consequence of the distribution involved? Explain.

(*c*) Use these data to estimate the mean and variance of the ozone level in this area.

14. In this exercise you will show that the most logical estimator for σ^2, namely, $\sum_{i=1}^{n}(X_i - \bar{X})^2/n$ is a biased estimator for σ^2 and tends to underestimate the true variance. Let $X_1, X_2, \ldots, X_n$ be a random sample of size n from a distribution with mean μ and variance σ^2.

(*a*) Show that $\sum_{i=1}^{n}(X_i - \bar{X})^2/n = (n-1)S^2/n$.

(*b*) Verify that $E[\sum_{i=1}^{n}(X_i - \bar{X})^2/n] \neq \sigma^2$, thus showing that this estimator is not an unbiased estimator for σ^2. Argue that it tends to underestimate σ^2.

(*c*) Consider the theoretical setting described in Exercise 12. Based on samples of size $n = 25$, what is $E[S^2]$? What is $E[\sum_{i=1}^{n}(X_i - \bar{X})^2/25]$?

Section 7.2

15. Suppose that when the experiment described in Example 7.2.1 is conducted, these data result:

$$x_1 = 13 \qquad x_3 = 15 \qquad x_5 = 17$$

$$x_2 = 12 \qquad x_4 = 10$$

Use the method of moments to estimate p, the probability that a seedling will survive the first winter.

16. Let $X_1, X_2, \ldots, X_m$ be a random sample of size m from a binomial distribution with parameters n, assumed to be known, and p. Show that the method of moments estimator for p is $\hat{p} = \bar{X}/n$.

17. Let $X_1, X_2, \ldots, X_n$ be a random sample from a Poisson distribution with parameter λs. Find the method of moments estimator for λs. Find the method of moments estimator for λ, the parameter underlying the Poisson process under observation.

18. In the study of traffic flow at an intersection a Poisson process with parameter λ is assumed. The basic unit of time assumed is 1 minute. These data are obtained on X, the number of vehicles arriving at the intersection during a 2-minute period:

2	5	1	3	2
3	0	8	2	5

Use these data to estimate λs, the average number of vehicles arriving during a two-minute period, and λ the average number arriving per minute. (Use the results of Exercise 17.)

19. Use the information obtained in Example 7.2.2 to find an estimator for σ^2, the variance of a gamma random variable. Is the estimator obtained unbiased for σ^2? *Hint:* Express M_1 and M_2 as arithmetic averages, and compare your result to that of Theorem 6.3.1.

20. An acid solution made by mixing a powder compound with water is used to etch aluminum. The pH of the solution, X, will vary due to slight variations in the amount of water used, the potency of the dry compound, and the pH of the water itself. Assume that X is gamma distributed with α and β unknown. From these data, estimate α, β, μ, and σ^2 using the method of moments:

1.2	2.0	1.6	1.8	1.1
2.5	2.1	2.6	2.2	1.7
1.5	1.7	2.0	3.0	1.8

21. Assume that the data of Exercise 13 are drawn from an exponential distribution with parameter β. Find the method of moments estimate for β. Use this to find the method of moments estimate for σ^2. Is this the same estimate as that obtained in Exercise 13(*c*)?

22. Assume that the burn time of a sparkler as described in Exercise 32 of Chap. 6 follows a gamma distribution with parameters α and β. Use the data of Exercise 32 to
(*a*) find an unbiased estimate for the average burn time.
(*b*) find the method of moments estimate for the average burn time.
(*c*) find an unbiased estimate for the variance in burn time.
(*d*) find the method of moments estimate for the variance in burn time.

23. Find the method of moments estimator for the parameter p of a geometric distribution.

24. Use the results of Exercise 23 and your data from Exercise 12(*d*) to find the method of moments estimate for p, the probability of rolling a sum of 7 on a single roll of a pair of fair dice. Compare your estimate to the true probability of $1/6$.

25. Using the method of moments estimator for p found in Exercise 23, find an estimator for σ^2 for the geometric distribution. Use this estimator to estimate σ^2 for your data from Exercise 12(*d*). Does this estimate differ from that found in Exercise 12(*d*)? If so, which estimate is closest to the true value of σ^2?

26. Let X be normal with mean μ and variance σ^2, both of which are unknown. Find the method of moments estimators for these parameters. Are the estimators obtained unbiased for their respective parameters? Explain.

27. Carbon dioxide is an odorless, colorless gas which constitutes about .035% by volume of the atmosphere. It affects the heat balance by acting as a one-way screen. It lets in the sun's heat to warm the oceans and the land but blocks some of the infrared heat that is radiated from the earth. This reflected heat is absorbed into the lower atmosphere, producing a greenhouse effect which causes the earth's surface to become warmer than it would be otherwise. Systematic measurements of CO_2 began in 1957 with Charles D. Keeling monitoring at Mauna Loa in Hawaii. (*McGraw-Hill Yearbook of Science and Technology*, 1983, p. 68.)
(*a*) Assume that these CO_2 readings (in ppm) are obtained:

319	338	337	339	328
325	340	331	341	336
330	330	321	327	337
320	343	350	322	334
326	349	341	338	332
339	335	338	333	334

Construct a stem-and-leaf diagram for these data using 31, 32, 32, 33, 33, 34, 34, 35 as stems. Graph leaves 0–4 on the first of each repeated stem and leaves 5–9 on the other. Is it reasonable to assume that the CO_2 level in the atmosphere is normally distributed? Explain.

(*b*) Estimate μ and σ^2 using the method of moments estimators.

(*c*) Find an unbiased estimate for σ^2.

28. Based on the data of Exercise 18, what is the maximum likelihood estimate for λ, the average number of vehicles arriving at an intersection per minute?

29. Based on the data of Exercise 27, what are the maximum likelihood estimates for the mean and variance of the atmospheric CO_2 level?

30. Let $X_1, X_2, X_3, \ldots, X_m$ be a random sample of size m from a binomial distribution with parameters n, assumed to be known, and p. Find the maximum likelihood estimator for p. Does it differ from the method of moments estimator found in Exercise 16?

31. Let W be an exponential random variable with parameter β unknown. Find the maximum likelihood estimator for β based on a sample of size n. Does it differ from the method of moments estimator?

32. A computer center employs consultants to answer users' questions. The center is open from 9 a.m. to 5 p.m. each weekday. Assume that calls arriving at the center constitute a Poisson process with unknown parameter λ calls per hour. To estimate λ, these observations were obtained on X, the number of calls arriving per hour:

8	6	12	15	12
4	9	7	20	10

(*a*) Find the maximum likelihood estimate for λ.

(*b*) Estimate the average time of arrival of the first call of the day. *Hint:* Consider Theorem 4.3.3.

33. A study of the noise level on takeoff of jet airplanes at a particular airport is studied. The random variable is X, the noise level in decibels of the plane as it passes over the first residential area adjacent to the airport. This random variable is assumed to have a gamma distribution with $\alpha = 2$ and β unknown.

(*a*) Find the maximum likelihood estimate for β based on a sample of size n.

(*b*) Use $\hat{\beta}$ to find an estimate for the mean value of X. Is this estimator unbiased for μ?

(*c*) Find the maximum likelihood estimate for β based on these data:

55	65	60	73	80
64	57	75	62	86
69	100	70	82	65
72	67	61	95	52

(*d*) Estimate the average decibel reading of these jets.

34. Computer terminals have a battery pack that maintains the configuration of the terminal. These packs must be replaced occasionally. Let X denote the life span in years of such a battery. Assume that X is exponentially distributed with unknown parameter β. Find the maximum likelihood estimate for β based on these data:

1.7	4.0	1.9	2.0	1.7
2.1	2.7	4.2	1.8	2.2
3.1	1.5	2.4	6.2	7.0
3.6	1.4	5.0	3.8	1.6

35. To estimate the proportion of defective microprocessor chips being produced by a particular maker, samples of five chips are selected at 10 randomly selected times during the day. These chips are inspected and X, the number of defective chips in each batch of size 5, is recorded. Assume that X is binomially distributed with $n = 5$ and p unknown. Use these data to find the maximum likelihood estimate for p:

1	0	1	2	0
0	0	0	1	0

36. A new material is being tested for possible use in the brake shoes of automobiles. These shoes are expected to last for at least 75,000 miles. Fifteen sets of four of these experimental shoes are subjected to accelerated life testing. The random variable X, the number of shoes in each group of four that fail early, is assumed to be binomially distributed with $n = 4$ and p unknown. Find the maximum likelihood estimate for p based on these data:

1	0	1	0	2	1	0	1
0	1	0	0	0	1	0	

If an early failure rate in excess of 10% is unacceptable from a business point of view, would you have some doubts concerning the use of this new material? Explain.

Section 7.3

37. In each part the moment generating function for a random variable X is given. Identify the family to which the random variable belongs and give the numerical values of pertinent distribution parameters.
(*a*) $m_X(t) = e^{2t+9t^2/2}$
(*b*) $m_X(t) = e^{8t^2}$
(*c*) $m_X(t) = .25e^t/(1 - .75e^t)$
(*d*) $m_X(t) = (.5 + .5e^t)^5$
(*e*) $m_X(t) = e^{6(e^t - 1)}$
(*f*) $m_X(t) = (1 - 3t)^{-5}$
(*g*) $m_X(t) = (1 - 2t)^{-8}$
(*h*) $m_X(t) = (1 - .5t)^{-1}$

38. (*a*) Let X be a random variable with moment generating function $m_X(t)$. Let $Y = \alpha + \beta X$. Show that $m_Y(t) = e^{\alpha t} m_X(\beta t)$. *Hint:* $m_Y(t) = E[e^{tY}] = E[e^{(\alpha+\beta X)t}]$.
(*b*) Let X be a normal random variable with mean 10 and variance 4. Find the moment generating function for the random variable $Y = 8 + 3X$. What is the distribution of Y?

39. Let $X_1, X_2, X_3, \ldots, X_n$ be a collection of independent random variables with moment generating functions $m_{X_i}(t)$ ($i = 1, 2, 3, \ldots, n$, respectively). Let $a_0, a_1,$

$a_2, \ldots, a_n$ be real numbers, and let

$$Y = a_0 + a_1X_1 + a_2X_2 + a_3X_3 + \cdots + a_nX_n$$

Show that the moment generating function for Y is given by

$$m_Y(t) = e^{a_0t} \prod_{i=1}^{n} m_{X_i}(a_it)$$

Note that this extends the result of Exercise 38(a) to more than one variable.

40. Let X_1 and X_2 be independent normal random variables with means 2 and 5 and variances 9 and 1, respectively. Let $Y = 3X_1 + 6X_2 - 8$. Use Exercise 39 to find the moment generating function for Y. What is the distribution of Y?

41. In this exercise you will prove that any linear combination of independent normally distributed random variables is also normally distributed. Let $X_1, X_2, X_3, \ldots, X_n$ be independent normal random variables with means μ_i and σ_i^2 ($i = 1, 2, 3, \ldots, n$, respectively). Let $a_0, a_1, a_2, \ldots, a_n$ be real numbers, and let

$$Y = a_0 + a_1X_1 + a_2X_2 + \cdots + a_nX_n$$

Use Exercise 39 to show that Y is normal with mean $\mu = a_0 + \sum_{i=1}^{n} a_i\mu_i$ and variance $\sigma^2 = \sum_{i=1}^{n} a_i^2\sigma_i^2$.

42. In this exercise you will prove that when sampling from a normal distribution, $\bar{X}$ is normally distributed. Let $X_1, X_2, X_3, \ldots, X_n$ be a random sample from a normal distribution with mean μ and variance σ^2. Use Exercise 41 to show that $\bar{X}$ is normal with mean μ and variance σ^2/n.

43. Let X_1 and X_2 be independent chi-squared random variables with 5 and 10 degrees of freedom, respectively. Show that $X_1 + X_2$ is a chi-squared random variable with 15 degrees of freedom.

44. In this exercise you will prove that the sum of a collection of independent chi-squared random variables also has a chi-squared distribution. Let $X_1, X_2, X_3, \ldots, X_n$ be independent chi-squared random variables with $\gamma_1, \gamma_2, \gamma_3, \ldots, \gamma_n$ degrees of freedom, respectively. Let

$$Y = X_1 + X_2 + X_3 + \cdots + X_n$$

Show that Y is a chi-squared random variable with γ degrees of freedom where $\gamma = \sum_{i=1}^{n} \gamma_i$.

45. It can be shown that the square of a standard normal random variable has a chi-squared distribution with $\gamma = 1$. That is, the random variable Z^2 follows a chi-squared distribution with 1 degree of freedom. Let $X_1, X_2, X_3, \ldots, X_n$ be a random sample from a normal distribution with mean μ and variance σ^2. Use Exercise 44 to show that

$$\sum_{i=1}^{n} \frac{(X_i - \mu)^2}{\sigma^2}$$

has a chi-squared distribution with n degrees of freedom.

46. Let X denote the time required to do a computation using an algorithm written in Fortran, and let Y denote the time required to do the same calculation using an algorithm written in Pascal. Assume that X is normally distributed with mean 10 seconds and standard deviation 3 seconds that Y is normally distributed with mean 9 seconds and standard deviation 4 seconds.

(*a*) What is the distribution of the random variable $X - Y$?

(*b*) Find the probability that a given calculation will run faster using Fortran than when using Pascal.

Section 7.4

47. As heat is added to a material its temperature rises. The heat capacity is a quantitative statement of the increase in temperature for a specified addition of heat. These data are obtained on X, the measured heat capacity of liquid ethylene glycol at constant pressure and 80°C. Measurements are in calories per gram degree Celsius.

.645	.654	.640	.627	.626
.649	.629	.631	.643	.633
.646	.630	.634	.631	.651
.659	.638	.645	.655	.624
.658	.658	.658	.647	.665

Past experience indicates that $\sigma = .01$.

(*a*) Evaluate $\bar{X}$ for these data thereby obtaining an unbiased point estimate for μ.

(*b*) Assume that X is normally distributed. Find a 95% confidence interval for μ.

(*c*) Would you expect a 90% confidence interval for μ based on these data to be longer or shorter than the interval of part (*b*)? Explain. Verify your answer by finding a 90% confidence interval on μ. *Hint:* Begin by sketching a curve similar to that shown in Fig. 7.3 with $1 - \alpha = .90$ and $\alpha/2 = .05$.

(*d*) Would you expect a 99% confidence interval for μ based on these data to be longer or shorter than the interval of part (*b*)? Explain. Verify your answer by finding a 99% confidence interval for μ.

48. The late manifestation of an injury following exposure to a sufficient dose of radiation is common. These data are obtained on the variable X, the time in days that elapses between the exposure to radiation and the appearance of peak erythema (skin redness):

16	12	14	16	13	9	15	7
20	19	11	14	9	13	11	3
8	21	16	16	12	16	14	20
7	14	18	14	18	13	11	16
18	16	11	13	14	16	15	15

(*a*) Even though the time at which the peak redness appears is recorded to the nearest day, time is actually a continuous random variable. Sketch a stem-and-leaf diagram for these data. Does the diagram lend support to the assumption that X is normally distributed?

(*b*) Evaluate $\bar{X}$ for these data.

(*c*) Assume that $\sigma = 4$ and find a 95% confidence interval on the mean time to the appearance of peak redness. Would you be surprised to hear a claim that $\mu = 17$ days? Explain, based on the confidence interval found in part (*b*).

49. When fission occurs, many of the nuclear fragments formed have too many neutrons for stability. Some of these neutrons are expelled almost instantaneously. These observations are obtained on X, the number of neutrons released during fission of plutonium-239:

3	2	2	2	2	3	3	3
3	3	3	3	4	3	2	3
3	2	3	3	3	3	3	1
3	3	3	3	3	3	3	3
3	3	2	3	3	3	3	3

(*a*) Is X normally distributed? Explain.

(*b*) Estimate the mean number of neutrons expelled during fission of plutonium-239.

(*c*) Assume that $\sigma = .5$. Find a 99% confidence interval on μ. What theorem justifies the procedure you used to construct this interval?

(*d*) The reported value of μ is 3.0. Do these data refute this value? Explain.

50. (*Central Limit Theorem.*) Consider an infinite population with 25% of the elements having the value 1, 25% the value 2, 25% the value 3, and 25% the value 4. If X is the value of a randomly selected item, then X is a discrete random variable whose possible values are 1, 2, 3, and 4.

(*a*) Find the population mean μ and population variance σ^2 for the random variable X.

(*b*) List all 16 possible distinguishable samples of size 2, and for each calculate the value of the sample mean. Represent the value of the sample mean $\bar{X}$ using a probability histogram (use one bar for each of the possible values for $\bar{X}$). Note that although this is a very small sample, the distribution of $\bar{X}$ does not look like the population distribution and has the general shape of the normal distribution.

(*c*) Calculate the mean and variance of the distribution of $\bar{X}$ and show that, as expected, they are equal to μ and σ^2/n, respectively.

REVIEW EXERCISES

51. Consider the random variable X with density given by

$$f(x) = (1 + \theta)x^{\theta} \qquad 0 < x < 1 \qquad \theta > -1$$

(*a*) Show that $\int_0^1 f(x)\,dx = 1$ regardless of the specific value chosen for θ.

(*b*) Find $E[X]$.

(*c*) Find the method of moments estimator for θ.

(*d*) Find the method of moments estimate for θ based on these data:

.5 .3 .1 .1 .2

(*e*) Find the maximum likelihood estimator for θ.

(*f*) Find the maximum likelihood estimate for θ based on the data of part (*d*). Does this value agree with the method of moments estimate?

52. Consider the random variable X with density given by

$$f(x) = 1/\theta \qquad 0 < x < \theta$$

(*a*) Find $E[X]$.

(*b*) Find the method of moments estimator for θ. Is this estimator unbiased for θ?
(*c*) Find the method of moments estimate for θ based on these data:

1	.5	1.4	2.0	.25

53. Studies have shown that the random variable X, the processing time required to do a multiplication on a new 3-D computer, is normally distributed with mean μ and standard deviation 2 microseconds. A random sample of 16 observations is to be taken.
(*a*) What is the distribution of $\bar{X}$?
(*b*) These data are obtained:

42.65	45.15	39.32	44.44
41.63	41.54	41.59	45.68
46.50	41.35	44.37	40.27
43.87	43.79	43.28	40.70

Based on these data, find an unbiased estimate for μ.
(*c*) Find a 95% confidence interval for μ. Would you be surprised to read that the average time required to process a multiplication on this system is 42.2 microseconds? Explain, based on the confidence interval.

54. Let X denote the unit price of a 3.5-inch floppy diskette. These observations are obtained from a random sample of 10 suppliers:

\$3.83	3.54	3.44	3.89	3.65
3.70	3.59	3.37	4.04	3.93

(*a*) Find an unbiased estimate for the mean price of these diskettes.
(*b*) Find an unbiased estimate for the variance in the price of these diskettes.
(*c*) Find the sample standard deviation. Is this an unbiased estimate for σ?
(*d*) Assume that X is normally distributed. Find the maximum likelihood estimate for σ^2. Does this agree with your answer to (*b*)?

55. (*Central Limit Theorem.*) In an attempt to approximate the proportion p of improperly sealed packages produced on an assembly line, a random sample of 100 packages is selected and inspected. Let

$$X_i = \begin{cases} 1 & \text{if the } i\text{th package selected is improperly sealed} \\ 0 & \text{otherwise} \end{cases}$$

(*a*) What is the distribution of X_i?
(*b*) Based on the Central Limit Theorem, what is the approximate distribution of $\bar{X}$?
(*c*) When the experiment is conducted, we observe five improperly sealed packages. Find a point estimate for the proportion of improperly sealed packages being produced on this assembly line.

56. (*Central Limit Theorem.*) In a study of the size of various computer systems the random variable X, the number of files stored, is considered. Past experience indicates that $\sigma = 5$. These data are obtained:

7	8	4	5	9	9
4	12	8	1	8	7
3	13	2	1	17	7
12	5	6	2	1	13
14	10	2	4	9	11
3	5	12	6	10	7

(*a*) Find an unbiased estimate for μ, the mean number of files per system.
(*b*) Based on the Central Limit Theorem, what is the approximate distribution of $\bar{X}$?
(*c*) Find an approximate 98% confidence interval on μ.
(*d*) In describing the size of such systems, an executive states that the average number of files exceeds 10. Does this statement surprise you? Explain.

57. Let X denote the time expended by a terminal user in a computing session (time from log on to log off). Assume that X is normally distributed with $\mu_X = 15$ minutes and $\sigma_X = 4$ minutes. Let Y denote the time required to access the system. Assume that Y is normally distributed with mean 1.5 minutes and $\sigma_Y = .5$ minutes. Assume that X and Y are independent.
(*a*) Find $m_X(t)$ and $m_Y(t)$.
(*b*) The random variable $T = X + Y$ denotes the total time required by the user to run his job. Find the moment generating function for T.
(*c*) What is the distribution of T?
(*d*) Find the probability that the total time required exceeds 20 minutes.

58. Let $X_1, X_2, \ldots, X_{100}$ be a random sample of size 100 from a gamma distribution with $\alpha = 5$ and $\beta = 3$.
(*a*) Find the moment generating function for $Y = \sum_{i=1}^{100} X_i$.
(b) What is the distribution of Y?
(*c*) Find the moment generating function for $\bar{X} = Y/n$.
(*d*) What is the distribution of $\bar{X}$?
(*e*) Use the Central Limit Theorem to approximate the probability that $\bar{X}$ is at most 14.

59. Consider the random variable X with density given by

$$f(x) = (1/\theta^2)xe^{-x/\theta} \qquad x > 0 \qquad \theta > 0$$

(*a*) What is the distribution of X?
(*b*) What is $E[X]$?
(*c*) Find the method of moments estimator for θ.
(*d*) Find the maximum likelihood estimator for θ based on a random sample of size n. Does this estimator differ from that found in part (*c*)?
(*e*) Estimate θ based on these data:

3	5	2	3	4
1	4	3	3	3

(*f*) Are the estimators found in parts (*c*) and (*d*) unbiased estimators for θ?

60. Let X be normally distributed with mean 2 and variance 25.
(*a*) What is the distribution of the random variable $(X - 2)/5$?
(*b*) What is the distribution of the random variable $[(X - 2)/5]^2$?
(*c*) Let $X_1, X_2, X_3, \ldots, X_{10}$ represent a random sample from the distribution of X. What is the distribution of the random variable

$$\sum_{i=1}^{10} \left(\frac{X_i - 2}{5} \right)^2$$

61. (*Central Limit Theorem.*) In this problem you will use the Central Limit Theorem to justify the normal approximation to the Poisson distribution given earlier. That is, you will show that a Poisson random variable X with parameter λs can be approximated using a normal random variable with mean and variance λs. To do so, let $Y_1, Y_2, Y_3, \ldots, Y_n$ be a random sample of size n from a Poisson distribution with parameter $\lambda s/n$.
(*a*) Use moment generating function techniques to show that

$$X = \sum_{i=1}^{n} Yi$$

has a Poisson distribution with parameter λs.
(*b*) Use the Central Limit Theorem to find the approximate distribution of $\bar{Y}$.
(*c*) Note that $n\bar{Y} = X$. Use this observation to argue that X is approximately normally distributed with mean λs and variance λs.

62. (*Central Limit Theorem.*) Consider the experiment of tossing a fair die once. Let X denote the number that occurs. Theoretically, X follows a discrete uniform distribution.
(*a*) Find the theoretical density, mean, and variance for X.
(*b*) Now consider an experiment in which the die is tossed 20 times and the results averaged. By the Central Limit Theorem, what is the theoretical mean and variance for the random variable $\bar{X}$?
(*c*) Perform the experiment of part (*b*) 25 times and record the value of $\bar{X}$ each time. (You will toss the die 500 times and obtain a data set that consists of 25 averages.) What shape should the stem-and-leaf diagram for these data assume? Explain. Construct a stem-and-leaf diagram for your data. Did the diagram take the shape that you expected?
(*d*) Approximately what value would you expect to obtain if you averaged the data of part (*c*)? Average your 25 observations on $\bar{X}$. Did the result come out as expected?
(*e*) Approximately what value would you expect to obtain if you found the sample variance for the data of part (*c*)? Explain. Find s^2 for your 25 observations on $\bar{X}$. Did the result come out as expected?
(*f*) If you were to construct 95% confidence intervals on μ based on each of the values of $\bar{X}$ found in part (*c*), approximately how many of them would you expect to contain the true value of μ? From your data, can you find an example of a confidence interval that does contain μ? of a confidence interval that does not contain μ?

CHAPTER 8

INFERENCES ON THE MEAN AND VARIANCE OF A DISTRIBUTION

One of the modern-day "miracles" is the rise of Japan's industrial strength after World War II. Much of this success has been attributed to the work of the American statistician W. Edwards Deming. This man not only helped the Japanese to implement the methods of statistical process control, but he also developed and taught his system of total quality management (TQM). TQM is a management system that is based on Deming's 14 points. One of the aims of TQM is to reduce variability. That is, with respect to a process, the aim is to produce goods that not only satisfy some target average value, but that also do so consistently. To achieve this, random variation must be reduced at every stage of the production process from the procurement of raw material through the marketing and servicing of the finished product. For this reason, there is interest in both the mean (the target value) and the variance of the random variable involved. In this chapter we present some statistical techniques that can be used to draw conclusions about these population parameters based on information obtained from samples.

We have seen how to estimate both the mean and the variance of a distribution via point estimation. We have also seen how to generate a confidence interval for the mean of a normal distribution when its variance is assumed to be *known*. Unfortunately, in most statistical studies, the assumption that σ^2 is known is unrealistic. If it is necessary to estimate the mean of a distribution, then its variance is usually unknown also. In this chapter we turn our attention to the problem of making inferences on the mean and variance of a distribution when both of these parameters are assumed to be unknown. We begin by considering the construction of a confidence interval for σ^2.

8.1 INTERVAL ESTIMATION OF VARIABILITY

In Theorem 7.1.3 we showed that the statistic S^2 is an unbiased estimator for σ^2. To obtain a $100(1-\alpha)\%$ confidence interval for σ^2, we need a random variable whose expression involves σ^2 and whose probability distribution is known. In Exercise 45, Chap. 7, we showed that the random variable $\sum_{i=1}^n (X_i - \mu)^2/\sigma^2$ has a chi-squared distribution with n degrees of freedom. The next theorem shows that if the population mean μ is replaced by the sample mean $\bar{X}$, the resulting random variable $\sum_{i=1}^n (X_i - \bar{X})^2/\sigma^2$ follows a chi-squared distribution with $n-1$ degrees of freedom. This theorem provides the random variable needed to construct a confidence interval for σ^2.

Theorem 8.1.1 [Distribution of $(n-1)S^2/\sigma^2$]. Let $X_1, X_2, X_3, \ldots, X_n$ be a random sample from a normal distribution with mean μ and variance σ^2. The random variable

$$(n-1)S^2/\sigma^2 = \sum_{i=1}^{n} (X_i - \bar{X})^2/\sigma^2$$

has a chi-squared distribution with $n-1$ degrees of freedom.

Proof. This argument does not constitute a rigorous proof of our theorem. However, it does suggest that the distribution of the random variable $(n-1)S^2/\sigma^2$ is as stated. We begin by rewriting the random variable as a difference between two chi-squared random variables:

$$(n-1)S^2/\sigma^2 = \sum_{i=1}^{n} \frac{(X_i - \bar{X})^2}{\sigma^2} = \sum_{i=1}^{n} \frac{[(X_i - \mu) - (\bar{X} - \mu)]^2}{\sigma^2}$$

$$= \sum_{i=1}^{n} \frac{(X_i - \mu)^2}{\sigma^2} - 2(\bar{X} - \mu) \sum_{i=1}^{n} \frac{(X_i - \mu)}{\sigma^2} + \frac{n(\bar{X} - \mu)^2}{\sigma^2}$$

$$= \sum_{i=1}^{n} \frac{(X_i - \mu)^2}{\sigma^2} - 2(\bar{X} - \mu) \frac{\left(\sum_{i=1}^{n} X_i - n\mu\right)}{\sigma^2} + \frac{n(\bar{X} - \mu)^2}{\sigma^2}$$

$$= \sum_{i=1}^{n} \frac{(X_i - \mu)^2}{\sigma^2} - \frac{2n(\bar{X} - \mu)^2}{\sigma^2} + \frac{n(\bar{X} - \mu)^2}{\sigma^2}$$

$$= \sum_{i=1}^{n} \frac{(X_i - \mu)^2}{\sigma^2} - \left(\frac{\bar{X} - \mu}{\sigma/\sqrt{n}}\right)^2$$

We now see that

$$(n-1)S^2/\sigma^2 + \left(\frac{\bar{X}-\mu}{\sigma/\sqrt{n}}\right)^2 = \sum_{i=1}^{n} \frac{(X_i-\mu)^2}{\sigma^2}$$

Note that the random variable $(\bar{X}-\mu)/(\sigma/\sqrt{n})$ is standard normal. By Exercise 45 of Chap. 7 $[(\bar{X}-\mu)/(\sigma/\sqrt{n})]^2$ has a chi-squared distribution with 1 degree of freedom, and $\sum_{i=1}^{n}[(X_i-\mu)^2/\sigma^2]$ has a chi-squared distribution with n degrees of freedom. Since the sum of independent chi-squared random variables is also a chi-squared random variable (see Exercise 44 of Chap. 7), it is logical to assume that the random variable $(n-1)S^2/\sigma^2$ has a chi-squared distribution with $n-1$ degrees of freedom as claimed.

To use the random variable $(n-1)S^2/\sigma^2$ to derive a $100(1-\alpha)\%$ confidence interval on σ^2, we first partition the X^2_{n-1} curve as shown in Fig. 8.1. Remember that in our notational convention, the subscript associated with a point denotes the area to the *right* of the point. If we partition the chi-squared curve so that $(1-\alpha)\%$ of the area is in the center of the curve, then the missing α is divided in half. Thus the right-hand chi-squared point has $\alpha/2\%$ of the area to its right; it is denoted by $\chi^2_{\alpha/2}$. Since the chi-squared distribution is never negative, the left-hand chi-squared point is not the negative of the right-hand point as was the case in the Z-type confidence interval on the mean developed in Sec. 7.4. Rather, it is simply a point with $\alpha/2$ area to its left and $1-\alpha/2$ area to its right; it is denoted by $\chi^2_{1-\alpha/2}$. To derive the confidence interval, we begin by giving a probability statement based on Fig. 8.1 that can be set equal to $1-\alpha$. It is evident that

$$P\left[\chi^2_{1-\alpha/2} \le (n-1)S^2/\sigma^2 \le \chi^2_{\alpha/2}\right] = 1-\alpha$$

To find the lower and upper bounds for the confidence interval, we isolate σ^2

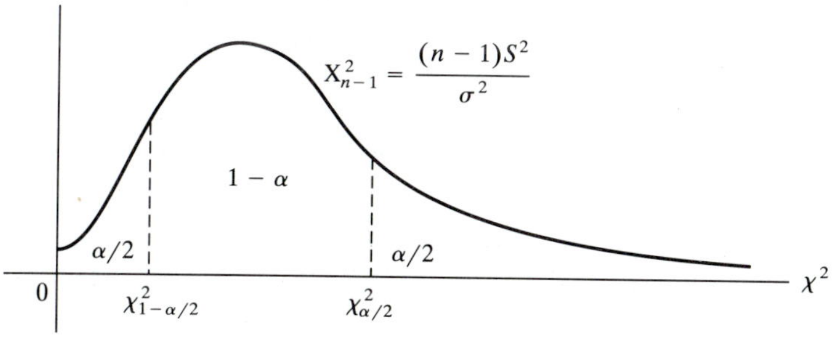

FIGURE 8.1
Partition of the X^2_{n-1} curve needed to derive a $100(1-\alpha)\%$ confidence interval on σ^2.

in the center of the inequality by inverting each term and solving for σ^2.

$$P\left[1/\chi^2_{\alpha/2} \le \sigma^2/(n-1)S^2 \le 1/\chi^2_{1-\alpha/2}\right] = 1 - \alpha$$

or

$$P\left[\frac{(n-1)S^2}{\chi^2_{\alpha/2}} \le \sigma^2 \le \frac{(n-1)S^2}{\chi^2_{1-\alpha/2}}\right] = 1 - \alpha$$

The desired confidence bounds can be read from the latter inequality and are given in Theorem 8.1.2.

Theorem 8.1.2 [100(1 − α)% confidence interval on σ^2]. Let $X_1, X_2, X_3, \ldots, X_n$ be a random sample of size n from a normal distribution with mean μ and variance σ^2. The lower and upper bounds, L_1 and L_2, respectively, for a 100(1 − α)% confidence interval on σ^2, are given by

$$L_1 = (n-1)S^2/\chi^2_{\alpha/2} \qquad \text{and} \qquad L_2 = (n-1)S^2/\chi^2_{1-\alpha/2}$$

As one would suspect, to obtain the bounds for a 100(1 − α)% confidence interval on the standard deviation of a normal random variable, we take the nonnegative square root of the bounds given in Theorem 8.1.2.

Example 8.1.1. In computing, "workload" is defined as a collection of processor and input-output (I/O) resource requests during a particular period of time. Workloads are compared via a measure called *relative I/O content*. The average commercial batch MVS installation provides the base for this measure and is given a relative I/O content rating of 1. Other installations are rated relative to this base. These observations on the relative I/O content for a large consulting firm over randomly selected 1-hour periods are obtained (based on "Processor, I/O Path and DASD Configuration Capacity" J. B. Major, *IBM Systems Journal*, vol. 20, no. 1, 1981, pp. 63–85):

3.4	3.6	4.0	0.4	2.0
3.0	3.1	4.1	1.4	2.5
1.4	2.0	3.1	1.8	1.6
3.5	2.5	1.7	5.1	.7
4.2	1.5	3.0	3.9	3.0

Let us construct a 95% confidence interval on the standard deviation of the relative I/O content for this installation. The stem-and-leaf diagram for these data is shown in Fig. 8.2. This diagram does not suggest a serious departure from normality. The partition of the X^2_{24} curve needed to construct the confidence interval is shown in Fig. 8.3. The values of s and s^2 obtained via a statistical

0	47
1	457486
2	0505
3	405611090
4	201
5	1

FIGURE 8.2
Stem-and-leaf diagram of the relative I/O content of the consulting firm of Example 8.1.1.

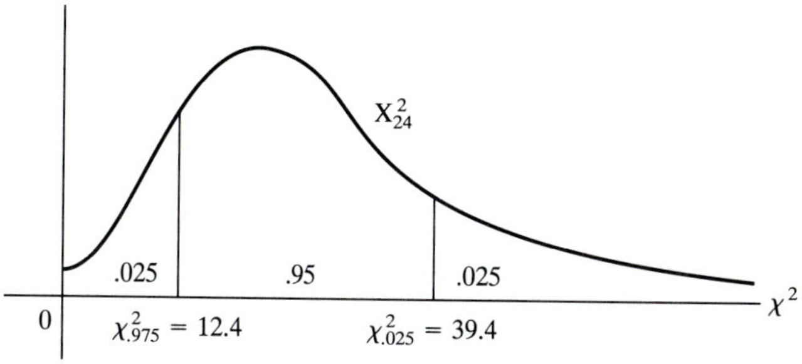

FIGURE 8.3
Partition of the X^2_{24} curve needed to construct a 95% confidence interval on the variance in relative I/O content of the consulting firm of Example 8.1.1.

calculator are $s = 1.186381052$ and $s^2 = 1.4075$. Since the sample variance is reported to two more decimal places than the data, $s^2 = 1.408$. The sample standard is reported to one more decimal place than the data. Here $s = 1.19$.

The bounds for a 95% confidence interval on σ^2 are

$$L_1 = (n-1)s^2/\chi^2_{.025} = 24(1.408)/39.4 = .858$$

$$L_2 = (n-1)s^2/\chi^2_{.975} = 24(1.408)/12.4 = 2.725$$

The bounds for a 95% confidence interval on σ are

$$L_1 = \sqrt{.858} \doteq .926$$

$$L_2 = \sqrt{2.725} \doteq 1.65$$

Thus we can say that we are 95% confident that the true variance in the relative I/O content at this consulting firm lies between .858 and 2.725; we are 95% confident that the true standard deviation lies between .926 and 1.65.

8.2 ESTIMATING THE MEAN AND THE STUDENT-*t* DISTRIBUTION

Note that to obtain a point estimate for a population mean μ, it is not necessary to know the population variance; the sample mean $\bar{X}$ provides an unbiased estimator for μ regardless of the value of σ^2. However, the bounds for a $100(1-\alpha)\%$ confidence interval on μ given in Sec. 7.4 are $\bar{X} \pm z_{\alpha/2}\sigma/\sqrt{n}$. It is assumed that even though the population mean is unknown, the population variance is known. Practically speaking, this assumption is not very realistic. In most instances when a statistical study is being conducted, it is being done for the first time; there is no way to know prior to the study either the mean or the variance of the population of interest. We consider in this section the more realistic problem of constructing a confidence interval on a population mean when the population variance is assumed to be *unknown*.

To derive a general formula for a $100(1-\alpha)\%$ confidence interval on μ under these circumstances, it is natural to begin by considering the random variable used earlier, namely,

$$\frac{\bar{X} - \mu}{\sigma/\sqrt{n}}$$

There are two problems to overcome:

1. The value of σ is not known and must be estimated.
2. The distribution of the random variable obtained by replacing σ by an estimator is not known.

The first problem is easy to overcome. We shall use the sample standard deviation S as an estimator for σ. The second problem is a little more difficult to solve. When we replace σ by its estimator S, the random variable $(\bar{X} - \mu)/(S/\sqrt{n})$ results. It can be shown that the distribution of this random variable is no longer standard normal. Rather, when sampling from a normal distribution, it follows what is called a *Student-t*, or simply a T distribution. This distribution was first described by W. S. Gosset in 1908. He used the pen name "Student" because his employers, an Irish brewery, did not want their competitors to know that they were using statistical methods in their work. We pause briefly to consider this distribution.

The T Distribution

Definition 8.2.1 (T Distribution). Let Z be a standard normal random variable and let X^2_γ be an independent chi-squared random variable with γ degrees of freedom. The random variable

$$T = \frac{Z}{\sqrt{X^2_\gamma/\gamma}}$$

is said to follow a T distribution with γ degrees of freedom.

This definition implies that to show that a random variable follows a T distribution, we must show that it can be written as a ratio of a standard normal random variable to the square root of an independent chi-squared random variable divided by its degrees of freedom.

We note here the characteristics of T distributions that will be useful in the work that follows.

1. There are infinitely many T distributions, each identified by one parameter γ, called *degrees of freedom*. This parameter is always a positive integer. The notation T_γ denotes a T random variable with γ degrees of freedom.

2. Each T random variable is continuous. The density for a T random variable with γ degrees of freedom is given by

$$f(t) = \frac{\Gamma(\gamma+1)/2}{\Gamma(\gamma/2)\sqrt{\pi\gamma}}\left(1+\frac{t^2}{\gamma}\right)^{-(\gamma+1)/2} \qquad -\infty < t < \infty$$

3. The graph of the density of a T_γ random variable is a symmetric bell-shaped curve centered at 0.
4. The parameter γ is a shape parameter in the sense that as its value increases, the variance of the random variable T_γ decreases. Thus as the value of γ increases, the bell-shaped curve associated with T_γ becomes more compact.
5. As the number of degrees of freedom increases, the bell-shaped curve associated with the T_γ random variable approaches the standard normal curve.

These ideas are illustrated in Fig. 8.4.

A partial summary of the cumulative distribution for selected values of γ is given in Table VI of App. A. The table is read just as the chi-squared table is read. That is, the degrees of freedom are listed as row headings, pertinent probabilities are listed as column headings, and the points associated with those probabilities are listed in the body of the table. We use our previous convention

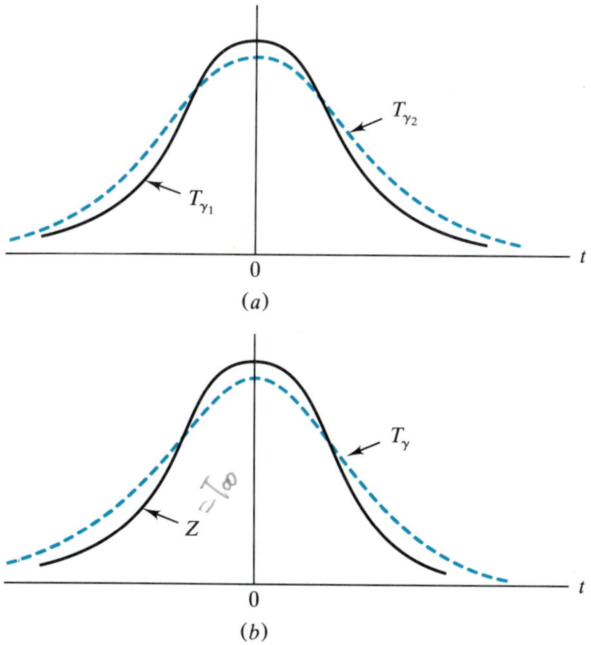

FIGURE 8.4
(a) Typical relationship between two T curves with $\gamma_1 > \gamma_2$; (b) typical relationship between a T curve and the standard normal curve.

of denoting by t_r the point associated with the T_γ curve such that the area to the right of the point is r.

Example 8.2.1. Consider the random variable T_{10}.

1. From Table VI of App. A, $P[T_{10} \le 1.372] = F(1.372) = .90$. By our notational convention, $t_{.10} = 1.372$. [See Fig. 8.5(*a*).]
2. Due to the symmetry of the T curve, $t_{.90} = -t_{.10} = -1.372$.
3. The point t such that $P[-t \le T_{10} \le t] = .95$ is $t_{.025} = 2.228$. [See Fig. 8.5(*b*).]

The last row in Table VI of App. A is labeled ∞. The points listed in that row are actually points associated with the standard normal curve. Note that as r increases, the values in each column of the table approach the value listed in the last row. This occurs because, for large values of r, the graph of the density for the T_r random variable, for all practical purposes, coincides with that of the standard normal or Z curve. For small values of r there is quite a bit of difference between the two curves.

Let us now show that the random variable $(\bar{X} - \mu)/(S/\sqrt{n})$ follows a T distribution as claimed. The proof of this theorem depends on a result that is beyond the scope of this discussion mathematically. In particular, it can be shown that when sampling from a normal distribution, the sample mean $\bar{X}$ and the sample standard deviation S are independent. This result is not surprising. It says simply that knowledge of the center of location of a normal random variable does not contribute to knowledge of its variability. The next theorem

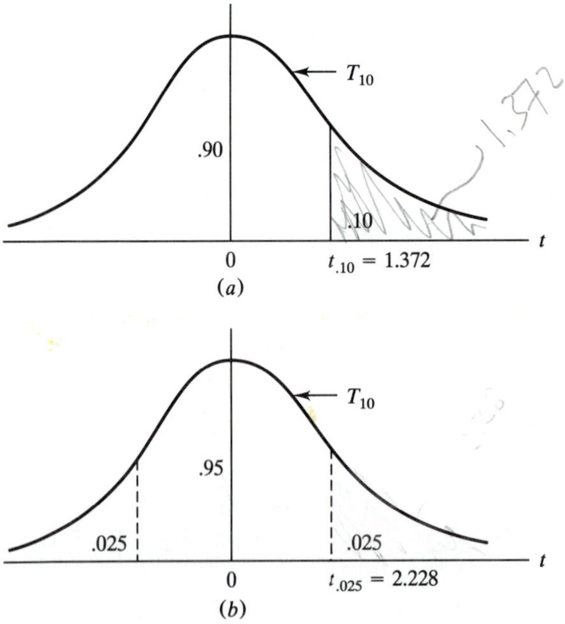

FIGURE 8.5
(*a*) $P[T_{10} \le 1.372] = .90$;
(*b*) $P[-2.228 \le T_{10} \le 2.228] = .95$.

provides the basis for the construction of a $100(1-\alpha)\%$ confidence interval on μ when σ^2 is assumed to be unknown.

Theorem 8.2.1. Let $X_1, X_2, X_3, \ldots, X_n$ be a random sample from a normal distribution with mean μ and variance σ^2. The random variable

$$\frac{\bar{X}-\mu}{S/\sqrt{n}}$$

follows a T distribution with $n-1$ degrees of freedom.

Proof. We shall show that the random variable $(\bar{X}-\mu)/(S/\sqrt{n})$ can be written as the ratio of a standard normal random variable to the square root of an independent chi-squared random variable divided by its degrees of freedom. By Theorem 7.3.4, $\bar{X}$ is normal with mean μ and variance σ^2/n. Standardizing, $(\bar{X}-\mu)/(\sigma/\sqrt{n})$ is standard normal. By Theorem 8.1.1, $(n-1)S^2/\sigma^2$ is a chi-squared random variable with $n-1$ degrees of freedom. Consider the random variable

$$\frac{Z}{\sqrt{X_\gamma^2/\gamma}} = \frac{(\bar{X}-\mu)/(\sigma/\sqrt{n})}{\sqrt{(n-1)S^2/\sigma^2(n-1)}} = \frac{\bar{X}-\mu}{S/\sqrt{n}}$$

Since $\bar{X}$ and S are independent, this random variable follows a T distribution with $n-1$ degrees of freedom as claimed.

Confidence Interval on the Mean: Variance Estimated

It is now easy to determine the general form for a $100(1-\alpha)\%$ confidence interval on μ when σ^2 is unknown. We need only note that the two random variables

$$Z = \frac{\bar{X}-\mu}{\sigma/\sqrt{n}} \qquad \text{and} \qquad T_\gamma = \frac{\bar{X}-\mu}{S/\sqrt{n}}$$

have the same algebraic structure. Thus the algebraic argument given in Sec. 7.4 will hold with σ being replaced by S and $z_{\alpha/2}$ being replaced by $t_{\alpha/2}$. These substitutions result in Theorem 8.2.2.

Theorem 8.2.2 [$100(1-\alpha)\%$ Confidence interval on μ when σ^2 is unknown]. Let $X_1, X_2, X_3, \ldots, X_n$ be a random sample from a normal distribution with mean μ and variance σ^2. A $100(1-\alpha)\%$ confidence interval on μ is given by

$$\bar{X} \pm t_{\alpha/2} S/\sqrt{n}$$

$\gamma = n-1$

Example 8.2.2 illustrates the use of this theorem.

Example 8.2.2. Sulfur dioxide and nitrogen oxide are both products of fossil fuel consumption. These compounds can be carried long distances and converted to acid before being deposited in the form of "acid rain." These data are obtained on the sulfur dioxide concentration (in micrograms per cubic meter) in a Bavarian forest thought to have been damaged by acid rain (based on "Is Acid Deposition Killing West German Forests?" by Leslie Roberts, *Bioscience*, vol. 33, no. 5, May 1983, pp. 302–305):

52.7	43.9	41.7	71.5	47.6	55.1
62.2	56.5	33.4	61.8	54.3	50.0
45.3	63.4	53.9	65.5	66.6	70.0
52.4	38.6	46.1	44.4	60.7	56.4

A statistical calculator yields these values:

$$\bar{x} = 53.91666667 \qquad s = 10.07371382 \qquad s^2 = 101.4797102$$

Our rounding guidelines yield

$$\bar{x} \doteq 53.92\ \mu g/m^3 \qquad s \doteq 10.07\ \mu g/m^3 \qquad s^2 \doteq 101.480$$

The partition of the T_{23} curve needed to find a 95% confidence interval on the mean sulfur dioxide concentration in this forest is shown in Fig. 8.6. The confidence bounds for the interval are

$$\bar{x} \pm t_{\alpha/2} s/\sqrt{n} = 53.92 \pm 2.069(10.07)/\sqrt{24}$$

That is, we are 95% confident that the mean sulfur dioxide concentration in this forest lies in the interval [49.67, 58.17]. The average concentration of this compound in undamaged areas of the country is 20 $\mu g/m^3$. Since this value is not included in the above interval, there is evidence of an elevated sulfur dioxide concentration in the damaged forest.

Several things should be pointed out. First, the number of degrees of freedom involved in finding a confidence interval on μ when σ^2 is unknown is $n - 1$, the sample size minus 1. For large samples this value may not be listed in Table VI of App. A. In this case the last row in the table (∞) is used to find points of interest. Thus in the case of large samples we are in effect estimating a desired t point via a z point. As mentioned earlier, this is appropriate since for large samples the Z and T curves are virtually identical. Second, once again, a normality assumption has been made. We can check the validity of this assumption graphically using a stem-and-leaf diagram or a histogram. More precise methods for testing for normality are available. One procedure that can be used

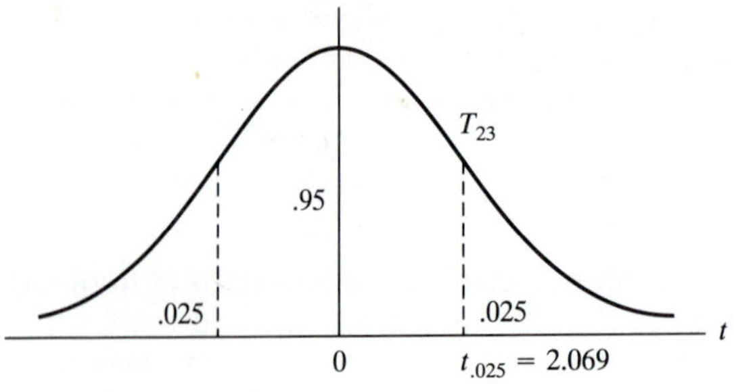

FIGURE 8.6
Partition of the T_{23} curve needed to construct a 95% confidence interval on the mean sulfur dioxide concentration in a Bavarian forest.

is given in Sec. 8.6. If there is reason to suspect that the variable under study has a distribution that is not normal and the sample size is small, then methods based on the T distribution may *not* be appropriate. Rather, some nonparametric technique should be employed. Some of these techniques are discussed in Sec. 8.7.

8.3 HYPOTHESIS TESTING

We have considered the basic ideas of estimation in some detail. Recall that in a typical estimation problem there is some population parameter, θ, whose value is to be approximated based on a sample. Usually, there is *no* preconceived notion concerning the actual value of this parameter. We are attempting simply to ascertain its value to the best of our ability. In contrast, when testing a hypothesis on θ, there is a preconceived notion concerning its value. This implies that two theories or hypotheses, are, in fact, involved in any statistical study of this sort: the hypothesis being proposed by the experimenter and the negation of this hypothesis. The former, denoted by H_1, is called the *alternative* or *research hypothesis*; the latter is denoted by H_0 and is called the *null hypothesis*. The purpose of the experiment is to decide whether the evidence tends to refute the null hypothesis. These three guidelines help in deciding how to state H_0 and H_1:

1. When testing a hypothesis concerning the value of some parameter θ, the statement of equality will always be included in H_0. In this way H_0 pinpoints a specific numerical value that could be the actual value of θ. This value is called the *null value* and is denoted by θ_0.
2. Whatever is to be detected or supported is the alternative hypothesis.
3. Since our research hypothesis is H_1, it is hoped that the evidence leads us to reject H_0 and thereby to accept H_1.

An example will help to clarify these ideas.

Example 8.3.1. Highway engineers have found that many factors affect the performance of reflective highway signs. One is the proper alignment of the automobile's headlights. It is thought that more than 50% of the automobiles on the road have misaimed headlights. If this contention can be supported statistically, then a new tougher inspection program will be put into operation. Let p denote the proportion of automobiles in operation that have misaimed headlights. Since we wish to support the statement that $p > .5$, this contention is taken as the alternative or research hypothesis, H_1. The null hypothesis is automatically the negation of H_1, namely, $p \leq .5$. Thus the two hypotheses are

$$H_0\colon p \leq .5$$

$$H_1\colon p > .5$$

not equal to,
H_1 → less than, statement
greater to,

Note that the statement of equality appears in the null hypothesis. This pinpoints the value .5 as a possible value for p; that is, the "null value" for p is $p_0 = .5$. Note also that if H_0 is rejected, then our research hypothesis is accepted and the new inspection program will be implemented.

Once a sample has been selected and the data have been collected, a decision must be made. The decision will be either to reject H_0 or to fail to do so. The decision is made by observing the value of some statistic whose probability distribution is known *under the assumption that the null value is the true value of* θ. Such a statistic is called a *test statistic*. If the test statistic assumes a value that is rarely seen when $\theta = \theta_0$ and tends to lend credence to the alternative hypothesis, then we reject H_0 in favor of H_1; if the value observed is a commonly occurring one under the assumption that $\theta = \theta_0$, then we do not reject the null hypothesis. This means that at the end of any study we shall be forced into exactly one of the following situations:

1. We shall have rejected H_0 when it was true and shall have committed what is known as a *Type I error*.
2. We shall have made the correct decision of rejecting H_0 when the alternative H_1 was true.
3. We shall have failed to reject H_0 when the alternative H_1 was true. In this case we shall have committed what is known as a *Type II error*.
4. We shall have made the correct decision of failing to reject H_0 when H_0 was true.

Example 8.3.2. In Example 8.3.1 we are testing

H_0: $p \leq .5$

H_1: $p > .5$ (majority of automobiles in operation have misaimed headlights)

If a Type I error is made, we shall have rejected H_0 when H_0 is true. Practically speaking, we shall have concluded that a majority of cars on the road have misaimed headlights when, in fact, this is not true. This error could lead to the implementation of an unnecessary inspection program. A Type II error occurs if we fail to reject H_0 when H_1 is true. In this case, the inspection program would not be implemented when, in fact, it is needed.

Note that regardless of what is done, an error is possible. Anytime H_0 is rejected, a Type I error might occur; anytime H_0 is not rejected, a Type II error might occur. There is no way to avoid this dilemma. The job of the statistician is to design methods for deciding whether or not to reject H_0 that keep the probabilities of making either error reasonably small.

Philosophically, there are two ways to determine whether or not to reject H_0. The first method, which we discuss in this section, is called *hypothesis testing*. This method has been used extensively in the past and is still used today. The second method, called *significance testing*, is becoming increasingly popular. It is discussed in the next section.

Hypothesis testing involves a procedure in which the values of the test statistic that lead to rejection of the null hypothesis are set before the experiment is conducted. These values constitute what is called the *critical*, or *rejection*, *region* for the test. The probability that the observed value of the test statistic will fall into this region by chance even though $\theta = \theta_0$ is called *alpha* (α), *the size of the test* or the *level of significance of the test*. If this occurs, a Type I error is committed. That is, in a hypothesis testing study, α is the probability of committing a Type I error. The next example illustrates the use of this method.

Example 8.3.3. To test the hypothesis of Example 8.3.1,

H_0: $p \le .5$

H_1: $p > .5$ (majority of automobiles in operation have misaimed headlights)

a random sample of 20 cars is selected and the headlights are tested. Let us design a test so that α, the probability of rejecting H_0 when p is equal to the null value of .5, is about .05. The test statistic that we shall use is X, the number of cars in the sample with misaimed headlights. If p is, in fact, equal to the null value, then X is binomial with $n = 20$, $p = .5$, and $E[X] = np = 10$. Thus if $p = .5$, then, on the average, 10 of every 20 cars tested will have misaimed headlights; if H_1 is true, this average value will be higher than 10. Logically, we should reject H_0 if the observed value of the test statistic X is somewhat larger than 10. Note from Table I of App. A that

$$
\begin{aligned}
P[X \ge 14|p = .5] &= 1 - P[X < 14|p = .5] \\
&= 1 - P[X \le 13|p = .5] \\
&= 1 - .9423 \\
&= .0577
\end{aligned}
$$

Let us agree to reject H_0 in favor of H_1 if the observed value of the test statistic, X, is 14 or greater. In this way we have split the possible values of X into two sets: $C = \{14, 15, 16, 17, 18, 19, 20\}$ and $C' = \{0, 1, 2, \ldots, 13\}$. If the observed value of X lies in C, we reject H_0 and conclude that the majority of cars in operation have misaimed headlights. The set of values of the test statistic that leads to rejection of the null hypothesis C is the *critical*, or *rejection*, *region* for the test. We chose C so that the probability that the test statistic will fall into C by chance, even though $p = .5$, is .0577. That is, we designed the test so that the probability of committing a Type I error (α) is approximately .05 as desired.

There is one point to note. In the previous example we use the null value $p_0 = .5$ to determine the critical region for the test even though the null hypothesis allows for values of p that are less than .5. It is safe to do this since values of X that are too large to occur by chance when $p = .5$ are also certainly too large to occur by chance when $p < .5$. That is, any value of X that leads us to reject .5 as a reasonable value for p also leads us to reject any value less than .5. (See Exercise 29.)

It is possible that the observed value of the test statistic does not fall into the rejection region even though H_0 is not true and should be rejected. If this occurs, a Type II error will be committed. The probability of this occurring is called *beta* (β). Beta is a little harder to handle than alpha, which can be dictated by the experimenter. For a particular test, β depends on the alternative. That is, β can be found only if a particular value of the alternative is specified. To illustrate, let us find β for the test designed in Example 8.3.3.

Example 8.3.4. The critical region for the test of Example 8.3.3 is $C = \{14, 15, 16, 17, 18, 19, 20\}$. Suppose that, unknown to the researcher, the true proportion of cars with misaimed headlights is .7. What is the probability that our test, as designed, is unable to detect this situation? To answer this question, we calculate β, the probability that H_0 will not be rejected given that $p = .7$. By definition

$$\begin{aligned} \beta &= P[\text{Type II error}] \\ &= P[\text{fail to reject } H_0 | p = .7] \\ &= P[X \text{ is not in the critical region} | p = .7] \\ &= P[X \leq 13 | p = .7] = .3920 \qquad \text{(Table I, App. A)} \end{aligned}$$

That is, for the test as designed there is not a very high probability that we shall be able to distinguish between $p = .5$ and $p = .7$. Beta is a function of the alternative in that if p is changed from .7 to .8, then β will change also. In this case

$$\beta = P[X \leq 13 | p = .8] = .0867$$

Note that as the difference between the null value of .5 and the alternative value of p increases, β decreases.

Remember that the hypothesis-testing procedure entails deciding on the level of significance (α) before the data are gathered and the test statistic is evaluated. That is, it involves presetting α. There are several reasons for wanting to do this. It gives a clear-cut way of making a decision. Once α is set, the critical region for the test is fixed also. If the observed value of the test statistic falls into this region, we reject H_0; otherwise we do not. There is no room for debate after the data are gathered. Hence there can be no charge that the statisticians are manipulating the results to suit themselves. In addition, if the consequences of making a Type I error are very serious, then by presetting α we are able to specify *before the fact* exactly how large a risk we are willing to tolerate. The language underlying hypothesis testing is summarized in Fig. 8.7.

	Actual situation	
Decision	H_0 true	H_1 true
Reject H_0	Type I error (probability α)	Correct decision
Fail to reject H_0	Correct decision	Type II error (probability β)

FIGURE 8.7

8.4 SIGNIFICANCE TESTING

In the last section we considered a method for deciding whether or not to reject a null hypothesis, called *hypothesis testing*. In this section we consider another method for doing so. This method, called *significance testing*, is coming into widespread use. This is due to its logical appeal and to the increasing use of computer packages in analyzing statistical data.

To understand why significance testing is so appealing, let us point out a bothersome aspect of hypothesis testing that might have occurred to you already. It is easy to spot the problem with a simple example. Suppose that we want to test

$$H_0: p \leq .1$$
$$H_1: p > .1$$

based on a sample of size 20. The test statistic is X, the number of "successes" that are observed in the 20 trials. Since the null value is $p_0 = .1$, when $p = p_0$ the test statistic follows a binomial distribution with $E[X] = np_0 = 20(.1) = 2$. Values of X somewhat larger than 2 tend to lend credence to the alternative hypothesis. Suppose that we want α to be "very small," so we define the critical region to be $C = \{9, 10, 11, \ldots, 20\}$. For this test,

$$\begin{aligned}
\alpha &= P[\text{Type I error}] \\
&= P[\text{reject } H_0 | p = p_0] \\
&= P[X \text{ is in the critical region} | p = .1] \\
&= P[X \geq 9 | p = .1] \\
&= 1 - P[X < 9 | p = .1] \\
&= 1 - P[X \leq 8 | p = .1] \\
&= 1 - .9999 \\
&= .0001
\end{aligned}$$

This is indeed a "very small value"! Now suppose that we conduct our test and observe 8 "successes." Via our rather rigid rules for hypothesis testing, we are unable to reject H_0 since 8 does not lie in the critical region. However, a little thought should make you a bit uneasy with this decision! Note that 8 is very close to 9, our rather arbitrarily selected lower boundary for the critical region. Let us see what the chances are of obtaining a value of 8 or more when $p = .1$:

$$\begin{aligned}
P[X \geq 8 | p = .1] &= 1 - P[X < 8 | p = .1] \\
&= 1 - P[X \leq 7 | p = .1] \\
&= 1 - .9996 \\
&= .0004
\end{aligned}$$

This probability is certainly also "very small." It is hard to imagine a situation in which we would be willing to tolerate 1 chance in 10,000 of making a Type I error but would declare vehemently that 4 chances in 10,000 of making such an error is much too large to risk! There is so little difference between these probabilities that it seems a bit silly to insist that we adhere rigidly to our original cutoff point of 9.

The problem just demonstrated can be avoided by performing what is called a *significance test* rather than a hypothesis test. This method of deciding whether or not to reject H_0 entails setting up H_0 and H_1 exactly as before. However, we do not then preset α and specify a rigid critical region. Rather, we evaluate the test statistic and then determine the probability of observing a value of the test statistic at least as extreme as the value noted under the assumption that $\theta = \theta_0$. This probability is referred to by a variety of names, including the *critical level*, the *descriptive level of significance*, and the *probability*, or *P value* of the test. We use the term "P value" in this text. Note that the P value is the smallest level at which we could have preset α and still have been able to reject H_0. We reject H_0 if we consider this P value to be small.

Example 8.4.1. Automotive engineers are using more and more aluminum in the construction of automobiles in hopes of reducing the cost and improving gas mileage. For a particular model the number of miles per gallon obtained on the highway currently has a mean of 26 mpg with a standard deviation of 5 mpg. It is hoped that a new design, which utilizes more aluminum, will increase the mean mileage rating. Assume that σ is not affected by this change. Since our research hypothesis is taken as the alternative hypothesis, we are testing

$$H_0: \mu \le 26$$

$$H_1: \mu > 26 \qquad \text{(the new design increases gas mileage on the highway)}$$

Since the sample mean is an unbiased estimator for the population mean, a logical test statistic is $\bar{X}$. Let us agree to reject H_0 in favor of H_1 if the observed value of the sample mean is "somewhat larger" than 26. By "somewhat larger" we mean too large to have reasonably occurred by chance if the true mean highway mileage is still 26 mpg. These data are obtained during road testing:

33.8	24.3	18.8	23.7	25.3	29.6
24.9	31.5	34.4	28.0	20.5	36.7
30.3	33.5	27.4	27.6	22.5	30.7
28.6	27.1	28.8	16.5	32.7	25.2
33.1	37.5	25.1	34.5	29.5	26.8
30.0	28.4	25.6	19.8	28.9	27.7

The sample mean for these data is $\bar{x} = 28.04$ mpg. This value is larger than the null value for μ of 26 mpg. To see if there is enough difference to cause us to reject H_0, we find the P value for the test. That is, we compute the probability of observing a sample mean of 28.04 or larger if $\mu = 26$ and $\sigma = 5$. This is done by noting if $\mu = 26$ and $\sigma = 5$, then the test statistic $\bar{X}$ is, by the Central Limit Theorem (Theorem 7.4.2), at least approximately normally distributed with mean

$\mu = 26$ and standard deviation $\sigma/\sqrt{n} = 5/6$. Therefore

$$P[\bar{X} \geq 28.04 | \mu = 26, \sigma = 5] = P\left[\frac{\bar{X} - 26}{(5/6)} \geq \frac{28.04 - 26}{(5/6)}\right]$$
$$\doteq P[Z \geq 2.45]$$
$$= 1 - P[Z \leq 2.45]$$
$$= 1 - .9929 \qquad \text{(Table V, App. A)}$$
$$= .0071$$

There are two explanations for this very small probability. The null hypothesis is true and we have observed a very rare sample that *by chance* has a large sample mean; the null hypothesis is not true and the new process has, in fact, resulted in a higher mean mileage rating. We prefer the latter explanation! That is, we shall reject H_0 and report that the P value of our test is .0071.

Significance testing is an interesting and exciting concept. There are still some questions to be answered concerning its proper use. One question still to be resolved is, "How do we compute a P value for a two-tailed test?" If the distribution of the test statistic is symmetric, as it is for a Z or T statistic, then it is logical to double the apparent one-tailed P value. If the distribution is not symmetric, as with a chi-squared statistic, then presumably the two-tailed P value is nearly double the one-tailed value. This is only one of several proposed solutions to the problem, but it is the convention that we shall use. The reader is referred to [17] for an excellent discussion of the problem.

8.5 HYPOTHESIS AND SIGNIFICANCE TESTS ON THE MEAN

One of the most commonly encountered problems is that of testing a hypothesis concerning the value of the mean. We have seen how this can be done if it is assumed that σ^2 is known. Since this assumption is usually not valid, we turn our attention to a method that can be used to test hypotheses concerning μ when σ^2 is unknown and must be estimated from the data at hand. Consider these examples.

Example 8.5.1. The maximum acceptable level for exposure to microwave radiation in the United States is an average of 10 microwatts per square centimeter. It is feared that a large television transmitter may be polluting the air nearby by pushing the level of microwave radiation above the safe limit. Since our research hypothesis is taken as the alternative, we are testing

$$H_0: \mu \leq 10$$
$$H_1: \mu > 10 \qquad \text{(unsafe)}$$

Example 8.5.2. Design engineers are working on a low-effort steering system that can be used in vans modified to fit the needs of disabled drivers. The old-type steering system required a force of 54 ounces to turn the van's 15-inch-diameter steering wheel. It is hoped that the new design will reduce the average force required to turn the wheel. In this case we are testing

H_0: $\mu \geq 54$

H_1: $\mu < 54$ (new system requires less force to operate than the old)

(*Design News*, April 1983, pp. 14–16.)

Example 8.5.3. A computer system currently has 10 terminals and uses a single printer. The average turnaround time for the system is 15 minutes. Ten new terminals and a second printer are added to the system. We want to determine whether or not the mean turnaround time is affected. To decide, we want to test

H_0: $\mu = 15$

H_1: $\mu \neq 15$ (the new equipment has an impact on turnaround time)

As you can see, a hypothesis on μ can take one of three general forms. With μ_0 denoting the null value of the mean, these are:

I	II	III
H_0: $\mu \leq \mu_0$	H_0: $\mu \geq \mu_0$	H_0: $\mu = \mu_0$
H_1: $\mu > \mu_0$	H_1: $\mu < \mu_0$	H_1: $\mu \neq \mu_0$
Right-tailed test	Left-tailed test	Two-tailed test

Form I is called a *right-tailed test* because when a hypothesis of this form is tested, the natural critical region is the upper- (or right-) tail region of the distribution of the test statistic. This point is explained in Example 8.5.4. Similarly form II is a left-tailed test because the natural critical region is the lower- (or left-) tail region of the appropriate distribution. In a two-tailed test the critical region consists of both the lower- and upper-tail regions of the distribution of the test statistic. This is easy to remember because in a one-sided test, forms I and II, the inequality in the *alternative* hypothesis points toward the critical region.

There is one general statement to keep in mind when you test a hypothesis on any parameter:

To test a hypothesis on a parameter θ, you must find a statistic whose probability distribution is known at least approximately under the assumption that $\theta = \theta_0$.

This statistic will serve as a test statistic. In the case at hand such a statistic is easy to find. From the discussion of Sec. 8.2 we know that if X is normal, the statistic $(\bar{X} - \mu_0)/(S/\sqrt{n})$ follows a T_{n-1} distribution. Tests based on this statistic are commonly called *T tests*.

Tests of hypotheses on μ are actually conducted by testing H_0: $\mu = \mu_0$ against one of the alternatives $\mu > \mu_0$, $\mu < \mu_0$ or $\mu \neq \mu_0$. It is safe to do this for reasons analogous to those discussed in Sec. 8.3. In particular, values of the test statistic that lead us to reject μ_0 and to conclude that $\mu > \mu_0$ will also lead us to reject any value less than μ_0; values of the test statistic that lead us to reject μ_0 and to conclude that $\mu < \mu_0$ will also lead us to reject any value greater than μ_0. For this reason, many statisticians prefer to express the three forms as

I		II		III	
	$H_0: \mu = \mu_0$		$H_0: \mu = \mu_0$		$H_0: \mu = \mu_0$
	$H_1: \mu > \mu_0$		$H_1: \mu < \mu_0$		$H_1: \mu \neq \mu_0$
	Right-tailed test		Left-tailed test		Two-tailed test

This emphasizes the fact that when performing a hypothesis test on μ, α is computed assuming that $\mu = \mu_0$; when performing a significance test on μ, the P value is computed under the assumption that $\mu = \mu_0$. We shall follow this notational convention in the remainder of this text.

Example 8.5.4. To determine whether a large television transmitter is polluting the nearby air (see Example 8.5.1), we intend to test

$$H_0: \mu = 10$$
$$H_1: \mu > 10$$

A sample of 25 readings is to be obtained at randomly selected times over a 1-week period. Our test statistic, $(\bar{X} - 10)/(S/\sqrt{25})$, follows a T_{24} distribution if H_0 is true. Since $\bar{X}$ is an unbiased estimator for the mean, we expect the observed value of $\bar{X}$ to be close to 10 if H_0 is true. This forces the numerator of the test statistic, $(\bar{X} - 10)$, to be small, causing the observed value of the test statistic to be small also. However, if H_1 is true, we expect $\bar{X}$ to be larger than 10, forcing $\bar{X} - 10$ to be large and *positive*. This in turn results in a *large positive* value for the test statistic. Hence logically we should reject H_0 in favor of H_1 whenever the observed value of the test statistic is positive and too large to have reasonably occurred by chance. Thus the natural critical region for the test is the right-tail, or upper, region of the T_{24} distribution. To decide how large a value is needed in order to reject H_0, let us preset α. If we make a Type I error, we shall shut down the transmitter unnecessarily; if we make a Type II error, we shall fail to detect a potential health hazard. We want α small but not so small as to force β to be extremely large. Let us choose α to be .1. The critical point for the test, read from Table VI of App. A and shown in Fig. 8.8, is 1.318. We shall reject H_0 in favor of H_1 if the observed value of the test statistic is 1.318 or larger. When the

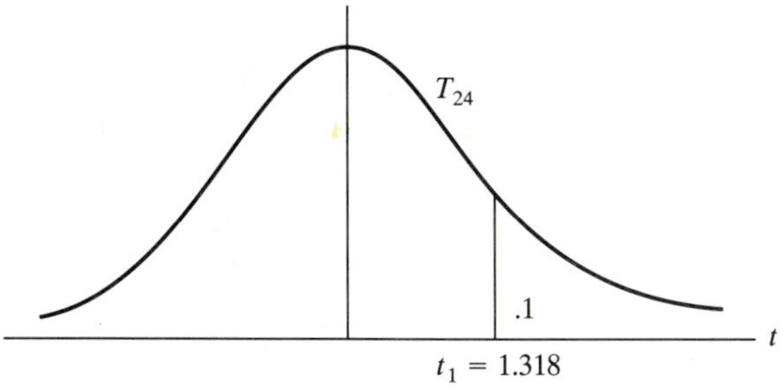

FIGURE 8.8
Critical region for an $\alpha = .1$ level right-tailed test ($n = 25$).

experiment is conducted, it is found that $\bar{x} = 10.3$ and $s = 2$. The observed value of the test statistic is

$$(\bar{x} - 10)/(s/\sqrt{25}) = (10.3 - 10)/(2/5) = .75$$

Since this value falls below the critical point of 1.318, we are unable to reject H_0. These data do not support the contention that the transmitter is forcing the average microwave level above the safe limit.

It should be pointed out that, in practice, it is not really necessary to find a critical point even if α has been preset. Rather, we can simply always evaluate the test statistic and find the P value. If the P value is at most equal to the preset α, then H_0 can be rejected at that α level. For instance, in the previous example α was preset at .05. The observed value of the test statistic is .75. This value and the associated P value is pictured in Fig. 8.9(a). From the T table with 24 degrees of freedom we see that this value lies between .685 and 1.318. [See Fig. 8.9(b).] The area to the right of .685 is .25; the P value is clearly smaller than .25. The area to the right of 1.318 is .10; the P value is larger than this. By combining these results, we can conclude that $.10 < P < .25$. Since this P value exceeds the preset α of .05, H_0 cannot be rejected at this level. This technique is especially useful now as the most serious data analysis is done using one of the commercially available computer packages. These packages typically

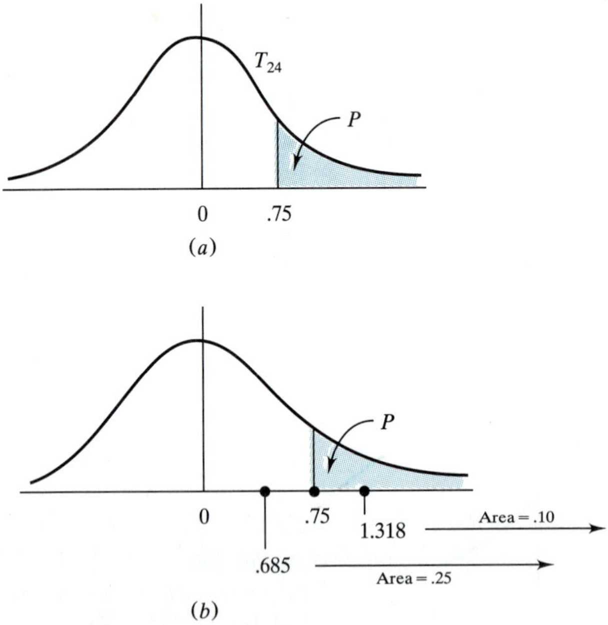

FIGURE 8.9
(a) Since the test is right-tailed, P = shaded area = area to the right of .75; (b) since P is smaller than the area to the right of .685 and larger than the area to the right of 1.318, $.10 < P < .25$.

report a P value automatically. To conduct a test in which α is preset, compare the reported P value to α. If $P \le \alpha$, then H_0 can be rejected at the α level of significance; otherwise it cannot be rejected. Warning! Most packages report two-tailed P values. If you are conducting a one-tailed test, the true P value is half that reported by the computer. Check the package documentation to be sure that you understand exactly what probability is being given.

The next example illustrates the use of significance testing in testing a two-tailed hypothesis.

Example 8.5.5. In studying the effect of adding 10 new terminals and one printer to an existing computer system (see Example 8.5.3), we are testing

$$H_0\colon \mu = 15$$

$$H_1\colon \mu \neq 15 \qquad \text{(the new equipment has an impact on turnaround time)}$$

Since we do not have any *preconceived* notion as to whether the new equipment increases or decreases the mean turnaround time, we are conducting a two-tailed test. We shall reject H_0 in favor of H_1 if the observed value of the test statistic is too large in either the positive or negative sense to have occurred by chance. When the data are gathered, a sample of size 30 yields $\bar{x} = 14.0$ and $s = 3$. The observed value of the test statistic is

$$(\bar{x} - 15)/(s/\sqrt{30}) = (14 - 15)/(3/\sqrt{30}) \doteq -1.83$$

From Table VI of App. A we see that

$$P[T_{29} \le -1.699] = .05 \qquad \text{and} \qquad P[T_{29} \le -2.045] = .025$$

Since -1.83 lies between -1.699 and -2.045, the probability of observing a value as large in the negative sense as that observed lies between .025 and .05. However, we were running a two-tailed test. This means that the P value of the test is the probability of observing a value as extreme as that observed in either the *positive or the negative* sense. That is, the P value is assumed to be double that computed above. We can report that for this test $.05 < P < .1$. Since this probability is still small, we reject H_0 and conclude that the new equipment does affect the mean turnaround time.

It should be emphasized that the statistic $(\bar{X} - \mu_0)/(S/\sqrt{n})$ follows the T_{n-1} distribution if X is *normal*. If X is not normal, then care must be taken. It has been found that for samples of moderate to large size ($n \ge 25$), violating this assumption does not seriously affect the distribution of the test statistic in that the probability of committing a Type I and a Type II error is not appreciably changed [6]. This property is called *robustness*. However, if the sample size is small, then T tests should not be run on nonnormal data. Rather, some nonparametric method should be used. Several of these are presented in

Sec. 8.7. We present a quick method for testing for normality in Sec. 8.6 so that you can be sure that the statistic used to test a hypothesis is appropriate.

8.6 TESTING FOR NORMALITY AND HYPOTHESIS TESTS ON THE VARIANCE (OPTIONAL)

As has been pointed out, the methods discussed in the previous sections of this chapter assume that sampling is from a normal distribution. For many years after the discovery of the normal curve, practitioners felt that virtually every random variable was at least approximately normally distributed. As more and more data became available it became evident that this was not true. However, the statistical tools developed by Fisher, Pearson, and "Student," which presuppose normality, were so appealing to researchers in a wide variety of fields that they were eager to adopt them. To laymen, unable to follow the mathematical derivations of these techniques, the normality "assumption" was thought to be unimportant, a law of nature, or at best satisfied due to some sophisticated mathematical magic. The situation is best described by the words of Lippman in a remark to Poincare (1912) [6]:

> Everyone believes in it [normal law of errors] however, said Monsieur Lippman to me one day, for the experimenters fancy that it is a theorem in mathematics and the mathematicians that it is an experimental fact.

Thus far we have done only a rough check of the normality assumption using the stem-and-leaf diagram. If the stem-and-leaf diagram assumes a definite bell shape, then the normality assumption is probably reasonable. However, if there is a doubt in your mind concerning its validity, then you should test to see if there is statistical evidence that the data are drawn from a nonnormal distribution. Several methods are available for doing so. Here we consider a graphical method that is especially useful when sample sizes are small. A nongraphical procedure for use with large samples is in SAS supplement III. If you wish to skip the discussion here and go on to the subsection on hypothesis tests on the variance you can do so. The exercises for this section are flexible enough to allow this.

Testing for Normality

The method for detecting nonnormality that we consider here, called *the Lilliefors test for normality*, was developed by H. W. Lilliefors in the late 1960s. Although it can be used with large samples, it is most helpful when samples are relatively small. The test basically compares the observed relative cumulative frequency distribution of the sample to that of the standard normal distribution.

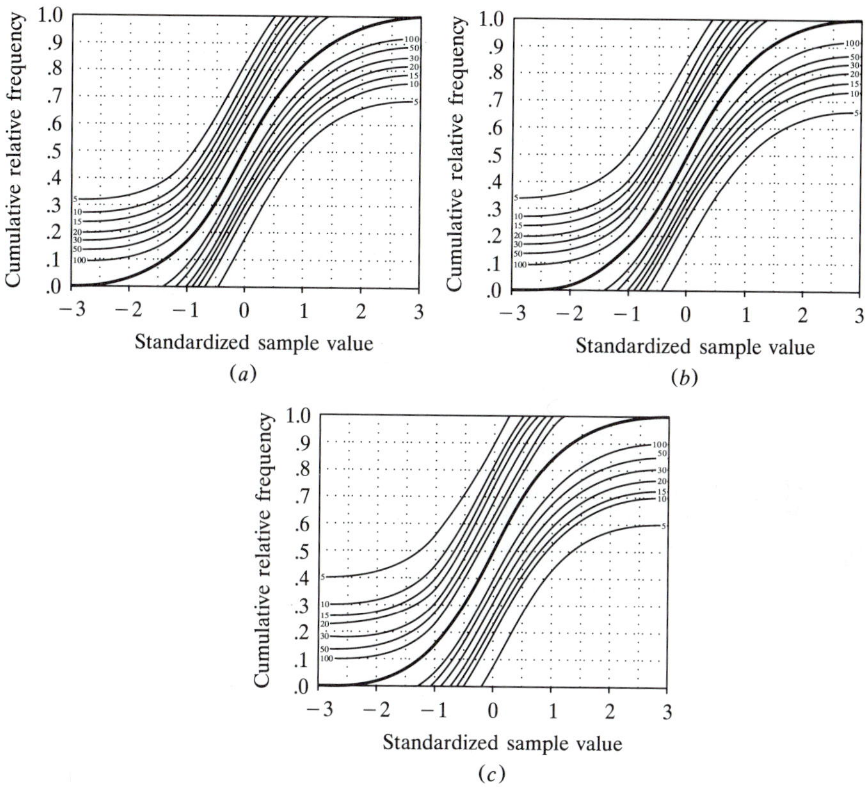

FIGURE 8.10
(a) 90% Lilliefors bounds for normal samples ($\alpha = .1$); (b) 95% Lilliefors bounds for normal samples ($\alpha = .05$); (c) 99% Lilliefors bounds for normal samples ($\alpha = .01$).

This is done by graphing the observed distribution on a Lilliefors graph. Figure 8.10 gives the Lilliefors graphs needed to test

$$H_0\text{: data are from a normal distribution}$$

$$H_1\text{: data are not from a normal distribution}$$

for various significance levels. The heavy curve in the center of the graph is the cumulative distribution for the standard normal curve. Curves to either side represent the Lilliefors bounds for the sample sizes indicated. If the observed relative cumulative frequency falls outside the bounds given for the specified sample size, then H_0 is rejected and it is concluded that the data are not from a normal distribution. The use of this technique is illustrated in Example 8.6.1.

Example 8.6.1. One random variable studied while designing the front-wheel-drive half-shaft of a new model automobile is the displacement (in millimeters) of the constant velocity (CV) joints. With the joint angle fixed at 12°, 20 simulations were

TABLE 8.1

Observation	Standardized observation	Relative cumulative frequency
1.1	−1.62	.05
1.3	−1.48	.10
1.4	−1.41	.15
1.5	−1.34	.20
1.9	−1.06	.25
2.5	−.63	.30
2.6	−.56	.35
3.2	−.13	.40
3.5	.08	.45
3.7	.22	.50
3.7	.22	.50
3.9	.36	.60
4.1	.50	.65
4.2	.57	.70
4.2	.57	.70
4.4	.72	.80
4.6	.86	.85
4.8	1.00	.90
4.9	1.07	.95
6.2	1.99	1.00

conducted. These data result ("Closed Loop," *Magazine of Mechanical Testing*, June 1980, p. 18):

6.2	1.9	4.4	4.9	3.5
4.6	4.2	1.1	1.3	4.8
4.1	3.7	2.5	3.7	4.2
1.4	2.6	1.5	3.9	3.2

For these data $\bar{x} = 3.39$ and $s = 1.41$. Since we are to compare the observed relative cumulative frequency distribution to that of the *standard* normal distribution, we first "standardize" these observations by subtracting $\bar{x}$ and dividing by s. The standardized observations are then ordered from the smallest to the largest and the relative cumulative frequency for each observation is found. The results of these calculations are shown in Table 8.1. To use these data to test

$$H_0\text{: data are from a normal distribution}$$

$$H_1\text{: data are not from a normal distribution}$$

at the $\alpha = .05$ level we graph the observed relative cumulative frequency on the Lilliefors graph of Fig. 8.10(*b*). The result is given in Fig. 8.11. Since the graph of the observed relative cumulative frequency does not fall outside the bands labeled 20, the size of our sample, we are unable to reject H_0. We have no evidence that the data are drawn from a nonnormal distribution.

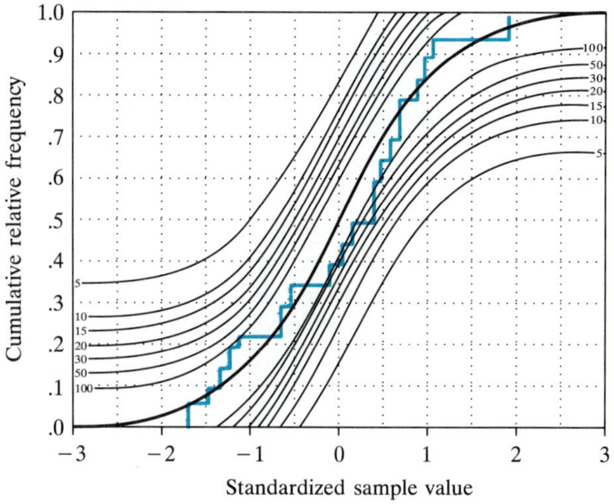

FIGURE 8.11
Test for normality of displacement of the CV joint at the $\alpha = .05$ level.

Testing Hypotheses on the Variance

We now turn our attention to testing hypotheses on the value of σ^2 or σ. These tests take the same general form as tests on the mean. These are summarized below with σ_0^2 denoting the null value of the population variance.

I	II	III
$H_0: \sigma^2 = \sigma_0^2$	$H_0: \sigma^2 = \sigma_0^2$	$H_0: \sigma^2 = \sigma_0^2$
$H_1: \sigma^2 > \sigma_0^2$	$H_1: \sigma^2 < \sigma_0^2$	$H_1: \sigma^2 \neq \sigma_0^2$
Right-tailed test	Left-tailed test	Two-tailed test

The test statistic used to test each of these is $(n-1)S^2/\sigma_0^2$. When sampling from a normal distribution, this statistic is known to follow a chi-squared distribution with $n-1$ degrees of freedom provided $\sigma^2 = \sigma_0^2$. As expected, the critical regions for right- and left-tailed tests are the upper- and lower-tail regions of the X_{n-1}^2 distribution respectively; the critical region for the two-tailed test consists of both the upper- and lower-tail regions of the distribution.

Example 8.6.2. Engineers designing the front-wheel-drive half-shaft described in Example 8.6.1 claim that the standard deviation in the displacement of the CV shaft is less than 1.5 millimeters. The estimated standard deviation based on the given 20 observations is 1.41 millimeters. Do these data support the contention of the engineers? To answer this question, we test

$$H_0: \sigma = 1.5$$

$$H_1: \sigma < 1.5$$

This is equivalent to testing

$$H_0: \sigma^2 = (1.5)^2$$

$$H_1: \sigma^2 < (1.5)^2$$

The observed value of the test statistic is

$$\frac{(n-1)s^2}{\sigma_0^2} = \frac{19(1.41)^2}{(1.5)^2} = 16.79$$

Since the test is left-tailed, we reject H_0 if this value is too small to have occurred by chance when H_0 is true. From the chi-squared table we see that

$$P[X_{19}^2 \le 14.6] = .25 \qquad \text{and} \qquad P[X_{19}^2 \le 18.3] = .50$$

Since the observed value of the test statistic, 16.79, lies between 14.6 and 18.3, the P value of the test lies between .25 and .50. Since this P value is rather large, we are unable to reject H_0. These data are not sufficient to allow us to claim that $\sigma < 1.5$ millimeters.

Recall that when sample sizes are moderate to large ($n \ge 25$), the T statistic can be used to make inferences on μ even though the normality assumption may be violated. It is when sample sizes are small that this becomes a serious problem. Unfortunately, the same cannot be said concerning the use of the X_{n-1}^2 statistic for making inferences on σ^2 and σ. For this reason, when constructing confidence intervals on σ^2 or testing hypotheses on the value of this parameter, a check for normality must be made. If the data are nonnormal then these methods should not be used.

8.7 ALTERNATIVE NONPARAMETRIC METHODS

We have seen how to use the Z and T statistics to test hypotheses concerning the mean of a normal distribution. The procedures presented assume that either we are sampling from a normal distribution or sample sizes are large enough so that deviations from the normality assumption do not seriously affect our results. In reality, experimenters often obtain data for which it is clearly unreasonable to assume an underlying normal distribution and for which sample sizes are small. When this occurs, usually the experimenter is advised to use a "nonparametric" test for location rather than the usual Z or T test. In this section we examine the meaning of the term "nonparametric" test. We also present some nonparametric alternatives for the usual Z and T tests for location.

The terms "nonparametric" and "distribution free" are often used interchangeably. When we use the term "nonparametric test," we shall mean a test with the property that no assumption is being made concerning the specific distribution from which the sample is drawn. Although we usually assume that the distribution is continuous, we do not have to specify the family to which the random variable under study belongs. In particular, we shall no longer have to assume that the random variable being studied is normally distributed. Hence nonparametric methods are applicable to a larger class of distributions than their normal theory analogs.

When comparing two statistical procedures designed to test essentially the same thing, we look at two characteristics: the probability of committing a Type I error and the power of the test. We want α to be small but at the same time we want a high probability of rejecting a false null hypothesis. Typically, for a fixed α level the normal theory procedures are more powerful than their nonparametric counterparts *when the assumptions underlying the normal theory test are met*. However, studies have shown that when these assumptions are not met, the use of normal theory procedures leads to tests that are approximate in the sense that the apparent α level is suspect. For example, if we run a chi-squared test for variance on data that is far from normal at an apparent α level of .05, the actual probability of rejecting a true null hypothesis may be far from .05. In some cases the approximations are excellent, but in others they are so bad as to be completely unacceptable. In any case, using a normal theory procedure in situations in which the normal theory assumptions are not valid is dangerous. In such cases we turn to nonparametric procedures. These methods are usually superior for analyzing data when the normal theory assumptions are not met; they compare very favorably to the normal theory tests even when the normal theory assumptions are met. The safe course of action is to follow the advice: when in doubt use a nonparametric test!

In this section we shall discuss the sign test and the Wilcoxon signed-rank test, both of which can be used to test for location in the form of population medians.

Sign Test for Median

Recall that for a continuous distribution the median for a random variable X is defined to be the value M such that

$$P(X < M) = P(X > M) = 1/2$$

That is, the median is the 50th percentile of the distribution. For a symmetric distribution such as the normal, the population mean and median are identical. We shall see that the sign test is simply a form of the binomial test, which was discussed in Sec. 8.3. Let X denote a continuous random variable with median M and let $X_1, X_2, \ldots, X_n$ denote a random sample of size n from this unspecified distribution. If M_0 denotes the hypothesized value of the population median, then the usual forms of the hypothesis to be tested can be stated as follows:

H_0: $M = M_0$	H_0: $M = M_0$	H_0: $M = M_0$
H_1: $M > M_0$	H_1: $M < M_0$	H_1: $M \neq M_0$
Right-tailed test	Left-tailed test	Two-tailed test

Under the assumption of a continuous distribution each of the differences $X_i - M_0$ has probability 1/2 of being positive, probability 1/2 of being negative, and probability 0 of being zero.

Let Q_+ denote the number of positive differences obtained. If H_0 is true, Q_+ is binomially distributed with parameters n and 1/2 and the expected value of Q_+ is $n/2$. That is, if H_0 is true, half the differences should be positive and the rest is negative. Note that in running a left-tailed test we want to detect a situation in which the true median M lies below the hypothesized median M_0. If this is true, we expect more than half the differences to be negative. This creates fewer positive differences than expected. Thus a logical procedure is to reject H_0: $M = M_0$ in favor of H_1: $M < M_0$ if the observed value of Q_+ is too small to have occurred by chance. In conducting a right-tailed test, the situation is reversed. In this case we reject H_0: $M = M_0$ in favor of H_1: $M > M_0$ if the observed value of Q_-, the number of negative differences obtained, is too small to have occurred by chance. A two-tailed test is conducted by rejecting H_0: $M = M_0$ in favor of H_1: $M \neq M_0$ if the smaller of Q_+ and Q_- is too small to have occurred by chance. The next example illustrates the use of the sign test.

Example 8.7.1. A standard method for completing a task on an assembly line yields a median completion time of 55 seconds. A new procedure is developed that should reduce the median time required. We want to test

$$H_0: M = 55$$

$$H_1: M < 55$$

To do so, 15 subjects are asked to complete the task, and these observations are obtained on the random variable X, the time required:

35	65	48	40	70	50	58	36
47	41	49	39	34	33	31	

The stem-and-leaf diagram for these data is shown in Fig. 8.12. Note that the diagram does suggest that X is not normally distributed. Since the sample size is rather small, we shall test for location using the nonparametric sign test. The test is left-tailed. Hence the test statistic is Q_+, the number of positive differences obtained when 55 is subtracted from each observation. From the stem-and-leaf diagram it is easy to see that only three observations exceed 55. Thus the observed value of the test statistic Q_+ is 3. The P value of the test is found by computing the probability of seeing a value this small or smaller under the assumption that Q_+ is binomially distributed with $n = 15$ and $p = 1/2$. From Table I of App. A, $P = P[Q_+ \leq 3 | n = 15, p = 1/2] = .0176$. Since this P value is small, we

3	569431
4	80719
5	08
6	5
7	0

FIGURE 8.12
Stem-and-leaf diagram for the time required to complete a task on an assembly line: diagram suggests a nonnormal population.

reject H_0. We do have strong statistical evidence that the new procedure reduces the median time required to complete the task.

Since we assume that the underlying distribution is continuous, theoretically zero differences should not occur when conducting a sign test. However, as you might guess, sometimes zeros do occur in practice. These occur for various reasons but the primary problem is the lack of instruments capable of precise measurement of continuous phenomena such as time, length, speed, and volume. Treatment of zero differences has been considered extensively. Various recommendations as to how to treat these differences have resulted. These are our recommendations:

1. Assign to the zero differences the algebraic sign least conducive to the rejection of the null hypothesis. Thus for a left-tailed test we would consider zero differences to be positive; for a right-tailed test they would be considered to be negative. In a two-tailed test we assign to zero differences the algebraic sign of the less frequently occurring difference. For example, if one observed 3 negative signs, 15 positive signs, and 6 zeros in running a two-tailed test, then the 6 zeros would all be treated as though they were negative. This procedure makes sense because a zero difference supports the null hypothesis that $M = M_0$. The suggested technique gives the null hypothesis the benefit of the doubt by making it harder to reject H_0.
2. If the number of zeros is small relative to the sample size n, discard these differences and reduce the sample size accordingly.

Occasionally a situation arises in which the differences $X_i - M_0$ are such that we can observe the algebraic sign of each difference but not its magnitude. In this case, the sign test is about the only choice available for testing location. Exercise 53 is an example of this type of problem. Usually, the actual numerical value of the differences can be obtained. Unfortunately, the sign test does not make use of this additional information. It treats a negative difference of $-.1$ in exactly the same way as it does a negative difference of -1000. For data in which the actual differences can be found, a second nonparametric test is available for testing for location. This test, the Wilcoxon signed-rank test, makes use of both the sign and magnitude of the observed differences $X_i - M_0$.

Wilcoxon Signed-Rank Test

In this test we assume that $X_1, X_2, \ldots, X_n$ is a random sample of size n from a continuous distribution that is symmetric about an unknown median M. Consider the set of differences $X_i - M_0$, $i = 1, 2, 3, \ldots, n$, where M_0 is the hypothesized median of the distribution from which the sample is drawn. The null hypothesis to be tested is H_0: $M = M_0$ versus the usual alternatives H_1: $M > M_0$, H_1: $M < M_0$, or H_1: $M \neq M_0$. If H_0 is true, the differences $X_i - M_0$ are drawn from a distribution that is symmetric about zero. It is assumed that the differences are such that the magnitude as well as the algebraic sign of each can be obtained. To conduct the test, we form the set of n absolute differences $|X_i - M_0|$. These are then ranked from 1 to n in order of absolute magnitude with the smallest absolute difference receiving a rank of 1. These ranks, which we denote by $R_1, R_2, \ldots, R_n$, are then assigned the algebraic sign of the difference score that generated the rank. If H_0 is true, then each rank is just as likely to be assigned a positive sign as a negative one. Consider the statistics

$$W_+ = \sum_{\substack{\text{all}\\ \text{positive}\\ \text{ranks}}} R_i \qquad \text{and} \qquad |W_-| = \sum_{\substack{\text{all}\\ \text{negative}\\ \text{ranks}}} |R_i|$$

If H_0 is true, then we should expect W_+ and $|W_-|$ to be approximately equal. If $M > M_0$, then W_+ would tend to be too large and $|W_-|$ too small. Similarly, if $M < M_0$, we would expect the reverse to be true. Hence, we define our test statistic to be $W = \min(W_+, |W_-|)$. The exact distribution of W has been tabled for various values of the sample size n and significance level α. One such table is Table VIII of App. A. Using this table, we reject H_0 if the observed value of W is less than or equal to the stated critical value.

In practice, ties in the difference scores $X_i - M_0$ can occur. If ties occur, the values for each tied group should be given the midrank of the group. For example, suppose that we observe difference scores of 3, -3, 3 which should occupy ranks 8, 9, and 10. We would assign each of the three values a rank of 9 and then assign the next largest difference score a rank of 11. Example 8.7.2 illustrates the idea.

Example 8.7.2. The melting point for a new lightweight material designed for use in automobile interiors is being investigated. It is known that due to impurities in the material, the melting point is a random variable uniformly distributed over a small temperature interval. It is thought that the median melting point is less than 120°C. Do these data support this contention?

115.1	117.8	116.5	121.0
120.3	119.0	119.8	118.5

We are testing

$$H_0\colon M = 120$$
$$H_1\colon M < 120$$

We first subtract 120 from each observation and then find the absolute value of each difference.

x_i	115.1	120.3	117.8	119.0	116.5	119.8	121.0	118.5
$x_i - 120$	-4.9	.3	-2.2	-1.0	-3.5	$-.2$	1.0	-1.5
$\lvert x_i - 120 \rvert$	4.9	.3	2.2	1.0	3.5	.2	1.0	1.5

We next rank these absolute differences from 1 to 8. Note that the value 1.0 occurs twice in what would normally be positions 3 and 4. We assign a rank of 3.5 to each of these values. The algebraic sign attached to each rank is the same as that of the difference that generated the rank.

$\lvert x_i - 120 \rvert$	4.9	.3	2.2	1.0	3.5	.2	1.0	1.5
Rank	8	2	6	3.5	7	1	3.5	5
Signed rank	−8	2	−6	−3.5	−7	−1	3.5	−5

For these data

$$W_+ = \sum_{\substack{\text{all}\\ \text{positive}\\ \text{ranks}}} R_i = 2 + 3.5 = 5.5$$

$$|W_-| = \sum_{\substack{\text{all}\\ \text{negative}\\ \text{ranks}}} |R_i| = 8 + 6 + 3.5 + 7 + 1 + 5 = 30.5$$

Since the test is a left-tailed test, the test statistic is W_+. We reject H_0 if the observed value of this statistic is too small to have occurred by chance. From Table VIII of App. A with $n = 8$ we see that we can reject H_0 at the $\alpha = .05$ level (critical point = 6), but we are unable to reject at $\alpha = .025$ (critical point = 4). Thus the P value of the test lies between .025 and .05. Since this P value is fairly small, we reject H_0 and conclude that the median melting point of this material is below 120°C.

If the sample size n exceeds values given in Table VIII of App. A, a large sample normal approximation may be used. The statistic

$$\frac{W - E[W]}{\sqrt{\text{Var } W}}$$

is approximately distributed as a standard normal random variable with

$$E[W] = \frac{n(n+1)}{4}$$

and

$$\text{Var } W = \frac{n(n+1)(2n+1)}{24}$$

Exercise 57 illustrates this approximation procedure.

The Wilcoxon signed-rank test is almost as sensitive to departures from the null hypothesis as the normal theory T test even when the underlying distribution is normal. For other symmetric distributions the signed-rank test is usually more powerful than the T test. Hence this test should be considered a strong competitor to the T test for practical problems. This is particularly true for small samples where violations of the normal theory tests assumptions are of greatest concern.

Note that although a Wilcoxon signed-rank test does not assume normality, it does assume symmetry. Procedures have been developed to test the validity of this assumption. One such test is given in [20].

CHAPTER SUMMARY

In this chapter we considered confidence interval estimation of the variance and standard deviation of a normal distribution. We also considered interval estimation of a mean when the population variance is unknown. This procedure entails the use of the Student-*t* or *T* distribution. We discussed this new continuous distribution in detail and saw that its properties are similar to those of the *Z* or standard normal distribution. In particular, we saw that for large sample sizes *t* points are well approximated by *z* points.

We next turned our attention to methods used in testing a statistical hypothesis. We found that we are always dealing with two hypotheses, the null hypothesis H_0 and its alternative H_1. The point of view of the researcher is stated as the alternative hypothesis. Thus we hope that our data will allow us to reject H_0, thereby accepting H_1. We design our tests in such a way so that we always know the probability of rejecting a true null hypothesis. We found that we are always subject to error when testing a hypothesis. If we reject a true null hypothesis, we commit a Type I error; if we fail to reject a false null hypothesis, a Type II error is committed. Two methods were described for deciding whether or not to reject H_0. The first method is referred to as hypothesis testing. In conducting a hypothesis test, we preset α. This is done by setting up a rejection or critical region prior to data collection. We reject H_0 if the observed value of the test statistic falls into this critical region. The second method for deciding whether to reject H_0 is called significance testing. Here no critical region is set prior to data gathering. Rather, we evaluate the test statistic and find the probability or *P* value of the test. The *P* value is the probability of observing a value of the test statistic as unusual or more unusual than that observed if the null value of the parameter θ is correct. Thus the *P* value is the smallest value at which we could have preset α and still have been able to reject H_0. We reject H_0 if the *P* value is deemed to be small. There are advantages and disadvantages to each method. You should be familiar with both as they are both used extensively.

We considered in some detail what are commonly called *T tests*. These are tests specifically designed to test a hypothesis on the mean of a normal distribution. We saw that these tests require that sampling be from a normal distribution and that this restriction is especially important for small samples. In Sec. 8.6 a method was given for testing for normality; in Sec. 8.7 we presented some nonparametric alternatives to the *T* test if the normality assumption appears to be invalid. Nonparametric tests are tests that make no assumption as to the family of distribution from which sampling is done.

Finally, we considered a method for testing a hypothesis on the variance or standard deviation of a normal distribution.

We introduced and discussed many new important terms and concepts that you should know. Some of these are:

Student-t distribution	Null hypothesis
Alternative hypothesis	Research hypothesis
Null value	Test statistic
Type I error	Type II error
α	β
Power	Size of test
Critical or rejection region	Level of significance
Significance test	Hypothesis test
Probability or P value	Critical level
Descriptive level of significance	Right-tailed test
Left-tailed test	Two-tailed test
Nonparametric test	Median

EXERCISES

Section 8.1

1. When programming from a terminal, one random variable of concern is the response time in seconds. These data are obtained for one particular installation:

1.48	1.26	1.52	1.56	1.48	1.46
1.30	1.28	1.43	1.43	1.55	1.57
1.51	1.53	1.68	1.37	1.47	1.61
1.49	1.43	1.64	1.51	1.60	1.65
1.60	1.64	1.51	1.51	1.53	1.74

(*a*) Construct a stem-and-leaf diagram. Does the assumption of normality appear reasonable?
(*b*) Find the unbiased point estimate for σ^2.
(*c*) Find a 95% confidence interval on σ^2.
(*d*) Find a 95% confidence interval on σ.
(*e*) Would you be surprised to hear the director of this installation claim that the standard deviation in response time is more than .2 second? Explain.

2. Highway engineers have found that the ability to see and read a sign at night depends in part on its "surround luminance." That is, it depends on the light intensity near the sign. These data are obtained on the surround luminance (in candela per square meter) of 30 randomly selected highway signs in a large metropolitan area. (Based on "Use of Retroreflectors in the Improvement of Nighttime Highway Visibility," H. Waltman, *Color*, pp. 247–251.)

10.9	1.7	9.5	2.9	9.1	3.2
9.1	7.4	13.3	13.1	6.6	13.7
1.5	6.3	7.4	9.9	13.6	17.3
3.6	4.9	13.1	7.8	10.3	10.3
9.6	5.7	2.6	15.1	2.9	16.2

(*a*) Find the sample variance for these data.
(*b*) Assume that the data are drawn from a normal distribution. Find a 90% confidence interval on the variance in the surround luminance in this area.
(*c*) Find a 90% confidence interval on the standard deviation in surround luminance.
(*d*) The normal probability rule (Sec. 4.5) implies that a normal random variable will lie within two standard deviations of its mean with probability .95. Use $\bar{X}$ and S to estimate the mean and standard deviation of the surround luminance in this area. Would it be unusual for the surround luminance for a randomly selected sign to exceed 18 cd/m^2? Explain.

3. X-ray microanalysis has become an invaluable method of analysis. With the electron microprobe, both quantitative and qualitative measures can be taken and analyzed statistically. One method for analyzing crystals is called the two-voltage technique. These measurements are obtained on the percentage of potassium present in a commercial product which theoretically contains 26.6% potassium by weight (based on "Quantitative Electron Probe Analysis of Low Atomic Number Samples with Irregular Surfaces" Klara Kiss, *Applied Spectroscopy*, February 1983, pp. 19–24):

21.9	23.4	22.1	22.1	24.7	24.6
24.0	24.1	24.2	26.5	23.8	25.3
24.8	24.8	24.5	27.8	24.9	
27.2	25.1	25.5	23.7	26.5	
22.0	26.7	25.2	23.1	25.4	

(*a*) Check the reasonableness of the normality assumption by constructing a stem-and-leaf diagram for these data.
(*b*) Find the sample variance for these data.
(*c*) Find a 99% confidence interval for σ^2.
(*d*) Find a 99% confidence interval for σ. Note that this confidence interval is fairly long. Suggest a way to improve the interval estimate for σ based on these data. Try your suggestion to see if the new estimate is more informative than that given by the 99% confidence interval.

***4.** (*One-sided confidence interval on σ^2.*) Since variance is a measure of consistency, it is usually hoped that σ^2 will be small. For this reason, it is sometimes useful to construct what is called a *one-sided* confidence interval for σ^2. That is, we want to find an interval of the form $(0, L]$, where L is a statistic with the property that $P[\sigma^2 \leq L] \doteq 1 - \alpha$. The derivation of such an interval is similar to that of the two-sided interval and is based on the diagram of Fig. 8.13. Use Fig. 8.13 and Theorem 8.1.1 to show that the upper bound for the one-sided confidence interval described is given by $L = (n - 1)S^2/\chi^2_{1-\alpha}$.

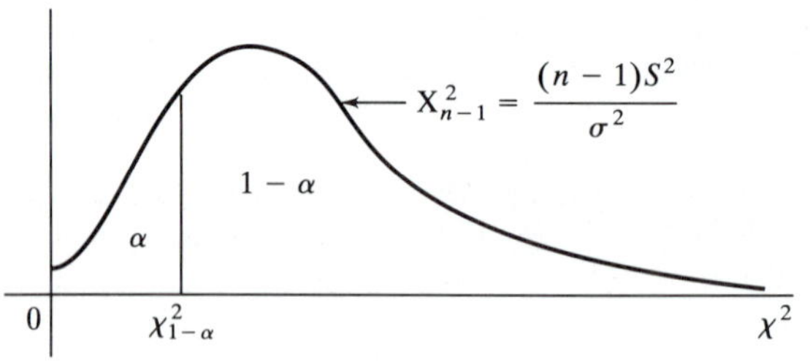

FIGURE 8.13
Partition of the X^2_{n-1} curve needed to derive a $100(1 - \alpha)\%$ "one-sided" confidence interval on σ^2.

***5.** Robotic technology is an area of rapid growth. It is reported that 315,000 industrial robots will be in use in American industry by the year 1995. One important feature of a robot is its accuracy. In a study of a particular robot used to apply adhesive to a specified location these data are obtained on the error (in inches) in the placement of the adhesive (based on "Robotics Growth—No End in Sight," D. Hegland, *Production Engineering*, April 1983, pp. 46–51):

.001	.002	.003	.002	.002
.007	.003	.004	.003	.006
.006	.003	.005	.004	.004
.001	.008	.001	.004	.003
.001	.003	.003	.005	.006

(*a*) Construct a stem-and-leaf diagram. Does the assumption that the placement error is normally distributed appear reasonable?
(*b*) Find the sample variance for these data.
(*c*) Use Exercise 4 to find 90% one-sided confidence intervals on σ^2 and σ.
(*d*) This robot is acceptable if its standard deviation does not exceed .005 inch. Does this criteria appear to be met? Explain.

***6.** In Theorem 7.1.3 we showed that the sample variance is an unbiased estimator for σ^2 regardless of the distribution of the random variable X. If X is normal, this property is obtained more easily by making use of Theorem 8.1.1 and the properties of the chi-squared distribution given in Sec. 4.3. Use these results to show that for a normal random variable X, $E[S^2] = \sigma^2$ and $\text{Var}\, S^2 = 2\sigma^4/(n-1)$.

***7.** (*Normal Approximation to* χ^2.) Note that the chi-squared table lists values of γ from 1 to 30. Therefore it can be used for sample sizes from 2 to 31. For samples larger than this, chi-squared points can be approximated by the formula

$$\chi_r^2 \doteq (1/2)\left[z_r + \sqrt{2\gamma - 1}\right]^2$$

For example, for a sample of size 50, the point $\chi^2_{.025}$ is given by

$$\begin{aligned}\chi^2_{.025} &\doteq (1/2)\left[z_{.025} + \sqrt{2\gamma - 1}\right]^2 \\ &= (1/2)\left[1.96 + \sqrt{2(49) - 1}\right]^2 \\ &\doteq 69.72\end{aligned}$$

(*a*) For a sample of size 100, approximate the points $\chi^2_{.05}$ and $\chi^2_{.95}$.
(*b*) Recent research indicates that heating and cooling commercial buildings with groundwater-source heat pumps is economically sound. The crucial random variable being studied is the water temperature. A sample of 150 wells in the state of California yields a sample standard deviation of 7.5° F. Find a 95% confidence interval on the standard deviation in temperature of wells in California. (Based on data appearing in the *Consulting Engineer*, May 1983, p. 18.)

***8.** In pouring glass for use in automobile windshields uniformity of thickness is desirable to prevent distortion. Find a 95% one-sided confidence interval on the standard deviation in thickness if a sample of 100 windshields yields a sample standard deviation of 0.01 inch.

Section 8.2

9. Use the T table to find each of these points:
(a) $t_{.05}(\gamma = 8)$;
(b) $t_{.95}(\gamma = 8)$;
(c) $t_{.975}(\gamma = 12)$;
(d) $t_{.025}(\gamma = 12)$;
(e) $t_{.05}(\gamma = 121)$;
(f) $t_{.05}(\gamma = 150)$;
(g) Point t such that $P[-t \leq T_{25} \leq t] = .90$;
(h) Point t such that $P[-t \leq T_{25} \leq t] = .95$;
(i) Point t such that $P[T_{15} \geq t] = .05$;
(j) Point t such that $P[T_{20} \geq t] = .10$;
(k) Point t such that $P[T_{16} \leq -t] = .05$;
(l) Point t such that $P[T_{30} \leq -t] = .10$.

10. The "supergopher" is a device invented to drill through arctic pack ice. It is a cone-shaped apparatus 5 feet high, 4 feet wide, and wound with a copper coil. Water heated to 180° F is pumped through the coil. This allows the gopher to melt a vertical round shaft through the ice. Let X denote the distance or depth that the gopher can drill per hour. These data are obtained on 10 test holes (depth is in feet):

2.0	1.7	2.6	1.5	1.4
2.1	3.0	2.5	1.8	1.4

(a) Use these data to find $\bar{x}$, s^2, and s.
(a) Find a 90% confidence interval on the average distance that can be drilled in an hour. (Based on information from "The Lost Squadron," by Steven Petrow, *LIFE*, December, 1992.)

11. Metal conduits or hollow pipes are used in electrical wiring. In testing 1-inch pipes, these data are obtained on the outside diameter (in inches) of the pipe:

1.281	1.293	1.287	1.286
1.288	1.293	1.291	1.295
1.292	1.291	1.290	1.296
1.289	1.289	1.286	1.291
1.291	1.288	1.289	1.286

(a) Find $\bar{x}$, s^2, and s for this sample.
(b) Assume that sampling is from a normal distribution. Find a 95% confidence interval on the mean outside diameter of pipes of this type.
(c) The makers of this type of pipe claim that the mean outside diameter is 1.29 inches. Does the confidence interval lead you to suspect this reported figure? Explain.

12. Lightweight hand-held, laser rangefinders are now used by civil engineers in hydrographic surveys. In testing one brand of rangefinder these data are obtained on the error (in meters) made in locating an object at a distance of 500 meters (*Civil Engineering*, February 1983, p. 52.):

−.10	−.02	.10	−.03	.09
.01	−.05	.05	−.06	.01
.03	.06	.02	−.07	.03

(*a*) Find point estimates for the mean and standard deviation in the error made by the laser.

(*b*) Assume that these measurement errors are normally distributed. Find a 90% confidence interval on the mean measurement error.

(*c*) A competitor claims that this particular model, on the average, overestimates the distance by at least .05 meter. Is there reason to doubt the claim based on the observed data? Explain.

*(*d*) Based on the normal probability rule (Sec. 4.5), would you consider it unusual for a single measurement error to be in excess of .15 meter? Explain.

13. One of the classic problems of operations research is the vehicle routing problem (VRP). This problem entails studying a system consisting of a given number of customers with known locations and demand for a commodity who are being supplied from a single depot by a number of vehicles with known capacity. The object of the study is to route the vehicles in such a way that the total distance traveled is minimized. The characteristics of a new algorithm, written in Pascal, are being investigated. These data are obtained on the cpu time required to solve the problem (*European Journal of Operational Research*, April 1983, pp. 388–393):

2.0	1.4	3.5	2.3	3.2	3.6
.1	3.5	2.2	2.1	2.4	1.5
2.2	2.3	2.7	1.9	1.7	1.8
3.1	1.5	1.5	2.6	2.8	2.5
2.5	3.9	.8	1.8	3.3	3.7

(*a*) Estimate the mean and standard deviation in the time required to solve a problem via this algorithm.

(*b*) Find a 99% confidence interval on the mean time required to solve a problem.

(*c*) The best algorithm known to date for solving the problems studied is written in Algol and requires an average of 6.6 seconds cpu time. The solutions obtained are equivalent. Does the new algorithm appear to be more efficient than the old with respect to computing time? Explain.

14. To estimate the average number of pounds of copper recovered per ton of ore mined, a sample of 150 tons of ore is monitored. A sample mean of 11 pounds with a sample standard deviation of 3 pounds is obtained. Construct a 95% confidence interval on the mean number of pounds of copper recovered per ton of ore mined.

15. A certain amount of natural gas is produced with each barrel of crude oil. This gas escapes from the oil near the top of the well pipe. In an attempt to estimate the amount of natural gas available from wells in Kuwait these data are obtained on X, the number of cubic feet of gas obtained per barrel of crude oil (based on information found in "The Oil/Gas Separator: A New Cap for Quenching Oil Well Fires," *Energy and Technology*, December 1991, p. 1):

290	610	790	670	770
420	600	350	800	920
410	810	620	560	550
610	510	390	480	630
470	380	550	570	730
680	530	650	1000	720

(*a*) Construct a stem-and-leaf diagram for these data. Does the normality assumption that underlies the T procedures appear to be met? Explain.
(*b*) Construct a boxplot for these data. Are any data points flagged as outliers?
(*c*) Find a 99% confidence interval for the average volume of natural gas produced per barrel of crude oil by wells in Kuwait.
(*d*) If we wanted an interval based on these data that was shorter than the one found in part (*c*), what could be done to accomplish this?

16. Surface finishing for corrosion protection is usually the last manufacturing process that takes place before the sale or assembly of metal parts used in such things as automobiles and electrical appliances. This process is often done in shops that specialize in this procedure. A new technique for applying bright zinc plating to steel is being tested. The variable under study is the thickness of the resulting coating in microns. These data are obtained on 25 test strips (based on figures found in "The Cinderella of Manufacturing," D. J. C. Hemsley, *Professional Engineering*, vol. 5, no. 7, July/August 1992, pp. 18–20):

6.4	8.3	7.9	7.5	6.9
8.5	7.0	7.4	7.2	6.8
7.1	8.1	7.5	7.7	8.5
7.8	7.3	8.4	8.0	7.8
7.5	7.8	7.6	8.4	9.9

(*a*) Construct a double stem-and-leaf diagram for these data. Comment on the possible distribution of this random variable.
(*b*) Construct a boxplot for these data, and identify the data point that is flagged as an outlier.
(*c*) Suppose that upon investigation it is found that the unusual data point discovered via your boxplot was for a strip that was inadvertantly left in the coating solution longer than the procedure specified. What should be done with this data point?
(*d*) Construct a 95% confidence interval on the average thickness of the coating obtained via the new process.
(*e*) Would you be surprised to hear a claim that this average is 7.7 microns? Explain based on your confidence interval.

***17.** (*One-sided confidence interval on* μ.) A "one-sided" confidence interval can be used to approximate the maximum or minimum value of a population mean.
(*a*) An interval of the form $(-\infty, L]$ such that $P[\mu \leq L] \doteq 1 - \alpha$ allows us to place bounds on the maximum value of the population mean. Use Fig. 8.14 to show that $L = \bar{X} + t_\alpha S/\sqrt{n}$.

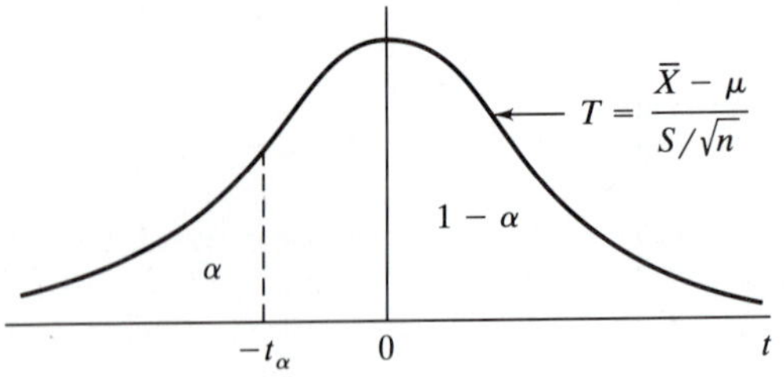

FIGURE 8.14

(*b*) An interval of the form $[L, \infty)$ allows us to place bounds on the minimum feasible value of the population mean. Show that $L = \bar{X} - t_\alpha S/\sqrt{n}$.

*__18.__ These data are obtained on the total nitrogen concentration (in ppm) of water drawn from a lake being considered for use as a source of drinking water for a locality:

.042	.023	.049	.036	.045	.025
.048	.035	.048	.043	.044	.055
.045	.052	.049	.028	.025	.039
.023	.045	.038	.035	.026	.059

Find a 95% one-sided confidence interval on the largest feasible value for μ. To be acceptable as a source of drinking water, the mean nitrogen content must lie below .07 ppm. Does this lake appear to meet this criterion? Explain.

*__19.__ (*Sample size required to estimate* μ.) Three factors determine the length of a confidence interval on μ. These are the confidence desired, the variability in the data, and the sample size. In an undesigned experiment it is possible that the resulting confidence interval is so long that it is almost useless. If σ is known or can be estimated from a small preliminary or "pilot" study, then it is possible to design an experiment in such a way that the resulting confidence interval will be short enough to be useful. This is done by selecting the sample size carefully.

(*a*) Let d denote the distance between $\bar{X}$, the center of the confidence interval and $\bar{X} + z_{\alpha/2}\sigma/\sqrt{n}$, the upper confidence bound. Thus $d = z_{\alpha/2}\sigma/\sqrt{n}$. Note that the confidence interval itself is of length $2d$. Solve this equation for n to show that the sample size required to estimate μ to within d units with $100(1-\alpha)\%$ confidence is

$$n \doteq \frac{(z_{\alpha/2})^2\sigma^2}{d^2} \qquad \sigma \text{ known}$$

$$n \doteq \frac{(z_{\alpha/2})^2\hat{\sigma}^2}{d^2} \qquad \sigma \text{ unknown}$$

(*b*) Reading digital displays in bright light poses a problem. Engineers want to design a filter to maximize both the luminance (brightness) and the chrominance (color) contrast. To do so, they intend to estimate the average number of footcandles in the cockpit of commercial airliners where the filter will be used. A preliminary pilot study is run and an estimated standard deviation of 500 footcandles is obtained. How large a sample is needed to estimate μ to within 50 footcandles with 95% confidence? ("LED's Have a Place in the Sun," M. Christiansen, *Design News*, April 1983, pp. 41–52.)

(*c*) To determine whether or not the copper ore in a particular area is pure enough for open pit mining to be feasible, mining engineers must estimate the average grade of the ore. Past experience with this type of ore indicates that the grade ranges from 1 to 4% copper. The normal probability rule and Exercise 25, Chap. 6, imply that a rough estimate of σ is $1/4$ of the range, or .75. How many test holes must be drilled to estimate μ to within .1% with 90% confidence?

*__20.__ A study is being designed to estimate the mean time required to assemble a panel of microprocessor chips for use in color television sets. An estimate of this mean is needed in order to set reasonable quotas for assembly line workers. A small pilot study is conducted and these data are obtained on the assembly time in minutes:

1.0	1.5	2.2	3.0	2.7
2.0	2.4	2.6	2.3	1.7

(*a*) Based on these data, estimate σ.
(*b*) How large a sample is required to estimate μ to within .2 minute with 99% confidence?

Section 8.3

21. In 1969 in the United States, on average, 8% of household waste was metal. Because of the increase in recycling efforts, it is hoped that this figure has been reduced. An experiment is run to verify this contention.
(*a*) Set up the appropriate null and alternative hypotheses for the experiment.
(*b*) Explain in a practical sense what has occurred if a Type I error has been committed.
(*c*) Explain in a practical sense what has occurred if a Type II error has been committed.
(*d*) Explain in a practical sense what it means to say that H_0 has been rejected at the $\alpha = .05$ level of significance.

22. The mean level of background radiation in the United States is .3 rem per year. It is feared that as a result of the increased use of radioactive materials, this figure has increased.
(*a*) Set up the appropriate null and alternative hypotheses to document this claim.
(*b*) Explain in a practical sense the consequences of making a Type I and a Type II error.

23. As mentioned in Chap. 1, an important aspect of the engineering sciences is model building. Once a theoretical model is devised to explain a physical phenomenon it must be tested to see that it yields results that are realistic. This testing is often done via computer simulation. In testing a model, we are testing

$$H_0\text{: model is credible}$$

$$H_1\text{: model is not credible}$$

(*a*) Explain in a practical sense what has occurred if a Type I error is committed. The probability of committing this error is referred to as the "model builder's risk." Do you see why this language is appropriate?
(*b*) Explain in a practical sense what has occurred if a Type II error is committed. The probability of committing an error of this type is called the "model user's risk." Does this seem appropriate?

24. (*Power.*) The probability of making the *correct* decision of rejecting H_0 when H_1 is true is called the *power* of the test. Thus power $= 1 - \beta$. Where does power fit into the chart of Fig. 8.7? What is the power of the test of Example 8.3.4 when $p = .7$? When $p = .8$?

25. Suppose we want to test

$$H_0\colon p \le .4$$

$$H_1\colon p > .4$$

based on a sample of size 15.

α → prob. of type error

(*a*) Find the critical region for an $\alpha \doteq .05$ level test.
(*b*) If when the data are gathered, $x = 11$, will H_0 be rejected?

26. Suppose we want to test

$$H_0\colon p \geq .7$$

$$H_1\colon p < .7$$

based on a sample of size 10.
(*a*) Find the critical region for an $\alpha \doteq .05$ level test.
(*b*) If when the data are gathered, $x = 5$, will H_0 be rejected?

27. It is a common practice to subject long-life items to larger than usual stress so that failure data can be obtained in a short amount of test time. Such tests are called accelerated life tests. Equipment used in computing makes use of metal oxide semiconductors (MOS). It is thought that "oxide short circuits" account for a majority of the early failures found in MOS integrated circuits. To verify this contention, a high-voltage screen test is applied to a number of circuits and 15 early failures are observed. Let X denote the number of failures due to oxide short circuits. (Based on "Testing for MOS IC Failure Modes," D. Edwards, *IEEE Transactions on Reliability*, April 1982, pp. 9–17.)
(*a*) Set up the appropriate null and alternative hypotheses.
(*b*) If H_0 is true and $p = .5$, what is the expected number of failures due to oxide short circuits in the 15 trials?
(*c*) Let us agree to reject H_0 in favor of H_1 if X is 11 or more. In this way we are presetting α at what level?
(*d*) Find β if $p = .6$; if $p = .7$; if $p = .8$; if $p = .9$.
(*e*) Find the power of the test if $p = .6$; if $p = .7$; if $p = .8$; if $p = .9$.
(*f*) If when the data are gathered, we observe 12 early failures that are due to oxide short circuits, will H_0 be rejected? What type error might be committed?
(*g*) If when the data are gathered, we observe 10 early failures due to oxide short circuits, will H_0 be rejected? What type error might be committed?

28. Quality and reliability are becoming important aspects of computer hardware and software. Past experience shows that the probability of failure during the first 1000 hours of operation for 16-kbit dynamic RAMs produced by a United States firm is .2. It is hoped that new technology and stricter quality controls have reduced this failure rate. To verify this contention, 20 systems will be monitored for 1000 hours and the number of failures will be recorded. (Based on "Software Quality Improvement," Y. Mizuno, *Computer*, March 1983, pp. 66–72.)
(*a*) Set up the appropriate null and alternative hypotheses.
(*b*) Explain in a practical sense the consequences of making a Type I and a Type II error.
(*c*) If H_0 is true and $p = .2$, what is the expected number of failures during the first 1000 hours in the 20 trials?
(*d*) Let us agree to reject H_0 in favor of H_1 if the observed number of failures, X, is at most 1. In this way we are presetting α at what level?
(*e*) Suppose that it is essential that the test be able to distinguish between a failure rate of .2 and a failure rate of .1. Find the probability that the test as designed will be unable to do so. That is, find β if $p = .1$. Find the power of the test if $p = .1$.

failed to rej. H_0 | $p=0.1$

(*f*) The results of part (*e*) indicate that the test as designed cannot distinguish well between $p = .1$ and $p = .2$. Keeping the sample size fixed at $n = 20$, can you suggest a way to modify the test that will lower β and to increase the power for detecting a failure rate of .1? Will α still be small enough to be acceptable? If not, can you suggest a way to redesign the experiment that will make both α and β low enough to be acceptable?

29. In Example 8.3.3 we test

$$H_0: p \leq .5$$

$$H_1: p > .5 \qquad \text{(majority of automobiles in operation have misaimed headlights)}$$

at the $\alpha = .0577$ level by agreeing to reject H_0 if at least 14 of the 20 cars sampled have misaimed headlights. We claim that values of X that are too large to occur by chance when $p = .5$ are also too large to occur by chance when $p < .5$. That is, if these values are rare when $p = .5$, they are even more rare when $p < .5$. To help see that this is true, find $P[X \geq 14]$ when $p = .4; .3; .2; .1$. Are each of these probabilities less than .0577 as expected?

30. A sample of size 9 from a normal distribution with $\sigma^2 = 25$ is used to test

$$H_0: \mu = 20$$

$$H_1: \mu = 28$$

The test statistic used is the sample mean, $\bar{X}$. Let us agree to reject H_0 in favor of H_1 if the observed value of $\bar{X}$ is greater than 25.

(*a*) If H_0 is true, what is the distribution of $\bar{X}$?
(*b*) In the diagram of Fig. 8.15, shade the region whose area is α.
(*c*) Find α. Remember that α is computed under the assumption that H_0 is true.
(*d*) If H_1 is true, what is the distribution of $\bar{X}$?
(*e*) In the diagram of Fig. 8.15, shade the region whose area is β. Remember that β is computed under the assumption that H_1 is true.
(*f*) Find β.
(*g*) Find the power of the test.
(*h*) If the sample size is increased, the standard deviation of $\bar{X}$ will decrease. What is the geometric effect of this on the two curves of Fig. 8.15?
(*i*) If the sample size is increased but the critical point is not changed, what will be the effect on α and β?

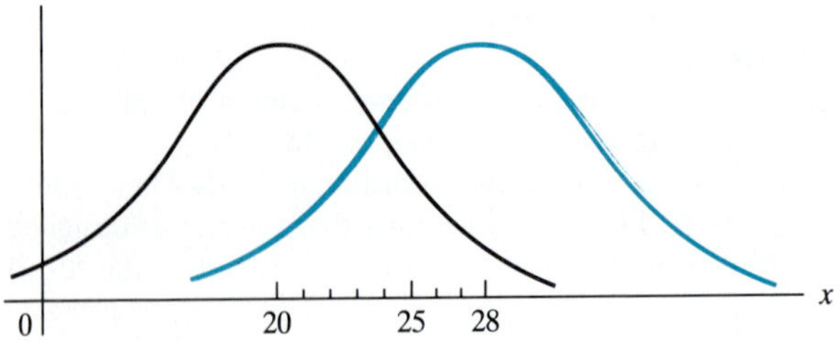

FIGURE 8.15

Section 8.4

31. Whenever a motorist encounters braking problems, especially an unpredictable pulling to one side, the villain is always held to be the brake pad. Trace elements, especially titanium, can combine with other elements to form minute particles of titanium carbonitride which alters the degree of friction between the pad and disc and leads to unequal wear. The percentage of titanium in a brake pad should not exceed 5%. A study is conducted to detect a situation in which the mean percentage of titanium in the brake pads being produced by a particular manufacturer exceeds 5%. (*Design Engineering*, February 1982, p. 24.)

(*a*) Set up the appropriate null and alternative hypotheses.

(*b*) Discuss the practical consequences of making a Type I and a Type II error.

(*c*) A sample of 100 brake pads yields a mean percentage of $\bar{x} = .051$. Assume that $\sigma = .008$. Find the P value for the test. Do you think that H_0 should be rejected? Explain. To what type error are you now subject?

32. The current particulate standard for diesel car emission is .6 g/mi. It is hoped that a new engine design has reduced the emissions to a level below this standard. (*Design Engineering*, February 1982, p. 13.)

(*a*) Set up the appropriate null and alternative hypotheses for confirming that the new engine has a mean emission level below the current standard.

(*b*) Discuss the practical consequences of making a Type I and a Type II error.

(*c*) A sample of 64 engines tested yields a mean emission level of $\bar{x} = .5$ g/mi. Assume that $\sigma = .4$. Find the P value of the test. Do you think that H_0 should be rejected? Explain. To what type error are you now subject?

33. It is thought that more than 15% of the furnaces used to produce steel in the United States are still open-hearth furnaces. To verify this contention, a random sample of 40 furnaces is selected and examined.

(*a*) Set up the appropriate null and alternative hypotheses required to support the stated contention.

(*b*) When the data are gathered, it is found that 9 of the 40 furnaces inspected are open-hearth furnaces. Use the normal approximation to the binomial distribution (Sec. 4.6) to find the P value for the test. Do you think that H_0 should be rejected? Explain. To what type error are you now subject?

34. It is known that defective items will be produced even on automated assembly lines. A particular process typically produces 5% defectives. If the proportion of defectives exceeds 5%, then the line must be shut down and adjusted.

(*a*) Set up the null and alternative hypotheses needed to detect a situation in which the proportion of defectives produced exceeds .05.

(*b*) Discuss the practical consequences of committing a Type I and a Type II error.

(*c*) A random sample of 100 items is selected and tested. Of these, 7 are found to be defective. Use the normal approximation to the binomial distribution to find the P value of the test. Do you think that H_0 should be rejected?

Section 8.5

35. Find the critical point(s) for conducting a hypothesis test on the mean with σ^2 unknown for

(*a*) a left-tailed test with $n = 25$; $\alpha = .05$

(*b*) a left-tailed test with $n = 150$; $\alpha = .10$

(*c*) a right-tailed test with $n = 20$; $\alpha = .025$

(*d*) a right-tailed test with $n = 16$; $\alpha = .01$
(*e*) a two-tailed test with $n = 20$; $\alpha = .10$
(*f*) a two-tailed test with $n = 30$; $\alpha = .05$

36. A new 8-bit microcomputer chip has been developed that can be reprogrammed without removal from the microcomputer. It is claimed that a byte of memory can be programmed in less than 14 seconds. (*Design News*, April 1983, p. 26.)
(*a*) Set up the null and alternative hypotheses needed to verify this claim.
(*b*) What is the critical point for an $\alpha = .05$ level test based on a sample of size 15?
(*c*) These data are obtained on X, the time required to reprogram a byte of memory:

11.6	14.7	12.9	13.3	13.2
13.1	14.2	15.1	12.5	15.3
13.3	13.4	13.0	13.8	12.3

Construct a stem-and-leaf diagram for these data. Does the normality assumption look reasonable?
(*d*) Test the null hypothesis. Can H_0 be rejected at the $\alpha = .05$ level? Interpret your result in a practical sense. To what type error are you now subject?

37. Ozone is a component of smog that can injure sensitive plants even at low levels. In 1979 a federal ozone standard of .12 ppm was set. It is thought that the ozone level in air currents over New England exceeds this level. To verify this contention, air samples are obtained from 30 monitoring stations set up across the region. ("Air Pollution Stress and Energy Policy," F. Bormann, *Ambio*, vol. XI, 1982, pp. 188–194.)
(*a*) Set up the appropriate null and alternative hypotheses for verifying the contention.
(*b*) What is the critical point for an $\alpha = .01$ level test based on a sample of size 30?
(*c*) When the data are analyzed, a sample mean of .135 and a sample standard deviation of .03 are obtained. Use these data to test H_0. Can H_0 be rejected at the $\alpha = .01$ level? What does this mean in a practical sense?
(*d*) What assumption are you making concerning the distribution of the random variable X, the ozone level in the air?

38. A model of Saudi Arabia's oil export strategy has been devised based on interviews with informed economists. The model is to be used to estimate the mean number of barrels of oil produced per day by this country. The usefulness of the model is to be partially checked by comparing the predicted mean for the year 1980 to its known value for that year, namely, 9.5 million barrels per day. ("Simulating Saudi Arabia's Oil Export Strategy," A. Picardi and A. Shorb, *Simulation*, January 1983, pp. 20–27.)
(*a*) Find the critical points for testing

$$H_0: \mu = 9.5$$

$$H_1: \mu \neq 9.5$$

at the $\alpha = .05$ level based on a sample of 50 simulations.
(*b*) For the data collected $\bar{x} = 9.8$ and $s = 1.2$. Test H_0. Can H_0 be rejected at the $\alpha = .05$ level? Based on these data, is there evidence that the model is not adequate? To what type error are you now subject?

39. A low-noise transistor for use in computing products is being developed. It is claimed that the mean noise level will be below the 2.5-dB level of products currently in use. (*Journal of Electronic Engineering*, March 1983, p. 17.)

(*a*) Set up the appropriate null and alternative hypotheses for verifying the claim.

(*b*) A sample of 16 transistors yields $\bar{x} = 1.8$ with $s = .8$. Find the P value for the test. Do you think that H_0 should be rejected? What assumption are you making concerning the distribution of the random variable X, the noise level of a transistor?

(*c*) Explain, in the context of this problem, what conclusion can be drawn concerning the noise level of these transistors. If you make a Type I error, what will have occurred? What is the probability that you are making such an error?

40. The Elbe River is important in the ecology of central Europe as it drains much of this region. Due to increased industrialization, it is feared that the mineral content in the soil is being depleted. This will be reflected in an increase in the level of certain minerals in the water of the Elbe. A study of the river conducted in 1982 indicated that the mean silicon level was 4.6 mg/l. ("Natural and Anthropogenic Flux of Major Elements from Central Europe." T. Paces, *Ambio*, vol. XI, November 1982, pp. 206–208.)

(*a*) Set up the null and alternative hypotheses needed to gain evidence to support the contention that the mean silicon concentration in the river has increased.

(*b*) A sample of size 28 yields $\bar{x} = 5.2$ and $s = 1.6$. Find the P value for the test. Do you think that H_0 should be rejected?

(*c*) What practical conclusion can be drawn from these data?

41. Coal-handling maintenance is a very young technology. The emission standard for coal-burning plants is 4.8 pounds SO_2/per million Btu/per 24-hour average. In an attempt to get emissions below this level engineers are experimenting with burning a blend of high- and low-sulfur coal. ("Upgrading and Maintaining Coal Handling," R. Rittenhouse, *Power Engineering*, March 1983, pp. 42–50.)

(*a*) Set up the null and alternative hypotheses needed to support the contention that the new mixture falls below the emission standard set by the government.

(*b*) Find the P value for the test if a sample of 200 readings yields a sample mean of 4.7 with a sample standard deviation of .5. Do you think that H_0 should be rejected? What does this mean in a practical sense?

42. Lasers are now used to detect structural movement in bridges and large buildings. These lasers must be extremely accurate. In laboratory testing of one such laser measurements of the error made by the device are taken. The data obtained are used to test

$$H_0: \mu = 0$$

$$H_1: \mu \neq 0$$

A sample of 25 measurements yields $\bar{x} = .03$ millimeter over 100 meters and $s = .1$. Find the P value for this two-tailed test. Do you think H_0 should be rejected? Interpret your result in a practical sense.

43. Clams, mussels, and other organisms that adhere to the water intake tunnels of electrical power plants are called *macrofoulants*. These organisms can, if left unchecked, inhibit the flow of water through the tunnel. Various techniques have been tried to control this problem, among them increasing the flow rate and coating the tunnel with Teflon, wax, or grease. In a year's time at a particular plant an unprotected tunnel accumulates a coating of macrofoulants that averages 5 inches in thickness over the length of the tunnel. A new silicone oil paint is being tested. It is hoped that this paint will reduce the amount of macrofoulants that adhere to the tunnel walls. The tunnel is cleaned, painted with the new paint, and put back into operation under normal working conditions. At the end of a year's time the thickness in inches of the macrofoulant coating is measured at 16 randomly selected locations within the tunnel. These data result (based on information from "Consider Non-fouling Coatings for Relief from Macrofouling, A. Christopher Gross, *Power*, October 1992, pp. 29–34):

4.2	4.5	4.1	4.6
4.4	4.0	4.7	4.3
5.0	6.2	3.6	4.5
5.1	3.5	3.0	2.8

(*a*) State the research hypothesis.

(*b*) Do these data support the contention that the new paint reduces the average thickness of the macrofoulants within this tunnel? Explain, based on the P value of the test.

(*c*) If α had been preset at .05, would H_0 have been rejected?

(*d*) Data in this problem are fictitious. Actually, the paint discussed in the journal article was much more effective than these data indicate. If these data had been real, do you think from a practical engineering point of view that the paint would be considered a major breakthrough in controlling the accumulation of macrofoulants? Explain.

44. Refineries, steel mills, food processing plants, and other industries separate oil and water using polyelectrolytes. These work better when pH is closely controlled. For example, chrome plating waste typically has a pH of 2.5. This wastewater must be neutralized before it is released into the environment. These data are obtained on the pH of wastewater samples that have been treated (based on a discussion found in "How to Choose a pH Measurement System," David M. Gray and Jeff Marshall, *Pollution Engineering*, November 1992, pp. 45–47):

6.2	6.5	7.6	7.7	7.0
7.0	7.2	6.8	7.5	8.1
7.1	7.0	7.1	7.8	8.5

Based on these data, is there evidence that the treatment process does not yield an average pH of 7 as desired? Explain based on the P value of the two-tailed test. If α had been preset at the .10 level, would H_0 have been rejected?

***45.** (*Confidence Intervals and Hypothesis Tests: A Connection.*) Confidence interval estimation on μ and hypothesis testing on μ, when σ^2 is unknown, are closely related. Both techniques make use of the T statistic $(\bar{X} - \mu)/(S/\sqrt{n})$. Show that if a given sample leads us to reject H_0: $\mu = \mu_0$ in favor of H_1: $\mu \neq \mu_0$ at the $\alpha = .05$ level, then a 95% confidence interval on μ based on the same data will *not* contain μ_0. *Hint:* Recall that H_0 is rejected if the observed value of the test statistic falls below the point $-t_{.025}$ or above the point $t_{.025}$.

***46.** (*Approximating sample sizes.*) In testing the hypothesis H_0: $\mu = \mu_0$, the experimenter can set α at any desired level. However, the value of β depends not only on the choice of α, but also on the difference between μ_0 and the alternative value μ_1.

The farther apart these values lie, the more likely it is that we shall be able to distinguish them from one another. In designing an experiment, we want to pick a sample size that gives us a high probability of rejecting H_0 when there is a real practical difference between μ_0 and μ_1. That is, we want β to be small. Choosing the appropriate size for a T test is not easy. The problem is due to the fact that when H_0 is not true, our test statistic no longer follows a T distribution. Rather, it has what is called a noncentral T distribution. Fortunately, tables have been constructed using this distribution that allow us to determine the proper sample size for testing H_0: $\mu = \mu_0$ for various values of α, β, and Δ, where $\Delta = |\mu_0 - \mu_1|/\sigma$ and σ is the standard deviation of X. Table VII of App. A is one such table. Its use is illustrated here.

Example. Let us test H_0: $\mu = 10$ versus H_1: $\mu > 10$ at the $\alpha = .05$ level. Assume that we want to be 90% sure of detecting a situation in which μ has gotten as large as 12. Assume also that a pilot study has been run and that $\hat{\sigma} = 4$. Here

$$\Delta \doteq |\mu_0 - \mu_1|/\hat{\sigma} = |10 - 12|/4 = .5$$

$$\alpha = .05 \qquad \text{and} \qquad \beta = .1$$

From Table VII of App. A we see that for a one-sided test with these characteristics we need a sample of size $n = 36$.

(*a*) A pilot study indicates that the standard deviation of a particular random variable X is 1.25. How large a sample is required to test

$$H_0: \mu = 20$$
$$H_1: \mu > 20$$

at the $\alpha = .05$ level and $\beta = .05$ level if it is important to be able to distinguish between $\mu = 20$ and $\mu = 21$?

(*b*) In Exercise 37 we tested

$$H_0: \mu = .12$$
$$H_1: \mu > .12$$

at the $\alpha = .01$ level based on a sample of size 30. From this study we see that $\hat{\sigma} = .03$. Suppose that a mean ozone level of .14 is so serious that we must have a probability of .95 of detecting the situation. Approximately how large a sample is required?

(*c*) In Exercise 39 we tested

$$H_0: \mu = 2.5$$
$$H_1: \mu < 2.5$$

A sample of size 16 yielded $s = .8$. Assume that the new transistors are not financially worth marketing unless they reduce to mean noise level to at most 2 dB. Approximately how large a sample is needed to distinguish between a mean of 2.5 and a mean of 2.0 if $\alpha = .025$ and $\beta = .05$?

Section 8.6

47. A new process for producing small precision parts is being studied. The process consists of mixing fine metal powder with a plastic binder, injecting the mixture into

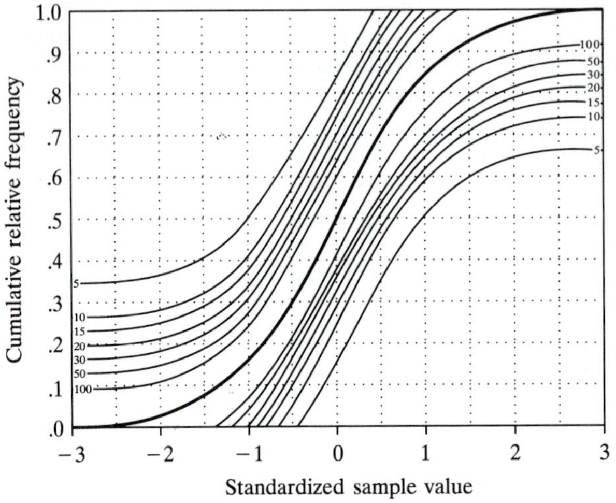

FIGURE 8.16
95% Lilliefors bounds for normal samples.

a mold, and then removing the binder with a solvent. These data are obtained on parts which should have a 1-inch diameter and whose standard deviation should not exceed .0025 inch:

1.0030	.9997	.9990	1.0054	.9991
1.0041	.9988	1.0026	1.0032	.9943
1.0021	1.0028	1.0002	.9984	.9999

For these data $\bar{x} = 1.00084$ and $s = .00282$.

(*a*) Use the Lilliefors graph of Fig. 8.16, to show that these data do not allow us to reject the normality assumption at the $\alpha = .05$ level.

(*b*) Test

$$H_0: \mu = 1$$
$$H_1: \mu \neq 1$$

at the $\alpha = .05$ level.

(*c*) Test

$$H_0: \sigma = .0025$$
$$H_1: \sigma > .0025$$

at the $\alpha = .05$ level.

48. Indoor natatoriums or swimming pools are noted for their poor acoustical properties. The goal is to design a pool in such a way that the average time that it takes a low-frequency sound to die is at most 1.3 seconds with a standard deviation of at most .6 second. Computer simulations of a preliminary design are conducted to see whether these standards are exceeded. These data are obtained on the time required for a low-frequency sound to die ("Acoustic Design in Natatoriums," R. Hughes and M. Johnson, *The Sound Engineering Magazine*, April 1983, pp. 34–36):

1.8	3.7	5.0	5.3	6.1	.5
2.8	5.6	5.9	2.7	3.8	5.9
4.6	.3	2.5	1.3	4.4	4.6
5.3	4.3	3.9	2.1	2.3	7.1
6.6	7.9	3.6	2.7	3.3	3.3

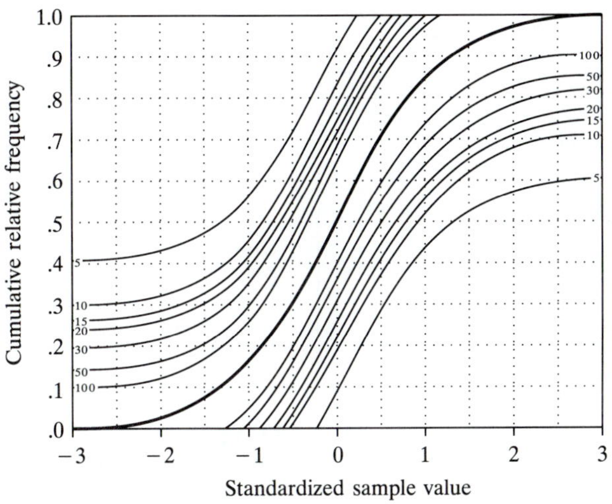

FIGURE 8.17
99% Lilliefors bounds for normal samples.

For these data $\bar{x} = 3.97$ and $s = 1.89$.

(*a*) Use the Lilliefors graph of Fig. 8.17 to show that these data do not allow us to reject the normality assumption at the $\alpha = .01$ level.

(*b*) Test

$$H_0: \mu = 1.3$$

$$H_1: \mu > 1.3$$

at the $\alpha = .01$ level.

(*c*) Test

$$H_0: \sigma = .6$$

$$H_1: \sigma > .6$$

at the $\alpha = .01$ level. Does it appear that the design specifications are being met?

49. Incompatibility is always a problem when working with computers. A new digital sampling frequency converter is being tested. It takes the sampling frequency from 30 to 52 kilohertz word lengths of 14 to 18 bits and arbitrary formats and converts it to the output sampling frequency. The conversion error is thought to have a standard deviation of less than 150 picoseconds. These data are obtained on the sampling error made in 20 tests of the device ("The Compatability Solution," K. Pohlmann, *The Sound Engineering Magazine*, April 1983, pp. 12–14):

133.2	−11.5	−126.1	17.9	139.4
−81.7	314.8	147.1	−70.4	104.3
56.9	44.4	1.9	−4.7	96.1
−57.3	−43.8	−95.5	−1.2	9.9

For these data $\bar{x} = 28.69$ and $s = 104.93$.

(*a*) Use the Lilliefors graph of Fig. 8.18 to show that these data do not allow us to reject the normality assumption at the $\alpha = .1$ level.

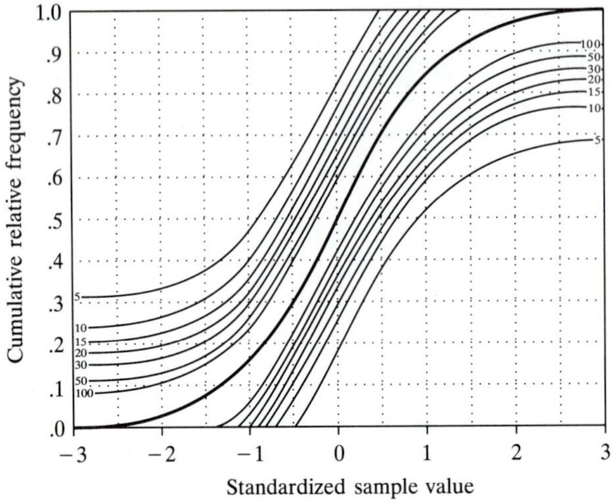

FIGURE 8.18
90% Lilliefors bounds for normal samples.

(*b*) Test

$$H_0\colon \mu = 0$$
$$H_1\colon \mu \neq 0$$

at the $\alpha = .1$ level.

(*c*) Test

$$H_0\colon \sigma = 150$$
$$H_1\colon \sigma < 150$$

at the $\alpha = .1$ level. Does the converter appear to be as accurate as claimed?

***50.** Consider the data of Example 6.2.1 concerning the life span of a lithium battery. Based on a stem-and-leaf diagram, we have reason to suspect that these data are not drawn from a normal distribution. Use the Lilliefors graph of Fig. 8.19 to test for normality at the $\alpha = .1$ level. Are our suspicions justified?

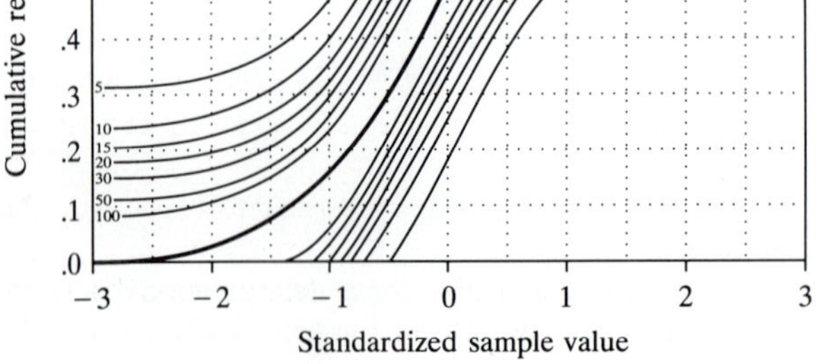

FIGURE 8.19
90% Lilliefors bounds for normal samples.

Section 8.7

51. In each case, use the sign test to decide whether H_0: $M = M_0$ will be rejected in favor of the stated alternative at the $\alpha = .05$ level based on the data given. Do not discard zeros.
(*a*) H_1: $M > M_0$; $n = 15$, $Q_+ = 13$, no zeros
(*b*) H_1: $M > M_0$; $n = 20$, $Q_+ = 15$, no zeros
(*c*) H_1: $M > M_0$; $n = 20$, $Q_+ = 15$, three zeros
(*d*) H_1: $M < M_0$; $n = 10$, $Q_+ = 1$, no zeros
(*e*) H_1: $M < M_0$; $n = 10$, $Q_+ = 1$, one zero
(*f*) H_1: $M \neq M_0$; $n = 15$, $Q_+ = 2$, no zeros
(*g*) H_1: $M \neq M_0$; $n = 15$, $Q_+ = 2$, one zero
In each case above, what is the P value of the test?

52. Engineers are designing the safety devices for use in a new amusement-park ride. They think that the median height of patrons of rides of this sort exceeds 68 inches. Based on the sign test, do these data support this contention? Support your answer by finding the P value of the conservative sign test.

Height in inches

65	73	72	71	68
74	74	66	68	69
70	66	72	67	73
69	70	73	70	74

53. Even with careful workmanship, digital scales may need some adjustment before being put into use. Unless there are systematic errors being made, the apparent zero of the scales before adjustment should fluctuate about true zero. That is, some scales should weigh a little heavy, whereas others should give readings that are a little light. Ten such scales are randomly selected and tested. These data are obtained on the accuracy of the zero reading:

heavy	light	heavy	heavy	heavy
light	light	light	heavy	heavy

Based on these data, can we reject H_0: $M = 0$ in favor of H_1: $M > 0$ at the $\alpha = .05$ level?

54. In Example 8.7.2 we were able to reject

$$H_0: M = 120$$

$$H_1: M < 120$$

at the $\alpha = .05$ level. If we had used the sign test, which ignores the magnitude of the difference scores, could we have rejected H_0 at the $\alpha = .05$ level? Explain by finding $P[Q_+ \leq 2 | n = 8 \text{ and } p = 1/2]$.

55. An experiment for treating tar sand wastewater was conducted to determine if a new treatment process removed more total organic carbon than a standard treatment process which is known to remove a median of 40 mg/l in a fixed detention time. Under the same experimental conditions the new process was replicated 10 times, yielding total organic carbon amounts removed of 38.8, 53.6, 39.0, 51.6, 40.1, 46.9, 40.9, 44.9, 41.0, and 43.2.

(*a*) What is $E[W]$?

(*b*) Using the signed-rank test, is there evidence that the new process removes significantly more total organic carbon than the standard process at the .05 level?

56. In an attempt to determine how many consultants are needed to answer questions of users at a computer center, these data are collected on X, the time in minutes required to answer a telephone inquiry:

1.5	1.0	5.0	1.9	3.0
1.3	2.1	1.7	6.5	4.2
6.3	5.6	5.1	2.5	6.9

(*a*) What is $E[W]$?

(*b*) Based on the signed-rank test, can we conclude that the median time required is less than 5 minutes? Explain, based on the P value of your test. (A zero score should be given the lowest rank and should be assigned the algebraic sign least conducive to rejecting the null hypothesis.)

***57.** A study of the expansion joints used in bridge beds is conducted. It is thought that these joints are expanding more than they were designed to expand, thus creating cracks in the pavement near the joint. The median design expansion at 95°F is 2 inches. Laboratory tests of 100 such joints are conducted at this temperature.

(*a*) What is $E[W]$?

(*b*) What is $\mathrm{Var}[W]$?

(*c*) Set up the appropriate null and alternative hypotheses.

(*d*) If $|W_-| = 1600$, can H_0 be rejected? Explain, based on the P value of the test.

REVIEW EXERCISES

58. A consumer group wants to estimate the mean cost of the base system for a personal computer with certain specifications. It is thought that these computers range in price from \$2390 to \$4000.

(*a*) How large a sample should be taken to estimate μ to within \$100 with 90% confidence?

(*b*) A random sample of size 50 yields these data (data in thousands of dollars):

2.43	2.86	2.74	2.75	2.69	2.64	2.91
2.89	3.18	3.00	3.21	3.07	3.72	3.24
3.17	3.57	3.37	3.56	3.30	2.32	3.09
2.99	3.20	3.25	3.70	3.45	2.82	2.88
2.71	3.25	2.86	2.93	3.45	3.11	3.86
2.96	3.00	2.88	3.19	3.56	3.21	3.33
3.39	3.14	2.90	3.49	3.02	3.56	2.87
2.32						

Construct a stem-and-leaf chart for these data. Use the digits 2 and 3 as stems 5 times each. Graph numbers beginning 2.0 and 2.1 on the first stem, those beginning 2.2 and 2.3 on the second stem, and so forth. Does the stem-and-leaf chart lead you to suspect that these data are not drawn from a distribution that is at least approximately normal?

(*c*) Find unbiased estimates for μ and σ^2 based on these data. Estimate σ. Is the estimate for σ unbiased?
(*d*) Find 90% confidence intervals on σ^2 and σ.
(*e*) Find a 90% confidence interval on μ.

59. Researchers are experimenting with a new compound used to bond Teflon to steel. The compounds currently in use require an average drying time of 3 minutes. It is thought that the new compound dries in a shorter length of time.
(*a*) Set up the null and alternative hypotheses needed to support the claim that the new compound dries faster than those currently in use.
(*b*) Discuss the practical consequences of making a Type I error; a Type II error.
(*c*) A pilot study shows that $\hat{\sigma} = .5$. Suppose that the new product is worth marketing if the average drying time can be shown to be 2.5 minutes or less. How large a sample is required to detect this situation with probability .95 with α set at .05?
(*d*) When the experiment is conducted, these data are obtained:

1.4	2.1	2.8	.9
2.4	1.7	3.7	2.7
2.6	1.9	2.8	2.8
2.2	2.2	3.4	1.9

Test the null hypothesis of part (*a*) at the $\alpha = .05$ level. Would you suggest marketing this new product?

60. It is thought that a majority of the procedures used in a statistical computer package run in less than .1 second. To verify this contention, a random sample of 20 programs which entail exactly one procedure is to be examined.
(*a*) Set up the appropriate null and alternative hypotheses needed to verify the claim.
(*b*) Let X denote the number of programs in which the procedure used runs in less than .1 second. Find the critical region for an $\alpha \doteq .025$ level test.
(*c*) When the test is conducted, 14 programs are found in which the procedure used runs in less than .1 second. Will H_0 be rejected? To what type error are you now subject?
(*d*) Find β if $p = .6$; if $p = .7$; if $p = .8$; if $p = .9$.
(*e*) Find the power of the test if $p = .6$; if $p = .7$; if $p = .8$; if $p = .9$.

61. Nickel powders are used in coatings used to shield electronic equipment from electromagnetic interference. It is thought that the mean size of the individual nickel particles in one such coating is less than 3 micrometers. Do these data support this contention? Explain, based on the P value of the appropriate test.

3.26	3.07	2.46	1.76
1.89	2.95	3.35	3.82
2.42	1.39	1.56	2.42
2.03	3.06	1.79	2.96

62. We want to test

$$H_0: \mu = 5$$
$$H_1: \mu > 5$$

based on a random sample of size 25. The sample standard deviation is 2 and the observed value of the sample mean is 5.5. What is the P value for the test?

63. The accuracy of a tank's artillery is obviously affected by target size, distance of the tank from the target, and other random factors such as wind and terrain. A series of tests is conducted on a standard target of size 2.3 by 3.4 meters. This is the average size of targets that NATO tanks are likely to encounter. Each target is 3000 meters from the tank. The angular measure from the center of the target to the point of impact on the target is given in mils. A mil is an angle of size 1/6400th of a 360° circle. (See Fig. 8.20.) The system aiming error that can be tolerated and still hit the target is investigated. These data are obtained. Note that zero denotes a direct center hit (no error); positive errors result in a high hit; negative errors in a low hit. (Based on information found in "Tank Gun Accuracy," Major Bruce Held and Master Sergeant Edward Sunoski, *Armor*, January 1993, p. 6.)

.8	−.3	.2	.2	0	0
−.1	.3	0	0	.2	.2
.1	0	0	0	.1	.1
−.1	−.1	−.1	.1	−.1	0
−.2	.1	.1	−.2	−.1	0
−.2	−.2	−.1	0	0	−.1
−.3					

(*a*) Sketch a stem-and-leaf diagram for these data. Are there any suspicious values in the data set?

(*b*) Sketch a boxplot for the data, and see if the one rather large data point qualifies as an outlier.

(*c*) It is thought that the "hit" that occurred when the outlier was obtained was, in fact, not a hit at all but, rather, an error in coding the data. For this reason, the outlier will be dropped from the data set in future analyses. Use the data given to find a 95% confidence interval on the average system error that occurs in firings that result in a hit.

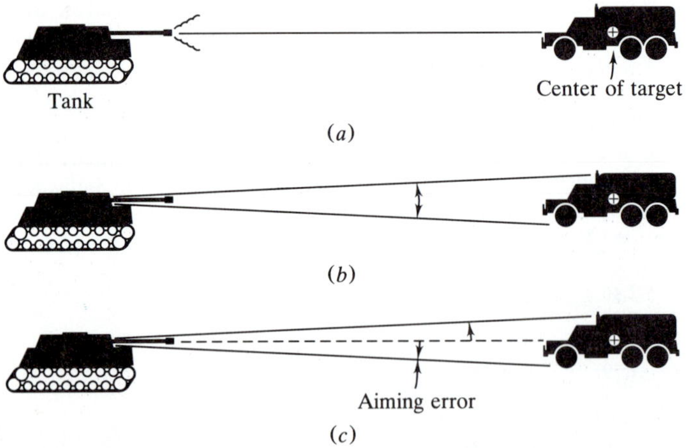

FIGURE 8.20
(*a*) A direct hit on the center of the target; (*b*) any shell fired within the window shown should hit the target; (*c*) angular measure from center of target to actual impact point = aiming error that still allows for a hit.

64. A pattern recognition device involves 3000 bits. Each bit can be either on (the bit value is 1) or off (the bit value is 0). These bits are either fixed or changeable. If they are fixed, they cannot be reversed by the programmer. If more than 95% of the bits are fixed, then the system will crash. Bits will be tested sequentially until a changeable bit is found. We want to detect a situation in which the system will crash. The test statistic is X, the number of bits sampled in order to obtain the first changeable bit. Note that if the system will crash, then the probability of finding a fixed bit exceeds .95 and the probability of finding a changeable bit is less than .05. The random variable X is approximately geometrically distributed. We are testing

$$H_0\colon\ p \geq .05 \qquad \text{(system will not crash)}$$

$$H_1\colon\ p < .05 \qquad \text{(system will crash)}$$

(*a*) Explain why X is only approximately geometrically distributed. That is, what geometric property is not strictly met?
(*b*) If H_0 is true, what is $E[X]$?
(*c*) If H_1 is true, would you expect X to exceed $E[X]$ or to be smaller than $E[X]$?
(*d*) Find the critical point for the test if you want α to be between .05 and .10.
(*e*) On a particular run 30 consecutive fixed bits are found. Is this evidence yet that the system will crash using the α level of part (*d*)? In this case, what type error is possible? In the context of this problem, what are the consequences of making this error?
(*f*) On a particular run 60 consecutive fixed bits are found. What conclusion can you draw from this? What type error is possible? In the context of this problem, what are the consequences of making this error? (Based on a study conducted in 1993–1994 by Eyal Schwartz, Department of Computer Science, Radford University.)

65. To study the cost effectiveness of energy-saving programs, data are gathered on the cost of such programs. These data are obtained on the cost of various programs per kilowatt hour of electricity saved.

Residential programs (cost in cents)

3	8	7	4	7	7
6	4	8	8	5	8
6	9	9	11	8	8
10	8	3	8	9	9
7	9	6	11	13	5
10	12	7	12	8	8
8	10	11	10	14	6
181					

Commercial and industrial programs (cost in cents)

1	3	3	2	3
3	2	4	5	3
4	3	3	6	
3	4	5	3	

(*a*) Construct stem-and-leaf diagrams for each data set. Comment on the likelihood that the normality assumption underlying T statistics is satisfied in each case.

(*b*) Construct a boxplot for each data set. Identify the extreme outlier in the residential data set. Suppose that upon investigation it is found that this data point was obtained in a very atypical program in an affluent region of the country. Since the program is so unusual, it is decided not to include it in tying to estimate the cost of programs that could be put in place in most areas of the country. Use the remaining data to construct a 95% confidence interval on the average cost of residential energy-saving programs currently in use. If the outlier had been included, what effect would it have in the confidence interval obtained?

(*c*) Find a 95% confidence interval on the variance of the cost of residential energy-saving programs. Do not use the extreme outlier in your calculation. If the outlier were used, what would be the effect on the length of the confidence interval obtained?

(*d*) Find 95% confidence intervals on the mean, variance, and standard deviation of the cost of energy-saving programs in the industrial and commercial sectors.

(*e*) If the true average cost of electricity to residential customers is 8 cents per kilowatt hour, is there reason to question the cost effectiveness of residential energy-saving programs? Explain.

(*f*) If the true average cost of electricity to industrial and commerical customers is 5 cents per kilowatt hour is there reason to question the cost effectiveness of commerical energy-saving programs? Explain. (Based on information found in "The Real Cost of Saving Electricity," *Technology Review*, February/March 1993, p. 12.)

COMPUTING SUPPLEMENT

II One-Sample T Tests and Confidence Intervals

If you have not read the computing supplement at the end of Chap. 6, do so now. We present here an SAS program that will generate the statistics needed to construct confidence intervals on μ, σ^2, or σ. It will also test a null hypothesis on the value of μ and print its P value. It utilizes PROC MEANS introduced in Chap. 6. Data used are the data of Exercise 36.

Statement	*Function*
DATA CHIP;	names data set
INPUT TIME;	names variable
STIME = TIME − 14;	creates a new random variable whose hypothesized mean is 0; the constant subtracted is the null value of μ
LINES;	signals beginning of data lines
11.6	
13.1	
13.3	data (one observation per line)
⋮	
12.3	
;	signals end of data

PROC MEANS MEAN VAR STD STDERR MAXDEC = 3; VAR TIME;	asks for the sample mean, variance, standard deviation, and standard error to be computed for the variable TIME
TITLE1 MAKING INFERENCES; TITLE2 ON THE MEAN;	titles the output
PROC MEANS T PRT MAXDEC = 3; VAR STIME;	asks for the observed value of the test statistic (T) and the P value (PRT) needed to test the null hypothesis H_0: $\mu = 14$

The printout of this program follows. Note that

① gives the observed value of $\bar{X}$;
② gives the observed value of S^2
③ gives the observed value of S;
④ gives the observed value of $S/\sqrt{n}$;
⑤ gives the observed value of the test statistic $(\bar{X} - 14)/(S/\sqrt{n})$ used to test H_0: $\mu = 14$;
⑥ gives the P value for a two-tailed test; the P value for a one-tailed test is half of this value or .0279.

MAKING INFERENCES
ON THE MEAN

VARIABLE	MEAN	VARIANCE	STANDARD DEVIATION	STD ERROR OF MEAN
TIME	13.447	1.056	1.027	0.265
	①	②	③	④

MAKING INFERENCES
ON THE MEAN

VARIABLE	T	PR > !T!
STIME	−2.09	0.0558
	⑤	⑥

III Testing for Normality

The procedure PROC UNIVARIATE generates a wide variety of summary statistics, including those generated by PROC MEANS. In addition, it can be used to construct a stem-and-leaf diagram for a data set and to test for normality. The test used is not the Lilliefors test. Rather, it uses the Shapiro-Wilk W Statistic or a modified Kolmogorov-Smirnov statistic. In either case the hypothesis of normality is rejected for small values of W. We illustrate its use by testing the data of Example 8.6.1 for normality.

Statement	*Function*
DATA AUTO;	names the data set
INPUT DISPLACE;	names the variable
LINES;	signals beginning of data lines
6.2	
4.6	
4.1	data (one observation per line)
⋮	
3.2	
;	signals end of data
PROC UNIVARIATE PLOT NORMAL;	asks for a stem and leaf plot and for a test for normality
TITLE1 IS THE DISTRIBUTION;	titles the output
TITLE2 NORMAL?;	

The printout of this program follows. Note that

① gives a stem-and-leaf plot for the data;
② gives the value of W, the statistic used to test for normality;
③ gives the P value for the test for normality. Since $P = .381$ is large, we are unable to reject the null hypothesis that the data are drawn from a normal distribution.
④ gives a box plot for the data.

IS THE DISTRIBUTION
NORMAL?

UNIVARIATE

VARIABLE = DISPLACE

MOMENTS				QUANTILES(DEF = 4)			
N	20	SUM WGTS	20	100% MAX	6.2	99%	6.2
MEAN	3.385	SUM	67.7	75% Q3	4.35	95%	6.135
STD DEV	1.40947	VARIANCE	1.98661	50% MED	3.7	90%	4.89
SKEWNESS	−0.143406	KURTOSIS	−0.670715	25% Q1	2.05	10%	1.31
USS	266.91	CSS	37.7455	0% MIN	1.1	5%	1.11
CV	41.6387	STD MEAN	0.315167			1%	1.1
T: MEAN = 0	10.7403	PROB > !T!	0.0001	RANGE	5.1		
SGN RANK	105	PROB > !S!	0.0001	Q3 − Q1	2.3		
NUM^ = 0	20			MODE	3.7		
W: NORMAL	0.947402 ②	PROB < W	0.381 ③				

EXTREMES

LOWEST	HIGHEST
1.1	4.4
1.3	4.6
1.4	4.8
1.5	4.9
1.9	6.2

```
STEM   LEAF                       ≠      BOX PLOT
    6  2                          1
    5                                        !
①   4  1224689                    7          !          ④
                                         +-----+
    3  25779                      5      *--+--*
    2  56                         2      +-----+
    1  13459                      5         !
       ----+----+----+----+
```

NORMAL PROBABILITY PLOT

```
6.5 +                                                   *
    !                                                    +
    !                                    *****  **       *
    !                             ******++
    !                      ++* *+
1.5 +         *    *  * * *
     +----+----+----+----+----+----+----+----+----+----+
         -2        -1        +0        +1        +2
```

CHAPTER 9

INFERENCES ON PROPORTIONS

In this chapter we discuss inferences on one proportion and the comparison of two proportions. As we have already seen, the binomial distribution can be used to test hypotheses on a proportion p when sample sizes are small. Here we see how to use the standard normal distribution to construct confidence intervals on p and test hypotheses concerning its value for large samples. We also begin our study of two sample problems by learning how to compare proportions based on samples drawn from two distinct populations.

9.1 ESTIMATING PROPORTIONS

The typical situation calling for the estimation of a proportion is as follows: There is a population of interest, a particular trait is being studied, and each member of the population can be classed as either having or failing to have the trait. We want to make inferences on p, the proportion of the population with the trait.

Example 9.1.1. Quality and reliability are important aspects of software. The smallest of bugs in computer software once foiled a space shuttle launch; in Japan a signal malfunction in an electronic telephone exchanger shut down phone lines for hours. To estimate the reliability of 16-kilobit (kbit) dynamic RAMs being produced by a particular company, a sample of size 100 is to be drawn and tested. We are interested in estimating p, the proportion of circuits that operate correctly during the first 1000 hours of operation. Here the population consists of all 16-kbit dynamic RAMs produced by the company; the trait being studied is the ability of the circuit to function correctly during the first 1000 hours of use. Each circuit either will have the trait, that is, it will operate correctly, or else it will not. ("Software Quality Improvement," Y. Mizuno, *Computer*, March 1983, pp. 66–72.)

To develop a logical point estimator for p, note that associated with a random sample of size n drawn from the population is a collection of n independent random variables $X_1, X_2, X_3, \ldots, X_n$, where

$$X_i = \begin{cases} 1 & \text{if the } i\text{th member of the sample has the trait} \\ 0 & \text{if the } i\text{th member of the sample does not have the trait} \end{cases}$$

For example, if we sample 100 circuits, we are dealing with a sample that would look something like this:

$$\begin{array}{lllll} x_1 = 1 & x_4 = 1 & & x_{98} = 1 \\ x_2 = 0 & x_5 = 0 & \cdots & x_{99} = 0 \\ x_3 = 0 & \vdots & & x_{100} = 0 \end{array}$$

In this case the first circuit sampled operates correctly during the first 1000 hours of use, so $x_1 = 1$; the second circuit does not operate correctly for this period of time, and so $x_2 = 0$, and so forth. Note that, in general, $X = \sum_{i=1}^{n} X_i$ gives the number of objects in the sample with the trait and that the statistic X/n gives the proportion of the sample with the trait. This statistic, called the *sample proportion*, is a logical point estimator for p:

Point estimator for p

$$\hat{p} = \frac{X}{n} = \frac{\text{number in samples with trait}}{\text{sample size}}$$

Example 9.1.2. Suppose that when the 100 tests mentioned in Example 9.1.1 are conducted, it is found that 91 of the 100 circuits tested perform properly during the first 1000 hours of operation. Thus 91 of the random variables $X_1, X_2, X_3, \ldots, X_{100}$ have value 1 and 9 assume the value 0. Based on these data, $\sum x_i = x = 91$ and

$$\hat{p} = x/n = 91/100 = .91$$

Confidence Interval on p

To develop a confidence interval on p, the distribution of $\hat{p}$ must be determined. This is accomplished by noticing that $\hat{p} = \sum X_i/n$ is actually nothing more than a very special sample mean. That is, $\hat{p} = \bar{X}$ is the average of the point binomial or zero-one random variables X_i. By the Central Limit Theorem $\hat{p}$ is approximately normally distributed with the same mean as the X_i's and with variance equal to $(\text{Var}\, X_i)/n$. The mean and variance of X_i is determined easily. Since $X_i = 1$ only if an object with the trait is sampled and the true proportion of objects in the sample with the trait is p, $P[X_i = 1] = p$. Consequently, $P[X_i = 0] = 1 - p$. The density for X_i is shown below:

x_i	1	0
$f(x_i)$	p	$1-p$

From this density it is easy to see that

$$E[X_i] = 1(p) + 0(1-p) = p$$

$$E[X_i^2] = 1^2(p) + 0^2(1-p) = p$$

and

$$\operatorname{Var} X_i = E[X_i^2] - (E[X_i])^2 = p - p^2 = p(1-p)$$

By the Central Limit Theorem we can conclude that $\hat{p}$ is approximately normally distributed with mean p and variance $p(1-p)/n$. Notice that we have just shown that $\hat{p}$ has the properties desirable in a point estimator. It is unbiased for p and has a small variance for large sample sizes.

To obtain a random variable that involves p whose distribution is known to serve as a starting point for the confidence interval derivation, standardize $\hat{p}$. The resulting random variable,

$$(\hat{p} - p)/\sqrt{p(1-p)/n}$$

follows an approximate Z distribution for large sample sizes. The partition of the standard normal curve shown in Fig. 9.1 is needed to derive the bounds for a $100(1-\alpha)\%$ confidence interval on p. From this diagram it can be seen that

$$P\left[-z_{\alpha/2} \le (\hat{p} - p)/\sqrt{p(1-p)/n} \le z_{\alpha/2}\right] \doteq 1 - \alpha$$

Isolating p in the middle of this inequality, we see that

$$P\left[\hat{p} - z_{\alpha/2}\sqrt{p(1-p)/n} \le p \le \hat{p} + z_{\alpha/2}\sqrt{p(1-p)/n}\right] \doteq 1 - \alpha$$

It appears that the confidence bounds for p are

$$\hat{p} \pm z_{\alpha/2}\sqrt{p(1-p)/n}$$

However, there is a problem here that has not been encountered before. The bounds for a confidence interval must be *statistics*. That is, they must be random variables whose expression contains no unknown parameters so that their

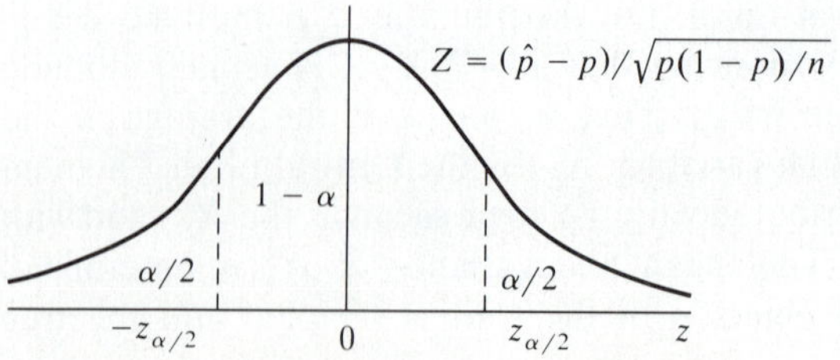

FIGURE 9.1
Partition of the Z curve needed to construct a $100(1-\alpha)\%$ confidence interval on p.

numerical value can be obtained from a sample. Unfortunately, as written, the above bounds are not statistics since the unknown parameter p appears in the expressions given. This means that we are attempting to use p to estimate p, a seemingly impossible situation! The problem can be overcome easily. The obvious method is to replace p by its unbiased estimator $\hat{p}$, to yield these bounds:

$$\hat{p} \pm z_{\alpha/2}\sqrt{\hat{p}(1-\hat{p})/n}$$

A legitimate question to ask is, "Since we are replacing the true standard deviation of $\hat{p}$ by an estimator for this standard deviation, should we switch from Z to T_{n-1} as was done when estimating μ?" The answer to this question lies in considering the sample size. The derivation of the confidence bounds is based on the Central Limit Theorem, which assumes that a large sample is available. Furthermore, in experimental settings in which this formula is to be applied the sample size is expected to be large enough so that there is very little difference between a z and a t point. Thus we shall write the confidence bounds as

Confidence interval on p

$$\hat{p} \pm z_{\alpha/2}\sqrt{\hat{p}(1-p)/n}$$

and use z points when the formula is applied. Confidence bounds for samples of size 1 through 30 have been developed based on the binomial distribution, and should be used for samples this small. Tables for these are found in [10]. The use of this method is illustrated in the next example.

Example 9.1.3. The point estimate for the proportion of 16-kbit dynamic RAMs that function correctly for at least 1000 hours based on a sample of size 100 is .91. From the standard normal table the point required to construct a 95% confidence interval on p is $z_{.025} = 1.96$. The bounds for the confidence interval are

$$\hat{p} \pm z_{\alpha/2}\sqrt{\hat{p}(1-\hat{p})/n}$$

or

$$.91 \pm 1.96\sqrt{.91(.09)/100}$$

$$.91 \pm .056$$

We can be approximately 95% confident that the true proportion of circuits that function correctly during the first 1000 hours of operation lies between .854 and .966. Converting to percentages, we can be approximately 95% confident that the true percentage of satisfactory circuits produced by this company lies between 85. and 96.6%. The word "approximately" is employed because we are approximating the distribution of $\hat{p}$ via the Central Limit Theorem and are also approximating p by $\hat{p}$ in finding the confidence bounds.

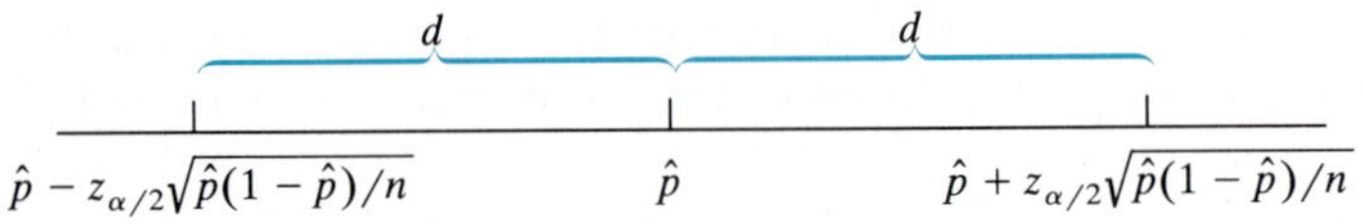

FIGURE 9.2
$100(1 - \alpha)\%$ confidence interval on p.

Sample Size for Estimating p

As when estimating a mean, it is possible that an experiment yields a confidence interval on p that is so long that it is virtually useless. This brings up one other important question."How large a sample should be selected so that $\hat{p}$ lies within a specified distance d of p with a stated degree of confidence?" There are two ways to answer this question. The first is applicable when an estimate of p based on some prior experiment is available. Consider the diagram of Fig. 9.2.

Since we are $100(1 - \alpha)\%$ sure that p lies in the interval shown, we are $100(1 - \alpha)\%$ sure that $\hat{p}$ and p differ by at most d, where d is given by

$$d = z_{\alpha/2}\sqrt{\hat{p}(1 - \hat{p})/n}$$

This equation is solved for n to obtain the following formula for finding the sample size needed to estimate p with a stated degree of accuracy and confidence when a prior estimate of p is available:

Sample size for estimating p, prior estimate available

$$n \doteq \frac{z_{\alpha/2}^2 \hat{p}(1 - \hat{p})}{d^2}$$

Example 9.1.4. How large a sample is required to estimate the proportion of 16-kbit dynamic RAMs that function properly during the first 1000 hours of use to within .01 (1 percentage point) with 95% confidence? We do have a prior estimate of p available, namely, $\hat{p} = .91$. By the above formula

$$n \doteq \frac{z_{\alpha/2}^2 \hat{p}(1 - \hat{p})}{d^2}$$

Since we want 95% confidence, the point $z_{\alpha/2} = z_{.025} = 1.96$. The maximum desired difference between $\hat{p}$ and p is $d = .01$. Substituting, we obtain

$$n \doteq \frac{(1.96)^2(.91)(.09)}{(.01)^2} \doteq 3147$$

To get the desired accuracy, we need substantially more data than we now have available!

The second method for determining sample size for estimating proportions is based on a result from elementary calculus. It can be shown (Exercise 9) that $\hat{p}(1 - \hat{p})$ will never exceed 1/4. Therefore this term can be replaced by 1/4 in the previous sample size formula to obtain the following formula for use when no prior estimate of p is available:

Sample size for estimating p, no prior estimate available

$$n \doteq \frac{z_{\alpha/2}^2}{4d^2}$$

This expression will be very useful to you since in most applications no prior estimate of p is available.

Example 9.1.5. A new method of precoating fittings used in oil, brake, and other fluid systems in heavy-duty trucks is being studied. How large a sample is needed to estimate the proportion of fittings that leak to within .02 with 90% confidence? Since no prior estimate of p is available,

$$n \doteq \frac{z_{\alpha/2}^2}{4d^2}$$

Here $z_{\alpha/2} = z_{.05} = 1.645$ and $d = .02$. Substituting, we have

$$n \doteq \frac{(1.645)^2}{4(.02)^2} \doteq 1692$$

(*Production Engineering*, March 1983, p. 66.)

It should be pointed out that sampling from a large finite population is usually done *without* replacement. Strictly speaking, the proportion of objects in the population with the given trait does vary from trial to trial. However, the change is so slight that its effect on our calculations is negligible. For this reason, the methods of this section can be used to study large populations even though the mathematical assumptions underlying the methods are not met completely.

9.2 TESTING HYPOTHESES ON A PROPORTION

When we have a preconceived idea of the value of a proportion or a percentage and we want statistical evidence to support our contention, we are in a hypotheses-testing situation. The hypotheses tested can assume any one of the usual three forms, depending on the purpose of the study. Let p_0 denote the null value of p. These forms are

I	II	III
$H_0: p = p_0$	$H_0: p = p_0$	$H_0: p = p_0$
$H_1: p > p_0$	$H_1: p < p_0$	$H_1: p \neq p_0$
Right-tailed test	Left-tailed test	Two-tailed test

In Sec. 8.3 we saw how to test these hypotheses for *small* samples. The test statistic used is X, the number of objects in the sample with the trait of interest. When the null hypothesis is true, this statistic has a binomial distribution with parameters n and p_0. When sample sizes are large, appropriate binomial tables usually are not available. In this case we must find another logical test statistic.

Consider the random variable used to generate the confidence bounds for p. That is, consider the statistic

$$(\hat{p} - p_0)\Big/\sqrt{p_0(1 - p_0)/n}$$

The statistic is a logical choice since, basically, it compares the unbiased point estimator for p, $\hat{p}$, to the null value p_0. Furthermore, if H_0 is true, then by the Central Limit Theorem this statistic has a standard normal distribution. Tests are conducted as you would expect. Namely, for a right-tailed test H_0 is rejected in favor of H_1 if the observed value of the test statistic is a large *positive* number; large *negative* numbers lead to rejection in a left-tailed test. In a two-tailed test H_0 is rejected for values of the test statistic that are too large in either the positive or the negative sense. These ideas are illustrated in the next example.

Example 9.2.1. The majority of faults on transmission lines are the result of external influences and are usually transitory. It is thought that more than 70% of all faults are caused by lightning. To gain evidence to support this contention we test

$$H_0: p = .7$$
$$H_1: p > .7$$

Data gathered over a year-long period show that 151 of 200 faults observed are due to lightning. The observed value of the test statistic is

$$(\hat{p} - p_0)\Big/\sqrt{p_0(1 - p_0)/n} = (151/200 - .7)\Big/\sqrt{.7(.3)/200}$$
$$\doteq 1.697$$

Since we are conducting a right-tailed test, we reject H_0 if this value, is unusually large. To decide whether 1.697 is a large positive value, we find the P value. From the standard normal table we see that $P[Z \geq 1.69] = .0455$ and $P[Z \geq 1.70] = .0446$. Since our observed value, 1.697, lies between 1.69 and 1.70, the P value lies between .0446 and .0455. There are two explanations for this small P value. The null hypothesis is true and we have just observed a rare event, one that occurs only about 4 times in every 100 trials; or the null hypothesis is not true and the true percentage of faults due to lightning exceeds 70%. The latter explanation seems more plausible, so we shall reject H_0 and conclude that $p > .7$. ("Standard Line Protection Packages," V. Narayen and F. Frey, *Brown Boveri Review*, February 1983, pp. 58–61.)

This method for testing hypothesis on p does assume that the sample size is "large." Following the guidelines given in Sec. 4.6, this is interpreted to mean that n and p_0 are such that $p_0 \leq .5$ and $np_0 > 5$ or $p_0 > .5$ and $n(1 - p_0) > 5$. These criteria are met in Example 9.2.1 since $p_0 = .7 > .5$ and $n(1 - p_0) = 200(.3) = 60 > 5$.

9.3 COMPARING TWO PROPORTIONS: ESTIMATION

The problem of comparing two proportions arises frequently in the engineering sciences. The general situation can be described as follows: There are two populations of interest, the same trait is studied in each population, each member of each population can be classed as either having the trait or failing to have it, and in each population the proportion having the trait is unknown. Random samples are drawn from each population. These samples are *independent* of one another in the sense that the objects drawn from one population do not determine in any way which objects are selected from the second population. Inferences are to be made on p_1, p_2, and $p_1 - p_2$, where p_1 and p_2 are the proportions in the first and second populations with the trait, respectively.

Example 9.3.1. A study is conducted to compare computer usage in Canadian business to that of businesses in the United States. Interest centers on the proportion of businesses in each country with an on-site mainframe computer. Here the two populations being studied are "businesses" in Canada and "businesses" in the United States. Remember that before sampling is done we must clearly specify what constitutes a "business." That is, we must clearly define the target populations. The trait under study is that of having an on-site computer; each business sampled either does or does not own such equipment. We draw a sample at random from each population. We use the sample data to compare the proportion of Canadian businesses with an on-site mainframe computer to that of businesses in the United States. (See Fig. 9.3.)

The problem of point estimation of the difference between two proportions is solved in the obvious way. We simply estimate p_1 and p_2 individually and take as our estimate for $p_1 - p_2$ the difference between the two. That is, our point estimator is

Point estimator for $p_1 - p_2$

$$\widehat{p_1 - p_2} = \hat{p}_1 - \hat{p}_2 = X_1/n_1 - X_2/n_2$$

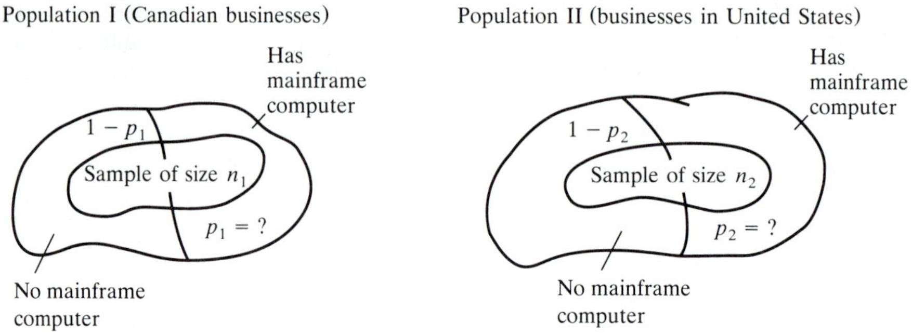

FIGURE 9.3
Independent samples drawn to estimate $p_1 - p_2$.

where n_1 and n_2 are the sizes of the samples drawn from the two populations and X_1 and X_2 are the number of objects, respectively, in the samples with the trait.

Example 9.3.2. Independent random samples of size 375 are selected from the population of Canadian businesses and from the population of businesses in the United States. It is found that 221 of the Canadian firms and 232 of the firms in the United States have mainframe computers. For these data

$$\hat{p}_1 = x_1/n_1 = 221/375 = .589$$
$$\hat{p}_2 = x_2/n_2 = 232/375 = .619$$
$$\widehat{p_1 - p_2} = \hat{p}_1 - \hat{p}_2 = .589 - .619 = -.03$$

(*Industrial Engineering*, February 1983, p. 6.)

Confidence Interval on $p_1 - p_2$

To extend the point estimator $\hat{p}_1 - \hat{p}_2$ to an interval estimator, we must pause to consider the probability distribution of this statistic. Its approximate distribution is given in Theorem 9.3.1.

Theorem 9.3.1. For large samples, the estimator $\hat{p}_1 - \hat{p}_2$ is approximately normal with mean $p_1 - p_2$ and variance $p_1(1 - p_1)/n_1 + p_2(1 - p_2)/n_2$.

Proof. We have shown in Sec. 9.1 that both $\hat{p}_1$ and $\hat{p}_2$ are approximately normal with means p_1 and p_2 and variances $p_1(1 - p_1)/n_1$ and $p_2(1 - p_2)/n_2$, respectively. Since the sum or difference of two normal random variables is normal (Exercise 41, Chap. 7), we can conclude that, the statistic $\hat{p}_1 - \hat{p}_2$ is at least approximately normally distributed. Furthermore, by the rules for expectation and the rules for variance

$$E[\hat{p}_1 - \hat{p}_2] = E[\hat{p}_1] - E[\hat{p}_2] = p_1 - p_2$$

and $$\text{Var}[\hat{p}_1 - \hat{p}_2] = \text{Var}\,\hat{p}_1 + \text{Var}\,\hat{p}_2 = p_1(1 - p_1)/n_1 + p_2(1 - p_2)/n_2$$

Note that Theorem 9.3.1 shows that the statistic $\hat{p}_1 - \hat{p}_2$ is an *unbiased* estimator for $p_1 - p_2$. To construct a $100(1 - \alpha)\%$ confidence interval on $p_1 - p_2$, we need a random variable whose expression involves this parameter and whose probability distribution is known at least approximately. This is easy to do via Theorem 9.3.1. We simply use the results of this theorem to standardize the statistic $\hat{p}_1 - \hat{p}_2$. In particular, we now know that the random variable

$$\frac{(\hat{p}_1 - \hat{p}_2) - (p_1 - p_2)}{\sqrt{p_1(1 - p_1)/n_1 + p_2(1 - p_2)/n_2}}$$

is at least approximately standard normal. Rather than repeat an algebraic argument given previously, let us consider three intervals that have been derived already and note their similarities.

Parameter being estimated	**Began derivation with**	**Distribution**	**Bounds**
$\mu(\sigma^2$ known)	$\dfrac{\bar{X} - \mu}{\sigma/\sqrt{n}}$	Z	$\bar{X} \pm z_{\alpha/2}\sigma/\sqrt{n}$
$\mu(\sigma^2$ unknown)	$\dfrac{\bar{X} - \mu}{S/\sqrt{n}}$	T	$\bar{X} \pm t_{\alpha/2}S/\sqrt{n}$
p	$\dfrac{\hat{p} - p}{\sqrt{p(1 - p)/n}}$	$\sim Z$	$\hat{p} \pm z_{\alpha/2}\sqrt{p(1 - p)/n}$

The algebraic structure of each of the beginning variables is the same and is of the form

$$\frac{\text{Estimator} - \text{parameter}}{D}$$

where D is either the standard deviation of the estimator or an estimator for this standard deviation. This is also the algebraic form assumed by the variable

$$\frac{(\hat{p}_1 - \hat{p}_2) - (p_1 - p_2)}{\sqrt{p_1(1 - p_1)/n_1 + p_2(1 - p_2)/n_2}} \sim Z$$

The confidence bounds in the previous cases took the form

$$\text{Estimator} \pm \text{probability point} \cdot D$$

Applying the notion to the case at hand, we find that the proposed confidence bounds for a confidence interval on $p_1 - p_2$ will be

$$(\hat{p}_1 - \hat{p}_2) \pm z_{\alpha/2}\sqrt{p_1(1 - p_1)/n_1 + p_2(1 - p_2)/n_2}$$

Once again there is a slight problem. The proposed bounds are not *statistics*. They include the unknown population proportions p_1 and p_2. As in the one sample case, this problem can be overcome by replacing the population proportions with their estimators $\hat{p}_1$ and $\hat{p}_2$. This leads to the following formula for finding confidence intervals on the difference between two population proportions:

Confidence interval on $p_1 - p_2$

$$(\hat{p}_1 - \hat{p}_2) \pm z_{\alpha/2}\sqrt{\hat{p}_1(1-\hat{p}_1)/n_1 + \hat{p}_2(1-\hat{p}_2)/n_2}$$

Example 9.3.3. The point estimate for the difference in the proportion of businesses in Canada and the proportion of businesses in the United States with on-site mainframe computers is $\hat{p}_1 - \hat{p}_2 = .589 - .619 = -.03$. A 95% confidence interval for this difference is

$$(\hat{p}_1 - \hat{p}_2) \pm z_{\alpha/2}\sqrt{\hat{p}_1(1-\hat{p}_1)/n_1 + \hat{p}_2(1-\hat{p}_2)/n_2}$$

or

$$-.03 \pm 1.96\sqrt{(.589)(.411)/375 + (.619)(.381)/375}$$

$$-.03 \pm .07$$

That is, we are 95% confident that the true difference in proportions lies in the interval $[-.10, .04]$. Note that since this interval contains the number 0, it is possible that there is really no difference in the two population proportions p_1 and p_2.

The question of determining the sample size needed to estimate the difference between two proportions with a stated degree of accuracy and confidence is more complex than in the one sample case. However, if samples of equal size are chosen from each population, then the problem can be solved just as in the one sample case. The procedure is outlined in Exercise 21.

9.4 COMPARING TWO PROPORTIONS: HYPOTHESIS TESTING

Sometimes problems arise in which it is theorized prior to the experiment that one proportion or percentage differs from another by a specified amount. The purpose of the experiment is to gain statistical support for the contention. These hypotheses take any one of these three forms, where $(p_1 - p_2)_0$ represents the null value of the difference in proportions:

I $H_0\colon p_1 - p_2 = (p_1 - p_2)_0$

$H_1\colon p_1 - p_2 > (p_1 - p_2)_0$

Right-tailed test

II $H_0\colon p_1 - p_2 = (p_1 - p_2)_0$

$H_1\colon p_1 - p_2 < (p_1 - p_2)_0$

Left-tailed test

III $H_0\colon p_1 - p_2 = (p_1 - p_2)_0$

$H_1\colon p_1 - p_2 \neq (p_1 - p_2)_0$

Two-tailed test

To test such hypotheses, a test statistic must be found. To derive such a statistic, consider the approximately standard normal random variable

$$\frac{(\hat{p}_1 - \hat{p}_2) - (p_1 - p_2)_0}{\sqrt{p_1(1 - p_1)/n_1 + p_2(1 - p_2)/n_2}}$$

that was used to construct confidence intervals on $p_1 - p_2$ in the previous section. This random variable is not a statistic since it contains the unknown population proportions p_1 and p_2. We again overcome this problem in the logical way. In particular, we replace p_1 and p_2 by their unbiased estimators $\hat{p}_1$ and $\hat{p}_2$ to obtain the approximately standard normal test statistic

$$\frac{(\hat{p}_1 - \hat{p}_2) - (p_1 - p_2)_0}{\sqrt{\hat{p}_1(1 - \hat{p}_1)/n_1 + \hat{p}_2(1 - \hat{p}_2)/n_2}}$$

This is a logical choice for a test statistic since it compares the estimated difference in proportions $\hat{p}_1 - \hat{p}_2$ with the hypothesized difference $(p_1 - p_2)_0$. If the hypothesized value is correct, then the estimated difference and the hypothesized difference should be close in value. This forces the numerator above to be close to zero and thus yields a small value for the test statistic. Large positive or large negative values of the test statistic indicate that the null hypothesis is not true and should be rejected in favor of an appropriate alternative.

Example 9.4.1. A corporation operates two foundries which are similar in size and which are engaged in the same production operations. An experimental safety program has been implemented at one location. Before expanding the program, the management wants to compare the proportion of workers injured during the trial period at the experimental site to that of its other plant. It is thought that the program is cost effective if these proportions differ by more than .05. We are testing

$$H_0\colon p_1 - p_2 = .05$$
$$H_1\colon p_1 - p_2 > .05$$

where p_1 and p_2 denote the proportions of injured workers at the control and experimental plants, respectively. Since making a Type I error is costly, let us preset α at .01. The critical point for this right-tailed test is $z_{.01} = 2.33$. When the trial period ends, it is found that 24 of the 263 workers at the control plant were injured whereas only 5 of the 250 workers at the experimental site received injuries. Based on these data,

$$\hat{p}_1 = 24/263 = .091 \qquad \hat{p}_2 = 5/250 = .020 \qquad \hat{p}_1 - \hat{p}_2 = .071$$

Is this difference large enough to allow us to conclude that the true difference in proportions exceeds .05? To decide, we evaluate the test statistic

$$\frac{(\hat{p}_1 - \hat{p}_2) - (p_1 - p_2)_0}{\sqrt{\hat{p}_1(1 - \hat{p}_1)/n_1 + \hat{p}_2(1 - \hat{p}_2)/n_2}} = \frac{.071 - .05}{\sqrt{(.091)(.909)/263 + (.02)(.98)/250}}$$
$$= 1.059$$

Since this value does not exceed the critical point of 2.33, we are unable to reject the null hypothesis at the $\alpha = .01$ level. We do not have the evidence that is felt necessary to justify expanding the safety program.

Pooled Proportions

Although the hypothesized difference $(p_1 - p_2)_0$ can be any value at all, the most commonly proposed value is zero. In this case the hypotheses considered previously compare p_1 and p_2 and take these forms:

I	II	III
H_0: $p_1 = p_2$	H_0: $p_1 = p_2$	H_0: $p_1 = p_2$
H_1: $p_1 > p_2$	H_1: $p_1 < p_2$	H_1: $p_1 \neq p_2$
Right-tailed test	Left-tailed test	Two-tailed test

Hypotheses of this sort can be tested via the previously developed test statistic with $(p_1 - p_2)_0$ set equal to zero. However, an alternative procedure is available. This alternative procedure, which is preferred by many statisticians, makes use of the fact that if H_0 is true, $\hat{p}_1$ and $\hat{p}_2$ are both estimators for the same proportion, which we denote by p. To see how to use this information, note that the variance of $\hat{p}_1 - \hat{p}_2$ is given by

$$p_1(1 - p_1)/n_1 + p_2(1 - p_2)/n_2$$

If H_0 is true, we can write this variance as

$$p(1 - p)/n_1 + p(1 - p)/n_2 = p(1 - p)(1/n_1 + 1/n_2)$$

We see that the random variable

$$\frac{\hat{p}_1 - \hat{p}_2}{\sqrt{p(1 - p)(1/n_1 + 1/n_2)}}$$

has a distribution that is approximately standard normal. We are now faced with the problem of estimating the unknown common population proportion p. Since $\hat{p}_1$ and $\hat{p}_2$ are both unbiased estimators for p, it makes sense to combine them in some way. We can simply average these estimators, but in so doing we ignore whatever differences might exist between the two sample sizes involved. To take these differences into account, we use a weighted average. Namely, we multiply each estimator by its corresponding sample size to obtain this "pooled" estimator for p:

Pooled estimator for p when $p_1 = p_2$

$$\hat{p} = \frac{n_1\hat{p}_1 + n_2\hat{p}_2}{n_1 + n_2}$$

The test statistic that results when p is replaced by $\hat{p}$ is

$$\frac{\hat{p}_1 - \hat{p}_2}{\sqrt{\hat{p}(1-\hat{p})(1/n_1 + 1/n_2)}}$$

The use of this statistic is demonstrated in our next example.

Example 9.4.2. Many consumers think that automobiles built on Mondays are more likely to have serious defects than those built on any other day of the week. To support this theory, a random sample of 100 cars built on Monday is selected and inspected. Of these, eight are found to have serious defects. A random sample of 200 cars produced on other days reveals 12 with serious defects. Do these data support the stated contention? To decide, we test

$$H_0\colon p_1 = p_2$$

$$H_1\colon p_1 > p_2$$

where p_1 denotes the proportion of cars with serious defects produced on Mondays. Estimates for p_1 and p_2 are

$$\hat{p}_1 = x_1/n_1 = 8/100 = .08 \qquad \text{and} \qquad \hat{p}_2 = x_2/n_2 = 12/200 = .06$$

The pooled estimate for the common population proportion is

$$\hat{p} = \frac{n_1\hat{p}_1 + n_2\hat{p}_2}{n_1 + n_2} = \frac{100(.08) + 200(.06)}{100 + 200}$$

$$= 20/300$$

$$= .066$$

The observed value of the test statistic is

$$\frac{\hat{p}_1 - \hat{p}_2}{\sqrt{\hat{p}(1-\hat{p})(1/n_1 + 1/n_2)}} = \frac{.08 - .06}{\sqrt{.066(.934)(1/100 + 1/200)}}$$

$$= .658$$

From the standard normal table we see that the probability of observing a value this large or larger is approximately .2546. That is, the P value is approximately .2546. Since this probability is large, we shall not reject H_0. We do not have sufficient statistical evidence to support the claim that cars built on Mondays are more likely to have serious defects than those built on other days.

Either one of the test statistics presented can be used to test H_0: $p_1 - p_2 = 0$ or H_0: $p_1 = p_2$, although the pooled statistic is preferable since it is thought to be more powerful. To test H_0: $p_1 - p_2 = (p_1 - p_2)_0$, where $(p_1 - p_2)_0 \neq 0$, pooling is not appropriate because $\hat{p}_1$ and $\hat{p}_2$ are estimating different proportions. In this case the first statistic presented is the proper test statistic.

Note that we are comparing proportions based on *independent* random samples drawn from two populations. In Chap. 14 we shall consider a method for comparing two proportions when the samples drawn are not independent.

CHAPTER SUMMARY

In this chapter, we considered methods that can be used to make inferences on a single proportion when sample sizes are large. We also saw how to determine the sample size required to estimate p to any desired degree of accuracy when we do and do not have prior estimates for p available.

We began our study of two sample problems by considering both point and interval estimation of the difference between two population proportions. The methods presented assume that samples are drawn independently. We also saw that H_0: $p_1 - p_2 = (p_1 - p_2)_0$ can be tested using as a test statistic the same random variable used to generate our confidence interval on $p_1 - p_2$, namely,

$$\frac{(\hat{p}_1 - \hat{p}_2) - (p_1 - p_2)_0}{\sqrt{\hat{p}_1(1 - \hat{p}_1)/n_1 + \hat{p}_2(1 - \hat{p}_2)/n_2}}$$

However, if $(p_1 - p_2)_0 = 0$, then a pooled procedure is preferable. This procedure makes use of the fact that if H_0 is true, $p_1 = p_2$. Since $\hat{p}_1$ and $\hat{p}_2$ are estimating the same thing, we pool them to form this estimator of the common population proportion p:

$$\hat{p} = \frac{n_1\hat{p}_1 + n_2\hat{p}_2}{n_1 + n_2} = \frac{X_1 + X_2}{n_1 + n_2}$$

Using this estimator, the test statistic used to test H_0: $p_1 - p_2 = 0$ is

$$\frac{\hat{p}_1 - \hat{p}_2}{\sqrt{\hat{p}(1 - \hat{p})(1/n_1 + 1/n_2)}}$$

We introduced the following term: Pooled estimator for p.

EXERCISES

Section 9.1

1. In order to be effective, reflective highway signs must be picked up by the automobile's headlights. To do so at long distances requires that the beams be on "high." A study conducted by highway engineers reveals that 45 of 50 randomly selected cars in a high-traffic-volume area have the headlights on low beam.
   (*a*) Find a point estimate for p, the proportion of automobiles in this type area that use low beams.
   (*b*) Find a 90% confidence interval on p.
   (*c*) How large a sample is required to estimate p to within .02 with 90% confidence?

2. A study of the electromechanical protection devices used in electrical power systems showed that of 193 devices that failed when tested, 75 were due to mechanical parts failures. ("Reliability of Protection Equipment in Operation," H. Hubensteiner, *Brown Boveri Review*, February 1983, pp. 111–114.)
   (*a*) Find a point estimate for p, the proportion of failures that are due to mechanical failures.
   (*b*) Find a 95% confidence interval on p.
   (*c*) How large a sample is required to estimate p to within .03 with 95% confidence?
3. In 1980 the Bureau of Labor Statistics conducted a study of 1000 minor eye injuries received by workers in the workplace. The study revealed that 600 of the workers involved were not wearing eye protection at the time of the injury. It also revealed that 900 of the injuries received could have been prevented through the proper use of protective eyewear. Assume that current conditions in the workplace have not changed substantially from those encountered in 1980 relative to the use of eye protection. (*Professional Safety*, March 1983, p. 12.)
   (*a*) Find a 90% confidence interval on the proportion of workers who receive minor eye injuries this year that will not be wearing eye protection at the time of the injury.
   (*b*) Find a 95% confidence interval on the proportion of minor eye injuries occurring this year that could be prevented through the proper use of protective eyewear.
4. A survey of companies using industrial robots showed that of 200 robots in use, 48 were used for loading and unloading. ("The Robots Are Coming," *Professional Safety*, P. Salem, March 1983, pp. 17–23.)
   (*a*) Find a 95% confidence interval on p, the proportion of industrial robots currently being used for loading and unloading.
   (*b*) Would you be surprised to hear someone claim that a majority of the robots in use are used for loading and unloading? Explain.
5. One problem associated with the use of the supersonic transport (SST) is the sonic boom. In the late 1960s and early 1970s preliminary tests were run over Oklahoma City, St. Louis, and other areas. After the tests were run a survey was to be conducted to estimate the percentage of people who felt that they could not live with the sonic booms. How large a sample should have been chosen to estimate this percentage to within 3 percentage points with 95% confidence?
6. The Environmental Protection Agency recently identified 30,000 waste dumping sites in the United States that were considered to be at least potentially dangerous. How large a sample is needed to estimate the percentage of these sites that do pose a serious threat to health to within 2 percentage points with 90% confidence?
7. It is said that "doctors bury their mistakes, architects cover them with ivy, engineers write long reports which never see the light of day" ("Foundation Failures," D. W. Joyce, *Civil Engineering*, April 1983, pp. 17–18). One area in which engineering mistakes are critical is dam-building. How large a sample is necessary to estimate the percentage of nonfederal earthen dams in the United States that are in need of immediate repair to within 1 percentage point with 90% confidence?
8. A market research study is to be conducted among users of a particular type of computer system. How many users should be sampled to estimate the percentage of users who plan to add terminals to within 4 percentage points with 90% confidence?

9. Consider the function $g(\hat{p}) = \hat{p}(1 - \hat{p})$.
(*a*) Find $g'(\hat{p})$.
(*b*) Find the critical point for g.
(*c*) Find $g''(\hat{p})$ and use this to argue that g assumes its maximum value at the critical point.
(*d*) What is the maximum value assumed by the function g?

Section 9.2

10. A poll of investment analysts taken earlier suggests that a majority of these individuals think that the dominant issue affecting the future of the solar energy industry is falling energy prices. A new survey is being taken to see if this is still the case. Let p denote the proportion of investment analysts holding this opinion. ("On the Way to Wall Street," C. Gadomski, *Solar Age*, May 1983, pp. 35–38.)
(*a*) Set up the appropriate null and alternative hypotheses.
(*b*) When the survey is conducted, 59 of the 100 analysts sampled agreed that the major issue is falling energy prices. Is this sufficient to allow us to reject H_0? Explain, based on the P value of the test.
(*c*) Interpret your results in the context of this problem.

11. A new computer network is being designed. The makers claim that it is compatible with more than 99% of the equipment already in use.
(*a*) Set up the null and alternative hypotheses needed to get evidence to support this claim.
(*b*) A sample of 300 programs is run, and 298 of these run with no changes necessary. That is, they are compatible with the new network. Can H_0 be rejected? Explain, based on the P value of the test.
(*c*) What practical conclusion can be drawn on the basis of your test?

12. It is thought that the no defect rate for 64-K-RAM devices produced in Japan is less than 8%.
(*a*) Set up the null and alternative hypotheses needed to support this claim.
(*b*) A sample of 64 of these devices is tested and 4 are found to have no defects. Can H_0 be rejected? Explain, based on the P value of the test. (*Computer*, March 1983, p. 17.)
(*c*) In the context of this problem, what conclusion can be drawn from your data?

13. It is thought that over 60% of the business offices in the United States have a mainframe computer as part of their equipment. (*Industrial Engineering*, February 1983, p. 6.)
(*a*) Set up the appropriate null and alternative hypotheses for supporting this claim.
(*b*) Find the critical point for an $\alpha = .05$ level test.
(*c*) When data are gathered, it is found that 233 of the 375 offices studied have mainframe computers. Can H_0 be rejected at the $\alpha = .05$ level? To what type error are you now subject?
(*d*) Explain, in the context of this problem, the practical consequences of making the error to which you are subject.

14. Opponents of the construction of a dam on the New River claim that less than half the residents living along the river are in favor of its construction. A survey is conducted to gain support for this point of view.
(*a*) Set up the appropriate null and alternative hypotheses.
(*b*) Find the critical point for an $\alpha = .1$ level test.

(*c*) Of 500 people surveyed, 230 favor the construction. Is this sufficient evidence to justify the claim of the opponents of the dam?

(*d*) To what type of error are you now subject? Discuss the practical consequences of making such an error.

15. A battery-operated digital pressure monitor is being developed for use in calibrating pneumatic pressure gauges in the field. It is thought that 95% of the readings it gives lie within .01 lb/in^2 of the true reading. In a series of 100 tests, the gauge is subjected to a pressure 10,000 lb/in^2. A test is considered to be a success if the reading lies within 10,000 $\pm$ .01 lb/in^2. We want to test

$$H_0: p = .95$$
$$H_1: p \neq .95$$

at the $\alpha = .05$ level.

(*a*) What are the critical points for the test?

(*b*) When the data are gathered, it is found that 98 of the 100 readings were successful. Can H_0 be rejected at the $\alpha = .05$ level? To what type error are you now subject? (*Nuclear News*, May 1983, p. 114.)

16. Power line noise, voltage variations, and power outages all can affect computer performance. When noise enters a television set, the result is static and snow; when noise enters a computer, errors can occur and circuits can be damaged. It is thought that more than 80% of all line disturbances at a particular computer site are noise.

(*a*) Set up the null and alternative hypotheses needed to verify this contention.

(*b*) Find the critical point for an $\alpha = .01$ level test.

(*c*) Of 150 line disturbances that occur during the study time, 133 are due to noise. Can H_0 be rejected at the $\alpha = .01$ level? Interpret your results in the context of this problem. (*Industrial Research and Development*, March 1983, p. 47.)

Section 9.3

17. A random sample of 500 workers engaged in research and development (R & D) during 1982 is selected. Of these, 178 earn over $36,000 per year. Of the 450 workers in R & D studied during 1983, 220 earn in excess of $36,000 per year. ("Salary Gains in R & D Continue, but at a Reduced Rate," R. Jones, *Industrial Research and Development*, March 1983, pp. 97–101.)

(*a*) Let p_1 and p_2 denote the proportion of workers engaged in research and development who earned over $36,000 per year in 1982 and 1983, respectively. Find point estimates for p_1, p_2, and $p_1 - p_2$.

(*b*) Find a 95% confidence interval for $p_1 - p_2$.

(*c*) Would you be surprised to hear someone claim that the proportion of R & D workers earning over $36,000 was the same in 1983 as it was in 1982? Explain, on the basis of the confidence interval of part (*b*).

18. Superplasticized concrete is formed by adding chemicals to conventional concrete to make it more fluid so that it can be placed more easily. Suppose that a sample of 50 new construction projects in the Dallas-Fort Worth area yields 15 that are using this type of concrete. A sample of 60 new projects in the Boston area also yields 15 using superplasticized concrete. ("Superplasticized Concrete Takes Off in Dallas," *Civil Engineering*, March 1983, pp. 39–42.)

(*a*) Let p_1 and p_2 denote the proportion of new construction projects in Dallas-Fort Worth and Boston, respectively, that are using superplasticized concrete. Find point estimates for p_1, p_2, and $p_1 - p_2$.

(*b*) Find a 95% confidence interval for $p_1 - p_2$.

(*c*) Would you be surprised to hear someone claim that the proportion of Dallas-Fort Worth projects using this type concrete is clearly larger than that in the Boston area? Explain, based on the confidence interval of part (*b*).

19. A study of the computer market is conducted. Random samples are drawn from among the users of the two leading mainframes. The purpose of the study is to estimate the proportion of users in each population that either do use or would like to use the small office system built by the mainframe supplier. These data result:

Type I	**Type II**
$n_1 = 200$	$n_2 = 190$
$x_1 = 62$	$x_2 = 76$

(*a*) Find point estimates for p_1, p_2, and $p_1 - p_2$.

(*b*) Find a 90% confidence interval on $p_1 - p_2$.

(*c*) Would you be surprised to hear someone claim that $p_1 = p_2$? Explain, based on the confidence interval of part (*b*).

20. The computer is expected to play an increasingly important role in crime control in the years to come. In 1983 the FBI had a noncomputerized Ident system containing the records of thousands of persons across the country. A random sample of 500 records shows that only 70% of these records include information on the disposition of the case. This is unfortunate since approximately 1/3 of all cases are eventually dismissed. If the dismissal is not a part of the record, then an innocent person could be stigmatized. ("When Computers Track Criminals," *Technology Review*, April 1983, p. 75.)

(*a*) Assume that a new computerized criminal history system is developed and implemented. A random sample of size 500 is selected from the cases recorded in the new system. It is found that 410 of these include information on the disposition of the case. Estimate the proportion of cases in the new system that include information on the disposition of the case.

(*b*) Estimate the difference in proportions between the old Ident system and the new computerized system. (Subtract in the order New − Ident.)

(*c*) Find a 95% confidence interval on the difference in proportions.

(*d*) Is it safe to say that the new system is superior to Ident in the sense that it contains more "disposition of case" information? Explain, based on the confidence interval of part (*c*).

21. (*Sample size for estimating* $p_1 - p_2$.) The difference between two population proportions, $p_1 - p_2$, is to be estimated based on independent random samples drawn from the respective populations. Each of the samples is each to be of size n. Show that in order to to estimate $p_1 - p_2$ to within d with $100(1 - \alpha)\%$ confidence, n is given by

Sample size for estimating $p_1 - p_2$, prior estimates for p_1 and p_2 available

$$n \doteq z_{\alpha/2}^2 \frac{[\hat{p}_1(1 - \hat{p}_1) + \hat{p}_2(1 - \hat{p}_2)]}{d^2}$$

or

Sample size for estimating $p_1 - p_2$, no prior estimates for p_1 and p_2 available

$$n \doteq \frac{z_{\alpha/2}^2}{2d^2}$$

22. What common sample size must we take from the populations of R & D workers in each of the years 1982 and 1983 to estimate $p_1 - p_2$ to within .02 with 90% confidence? Use the data of Exercise 17 to obtain estimates for p_1 and p_2.

23. What common sample size should be selected from the Ident files and the new computer files to estimate $p_1 - p_2$ to within .03 with 95% confidence? Use the data of Exercise 20 to obtain estimates for p_1 and p_2.

24. A study is to be conducted to estimate the difference in the proportions of defective items produced during two different shifts of assembly line workers. What common sample size should be used to estimate this difference to within .04 with 90% confidence?

25. Automotive engineers want to compare the performance of their new six-cylinder front-wheel-drive automobiles to their four-cylinder model. Let p_1 and p_2 denote the proportion of automobiles experiencing engine problems during the first 5000 miles of use for the two models, respectively. What common sample size should be used to estimate $p_1 - p_2$ to within .05 with 90% confidence?

Section 9.4

26. The use of optical fibers in telecommunications, the military, and industry is increasing rapidly. These fibers must be strong, durable, able to operate over a wide temperature range, and insensitive to radiation. Most fiber failures are due to a brittle fracture that grows into a complete crack. Two different fiber-drawing heat sources are being studied. These are carbon furnaces and CO_2 laser heating. A company currently uses a carbon furnace but will switch to laser heating if it can be shown that the latter method reduces the proportion of failures by more than .02. ("Optical Fiber Technology Research in the 1980s," F. Akers, P. W. Black, J. Irven, *Electrical Communications*, vol. 56, no. 4, 1981, pp. 331–337.)

(*a*) Let p_1 and p_2 denote the proportions of failures occurring using the carbon furnace and CO_2 laser heating, respectively. Set up the appropriate null and alternative hypotheses needed to support a move to the laser technique.

(*b*) Find the critical point for an $\alpha = .05$ level test.

(*c*) Of 100 test fibers produced using the carbon furnace, 5 failed whereas only 1 of the 100 fibers produced using the laser technique resulted in failure. Estimate p_1, p_2, and $p_1 - p_2$. Can H_0 be rejected at the $\alpha = .05$ level? Would you recommend that the company switch production methods?

(*d*) To what type error are you now subject? Discuss the practical consequences of making this error.

27. The cost of correcting a defect in a bipolar digital integrated circuit depends on when the defect is discovered. If it is discovered before it is integrated into a computer system, the cost may be only pennies. However, if it is not found until after the device is in the field it could cost thousands of dollars to repair. The electrical defect rate of two types of circuits produced by a particular company is being studied. It is suspected that the defect rate of their ALS circuits (advanced lower-power Schottky) is smaller than that of their LPS circuits (lower-power Schottky). (*EDN*, June 1983, p. 52.)

(*a*) Let p_1 and p_2 denote the proportions of ALS circuits and LPC circuits produced, respectively, that have electrical defects. Set up the null and alternative hypotheses needed to confirm their suspicions.

(*b*) What is the critical point for an $\alpha = .1$ level test?

(*c*) Two thousand circuits of each type are randomly selected and tested. It is found that three of the ALS and five of the LPS circuits have electrical defects. Estimate p_1, p_2, and $p_1 - p_2$. Based on these data, can H_0 be rejected at the $\alpha = .1$ level?

28. Today's diesel engines require smoother surface finishes and better consistency than in the past. Two types of abrasives are being tested for use on the microfinishers that are used to polish crankshafts. The first uses a paper and cloth abrasive; the second, a coated abrasive film. Both come on rolls that can tear, causing downtime and delay in the polishing process. It is thought that the proportion of rolls that tear is higher for the paper-cloth abrasive than for the abrasive film. However, since the abrasive film is the more expensive of the two, the difference in these proportions must exceed .10 in order for the abrasive film to be economical. ("Microfinishing Abrasive Film Meets Tighter Finishing Specifications," *Manufacturing Engineering*, May 1983, p. 65.)

(*a*) Set up the null and alternative hypotheses needed to support the contention that the abrasive film is economical. Let p_1 denote the proportion of rolls of the paper-cloth abrasive that tear during testing.

(*b*) What is the critical point for an $\alpha = .025$ level test.

(*c*) Fifteen of 50 rolls of the paper-cloth abrasive tear during testing, whereas only two of the 40 rolls of the abrasive film do so. Estimate p_1, p_2, and $p_1 - p_2$. Can H_0 be rejected at the $\alpha = .025$ level?

(*d*) To what type error are you now subject? Discuss the practical consequences of committing such an error.

29. Two types of metal detectors are in use in airports around the world. One is called a continuous wave detector, whereas the other is called a pulse field wave detector. Both devices are equally efficient at detecting large metal objects such as guns or knives. However, it is thought that the continuous wave detector tends to be less efficient in that it can be triggered more easily by objects such as coins, lipstick holders, and other small harmless metal objects. ("The Weapons Detector," C. Mann, *Technology Illustrated*, May 1982, pp. 80–82.)

(*a*) Let p_1 and p_2 denote the proportions of passengers that pass through the continuous wave and the pulse wave detectors, respectively, that trigger the device. Set up the null and alternative hypotheses needed to support the contention that the continuous wave detector will be triggered by a higher proportion of passengers than will the pulse wave device.

(*b*) Random samples of 175 passengers are observed passing through each of these types of devices. Of those passing through the continuous wave device, 113 triggered a warning. However, only 4 of those passing through the pulse field detector activated an alarm. Do you think that H_0 should be rejected? What is the P value of the test? What practical conclusion can be drawn from these data?

30. Shot peening is used to compress the surface area of metal parts to make them more resistant to fractures. It is done by bombarding the surface with small particles hurled at high velocity. Each time a particle hits, it puts a small dent in the surface and compresses the area directly beneath the surface. The bombardment continues until eventually the entire surface is compressed. Tests are conducted on a particular part to see if shot peening reduces the proportion of parts that fracture when put into use. These data result:

Not shot peened	**Shot peened**
$n_1 = 35$	$n_2 = 40$
number fractured = 7	number fractured = 3

Set up the appropriate null and alternative hypotheses. Based on these data, do you think that shot peening reduces the probability that a part will fracture when put into use? Explain, based on the P value of the test. ("Understanding Shot Peening: A Case History," G. Fett, *Modern Casting*, June 1983, pp. 29–31.)

31. Show that $\hat{p} = (X_1 + X_2)/(n_1 + n_2)$. That is, show that $\hat{p}$ can be found by combining the two samples into one and by finding the usual sample proportion for the new sample. Verify this numerically using the data of Exercise 30.

***32.** Let X_1 and X_2 denote the number of objects with the trait of interest in independently drawn random samples of sizes n_1 and n_2, respectively. Assume that these random variables are binomially distributed with parameters p_1 and p_2.

(*a*) Find the expected value of the pooled estimator $\hat{p}$.

(*b*) Show that if H_0: $p_1 = p_2$ is true, then $\hat{p}$ is an unbiased estimator for the common population proportion p.

REVIEW EXERCISES

33. A survey of mining companies is to be conducted to estimate p, the proportion of companies that anticipate hiring either graduating seniors or experienced engineers during the coming year.

(*a*) How large a sample is required to estimate p to within .04 with 94% confidence?

(*b*) A sample of size 500 yields 105 companies that plan to hire such engineers. Find a point estimate for p. Find a 94% confidence interval for p.

34. It is thought that the majority of the mining engineers that graduated in 1970 from U.S. schools are now employed in the coal mining industry.

(*a*) Set up the null and alternative hypotheses needed to gain statistical evidence to support this contention.

(*b*) A random sample of 50 of these individuals is selected and their current place of employment is determined. Twenty-six are working in the coal mining industry. Do you think that H_0 should be rejected? Explain, based on the P value of the test.

35. A procedure used to produce identical twins in cattle entails the microsurgical division of the embryo into two groups of cells followed by immediate embryo transfer. This procedure is thought to be more than 50% effective. ("Embryo

Transfer Technology for the Enhancement of Animal Reproduction," R. Mapletoft, *Biotechnology*, February 1984, pp. 149–159.)

(a) Set up the null and alternative hypotheses needed to support this claim.

(b) Find the critical point for an $\alpha = .05$ level test based on a sample of size 100.

(c) When the experiment is conducted, 55 of the transplants result in the birth of twins. Can H_0 be rejected at the $\alpha = .05$ level? Interpret your results in the context of this problem.

36. A programmable lighting control system is being designed. The purpose of the system is to reduce electricity consumption costs in buildings. The system eventually will entail the use of a large number of transceivers. Two types are being considered. In life testing these data are gathered on the number of transceiver failures for each type ("Effective Lighting Control," D. Peterson, *Lighting Design and Application*, February 1983, pp. 18–23):

Type I	Type II
$n_1 = 100$	$n_2 = 100$
$x_1 = 2$	$x_2 = 4$

(a) Find a point estimate for $p_1 - p_2$, the difference in the failure rates for the two types of transceivers.

(b) Find a 95% confidence interval for $p_1 - p_2$.

(c) Based on the interval of part (b), can we claim that $p_1 < p_2$? Explain.

(d) Is the interval found in part (b) short enough to give us a good idea of the actual value of $p_1 - p_2$? What common sample size is needed to estimate $p_1 - p_2$ to within .01 with 95% confidence?

37. One measure of quality and customer satisfaction is repeat business. A supplier of paper used for computer printouts sampled 75 customer accounts last year and found that 40 of these had placed more than one order during the year. A similar survey conducted at the end of the current year revealed that 35 of 50 customers ordered again. Do these data support the contention that there has been an increase in the proportion of repeat business over the 2-year period? Explain, based on the P value of your test.

38. A company is experimenting with a new method for etching circuits that should decrease the proportion of circuits that must be etched a second time. To be cost effective the difference in proportions between the old and new methods must exceed .1.

(a) Letting p_1 denote the proportion of circuits that must be redone using the old method, set up the null and alternative hypotheses required to show that the new method is cost effective.

(b) Find the critical point for $\alpha = .05$ level test of the hypothesis of part (a).

(c) These data are obtained on the number of circuits that must be reworked using each method:

Old	New
$n_1 = 25$	$n_2 = 50$
$x_1 = 4$	$x_2 = 2$

Can H_0 be rejected at the $\alpha = .05$ level? To what type error are you now subject? What are the practical consequences of making such an error?

39. One source of water pollution is gasoline leakage from underground storage tanks. A random sample of 100 gasoline stations is selected and the tanks are inspected. Twenty are found to have at least one leaking tank.

(*a*) Find a 95% confidence interval on the proportion of stations across the country with a leakage problem.

(*b*) Assume that there are approximately 375,000 stations in the United States. Find a 95% confidence interval on the number of stations with a leakage problem.

(*c*) How large a sample is required to estimate the proportion of stations with a leakage problem to within .02 with 95% confidence?

40. "The Desert Storm rules of engagement dictated that when an aircrew could not locate or positively identify their primary or secondary targets they were to return to base with their weapons." This rule was intended to minimize damage to civilian populations. During Desert Storm and Desert Shield 72,000 combat sortees were flown by allied forces. In 18,000 cases planes returned to base with their weapons. (Based on information taken from "Operations Law and the Rules of Engagement in Operations Desert Shield and Desert Storm," Lt. Col. John G. Humphries, *Airpower Journal*, Fall 1992, pp. 25–41.)

(*a*) Based on these data, find a point estimate for p, the proportion of combat missions which will return to base with their weapons in similar future engagements in which these rules of engagement are in force.

(*b*) Find a 95% confidence interval on p.

(*c*) Suppose that, in a future engagement, 10,000 combat missions are flown. Find a 95% confidence interval on the number of missions in which planes will return to base with their weapons.

CHAPTER 10

COMPARING TWO MEANS AND TWO VARIANCES

In this chapter, we continue the study of two sample problems by considering methods for comparing the means of two populations. This problem is considered under two different experimental conditions, namely when the samples drawn are independent and when the data are paired. These terms are explained in depth in the sections to come.

10.1 POINT ESTIMATION: INDEPENDENT SAMPLES

The general situation which we consider now is described as follows:

> There are two populations of interest, each with unknown mean. One random sample is drawn from the first population and one from the second in such a way that the objects selected from the first population have no bearing on those selected from the second. Samples selected in this way are said to be *independent* of one another. We want to estimate $\mu_1 - \mu_2$, the difference in population means, via a point estimator.

Example 10.1.1 illustrates this idea in a practical context.

Example 10.1.1. A study is conducted to compare the time required to inspect the wiring connections and insulation in two types of circuit breakers. Population I consists of all circuit breakers of the vacuum-interruptor type and population II

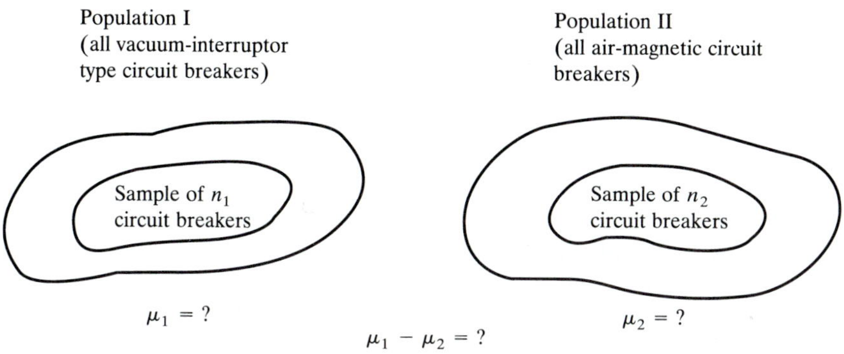

FIGURE 10.1
Independent samples of circuit breakers drawn from two different populations.

consists of all air-magnetic circuit breakers. A random sample is selected from each of these populations, and each circuit breaker chosen is inspected and the time in minutes required for the inspection is recorded. The samples are independent in the sense that the choice of a circuit breaker from population I has no effect whatsoever on the choice of circuit breakers from population II. We want to estimate $\mu_1 - \mu_2$, the difference in the mean times required to perform the inspection for the two populations. The study is visualized in Fig. 10.1. ("Vacuum-Interruptor CB's in High Voltage Applications," *EC + M*, July 1983, p. 34.)

The logical way to estimate $\mu_1 - \mu_2$ is to estimate each mean separately via its corresponding sample mean and then estimate $\mu_1 - \mu_2$ to be the difference between these sample means. That is, a logical point estimator for the difference in population means is the difference in sample means. Notationally, we are implying that

$$\widehat{\mu_1 - \mu_2} = \hat{\mu}_1 - \hat{\mu}_2 = \bar{X}_1 - \bar{X}_2$$

where $\bar{X}_1$ is the sample mean for the sample drawn from population I and $\bar{X}_2$ is the sample mean based on the independent sample drawn from population II.

Example 10.1.2. When the study of Example 10.1.1 is completed, these data result:

Vacuum-interruptor (I)				Air-magnetic (II)		
3.0	5.3	6.9	4.1	7.1	9.3	8.2
8.0	6.7	6.3		10.4	9.1	8.7
7.1	4.2	7.2		12.1	10.7	10.6
5.1	5.5	5.8		10.5	11.3	11.5

Based on these data,

$$\hat{\mu}_1 = \bar{x}_1 = 75.2/13 = 5.78 \text{ min}$$

$$\hat{\mu}_2 = \bar{x}_2 = 119.5/12 = 9.96 \text{ min}$$

The estimated difference in mean inspection times is

$$\widehat{\mu_1 - \mu_2} = \hat{\mu}_1 - \hat{\mu}_2 = \bar{x}_1 - \bar{x}_2 = 5.78 - 9.96 = -4.18$$

Based on these data, it appears that, on the average, the vacuum-interruptor circuit breaker can be inspected in about 4.18 minutes less time than the air-magnetic type breaker.

When finding confidence intervals for $\mu_1 - \mu_2$ or when testing a hypothesis concerning the value of this difference, it is necessary to know the distribution of the random variable $\bar{X}_1 - \bar{X}_2$. The next theorem pinpoints its distribution under the assumption that both samples are drawn from normal distributions. The theorem also shows that the estimator $\bar{X}_1 - \bar{X}_2$ is an unbiased estimator for $\mu_1 - \mu_2$. We shall use this theorem to motivate many of the statistical procedures presented later.

Theorem 10.1.1 (Distribution of $\bar{X}_1 - \bar{X}_2$). Let $\bar{X}_1$ and $\bar{X}_2$ be the sample means based on independent random samples of sizes n_1 and n_2 drawn from normal distributions with means μ_1 and μ_2 and variances σ_1^2 and σ_2^2, respectively. Then $\bar{X}_1 - \bar{X}_2$ is normal with mean $\mu_1 - \mu_2$ and variance $\sigma_1^2/n_1 + \sigma_2^2/n_2$.

Proof. In Theorem 7.3.4 we show that when sampling from a normal distribution, the sample mean is normal with mean μ and variance σ^2/n. Applying the result, we can conclude that $\bar{X}_1$ and $\bar{X}_2$ are normal with means μ_1 and μ_2 and variances σ_1^2/n_1 and σ_2^2/n_2, respectively. Exercise 41, Chap. 7, shows that any linear combination of independent normal random variables is normal. Since $\bar{X}_1$ and $\bar{X}_2$ are based on samples drawn independently from two populations, $\bar{X}_1$ and $\bar{X}_2$ are themselves independent. Applying Exercise 41, we can conclude that $\bar{X}_1 - \bar{X}_2$ is normal with mean $\mu_1 - \mu_2$ and variance $\sigma_1^2/n_1 + \sigma_2^2/n_2$ as claimed.

As in the one-sample case, because of the Central Limit Theorem, it is safe to assume that for large sample sizes $\bar{X}_1 - \bar{X}_2$ is at least approximately normal even if the samples are drawn from populations that are not themselves normal.

10.2 COMPARING VARIANCES: THE *F* DISTRIBUTION

There are two opinions as to the best way to compare the means of two normal populations. This is due to the fact that there are two distinct possibilities. These are

1. σ_1^2 and σ_2^2 are unknown and equal.
2. σ_1^2 and σ_2^2 are unknown and unequal.

One philosophy is that the experimental data or past experience should be used as a guide to determine the prevailing situation. Then one of two possible test statistics is chosen to compare means with the choice dependent on the perceived relationship between the population variances. A second philosophy disregards the relationship between the variances and uses the same test statistic to compare means in both cases. Since you will see both approaches used in research literature, we shall discuss them both. You can decide for yourself which you prefer.

The first philosophy mentioned requires that we develop a test for comparing the variances of two normal populations.

Hypothesis tests on the relationship between two variances can take any of the usual three forms depending on the purpose of the study. These forms are

I	II	III
$H_0: \sigma_1^2 = \sigma_2^2$	$H_0: \sigma_1^2 = \sigma_2^2$	$H_0: \sigma_1^2 = \sigma_2^2$
$H_1: \sigma_1^2 > \sigma_2^2$	$H_1: \sigma_1^2 < \sigma_2^2$	$H_1: \sigma_1^2 \neq \sigma_2^2$
Right-tailed test	Left-tailed test	Two-tailed test

To test any of these hypotheses, a test statistic must be developed. The statistic should be logical, but more importantly, it must be such that its probability distribution is known under the assumption that the null hypothesis is true. That is, its distribution must be known when it is assumed that the population variances are equal.

It is easy to find a logical statistic for comparing variances. Recall that the sample variances S_1^2 and S_2^2 are unbiased estimators for the population variances σ_1^2 and σ_2^2, respectively. Thus to compare σ_1^2 with σ_2^2, we simply compare S_1^2 with S_2^2. This is done not by looking at the difference of the two, but, rather, by looking at their ratio, S_1^2/S_2^2. If the null hypothesis is true and the population variances are really equal, then we expect S_1^2 and S_2^2 to be close in value, forcing S_1^2/S_2^2 to be close to 1. If the ratio is close to zero, then we naturally conclude that the population variances are not equal and that, in fact, $\sigma_1^2 < \sigma_2^2$. Conversely, if S_1^2/S_2^2 is much larger than 1, we also conclude that the population variances are different and, in this instance, that $\sigma_1^2 > \sigma_2^2$.

When we use the phrases *close to 0* and *much larger than 1*, we are speaking in terms of probabilities. That is, an observed value of the statistic is "close to 0" when it is too small to have reasonably occurred by chance if, in fact, the population variances are equal. Similarly, an observed value is "much larger than 1" if it is too large to have reasonably occurred by chance. To determine the probability of observing various values of the statistic S_1^2/S_2^2, we must know its probability distribution. We shall show that this statistic follows a distribution previously unencountered. In particular, if the population variances are equal, it follows what is called an F distribution. This distribution is defined in terms of a distribution previously studied, namely, the chi-squared distribution. In particular, any F random variable can be written as the ratio of two independent chi-squared random variables, each divided by their respective degrees of freedom. The formal definition of the F distribution is given in Definition 10.2.1.

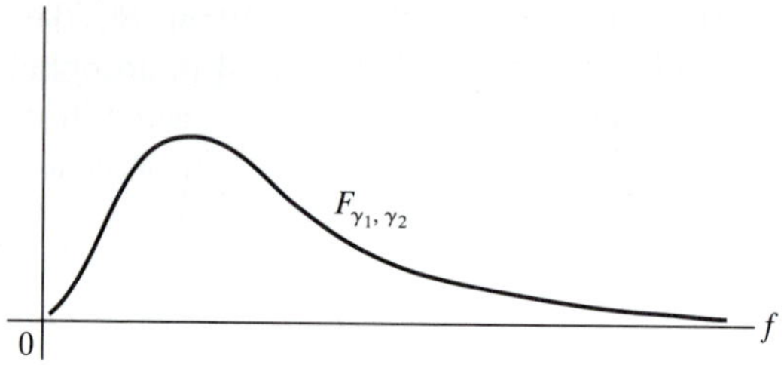

FIGURE 10.2
A typical F density.

Definition 10.2.1 (*F* **distribution**). Let $\mathbf{X}^2_{\gamma 1}$ and $\mathbf{X}^2_{\gamma 2}$ be independent chi-squared random variables with γ_1 and γ_2 degrees of freedom, respectively. The random variable

$$\frac{\mathbf{X}^2_{\gamma_1}/\gamma_1}{\mathbf{X}^2_{\gamma_2}/\gamma_2}$$

follows what is called an F distribution with γ_1 and γ_2 degrees of freedom.

The important properties of the family of F random variables are summarized below:

1. There are infinitely many F random variables, each identified by two parameters γ_1 and γ_2, called *degrees of freedom*. These parameters are always positive integers; γ_1 is associated with the chi-squared random variable of the numerator of the F random variable, and γ_2 is associated with the chi-squared random variable of the denominator. The notation F_{γ_1,γ_2} denotes an F random variable with γ_1 and γ_2 degrees of freedom.
2. Each F random variable is continuous.
3. The graph of the density of each F random variable is an asymmetric curve of the general shape shown in Fig. 10.2.
4. F random variables cannot assume negative values.

A partial summary of the cumulative distribution for F random variables with selected degrees of freedom is given in Table IX of App. A. In the table γ_1, the degrees of freedom for the numerator, appears as column headings; γ_2, the degrees of freedom for the denominator, appears as row headings. F points for degrees of freedom that exceed 120 may be approximated well via row or column 120. Once again, we use the notational convention of denoting the point of the F_{γ_1,γ_2} curve with area r to its right by f_r. Example 10.2.1 illustrates the use of Table IX.

Example 10.2.1. Consider $F_{10,15}$, the F random variable with 10 and 15 degrees of freedom.

(*a*) Find $P[F_{10,15} \leq 2.544]$. This probability can be read directly from Table IX. Simply scan the numbers in column 10 and row 15 until you locate 2.544. It can be seen that $P[F_{10,15} \leq 2.544] = F(2.544) = .95$

(*b*) Find $P[F_{10,15} > 2.059]$. Since the F distribution is continuous, this probability is $1 - F(2.059)$. From Table IX, $F(2.059) = .90$. Hence $P[F_{10,15} > 2.059] = .10$.

(*c*) Via our notational convention we can say that $f_{.05} = 2.544$ and $f_{.10} = 2.059$.

(*d*) Find $f_{.95}$. This is a point with 95% of the area under the graph of the density to its right and 5% of the area to its left. Since Table IX is a cumulative table, this value can be read directly from the table. Its value is $f_{.95} = .35149$.

We are now in a position to verify that our proposed statistic for testing $H_0\colon \sigma_1^2 = \sigma_2^2$ does indeed follow an F distribution when H_0 is true.

Theorem 10.2.1 (Distribution of S_1^2/S_2^2). Let S_1^2 and S_2^2 be sample variances based on independent random samples of sizes n_1 and n_2 drawn from normal populations with means μ_1 and μ_2 and variances σ_1^2 and σ_2^2, respectively. If $\sigma_1^2 = \sigma_2^2$, then the statistic S_1^2/S_2^2 follows an F distribution with $n_1 - 1$ and $n_2 - 1$ degrees of freedom.

Proof. We have already shown that the random variable $(n - 1)S^2/\sigma^2$ follows a chi-squared distribution with $n - 1$ degrees of freedom. (Theorem 8.1.1.) Applying this result here, we can conclude that the random variables $(n_1 - 1)S_1^2/\sigma_1^2$ and $(n_2 - 1)S_2^2/\sigma_2^2$ are chi-squared random variables with $n_1 - 1$ and $n_2 - 1$ degrees of freedom respectively. Furthermore, since sampling is independent, these chi-squared random variables are independent. By Definition 10.2.1 the random variable

$$\frac{\dfrac{(n_1 - 1)S_1^2/\sigma_1^2}{(n_1 - 1)}}{\dfrac{(n_2 - 1)S_2^2/\sigma_2^2}{(n_2 - 1)}} = \frac{\sigma_2^2 S_1^2}{\sigma_1^2 S_2^2}$$

follows an F distribution with $n_1 - 1$ and $n_2 - 1$ degrees of freedom. If $\sigma_1^2 = \sigma_2^2$, then the above ratio reduces to S_1^2/S_2^2 as desired.

Note that the degrees of freedom associated with the statistic S_1^2/S_2^2 are $n_1 - 1$ and $n_2 - 1$. That is, the number of degrees of freedom for the numerator is 1 less than the size of the sample drawn from population I; that of the denominator is 1 less than the size of the sample drawn from population II.

There are several things to realize concerning this F test.

1. Normality is assumed and the test is sensitive to violations of this assumption. If it appears from the stem-and-leaf diagram or a histogram that either population does not have at least an approximate bell shape, then the test should not be used.
2. The test for equality of variances performs best when sample sizes are equal. If they are very different and there is any doubt concerning the normality of the two sampled populations, then the test should not be used.

3. The test is not very powerful. That is, the null hypothesis that $\sigma_1^2 = \sigma_2^2$ will not be rejected fairly often when, in fact, the variances are different. To minimize this problem, it is suggested that the test be performed at a relatively high α level. (α levels as high as .20 are satisfactory.)

These restrictions on the use of the F test partially explain the preference of some statisticians for the second philosophy mentioned for comparing means.

Example 10.2.2. A study of two types of materials used in electrical conduits, tubes used to house electrical wires, is to be conducted. The purpose of the study is to compare the strength of one to the other. Strength is to be assessed by measuring the load in pounds required to crush a 6-inch piece of material to 40% of its original diameter. Two questions are posed. Each is to be answered statistically, based on information obtained from independently drawn samples of the two materials. The primary question is, "Does material I on the average withstand a heavier load than material II?" That is, "Is $\mu_1 > \mu_2$?" However, before this question can be answered, we want to consider the question, "Is $\sigma_1^2 = \sigma_2^2$?"

We wish first to test

$$H_0: \sigma_1^2 = \sigma_2^2$$

$$H_1: \sigma_1^2 \neq \sigma_2^2$$

These data are obtained:

Material I	Material II
$n_1 = 25$	$n_2 = 16$
$\bar{x}_1 = 380$ lb	$\bar{x}_2 = 370$ lb
$s_1^2 = 100$	$s_2^2 = 400$

The test is two-tailed. The null hypothesis should be rejected if the observed value of the test statistic is too large or too small to have occurred by chance when the population variances are equal. The number of degrees of freedom associated with the test statistic is $n_1 - 1 = 25 - 1 = 24$ and $n_2 - 1 = 16 - 1 = 15$. Let us test H_0 at the $\alpha = .20$ level. From Table IX of App. A, the critical points are

$$f_{.10}(24, 15, \text{df}) = 1.899 \qquad \text{and} \qquad f_{.90}(24, 15\ \text{df}) = .561$$

The observed value of the test statistic is $s_1^2/s_2^2 = 100/400 = .25$. Since $.25 < .561$, H_0 can be rejected at the .20 level. There is evidence that the population variances are not equal.

10.3 COMPARING MEANS: VARIANCES EQUAL (POOLED TEST)

Suppose that the primary objective of a study is to compare means and, after considering the information at hand, we have no reason to believe that population variances are unequal. In this case we can use a procedure called the *pooled, independent*, or *uncorrelated T test* to compare μ_1 to μ_2. The comparison can be done via confidence interval estimation or by means of a hypothesis or a significance test. We begin by developing the bounds for a $100(1-\alpha)\%$ confidence interval on the differences in population means.

Confidence Interval on $\mu_1 - \mu_2$: Pooled

It has been shown that $\bar{X}_1 - \bar{X}_2$ is an unbiased estimator for $\mu_1 - \mu_2$. To extend this point estimator to a confidence interval, once again, we must find a random variable whose expression involves the parameter of interest, in this case $\mu_1 - \mu_2$, whose distribution is known. Such a random variable is provided by Theorem 10.1.1. This theorem states that when normal populations are sampled, the random variable $\bar{X}_1 - \bar{X}_2$ is normal with mean $\mu_1 - \mu_2$ and variance $\sigma_1^2/n_1 + \sigma_2^2/n_2$. By standardizing this random variable, it can be concluded that the random variable

$$\frac{(\bar{X}_1 - \bar{X}_2) - (\mu_1 - \mu_2)}{\sqrt{\sigma_1^2/n_1 + \sigma_2^2/n_2}}$$

is standard normal. If the population variances have been compared and no difference has been detected, then we assume that they are equal. Let σ^2 denote this common population variance. That is, let $\sigma_1^2 = \sigma_2^2 = \sigma^2$. Substituting into the above expression, we conclude that

$$\frac{(\bar{X}_1 - \bar{X}_2) - (\mu_1 - \mu_2)}{\sqrt{\sigma^2(1/n_1 + 1/n_2)}}$$

is standard normal. Since σ^2 is unknown, it must be estimated from the data. This is done by a *pooled* sample variance. Note that we already have two unbiased estimators for σ^2, namely, S_1^2 and S_2^2. The idea is to pool, or combine, these estimators to form a single unbiased estimator for σ^2 in such a way that sample sizes are taken into account. It is natural to want to attach greater importance, or "weight," to the sample variance associated with the larger sample. The pooled variance, as defined now, does exactly this.

Definition 10.3.1 (Pooled variance). Let S_1^2 and S_2^2 be the sample variances based on independent samples of sizes n_1 and n_2, respectively. The *pooled variance*, denoted by S_p^2, is given by

$$S_p^2 = \frac{(n_1 - 1)S_1^2 + (n_2 - 1)S_2^2}{n_1 + n_2 - 2}$$

Note that we weight S_1^2 and S_2^2 by multiplying by $n_1 - 1$ and $n_2 - 1$, respectively. The more natural way to weight is to multiply by the corresponding sample sizes n_1 and n_2. We choose to weight in this somewhat odd way so that the random variable $(n_1 + n_2 - 2)S_p^2/\sigma^2$ will follow a chi-squared distribution. This is necessary so that the test statistic that we use to test for equality of means will follow a T distribution.

Example 10.3.1. Consider a sample variance $s_1^2 = 24$ based on a sample of size 16 and a second sample variance $s_2^2 = 20$ based on a sample of size 121. The value of the ratio s_1^2/s_2^2 is $24/20 = 1.20$. Based on these sample variances, the population variances σ_1^2, and σ_2^2 cannot be declared to be different even with α set at .2 ($f_{.1} = 1.545$). The pooled estimate for the common population variance is

$$\hat{\sigma}^2 = s_p^2 = \frac{(n_1 - 1)s_1^2 + (n_2 - 1)s_2^2}{n_1 + n_2 - 2}$$

$$= \frac{15(24) + 120(20)}{16 + 121 - 2}$$

$$= \frac{2760}{135} = 20.44$$

Note that this estimate is quite different from 22, the value obtained by ignoring sample sizes and arithmetically averaging s_1^2 and s_2^2.

To obtain a random variable that can be used to construct a $100(1 - \alpha)\%$ confidence interval on $\mu_1 - \mu_2$, we replace the unknown population variance σ^2 in the Z random variable

$$\frac{(\bar{X}_1 - \bar{X}_2) - (\mu_1 - \mu_2)}{\sqrt{\sigma^2(1/n_1 + 1/n_2)}}$$

by the pooled estimator S_p^2, to obtain the random variable

$$\frac{(\bar{X}_1 - \bar{X}_2) - (\mu_1 - \mu_2)}{\sqrt{S_p^2(1/n_1 + 1/n_2)}}$$

As in the one-sample case, replacing the population variance by its estimator does affect the distribution. The former random variable is a Z random variable; the latter has a T distribution with $n_1 + n_2 - 2$ degrees of freedom. The proof of this result is outlined in Exercise 18. The algebraic structure of this random variable is the same as that encountered previously, namely,

$$\frac{\text{Estimator} - \text{parameter}}{D}$$

Therefore the confidence interval on $\mu_1 - \mu_2$ takes the same general form as most of the intervals encountered previously. These bounds are given in Theorem 10.3.1.

Theorem 10.3.1 (Confidence interval on $\mu_1 - \mu_2$: Pooled variance). Let $\bar{X}_1$ and $\bar{X}_2$ be sample means based on independent random samples drawn from normal distributions with means μ_1 and μ_2, respectively, and common variance σ^2. Let S_p^2 denote the pooled sample variance. The bounds for a $100(1 - \alpha)\%$ confidence interval on $\mu_1 - \mu_2$ are

$$\left(\bar{X}_1 - \bar{X}_2\right) \pm t_{\alpha/2}\sqrt{S_p^2(1/n_1 + 1/n_2)}$$

where the point $t_{\alpha/2}$ is found relative to the $T_{n_1+n_2-2}$ distribution.

The proof of this theorem is outlined in Exercise 19. Its use is illustrated in Example 10.3.2.

Example 10.3.2. A study is conducted to estimate the difference in the mean occupational exposure to radioactivity in utility workers in the years 1973 and 1979. These data based on independent samples of workers for the 2 years are obtained:

1973	1979
$n_1 = 16$	$n_2 = 16$
$\bar{x}_1 = .94$ rem	$\bar{x}_2 = .62$ rem
$s_1^2 = .040$	$s_2^2 = .028$

We first check for equality of variances by testing

$$H_0\colon \sigma_1^2 = \sigma_2^2$$
$$H_1\colon \sigma_1^2 \neq \sigma_2^2$$

at the $\alpha = .2$ level. The critical points for this two-tailed test are

$$f_{.10}(15, 15 \text{ df}) = 1.972 \quad \text{and} \quad f_{.90}(15, 15 \text{ df}) = .507$$

The observed value of the test statistic is $s_1^2/s_2^2 = .040/.028 \doteq 1.43$. Since $1.43 \ngtr 1.972$ and $1.43 \nless .507$, we are unable to reject H_0. We, therefore, pool s_1^2 and s_2^2 and estimate σ^2 by

$$\hat{\sigma}^2 = s_p^2 = \frac{15(.040) + 15(.028)}{16 + 16 - 2} = .034$$

To compare means, let us find a 95% confidence interval on $\mu_1 - \mu_2$. The partition of the $T_{16+16-2} = T_{30}$ curve needed is shown in Fig. 10.3. The bounds for the confidence interval are

$$\left(\bar{x}_1 - \bar{x}_2\right) \pm t_{\alpha/2}\sqrt{s_p^2(1/n_1 + 1/n_2)} = (.94 - .62) \pm 2.042\sqrt{.034(1/16 + 1/16)}$$
$$= .32 \pm .13$$

We can be 95% confident that the difference in mean occupational exposure to radioactivity for the 2 years is between .19 and .45 rem. This interval does not contain the number 0 and is positive-valued throughout, an indication that the

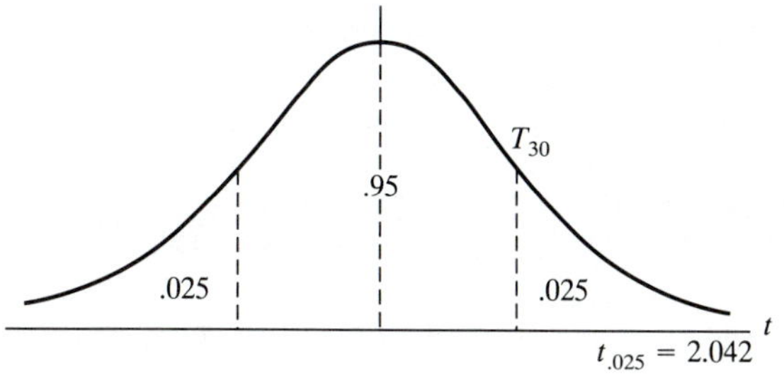

FIGURE 10.3
Partition of the T_{30} curve needed to obtain a 95% confidence interval on $\mu_1 - \mu_2$.

mean exposure in 1973 was, in fact, higher than in 1979. ("Estimating the Risks from Occupational Exposure," *Nuclear Engineering International*, February 1983, pp. 27–29.)

Pooled T Test

As in previous instances, the random variable used to derive confidence bounds for a parameter also serves as a test statistic for testing various hypotheses concerning the parameter. In this case the random variable

$$\frac{(\bar{X}_1 - \bar{X}_2) - (\mu_1 - \mu_2)_0}{\sqrt{S_p^2(1/n_1 + 1/n_2)}} = T_{n_1+n_2-2}$$

serves as a test statistic for testing any of the usual hypotheses, where $(\mu_1 - \mu_2)_0$ denotes the hypothesized difference in population means. The hypothesized difference can be any value whatsoever. However, the most commonly encountered hypothesized value is zero. In this case the purpose is to determine whether the population means differ and, if so, which is the larger. Such hypotheses take these forms:

I	II	III
H_0: $\mu_1 = \mu_2$	H_0: $\mu_1 = \mu_2$	H_0: $\mu_1 = \mu_2$
H_1: $\mu_1 > \mu_2$	H_1: $\mu_1 < \mu_2$	H_1: $\mu_1 \neq \mu_2$
Right-tailed test	Left-tailed test	Two-tailed test

We can distinguish between H_0 and H_1 by presetting α and performing a hypothesis test or by performing a significance test and then reporting its P value, leaving to the researcher the final decision of whether or not to reject H_0.

Example 10.3.3. The tensile strength of a material is the ability that the material possesses to resist deformation when a force or a load is applied to it. A study of the tensile strength of ductile iron annealed or strengthened at two different temperatures is conducted. It is thought that the lower temperature will yield the higher mean tensile strength. These data result:

1450°F	1650°F
$n_1 = 10$	$n_2 = 16$
$\bar{x}_1 = 18{,}900$ psi	$\bar{x}_2 = 17{,}500$ psi
$s_1^2 = 1600$	$s_2^2 = 2500$

We first test H_0: $\sigma_1^2 = \sigma_2^2$ to be sure that pooling is appropriate. The critical points for an $\alpha = .2$ level two-tailed test are $f_{.10}(9, 15 \text{ df}) = 2.086$ and $f_{.90}(9, 15 \text{ df}) = .427$. The observed value of the test statistic is $s_1^2/s_2^2 = 1600/2500 = .64$. Since $.64 \not> 2.086$ and $.64 \not< .427$, we are unable to reject H_0, and therefore we shall pool s_1^2 and s_2^2 to estimate the common population variance. In this case

$$s_p^2 = \frac{(n_1 - 1)s_1^2 + (n_2 - 1)s_2^2}{n_1 + n_2 - 2} = \frac{9(1600) + 15(2500)}{10 + 16 - 2} = 2162.5$$

The primary purpose of the study is to test

$$H_0: \mu_1 = \mu_2$$

$$H_1: \mu_1 > \mu_2$$

The observed value of the test statistic is

$$\frac{(\bar{x}_1 - \bar{x}_2) - (\mu_1 - \mu_2)_0}{\sqrt{s_p^2(1/n_1 + 1/n_2)}} = \frac{(18{,}900 - 17{,}500) - 0}{\sqrt{2162.5(1/10 + 1/16)}} = 74.68$$

Based on the $T_{10+16-2} = T_{24}$ distribution, the P value, the probability of observing a value of 74.68 or larger if $\mu_1 = \mu_2$, is less than .0005($t_{.0005} = 3.745$). We have very strong evidence that the mean tensile strength of iron annealed at 1450°F is higher than the mean strength of that annealed at 1650°F. ("Ductile Iron for Elevated Temperature Service," D. Torkington, *Foundry Management and Technology*, July 1983, pp. 24–30.)

10.4 COMPARING MEANS: VARIANCES UNEQUAL

If a difference is detected when the population variances are compared, then pooling is inappropriate. It is still possible to compare means using an approximate T statistic. Again, the desired statistic is found by modifying the Z random variable

$$\frac{(\bar{X}_1 - \bar{X}_2) - (\mu_1 - \mu_2)}{\sqrt{\sigma_1^2/n_1 + \sigma_2^2/n_2}}$$

in a logical way. Since now there is evidence that $\sigma_1^2 \neq \sigma_2^2$, each population variance is estimated separately; these estimates are *not* combined. Instead, the population variances in the Z random variable above are replaced by their respective estimators, S_1^2 and S_2^2, to obtain the random variable

$$\frac{(\bar{X}_1 - \bar{X}_2) - (\mu_1 - \mu_2)}{\sqrt{S_1^2/n_1 + S_2^2/n_2}}$$

As in the past, making this change results in a change in distribution from Z to an approximate T. This time, however, the number of degrees of freedom must be estimated from the data. Several methods have been suggested for doing this. Here we demonstrate the *Smith-Satterthwaite* procedure. According to this procedure, γ, the number of degrees of freedom, is given by

$$\gamma \doteq \frac{\left[S_1^2/n_1 + S_2^2/n_2\right]^2}{\dfrac{\left[S_1^2/n_1\right]^2}{n_1 - 1} + \dfrac{\left[S_2^2/n_2\right]^2}{n_2 - 1}}$$

The value for γ will not necessarily be an integer. If it is not, we round it *down* to the nearest integer. We round down rather than up in order to take a conservative approach. As the number of degrees of freedom associated with T random variables increase, the corresponding bell-shaped curves become more compact. Practically speaking, this means that, for example, the point $t_{.05}$ associated with the T_{10} curve (1.812) is a little larger than the point $t_{.05}$ associated with the T_{11} curve (1.796). If we can reject a null hypothesis based on the T_{10} distribution, it will also be rejected based on the T_{11} distribution. The converse does not necessarily hold.

The Smith-Satterthwaite procedure is illustrated in a significance testing context in the next example.

Example 10.4.1. In Example 10.2.2 we began a study of the load-bearing properties of two materials used in electrical conduits. The primary question posed was, "Is material I, on the average, better able to withstand a heavy load than material II?" That is, "Is $\mu_1 > \mu_2$?" These data were gathered:

Material I	Material II
$n_1 = 25$	$n_2 = 16$
$\bar{x}_1 = 380$ lb	$\bar{x}_2 = 370$ lb
$s_1^2 = 100$	$s_2^2 = 400$

We tested for equality of variances and found evidence that $\sigma_1^2 \neq \sigma_2^2$. Therefore to test

$$H_0: \mu_1 = \mu_2$$
$$H_1: \mu_1 > \mu_2$$

we do *not* pool s_1^2 and s_2^2. Rather, we use the Smith-Satterthwaite procedure. The degrees of freedom required are

$$\gamma \doteq \frac{\left[s_1^2/n_1 + s_2^2/n_2\right]^2}{\dfrac{\left[s_1^2/n_1\right]^2}{n_1 - 1} + \dfrac{\left[s_2^2/n_2\right]^2}{n_2 - 1}}$$

$$= \frac{[100/25 + 400/16]^2}{\dfrac{[100/25]^2}{25 - 1} + \dfrac{[400/16]^2}{16 - 1}}$$

$$= 19.86$$

This value is rounded *down* to 19. The observed value of the test statistic is

$$\frac{(\bar{x}_1 - \bar{x}_2) - (\mu_1 - \mu_2)_0}{\sqrt{s_1^2/n_1 + s_2^2/n_2}} = \frac{(380 - 370) - 0}{\sqrt{100/25 + 400/16}} = 1.857$$

Based on the T_{19} distribution, $t_{.05} = 1.729$ and $t_{.025} = 2.093$. Since the observed value of our test statistic lies between these two values, the P value of our test lies between .025 and .05. Since these values are relatively small, we can reject H_0 and conclude that material I is capable of withstanding heavier loads on the average than is material II.

The Smith-Satterthwaite procedure can be used to construct confidence bounds on $\mu_1 - \mu_2$ when the population variances are unequal. The derivation of these bounds is outlined in Exercise 27.

We mentioned earlier that there are two opinions concerning the best course of action when comparing two means. The first, which entails the use of a preliminary F test run at a high α level to decide whether or not to pool, has been demonstrated in the last two sections. It embraces a philosophy that might be called a "sometimes pool" point of view. The second philosophy makes use of a very nice property of the Smith-Satterthwaite procedure. Namely, recent simulation studies have shown that not only does it perform well when variances are unequal, but it yields results that are virtually equivalent to those obtained with the pooled T test when variances are equal. For this reason, there seems to be no real need to pool; simply use the Smith-Satterthwaite procedure in all cases. Neither philosophy is clearly "best." However, the latter may be the safer road to take as it avoids the pitfalls inherent in the use of the F test for variances. You are free to choose the procedure that appeals to you.

10.5 COMPARING MEANS: PAIRED DATA

In many instances problems arise in which two random samples are available but they are *not* independent; rather, each observation in one sample is naturally or by design paired with an observation in the other. To see what we mean, consider Example 10.5.1.

Example 10.5.1. One important aspect of computing is the cpu time required by a particular algorithm to solve a problem. A new algorithm is developed to solve zero-one multiple objective problems in linear programming. It is thought that the new algorithm will solve problems faster than the algorithm currently used. To obtain statistical evidence to support this research hypothesis, a number of problems will be selected at random. Each problem will be solved twice; once using the current algorithm and once using the newly developed one. Thus each test problem generates two observations, and we have two data sets. One data set represents a random sample of cpu times using the old algorithm; the other, represents a sample of cpu times for the new one. These data sets are *not* independent; they are based on the same problems solved by two different methods and so are paired by design. The idea is illustrated in Fig. 10.4. (*European Journal of Operational Research*, April 1983, p. 371.)

When pairing as we just illustrated occurs, the methods of Secs. 10.3 and 10.4 are no longer applicable. Rather, a procedure for answering the question, "What is $\mu_X - \mu_Y$?" must be developed that takes into account the fact that the observations are paired. This is done easily. Note that when data are paired, we can define a new random variable D by $D = X - Y$. The n differences $D_i = X_i - Y_i$, $i = 1, 2, 3, \dots, n$ constitute a set of observations on D; that is, they constitute a random sample of size n drawn from the population of differences.

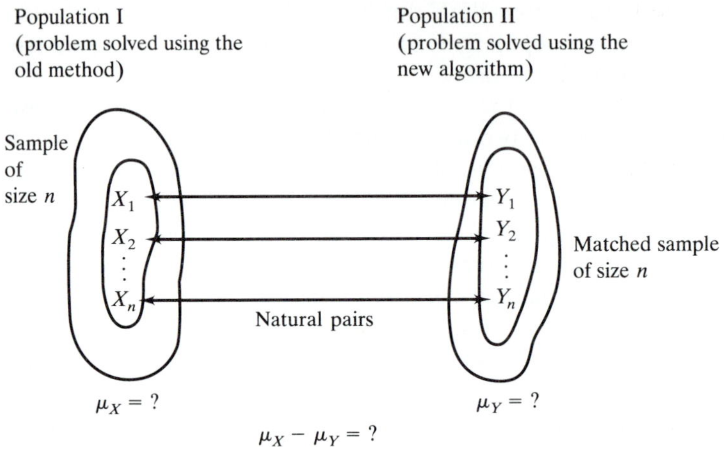

FIGURE 10.4
Matched or paired samples of cpu times drawn from two different populations.

Since, by the rules for expectation,

$$\mu_X - \mu_Y = E[X] - E[Y] = E[X - Y] = E[D] = \mu_D$$

the original question, "What is $\mu_X - \mu_Y$?" is equivalent to, "What is μ_D?" We are reduced from the original two-sample problem to the *one*-sample problem of making an inference on the mean of the population of differences. This problem is not new, and it can be handled using the methods of Chap. 8. In particular, the formula for the $100(1 - \alpha)\%$ confidence bounds on $\mu_X - \mu_Y = \mu_D$ is

Confidence bounds on $\mu_X - \mu_Y$ for paired data

$$\bar{D} \pm t_{\alpha/2} S_d/\sqrt{n}$$

In this formula $\bar{D}$ and S_d are the sample mean and sample standard deviation of the sample of difference scores, respectively, and $t_{\alpha/2}$ is the appropriate point relative to the T_{n-1} distribution.

Paired *T* Test

The null hypothesis $\mu_X = \mu_Y$ is equivalent to the hypothesis $\mu_D = 0$. The test statistic for testing this hypothesis based on the sample of difference scores is

$$\frac{\bar{D} - 0}{S_d/\sqrt{n}}$$

which follows a T distribution with $n - 1$ degrees of freedom if H_0 is true. The use of this statistic is now illustrated.

Example 10.5.2. When the experiment described in Example 10.5.1 is conducted, these data result:

cpu time, s

Program	Old (x)	New (y)	Difference $d = x - y$
1	8.05	.71	7.34
2	24.74	.74	24.00
3	28.33	.74	27.59
4	8.45	.77	7.68
5	9.19	.80	8.39
6	25.20	.83	24.37
7	14.05	.82	13.23
8	20.33	.77	19.56
9	4.82	.71	4.11
10	8.54	.72	7.82

For these data $\bar{d} = 14.409$ and $s_d = 8.653$. We want to test

$$H_0: \mu_X = \mu_Y$$
$$H_1: \mu_X > \mu_Y$$

This is equivalent to testing

$$H_0: \mu_D = 0$$
$$H_1: \mu_D > 0$$

The observed value of the test statistic is

$$\frac{\bar{d} - 0}{s_d/\sqrt{n}} = \frac{14.409 - 0}{8.653/\sqrt{10}} = 5.266$$

Based on the $T_{n-1} = T_9$ distribution, the P value for this test is less than .0005 ($t_{.0005} = 4.781$). Since this probability is very small, we reject H_0 and conclude that, on the average, the new algorithm is faster than the old.

In using the "paired T" procedures, it is assumed that the random variable $D = X - Y$ is at least approximately normally distributed. Note that in the case of a paired comparison we do not need to check for equality of variances. This is due to the fact that we are actually studying a single population, the population of differences. We are concerned only with the variance of this one population.

10.6 ALTERNATIVE NONPARAMETRIC METHODS

In Sec. 10.3 the T test for testing equality of population means for two independent samples was discussed. Under the assumptions of normally distributed random variables with equal but unknown population variances, this is the most powerful test for testing means. However, as one might expect, these rather restrictive assumptions are not always reasonable to assume in applications. For such situations an alternative nonparametric test is available which is almost as good as the T test even when all the necessary assumptions are met and may be considerably superior to the T test when the assumptions are clearly not met. If the sample sizes are reasonably large, the T test is quite robust. That is, the test is not very sensitive to departures from normality. However, for small samples, and particularly when the variances are unequal, the T test can lead to invalid conclusions. Under these circumstances a nonparametric test should be strongly considered as an alternative approach for testing equality of location for two populations. The most widely used test such as this one is the *Wilcoxon rank-sum* test.

Wilcoxon Rank-Sum Test

Let X and Y be continuous random variables. Let $X_1, X_2, \ldots, X_m$ and $Y_1, Y_2, \ldots, Y_n$ be independent random samples of size m and n from the underlying distribution of X and Y, respectively. For convenience we assume that the X sample represents the smaller sample, and hence $m \le n$. The null hypothesis to be tested is that the X and Y populations are identical. However, the test that we use is especially sensitive to differences in location. For this reason, the null hypothesis is usually stated in terms of equal population medians. Thus the three forms that hypotheses may take are:

H_0: $M_X = M_Y$	H_0: $M_X = M_Y$	H_0: $M_X = M_Y$
H_1: $M_X > M_Y$	H_1: $M_X < M_Y$	H_1: $M_X \neq M_Y$
Right-tailed test	Left-tailed test	Two-tailed test

To perform the test, the $m + n$ observations are pooled to form a single sample with the group identity of each observation retained. These observations are then ordered smallest to largest and ranked from 1 to $N = m + n$. If ties occur, each tied value receives the average group rank as in previous Wilcoxon procedures. The test statistic, denoted by W_m, is the sum of the ranks associated with the observations that originally constituted the smaller sample (X values). The logic behind this choice of test statistic is this. If the X population is located below the Y population, then the smaller ranks will tend to be associated with the X values. This produces a small value of W_m. If the reverse is true, then W_m will tend to be large. Thus, logically, we should reject H_0: $M_X = M_Y$ in favor of H_1: $M_X < M_Y$ for small values of W_m; we reject H_0: $M_X = M_Y$ in favor of H_1: $M_X > M_Y$ for large values of W_m. Upper and lower critical points for selected values of m, n, and α are found in Table X of App. A. Example 10.6.1 demonstrates the use of this table.

Example 10.6.1. An experiment is conducted on two brands of kerosene heaters. The manufacturer of brand A claims that his model will heat an 8-foot room from 60 to 70°F in less time than a competitor's brand B. Hence to test the manufacturer's claim, the following hypothesis is to be tested.

$$H_0: M_B = M_A$$
$$H_1: M_B > M_A$$

A random sample of 12 heaters is selected from brand B; an independent sample of 15 heaters is selected from brand A. The observations are time in seconds to raise the room temperature the 10° specified.

Brand B		Brand A	
69.3	52.6	28.6	30.6
56.0	34.4	25.1	31.8
22.1	60.2	26.4	41.6
47.6	43.8	34.9	21.1
53.2		29.8	36.0
48.1		28.4	37.9
23.2		38.5	13.9
13.8		30.2	

Ordering the pooled observations from the smallest to the largest, retaining group identity, we obtain the following corresponding ranks:

Observation	13.8	13.9	21.1	22.1	23.2	25.1	26.4	28.4	28.6
Brand	B	A	A	B	B	A	A	A	A
Rank	1	2	3	4	5	6	7	8	9

Observation	29.8	30.2	30.6	31.8	34.4	34.9	36.0	37.9	38.5
Brand	A	A	A	A	B	A	A	A	A
Rank	10	11	12	13	14	15	16	17	18

Observation	41.6	43.8	47.6	48.1	52.6	53.2	56.0	60.2	69.3
Brand	A	B	B	B	B	B	B	B	B
Rank	19	20	21	22	23	24	25	26	27

Brand B is the smaller sample ($m = 12$), and hence the test statistic W_m is

$$W_m = 1 + 4 + 5 + 14 + 20 + 21 + 22 + 23 + 24 + 25 + 26 + 27 = 212$$

From Table X, for $m = 12$, $n = m + 3 = 15$ and $\alpha = .05$, the critical value for a right-tailed test is 202. Since $W_m = 212 > 202$, we reject H_0 and conclude that brand A heaters do, in fact, raise the temperature in less time than brand B.

When the sample sizes m or n exceed the values in Table X, a large sample normal approximation can be used to test H_0. The test statistic is

$$\frac{W_m - E(W_m)}{\sqrt{\text{Var } W_m}}$$

This statistic is approximately distributed as a standard normal random variable, where

$$E(W_m) = [m(m + n + 1)/2]$$

and

$$\text{Var } W_m = mn(m + n + 1)/12$$

Several things should be pointed out concerning the Wilcoxon statistic. First, although the null hypothesis is stated in terms of medians, if the distributions of X and Y are symmetric, we are also testing equality of means. Thus for normal populations the Wilcoxon statistic is analogous to the normal theory T test for independent samples. Second, the Wilcoxon statistic can be used with data that cannot be measured but which can, nevertheless, be ranked. Examples of data of this sort are given in Exercises 39 and 40.

Wilcoxon Signed-Rank Test for Paired Observations

In Sec. 10.5 the T test for paired data was discussed. The signed-rank test for paired observations is the nonparametric analog when the normal assumptions are not met. This test is almost as good as the paired T test even when the underlying distribution is normal, and is usually preferred to the paired T test for other distributions.

We discussed the Wilcoxon signed-rank test for a single sample in Sec. 8.7. The corresponding test for paired data is a simple modification of the method given in Sec. 8.7. Here we let X and Y be continuous random variables that are assumed to have symmetric distributions. We want to test the hypothesis that the medians of these two distributions are equal. Thus our hypotheses take the form

H_0: $M_X = M_Y$	H_0: $M_X = M_Y$	H_0: $M_X = M_Y$
H_1: $M_X > M_Y$	H_1: $M_X < M_Y$	H_1: $M_X \neq M_Y$
Right-tailed test	Left-tailed test	Two-tailed test

Consider a random sample $(X_1, Y_1), (X_2, Y_2), \ldots, (X_n, Y_n)$ of paired observations on X and Y. We first form the differences $X_1 - Y_1, X_2 - Y_2, \ldots, X_n - Y_n$. If the null hypothesis is true, the population of difference scores is symmetric about 0. Thus to test H_0: $M_X = M_Y$, we test H_0: $M_{X-Y} = 0$. The test is performed exactly as before. We first order the absolute values of the differences from the smallest to the largest and rank them from 1 to n. Tied scores are assigned the average group rank. Each rank is assigned the sign of the difference that generated the rank. Once again, the test statistics used are

$$W_+ = \sum_{\substack{\text{all} \\ \text{positive} \\ \text{ranks}}} R_i \quad \text{and} \quad |W_-| = \sum_{\substack{\text{all} \\ \text{negative} \\ \text{ranks}}} |R_i|$$

Right-tailed tests are conducted via $|W_-|$, and left-tailed tests utilize W_+ as the test statistic. In each case we reject H_0 for values that are too small to have occurred by chance based on the critical points found in Table VIII of App. A.

The next example should refresh your memory of the Wilcoxon signed-rank procedure.

Example 10.6.2. An experiment is conducted to compare the amount of memory required to analyze a data set using the two leading statistical packages. These data are obtained:

Program	Package X	Package Y	Difference $X - Y$
1	512K	500K	12
2	650K	600K	50
3	890K	890K	0
4	410K	400K	10
5	1050K	1025K	25
6	1500K	1400K	100
7	600K	625K	−25
8	750K	710K	40

Let us test

$$H_0: M_X = M_Y$$
$$H_1: M_X \neq M_Y$$

at the $\alpha = .1$ level. To do so, we order the absolute values of the differences from the smallest to the largest and rank them from 1 to 8. We then assign to each rank the algebraic sign of the difference that generated the rank. The zero difference is assigned the algebraic sign that is least conductive to the rejection of H_0. In this case the zero difference is considered to be negative. We thus obtain these signed ranks:

$\lvert X - Y\rvert$	0	10	12	25	25	40	50	100
Rank	1	2	3	4.5	4.5	6	7	8
Signed rank	−1	2	3	−4.5	4.5	6	7	8

For these data

$$W_+ = 2 + 3 + 4.5 + 6 + 7 + 8 = 30.5$$
$$|W_-| = |-1| + |-4.5| = 5.5$$

For a two-tailed test the test statistic is W, the smaller of W_+ and $|W_-|$. From Table VIII of App. A we see that the critical point for a two-tailed test at the $\alpha = .1$ level is 6. Since $5.5 < 6$, we can reject H_0 and claim that there are differences in the medians of these two populations.

One other comment should be made. If the differences are such that we only know whether a difference is positive or negative, then the null hypothesis can be tested via the sign test. This idea was discussed in the one-sample context in Sec. 8.7. An example of data of this sort is given in Exercise 44.

CHAPTER SUMMARY

In this chapter we continued our study of two sample problems by learning how to compare the means, variances, and medians of two populations. We considered two different experimental settings. Namely, those problems in which independent samples are drawn from the two populations and problems in which data are paired.

To compare variances based on independent samples, it was necessary to introduce a new continuous distribution called the F distribution. Although some studies are designed specifically to compare variances, more often variances are compared as a first step in comparing means. If there is statistical evidence based on the F test that the population variances are not equal, then we use the Smith-Satterthwaite T procedure to compare means. Otherwise we can use either a pooled T procedure or the Smith-Satterthwaite procedure. The choice is yours. Each of these procedures assumes that sampling is from normal distributions. A nonparametric alternative to these tests was presented. This alternative procedure, called the Wilcoxon rank-sum test, does not require normality and no knowledge of population variances is necessary.

When data are paired, we do not need to consider the individual population variances. In this case we work with a population of difference scores. It is assumed that this population is normally distributed and inferences are made on the difference in population means via a one-sample "paired" T test. The Wilcoxon signed-rank test for paired data was introduced as a nonparametric test for location when it is evident that the population of difference scores is not normally distributed.

We introduced and defined important terms that you should know. These are:

F distribution
Pooled estimator for σ^2
Pooled T test
Smith-Satterthwaite test
Paired T test

EXERCISES

Section 10.1

1. A firm receives integrated circuits in lots of 100 from two different suppliers. These data are obtained on the number of defective items found per lot.

Supplier I				Supplier II				
3	8	5	7	0	2	3	1	3
2	3	8	1	1	1	5	4	0
5	0	6	2					

Estimate μ_1, μ_2, and $\mu_1 - \mu_2$.

2. Many gold and silver deposits that were once considered uneconomical are now being exploited, thanks to improvements in methods for recovering precious metals from ore. In a study to compare the potential of two different open-pit gold mines ore samples are obtained from each mine. The mean number of ounces of gold recovered per ton of ore is .233 for the first mine and .127 for the second. Estimate the difference in the mean number of ounces of gold per ton of ore for these two mines. ("Leaching and Precipitation Technology for Gold and Silver Ore," *Engineering and Mining Journal*, June 1983, pp. 48–55.)

3. A press used to remove water from copper-bearing materials is being tested using two different types of filter plates. These data are obtained on the percentage of moisture remaining in the material after treatment.

Regular chamber (I)			**Diaphragm chamber (II)**		
8.10	8.16	8.16	7.58	7.65	7.69
7.96	7.98	7.93	7.66	7.67	7.67
7.97	8.08	8.06	7.58	7.62	7.65
8.02	7.87	7.94	7.65	7.58	7.71
7.82	8.11	7.92	7.63	7.54	
8.15	7.91	8.00	7.46	7.40	

Estimate μ_1, μ_2, and $\mu_1 - \mu_2$. ("How Pressure Dewaters Butte Concentrations," *Engineering and Mining Journal*, June 1983, pp. 60–73.)

4. Let $\bar{X}_i$ be the sample mean based on a sample of size 25 drawn from a normal distribution with mean 8 and variance 16. Let $\bar{X}_2$ be the sample mean based on a sample of size 36 drawn from a normal distribution with mean 5 and variance 9. What is the distribution of each of these random variables?

(*a*) $\bar{X}_1$

(*b*) $\bar{X}_2$

(*c*) $(\bar{X}_1 - 8)/(4/5)$

(*d*) $(\bar{X}_2 - 5)/(3/6)$

(*e*) $\bar{X}_1 - \bar{X}_2$

(*f*) $\dfrac{(\bar{X}_1 - \bar{X}_2) - (8 - 5)}{\sqrt{16/25 + 9/36}}$

Section 10.2

5. Use Table IX of App. A to find each of the following:

(*a*) $P[F_{10,9} \le 2.416]$ (*b*) $f_{.1}(10, 9 \text{ df})$

(*c*) $f_{.9}(10, 9 \text{ df})$ (*d*) $P[F_{30,30} \le .622]$

(*e*) $f_{.9}(30, 30 \text{ df})$ (*f*) $f_{.05}(30, 30 \text{ df})$

(*g*) $P[F_{20,4} \ge 5.803]$ (*h*) $f_{.05}(20, 4 \text{ df})$

(*i*) $f_{.95}(20, 4 \text{ df})$ (*j*) $P[F_{20,120} \ge 1.659]$

(*k*) $f_{.05}(20, 150 \text{ df})$ (*l*) $f_{.95}(20, 150 \text{ df})$

6. In each part of Fig. 10.5, find the points *a* and *b* indicated.

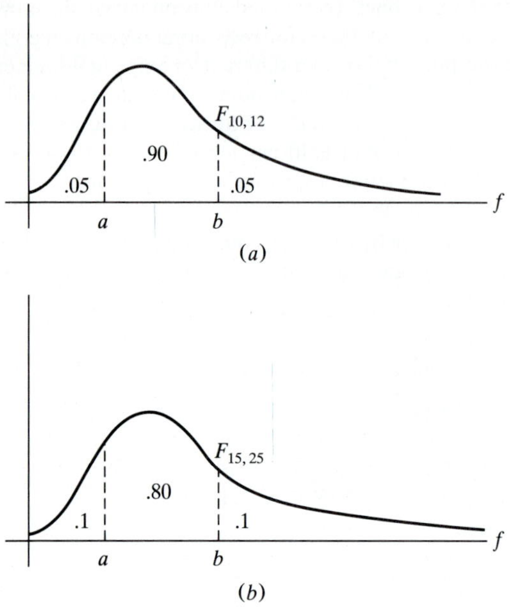

FIGURE 10.5

7. In each case, test for equality of variances at the indicated level.
   (a) $n_1 = 10$ $n_2 = 8$ $\alpha = .20$
   $s_1^2 = .25$ $s_2^2 = .05$
   (b) $n_1 = 13$ $n_2 = 20$ $\alpha = .10$
   $s_1^2 = 4$ $s_2^2 = 2$
   (c) $n_1 = 121$ $n_2 = 150$ $\alpha = .10$
   $s_1^2 = 6$ $s_2^2 = 4.4$
8. The cost of repairing a fiberoptic component may depend on the stage of production at which it fails. These data are obtained on the cost of repairing parts that fail when installed in the system and on the cost of repairing parts that fail after the system is installed in the field. ("Strategies for Active Component Testing," J. Hayes, *Laser Focus*, July 1983, pp. 93–98.)

System failure	Field failure
$n_1 = 21$	$n_2 = 25$
$\bar{x}_1 = \$65$	$\bar{x}_2 = \$120$
$s_1^2 = 25$	$s_2^2 = 100$

   (a) It is thought that the variance in cost of repairs made in the field is larger than the variance in cost of repairs made when the component is placed into the system. Set up the null and alternative hypotheses needed to gain statistical evidence to support this contention.
   (b) Find the critical point for an $\alpha = .10$ level test of the hypothesis of part (a). Use the given data to test H_0.

9. A study of the sodium content in a 6-fluid ounce serving of a soft drink is conducted. These data are obtained on various types of ginger ales and cola drinks.

Ginger ale	**Cola**
$n_1 = 10$	$n_2 = 10$
$\bar{x}_1 = 9.6$	$\bar{x}_2 = 9.9$
$s_1^2 = 10.89$	$s_2^2 = 11.90$

(*a*) It is thought that the variability in the sodium content in ginger ales is smaller than that of colas. Set up the null and alternative hypotheses needed to support this contention.

(*b*) Find the critical point for an $\alpha = .1$ level test for the hypothesis of part (*a*). Use the given data to test H_0.

10. (*Confidence interval on* σ_1^2/σ_2^2.) Show that the lower and upper bounds for a $100(1-\alpha)\%$ confidence interval on σ_1^2/σ_2^2 are given by

$$L_1 = \frac{S_1^2}{S_2^2}\,\frac{1}{f_{\alpha/2}(n_1-1, n_2-1 \text{ df})}$$

and

$$L_2 = \frac{S_1^2}{S_2^2}\,\frac{1}{f_{1-\alpha/2}(n_1-1, n_2-1 \text{ df})}$$

respectively. *Hint:* Start by noting that the random variable $\sigma_2^2 S_1^2/\sigma_1^2 S_2^2$ follows an F distribution with $n_1 - 1$ and $n_2 - 1$ degrees of freedom. Therefore

$$P\left[f_{1-\alpha/2}(n_1-1, n_2-1 \text{ df}) \le \frac{\sigma_2^2 S_1^2}{\sigma_1^2 S_2^2} \le f_{\alpha/2}(n_1-1, n_2-1 \text{ df})\right] \doteq 1-\alpha$$

Isolate σ_1^2/σ_2^2 in the middle of this inequality to find L_1 and L_2. If $\sigma_1^2 = \sigma_2^2$, what number is expected to lie within the confidence interval?

11. A study is conducted to compare the variability in the number of hours that a rechargable flashlight will operate after its battery has been fully charged. These data are obtained for two different brands of batteries.

Brand X	**Brand** Y
$n_1 = 25$	$n_2 = 21$
$s_1^2 = .021$	$s_2^2 = .018$

(*a*) Find a 90% confidence interval on σ_1^2/σ_2^2.

(*b*) Note that if $\sigma_1^2 = \sigma_2^2$, then $\sigma_1^2/\sigma_2^2 = 1$. If 1 does not lie in a particular confidence interval, then there is some evidence that the population variances are not equal. Does the interval of part (*a*) suggest that these population variances are not equal? Explain.

Section 10.3

12. (*a*) Let $s_1^2 = 42$, $s_2^2 = 37$, $n_1 = 10$, $n_2 = 14$. Find s_p^2.

(*b*) Let $s_1^2 = 28$, $s_2^2 = 30$, $n_1 = 20$, $n_2 = 20$. Find s_p^2. Do not use your calculator!

(*c*) Let $s_1^2 = 20$, $s_2^2 = 40$, $n_1 = 10$, $n_2 = 50$. Find s_p^2. Why is s_p^2 closer in value to s_2^2 than to s_1^2?

13. A study of report writing by engineers is conducted. A scale that measures the intelligibility of engineers' English is devised. This scale, called an "index of confusion," is devised so that low scores indicate high readability. These data are obtained on articles randomly selected from engineering journals and from unpublished reports written in 1979. ("Engineers' English," W. H. Emerson, *CME*, June 1983, pp. 54–56.)

Journals				**Unpublished reports**			
1.79	1.75	1.67	1.65	2.39	2.51	2.86	2.14
1.87	1.74	1.94		2.56	2.29	2.49	
1.62	2.06	1.33		2.36	2.58	2.33	
1.96	1.69	1.70		2.62	2.41	1.94	

(*a*) Test H_0: $\sigma_1^2 = \sigma_2^2$ at the $\alpha = .2$ level to be sure that pooling is appropriate.
(*b*) Find s_p^2.
(*c*) Find a 90% confidence interval on $\mu_1 - \mu_2$.
(*d*) Does there appear to be a difference between μ_1 and μ_2? Explain, based on the confidence interval of part (*c*).

14. To decide whether or not to purchase a new hand-held laser scanner for use in inventorying stock, tests are conducted on the scanner currently in use and on the new scanner. These data are obtained on the number of 7-inch bar codes that can be scanned per second:

New	**Old**
$n_1 = 61$	$n_2 = 61$
$\bar{x}_1 = 40$	$\bar{x}_2 = 29$
$s_1^2 = 24.9$	$s_2^2 = 22.7$

(*a*) Test H_0: $\sigma_1^2 = \sigma_2^2$ at the $\alpha = .2$ level to be sure that pooling is appropriate.
(*b*) Find s_p^2.
(*c*) Find a 95% confidence interval on $\mu_1 - \mu_2$.
(*d*) Does the new laser appear to read more bar codes per second on the average? Explain.
(*e*) Since the number of bar codes that can be scanned per second is discrete, we have not satisfied the normality requirement. What theorem justifies the use of the pooled T procedure in this case?

15. Environmental testing is an attempt to test a component under conditions that closely simulate the environment in which the component will be used. An electrical component is to be used in two different locations in Alaska. Before environmental testing can be conducted, it is necessary to determine the soil composition in these localities. These data are obtained on the percentage of SiO_2 by weight of the soil. ("Environmental Synergisms in the Performance Assessment of Armament Electrical Connectors," K. Wittendorfer, *Journal of Environmental Sciences*, May 1983, pp. 19–28.)

Anchorage	Kodiak
$n_1 = 10$	$n_2 = 16$
$\bar{x}_1 = 64.94$	$\bar{x}_2 = 57.06$
$s_1^2 = 9$	$s_2^2 = 7.29$

(*a*) Test H_0: $\sigma_1^2 = \sigma_2^2$ at the $\alpha = .2$ level.
(*b*) Find s_p^2.
(*c*) Find a 99% confidence interval on $\mu_1 - \mu_2$.
(*d*) Based on the interval of part (*c*), does there appear to be a difference between μ_1 and μ_2? Explain.

16. Show that $E[S_p^2] = \sigma^2$, thus proving that the pooled variance is an unbiased estimator for the common population variance.

17. Show that the random variable $(n_1 + n_2 - 2)S_p^2/\sigma^2$ has a chi-squared distribution with $n_1 + n_2 - 2$ degrees of freedom. *Hint:* Recall that $(n - 1)S^2/\sigma^2$ follows a chi-squared distribution with $n - 1$ degrees of freedom (Theorem 8.1.1), and apply Exercise 44, Chap. 7.

18. Show that the random variable

$$\frac{(\bar{X}_1 - \bar{X}_2) - (\mu_1 - \mu_2)}{\sqrt{S_p^2(1/n_1 + 1/n_2)}}$$

follows a T distribution with $n_1 + n_2 - 2$ degrees of freedom. *Hint:* Show that this random variable can be expressed as the ratio of a Z random variable divided by the square root of a chi-squared random variable with $n_1 + n_2 - 2$ degrees of freedom divided by its degrees of freedom, and apply Definition 8.2.1.

19. (*Confidence interval on $\mu_1 - \mu_2$: Variances equal.*) Show that the lower and upper bounds for a $100(1 - \alpha)\%$ confidence interval on $\mu_1 - \mu_2$ when $\sigma_1^2 = \sigma_2^2$ are given by

$$(\bar{X}_1 - \bar{X}_2) \pm t_{\alpha/2}\sqrt{S_p^2(1/n_1 + 1/n_2)}$$

as claimed in Theorem 10.3.1. *Hint:* Start by noting that $P[-t_{\alpha/2} \le T_{n_1+n_2-2} \le t_{\alpha/2}] \doteq 1 - \alpha$ and that in particular,

$$P\left[-t_{\alpha/2} \le \frac{(\bar{X}_1 - \bar{X}_2) - (\mu_1 - \mu_2)}{\sqrt{S_p^2(1/n_1 + 1/n_2)}} \le t_{\alpha/2}\right] \doteq 1 - \alpha$$

Isolate $\mu_1 - \mu_2$ in the middle of this inequality to find L_1 and L_2.

Section 10.4

20. Calculate the number of degrees of freedom for a Smith-Satterthwaite procedure based on these data:
(*a*) $n_1 = 9$ $\quad n_2 = 16$
$s_1^2 = 38.07$ $\quad s_2^2 = 16.89$
(*b*) $n_1 = 25$ $\quad n_2 = 25$
$s_1^2 = .42$ $\quad s_2^2 = 1$

21. Strontium-90, a radioactive element produced by nuclear testing, is closely related to calcium. In dairy lands strontium-90 can make its way into milk via the grasses eaten by dairy cows. It becomes concentrated in the bones of those who drink the milk. In 1959 a study was conducted to compare the mean concentration of strontium-90 in the bones of children to that of adults. It was thought that the level in children was higher because the substance was present during their formative years.

(*a*) Set up the null and alternative hypotheses needed to verify this contention.

(*b*) Based on these data, is pooling appropriate?

Children	Adults
$n_1 = 121$	$n_2 = 61$
$\bar{x}_1 = 2.6$ picocuries per gram	$\bar{x}_2 = .4$ picocurie per gram
$s_1^2 = 1.44$	$s_2^2 = .0121$

(*c*) Test the null hypothesis of part (*a*). Can H_0 be rejected? Explain, based on the P value of your test. What practical conclusion can be drawn from these data?

22. Water and other nonaqueous volatiles are present in differing concentrations in coal from different seams. To measure the percentage by weight of these substances for a particular seam, readings are taken at two different temperatures. These data result ("Determination of Moisture and Low Temperature Volatiles in Solid Fuels," D. Jenke and M. Hannifer, *American Laboratory*, July 1983, pp. 38–45):

Water

105°C			160°C		
15.11	15.30	15.44	15.14	15.33	15.40
15.23	15.32	15.48	15.28	15.34	15.77
15.27	15.37	15.36	15.26	15.38	15.52

Nonaqueous volatiles

105°C			160°C		
.343	.601	.676	1.533	1.780	1.625
.481	.543	.541	1.190	1.636	1.692
.475	.108	.106	2.015	1.464	1.991

(*a*) Use the water data to test

$$H_0: \mu_1 = \mu_2$$

$$H_1: \mu_1 \neq \mu_2$$

at the $\alpha = .05$ level. Does the temperature at which the readings are taken appear to affect the mean reading of the water concentration of the coal? Explain. Be ready to defend your choice of test statistic.

(*b*) Use the nonaqueous volatiles data to test

$$H_0: \mu_1 = \mu_2$$

$$H_1: \mu_1 \neq \mu_2$$

at the $\alpha = .05$ level. Does the temperature at which the readings are taken appear to affect the mean reading of the concentration of nonaqueous volatiles in the coal? Explain.

23. It is thought that the gas mileage obtained by a particular model of automobile will be higher if unleaded premium gasoline is used in the vehicle rather than regular unleaded gasoline. To gather evidence to support this contention, 10 cars are randomly selected from the assembly line and tested using a specified brand of premium gasoline; 10 others are randomly selected and tested using the brand's regular gasoline. Tests are conducted under identical controlled conditions. These data result:

Premium		Regular	
35.4	31.7	29.7	34.8
34.5	35.4	29.6	34.6
31.6	35.3	32.1	34.8
32.4	36.6	35.4	32.6
34.8	36.0	34.0	32.2

(*a*) Set up the null and alternative hypotheses needed to compare the mean mileage for these two gasolines.

(*b*) Decide whether or not to reject H_0 via a significance test. What is the approximate P value of the test? Be ready to defend your choice of test statistic.

(*c*) Interpret your results in the context of this problem.

24. A new coal liquefaction process is being studied. It is claimed that the new process results in a higher yield of distillate synthetic fuel than the current process. These observations are obtained on the number of kilograms of distillate synthetic fuel produced per kilogram of hydrogen consumed in the process ("Liquefaction Process Promises Better Efficiency," *Modern Power Systems*, May 1983, p. 13):

New				Old			
16.4	12.8	15.4	17.0	11.1	10.5	10.9	10.1
17.7	12.2	18.7		12.8	13.2	12.6	
15.9	14.7	19.1		12.1	14.5	15.6	
11.3	14.1	16.5		14.2	15.3	14.2	

(*a*) Set up the null and alternative hypotheses needed to support the stated claim.

(*b*) Since putting the new process into production is very expensive, a Type I error is costly. To compensate for this, test the null hypothesis of part (*a*) at the $\alpha = .01$ level. Would you recommend that the new process be used? Explain.

25. A study is conducted to compare the tensile strength of two types of roof coatings. It is thought that, on the average, butyl coatings are stronger than acrylic coatings. These data are gathered ("Urethane Foam—An Efficient Alternative," J. Collett, *Military Engineer*, August 1983, pp. 348–350):

Tensile strength, lb / in²

Acrylic				Butyl			
246.3	247.7	287.5	248.3	340.7	263.4	272.6	271.4
255.0	246.3	284.6	243.7	270.1	341.6	332.6	303.9
245.8	214.0	268.7	276.7	371.6	307.0	362.2	324.7
250.7	242.7	302.6	254.9	306.6	319.1	358.1	360.1

(*a*) Set up the null and alternative hypotheses needed to verify the research hypothesis.
(*b*) Is pooling appropriate? Explain.
(*c*) Test the null hypothesis of part (*a*). Can H_0 be rejected? Explain, based on the P value of your test. What practical conclusion can be drawn in this case?

26. It is thought that the application of a plasma coating that contains submicron particles of tungsten carbide will reduce wear to rotary valves used in the pulp and paper industry. Tests are conducted to compare the wear in coated and uncoated valves. These data are gathered on the wear of the part in millimeters over the test period:

Coated		Uncoated		
.075	.099	.095	.074	.104
.078	.082	.096	.149	.052
.092	.088	.136	.081	
.078	.072	.156	.099	

(*a*) Set up the null and alternative hypotheses needed to support the contention that the plasma coating on the average reduces the wear in these valves.
(*b*) Based on these data, is pooling appropriate?
(*c*) Test the null hypothesis of part (*a*). Does it appear that the coating is effective in reducing wear? Explain, based on the P value of your test.

27. (*Confidence interval on $\mu_1 - \mu_2$: Variances unequal.*) Show that the lower and upper bounds for a 100(1 − α)% confidence interval on $\mu_1 - \mu_2$ when $\sigma_1^2 \neq \sigma_2^2$ are given by

$$(\bar{X}_1 - \bar{X}_2) \pm t_{\alpha/2}\sqrt{S_1^2/n_1 + S_2^2/n_2}$$

where γ, the degrees of freedom associated with the T distribution, is given by the Smith-Satterthwaite formula. *Hint:* Start by noting that

$$P[-t_{\alpha/2} \le T_\gamma \le t_{\alpha/2}] \doteq 1 - \alpha$$

and that in particular

$$P\left[-t_{\alpha/2} \le \frac{(\bar{X}_1 - \bar{X}_2) - (\mu_1 - \mu_2)}{\sqrt{S_1^2/n_1 + S_2^2/n_2}} \le t_{\alpha/2}\right] \doteq 1 - \alpha$$

Isolate $\mu_1 - \mu_2$ in the middle of this inequality to find L_1 and L_2.

28. A manufacturer of power-steering components buys hydraulic seals from two sources. Samples are selected from among the seals obtained from these two suppliers and each seal is tested to determine the amount of pressure that it can withstand. These data result:

Supplier I	Supplier II
$n_1 = 10$	$n_2 = 10$
$\bar{x}_1 = 1350$ lb/in^2	$\bar{x}_2 = 1338$ lb/in^2
$s_1^2 = 100$	$s_2^2 = 29$

(*a*) We want to find a 95% confidence interval on $\mu_1 - \mu_2$. Is pooling appropriate?
(*b*) Construct a 95% confidence interval on $\mu_1 - \mu_2$.
(*c*) Is there evidence based on the confidence interval that, on the average, the seals from supplier I can withstand higher pressures than those from supplier II? Explain.

29. Aseptic packaging of juices is a method of packaging that entails rapid heating followed by quick cooling to room temperature in an air-free container. Such packaging allows the juices to be stored unrefrigerated. Two machines used to fill aseptic packages are compared. These data are obtained in the number of containers that can be filled per minute:

Machine I	Machine II
$n_1 = 25$	$n_2 = 25$
$\bar{x}_1 = 115.5$	$\bar{x}_2 = 112.7$
$s_1^2 = 25.2$	$s_2^2 = 7.6$

(*a*) Is pooling appropriate?
(*b*) Find a 90% confidence interval on $\mu_1 - \mu_2$.
(*c*) Is there evidence based on the confidence interval that machine I is faster, on the average, than machine II? Explain.

30. These data are obtained on the power output in kilowatts of two new diesel motors for small cars:

Direct fuel injection				Indirect fuel injection			
38.5	38.2	39.2	38.5	38.9	38.3	38.4	39.0
38.9	38.0	39.1	39.1	37.7	37.2	38.4	
37.4	37.6	39.0	38.0	38.2	37.0	37.9	
39.0	37.7	38.1	37.4	38.2	37.5	39.7	

Construct a 95% confidence interval on $\mu_1 - \mu_2$ using the appropriate T procedure. Based on your confidence interval, does there appear to be a difference in the mean power of these two engines? Explain.

Section 10.5

31. Information about ocean weather can be extracted from radar returns with the aid of a special algorithm. A study is conducted to estimate the difference in wind speed as measured on the ground and via the Seasat satellite. To do so, wind speeds are measured using the two methods simultaneously at 12 specified times. These data result ("Mapping Ocean Winds by Radar," *NASA Tech Briefs*, Fall 1982, p. 27):

Windspeed, m / s

Time	Ground (x)	Satellite (y)	Differences $(d = x - y)$
1	4.46	4.08	
2	3.99	3.94	
3	3.73	5.00	
4	3.29	5.20	
5	4.82	3.92	
6	6.71	6.21	
7	4.61	5.95	
8	3.87	3.07	
9	3.17	4.76	
10	4.42	3.25	
11	3.76	4.89	
12	3.30	4.80	

(*a*) Find the difference scores for the above data subtracting in the order indicated.
(*b*) Find $\bar{d}$ and s_d.
(*c*) Find a 95% confidence interval on the mean difference in measurements taken by these methods. Based on this interval, is there reason to believe that, on the average, the satellite measurements differ from those taken on the ground? Explain.

32. A study is conducted to estimate the average difference in the cost of analyzing data using two different statistical packages. To do so, 15 data sets are used. Each is analyzed by each package and the cost of the analysis is recorded. These observations result:

Program	Package I	Package II	Program	Package I	Package II
	\$	\$		\$	\$
1	.26	.29	9	.19	.33
2	.24	.32	10	.25	.33
3	.26	.24	11	.29	.33
4	.22	.33	12	.25	.28
5	.25	.28	13	.23	.30
6	.23	.27	14	.20	.24
7	.18	.25	15	.25	.32
8	.25	.26			

(*a*) Find the set of difference scores subtracting in the order package I minus package II.
(*b*) Find $\bar{d}$ and s_d.

(*c*) Find a 90% confidence interval on the mean difference in the cost of running a data analysis using the two packages.

33. Post Three Mile Island regulations require provisions by which people within 10 miles of a nuclear power plant can be notified promptly in the event of a general nuclear emergency. In a study of one such system the sound level at 69 locations within 10 miles of the plant is first simulated and then field tested. Subtracting in the order measured siren level minus simulated siren level, it is found that $\bar{d} = .04$ decibels and $s_d = 2.43$. Find a 95% confidence interval on the mean difference between the actual siren level and the simulated level. Based on this interval, is there reason to suspect that a difference exists? Explain. ("Prompt Notification Siren System Design," *Power Engineering*, March 1983.)

34. A new method for measuring the concentration of Pu^{239} based on the registration of α-particles and fission-fragment tracks is studied. Test solution media of various concentrations are obtained and each is split into two portions. The concentration of the first portion is determined using the new method; the second portion is measured using the standard technique. It is thought that the new procedure tends to give a higher average reading than the standard techniques. Do the following data support this research hypothesis? Explain, based on the *P* value of your test. ("Analysis of Uranium and Plutonium in Binary Mixtures Using the Technique of Track Registration from Solutions," N. K. Chaudhuri and V. Natarajan, *Nuclear Tracks and Radiation Measurements*, June 1982, pp. 109–113.)

Concentration of Pu^{239} (μ / ml)

Sample number	New	Old
1	3.78	3.35
2	3.58	3.60
3	3.77	3.41
4	3.82	3.69
5	3.67	3.48
6	3.66	3.50
7	3.48	3.33
8	3.63	3.64
9	3.88	3.65
10	3.53	3.64

35. Highway engineers studying the effects of wear on dual-lane highways suspect that more cracking occurs in the travel lane of the highway than in the passing lane. To verify this contention, 30 one-hundred-feet-long test strips are selected, paved, and studied over a period of time. It is found that the mean difference in the number of major cracks is 4.5 with a sample standard deviation of 8.1. Do these data support the research hypothesis? Explain, based on the *P* value of the test.

36. Two different Fortran compilers are compared for efficiency. The comparison is done by running 25 randomly selected programs using each compiler. These data on the compile time in seconds are obtained:

Program	Compiler I	Compiler II	Program	Compiler I	Compiler II
1	3.76	4.28	14	4.02	4.25
2	4.78	3.89	15	4.34	4.28
3	4.66	3.30	16	4.10	3.35
4	3.38	3.53	17	3.25	3.83
5	2.52	2.73	18	4.52	3.82
6	3.46	3.19	19	4.24	3.75
7	4.19	3.34	20	5.33	3.66
8	4.15	4.71	21	3.84	4.14
9	3.61	4.21	22	3.95	3.68
10	3.91	3.76	23	4.32	3.09
11	4.47	3.62	24	4.22	3.12
12	3.53	3.26	25	4.25	3.97
13	4.14	2.87			

(*a*) It is thought that the second compiler is the faster of the two. Set up the null and alternative hypotheses needed to support this contention.

(*b*) What is the critical point for an $\alpha = .05$ level test of this hypothesis? Can H_0 be rejected at the .05 level? Interpret your results in the context of this problem.

Section 10.6

37. A study is conducted to determine the effect of acid rain and other industrial pollutants on lake water. Random samples are drawn from 10 lakes in a heavily industrialized area and from eight lakes in a primitive forest area. These data are obtained on the pH of the water:

Industrial area (I)		Primitive area (P)	
6.9	7.0	7.0	6.8
6.2	6.5	6.9	7.1
6.3	6.6	6.7	7.0
5.9	5.5	7.1	7.2
6.0	7.3		

At the $\alpha = .025$ level, can we claim that the pH of the water in the industrialized area tends to be lower than that in the primitive area?

38. Polychlorinated biphenyls (PCB) are worldwide environmental contaminants of industrial origin that are related to DDT. They are being phased out in the United States, but they will remain in the environment for many years. An experiment is run to study the effects of PCB on the reproductive ability of screech owls. The purpose is to compare the shell thickness of eggs produced by birds exposed to PCB to that of birds not exposed to the contaminant. It is thought that shells of the former group will be thinner than those of the latter. Do these data support this research hypothesis? Explain.

Shell thickness, mm

Exposed to PCB (E)		Free of PCB (F)	
.21	.226	.22	.27
.223	.215	.265	.18
.25	.24	.217	.187
.19	.136	.256	.23
.20		.20	

39. An automobile manufacturer is experimenting with a new type of paint designed to resist corrosion. Five automobile hoods are painted with the new paint; seven are painted using the old mixture. All hoods are subjected to identical accelerated life testing. At the end of the testing period an impartial judge is asked to rank the hoods from 1 to 12, with lower ranks indicating less corrosion. These data result (N = new, O = old):

Rank	1	2	3	4	5	6	7	8	9	10	11	12
Type	N	O	N	N	N	O	N	O	O	O	O	O

At the $\alpha = .05$ level, can we reject H_0: $M_N = M_O$ and conclude that the new paint resists corrosion better than the old?

40. A study is conducted to compare a new drill tip to be used in drilling oil wells to the drill tips currently in use. Four new tips (N) are field tested. The length of time each is usable is recorded. After comparison with file data on the old tips (O), these data are obtained (a lower rank indicates a longer lasting drill tip):

Rank	1	2	3	4	5	6	7	8	9
Type	N	O	O	N	N	N	O	O	O

At the $\alpha = .05$ level, can we claim that the new drill tip tend to last longer than the old ones?

41. Manufacturers of brand A mainframe computers claim that maintenance costs are lower for their equipment than for that of their nearest competitor. Before purchasing brand A, a company makes an independent investigation of this claim. Samples of repair records are obtained from users of the two types of equipment. These data result:

Brand A	**Competitor**
$m = 75$	$n = 100$
$W_m = 5937$	

(*a*) What is $E[W_m]$?
(*b*) What is Var W_m?
(*c*) Do the data support the claim of the makers of brand A equipment? Explain, based on the P value of your test.

42. A study of visual and auditory reaction time is conducted for a group of college basketball players. Visual reaction time is measured by the time needed to respond to a light signal, and auditory reaction time is measured by the time needed to respond to the sound of an electric switch. Fifteen subjects were measured with time recorded to the nearest millisecond:

Subject	Visual	Auditory
1	161	157
2	203	207
3	235	198
4	176	161
5	201	234
6	188	197
7	228	180
8	211	165
9	191	202
10	178	193
11	159	173
12	227	137
13	193	182
14	192	159
15	212	156

Is there evidence that the visual reaction time tends to be slower than the auditory reaction time?

43. A firm has two possible sources for its computer hardware. It is thought that supplier X tends to charge more than supplier Y for comparable items. Do these data support this contention at the $\alpha = .05$ level?

Item	Price (X), $	Price (Y), $
1	6000	5900
2	575	580
3	15,000	15,000
4	150,000	145,000
5	76,000	75,000
6	5,650	5,600
7	10,000	9,975
8	850	870
9	900	890
10	3,000	2,900

Would the sign test have yielded the same results? If not, explain the discrepancy.

44. An experiment was conducted to compare the appearance of two types of paint on houses after normal exposure for a period of 2 years. Twenty pairs of similarly constructed homes were selected and brand A paint was applied to one house of each pair and brand B was applied to the other member of each pair. After 2 years a paint expert was asked to judge the appearances of the two brands for each pair of houses. $A > B$ and $B > A$ denotes the order of preferred appearance of A preferred to B and B preferred to A, respectively. The outcome of the experiment is given below.

Pair		Pair		Pair		Pair	
1	$A > B$	6	$B > A$	11	$A > B$	16	$A > B$
2	$A > B$	7	$B > A$	12	$A > B$	17	$B > A$
3	$B > A$	8	$A > B$	13	$A > B$	18	$A > B$
4	$A > B$	9	$B > A$	14	$B > A$	19	$A > B$
5	$A > B$	10	$A > B$	15	$A > B$	20	$A > B$

Using an appropriate test, test the hypothesis that the two brands of paint are equally preferred in terms of appearance after 2 years.

REVIEW EXERCISES

45. Researchers are experimenting with the use of microprocessors to help reduce fuel and power consumption in furnaces used to process magnetite ore. A particular system is designed to maintain gas flow through the machine in such a way as to ensure that sufficient heat is available to raise the raw ore pellets to 1300°C. A study is conducted to compare the temperature setting needed to accomplish this using the computerized system to that setting needed using the conventional method. It is thought that the computerized system will result in a lower average required setting with a smaller variability in settings than the conventional system. ("Reduction in Fuel and Power Consumption in a Traveling Grade Pelletizing Furnace through Computer Control," R. C. Corson, *Mining Engineering*, September 1983, pp. 1309–1312.)

(*a*) We are interested in testing two null hypotheses. State these null hypotheses and their alternatives.

(*b*) Sample runs yield these data:

Computerized	Conventional
$n_1 = 25$	$n_2 = 25$
$\bar{x}_1 = 733°C$	$\bar{x}_2 = 822°C$
$s_1 = 10°C$	$s_2 = 50°C$

Assuming normality, test the null hypotheses of part (*a*). Be ready to defend your choice of test statistics. Do these data support the two contentions stated concerning the computerized system? Explain.

46. Dross is scum that forms on the surface of molten metal during processing. A new technique is being developed to reduce the formation of this substance. To be profitable, the reduction must amount to an average of more than 15 kilograms per ton over the current method. ("Metallurgy of Dross Formation on Aluminum Melts," S. Freti, J. Bornand and K. Buxmann, *Light Metal Age*, June 1982 pp. 12–16.)

(*a*) Set up the null and alternative hypotheses needed to support the contention that the new process will be profitable.

(*b*) Trial runs produce these data:

Old method	New method
$n_1 = 10$	$n_2 = 10$
$\bar{x}_1 = 20$ kg/t	$\bar{x}_2 = 1$ kg/t
$s_1 = 2.5$ kg/t	$s_2 = .5$ kg/t

Assuming normality, test the null hypothesis of part (*a*). Be ready to defend your choice of test statistics. Does it appear that the new process will be profitable? Explain.

47. A composite of 6/6 nylon and steel is being studied for possible use in cam gears. Sixteen gears of different types are produced and the noise level obtained using these gears is compared to that of an identical gear made of cast iron. These data are obtained:

Noise level in dB

Gear	Cast iron	Composite
1	75	74
2	90	88
3	80	81
4	60	60
5	110	107
6	95	92
7	93	90
8	88	84
9	70	66
10	65	64
11	91	86
12	100	97
13	85	83
14	50	44
15	62	60
16	67	64

(*a*) Construct a stem-and-leaf diagram for the differences in the noise levels for the two gears. Subtract in the order cast iron minus composite. Does it appear that these differences are approximately normally distributed?

(*b*) Find a 95% confidence interval on the mean difference in the reduction in the noise level. Does it appear that gears made from the composite have a lower average decibel level than those made from cast iron? Explain.

48. Chains have long been used in kilns in cement plants to help reduce heat consumption. A study is conducted to determine if chains will have the same effect when using cheaper raw materials with high sulfur and chlorine content. The purpose of the study is to estimate the difference in specific heat consumption in kilns with and without the use of chains. Independent samples of sizes 14 and 16, respectively, are used in the study. These data result:

Without chains	With chains
$n_1 = 16$	$n_2 = 14$
$\bar{x}_1 = 6150$ kJ/kg	$\bar{x}_2 = 5250$
$s_1 = 80$ kJ/kg	$s_2 = 75$

Find a 95% confidence interval on $\mu_1 - \mu_2$. Be ready to defend your choice of the confidence bounds used. Does it appear that the chains are effective? Explain. ("Kiln Systems for Difficult Raw Materials and Fuels," H. Herchenbach, *Cement World*, December 1983, vol. 14, pp. 362–369.)

49. It is thought that the heat loss in glass pipes is smaller than that in steel pipes of the same size. To verify this contention, nine pairs of pipes of assorted diameters are obtained. Various liquids at identical starting temperatures are run through 50-meter segments of each type of pipe and the heat loss is measured in each case. These data result:

Heat loss (in C°)

Pair	Steel	Glass
1	4.6	2.5
2	3.7	1.3
3	4.2	2.0
4	1.9	1.8
5	4.8	2.7
6	6.1	3.2
7	4.7	3.0
8	5.5	3.5
9	5.4	3.4

Assuming normality, can we conclude that the mean heat loss is higher in steel pipes than in those made of glass? Explain, based on the P value of your test.

50. A study is conducted to compare the total printing time in seconds of two brands of laser printers on various tasks. Data below are for the printing of charts. (Based on information found in *MACWORLD*, March 1993, p. 1980.)

Task	Brand 1 time	Brand 2 time
1	21.8	36.5
2	22.6	35.2
3	21.0	36.2
4	19.7	34.0
5	21.9	36.4
6	21.6	36.1
7	22.5	37.5
8	23.1	38.0
9	22.2	36.3
10	20.1	35.9
11	21.4	35.7
12	20.5	34.9
13	22.7	37.1
14	20.5	34.2
15	21.3	35.4

(*a*) Estimate the average difference in printing time for these two lasers.
(*b*) Find a 95% confidence interval on the average difference in printing times.
(*c*) Based on the confidence interval found in part (*b*), would you be surprised to hear a claim that these two printers are equally fast in printing charts? Explain.

51. Two drugs, amantadine (A) and rimantadine (R), are being studied for use in combatting the influenza virus. A single 100-milligram dose is administered orally to healthy adults. The variable studied is T_{max}, the time in minutes required to reach maximum plasma concentration. These data are obtained (based on information found in "Drug Therapy", Gordon Douglas, Jr., *New England Journal of Medicine*, vol. 322, February 1990, pp. 443–449):

$T_{max}(A)$			$T_{max}(R)$		
105	123	12.4	221	227	280
126	108	134	261	264	238
120	112	130	250	236	240
119	132	130	230	246	283
133	136	142	253	273	516
145	156	170	256	271	
200					

(*a*) Construct a box plot for each data set and identify outliers.

(*b*) Assume that the outlier 12.4 of set A is the result of a misplaced decimal point. Replace the outlier 12.4 with the true value 124. Test for equality of variances at the $\alpha = .20$ level.

(*c*) Construct a 95% confidence interval on the difference in the average time required to reach maximum plasma concentration for these two drugs.

(*d*) Based on the confidence interval of part (*c*), can it be concluded that there is a difference in means? Explain.

52. A study is conducted to help understand the effect of smoking on sleep patterns. The random variable considered is X, the time in minutes that it takes to fall asleep. Samples of smokers and nonsmokers yield these observations on X:

Nonsmokers						Smokers					
17.2	19.7	18.1	15.1	18.3	17.6	15.1	20.5	17.7	21.3	16.0	24.8
16.2	19.9	19.8	23.6	24.9	20.1	16.8	21.2	18.1	22.1	15.9	25.2
19.8	22.6	20.0	24.1	25.0	21.4	22.8	22.4	19.4	25.2	18.3	25.0
21.2	18.9	22.1	20.6	23.3	20.2	25.8	24.1	15.0	24.1	21.6	16.3
21.1	16.9	23.0	20.1	17.5	21.3	24.3	25.7	15.2	18.0	23.8	17.9
21.8	22.1	21.1	20.5	20.4	20.7	23.2	25.1	16.1	17.2	24.9	19.9
19.5	18.8	19.2	22.4	19.3	17.4	15.7	15.3	19.9	23.1	23.0	25.1

(*a*) Construct a stem-and-leaf diagram for each of these data sets. Use the integers from 15 to 25 inclusive as stems.

(*b*) Would you be surprised to hear someone claim that there is no difference in the distribution of X for the two groups? Explain.

(*c*) Perform any statistical tests that you believe are appropriate to detect differences that might exist. They can be either normal theory or nonparametric.

53. Recycling has become important as landfills become harder to obtain. A study of white paper disposal is conducted. These data are obtained on the amount of white paper thrown out per year by bank employees and employees in other businesses. (Data are in hundreds of pounds.)

Bank employees		Other Businesses	
3.1	2.6	6.9	5.3
2.9	2.0	6.4	5.2
3.8	3.2	4.7	5.1
3.3	2.4	4.3	5.9
2.7	2.3	5.1	5.8
3.0	3.1	6.3	4.9
2.8	2.1	5.9	4.8
2.5	3.4	5.4	4.0
2.0	1.9	5.2	4.0
2.9	2.5	5.7	5.2
2.1	2.5	6.2	5.0
2.7	2.3	4.2	4.1
2.2	2.3	5.0	3.9
1.8	1.5	3.7	3.7
1.9	1.2	5.1	3.4
1.7	1.7		

(Based on information found in "White Paper Recycling," *MIT Technology*, *Review*, August/September 1992, p. 20.)

Do these data support the contention that, on the average, bank employees dispose of more white paper per year than do employees in other businesses? Explain by conducting appropriate statistical tests (either normal theory or nonparametric). Be ready to defend your choice of tests.

COMPUTING SUPPLEMENT

IV Comparing Means and Variances

The procedure PROC TTEST can be used to test for equality of both variances and means by utilizing either the pooled T test or the Smith-Satterthwaite procedure. This procedure is illustrated by using the data of Exercise 26 to test H_0: $\mu_1 = \mu_2$ versus H_1: $\mu_1 < \mu_2$.

Statement	*Function*
DATA PLASMA;	names data set
INPUT GROUP $ WEAR @@;	indicates two variables named GROUP and WEAR to be entered; $ indicates that the variable GROUP is not numeric; @@ indicates that there will be more than one observation entered per line
LINES;	indicates that data follow
C .075 C .099	
C .078 C .082	

C .092 C .088	data coded so that *C* represents
C .078 C .072	coated material and *U* represents
U .095 U .074 U .104	uncoated material
U .096 U .149 U .052	
U .136 U .081	
U .156 U .099	
;	signals end of data
PROC TTEST;	calls for a *T* test to be run
CLASS GROUP;	names variable that identifies the two groups (populations) being compared
TITLE1 TESTING FOR EQUALITY;	titles output
TITLE2 OF;	
TITLE3 MEANS AND VARIANCES;	

The output of this program is as follows:

```
                         TESTING FOR EQUALITY
                                  OF
                        MEANS AND VARIANCES

                          TTEST PROCEDURE

VARIABLE: WEAR

GROUP  N    MEAN       STD DEV    STD ERROR   MINIMUM     MAXIMUM

C      8  0.08300000  0.00924276  0.00326781  0.07200000  0.09900000
H     10  0.10420000  0.03342587  0.01057019  0.05200000  0.15600000

VARIANCES          T             DF PROB > !T!

UNEQUAL       -1.9162 ③     10.7 ④     0.0825 ⑤
EQUAL         -1.7320       16.0       0.1025

FOR HO: VARIANCES ARE EQUAL, F' = 13.08 WITH 9 AND 7 DF

                        ①       PROB > F' = 0.0027
                                       ②
```

The value of the *F* statistic used to compare variances is 13.08. This value is shown in ①. The *P* value for the two-tailed *F* test is .0027. This is the *P* value listed in ②. Since this value is small, we conclude that $\sigma_c^2 \neq \sigma_u^2$, and use the Smith-Satterthwaite procedure to compare means. The *T* statistic and its corresponding degrees of freedom are shown in ③ and ④, respectively. The *P* value for the two-tailed test is .0825. This value is shown in ⑤. The one-tailed *P* value is .0825/2.

V Paired *T* Test

The SAS procedure PROC MEANS can be used to run a paired *T* test. The method is illustrated by using the data of Example 10.5.2.

Statement	*Function*
DATA CPU;	names data set
INPUT OLD NEW;	indicates two variables, OLD and NEW, will be used; each line will contain a value for OLD followed by a value for NEW with at least one blank between
DIFF = OLD − NEW;	Forms a new variable named DIFF that is difference between OLD value and paired NEW value
LINES;	indicates that data follows
8.05 .71	data
24.74 .74	
⋮	
8.54 .72	
;	signals end of data
PROC MEANS MEAN STD STDERR T PRT;	asks for mean, standard deviation, and standard error ($S_d/\sqrt{n}$) to be printed for variable DIFF; asks for test of H_0: $\mu_{\text{DIFF}} = 0$ and P value to be reported
VAR DIFF;	
TITLE PAIRED T TEST;	titles output

The output of this program is given below:

PAIRED T TEST

VARIABLE	MEAN	STANDARD DEVIATION	STD ERROR OF MEAN	T	PR > !T!
DIFF	14.40900000	8.65276635	2.73624497	5.27	0.0005
	④	③	⑤	①	②

These quantities are given on the printout:

① The observed value of the test statistic;
② The P value for a two-tailed test;
③ The observed value of S_d;
④ The observed value of $\bar{D}$;
⑤ The observed value of $S_d/\sqrt{n}$.

CHAPTER 11

SIMPLE LINEAR REGRESSION AND CORRELATION

We introduced the idea of regression in the theoretical sense in Chap. 5. There we assumed that both X and Y were random variables. We used the theoretical densities to find the graph of $\mu_{Y|x}$, the mean value of Y given that X has assumed the value x. That is, we graphed the mean of Y as a function of x.

In this chapter we study a similar problem. However, there is one important difference that should be noted. We shall now assume that the variable X is *not* a random variable. Rather, it is a mathematical variable—an entity that can assume different values but whose value at the time under consideration is not determined by chance. To illustrate, suppose that we are developing a model to describe the temperature of the water off the continental shelf. Since the temperature depends in part on the depth of the water, two variables are involved. These are X, the water depth, and Y, the water temperature. We are not interested in making inferences on the depth of the water. Rather, we want to describe the behavior of the water temperature under the assumption that the depth of the water is known precisely in advance. Even if the depth of the water is fixed at some value x, the water temperature will still vary due to other random influences. For example, if several temperature measurements are taken at various places each at a depth of $x = 1000$ feet, these measurements will vary in value. For this reason, we must admit that for a given x we are really dealing with a "conditional" random variable, which we denote by $Y|x$ (Y given that $X = x$). This conditional random variable has a mean denoted by $\mu_{Y|x}$. It is obvious that the average temperature of ocean water depends in part on the depth of the water; we do not expect the average temperature at $x = 1000$ feet to be the same as that at $x = 5000$ feet. That is, it is reasonable to assume that $\mu_{Y|x}$ is a function of x. We call the graph of this function the *curve of regression of* Y *on* X. Since we assume that the value of X is known in advance and that the value assumed by Y depends in part on the particular value of X under consideration, Y is called the *dependent* or *response* variable. The variable X whose value is used to help predict the behavior of $Y|x$ is called the *independent* or *predictor* variable or the *regressor*.

Our immediate problem is to estimate the form of $\mu_{Y|x}$ based on data obtained at some selected values $x_1, x_2, x_3, \ldots, x_n$ of the predictor variable X. The actual values used to develop the model are not overly important. If a functional relationship exists, it should become apparent regardless of which X values are used to discover it. However, to be of practical use, these values should represent a fairly wide range of possible values of the independent variable X. Sometimes the values used can be preselected. For example, in studying the relationship between water temperature and water depth, we might know that our model is to be used to predict water temperature for depths from 1000 to 5000 feet. We can choose to measure water temperatures at any depths that we wish within this range. For example, we might take measurements at 1000-foot increments. In this way we preset our X values at $x_1 = 1000$, $x_2 = 2000$, $x_3 = 3000$, $x_4 = 4000$, and $x_5 = 5000$ feet. When the X values used to develop the regression equation are preselected, the study is said to be *controlled*. Sometimes the X values used to develop the equation are chosen via some random mechanism. For example, in studying the effect of air quality on the pH of rainwater, we shall be forced to select a sample of days, record the air quality reading for the day, and measure the pH of the rainwater. In this case the values of X used to develop the regression equation are not preselected by the researcher. They do represent a set of typical X values. Studies of this sort are called *observational studies*. Regardless of how the X values for study are selected, our random sample is properly viewed as taking the form

$$\{(x_1, Y|x_1), (x_2, Y|x_2), (x_3, Y|x_3), \ldots, (x_n, Y|x_n)\}$$

Note that the first member of each ordered pair denotes a value of the independent variable X; it is a *real number*. The second member of each pair is a random variable.

In this chapter we learn to estimate the curve of regression of Y on X when the regression is considered to be linear. In this case the equation for $\mu_{Y|x}$ is given by

$$\mu_{Y|x} = \beta_0 + \beta_1 x$$

where β_0 and β_1 denote real numbers.

Much of the theory behind the techniques presented depends on linear algebra. For this reason, we cannot prove some of the results based on material from this text. Where it is possible to verify results we shall do so.

11.1 MODEL AND PARAMETER ESTIMATION

Description of Model

Recall from elementary algebra that the equation for a straight line is $y = b + mx$, where b denotes the y intercept and m denotes the slope of the line. In the simple linear regression model

$$\mu_{Y|x} = \beta_0 + \beta_1 x$$

β_0 denotes the intercept and β_1 the slope of the regression line. To estimate the regression line, we must find a logical way to estimate the theoretical parameters β_0 and β_1. To understand how this is done, we first rewrite our model in an alternative form.

In conducting a regression study, we shall be observing the variable X at n points $x_1, x_2, x_3, \ldots, x_n$. These points are assumed to be measured without error. When they are preselected by the experimenter, we say that the study is a *controlled* study; when they are observed at random, then the study is called an *observational* study. Both situations are handled in the same way mathematically. In either case we shall be concerned with the n random variables $Y|x_1, Y|x_2, Y|x_3, \ldots, Y|x_n$. Recall that a random variable varies about its mean value. Let E_i denote the random difference between $Y|x_i$ and its mean, $\mu_{Y|x_i}$. That is, let

$$E_i = Y|x_i - \mu_{Y|x_i}$$

Solving this equation for $Y|x_i$, we conclude that

$$Y|x_i = \mu_{Y|x_i} + E_i$$

In this expression it is assumed that the random difference E_i has mean 0. Since we are assuming that the regression is linear, we can conclude that $\mu_{Y|x_i} = \beta_0 + \beta_1 x_i$. Substituting, we see that

$$Y|x_i = \beta_0 + \beta_1 x_i + E_i$$

It is customary to drop the conditional notation and to denote $Y|x_i$ by Y_i. Thus an alternative way to express the simple linear regression model is

$$Y_i = \beta_0 + \beta_1 x_i + E_i \tag{11.1}$$

where E_i is assumed to be a random variable with mean 0.

Our data consist of a collection of n pairs (x_i, y_i), where x_i is an observed value of the variable X and y_i is the corresponding observation for the random variable Y. The observed value of a random variable usually differs from its mean value by some random amount. This idea is expressed mathematically by writing

$$y_i = \beta_0 + \beta_1 x_i + \varepsilon_i \tag{11.2}$$

In this equation ε_i is a realization of the random variable E_i that appears in the alternative model for simple linear regression, Eq. (11.1).

In a regression study it is useful to plot the data points in the xy plane. Such a plot is called a *scattergram*. We do not expect these points to lie exactly in a straight line. However, if linear regression is applicable, then they should exhibit a linear trend. These theoretical ideas are illustrated in Fig. 11.1 in the context of our water temperature study. Note that since we do not know the true values for β_0 and β_1, we shall not know the true value for ε_i, the vertical distance from the point (x_i, y_i) to the true regression line.

Once β_0 and β_1 have been approximated from the available data, we can replace these theoretical parameters by their estimated values in the regression model. Letting b_0 and b_1 denote the estimates for β_0 and β_1, respectively, the estimated line of regression takes the form

$$\hat{\mu}_{Y|x} = b_0 + b_1 x$$

Just as the data points do not all lie on the theoretical line of regression, they also do not all lie on this estimated regression line. If we let e_i denote the vertical distance from a point (x_i, y_i) to the estimated regression line, then each data point satisfies the equation

$$y_i = b_0 + b_1 x_i + e_i$$

The term e_i is called the *residual*. Figure 11.2 illustrates this idea and points out the difference between ε_i and e_i graphically.

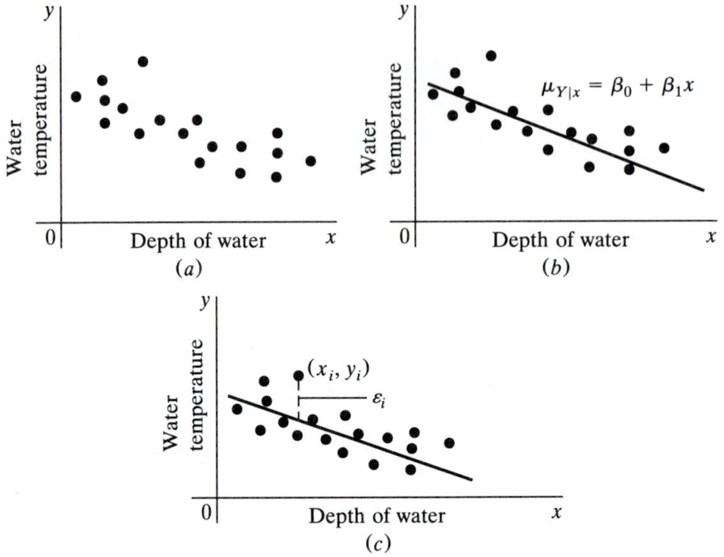

FIGURE 11.1

(a) A scattergram of hypothetical data on depth of water (x) versus its temperature (y)—the data exhibits a linear trend indicating that linear regression is reasonable; (b) theoretical and unknown line of regression passes through the data points; (c) ε_i is the distance from y_i to its mean value, $\mu_{Y|x_i}$.

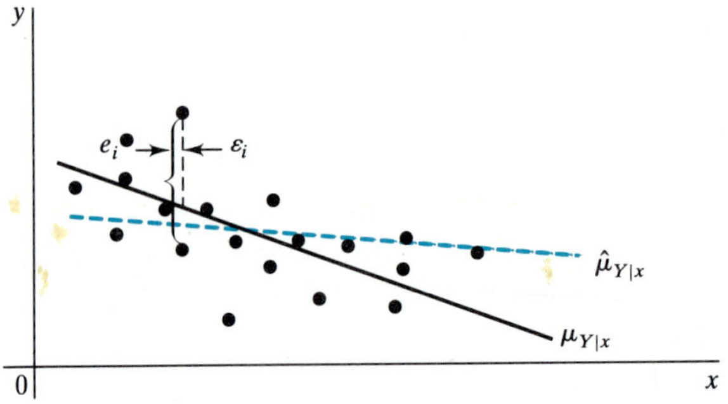

FIGURE 11.2

ε_i is the vertical distance from the point (x_i, y_i) to the true regression line $\mu_{Y|x} = \beta_0 + \beta_1 x$; e_i is the vertical distance from the point (x_i, y_i) to the estimated regression line $\hat{\mu}_{Y|x} = b_0 + b_1 x$.

Least-Squares Estimation

The parameters β_0 and β_1 are estimated by the method of *least squares*. The reasoning behind this method is quite simple. From the many straight lines that can be drawn through a scattergram we wish to pick the one that "best fits" the data. The fit is "best" in the sense that the values of b_0 and b_1 chosen are those that minimize the sum of the squares of the residuals. In this way we are

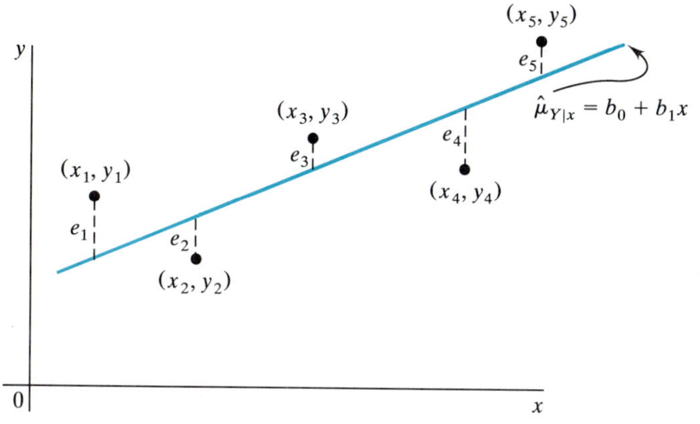

FIGURE 11.3
The least-squares procedure minimizes the sum of the squares of the residuals e_i.

essentially picking the line that comes as close as it can to all data points simultaneously. For example, if we consider the sample of five data points shown in Fig. 11.3, then the least-squares procedure selects that line which causes $e_1^2 + e_2^2 + e_3^2 + e_4^2 + e_5^2$ to be as small as possible.

The residuals are squared before summing for a very practical reason. Notice that the residual for a data point that lies above the estimated regression line is positive; for a point that lies below the line the residual is negative. If the residuals themselves are summed, the negative and positive values will counteract one another and the sum will always be 0. You are asked to verify this fact in Exercise 5.

The general derivation of the least-squares estimates for β_0 and β_1 depends on the minimization technique studied in elementary calculus. In particular, we shall express the sum of squares of the residuals as a function of the two variables b_0 and b_1, differentiate this function with respect to these variables, set these derivatives equal to 0, and solve the resulting equations for b_0 and b_1. Before presenting the derivation, let us note that the residual e_i is sometimes called the residual *error*. For this reason the sum of squares of the residuals often is called the *error sum of squares* and is denoted by SSE (sum of squares error). Since the word "error" tends to suggest that a mistake has been made, this language is somewhat misleading. However, it is recognized widely, and so we shall adhere to its use.

The sum of squares of the errors about the estimated regression line is given by

$$\text{SSE} = \sum_{i=1}^{n} e_i^2 = \sum_{i=1}^{n} (y_i - b_0 - b_1 x_i)^2$$

Differentiating SSE with respect to b_0 and b_1, we obtain

$$\frac{\partial \text{SSE}}{\partial b_0} = -2\sum_{i=1}^{n}(y_i - b_0 - b_1 x_i)$$

$$\frac{\partial \text{SSE}}{\partial b_1} = -2\sum_{i=1}^{n}(y_i - b_0 - b_1 x_i)x_i$$

We now set these partial derivatives equal to 0 and use the rules of summation to obtain the equations

$$nb_0 + b_1\sum_{i=1}^{n} x_i = \sum_{i=1}^{n} y_i$$

$$b_0\sum_{i=1}^{n} x_i + b_1\sum_{i=1}^{n} x_i^2 = \sum_{i=1}^{n} x_i y_i$$

These equations are called the *normal* equations. They can be solved easily to obtain these estimates for β_0 and β_1:

Least-squares estimates for β_0 and β_1

$$b_1 = \frac{n\sum_{i=1}^{n} x_i y_i - \left(\sum_{i=1}^{n} x_i\right)\left(\sum_{i=1}^{n} y_i\right)}{n\sum_{i=1}^{n} x_i^2 - \left(\sum_{i=1}^{n} x_i\right)^2}$$

$$b_0 = \bar{y} - b_1\bar{x}$$

Before illustrating these ideas, let us point out a very practical aspect of regression that we have not yet mentioned. Namely, even though the regression equation actually estimates the mean value of Y for a given value x, it is used extensively to estimate the value of Y itself. Common sense tells us that a logical choice for the predicted value of Y for a given value x is its estimated average value $\hat{\mu}_{Y|x}$. For example, if asked to predict the ocean water temperature at a depth of 1000 feet, a logical choice is the average temperature at this depth. To emphasize this use of the estimated regression line, we rewrite it in the form

$$\hat{y} = \hat{\mu}_{Y|x} = b_0 + b_1 x$$

Example 11.1.1. Since humidity influences evaporation, the solvent balance of water-reducible paints during sprayout is affected by humidity. A controlled study is conducted to examine the relationship between humidity (X) and the extent of solvent evaporation (Y). Knowledge of this relationship will be useful in that it will allow the painter to adjust his or her spraygun setting to account for humidity. These data are obtained (based on "Evaporation During Sprayout of a Typical

Water-Reducible Paint at Various Humidities," *Journal of Coating Technology*, Vol. 65, 1983):

Observation	(x) Relative humidity, (%)	(y) Solvent evaporation, (%) wt
1	35.3	11.0
2	29.7	11.1
3	30.8	12.5
4	58.8	8.4
5	61.4	9.3
6	71.3	8.7
7	74.4	6.4
8	76.7	8.5
9	70.7	7.8
10	57.5	9.1
11	46.4	8.2
12	28.9	12.2
13	28.1	11.9
14	39.1	9.6
15	46.8	10.9
16	48.5	9.6
17	59.3	10.1
18	70.0	8.1
19	70.0	6.8
20	74.4	8.9
21	72.1	7.7
22	58.1	8.5
23	44.6	8.9
24	33.4	10.4
25	28.6	11.1

Summary statistics for these data are

$$n = 25 \qquad \sum x = 1314.90 \qquad \sum y = 235.70$$

$$\sum x^2 = 76{,}308.53 \qquad \sum y^2 = 2286.07 \qquad \sum xy = 11{,}824.44$$

To estimate the simple linear regression line, we estimate the slope β_1 and intercept β_0. These estimates are

$$\hat{\beta}_1 = b_1 = \frac{n\sum xy - \left[\left(\sum x\right)\left(\sum y\right)\right]}{n\sum x^2 - \left(\sum x\right)^2}$$

$$= \frac{25(11{,}824.44) - [(1314.90)(235.70)]}{25(76{,}308.53) - (1314.90)^2}$$

$$= -.08$$

$$\hat{\beta}_0 = b_0 = \bar{y} - b_1\bar{x}$$

$$= 9.43 - (-.08)(52.60)$$

$$= 13.64$$

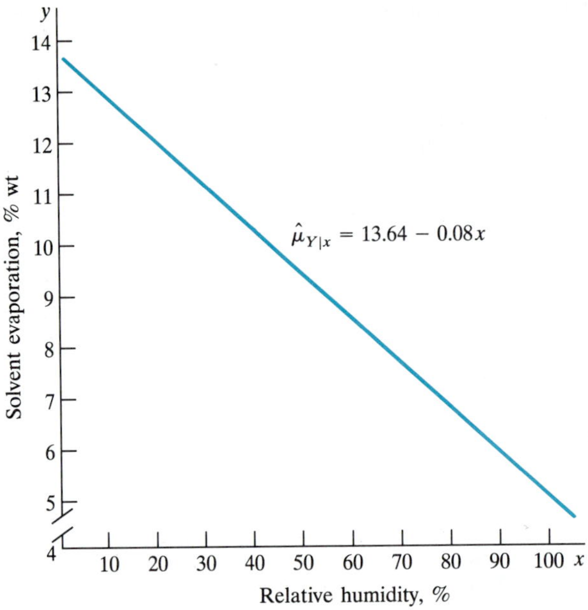

FIGURE 11.4
A graph of the estimated line of regression of Y, the extent of evaporation, on X, the relative humidity.

Hence the estimated regression equation is

$$\hat{\mu}_{Y|x} = \hat{y} = 13.64 - .08x$$

The graph of this equation is shown in Fig. 11.4. To predict the extent of solvent evaporation when the relative humidity is 50%, we substitute the value 50 for x in the equation

$$\hat{y} = 13.64 - .08x$$

to obtain $\hat{y} = 13.64 - .08(50) = 9.64$. That is, when the relative humidity is 50%, we predict that 9.64% of the solvent, by weight, will be lost due to evaporation.

Recall from elementary calculus that the slope of a line gives the change in y for a unit change in x. If the slope is positive, then as x increases so does y; as x decreases, so does y. If the slope is negative, things operate in reverse. An increase in x signals a decrease in y, whereas a decrease in x yields an increase in y. In the previous example the slope is $-.08$. If the relative humidity increases by 1 percentage point, then the mean solvent evaporation decreases by .08. If the relative humidity decreases by 3 percentage points, then the mean solvent evaporation should increase by $3(.08) = .24$.

We end this section with a word of caution. A given data set gives evidence of linearity only over those values of X spanned by the data set. For values of X beyond those covered there is no evidence of linearity. Thus it is dangerous to use an estimated regression line to predict values of Y corresponding to values of X that lie far beyond the range of the X values included in the data set.

11.2 PROPERTIES OF LEAST-SQUARES ESTIMATORS

For a given set of observations on (X, Y) the method of least squares yields estimates b_0 and b_1 for β_0 and β_1, the intercept and slope of the true regression line, respectively. Since the values obtained for b_0 and b_1 vary from data set to data set, it is evident that they are actually observed values of random variables, which we denote by B_0 and B_1. These random variables are estimators for β_0 and β_1 and are given by

Least-squares estimators for β_0 and β_1

$$B_1 = \hat{\beta}_1 = \frac{n \sum_{n=1}^{n} x_i Y_i - \left(\sum_{i=1}^{n} x_i\right)\left(\sum_{i=1}^{n} Y_i\right)}{n \sum_{i=1}^{n} x_i^2 - \left(\sum_{i=1}^{n} x_i\right)^2}$$

$$B_0 = \hat{\beta}_0 = \bar{Y} - B_1 \bar{x}$$

In this section we derive the mathematical properties of these estimators. Knowledge of these properties will allow us to find confidence intervals on β_0, β_1, $\mu_{Y|x}$, and $Y|x$ as well as to test hypotheses on the values of β_0 and β_1.

Recall that one way to express the simple linear regression model is

$$Y_i = \beta_0 + \beta_1 x_i + E_i$$

where E_i is assumed to be a random variable with mean 0. To determine the properties of B_0 and B_1, we must make certain other assumptions concerning E_i. In particular, we assume that $E_1, E_2, E_3, \ldots, E_n$ is a random sample from a distribution that is normal with mean 0 and variance σ^2. We express this by writing

$$E_i \sim N(0, \sigma^2)$$

Note that this implies that the random variables $E_1, E_2, E_3, \ldots, E_n$ are independent. Since our model expresses Y_i as a linear function of E_i, the assumptions concerning $E_1, E_2, E_3, \ldots, E_n$ impose some restrictions on the random variables

$Y_1, Y_2, Y_3, \ldots, Y_n$. Namely, we are assuming the following:

Model assumptions: Simple linear regression

1. The random variables Y_i are independently and, normally distributed.
2. The mean of Y_i is $\beta_0 + \beta_1 x_i$.
3. The variance of Y_i is σ^2.

We express these assumptions by writing

$$Y_i \sim N(\beta_0 + \beta_1 x_i, \sigma^2)$$

Notice that σ^2 is a measure of the variability of the responses about the true regression line. These assumptions are demonstrated in Fig. 11.5. Note that the mean values of $Y_1, Y_2, Y_3, \ldots, Y_n$ may differ but that each is assumed to have the same variance. Thus the associated normal curves may differ in location but all of them have the same shape.

Before using the assumptions just made to determine the distribution of B_0 and B_1, we pause to state some results that will make our work simpler. These results can be verified easily by applying the rules governing the behavior of the summation symbol.

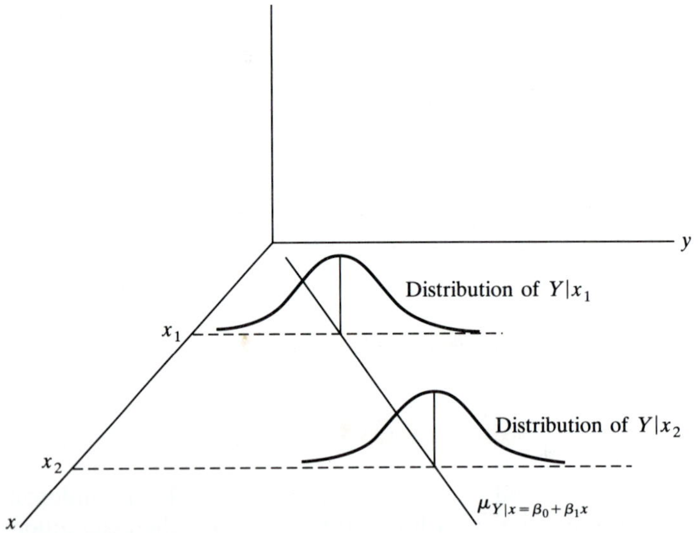

FIGURE 11.5
For each i, Y_i is normally distributed with mean $\mu_{Y|x_i} = \beta_0 + \beta_1 x_i$ and variance σ^2.

Some properties of summation:

1. $\sum_{i=1}^{n} (x_i - \bar{x}) = 0$

2. $\sum_{i=1}^{n} (x_i - \bar{x})(Y_i - \bar{Y}) = \sum_{i=1}^{n} (x_i - \bar{x})Y_i$

3. $\sum_{i=1}^{n} (x_i - \bar{x})(Y_i - \bar{Y}) = \left(n \sum_{i=1}^{n} x_i Y_i - \sum_{i=1}^{n} x_i \sum_{i=1}^{n} Y_i\right) \Big/ n$

4. $\sum_{i=1}^{n} (x_i - \bar{x})^2 = \sum_{i=1}^{n} (x_i - \bar{x})x_i$

5. $\sum_{i=1}^{n} (x_i - \bar{x})^2 = \left[n \sum_{i=1}^{n} x_i^2 - \left(\sum_{i=1}^{n} x_i\right)^2\right] \Big/ n$

Distribution of B_1

To develop confidence intervals or test hypotheses or the slope of a regression line, we need to know the distribution of B_1, the estimator for this slope. We shall show that the model assumptions on Y_i ensure the B_1 is normally distributed with $E[B_1] = \beta_1$ and $\operatorname{Var} B_1 = \sigma^2/\sum_{i=1}^{n}(x_i - \bar{x})^2$. Notice that this implies that the least-squares estimator for β_1 is an unbiased estimator for this parameter.

To derive the distribution of B_1, we first use properties 2, 3, and 5 above to rewrite the estimators as shown:

$$B_1 = \frac{n \sum_{i=1}^{n} x_i Y_i - \left(\sum_{i=1}^{n} x_i\right)\left(\sum_{i=1}^{n} Y_i\right)}{n \sum_{i=1}^{n} x_i^2 - \left(\sum_{i=1}^{n} x_i\right)^2}$$

$$= \frac{\sum_{i=1}^{n} (x_i - \bar{x})(Y_i - \bar{Y})}{\sum_{i=1}^{n} (x_i - \bar{x})^2}$$

$$= \frac{\sum_{i=1}^{n} (x_i - \bar{x})Y_i}{\sum_{i=1}^{n} (x_i - \bar{x})^2}$$

Letting

$$c_j = \frac{(x_j - \bar{x})}{\sum_{i=1}^{n} (x_i - \bar{x})^2} \qquad j = 1, 2, 3, \ldots, n$$

we have expressed B_1 in the form

$$B_1 = c_1Y_1 + c_2Y_2 + \cdots + c_nY_n$$

That is, we have expressed B_1 as a linear function of the independent normal random variables $Y_1, Y_2, \ldots, Y_n$. Since any linear function of independent normal random variables is normally distributed (see Exercise 41, Chap. 7), we can conclude that B_1 is normal. Using the rules for expectation, we see that

$$E[B_1] = E[c_1Y_1 + c_2Y_2 + \cdots + c_nY_n]$$

$$= E\left[\frac{(x_1 - \bar{x})Y_1 + (x_2 - \bar{x})Y_2 + \cdots + (x_n - \bar{x})Y_n}{\sum_{i=1}^{n} (x_i - \bar{x})^2}\right]$$

$$= \frac{\sum_{i=1}^{n} (x_i - \bar{x})E[Y_i]}{\sum_{i=1}^{n} (x_i - \bar{x})^2}$$

For each i, $E[Y_i] = \beta_0 + \beta_1 x_i$. Substituting, we see that

$$E[B_1] = \frac{\sum_{i=1}^{n} (x_i - \bar{x})(\beta_0 + \beta_1 x_i)}{\sum_{i=1}^{n} (x_i - \bar{x})^2}$$

$$= \frac{\sum_{i=1}^{n} (x_i - \bar{x})\beta_0 + \beta_1 \sum_{i=1}^{n} (x_i - \bar{x})x_i}{\sum_{i=1}^{n} (x_i - \bar{x})^2}$$

By summation properties 1 and 4,

$$E[B_1] = \beta_1$$

This result shows that B_1 is an unbiased estimator for β_1. We apply the rules of variance to find Var B_1 as follows:

$$\operatorname{Var} B_1 = \operatorname{Var}\left[\frac{\sum_{i=1}^{n}(x_i - \bar{x})Y_i}{\sum_{i=1}^{n}(x_i - \bar{x})^2}\right]$$

$$= \left[\frac{1}{\sum_{i=1}^{n}(x_i - \bar{x})^2}\right]^2 \operatorname{Var} \sum_{i=1}^{n}(x_i - \bar{x})Y_i$$

$$= \left[\frac{1}{\sum_{i=1}^{n}(x_i - \bar{x})^2}\right]^2 \sum_{i=1}^{n} \operatorname{Var}(x_i - \bar{x})Y_i$$

$$= \left[\frac{1}{\sum_{i=1}^{n}(x_i - \bar{x})^2}\right]^2 \sum_{i=1}^{n}(x_i - \bar{x})^2 \operatorname{Var} Y_i$$

Since Var Y_i is assumed to be σ^2 for each i, we can substitute to obtain

$$\operatorname{Var} B_1 = \left[\frac{1}{\sum_{i=1}^{n}(x_i - \bar{x})^2}\right]^2 \sum_{i=1}^{n}(x_i - \bar{x})^2\sigma^2$$

$$= \frac{\sigma^2}{\sum_{i=1}^{n}(x_i - \bar{x})^2}$$

These results are summarized by writing

$$B_1 \sim N\left(\beta_1, \sigma^2 \middle/ \sum_{i=1}^{n}(x_i - \bar{x})^2\right)$$

Distribution of B_0

Confidence intervals on the intercept of the regression line and hypothesis tests on this parameter are based on knowledge of the distribution of B_0, the estimator for this intercept. We shall show that this estimator is normally distributed with $E[B_0] = \beta_0$ and $\operatorname{Var} B_0 = \sigma^2 \sum_{i=1}^n x_i^2 / n \sum_{i=1}^n (x_i - \bar{x})^2$. Once again, the least-squares estimator for β_0 is an unbiased estimator for this parameter.

To derive the distribution of the estimator B_0, we note first that it can be shown that $\bar{Y}$ and B_1 are independent. (See Exercise 14.) Since

$$B_0 = \bar{Y} - B_1 \bar{x}$$

B_0 is a linear function of independent normal random variables and therefore is itself normally distributed. Using the rules of expectation, we see that

$$\begin{aligned}
E[B_0] &= E\left[\bar{Y} - B_1 \bar{x}\right] \\
&= E\left[(Y_1 + Y_2 + \cdots + Y_n)/n - B_1 \bar{x}\right] \\
&= (E[Y_1] + E[Y_2] + \cdots + E[Y_n])/n - \bar{x}E[B_1] \\
&= [(\beta_0 + \beta_1 x_1) + (\beta_0 + \beta_1 x_2) + \cdots + (\beta_0 + \beta_1 x_n)]/n - \bar{x}E[B_1] \\
&= \left(n\beta_0 + \beta_1 \sum_{i=1}^n x_i\right) \Big/ n - \bar{x}E[B_1] \\
&= \beta_0 + \bar{x}\beta_1 - \bar{x}\beta_1 \\
&= \beta_0
\end{aligned}$$

This result shows that B_0 is an unbiased estimator for β_0. The variance of B_0 is given by

$$\begin{aligned}
\operatorname{Var} B_0 &= \operatorname{Var}\left(\bar{Y} - B_1 \bar{x}\right) \\
&= \operatorname{Var} \bar{Y} + \bar{x}^2 \operatorname{Var} B_1
\end{aligned}$$

Note that

$$\begin{aligned}
\operatorname{Var}(\bar{Y}) &= \operatorname{Var}(Y_1 + Y_2 + \cdots + Y_n)/n \\
&= \frac{\operatorname{Var} Y_1 + \operatorname{Var} Y_2 + \cdots + \operatorname{Var} Y_n}{n^2} \\
&= \frac{n\sigma^2}{n^2} = \frac{\sigma^2}{n}
\end{aligned}$$

By substituting, we see that

$$\operatorname{Var} B_0 = \frac{\sigma^2}{n} + \frac{\bar{x}^2\sigma^2}{\sum_{i=1}^{n}(x_i - \bar{x})^2}$$

$$= \frac{\sigma^2 \sum_{i=1}^{n}(x_i - \bar{x})^2 + n\bar{x}^2\sigma^2}{n\sum_{i=1}^{n}(x_i - \bar{x})^2}$$

$$= \frac{\sigma^2\left[\dfrac{n\sum_{i=1}^{n}x_i^2 - \left(\sum_{i=1}^{n}x_i\right)^2}{n} + \dfrac{\left(\sum_{i=1}^{n}x_i\right)^2}{n}\right]}{n\sum_{i=1}^{n}(x_i - \bar{x})^2}$$

$$= \frac{\sum_{i=1}^{n}x_i^2}{n\sum_{i=1}^{n}(x_i - \bar{x})^2}\sigma^2$$

To summarize, we have shown that

$$B_0 \sim N\left[\beta_0, \frac{\sum_{i=1}^{n}x_i^2}{n\sum_{i=1}^{n}(x_i - \bar{x})^2}\sigma^2\right]$$

Estimator of σ^2

To test hypotheses and construct confidence intervals on various parameters, we must estimate the unknown variance σ^2. Recall that σ^2 denotes the variability of each of the random variables Y_i about the *true* regression line. To estimate this variability, we use information concerning the variability of the data points about the *fitted* regression line.

Since the residual measures the unexplained or random deviation of a data point from the estimated line of regression, the residuals are used to estimate σ^2. That is, our estimate makes use of SSE, the sum of the squares of the residuals. In particular, we shall estimate σ^2 by

$$S^2 = \hat{\sigma}^2 = \text{SSE}/(n-2)$$

We divide SSE by $n - 2$ so that the estimate will be unbiased for σ^2. (See Exercise 13.)

Summary of Theoretical Results

Before closing this section, let us introduce some notation that will make the results obtained here easier to remember. Namely, we shall denote $\sum_{i=1}^n (x_i - \bar{x})^2$ by S_{xx}. The symbol S_{yy} will denote $\sum_{i=1}^n (y_i - \bar{y})^2$ or $\sum_{i=1}^n (Y_i - \bar{Y})^2$. Whether we are dealing with the random variables Y_i or their observed values y_i should be clear from the context in which the symbol is used. Similarly, S_{xy} will denote either $\sum_{i=1}^n (x_i - \bar{x})(y_i - \bar{y})$ or $\sum_{i=1}^n (x_i - \bar{x})(Y_i - \bar{Y})$ and SSE will denote $\sum_{i=1}^n (y_i - b_0 - b_1 x_i)^2$ or $\sum_{i=1}^n (Y_i - B_0 - B_1 x_i)^2$. This notation can be used to rewrite the error sum of squares as follows:

$$\begin{aligned}
\text{SSE} &= \sum_{i=1}^n (Y_i - B_0 - B_1 x_i)^2 \\
&= \sum_{i=1}^n \left(Y_i - \bar{Y} + B_1 \bar{x} - B_1 x_i\right)^2 \\
&= \sum_{i=1}^n \left[\left(Y_i - \bar{Y}\right) - B_1(x_i - \bar{x})\right]^2 \\
&= \sum_{i=1}^n (Y_i - \bar{Y})^2 - 2B_1 \sum_{i=1}^n (x_i - \bar{x})(Y_i - \bar{Y}) + B_1^2 \sum_{i=1}^n (x_i - \bar{x})^2 \\
&= S_{yy} - 2B_1 S_{xy} + B_1^2 S_{xx}
\end{aligned}$$

Note that

$$B_1 = \frac{\sum_{i=1}^n (x_i - \bar{x})(Y_i - \bar{Y})}{\sum_{i=1}^n (x_i - \bar{x})^2} = \frac{S_{xy}}{S_{xx}}$$

By substituting, we see that

$$\begin{aligned}
\text{SSE} &= S_{yy} - 2B_1 S_{xy} + B_1 \frac{S_{xy}}{S_{xx}} S_{xx} \\
&= S_{yy} - B_1 S_{xy}
\end{aligned}$$

Let us summarize the theoretical results that we have obtained in this section. We have shown that

1. $\sum_{i=1}^{n}(x_i - \bar{x})^2 = S_{xx}$
2. $\sum_{i=1}^{n}(Y_i - \bar{Y})^2 = S_{yy}$
3. $\sum_{i=1}^{n}(x_i - \bar{x})(Y_i - \bar{Y}) = S_{xy}$
4. $B_1 = S_{xy}/S_{xx}$ is an unbiased estimator for β_1. This estimator is normally distributed with variance $\sigma_{B_1}^2 = \sigma^2/S_{xx}$.
5. $B_0 = \bar{Y} - B_1\bar{x}$ is an unbiased estimator for β_0. This estimator is normally distributed with variance $\sigma_{B_0}^2 = (\sum_{i=1}^{n} x_i^2 \sigma^2)/nS_{xx}$.
6. $S^2 = \text{SSE}/(n - 2)$ is an unbiased estimator for σ^2.

11.3 CONFIDENCE INTERVAL ESTIMATION AND HYPOTHESIS TESTING

In the previous sections we considered point estimation procedures for the parameters associated with the simple linear regression model. We showed that the estimators given are unbiased. With this information alone, we can estimate a regression line from a sample of paired observations (x_i, y_i) and predict the value of Y or estimate the mean value of Y for a given value x. As in the past, we do not end our study with point estimation. We continue by developing pertinent confidence intervals and by learning how to test hypotheses on the model parameters. In this section we consider these topics:

1. Hypothesis testing and confidence interval estimation on the slope of the regression line
2. Hypothesis testing and confidence interval estimation on the intercept of the regression line
3. Confidence interval estimation on the mean value of Y for a given value x
4. Prediction interval estimation on the value of Y itself for a given value x

We consider these ideas in the order listed.

Inferences about Slope

One of the first questions that a scientist wants to answer is, "Is the regression 'significant'?" The term "significant" regression as used here means that there is sufficient statistical evidence to conclude that the slope of the true regression line is not zero. Note that if $\beta_1 = 0$, then our regression model is

$$Y_i = \beta_0 + E_i$$

This implies that the variation in Y is due solely to random fluctuations about the line $y = \beta_0$. If $\beta_1 \neq 0$, then at least some of the variation in Y is explained by the fact that Y is being observed at different x values. In the latter case our regression model is helpful in estimating $\mu_{Y|x}$ and predicting $Y|x$.

To develop a test statistic for testing H_0: $\beta_1 = 0$, we reconsider B_1, the point estimator for β_1. Recall that

$$B_1 \sim N(\beta_1, \sigma^2/S_{xx})$$

By standardizing, we can conclude that the random variable

$$(B_1 - \beta_1)\Big/\left(\sigma\big/\sqrt{S_{xx}}\right)$$

is standard normal. It can be shown that the random variable $(n-2)S^2/\sigma^2 = \text{SSE}/\sigma^2$ has a chi-squared distribution with $n - 2$ degrees of freedom and that B_1 and S^2 are independent [19]. By applying Definition 8.2.1, the definition of a T random variable, we can conclude that the random variable

$$\frac{(B_1 - \beta_1)\Big/\left(\sigma\big/\sqrt{S_{xx}}\right)}{\sqrt{(n-2)S^2/\sigma^2(n-2)}} = \frac{B_1 - \beta_1}{S\big/\sqrt{S_{xx}}}$$

has a T distribution with $n - 2$ degrees of freedom. If $\beta_1 = 0$, then this random variable becomes

$$\frac{B_1}{S\big/\sqrt{S_{xx}}}$$

This statistic serves as the test statistic for testing any of the usual three hypotheses:

H_0: $\beta_1 = 0$	H_0: $\beta_1 = 0$	H_0: $\beta_1 = 0$
H_1: $\beta_1 > 0$	H_1: $\beta_1 < 0$	H_1: $\beta_1 \neq 0$
Right-tailed test	Left-tailed test	Two-tailed test

The null hypothesis is rejected for large positive values of the test statistic in conducting a right-tailed test; large negative values lead to rejection of H_0 in a left-tailed test. In a two-tailed test H_0 is rejected for large values in either the positive or negative direction.

One other point needs to be made. We have considered the null value to be 0 because this is the value most often encountered in practice. We can test H_0: $\beta_1 = \beta_1^0$, where β_1^0 denotes any hypothesized value for the slope of the regression line. The test statistic for this generalized null hypothesis is

$$T_{n-2} = \frac{(B_1 - \beta_1^0)}{S/\sqrt{S_{xx}}}$$

Example 11.3.1. In Example 11.1.1 we estimated the regression equation of Y, the extent of solvent evaporation while spray painting, on X, the relative humidity, to be

$$\hat{\mu}_{Y|x} = 13.64 - .08x$$

We now determine whether the regression is significant. That is, we test

$$H_0: \beta_1 = 0$$

$$H_1: \beta_1 \neq 0$$

Summary statistics for the data given previously are

$$n = 25 \qquad \sum x = 1314.90 \qquad \sum y = 235.70$$

$$\sum x^2 = 76{,}308.53 \qquad \sum y^2 = 2286.07 \qquad \sum xy = 11{,}824.44$$

For these data

$$\begin{aligned} S_{xx} &= \left[n\sum x^2 - \left(\sum x\right)^2\right]\Big/n \\ &= \left[25(76{,}308.53) - (1{,}314.90)^2\right]\Big/25 \\ &= 7{,}150.05 \end{aligned}$$

$$\begin{aligned} S_{yy} &= \left[n\sum y^2 - \left(\sum y\right)^2\right]\Big/n \\ &= \left[25(2{,}286.07) - (235.70)^2\right]\Big/25 \\ &= 63.89 \end{aligned}$$

$$\begin{aligned} S_{xy} &= \left[n\sum xy - \sum x \sum y\right]\Big/n \\ &= [25(11{,}824.44) - (1{,}314.90)(235.70)]/25 \\ &= -572.44 \end{aligned}$$

Using these data, we obtain

$$\begin{aligned} \text{SSE} &= S_{yy} - b_1 S_{xy} \\ &= 63.89 - (-.08)(-572.44) \\ &= 18.09 \end{aligned}$$

Hence

$$s^2 = \text{SSE}/(n-2)$$
$$= 18.09/23$$
$$= .79$$

The observed value of the $T_{n-2} = T_{23}$ test statistic is

$$t = \frac{b_1}{s/\sqrt{S_{xx}}}$$

$$= \frac{-.08}{\sqrt{.79}/\sqrt{7{,}150.05}}$$

$$= -7.62$$

From Table VI of App. A we see that $P[T_{23} \leq -7.62] < .0005$. Since this is a two-tailed test, $P < 2(.0005) = .001$. We can reject H_0 and conclude that the slope of the true regression line is not zero. That is, knowledge of the x value does help in estimating $\mu_{Y|x}$ and predicting $Y|x$.

The random variable

$$T_{n-2} = \frac{B_1 - \beta_1}{S/\sqrt{S_{xx}}}$$

is of the form

$$\frac{\text{Estimator} - \text{parameter}}{D}$$

where D is the estimator for the standard deviation of B_1. This is the same algebraic structure encountered several times in the past. (See Secs. 9.3 and 10.3.) The resulting confidence interval for β_1 assumes the familiar form

$$\text{Estimator} \pm \text{probability point} \cdot D$$

In this case the confidence interval is

Confidence interval on β_1, the slope of the regression line

$$B_1 \pm t_{\alpha/2} S/\sqrt{S_{xx}}$$

where $t_{\alpha/2}$ is the appropriate point based on the T_{n-2} distribution.

Inferences about Intercept

Hypothesis tests on β_0, the intercept of the true regression line, are conducted by noting that since

$$B_0 \sim N(\beta_0, \sigma^2 \sum x^2 / nS_{xx})$$

the random variable

$$\frac{B_0 - \beta_0}{\left(\sigma\sqrt{\sum x^2}\right)\Big/\sqrt{n \cdot S_{xx}}}$$

is standard normal. It can be shown that B_0 and S^2 are independent. Thus the random variable

$$\frac{(B_0 - \beta_0)\Big/\left(\sigma\sqrt{\sum x^2}\Big/\sqrt{nS_{xx}}\right)}{\sqrt{(n-2)S^2/\sigma^2(n-2)}} = \frac{B_0 - \beta_0}{\left(\dfrac{S\sqrt{\sum x^2}}{\sqrt{nS_{xx}}}\right)}$$

follows a T distribution with $n - 2$ degrees of freedom.

The test statistic for testing H_0: $\beta_0 = 0$ is

$$\frac{B_0}{\left(\dfrac{S\sqrt{\sum x^2}}{\sqrt{nS_{xx}}}\right)}$$

Confidence intervals on the value of β_0 are found as follows:

Confidence interval on β_0, the intercept of the regression line

$$B_0 \pm t_{\alpha/2} \frac{S\sqrt{\sum x^2}}{\sqrt{nS_{xx}}}$$

where $t_{\alpha/2}$ is the appropriate point based on the T_{n-2} distribution.

The next example illustrates the use of these confidence intervals.

Example 11.3.2. We continue the analysis of the data on the extent of solvent evaporation during spray painting and relative humidity by finding confidence intervals on β_0 and β_1. These summary statistics, found earlier, are needed:

$$s^2 = .79 \qquad \sum x^2 = 76{,}308.53 \qquad b_0 = 13.64$$

$$S_{xx} = 7150.05 \qquad b_1 = -.08 \qquad n = 25$$

A 99% confidence interval on the slope of the regression line is given by

$$b_1 \pm t_{.005} s \Big/ \sqrt{S_{xx}} \qquad \text{or} \qquad -.08 \pm 2.807\sqrt{.79} \Big/ \sqrt{7150.05}$$

The point $t_{.005}$ is based on the $T_{n-2} = T_{23}$ distribution. Completing the calculations, we see that we can be 99% confident that the slope of the true regression line lies in the interval $[-.109, -.051]$. Note that this interval does not contain 0. This is expected since we rejected $H_0\colon \beta_1 = 0$ in our last example.

A 90% confidence interval on the intercept of the regression line is given by

$$b_0 \pm t_{.05} s \sqrt{\sum x^2} \Big/ \sqrt{n S_{xx}} \qquad \text{or} \qquad 13.64 \pm 1.714\sqrt{.79}\sqrt{76{,}308.53} \Big/ \sqrt{25(7150.05)}$$

We can be 90% confident that the true regression line crosses the y axis between the points $y = 12.64$ and $y = 14.64$.

Inferences about Estimated Mean

In addition to finding a point estimate for $\mu_{Y|x}$, the mean value of Y for a given x value, it is useful to be able to obtain a confidence interval on this parameter. To do so, we consider the distribution of the point estimator for $\mu_{Y|x}$ by rewriting this estimator in the form

$$\begin{aligned}\hat{\mu}_{Y|x} &= B_0 + B_1 x \\ &= \bar{Y} - B_1\bar{x} + B_1 x \\ &= \bar{Y} + B_1(x - \bar{x})\end{aligned}$$

Since $\bar{Y}$ and B_1 are both normally distributed and independent, $\hat{\mu}_{Y|x}$ is normal. In Exercise 12 we found that this estimator is unbiased for $\mu_{Y|x}$. The only other information needed is its variance. Using the rules for variance, see that

$$\begin{aligned}\operatorname{Var}(\hat{\mu}_{Y|x}) &= \operatorname{Var}\left[\bar{Y} + B_1(x - \bar{x})\right] \\ &= \operatorname{Var}\bar{Y} + (x - \bar{x})^2 \operatorname{Var} B_1 \\ &= \sigma^2/n + (x - \bar{x})^2 \sigma^2/S_{xx} \\ &= \left[1/n + \frac{(x - \bar{x})^2}{S_{xx}}\right]\sigma^2\end{aligned}$$

To summarize, we can conclude that

$$\hat{\mu}_{Y|x} \sim N\left\{\mu_{Y|x}, \left[1/n + \frac{(x-\bar{x})^2}{S_{xx}}\right]\sigma^2\right\}$$

By the standardization process it can be shown that

$$\frac{\hat{\mu}_{Y|x} - \mu_{Y|x}}{\sigma\sqrt{\dfrac{1}{n} + \dfrac{(x-\bar{x})^2}{S_{xx}}}}$$

is standard normal. Dividing by $\sqrt{(n-2)S^2/\sigma^2(n-2)} = S/\sigma$, we find that the random variable

$$\frac{\hat{\mu}_{Y|x} - \mu_{Y|x}}{S\sqrt{\dfrac{1}{n} + \dfrac{(x-\bar{x})^2}{S_{xx}}}}$$

follows a T distribution with $n-2$ degrees of freedom. Since the random variable is of the same algebraic form as those encountered earlier, confidence intervals on $\mu_{Y|x}$ are found using this formula:

Confidence interval on $\mu_{Y|x}$, the mean value of Y when $X = x$

$$\hat{\mu}_{Y|x} \pm t_{\alpha/2} S\sqrt{\frac{1}{n} + \frac{(x-\bar{x})^2}{S_{xx}}}$$

where $t_{\alpha/2}$ is the appropriate point based on the T_{n-2} distribution.

This formula can be used to construct what is called a *confidence band* about the estimated regression line. To do so, one simply constructs $100(1-\alpha)\%$ confidence intervals at several selected points and then joins the endpoints of these intervals with a smooth curve. The true regression line should lie within the band. Figure 11.6 illustrates this idea.

Inferences about Single Predicted Value

One of the primary uses of the estimated regression line is to predict the value of Y itself for a specified value x. We know that the point estimator for $Y|x$ is

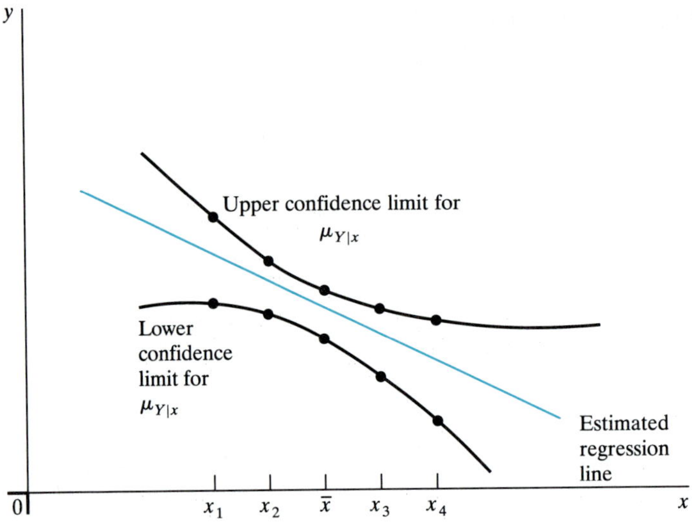

FIGURE 11.6
95% confidence band on $\mu_{Y|x}$.

the same as the point estimator for $\mu_{Y|x}$, namely,

$$\hat{Y}|x = \hat{\mu}_{Y|x} = B_0 + B_1 x$$

Note that $Y|x$ is a random variable, not an unknown constant. When we ask for a "prediction interval" on $Y|x$, we are asking for two statistics L_1 and L_2 with the property that

$$P[L_1 \leq Y|x \leq L_2] \doteq 1 - \alpha$$

That is, we are asking for two statistics that will trap the observed value of $Y|x$ between them $(1 - \alpha)100\%$ of the time. To find these statistics, we use the guideline for constructing a confidence interval given in Chap. 7. This guideline requires that we find a random variable whose expression involves $Y|x$ and whose distribution we know. Recall that

$$\hat{\mu}_{Y|x} \sim N\left\{\mu_{Y|x}, \left[\frac{1}{n} + \frac{(x - \bar{x})^2}{S_{xx}}\right]\sigma^2\right\}$$

and that, via our model assumptions

$$Y|x \sim N(\mu_{Y|x}, \sigma^2)$$

It can be shown that the random variable $\hat{Y}|x - Y|x$ is normally distributed

[19]. Using the rules for expectation, we obtain

$$E\left[\hat{Y}|x - Y|x\right] = E\left[\hat{Y}|x\right] - E[Y|x]$$
$$= \mu_{Y|x} - \mu_{Y|x} = 0$$

Similarly, the rules for variance are used to show that

$$\operatorname{Var}\left[\hat{Y}|x - Y|x\right] = \operatorname{Var}\hat{Y}|x + \operatorname{Var}Y|x$$
$$= \left[\frac{1}{n} + \frac{(x-\bar{x})^2}{S_{xx}}\right]\sigma^2 + \sigma^2$$
$$= \left[1 + \frac{1}{n} + \frac{(x-\bar{x})^2}{S_{xx}}\right]\sigma^2$$

In conclusion, it can be seen that

$$(\hat{Y}|x - Y|x) \sim N\left\{0, \left[1 + \frac{1}{n} + \frac{(x-\bar{x})^2}{S_{xx}}\right]\sigma^2\right\}$$

In this case standardization and division by S/σ results in the T random variable

$$T_{n-2} = \frac{\hat{Y}|x - Y|x}{S\sqrt{1 + \frac{1}{n} + \frac{(x-\bar{x})^2}{S_{xx}}}}$$

The algebraic structure of this random variable parallels that seen earlier. For this reason, we can conclude that a 100(1 − α)% "prediction interval" on $Y|x$ is given by

Prediction interval on $Y|x$, the value of Y when $X = x$

$$\hat{Y}|x \pm t_{\alpha/2}S\sqrt{1 + \frac{1}{n} + \frac{(x-\bar{x})^2}{S_{xx}}}$$

where $t_{\alpha/2}$ is the appropriate point based on the T_{n-2} distribution.

By evaluating the prediction limits at several x values, we can construct a prediction band on $Y|x$. Note that the confidence limits for $\mu_{Y|x}$ and $Y|x$ are similar. The difference is that the former entails the term

$$\sqrt{\frac{1}{n} + \frac{(x - \bar{x})^2}{S_{xx}}}$$

whereas the corresponding term in the latter is a little larger, namely,

$$\sqrt{1 + \frac{1}{n} + \frac{(x - \bar{x})^2}{S_{xx}}}$$

This is to be expected since we should be able to estimate an average response more precisely than we can predict an individual observation. Graphically, the confidence band on $\mu_{Y|x}$ will be contained in the corresponding prediction band for $Y|x$. This idea is illustrated in Fig. 11.7.

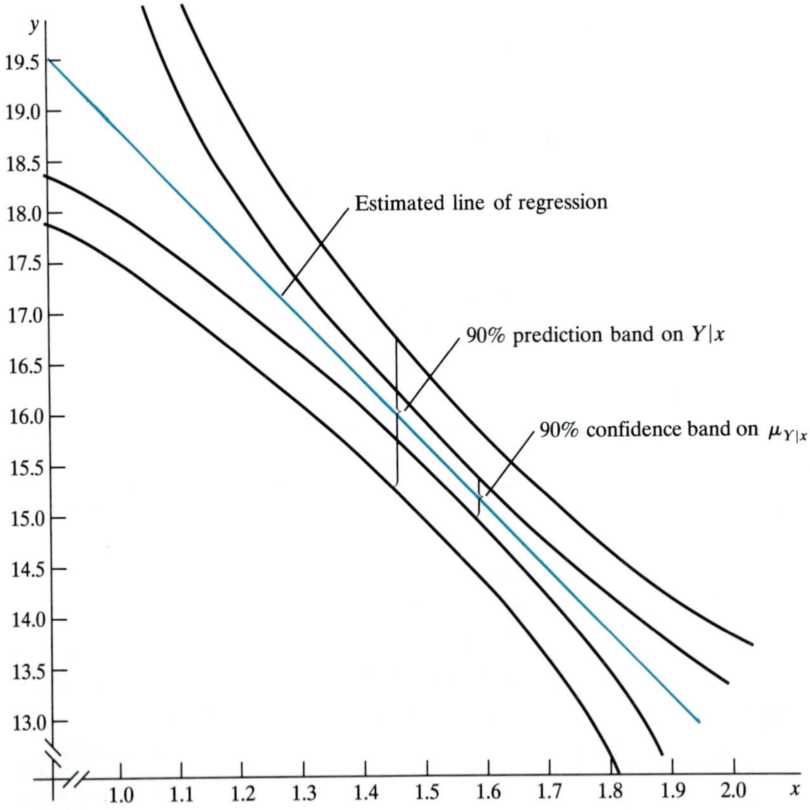

FIGURE 11.7
Relative positions of a 90% confidence band on $\mu_{Y|x}$ and a 90% prediction band on $Y|x$.

The next example should demonstrate clearly the difference between these two types of intervals.

Example 11.3.3. An investigation is conducted to study gasoline mileage in automobiles when used exclusively for urban driving. Ten properly tuned and serviced automobiles manufactured during the same year are used in the study. Each automobile is driven for 1000 miles and the average number of miles per gallon (mi/gal) obtained (Y) and the weight of the car in tons (X) are recorded. These data result:

Car number	1	2	3	4	5	6	7	8	9	10
Miles per gallon (y)	17.9	16.5	16.4	16.8	18.8	15.5	17.5	16.4	15.9	18.3
Weight in tons (x)	1.35	1.90	1.70	1.80	1.30	2.05	1.60	1.80	1.85	1.40

Summary statistics for these data are

$$n = 10 \qquad \sum x^2 = 28.6375 \qquad \sum y^2 = 2900.46 \qquad S_{xx} = .581 \qquad S_{xy} = -2.345$$

$$\sum x = 16.75 \qquad \sum y = 170.0 \qquad \sum xy = 282.405 \qquad S_{yy} = 10.46$$

$$\begin{aligned} \hat{\beta}_1 = b_1 &= \frac{n\sum xy - \sum x \sum y}{n\sum x^2 - \left(\sum x\right)^2} \\ &= \frac{10(282.405) - (16.75)(170.0)}{10(28.6375) - (16.75)^2} \\ &= -4.03 \\ \hat{\beta}_0 = b_0 &= \bar{y} - b_1\bar{x} \\ &= 17.0 - (-4.03)(1.675) \\ &= 23.75 \end{aligned}$$

The estimated line of regression is

$$\hat{\mu}_{Y|x} = b_0 + b_1 x = 23.75 - 4.03x$$

The reader can verify that H_0: $\beta_1 = 0$ can be rejected with $P < .0001$. Thus the regression is significant; the model is useful in predicting gasoline mileage based on automobile weight. Suppose that we are interested in all cars weighing 1.7 tons. The estimated average mileage for these cars is

$$\hat{\mu}_{Y|x=1.7} = 23.75 - 4.03(1.7) = 16.899 \text{ mi/gal}$$

This estimate is not very useful without some idea of its accuracy. To pinpoint the accuracy, we construct a 90% confidence interval on $\mu_{Y|x=1.7}$. To do so, we must compute SSE and s^2 for these data:

$$\begin{aligned} \text{SSE} &= S_{yy} - b_1 S_{xy} \\ &= 10.46 - (-4.03)(-2.345) \\ &= 1.01 \\ s^2 &= \text{SSE}/(n-2) = 1.01/8 = .126 \end{aligned}$$

A 90% confidence interval on $\mu_{Y|x}$ is

$$\hat{\mu}_{Y|x} \pm t_{\alpha/2} s \sqrt{\frac{1}{n} + \frac{(x - \bar{x})^2}{S_{xx}}}$$

or

$$16.899 \pm 1.86\sqrt{.126}\sqrt{\frac{1}{10} + \frac{(1.7 - 1.675)^2}{.581}}$$

$$16.899 \pm .21$$

We can be 90% confident that the average gas mileage for cars weighing 1.7 tons lies between 16.689 and 17.109 mi/gal.

To predict the gas mileage for a single car weighing 1.7 tons, we use the interval

$$\hat{y}|x \pm t_{\alpha/2} s \sqrt{1 + \frac{1}{n} + \frac{(x - \bar{x})^2}{S_{xx}}}$$

For these data this interval is

$$16.899 \pm 1.86\sqrt{.126}\sqrt{1 + \frac{1}{10} + \frac{(1.7 - 1.675)^2}{.581}}$$

$$16.899 \pm .69$$

We can be 90% confident that the gas mileage for any individual automobile weighing 1.7 tons lies between 16.209 and 17.589 mi/gal. As expected, the prediction interval used to predict the gas mileage for a single auto is wider than that used to predict the average mileage for a group of automobiles.

We should note here that the width of a confidence or prediction band is a function of x. To see why this is true, consider the formula for constructing a prediction interval of $Y|x$. The width of the interval is determined in part by the term

$$\sqrt{1 + \frac{1}{n} + \frac{(x - \bar{x})^2}{S_{xx}}}$$

It is evident that this term is smallest when $x = \bar{x}$. Hence we can predict the value of Y more precisely for values of x that are near the average value $\bar{x}$. This fact is evident graphically in the confidence bands shown in Fig. 11.7. In this figure the bands are narrowest at $\bar{x} = 1.675$.

11.4 REPEATED MEASUREMENTS AND LACK OF FIT

When we fit a straight line to a set of paired observations via the least-squares procedure, we are assuming at the outset that linear regression is appropriate. The reasonableness of this assumption can be checked visually via a scattergram. Unfortunately, two people can view the same scattergram differently; it might appear to exhibit a linear trend to one but not to the other! We need an analytic method to test the appropriateness of the linear regression model. In this section we present a statistical method for detecting model "lack of fit." The method is based on an examination of the residuals—the differences between the observed values of the dependent variable Y and the values predicted for Y via the estimated regression line.

The residual or error sum of squares, SSE, may be large either because Y exhibits a high variability naturally or because the assumed model is inappropriate. The method used to detect model lack of fit entails partitioning SSE into two components attributable to these sources of error. The portion attributable to natural variability in Y is called *pure* or *experimental error*; that attributable to inappropriateness of the model is called *error due to lack of fit*. If the model is appropriate, then logically we expect most of SSE to be pure error; if the model is inappropriate, then a large portion of SSE should be attributed to lack of fit. Our test is to determine the portion of SSE due to lack of fit and to reject our model if this appears to be too large to have occurred by chance.

To measure pure error, we must have available what are called repeated or replicated measurements. That is, at one or more points x_i $(i = 1, 2, \dots, k)$ we must have at least two observations on Y. Let Y_{ij} denote the jth observation on Y at point x_i $(j = 1, 2, 3, \dots, n_i)$. Using this notation, the data layout for our experiment is as shown in Table 11.1. Note that the total number of observations is

$$n = n_1 + n_2 + n_3 + \cdots + n_k = \sum_{i=1}^{k} n_i$$

Recall that we measure the natural variability in a random variable by considering its deviation about its mean. For each $i = 1, 2, 3, \dots, k$, we can view $Y_{i1}, Y_{i2}, Y_{i3}, \dots, Y_{in_i}$ as a random sample of size n_i from the distribution of the random variable $Y|x_i$. An unbiased estimator for $\mu_{Y|x_i}$ is the sample mean $\bar{Y}_i$ where

$$\bar{Y}_i = \sum_{j=1}^{n_i} Y_{ij}/n_i$$

TABLE 11.1

Value of x				
x_1	x_2	x_3	$\cdots$	x_k
Y_{11}	Y_{21}	Y_{31}		Y_{k1}
Y_{12}	Y_{22}	Y_{32}		Y_{k2}
Y_{13}	Y_{23}	Y_{33}		Y_{k3}
$\vdots$	$\vdots$	$\vdots$		$\vdots$
Y_{1n_1}	Y_{2n_2}	Y_{3n_3}		Y_{kn_k}

The statistic

$$\sum_{j=1}^{n_i} \left(Y_{ij} - \bar{Y}_i\right)^2$$

measures the natural variability of Y at the point x_i and is called an internal sum of squares.

To obtain a measure of the natural variability in Y over all x values, we pool the k internal sums of squares to form the statistic

$$\sum_{i=1}^{k} \sum_{j=1}^{n_i} \left(Y_{ij} - \bar{Y}_i\right)^2$$

This statistic is called the *sum of squares due to pure error*, denoted by SSE_{pe}. It is left to the reader to argue that the random variable

$$\text{SSE}_{\text{pe}}/\sigma^2$$

follows a chi-squared distribution with $n - k$ degrees of freedom (see Exercise 40).

The portion of SSE due to lack of fit is denoted by SSE_{lf}. It is found by subtraction. That is,

$$\text{SSE}_{\text{lf}} = \text{SSE} - \text{SSE}_{\text{pe}}$$

Since SSE/σ^2 has a chi-squared distribution with $n - 2$ degrees of freedom, it is reasonable to conclude that $\text{SSE}_{\text{lf}}/\sigma^2$ follows a chi-squared distribution with $(n - 2) - (n - k) = k - 2$ degrees of freedom.

To detect lack of fit, we test

H_0: the linear regression model is appropriate

H_1: the linear regression model is not appropriate

The test statistic used is the ratio

$$\frac{\text{SSE}_{\text{lf}}/(k-2)\sigma^2}{\text{SSE}_{\text{pe}}/(n-k)\sigma^2} = \frac{\text{SSE}_{\text{lf}}/(k-2)}{\text{SSE}_{\text{pe}}/(n-k)}$$

This statistic follows an F distribution with $k - 2$ and $n - k$ degrees of freedom. Note that a poor fit will be reflected in an inflated value for SSE_{lf} and a large F value. We reject H_0 for values of the F ratio that are too large to have occurred by chance.

The test for lack of fit is illustrated in the next example.

Example 11.4.1. Consider these data on X, the temperature, in degrees centigrade (° C), at which a chemical reaction is conducted, and Y, the percentage yield obtained:

Value of X

30	40	50	60	70
13.7	15.5	18.5	17.7	15.0
14.0	16.0	20.0	18.1	15.6
14.6	17.0	21.1	18.5	16.5

For these data $k = 5$, $n_i = 3$ for $i = 1, 2, 3, 4, 5$ and

$$n = \sum_{i=1}^{5} n_i = 15$$

The internal sum of squares for error at $x = 30$ is

$$\sum_{j=1}^{3} (y_{1j} - \bar{y}_1)^2 = \sum_{j=1}^{3} (y_{1j} - 14.1)^2 = .42$$

The sum of squares for pure error is found by computing the internal error sum of squares for each x value and then summing these values. That is,

$$\text{SSE}_{\text{pe}} = \sum_{i=1}^{5} \sum_{j=1}^{3} (y_{ij} - \bar{y}_i)^2 = 6.453$$

For these data $S_{yy} = 66.437$, $S_{xy} = 154$, and $b_1 = .051$. The total error sum of squares, SSE, is given by

$$\text{SSE} = S_{yy} - b_1 S_{xy} = 58.583$$

The sum of squares for lack of fit is found by subtraction. It is given by

$$\begin{aligned} \text{SSE}_{\text{lf}} &= \text{SSE} - \text{SSE}_{\text{pe}} \\ &= 58.583 - 6.453 = 52.13 \end{aligned}$$

The observed value of the $F_{k-2, n-k} = F_{3,10}$ statistic used to test

H_0: the linear regression model is appropriate

H_1: the linear regression model is not appropriate

is

$$\begin{aligned} F &= \frac{\text{SSE}_{\text{lf}}/(k-2)}{\text{SSE}_{\text{pe}}/(n-k)} \\ &= \frac{52.13/3}{6.453/10} \\ &= 26.928 \end{aligned}$$

Based on the $F_{3,10}$ distribution, we can reject H_0 with $P < .05$ (Table IX of App. A).

There is evidence that a linear regression model is not appropriate.

A word of warning is in order. The least-squares procedure can be used to fit a straight line to any set of paired observations. This line can be used to predict the value of Y for a given x value. However, these predictions probably will be useless if the linear regression model is inappropriate. It is your responsibility as a researcher to find a satisfactory model. Some techniques for choosing an alternative model are considered in Chap. 12.

11.5 RESIDUAL ANALYSIS

Recall that to construct confidence intervals on β_0, β_1, and $\mu_{Y|x}$ and prediction intervals on $Y|x$ or to test hypotheses concerning β_0 and β_1, we need to make some model assumptions. When the simple linear regression model is written as

$$Y_i = \beta_0 + \beta_1 x_i + E_i$$

then the model assumptions are expressed as assumptions concerning the behavior of the random variables $E_1, E_2, \ldots, E_n$. In particular, it is assumed that these random variables are independent, normally distributed random variables with mean 0 and common variance σ^2. Before a fitted regression line is used to make predictions in practice, an effort should be made to check the validity of these assumptions. In this section we present some graphical techniques that can be used to do so. These procedures utilize the residuals, whose behavior under ideal conditions should mirror that of the random variables E_i.

Residual Plots

Recall that e_i, the ith residual, is the vertical distance from the ith data point to the fitted regression line. Therefore

$$e_i = y_i - [b_0 + b_1 x_i]$$

For a given value x_i the predicted response is found by substitution into the regression equation. That is, $\hat{y}_i = b_0 + b_1 x_i$. It can be seen that the ith residual can be written as

$$e_i = y_i - \hat{y}_i$$

The ith residual is the difference between the ith observed response and its predicted value. To check the model assumptions, we construct residual plots. A residual plot is a scattergram of the points (x_i, e_i). In such a plot we are plotting the regressor value (horizontal axis) versus the residual value (vertical axis). A

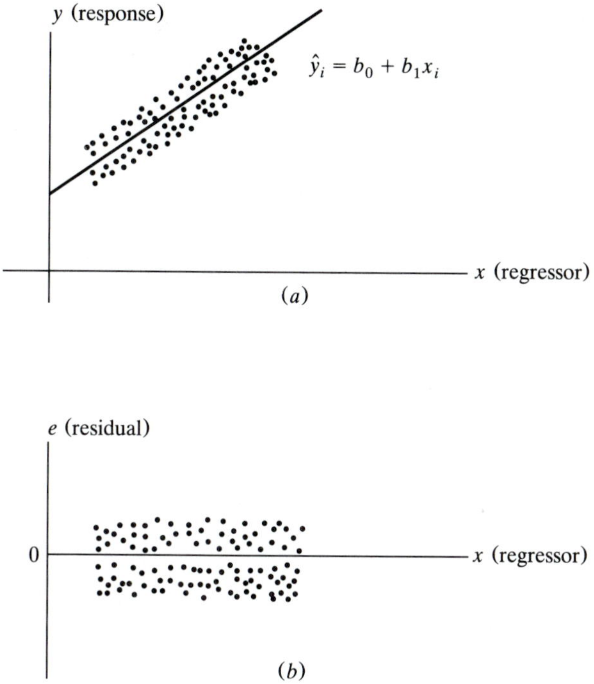

FIGURE 11.8
(a) The scattergram of a data set for which simple linear regression is appropriate. Points exhibit a linear trend with uniform spread about the estimated line of regression; (b) a residual plot for the ideal case. Residuals scatter randomly about 0 with a uniform spread.

residual plot can be used to help answer two questions:

1. Do the model assumptions underlying simple linear regression appear to be met?
2. If the model assumptions do not appear to be met, then which assumptions fail?

Since residual plots are useful in pinpointing or diagnosing problems that might exist, they are sometimes referred to as "diagnostic tools."

In Figure 11.8(a) shows a plot of a data set for which simple linear regression is appropriate. The data points exhibit an upward linear trend; they cluster tightly about the estimated line of regression; and the spread of the data points at each value of the regressor is about the same. The residual plot associated with an ideal data set of this sort is shown in Fig. 11.8(b). Notice that the residual plot depicts a set of points that scatter randomly about 0. The fact that these points vary about 0 is to be expected since it has been shown that the average value of the residuals is always 0. However, notice also that the spread of the residuals is about the same throughout the plot. This is expected whenever the assumption of a common variance σ^2 is valid.

In Fig. 11.9(a) a data set that signals trouble is shown. Notice that although there is an upward linear trend, the spread of the responses appears to

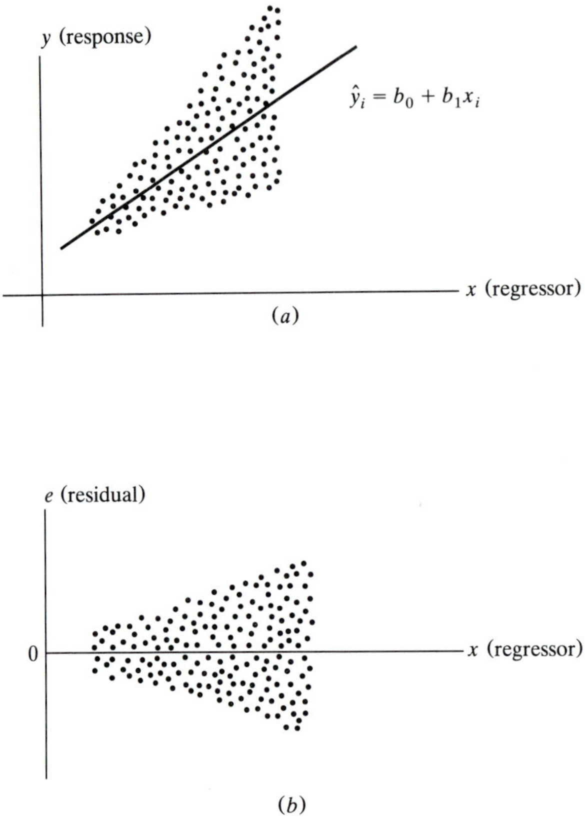

FIGURE 11.9
(*a*) A data set that signals that the common variance assumption is probably not valid; (*b*) a residual plot that throws doubt on the validity of the common variance assumption. Residuals scatter randomly about 0, but the spread of the points is not uniform.

increase as x increases. This is an indication that the common variance assumption might not be met. That is, the variance in response for small values of the regressor seems to be different from that for large values of x. How is this problem seen on a residual plot? Probably just as you suspect—the residual plot will show a random scatter about 0 with the spread of the points increasing as the value of x becomes larger. See Fig. 11.9(*b*).

Two other problems can be spotted with the help of a residual plot. They are model misspecification and gaps in the data. Model misspecification occurs when we try to fit a straight line to data that is not linear; gaps in the data occur due to poor experimental design or perhaps due to the loss of some data during experimentation. Both problems make the use of a linear prediction equation risky at best. Fig. 11.10(*a*) illustrates a set of data that is clearly not linear together with a "regression line" that has nevertheless been forced through the data. Figure 11.10(*b*) shows how this error is seen on a residual plot. Notice that the residuals do vary about 0 but there appears to be a pattern in the residuals. The scatter is not random. This lack of randomness is what we are looking for

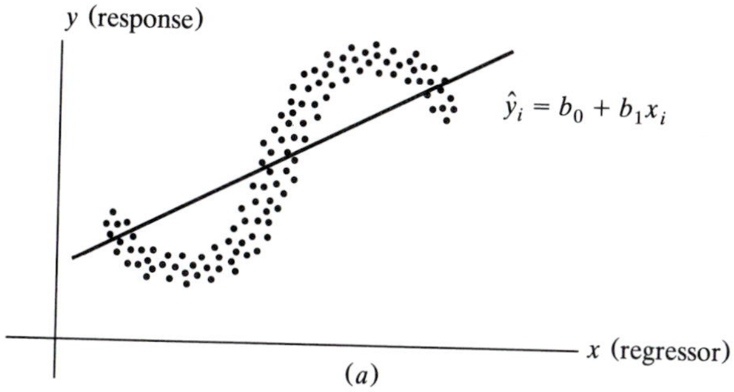

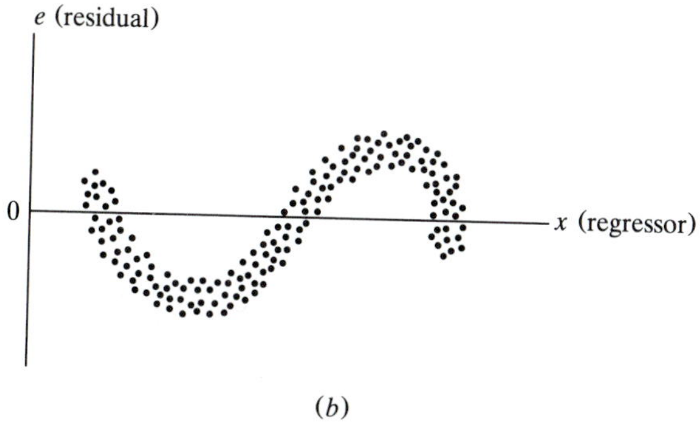

FIGURE 11.10
(a) A data set for which the simple linear regression model is inappropriate; (b) a residual plot in which a linear prediction equation has been fitted to a set of data points that do not exhibit a linear trend. The residuals exhibit a pattern rather than a random scatter about 0.

since it signals the possibility that simple linear regression does not adequately describe the relationship between the regressor and the response.

In Fig. 11.11(a) a data set that contains gaps with respect to the regressor values is shown. Even though it is possible to fit a line to these data, to do so is risky. We are assuming that the linear trend suggested by the responses for low and high values of the regressor continues in the midrange of x. We really have no evidence that this is the case and it is not appropriate to assume this. Figure 11.11(b) shows the residual plot for a data set with a definite gap in the data.

Residual plots are helpful in spotting potential problems. However, they are not always as easy to interpret as are those given in Fig. 11.8 through 11.11. Since patterns are hard to spot with small data sets except in extreme cases, residual plots are most useful with fairly large collections of data. Furthermore, to get a clear picture of the validity of the variance assumption, we should design experiments in such a way that multiple observations are taken at each distinct value of the regressor.

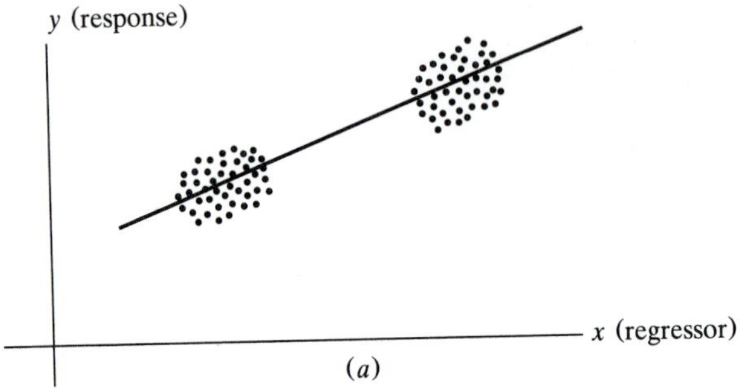

(a)

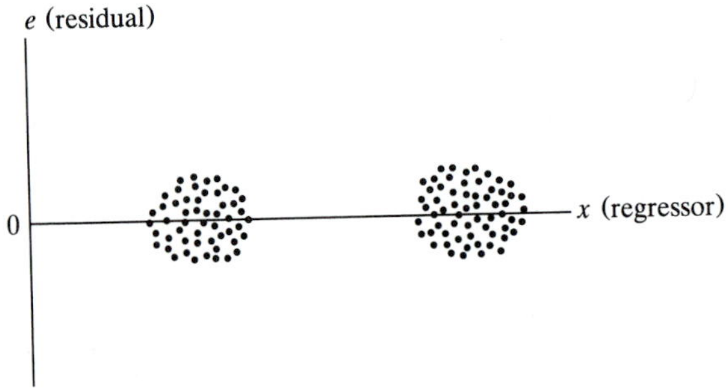

(b)

FIGURE 11.11
(a) A data set with a midrange gap. Simple linear regression is risky. (b) A residual plot showing gaps in the regressor values.

Checking for Normality: Stem-and-Leaf Plots and Boxplots

One model assumption that has not been investigated yet is that of normality. This assumption can be tested graphically via the Lilliefors test described in Sec. 8.6 or analytically by using SAS as shown in Computing Supplement III (page 313). Here we consider two visual checks that do not require formal testing. They work best for fairly large samples since they both require that some value judgments be made concerning shape. Such judgments are hard to make with small sample because patterns do not appear in such data sets unless the violations in the assumptions are extreme.

The first visual technique is one that should come to mind immediately. Namely, construct a stem-and-leaf diagram of the residuals as explained in Chap. 6. If the normality assumption is valid, we expect the plot to exhibit the approximate bell shape indicative of a normal curve.

The boxplot can also be used as a diagnostic tool. It allows us to check for possible violations of the normality assumption and also to detect the presence of outliers. A boxplot for residuals obtained when the normality assumption is valid should be symmetric with the median line near or at 0. Outliers are not expected since these values are extremely rare whenever the random variable involved is normally distributed. (See Exercise 26 of Chap. 6.) Thus asymmetry or the presence of outliers signals that the normality assumption might not hold.

Outliers are very troublesome in regression studies. They can greatly influence the regression line in that the line tends to be pulled toward the outlier. This can cause the fitted line not to pass through the center of the bulk of the data as is desired. If outliers are detected via a boxplot, then they must be investigated. If the data point is found to be an error or suspect in some way, then it should not be used in the analysis.

Example 11.5.1 illustrates the use of residual and stem-and-leaf plots in a regression study. In this example you will see that the plots associated with real data are not always as easy to interpret as you would like.

Example 11.5.1. Consider the problem described in Example 11.1.1 in which we found an equation by which the extent of solvent evaporation (Y) can be predicted based on knowledge of the humidity (X). The scattergram for the data given in Example 11.1.1 is shown in Fig. 11.12(a). The line shown in the picture is the graph of the estimated line of regression. Its equation is given by

$$\hat{\mu}_{Y|x} = 13.64 - .08x$$

There is a linear trend to the data and the data points appear to lie reasonably close to the estimated line of regression. Figure 11.12(b) shows the residual plot. Notice that this plot does not reveal a pattern that might indicate that the linear model is not appropriate; the points do appear to scatter randomly about 0 as desired. There are no obvious differences in spread as x increases. However, the experiment was not designed with multiple observations at distinct values of the regressor, so this assumption is not easy to verify. To check for normality, we construct a stem-and-leaf plot of the residuals. The residuals are given in Table 11.2 and the stem-and-leaf plot for these residuals is shown in Fig. 11.13. In the plot we use double stems, with the "leaf" being the first decimal place of the residual. For example, the value .18727 is graphed as 0|1 on the "low" 0 stem. Does this plot give convincing evidence of normality? Does it clearly exhibit the bell-shape characteristic of a normal curve? The answer to these questions is probably "no." The shape is rather nondescript. To determine whether there is enough evidence to indicate that the normality assumption is probably not valid, a formal test is needed. Figure 11.14 gives the SAS printout for PROC UNIVARIATE (see Computing Supplement III). This printout includes the observed value of the statistic W used to test

H_0: data are from a normal distribution

H_1: data are from a nonnormal distribution

The value of the statistic, .958717, is shown at (1); its P value, .4028, is shown at (2). Since this P value is large, H_0 should not be rejected. Based on these

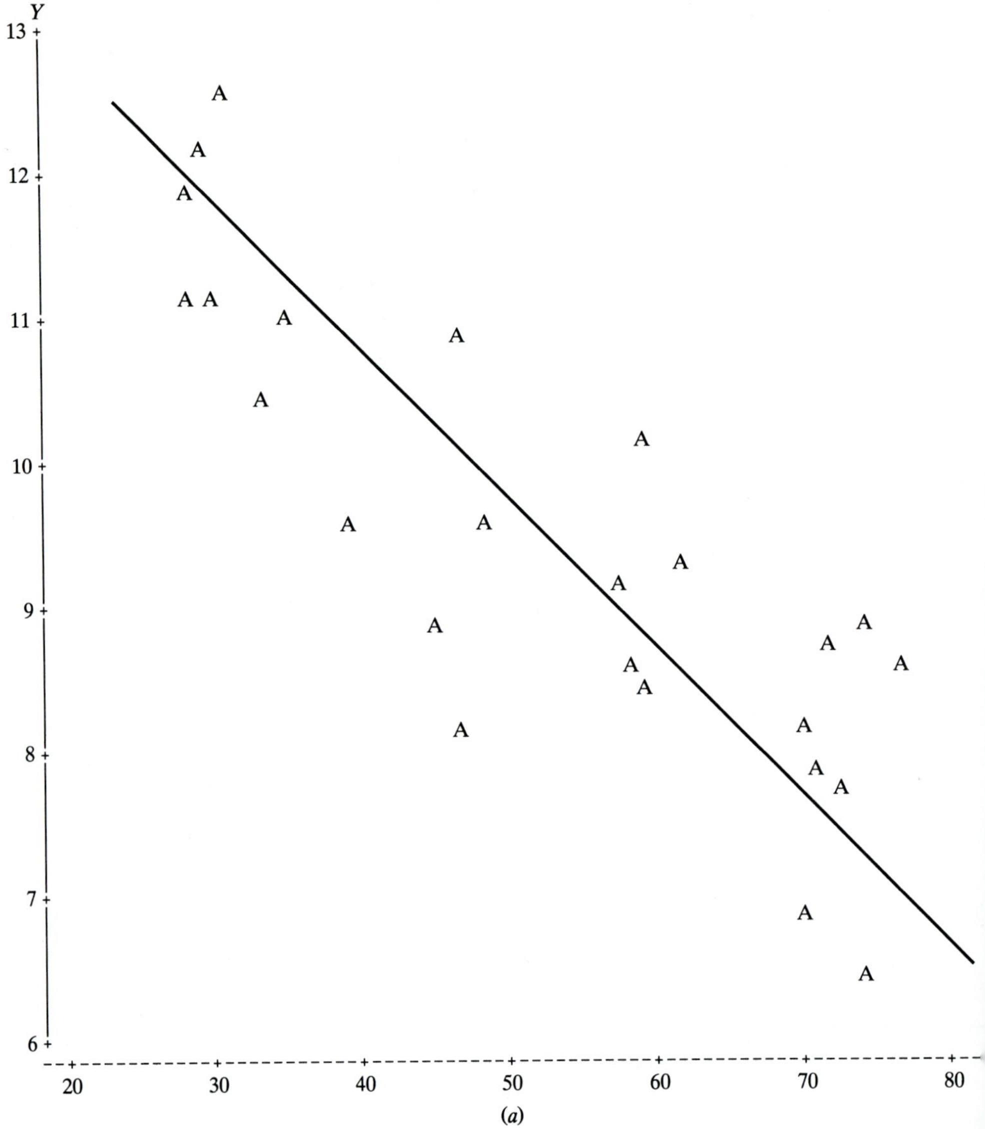

FIGURE 11.12
(a) Scattergram of data with an estimated line of regression.

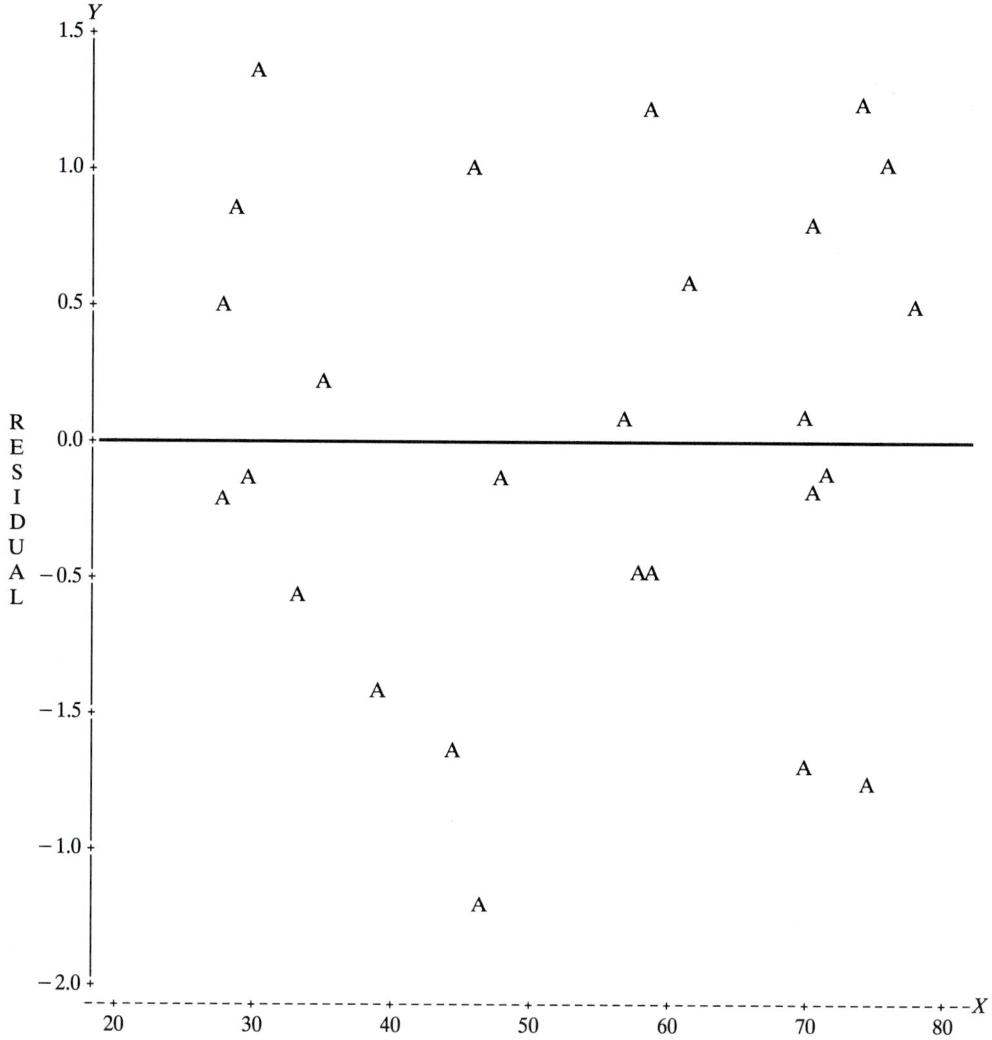

(b)

FIGURE 11.12 (*continued*)
(*b*) A residual plot for the data of Examples 11.1.1 and 11.5.1. Residuals show no obvious pattern that would signal model misspecification.

residuals, there is no reason to suspect that the residuals do not follow a normal distribution. Notice that SAS gives a stem-and-leaf diagram that is different from that given in Fig. 11.13. Remember that there are no set rules in defining leaves. The SAS plot has defined a "leaf" by first rounding the residual to one decimal place and then plotting the residuals. Does this clarify the question of normality? Probably not. There is still no clearly defined bell visible.

TABLE 11.2

Observed response (Y), predicted response (Predict.), and residual (Resid.) for the 25 data points of Examples 11.1.1 and 11.5.1

Observation number	Y	Predict.	Resid.
1	11.0	10.8127	0.18727
2	11.1	11.2611	−0.16107
3	12.5	11.1730	1.32700
4	8.4	8.9313	−0.53130
5	9.3	8.7231	0.57685
6	8.7	7.9305	0.76945
7	6.4	7.6824	−1.28236
8	8.5	7.4982	1.00178
9	7.8	7.9786	−0.17858
10	9.1	9.0354	0.06462
11	8.2	9.9241	−1.72406
12	12.2	11.3251	0.87488
13	11.9	11.3892	0.51084
14	9.6	10.5085	−0.90850
15	10.9	9.8920	1.00797
16	9.6	9.7559	−0.15593
17	10.1	8.8913	1.20873
18	8.1	8.0346	0.06537
19	6.8	8.0346	−1.23463
20	8.9	7.6824	1.21764
21	7.7	7.8665	−0.16650
22	8.5	8.9873	−0.48735
23	8.9	10.0682	−1.16816
24	10.4	10.9648	−0.56484
25	11.1	11.3491	−0.24913

1	3	0	0	2	2	
0	5	7	8	5		
0	1	0	0			
−0	1	1	1	1	4	2
−0	5	9	5			
−1	2	2	1			
−1	7					

FIGURE 11.13

A stem-and-leaf diagram for the residuals of Table 11.2. The diagram uses double stems, with a "leaf" being the first decimal place of the residual.

SAS

UNIVARIATE PROCEDURE

Variable = RESID Residual

Moments

	N	25	Sum Wgts	25	
	Mean	0	Sum	0	
	Std Dev	0.867485	Variance	0.752531	
	Skewness	−0.17648	Kurtosis	−0.84038	
	USS	18.06074	CSS	18.06074	
	CV	.	Std Mean	0.173497	
	T:Mean = 0	0	Prob > |T|	1.0000	
	Sgn Rank	−0.5	Prob > |S|	0.9896	
	Num ^= 0	25			
①	W:Normal	0.958717	Prob < W	0.4028	②

Quantiles (Def = 5)

	100% Max	1.326999	99%	1.326999
⑤	75% Q3	0.769453	95%	1.217641
③	50% Med	−0.15593	90%	1.208726
④	25% Q1	−0.5313	10%	−1.23463
	0% Min	−1.72406	5%	−1.28236
			1%	−1.72406
	Range	3.051055		
	Q3 − Q1	1.300757		
	Mode	−1.72406		

Extremes

Lowest	Obs	Highest	Obs
−1.72406(	11)	1.00178(	8)
−1.28236(	7)	1.007969(	15)
−1.23463(	19)	1.208726(	17)
−1.16816(	23)	1.217641(	20)
−0.9085 (	14)	1.326999(	3)

```
Stem Leaf                       #          Boxplot
   1 00223                      5             |
   0 5689                       4          +-----+
   0 112                        3          |  +  |
  -0 22222                      5      ⑥   *-----*
  -0 9655                       4          +-----+
  -1 322                        3             |
  -1 7                          1
     ----+----+----+----+
```

FIGURE 11.14
An analytic test for normality.

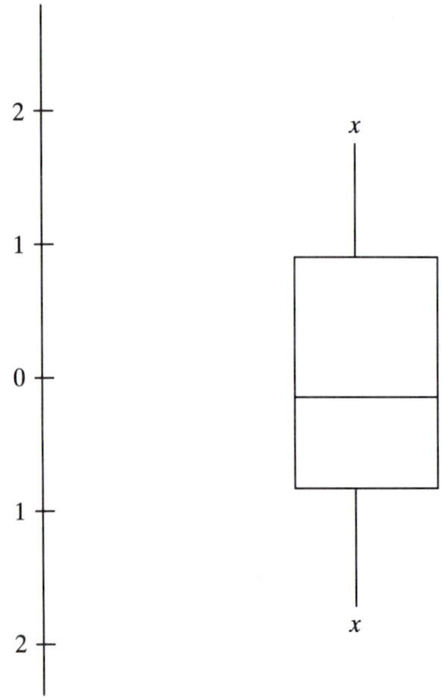

FIGURE 11.15
A boxplot for the residuals of Table 11.2 based on the stem-and-leaf plot of Fig. 11.13. No outliers are detected. Although the plot does not exhibit perfect symmetry, it is not skewed enough to reject the normality assumption.

In the previous example interest centered on checking for normality. In the next example the boxplot is used to check for symmetry and the presence of outliers. A definite lack of symmetry or the presence of outliers are both signals of possible violation of the normality assumption.

Example 11.5.2. To construct a quick boxplot for the residuals given in Table 11.2, we again retain only the first decimal place of the number as was done in constructing the stem-and-leaf plot shown in Fig. 11.13. For these data $n = 25$, the median location is $(n + 1)/2 = 13$, and the median value is $-.1$. Quartile locations are at $(13 + 1)/2 = 7$. The quartile values are $q_1 = -.5$ and $q_3 = .7$. The interquartile range is $q_3 - q_1 = 1.2$. Inner fences are located at

$$f_1 = q_1 - 1.5iqr = -.5 - 1.5(1.2) = -2.3$$

and
$$f_3 = q_3 + 1.5iqr = .7 + 1.5(1.2) = 2.5$$

Since no residual values lie beyond the inner fences, the set of residuals does not contain any outliers. This is good news, for outliers are very rare (occurring with probability .007) when sampling from a normal distribution. The boxplot for the residuals is shown in Fig. 11.15. Notice that it does not exhibit the perfect symmetry expected from a normal distribution, but it is not skewed enough to signal a clear violation of the normality assumption. Values for the median, q_1, and q_3 based on the actual residual values are given by SAS in Fig. 11.14 at (3), (4), and

(5), respectively. The SAS boxplot is shown at (6). The + in the box is at 0, the average value of the residuals. In a perfectly symmetric plot, the ideal plot, this + would coincide with the median. Since our data set is not perfect, as is usually the case with real data, these two values differ slightly.

Regression is an art as well as a science. Real-life data sets are seldom perfect. You will be called upon to make some value judgments as to the appropriateness of linear regression. The tools presented in this section will help you to make these judgments. In Sec. 12.8 some suggestions are made as to how to handle data that violate variance assumptions described in this section.

11.6 CORRELATION

Thus far in this chapter we have considered problems related to simple linear regression. Our primary problem has been to express the *mean* value of a random variable Y as a linear function of a *nonrandom* variable X. In this section we continue the study of correlation presented in a theoretical context in Sec. 5.3. There are two important differences between the regression studies that we have been considering and the correlation studies that we shall consider now. First, in a correlation study both X and Y must be *random variables*. Second, we are not looking for a linear relationship between X and the mean of Y; rather, we are trying to measure the strength of the linear relationship that exists between X and Y itself.

The theoretical parameter used to measure the linear relationship between X and Y is the Pearson coefficient of correlation ρ. This parameter is defined by

Pearson correlation coefficient

$$\rho = \frac{\text{Cov}(X, Y)}{\sqrt{(\text{Var}\, X)(\text{Var}\, Y)}}$$

The parameter ρ assumes values between -1 and 1 inclusive. Values of 1 or -1 indicate perfect positive or negative linear relationships respectively. A value of 0 indicates no linear relationship. When this occurs, we say that X and Y are uncorrelated. Figure 11.16 illustrates the graphical interpretation of ρ.

Previously we found the theoretical value of ρ based on knowledge of the joint density function for X and Y. Unfortunately, these densities are seldom known in practice. For this reason, the job of the researcher is to estimate ρ based on a set $\{(x_i, y_i)\colon\ i = 1, 2, 3, \ldots, n\}$ of observations on the random variable (X, Y). It is easy to see how this can be done. We must estimate Var X, Var Y and Cov(X, Y). We shall use the maximum likelihood estimators for

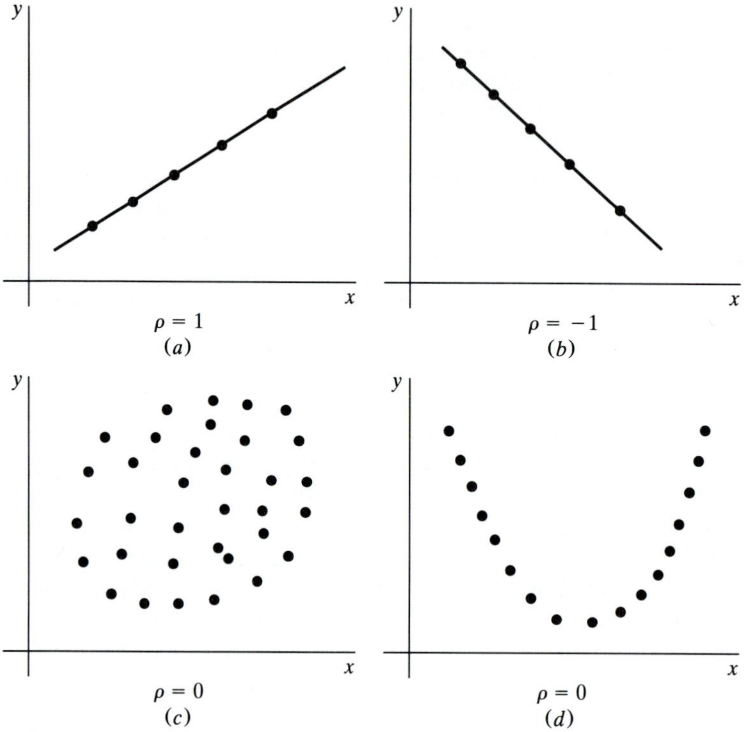

FIGURE 11.16
(a) $\rho = 1$, perfect positive relationship; (b) $\rho = -1$, perfect negative relationship; (c) $\rho = 0$, no relationship exists; (d) $\rho = 0$, a relationship exists but it is not linear.

variance. That is,

$$\widehat{\text{Var}\, X} = \sum_{i=1}^{n} (X_i - \bar{X})^2 \Big/ n = S_{xx}/n$$

$$\widehat{\text{Var}\, Y} = \sum_{i=1}^{n} (Y_i - \bar{Y})^2 \Big/ n = S_{yy}/n$$

To estimate Cov(X, Y), note that

$$\text{Cov}(X, Y) = E[(X - \mu_X)(Y - \mu_Y)]$$

We estimate Cov(X, Y) by averaging products analogous to that on the right-hand side of the above equation. Therefore

$$\widehat{\mathrm{Cov}(X,Y)} = \sum_{i=1}^{n} (X_i - \bar{X})(Y_i - \bar{Y})\Big/n = S_{xy}/n$$

When we combine these estimators, the estimator for ρ is given by

Estimator for ρ, the Pearson correlation coefficient

$$\hat{\rho} = R = \frac{S_{xy}}{\sqrt{S_{xx}S_{yy}}}$$

Many calculators will compute $\hat{\rho}$ for you automatically. If you have such a calculator, it should be used. Otherwise the following computational formula is useful:

Computational formula for r, the estimated Pearson correlation coefficient

$$\hat{\rho} = .r = \frac{n\Sigma xy - \Sigma x \Sigma y}{\sqrt{\left[n\Sigma x^2 - (\Sigma x)^2\right]\left[n\Sigma y^2 - (\Sigma y)^2\right]}}$$

Example 11.6.1. In studying the effect of sewage effluent on a lake, researchers take measurements of the nitrate concentration of the water. An older manual method has been used to monitor this variable. However, a new automated method has been devised. If a high positive correlation exists between the measurements taken by using the two methods, then the automated method will be put into routine use. These data are obtained on the nitrate concentration in micrograms of nitrate per liter of water:

x **(manual)**	y **(automated)**
25	30
40	80
120	150
75	80
150	200
300	350
270	240
400	320
450	470
575	583

The scattergram for these data is shown in Fig. 11.17. Since these points exhibit a

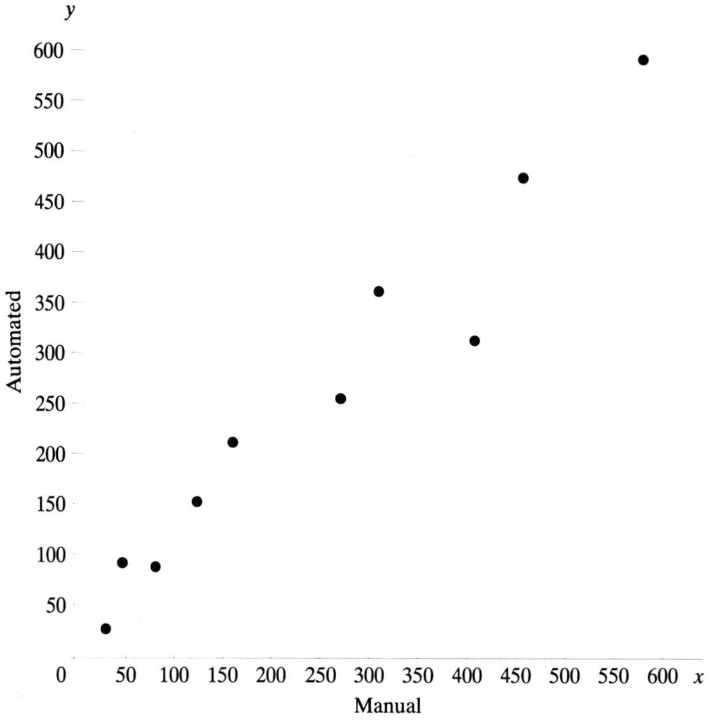

FIGURE 11.17
A scattergram of manual readings versus automated readings.

fairly well-defined increasing trend, we expect r to be positive and close in value to 1. Summary statistics for these data are

$$n = 10 \qquad \sum x^2 = 900{,}775 \qquad \sum y^2 = 919{,}489$$

$$\sum x = 2405 \qquad \sum y = 2503 \qquad \sum xy = 902{,}475$$

$$S_{xx} = 322{,}372.5 \qquad S_{xy} = 300{,}503.5$$

$$S_{yy} = 292{,}988.1$$

The estimated correlation between X and Y is

$$\hat{\rho} = r = \frac{S_{xy}}{\sqrt{S_{xx}S_{yy}}}$$

$$= \frac{300{,}503.5}{\sqrt{(322{,}372.5)(292{,}988.1)}}$$

$$\doteq .978$$

As expected, there appears to be a strong positive linear relationship between X and Y.

Interval Estimation and Hypothesis Tests on ρ

It is almost always possible to develop a logical point estimator for a parameter θ based on its definition alone. However, before confidence intervals can be constructed or hypothesis tests conducted, it is usually necessary to make some assumptions concerning the distribution of the random variable under study. This is true here. We have a logical point estimator for ρ. To draw statistical inferences concerning its value, we must assume a probability distribution for the two-dimensional random variable (X, Y). The distribution assumed is the *bivariate normal* distribution. The joint density for such a random variable is given by

Bivariate normal density

$$f(x, y) = k \exp\left\{ -\frac{1}{2(1-\rho^2)} \left[\left(\frac{x-\mu_X}{\sigma_X}\right)^2 - 2\rho\left(\frac{x-\mu_X}{\sigma_X}\right)\left(\frac{y-\mu_Y}{\sigma_Y}\right) + \left(\frac{y-\mu_Y}{\sigma_Y}\right)^2 \right] \right\}$$

where $k = \dfrac{1}{2\pi\sigma_X\sigma_Y\sqrt{1-\rho^2}}$

$$-\infty < x < \infty$$
$$-\infty < y < \infty$$
$$-\infty < \mu_X < \infty$$
$$-\infty < \mu_Y < \infty$$
$$\sigma_X > 0$$
$$\sigma_Y > 0$$
$$-1 < \rho < 1$$

This distribution has many interesting theoretical properties. Among them are the following:

1. The marginal distributions for both X and Y are normal. The parameters μ_X, μ_Y, σ_X, and σ_Y that appear in the expression for $f(x, y)$ are the means and standard deviations for X and Y respectively.
2. The parameter ρ that appears in the expression for $f(x, y)$ is the correlation coefficient between X and Y.
3. If $\rho = 0$, then X and Y are independent.
4. The curves of regression of X on Y and Y on X are both linear. The latter is given by

$$\mu_{Y|x} = \mu_Y + \rho\frac{\sigma_Y}{\sigma_X}(x - \mu_X)$$

Although we shall not be overly concerned with these theoretical properties, they will make it easier to understand the relationship between correlation and regression.

In assuming that (X, Y) has a bivariate normal distribution, we are assuming a linear regression model. That is, we are assuming that

$$\mu_{Y|x} = \beta_0 + \beta_1 x$$

where $\beta_1 = (\sigma_Y/\sigma_X)\rho$. Since σ_Y and σ_X are both positive, it is easy to see that the slope of the regression line and the correlation coefficient have the same algebraic sign. It is also easy to see that $\rho = 0$ if and only if $\beta_1 = 0$. Thus to test H_0: $\rho = 0$ against any one of the usual alternatives, we use the same test statistic as that used earlier to test H_0: $\beta_1 = 0$, namely, $B_1/(S/\sqrt{S_{xx}})$. Since we shall have a point estimate for ρ available when we test H_0: $\rho = 0$, it is convenient to express our test statistic in the alternative form

$$T_{n-2} = \frac{R\sqrt{n-2}}{\sqrt{1-R^2}}$$

(see Exercise 52).

We illustrate the use of this statistic in the next example.

Example 11.6.2. In our previous example we estimated the correlation between X, the manual nitrate reading, and Y, the automated reading by $r = .978$. Although intuition certainly leads us to suspect that we have strong evidence that $\rho \neq 0$, we must remember that the sample size is small with $n = 10$. For this reason, we should test

$$H_0: \rho = 0$$

$$H_1: \rho \neq 0$$

The observed value of the test statistic

$$T_{n-2} = \frac{R\sqrt{n-2}}{\sqrt{1-R^2}}$$

is

$$t = \frac{.978\sqrt{10-2}}{\sqrt{1-(.978)^2}} = 13.26$$

Based on the T_8 distribution, the null hypothesis can be rejected with $P < .0005$. We do have strong evidence that $\rho \neq 0$.

The exact distribution of R depends on the true value of ρ. Furthermore, for large values of ρ this distribution is decidedly nonnormal. Fortunately, there exists a simple change of variable that results in a random variable whose distribution is approximately normal. In particular, it can be shown that when

(X, Y) has a bivariate normal distribution, then the random variable

$$\frac{1}{2}\ln\left(\frac{1+R}{1-R}\right)$$

is approximately normally distributed with

$$\mu = \frac{1}{2}\ln\left(\frac{1+\rho}{1-\rho}\right) \qquad \text{and} \qquad \sigma^2 = \frac{1}{n-3}$$

This result, due to R. A. Fisher, was first published in 1921. Standardizing, we can conclude that the random variable

$$Z = \frac{\frac{1}{2}\ln\left(\frac{1+R}{1-R}\right) - \frac{1}{2}\ln\left(\frac{1+\rho}{1-\rho}\right)}{\sqrt{\frac{1}{n-3}}}$$

is approximately standard normal. To develop confidence bounds on ρ we note that

$$P\left[-z_{\alpha/2} \le \frac{\frac{1}{2}\ln\left(\frac{1+R}{1-R}\right) - \frac{1}{2}\ln\left(\frac{1+\rho}{1-\rho}\right)}{\sqrt{\frac{1}{n-3}}} \le z_{\alpha/2}\right] \doteq 1 - \alpha$$

Although the algebraic argument is a bit messy, this inequality can be solved for ρ to obtain these bounds for a $100(1-\alpha)\%$ confidence interval on ρ:

Confidence interval on ρ, the Pearson correlation coefficient

$$\text{Lower bound} = \frac{(1+R) - (1-R)\exp\left(2z_{\alpha/2}/\sqrt{n-3}\right)}{(1+R) + (1-R)\exp\left(2z_{\alpha/2}/\sqrt{n-3}\right)}$$

$$\text{Upper bound} = \frac{(1+R) - (1-R)\exp\left(-2z_{\alpha/2}/\sqrt{n-3}\right)}{(1+R) + (1-R)\exp\left(-2z_{\alpha/2}/\sqrt{n-3}\right)}$$

To see how to evaluate these bounds, consider the next example.

Example 11.6.3. We know that a point estimate for ρ, the correlation between the manual nitrate reading and the automated reading, is .978. To find a 95% confidence interval on ρ, we first note that $z_{.025} = 1.96$ and $n = 10$. The lower bound for the confidence interval is

$$\frac{(1 + r) - (1 - r)\exp\left(2z_{\alpha/2}\big/\sqrt{n - 3}\right)}{(1 + r) + (1 - r)\exp\left(2z_{\alpha/2}\big/\sqrt{n - 3}\right)}$$

$$= \frac{(1 + .978) - (1 - .978)\exp\left(2(1.96)\big/\sqrt{7}\right)}{(1 + .978) + (1 - .978)\exp\left(2(1.96)\big/\sqrt{7}\right)}$$

$$= \frac{(1 + .978) - .022(4.4)}{(1 + .978) + .022(4.4)}$$

$$= \frac{1.881}{2.075} \doteq .907$$

Substituting, we find that the upper bound is .995. We are 95% confident that the true value of the correlation coefficient lies in the interval [.907, .995]. Since we rejected H_0: $\rho = 0$, it is not surprising that 0 is not in this interval.

Although the usual null hypothesis concerning ρ is H_0: $\rho = 0$, other null values can be tested via the Fisher transformation. Letting ρ_0 denote any null value for ρ, we see that the Z statistic

$$Z = \frac{\dfrac{1}{2}\ln\left(\dfrac{1 + R}{1 - R}\right) - \dfrac{1}{2}\ln\left(\dfrac{1 + \rho_0}{1 - \rho_0}\right)}{\sqrt{\dfrac{1}{n - 3}}}$$

serves as the test statistic for testing H_0: $\rho = \rho_0$.

Coefficient of Determination

Strictly speaking, one should not use the techniques of simple linear regression presented in this chapter and the correlation techniques given here on the same data set. The former assumes that X is not a random variable; the latter requires that it be a random variable. Even so, R can be useful in a regression study. As we shall show, it is an indicator of the adequacy of the simple linear regression model. To see why this is true, note that

$$\text{SSE} = S_{yy} - B_1 S_{xy}$$

Dividing each side of this equation by S_{yy} and replacing B_1 with S_{xy}/S_{xx}, we see that

$$\frac{\text{SSE}}{S_{yy}} = 1 - \frac{S_{xy}^2}{S_{xx}S_{yy}}$$

Since $R = S_{xy}/\sqrt{S_{xx}S_{yy}}$, we may conclude that

$$\frac{\text{SSE}}{S_{yy}} = 1 - R^2$$

or that $R^2 = 1 - \text{SSE}/S_{yy}$. This equation can be rewritten as

$$R^2 = \frac{S_{yy} - \text{SSE}}{S_{yy}}$$

Since S_{yy} measures the total variability in Y and SSE measures the random variability in Y about the estimated regression line, S_{yy} − SSE measures the variability in Y explained by the linear regression model. The random variable R^2 represents the proportion of the variability in Y explained by the model. When this proportion is multiplied by 100%, we obtain a statistic called the *coefficient of determination*. If R lies close to 1 or −1, then R^2 will also be close to 1 yielding a coefficient of determination near 100%. When R is near 0, then the coefficient of determination is also near 0. Thus the relative size of $R^2 \times 100\%$ is a good descriptive measure of the adequacy of the model.

Although there are no hard and fast rules concerning the interpretation of R and R^2, the charts given in Fig. 11.18 are useful. Keep in mind the fact that the interpretation of these statistics is somewhat subject matter dependent. An R^2 value of 50% might be considered very large in a social science setting where human subjects are involved; however, the same figure could be considered very small in a designed engineering experiment. The interpretation of R and R^2 must be left to the discretion of the subject matter expert.

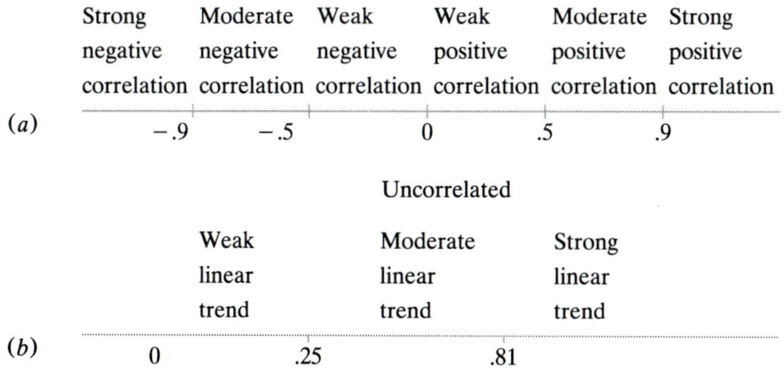

FIGURE 11.18
(*a*) A suggested interpretation of R; (*b*) a suggested interpretation of R^2.

CHAPTER SUMMARY

In this chapter we have considered most of the important aspects of simple linear regression and correlation. A verbal and mathematical description of the regression model was given along with the least-squares method for estimating the slope and intercept parameters of the model. We saw that under minimal assumptions these estimators were unbiased. When the random error E_i was assumed to be normally distributed, the distribution of $Y|x_i$, $\hat{\beta}_0$, and $\hat{\beta}_1$ was given. Utilizing these distributional properties, we considered methods for testing hypotheses and estimating confidence intervals about the slope β_1 and the intercept β_0. We carefully distinguished between predicting the mean response of the dependent variable Y at a fixed value of the independent variable x and predicting a single value of the dependent variable Y at x. Methods were given for constructing confidence intervals for both cases and we observed that prediction of the mean led to a shorter confidence interval (more precise) than for a single value. Finally for regression, when multiple measurements of the dependent variable Y are observed at values of the independent variable x, we considered a procedure which enables us to test the model for a lack of linear fit.

In addition to simple linear regression, we considered the Pearson correlation coefficient. Methods were given for estimating the true correlation, testing hypothesis about the correlation, and estimating confidence intervals for the correlation coefficient ρ.

We also introduced and defined terms that you should know. These are:

Linear regression	Least-squares estimation
Least-squares properties	Independent variable
Dependent variable	Slope of regression
Intercept of regression	Lack of fit
Pure error	Experimental error
Residual error	Pearson correlation
Significant correlation	Bivariate normal distribution
Observational study	SSE
Designed study	Coefficient of determination
Scattergram	Response variable
Residual	Regressor
	Predictor variable

EXERCISES

Section 11.1

1. Consider the following observations on the independent variable X and the dependent variable Y:

x	y	x	y
80	5.0	100	6.7
75	4.7	115	4.2
75	4.5	110	5.5
70	3.2	105	6.4
65	2.0	105	6.8
100	6.5	100	7.2
95	6.0	130	2.0
90	5.5	125	2.9
85	5.2	120	3.8

(*a*) Plot the scattergram for these data.

(*b*) Does it appear reasonable that a linear regression could be used for these data?

(*c*) Sketch, by eye, a linear regression line on the scattergram in part (*a*).

2. For each of the three following data sets, plot a scattergram and subjectively state whether it appears that a linear regression will (i) fit the data well, (ii) give only a fair fit, or (iii) fit the data poorly:

(*a*)

x	5	15	25	35	45	50
y	10	18	20	25	32	45

(*b*)

x	5	10	20	30	40	50
y	15	22	32	35	30	15

(*c*)

x	10	15	20	30	40	50
y	40	35	30	22	14	7

3. The normal equations were given in this section. Solve the normal equations for b_0 and b_1 and show that your solution can be written in the form given as the least-squares estimates for β_0 and β_1.

4. Consider any arbitrary data set $(x_1, y_1), (x_2, y_2), \ldots, (x_n, y_n)$. Let $\bar{x}$ and $\bar{y}$ denote the respective sample means for the independent variable X and the dependent variable Y. For the estimated linear regression equation $\hat{\mu}_{Y|x} = b_0 + b_1 x$, show that the point $(\bar{x}, \bar{y})$ always lies on the estimated regression line.

5. Verify that $\Sigma e_i = 0$. *Hint:* Write e_i as $y_i - (b_0 + b_1 x_i)$ and remember that $b_0 = \bar{y} - b_1 \bar{x}$.

6. For each of the data sets of Exercise 2, estimate β_0 and β_1. Find the residuals in each case and verify that, apart from round-off error, the residuals sum to 0.

7. The relationship between energy consumption and household income was studied in *Energy Policy*, September 1982. The following data are representative of household income X (in units of \$1000/year) and energy consumption Y (in units of 10^8 Btu/year).

Energy consumption (y)	Household income (x)
1.8	20.0
3.0	30.5
4.8	40.0
5.0	55.1
6.5	60.3
7.0	74.9
9.0	88.4
9.1	95.2

(*a*) Plot a scattergram of these data.
(*b*) Estimate the linear regression equation $\mu_{Y|x} = \beta_0 + \beta_1 x$.
(*c*) If $x = 50$ (household income of $50,000), estimate the average energy consumed for households of this income. What would your estimate be for a single household?
(*d*) How much would you expect the change in consumption to be if any household income increases $2000/year (2 units of $1000)?
(*e*) How much would you expect consumption to change if any household income decreases $2000/year?

8. Consider the data in Exercise 7.
(*a*) Write the normal equations for these data.
(*b*) Solve the normal equations for b_0 and b_1 and verify that your results are the same as those you obtained in part (*b*) of Exercise 7.

9. Connectors used in computers are subject to simultaneous multidimensional stresses such as high temperatures and mechanical stresses. A study is conducted to identify and quantify interface stresses. Experiments are conducted to investigate the relationship between pitch and connector length. These data are obtained:

Connector length, x (inches)	Pitch (millimeters)
.150	3.81
.100	2.54
.098	2.50
.079	2.00
.050	1.27
.040	1.02
.039	1.00
.032	0.80
.020	0.50
.016	0.40
.010	0.25
.005	0.13

(Based on information from "Interaction of Multichip Module Substrates with High Density Connectors," Yakov Belopolsy, *IEEE MICRO*, April 1993, pp. 36–43.)

(*a*) Plot a scattergram for these data.
(*b*) Estimate the regression line.
(*c*) Calculate the residuals and show that, apart from round-off error, they sum to 0.
(*d*) Estimate the average pitch for all connectors of length 0.03 inch (in.).
(*e*) Estimate the pitch of a particular connector of length 0.03 in.
(*f*) By how much would you expect the pitch to change if the connector length increased by 0.1 in.? By .05 in.?
(*g*) Would it be reasonable to expect to be able to use the estimated regression line to predict the pitch well for connectors of length .175 in.? .39 in.? 1.0 in.? Explain.

10. A particular type of power brush is a wheel made of wire strands extending outward around a hub. It is used for many purposes such as finishing aluminum bicycle rims, producing a matte finish on plastic, and removing burrs from gear teeth. The shorter the wire length and the coarser the wire, the more severe is the buffing action. A study is conducted to develop a chart for suggested use of the wheel. Tests are conducted on a 2-in.-brush-diameter wheel. These data are obtained:

x (rpm × 1000)	y (surface ft / min covered in removing burrs)
1.0	525, 520, 527
1.5	785, 780, 790
1.75	915, 900, 922
2.5	1300, 1295, 1310
3.0	1575, 1565, 1582
4.0	2100, 2110, 2090
6.0	3125, 3120, 3133
10.0	5250, 5256, 5245

(Based on information found in "Power Brushes for Surface Finishing," Kail M. Weiler, *Industrial Finishing Coatings Manufacturing and Application*, May 1993, pp. 36–37.)

(*a*) Sketch a scattergram for these data.
(*b*) Estimate the regression line.
(*c*) Estimate the surface feet per minute that can be covered when a wheel of this sort is used at 3450 revolutions per minute (rpm).

Section 11.2

11. Verify the following summation properties.

(*a*) $\sum_{i=1}^{n} (x_i - \bar{x}) = 0.$

(*b*) $\sum_{i=1}^{n} (x_i - \bar{x})(Y_i - \bar{Y}) = \sum_{i=1}^{n} (x_i - \bar{x})Y_i.$

(*c*) $\sum_{i=1}^{n} (x_i - \bar{x})(Y_i - \bar{Y}) = \left(n \sum_{i=1}^{n} x_i Y_i - \sum_{i=1}^{n} x_i \sum_{i=1}^{n} Y_i\right) \Big/ n.$

(*d*) $\sum_{i=1}^{n}(x_i - \bar{x})^2 = \sum_{i=1}^{n}(x_i - \bar{x})x_i.$

(*e*) $\sum_{i=1}^{n}(x_i - \bar{x})^2 = \left[n\sum_{i=1}^{n}x_i^2 - \left(\sum_{i=1}^{n}x_i\right)^2\right]\Big/ n.$

12. The estimator of the true mean of the dependent variable Y was given by $\hat{\mu}_{Y|x} = B_0 + B_1 x$. Show that $E(\hat{\mu}_{Y|x}) = \mu_{Y|x}$, and hence that $\hat{\mu}_{Y|x}$ is an unbiased estimator, for $\mu_{Y|x}$.

13. The proof that $S^2 = \text{SSE}/(n-2)$ is an unbiased estimator for σ^2 is tricky. The steps in the proof are outlined below:

(*a*) Show that $\text{SSE} = S_{yy} - S_{xx}B_1^2$.

(*b*) Show that $\text{SSE} = \sum_{i=1}^{n} Y_i^2 - n\bar{Y}^2 - S_{xx}B_1^2.$

(*c*) Show that $E[\text{SSE}] = \sum_{i=1}^{n} E[Y_i^2] - nE[\bar{Y}^2] - S_{xx}E[B_1^2].$

(*d*) Show that
$$E[Y_i^2] = \text{Var } Y_i + (E[Y_i])^2 = \sigma^2 + (\beta_0 + \beta_1 x_i)^2$$
$$E[\bar{Y}^2] = \text{Var } \bar{Y} + (E[\bar{Y}])^2 = \sigma^2/n + (\beta_0 + \beta_1 \bar{x})^2$$
$$E[B_1^2] = \text{Var } B_1 + (E[B_1])^2 = \sigma^2/S_{xx} + \beta_1^2.$$

(*e*) Substitute and simplify to show that $E[\text{SSE}] = (n-2)\sigma^2$.

(*f*) Show that $E[S^2] = \sigma^2$.

14. Recall the estimators for the parameters μ_Y and β_1 are denoted by $\bar{Y}$ and B_1, respectively. Prove that $\text{Cov}(\bar{Y}, B_1) = 0$. *Hint:* It is assumed that Y_i and Y_j are uncorrelated. Write $\bar{Y}$ and B_1 as linear combinations of Y_i ($\bar{Y} = \sum_{i=1}^{n} a_i Y_i$ and $B_1 = \sum_{i=1}^{n} c_i Y_i$). Note that $a_i = 1/n$ and $c_i = (x_i - \bar{x})/\sum_{i=1}^{n}(x_i - \bar{x})^2$.

15. Suppose that the *true* regression equation is known to be $\mu_{Y|x} = 10 + 2.5x$. Under the assumption of normality we have seen that the estimator for β_1, B_1 is also normally distributed with mean β_1 and variance $\sigma^2/\sum_{i=1}^{n}(x_i - \bar{x})^2$. Suppose that it is also known that $\text{Var } B_1 = 1.2$. For a sample of 25 observations (x_i, y_i), find the probability that the estimate of β_1 will be greater than 3.5.

Section 11.3

The following data are representative of a study in *Engineering Costs and Production Economics*, 1983, p. 137. In production flow-shop problems, performance is often evaluated by minimum make-span, the total elapsed time from starting the first job on the first machine until the last job is completed on the last machine. For a particular flow-shop the make-span was evaluated with respect to the number of jobs to be done. Let the independent variable X denote the number of jobs and the dependent variable Y denote the make-span (in standardized units):

Number of jobs (x)	4	5	6	7	8	9
Make-span (y)	3.75	4.90	4.88	7.2	7.3	9.1
	10	11	12	13	14	15
	9.0	11.9	11.5	14.1	13.9	17.5

Refer to these data for Exercises 16 through 18.

16. (*a*) Estimate the linear regression equation $\mu_{Y|x} = \beta_0 + \beta_1 x$.
(*b*) Plot the estimated regression equation.

17. Test for a significant linear regression at the $\alpha = .05$ level of significance.

18. (*a*) At $x = \bar{x}$, compute a 95% confidence interval for $\mu_{Y|x}$ and verbally explain the answer.
(*b*) At $x = 12$, compute a 95% confidence interval for $\mu_{Y|x}$ and verbally explain the answer.
(*c*) How do you explain the different widths of the intervals in parts (*a*) and (*b*)?

Refer to the data in Exercise 7 for Exercises 19 through 22.

19. Test $H_0: \beta_0 = 2$ at the .01 level of significance.

20. Calculate a 95% confidence interval for the true intercept β_0.

21. Test for significant linear regression; that is, test $H_0: \beta_1 = 0$ at the .05 level.

22. (*a*) If $x = 50$, estimate Y, a single predicted value of Y when $x = 50$.
(*b*) Calculate a 95% prediction interval for $Y|x = 50$ and interpret your answer.

Let x denote the number of lines of executable SAS code and let Y denote the execution time in seconds. Use the following summary information to do Exercises 23 through 26.

$$n = 10 \qquad \sum_{i=1}^{10} x_i = 16.75 \qquad \sum_{i=1}^{10} y_i = 170 \qquad \sum_{i=1}^{10} x_i^2 = 28.64$$

$$\sum_{i=1}^{10} y_i^2 = 2898 \qquad \sum_{i=1}^{10} x_i y_i = 285.625$$

23. Estimate and plot the line of regression.

24. (*a*) Estimate Var $Y_i = \sigma^2$.
(*b*) Estimate the standard deviation of B_1.
(*c*) Estimate the standard deviation of B_0.

25. Test the hypothesis $\beta_1 = 0$ at the .01 level and verbally state the conclusion.

26. Test the hypothesis $\beta_0 = 25$ at the .05 level and discuss the conclusion in the context of the problem.

27. If significant regression is found, estimate the average time required to run a SAS program with 15 lines of executable code.

28. If regression is not significant, what does this mean mathematically? Can you think of a practical reason from a computing standpoint that regression might not be significant in this case?

29. The following data are representative of information in *Energy Policy*, March 1983. The data represent carbon dioxide (CO_2) emissions from coal-fired boilers (in units of 1000 tons) over a period of years between 1965 and 1977. The independent variable (year) has been standardized to yield the following table.

Year (x)	0	5	8	9	10	11	12
CO_2 emission (y)	910	680	520	450	370	380	340

(a) Estimate the linear regression equation $\mu_{Y|x} = \beta_0 + \beta_1 x$.
(b) Is there a significant linear trend in CO_2 emission over this time span? That is, test H_0: $\beta_1 = 0$ at the .01 level of significance.
(c) Would it be wise to use the estimated regression line to estimate the average CO_2 emissions from coal-fired boilers for the year 2000? Explain.

The following data represent the known weights of calcium oxide (CaO) from nine different samples and the corresponding weights determined by a standard chemical procedure. The known weight is treated as the independent variable X.

CaO present (x)	3.0	7.0	11.5	15.0	19.0
CaO found (y)	2.7	6.8	11.1	14.6	18.8
	24.0	30.0	35.0	39.0	39.0
	23.5	29.7	34.5	38.4	38.5

Use these data to do Exercises 30 through 32.

30. Find the linear regression line used to estimate $\mu_{Y|x}$, the average weight of CaO found for a known weight x.

31. Compute an unbiased estimate of the variance of Y about the true linear regression line.

32. (a) If $x = 15$, estimate $\mu_{Y|x}$.
(b) Compute and interpret a 90% confidence interval for $\mu_{Y|x}$ when $x = 15$.

33. An experiment was completed to study the relationship between concentrations of estrone in saliva and in free plasma. The following data were obtained:

Subject	**Estrone in saliva** (x)	**Estrone in free plasma** (y)
1	7.4	30.0
2	7.5	25.0
3	8.5	31.5
4	9.0	27.5
5	9.0	39.5
6	11.0	38.0
7	13.0	43.2
8	14.0	49.0
9	14.5	55.0
10	16.0	48.5

(a) Plot a scattergram of the data.
(b) Estimate the line of regression of Y on X.
(c) If the estrone level is 12.1, predict the level of estrone in free plasma.
(d) Test for a significant linear regression at the .10 level.

Section 11.4

A study reported in the *Journal of Coatings Technology*, vol. 55, 1983, considered the ability to predict cracking of latex paints on exposed wood surfaces based on accelerated cracking tests. Representative data on accelerated crack rating (x) and exposure crack rating (y) are given below.

Accelerated crack rating (x)	Exposure crack rating (y)
2.0	1.9
2.0	2.3
3.0	2.7
3.0	3.9
4.0	3.0
4.0	4.2
5.0	3.1
5.0	4.8
6.0	4.8
6.0	6.7
7.0	5.1
7.0	6.4

Refer to these data for Exercises 34 and 35.

34. (*a*) Plot the data in a scattergram.
(*b*) Estimate the line of regression for predicting exposure crack rating from accelerated crack rating.
(*c*) Estimate the average exposure crack rating if the accelerated crack rating is 4.5.

35. (*a*) Test for lack of linear fit at the .05 level of significance.
(*b*) Can we conclude that a linear regression equation adequately fits the data?

Many chemicals dissolve in water at different rates depending on water temperature. This phenomenon was studied for a certain chemical with the experimental data given below. The dependent variable y denotes the amount [in grams per liter (g/l)] of the chemical dissolved and x denotes the temperature [in degrees Celsius (°C)] of the water:

x (C°)	y (g / l)
0	2.1, 2.8, 3.1
10	4.5, 4.8, 5.4
20	6.1, 8.2, 9.0
30	11.2, 12.0, 12.5
40	13.8, 15.2, 15.9
50	17.0, 18.1, 18.8

Use these data for Exercises 36 through 38.

36. Plot a scattergram.

37. (*a*) Estimate the true linear regression $\mu_{Y|x} = \beta_0 + \beta_1 x$.
(*b*) Estimate $\mu_{Y|x}$ when the temperature (x) is 35°C.

38. Test for adequacy of fit for linear regression at the .05 level of significance.

39. If the random variable Y_i follows a normal distribution with mean μ_Y and variance σ^2, what is the distribution of the random variable

$$\sum_{i=1}^{n} \left(Y_{ij} - \bar{Y}_i\right)^2 / \sigma^2$$

Hint: See Theorem 8.1.1.

40. Show that the random variable SSE_{pe}/σ^2 follows a chi-squared distribution with $n - k$ degrees of freedom when Y_i follows a normal distribution with mean μ_Y and variance σ^2. *Hint:* See Exercise 44, Chap. 7.

41. Consider the data of Exercise 10. Test for adequacy of fit.

Section 11.5

42. Reconsider Exercise 1. Even though the scattergram of the data suggests that simple linear regression is not appropriate, a straight line can be forced through the data via least squares.
(*a*) Estimate b_0 and b_1 to force a line through the data of Exercise 1.
(*b*) Use the estimated line of regression to find $\hat{y}_i$ for $i = 1$ to 18.
(*c*) Find the 18 residuals for these data and verify that, apart from round-off error, these residuals sum to 0.
(*d*) Form a residual plot and notice that it does not exhibit the ideal pattern expected when simple linear regression is appropriate.
(*e*) Can you suggest an equation that would probably describe the pattern seen in the raw data much better than does a straight line?
(*f*) Sketch and interpret the boxplot for the residuals.

43. Sketch residual plots for each of the data sets given in Exercise 2. (The residuals were found in Exercise 6.) Which, if any, of the plots suggest that the assumptions underlying simple linear regression are not met?

44. Consider the residuals found in Exercise 9.
(*a*) Sketch and interpret the residual plot.
(*b*) Sketch and interpret the boxplot of the residuals.

45. In the earliest stages of the development of electronic technology solders were used to assemble components. Due to the fact that they are applied hot, stress such as creeping, distortion, and metal fatigue can result. A study of the use of amalgams as alternatives to solder is conducted. An amalgam is an alloy between a liquid metal and a powder formed at room temperature. These data are obtained on the curing time in minutes (x) and the hardness rating (y) of a gallium/nickel/copper amalgam:

x **(curing time)**	y **(hardness in durometers,** D**)**
5	9, 9, 10
1500	68, 70, 72
1800	82, 80, 83
2000	87, 86, 86
3500	91, 90, 90
4200	91, 91, 92
5800	95, 96, 93

(Based on information from "Amalgams for Improved Electronics Interconnection," Colin A. MacKay, *IEEE MICRO*, April 1993, pp. 46–58.)

(*a*) Sketch a scattergram for these data.
(*b*) Even though simple linear regression is not appropriate, force a regression line through the data. Graph this line on the scattergram.
(*c*) Estimate each residual visually and sketch a rough residual plot. Discuss the plot.
(*d*) If you have SAS or some other computer software available, find the exact values of the residuals and form a residual plot by computer. (See Computing Supplement X.)

46. Figure 11.19 shows residual plots for various data sets. In each case, identify any model assumptions that might be violated.

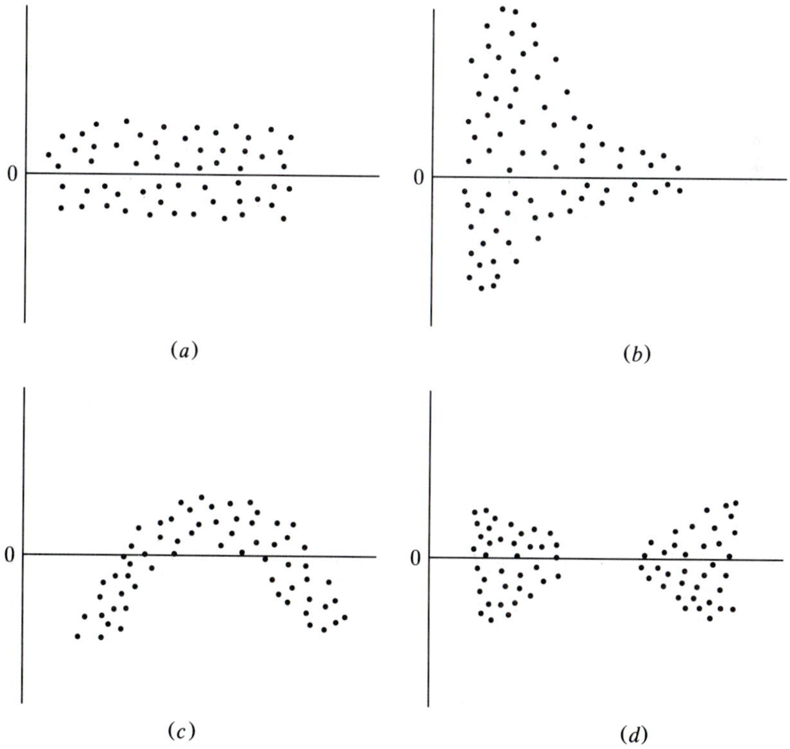

FIGURE 11.19
Residual plots.

Section 11.6

Pesticides used in food production can be found in food consumed by humans. A study focusing on chickens exposed to malaoxon was conducted. The chickens were also exposed to a liver enzyme inducer to determine whether liver detoxification of the pesticide is affected. Data were reported as a percentage of normal pesticide detoxification (y) and percentage of normal liver enzyme levels (x).

Enzyme level (x)	Detoxification level (y)
95	108
110	126
118	102
124	121
145	118
140	155
185	158
190	178
205	159
222	184

Refer to these data for Exercises 47 through 50.

47. (*a*) Plot a scattergram of the data.
(*b*) Estimate ρ, the correlation between X and Y.

48. Test the null hypothesis that X and Y are uncorrelated at the 0.10 level. That is, test H_0: $\rho = 0$. Discuss your conclusion.

49. Find a 90% confidence interval on ρ.

50. Test H_0: $\rho = .8$ at the $\alpha = .05$ level of significance.

51. These data are obtained in the random variables x, the percentage copper of a sample, and its Rockwell hardness rating y:

x	y
.01	58.0
.03	66.0
.01	55.0
.02	63.2
.10	58.3
.08	57.9
.12	69.3
.15	70.1
.10	65.2
.11	62.3

(*a*) Plot a scattergram of these data.
(*b*) Find a point estimate for ρ.
(*c*) Find a 95% confidence interval for ρ and discuss your conclusion.

52. Show that $B_1/(S/\sqrt{S_{xx}}) = R\sqrt{n-2}/\sqrt{1-R^2}$. *Hint:* Use the fact that $B_1 = S_{xy}/S_{xx}$ and $R = S_{xy}/\sqrt{S_{xx}S_{yy}}$ to show that $B_1/(S/\sqrt{S_{xx}}) = \sqrt{S_{yy}}R/S$. Then use the fact that $\text{SSE} = S_{yy} - B_1 S_{xy}$ and $S^2 = \text{SSE}/(n-2)$.

53. Show that if the random variable (X, Y) has a bivariate normal distribution with $\rho = 0$, then X and Y are independent.

54. Does a correlation coefficient of zero always imply that X and Y are independent?

55. Show that if the random variable (X, Y) has a bivariate normal distribution, then the point (μ_X, μ_Y) lies on the true line of regression of Y on X.

56. Find and interpret the coefficient of determination for the data sets of Exercise 2.
57. Find and interpret the coefficient of determination for the data of Exercise 7.
58. Find and interpret the coefficient of determination for the data of Exercise 10.
59. Find and interpret the coefficient of determination for the data of Exercise 1.

REVIEW EXERCISES

Investigators at an interactive graphics installation designed an experiment to study operator performance as a function of the length of time worked. The independent variable (fixed by the experimenter) was the length of time worked [in hours (h)]. The dependent variable was the number of commands per hour. Fifteen operators of comparable training were used with three operators randomly selected to work for each of the five lengths of time in the experiment. The study yielded the following data:

Length of work, h (x)	**Number of commands** (y)
1	136, 143, 139
2	165, 169, 173
3	168, 173, 176
4	170, 169, 176
5	165, 170, 172

Use these data for Exercises 60 and 61.

60. (*a*) Plot the data.
(*b*) Estimate and plot the curve of regression $\mu_{Y|x} = \beta_0 + \beta_1 x$.

61. (*a*) Test for lack of linear fit to the data at the 5% level of significance. Does a linear regression curve adequately fit the data?
(*b*) Do the plot in Exercise 60 and the test for lack of linear fit seem to agree with each other?

The following data represent the fuel gas temperature [in degrees Fahrenheit (°F)] and unit heat rate [in Btus per kilowatt hour (Btu/kWh)] for a combustion turbine to be used in coal gasification.

Gas temperature, °F (x)	**Heat, Btu / kWh Units of 100** (y)
100	99.1
150	98.5
200	98.2
250	98.0
300	97.8
350	97.6
400	97.5
450	97.0
500	96.8

Use these data for Exercises 62 through 66.

62. Estimate the regression curve $\mu_{Y|x} = \beta_0 + \beta_1 x$.

63. Test H_0: $\beta_1 = 0$ versus H_1: $\beta_1 < 0$. Use $\alpha = .05$.

64. Estimate the coefficient of determination as a measure of goodness of fit of the linear regression curve.

65. Calculate a 90% confidence interval on β_0 and discuss in the context of the data.

66. Calculate a 95% confidence interval on β_1 and discuss in the context of the data.

An engineer wishes to investigate the recovery of heat normally lost to the environment in the form of exhaust gases from furnaces. Her experiment is designed by fixing flow speed past heat pipes [in meters per second (m/sec)] and then measuring the recovery ratio. The study yielded the following data:

Flow speed, m / sec (x)	**Recovery ratio** (y)
1	.740
1.5	.745
2	.718
2.5	.678
3	.652
3.5	.627
4	.607
4.5	.507
5	.545

Refer to these data for Exercises 67 through 69.

67. Estimate the curve of regression $\mu_{Y|x} = \beta_0 + \beta_1 x$.

68. Test for significant regression at the .05 level.

69. (*a*) If the flow speed is fixed at 3.25 (m/sec), predict $\mu_{Y|x=3.25}$ and $Y|x = 3.25$.

(*b*) Calculate and interpret the 95% confidence interval on $\mu_{Y|x=3.25}$.

(*c*) Calculate and interpret the 95% prediction interval on $Y|x = 3.25$.

(*d*) How do you explain the differing widths of the intervals calculated in parts (*b*) and (*c*)?

In studying the effect of air quality on a lake, the experimenter takes observations on the pH of the water and the air quality as measured on an air quality index. The index goes from 0 to 100 with larger numbers representing high pollution. These data are obtained:

pH (x)	4.5	4.1	4.8	4.0	5.0	6.0	3.5	4.9	3.2	6.1
Air quality (y)	40	50	30	60	20	10	70	30	85	15

Refer to these data for Exercises 70 through 72.

70. (*a*) Plot the data on the *xy* plane.
(*b*) Estimate the correlation coefficient ρ.

71. Test for a significant negative correlation at the .05 level of significance.

72. Calculate and interpret the 90% confidence interval on ρ.

73. Suppose that a set of 10 pairs of data (x, y) yield an estimated correlation of $r = .3$.
(*a*) Give the approximate smallest P value for testing $H_0: \rho = 0$ versus $H_1: \rho \neq 0$.
(*b*) Suppose again that $r = .3$, but for $n = 50$ observations. What is the approximate smallest P value for testing the same hypothesis as given in part (*a*)? How do you explain the difference between P values in parts (*a*) and (*b*)?

74. (*a*) What relationship does the size (in absolute value) of the correlation coefficient have to the slope of the linear regression line of y on x.
(*b*) What relationship does the size (in absolute value) of the correlation coefficient have to the closeness of the points to the linear regression line of y on x.

75. When processing flow-shops involve semiautomatic or manual operators, processing times can be regarded as random variables. An investigator decided to study the correlation between make-spans (the time elapsed until the last job is completed on the last machine) for two different systems. The study yielded the following bivariate observations for a random selection of 10 sets of jobs:

Job	System 1 (x)	System 2 (y)
1	4.1	3.9
2	5.0	5.1
3	4.9	5.0
4	5.3	4.9
5	13.5	13.3
6	12.0	13.2
7	19.2	21.3
8	10.0	9.1
9	24.1	23.0
10	6.9	8.1

(*a*) Estimate the Pearson correlation coefficient.
(*b*) Compute a 95% confidence interval on the true correlation ρ.
(*c*) Test for a significant correlation at the .05 level. Do parts (*b*) and (*c*) tend to agree?
(*d*) Calculate the coefficient of determination. Explain its meaning.

76. Carbon dioxide is known to have a critical effect on microbiological growth. Small amounts of CO_2 stimulate the growth of many organisms, while high concentrations inhibit the growth of most. The latter effect is used commercially when perishable food products are stored. A study is conducted to investigate the effect of CO_2 on the growth rate of *Pseudomonas fragi*, a food spoiler. Carbon dioxide is administered at five different atmospheric pressures. The response noted is the percentage change in cell mass after a 1-hour growing time. Ten cultures are used at each level. The following data are found:

Factor level (CO_2 pressure in atmospheres)

0.0	.083	.29	.50	.86
62.6	50.9	45.5	29.5	24.9
59.6	44.3	41.1	22.8	17.2
64.5	47.5	29.8	19.2	7.8
59.3	49.5	38.3	20.6	10.5
58.6	48.5	40.2	29.2	17.8
64.6	50.4	38.5	24.1	22.1
50.9	35.2	30.2	22.6	22.6
56.2	49.9	27.0	32.7	16.8
52.3	42.6	40.0	24.4	15.9
62.8	41.6	33.9	29.6	8.8

Conduct a regression study and write a report that summarizes the results of all tests and graphical tools that you used in the analysis.

77. Below are the predicted values and the residuals for Example 11.4.1. Construct a residual plot and discuss its implications. Does the plot lead you to the same conclusion as that of the formal test conducted in the example?

X	Y	PREDICT.	RESID.
30	13.7	15.7600	−2.06000
30	14.0	15.7600	−1.76000
30	14.6	15.7600	−1.16000
40	15.5	16.2733	−0.77333
40	16.0	16.2733	−0.27333
40	17.0	16.2733	0.72667
50	18.5	16.7867	1.71333
50	20.0	16.7867	3.21333
50	21.1	16.7867	4.31333
60	17.7	17.3000	0.40000
60	18.1	17.3000	0.80000
60	18.5	17.3000	1.20000
70	15.0	17.8133	−2.81333
70	15.6	17.8133	−2.21333
70	16.5	17.8133	−1.31333

COMPUTING SUPPLEMENT

VI Scattergrams, Estimation of β_0 and β_1, and the Use of the Regression Line for Predictions

The SAS program PROC PLOT is used to plot a scattergram of a data set. PROC GLM is then used to estimate the regression line and to predict values of Y for specified values of X. We illustrate by using the data given in Example 11.1.1.

IS REGRESSION LINEAR?

PLOT OF Y*X   LEGEND: A = 1 OBS, B = 2 OBS, ETC

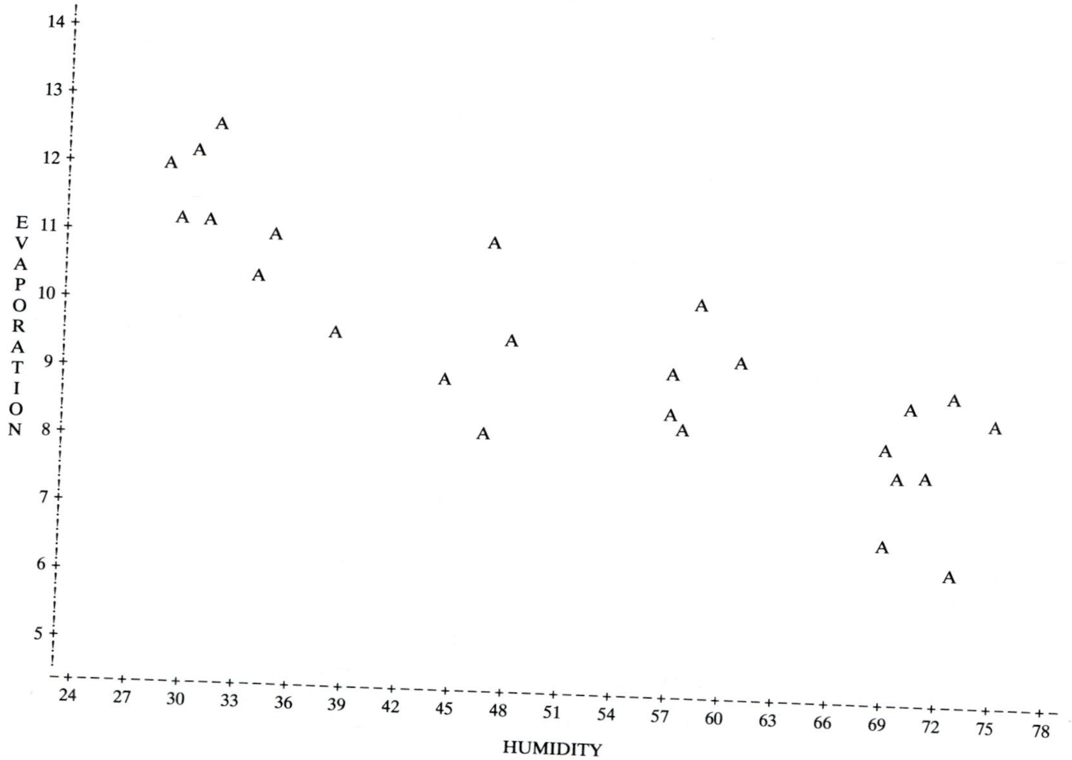

NOTE: 1 OBS HAD MISSING VALUES

Statement	*Function*
DATA PAINT;	names data set
INPUT X Y;	each line will contain the value of X and its corresponding Y value
LINES;	data follows
35.3 11.0 29.7 11.1 ⋮ 28.6 11.1	data
50.0	allows SAS to predict Y when x = 50
;	signals end of data
PROC PLOT;	asks for scattergram
PLOT $Y * X$;	
LABEL X = HUMIDITY Y = % EVAPORATION;	labels variables
TITLE IS REGRESSION LINEAR?;	titles output
PROC GLM;	asks for regression analysis to be conducted
MODEL $Y = X$;	specifies that simple linear regression is to be used; variable on the left is identified as the dependent or response variable; variable on the right is the independent variable
OUTPUT OUT = NEW P = PREDICT;	forms a new data set, NEW, that contains the predicted values of Y
TITLE1 ESTIMATED LINE; TITLE2 OF REGRESSION;	titles output
PROC PRINT;	prints the predicted values of Y
TITLE PREDICTIONS;	titles output

Note the downward linear trend exhibited in the scattergram. This is consistent with the estimated slope $\hat{\beta}_1$, which is slightly negative and given by ①. The estimated intercept, $\hat{\beta}_0$, is given by ②. The predicted value of Y when $x = 50$ is shown in ③.

ESTIMATED LINE
OF REGRESSION

GENERAL LINEAR MODELS PROCEDURE

DEPENDENT VARIABLE INFORMATION

NUMBER OF OBSERVATIONS IN DATA SET = 26

NOTE: ALL DEPENDENT VARIABLES ARE CONSISTENT WITH RESPECT TO THE PRESENCE OR ABSENCE OF MISSING VALUES. HOWEVER, ONLY 25 OBSERVATIONS IN DATA SET CAN BE USED IN THIS ANALYSIS

ESTIMATED LINE
OF REGRESSION

GENERAL LINEAR MODELS PROCEDURE

DEPENDENT VARIABLE: Y

SOURCE	DF	SUM OF SQUARES	MEAN SQUARE	F VALUE
MODEL	1	45.82966081	45.82966081	58.36
ERROR	23	18.06073919	0.78524953	PR > F
CORRECTED TOTAL	24	63.89040000		0.0001

R-SQUARE	C.V.	ROOT MSE	Y MEAN
0.717317	9.3991	0.88614306	9.42800000

SOURCE	DF	TYPE I SS	F VALUE	PR > F
X	1	45.82966081	58.36	0.0001

SOURCE	DF	TYPE III SS	F VALUE	PR > F
X	1	45.82966081	58.36	0.0001

PARAMETER	ESTIMATE	T FOR HO: PARAMETER = 0	PR > !T!	STD ERROR OF ESTIMATE
INTERCEPT	13.63886687 ②	23.56	0.0001	0.57898306
X	−0.08006059 ①	−7.64	0.0001	0.01047971

PREDICTIONS

OBS	X	Y	PREDICT
1	35.3	11.0	10.8127
2	29.7	11.1	11.2611
3	30.8	12.5	11.1730
4	58.8	8.4	8.9313
5	61.4	9.3	8.7231
6	71.3	8.7	7.9305
7	74.4	6.4	5.6824
8	76.7	8.5	7.4982
9	70.7	7.8	7.9786
10	57.5	9.1	9.0354
11	46.4	8.2	9.9241
12	28.9	12.2	11.3251
13	28.1	11.9	11.3892
14	39.1	9.6	10.5085
15	46.8	10.9	9.8920
16	48.5	9.6	9.7559
17	59.3	10.1	8.8913
18	70.0	8.1	8.0346
19	70.0	6.8	8.0346
20	74.4	8.9	7.6824
21	72.1	7.7	7.8665
22	58.1	8.5	8.9873
23	44.6	8.9	10.0682
24	33.4	10.4	10.9648
25	28.6	11.1	11.3491
26	50.0		9.6358 ③

VII Testing H_0: $\beta_1 = 0$

The printout shown below was produced by the program in our previous computing supplement. It is the analysis of the data of Example 11.1.1. The statistics needed to test

$$H_0: \beta_1 = 0$$

$$H_1: \beta_1 \neq 0$$

are given on this printout. In particular b_1 is given by ①, the standard error of this estimate $s/\sqrt{S_{xx}}$ is given by ②. The observed value of the test statistic is shown in ③, and its two-tailed P value is given by ④. Compare these results with those computed by hand in Example 11.3.1.

ESTIMATED LINE
OF REGRESSION

GENERAL LINEAR MODELS PROCEDURE

DEPENDENT VARIABLE: Y

SOURCE	DF	SUM OF SQUARES	MEAN SQUARE	F VALUE
MODEL	1	45.82966081	45.82966081	58.36
ERROR	23	18.06073919	0.78524953	PR > F
CORRECTED TOTAL	24	63.89040000		0.0001

R-SQUARE	C.V.	ROOT MSE	Y MEAN
0.717317	9.3991	0.88614306	9.42800000

SOURCE	DF	TYPE I SS	F VALUE	PR > F
X	1	45.82966081	58.36	0.0001

SOURCE	DF	TYPE III SS	F VALUE	PR > F
X	1	45.82966081	58.36	0.0001

PARAMETER	ESTIMATE	T FOR HO: PARAMETER = 0	PR > !T!	STD ERROR OF ESTIMATE
INTERCEPT	13.63886687	23.56	0.0001	0.57898306
X	−0.08006059 ①	−7.64 ③	0.0001 ④	0.01047971 ②

VIII Confidence Intervals on $\beta_0, \beta_1, \mu_{Y|x}, Y|x$

SAS can be used to generate a confidence interval on $\mu_{Y|x}$ or a prediction interval on $Y|x$. It also provides the statistics needed to construct a confidence interval on β_0 or β_1. We illustrate by using the data of Example 11.3.3.

Statement	*Function*
DATA CAR;	names data set
INPUT WEIGHT MILES;	names variables
LINES;	signals that data follows
1.35 17.9	
1.90 16.5	
⋮	data
1.40 18.3	
2.00 .	allows SAS to estimate the mileage of a car weighing 2.0 tons;
;	signals end of data
PROG GLM;	asks for regression analysis to be conducted
MODEL MILES = WEIGHT/ P CLM ALPHA = .10;	specifies that simple linear regression is to be used; variable on the left is identified as the dependent or response variable; variable on the right is the independent variable; P generates predicted values for Y at each x listed; CLM ALPHA = .10 generates a 90% confidence interval on $\mu_{Y\|x}$.
TITLE CONFIDENCE INTERVALS;	titles output

The printout gives these statistics:

① point estimate for β_1, the slope of the regression line;
② point estimate for β_0, the intercept of the regression line;
③ value of the test statistic used to test H_0: $\beta_1 = 0$;
④ P value for the test of significance; since $P < .0001$, we can reject H_0: $\beta_1 = 0$ and conclude that knowledge of x is useful in predicting Y;
⑤ $s\sqrt{\sum x^2}/\sqrt{nS_{xx}}$ used in constructing a $100(1-\alpha)\%$ confidence interval on β_0;
⑥ $s/\sqrt{S_{xx}}$ used in constructing a $100(1-\alpha)\%$ confidence interval on β_1;
⑦ $s^2 = \text{SSE}/(n-2)$ (this differs slightly from the value obtained in Example 11.3.3 due to round-off error);
⑧ $s = \sqrt{s^2}$;
⑨ point estimate for $\mu_{Y|x=1.7}$ and for $Y|x = 1.7$;
⑩ lower 90% confidence bound on $\mu_{Y|x=1.7}$ (this differs slightly from the value obtained in Example 11.3.3 due to round-off error);
⑪ upper 90% confidence bound on $\mu_{Y|x=1.7}$.

To obtain a 90% prediction interval on $Y|x$, replace the keyword CLM with the word CLI. To obtain a 99% or a 95% confidence interval, replace the words ALPHA = .10 with ALPHA = .01 or ALPHA = .05, respectively. The weight 1.7 was a member of the given data set. However, SAS can predict the mileage of a car with any weight whatsoever. For example, to predict the mileage of a car weighing 2.00 tons, a weight that is not in the data set, we simply place 2.00 · as the last data line. The results are shown as

⑫ point estimate for $\mu_{Y|x=2.0}$ and for $Y|x = 2.0$;

⑬ lower 90% confidence bound on $\mu_{Y|x=2.0}$.

⑭ upper 90% confidence bound on $\mu_{Y|x=2.0}$.

CONFIDENCE INTERVALS

GENERAL LINEAR MODELS PROCEDURE

DEPENDENT VARIABLE: MILES

SOURCE	DF	SUM OF SQUARES	MEAN SQUARE	F VALUE
MODEL	1	9.46068817	9.46068817	75.74
ERROR	8	0.99931183	0.12491398 ⑦	PR > F
CORRECTED TOTAL	9	10.46000000		0.0001

R-SQUARE	C.V.	ROOT MSE	MILES MEAN
0.904463	2.0790	0.35343172 ⑧	17.00000000

SOURCE	DF	TYPE I SS	F VALUE	PR > F
WEIGHT	1	9.46048817	75.74	0.0001

SOURCE	DF	TYPE III SS	F VALUE	PR > F
WEIGHT	1	9.46068817	75.74	0.0001

PARAMETER	ESTIMATE	T FOR HO: PARAMETER = 0	PR > !T!	STD ERROR OF ESTIMATE
INTERCEPT	23.75763441 ②	30.28	0.0001	⑤ 0.78449754
WEIGHT	−4.03440860 ①	−8.70 ③	0.0001 ④	⑥ 0.46357930

OBSERVATION	OBSERVED	PREDICTED RESIDUAL	LOWER 90% CLM UPPER 90% CLM
1	17.90000000	18.31118280 −0.41118280	17.96234155 18.66002404
2	16.50000000	16.09225806 0.40774194	15.80797447 16.37654166
3	16.40000000	16.89913978 ⑨ −0.49913978	16.69019039 ⑩ 17.10808917 ⑪
4	16.80000000	16.49569892 0.30430108	16.26158988 16.72980796
5	18.80000000	18.51290323 0.28709677	18.12858521 18.89722124
6	15.50000000	15.48709677 0.01290323	15.10277876 15.87141479
7	17.50000000	17.30258065 0.19741935	17.08492133 17.52023996

8	16.40000000	16.49569892	16.26158988
		−0.09569892	16.72980796
9	15.90000000	16.29397849	16.03716292
		−0.39397849	16.55079407
10	18.30000000	18.10946237	17.79419133
		0.19053763	18.42473340
11*		15.68881720 ⑫	15.33997596 ⑬
			16.03765845 ⑭

*OBSERVATION WAS NOT USED IN THIS ANALYSIS

SUM OF RESIDUALS	0.00000000
SUM OF SQUARED RESIDUALS	0.99931183
SUM OF SQUARED RESIDUALS − ERROR SS	−0.00000000
PRESS STATISTIC	1.53843187
FIRST ORDER AUTOCORRELATION	−0.48604150
DURBIN-WATSON D	2.76656568

IX Testing for Lack of Fit

We illustrate the use of SAS in detecting lack of fit by using the data of Example 11.4.1.

Statement	*Function*
DATA CHEMICAL;	names data set
INPUT X Y;	names variables
A = X;	groups values according to the value of the variable X;
LINES;	signals that data follows
30 13.7	
30 14.0	
⋮	data
70 16.5	
;	signals end of data
PROC GLM;	calls for a regression analysis to be conducted
CLASSES A;	identifies the group name
MODEL Y = X A;	
TITLE1 TESTING FOR LACK;	titles output
TITLE2 OF FIT;	

These quantities are shown on the printout:

① SSE_{pe}
② SSE_{lf}
③ F value used to detect lack of fit
④ P value of the test

TESTING FOR LACK
OF FIT

GENERAL LINEAR MODELS PROCEDURE

CLASS LEVEL INFORMATION

CLASS	LEVELS	VALUES
A	5	30 40 50 60 70

NUMBER OF OBSERVATIONS IN DATA SET = 15

TESTING FOR LACK
OF FIT

GENERAL LINEAR MODELS PROCEDURE

DEPENDENT VARIABLE: Y

SOURCE	DF	SUM OF SQUARES	MEAN SQUARE	F VALUE
MODEL	4	59.98400000	14.99600000	23.24
ERROR	10	6.45333333 ①	0.64533333	PR > F
CORRECTED TOTAL	14	66.43733333		0.0001

R-SQUARE	C.V.	ROOT MSE	Y MEAN
0.902866	4.7855	0.80332642	16.78666667

SOURCE	DF	TYPE I SS	F VALUE	PR > F
X	1	7.90533333	12.25	0.0057
A	3	52.078666667 ②	26.90 ③	0.0001 ④

SOURCE	DF	TYPE III SS	F VALUE	PR > F
X	0	0.00000000	.	.
A	3	52.07866667	26.90	0.0001

X Residual Plots

SAS can be used to construct residual plots, boxplots, and stem-and-leaf diagrams for use in detecting violations of the model assumptions of simple linear regression. Here we illustrate the use of PROC REG in the construction of residual plots. Boxplots and stem-and-leaf diagrams are found by using PROC UNIVARIATE as explained in Computing Supplement III. Data are those of Example 11.3.1.

Statement	*Function*
DATA CAR;	names data set
INPUT X Y;	names variables; X = weight, Y = miles;
LINES;	signals that data follows
1.35 17.9 1.90 16.5 ⋮ 1.40 18.3	data
;	signals end of data

PROC PLOT;	
PLOT MILES * WEIGHT;	asks for a scattergram of the data
PROC REG;	asks for a regression analyses to be conducted
MODEL Y = X;	identifies the variable Y (miles) as the response and X (weight), as the regressor in a simple linear regression model
OUTPUT OUT = NEW P = PREDICT R = RESID;	creates a new data set in which the predicted values are called PREDICT and the residuals are called RESID
PROC PRINT; VAR X Y PREDICT RESID;	asks for the original data as well as the predicted values and residuals to be printed
PROC PLOT; PLOT RESID * X;	produces the residual plot
PROC UNIVARIATE PLOT NORMAL; VAR RESID;	asks for statistics to be generated for the residuals requests a boxplot, stem-and-leaf plot, and a test of normality for this variable

The scattergram for the data is shown at ① It appears to indicate a downward linear trend. Predicted and residual values are given at ② and ③, respectively. Notice that the residuals sum to 0 as expected. The residual plot is given at ④. The stem-and-leaf plots are shown at ⑤ and ⑥, respectively, with the results of the normality test given at ⑦ and ⑧.

OBSERVED	X	Y	② PREDICT	③ RESID
1	1.35	17.9	18.3112	−0.41118
2	1.90	16.5	16.0923	0.40774
3	1.70	16.4	16.8991	−0.49914
4	1.80	16.8	16.4957	0.30430
5	1.30	18.8	18.5129	0.28710
6	2.05	15.5	15.4871	0.01290
7	1.60	17.5	17.3026	0.19742
8	1.80	16.4	16.4957	−0.09570
9	1.85	15.9	16.2940	−0.39398
10	1.40	18.3	18.1095	0.19054

SAS
UNIVARIATE PROCEDURE

Variable = RESID   Residual

Moments

	N	10	Sum Wgts	10	
	Mean	0	Sum	0	
	Std Dev	0.333219	Variance	0.111035	
	Skewness	−0.4567	Kurtosis	−1.4948	
	USS	0.999312	CSS	0.999312	
	CV	.	Std Mean	0.105373	
	T:Mean = 0	0	Prob > |T|	1.0000	
	Sgn Rank	−0.5	Prob > |S|	1.0000	
	Num ^= 0	10			
⑦	W:Normal	0.897654	Prob < W	0.1969	⑧

PLOT OF Y*X

```
    Y
    |              (1)
18.8 +
18.7 +
18.6 +
18.5 +
18.4 +
18.3 +         A
18.2 +
18.1 +
18.0 +
17.9 +    A
17.8 +
17.7 +
17.6 +
17.5 +                           A
17.4 +
17.3 +
17.2 +
17.1 +
17.0 +
16.9 +
16.8 +                                             A
16.7 +
16.6 +
16.5 +                                                      A
16.4 +                                    A        A
16.3 +
16.2 +
16.1 +
16.0 +
15.9 +                                                  A
15.8 +
15.7 +
15.6 +
15.5 +                                                                    A
     |
     -+--------+--------+--------+--------+--------+--------+--------+--------+-- X
     1.3      1.4      1.5      1.6      1.7      1.8      1.9      2.0      2.1
```

Quantiles (Def = 5)

100% Max	0.407742	99%	0.407742
75% Q3	0.287097	95%	0.407742
50% Med	0.10172	90%	0.356022
25% Q1	−0.39398	10%	−0.45516
0% Min	−0.49914	5%	−0.49914
		1%	−0.49914
Range	0.906882		
Q3 − Q1	0.681075		
Mode	−0.49914		

Extremes

Lowest	Obs	Highest	Obs
−0.49914(	3)	0.190538(	10)
−0.41118(	1)	0.197419(	7)
−0.39398(	9)	0.287097(	5)
−0.0957(	8)	0.304301(	4)
0.012903(	6)	0.407742(	2)

PLOT OF RESID * X

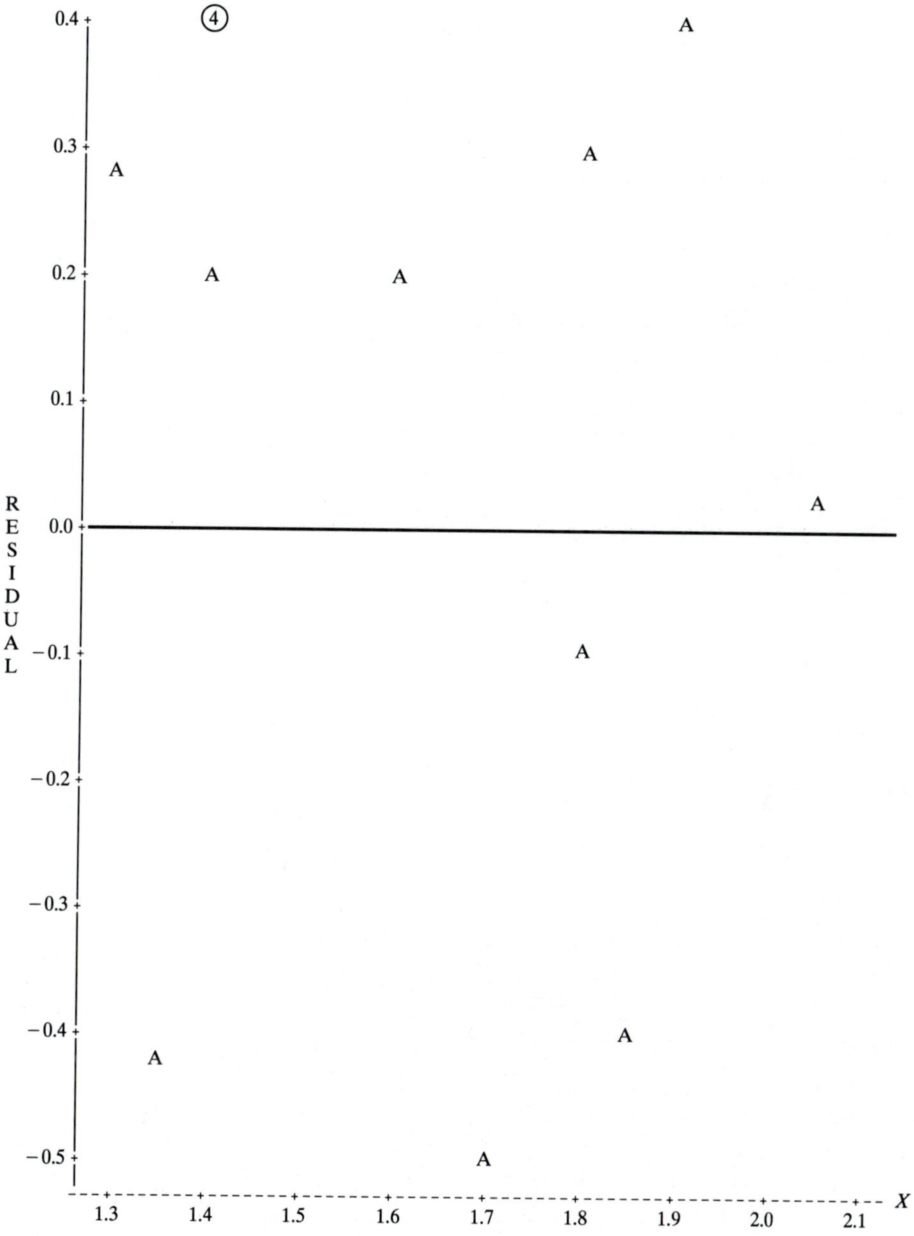

```
⑤ Stem          Leaf          #          ⑥ Boxplot
    4           1             1               |
    2           090           3           +-----+
    0           19            2           *--+--*
   -0           0             1           |     |
   -2           9             1           +-----+
   -4           01            2               |
      ----+----+----+----+
Multiply Stem. Leaf by 10** - 1
```

XI Correlation

The SAS procedure PROC CORR is used to estimate ρ. We illustrate the procedure the using the data of Example 11.5.1.

Statement	*Function*
DATA LAKE;	names data set
INPUT OLD NEW;	names variables
LABEL OLD = MANUAL;	labels variables
LABEL NEW = AUTOMATED;	
LINES;	signals that data follows
25 30	
40 80	
⋮	data
575 583	
;	signals end of data
PROC PLOT;	creates scattergram
PLOT NEW * OLD;	
PROC CORR;	asks for correlation analysis
TITLE CORRELATION;	titles output

The value of $\hat{\rho} = r$ is shown in ①.

CORRELATION

VARIABLE	N	MEAN	STD DEV	SUM	MINIMUM	MAXIMUM
OLD	10	240.500000	189.259522	2405.00000	25.000000	575.000000
NEW	10	250.300000	180.427917	2503.00000	30.000000	583.000000

CORRELATION COEFFICIENTS/PROB > !R! UNDER HO: RHO = 0/N = 10

	OLD	NEW
OLD	1.00000	0.97779 ①
MANUAL	0.0000	0.0001
NEW	0.97779	1.00000
AUTOMATED	0.0001	0.0000

SAS

PLOT OF NEW*OLD   LEGEND: A = 1 OBS, B = 2 OBS, ETC

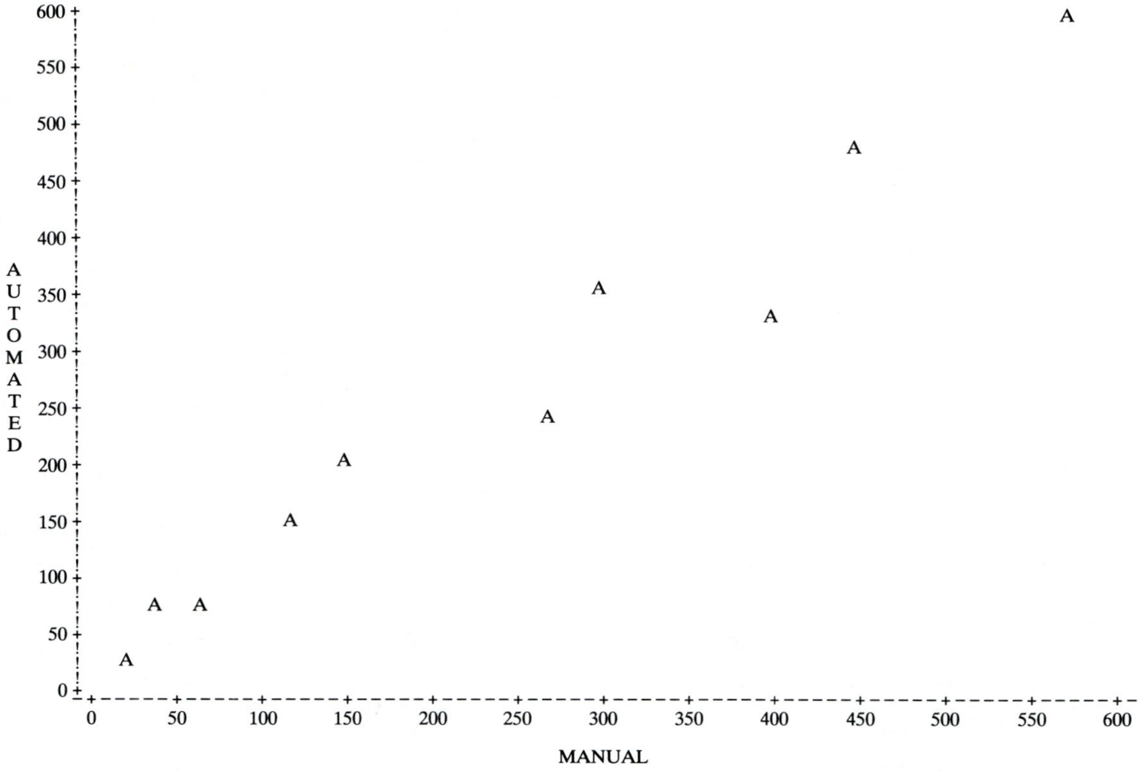

CHAPTER 12

MULTIPLE LINEAR REGRESSION MODELS

In the last chapter we studied the simple linear regression model. This model expresses the idea that the mean of a response variable Y depends on the value assumed by a single predictor variable X. In this chapter we extend the concepts studied earlier to cases in which the model becomes more complex. In particular, we distinguish between two basic models: the *polynomial model*, in which the single predictor variable can appear to a power greater than 1, and the *multiple linear regression model*, in which more than one distinct predictor variable can be used. The techniques employed in each case are similar and conceptually easy. However, it will soon become obvious that, except for the simplest cases, the analysis is too cumbersome to handle without the use of a computer. This should present no problem since computer software packages are available for the analysis of these models.

12.1 LEAST-SQUARES PROCEDURES FOR MODEL FITTING

In this section we develop the least-squares estimators for the parameters in both the polynomial and multiple regression models. Before introducing these models specifically, let us note that each of them is a special case of what is called the *general linear model*. These are models in which the mean value of a response variable Y is assumed to depend on the values assumed by one or more predictor variables. As in the case of simple linear regression, the predictor variables $X_1, X_2, \ldots, X_k$ are *not* treated as random variables. However, for a given set of numerical values for these variables $x_1, x_2, \ldots, x_k$, the response variable denoted by $Y|x_1, x_2, \ldots, x_k$ is assumed to be a random variable. The general linear model expresses the mean value of this conditional random variable as a function of $x_1, x_2, \ldots, x_k$. The model takes the following form:

General linear model

$$\mu_{Y|x_1, x_2, \ldots, x_k} = \beta_0 + \beta_1 x_1 + \beta_2 x_2 + \cdots + \beta_k x_k \tag{12.1}$$

In this model $x_1, x_2, \ldots, x_k$ denote known real numbers; $\beta_0, \beta_1, \ldots, \beta_k$ denote unknown parameters. Our task is to estimate the values of these parameters from a data set. The model is linear in the sense that it is linear in the *parameters* $\beta_0, \beta_1, \beta_2, \ldots, \beta_k$.

Example 12.1.1. Suppose that we want to develop an equation with which we can predict the gasoline mileage of an automobile based on its weight and the temperature at the time of operation. We might pose the model

$$\mu_{Y|x_1, x_2} = \beta_0 + \beta_1 x_1 + \beta_2 x_2$$

Here the response variable is Y, the mileage obtained. There are two independent or predictor variables. These are X_1, the weight of the car, and X_2, the temperature. The values assumed by these variables are denoted by x_1 and x_2, respectively. For example, we might want to predict the gas mileage for a car that weighs 1.6 tons when it is being driven in 85°F weather. Here $x_1 = 1.6$ and $x_2 = 85$. The unknown parameters in the model are β_0, β_1, and β_2. Their values are to be estimated from the data gathered.

It is possible to treat the polynomial and multiple regression models simultaneously from the mathematical standpoint. However, they differ enough in a practical sense to justify considering them separately. We begin with a description of the general polynomial model.

Polynomial Model of Degree p

The general polynomial regression model of degree p expresses the mean of the response variable Y as a polynomial function of one predictor variable X. It takes the form

$$\mu_{Y|x} = \beta_0 + \beta_1 x + \beta_2 x^2 + \cdots + \beta_p x^p \tag{12.2}$$

where p is a positive integer. If we let $x_1 = x$, $x_2 = x^2$, $x_3 = x^3, \ldots, x_p = x^p$,

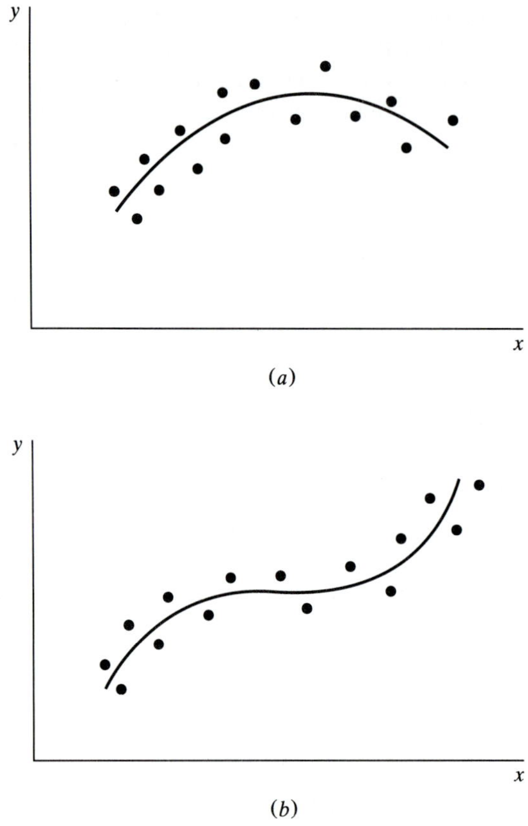

FIGURE 12.1
(a) Quadratic model:

$$\mu_{Y|x} = \beta_0 + \beta_1 x + \beta_2 x^2$$

(b) Cubic model:

$$\mu_{Y|x} = \beta_0 + \beta_1 x + \beta_2 x^2 + \beta_3 x^3$$

then the model can be rewritten in the general linear form as

$$\mu_{Y|x} = \beta_0 + \beta_1 x_1 + \beta_2 x_2 + \cdots + \beta_p x_p$$

Scattergrams are useful in determining when a polynomial model might be appropriate. The pattern shown in Fig. 12.1(a) suggests the quadratic model $\mu_{Y|x} = \beta_0 + \beta_1 x + \beta_2 x^2$; that of Fig. 12.1(b) points to the cubic model $\mu_{Y|x} = \beta_0 + \beta_1 x + \beta_2 x^2 + \beta_3 x^3$. Once we decide that a polynomial is appropriate, we are faced with the problem of estimating the parameters $\beta_0, \beta_1, \beta_2, \ldots, \beta_p$. To apply the method of least squares, we first express the polynomial model in the form

$$Y|x = \beta_0 + \beta_1 x + \beta_2 x^2 + \cdots + \beta_p x^p + \mathrm{E}$$

where $Y|x$ denotes the response variable when the predictor variable assumes the value x, and E denotes the random difference between $Y|x$ and its mean value, $\mu_{Y|x} = \beta_0 + \beta_1 x + \beta_2 x^2 + \cdots + \beta_p x^p$. A random sample of size n takes the form

$$\{(x_1, Y|x_1), (x_2, Y|x_2), \ldots, (x_n, Y|x_n)\}$$

where the first member of each ordered pair denotes a real number, and the second, a random variable. As in the case of simple linear regression, it is customary to drop the condition notation. The sample itself becomes

$$\{(x_1, Y_1), (x_2, Y_2), \ldots, (x_n, Y_n)\}$$

where for each $i = 1, 2, \ldots, n$,

$$Y_i = \beta_0 + \beta_1 x_i + \beta_2 x_i^2 + \cdots + \beta_p x_i^p + E_i$$

Once again, we assume that the random errors $E_1, E_2, \ldots, E_n$ are independent random variables, each with mean 0 and variance σ^2.

The estimated mean response, estimated value of Y for a given value of x, and estimated curve of regression are given by

$$\hat{y} = \hat{\mu}_{Y|x} = b_0 + b_1 x + b_2 x^2 + \cdots + b_p x^p$$

where $b_0, b_1, b_2, \ldots, b_p$ are the least-squares estimates for $\beta_0, \beta_1, \beta_2, \ldots, \beta_p$. To find these estimates, we minimize the sum of the squares of the residuals. Remember that a residual, e_i, is the difference between the observed response, y_i, and the estimated response when $x = x_i$, $\hat{y}_i = b_0 + b_1 x_i + b_2 x_i^2 + \cdots + b_p x_i^p$. We are therefore minimizing the expression

$$\text{SSE} = \sum_{i=1}^{n} e_i^2 = \sum_{i=1}^{n} [y_i - (b_0 + b_1 x_i + b_2 x_i^2 + \cdots + b_p x_i^p)]^2$$

This is done by finding the $p + 1$ partial derivatives

$$\frac{\partial \text{SSE}}{\partial b_0}, \frac{\partial \text{SSE}}{\partial b_1}, \frac{\partial \text{SSE}}{\partial b_2}, \ldots, \frac{\partial \text{SSE}}{\partial b_p}$$

These derivatives are then set equal to 0 to form a system of $p + 1$ normal equations. A little computation will show that these normal equations are given by

Normal equations, polynomial model

$$\begin{aligned}
b_0 n + b_1 \sum_{i=1}^{n} x_i + b_2 \sum_{i=1}^{n} x_i^2 + \cdots + b_p \sum_{i=1}^{n} x_i^p &= \sum_{i=1}^{n} y_i \\
b_0 \sum_{i=1}^{n} x_i + b_1 \sum_{i=1}^{n} x_i^2 + b_2 \sum_{i=1}^{n} x_i^3 + \cdots + b_p \sum_{i=1}^{n} x_i^{p+1} &= \sum_{i=1}^{n} x_i y_i \\
&\vdots \\
b_0 \sum_{i=1}^{n} x_i^p + b_1 \sum_{i=1}^{n} x_i^{p+1} + b_2 \sum_{i=1}^{n} x_i^{p+2} + \cdots + b_p \sum_{i=1}^{n} x_i^{2p} &= \sum_{i=1}^{n} x_i^p y_i
\end{aligned}$$

(12.3)

These equations are then solved simultaneously for the $p + 1$ unknowns $b_0, b_1, b_2, \ldots, b_p$ to find the least-squares estimates for the model parameters $\beta_0, \beta_1, \beta_2, \ldots, \beta_p$. Of course, this is easier said than done! You will find that even for moderate values of p, these calculations become cumbersome. To overcome the problem, we shall show you later how to express the model in matrix form. The solution can then be found by using standard matrix computer packages or any of the statistical software packages. We demonstrate these ideas with a very small hypothetical example.

Example 12.1.2. A study is conducted to develop an equation by which the unit cost of producing a new drug (Y) can be predicted based on the number of units produced (X). The proposed model is

$$\mu_{Y|x} = \beta_0 + \beta_1 x + \beta_2 x^2$$

This is a polynomial model of degree $p = 2$. Assume that these data are available:

Number of units produced (x)	**Cost in hundreds of dollars** (y)
5	14.0
5	12.5
10	7.0
10	5.0
15	2.1
15	1.8
20	6.2
20	4.9
25	13.2
25	14.6

The scattergram of these data is shown in Fig. 12.2. This scattergram does suggest

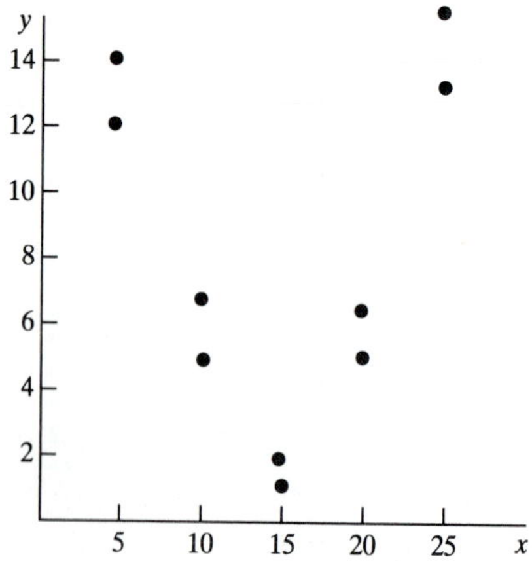

FIGURE 12.2
Scattergram of Y, unit cost in hundreds of dollars of producing a new drug, versus X, the number of units produced.

a quadratic model. The $p + 1 = 2 + 1 = 3$ normal equations are

$$nb_0 + b_1 \sum_{i=1}^{n} x_i + b_2 \sum_{i=1}^{n} x_i^2 = \sum_{i=1}^{n} y_i$$

$$b_0 \sum_{i=1}^{n} x_i + b_1 \sum_{i=1}^{n} x_i^2 + b_2 \sum_{i=1}^{n} x_i^3 = \sum_{i=1}^{n} x_i y_i$$

$$b_0 \sum_{i=1}^{n} x_i^2 + b_1 \sum_{i=1}^{n} x_i^3 + b_2 \sum_{i=1}^{n} x_i^4 = \sum_{i=1}^{n} x_i^2 y_i$$

For these data,

$n = 10$	$\sum x^2 = 2750$	$\sum x^4 = 1{,}223{,}750$	$\sum xy = 1228$
$\sum x = 150$	$\sum x^3 = 56{,}250$	$\sum y = 81.3$	$\sum x^2 y = 24{,}555$

Substituting, we have the normal equations

$$10b_0 + 150b_1 + 2750b_2 = 81.3$$

$$150b_0 + 2750b_1 + 56{,}250b_2 = 1{,}228$$

$$2750b_0 + 56{,}250b_1 + 1{,}223{,}750b_2 = 24{,}555$$

It takes quite a bit of time to solve even this small system by hand. The reader may wish to verify that $b_0 = 27.3$, $b_1 = -3.313$, and $b_2 = .111$. We shall let the computer do it for us later!

We continue our discussion by describing the multiple linear regression model.

Multiple Linear Regression Model

The multiple linear regression model expresses the mean of the response variable Y as a function of one or more distinct predictor variables $X_1, X_2, \ldots, X_k$. It takes the form

$$\mu_{Y|x_1, x_2, \ldots, x_k} = \beta_0 + \beta_1 x_1 + \beta_2 x_2 + \cdots + \beta_k x_k \qquad (12.4)$$

We note that this model differs conceptually from the polynomial model. In the polynomial model we dealt with one predictor variable that could appear to powers greater than 1. Here we deal with k distinct predictor variables, each of the first degree.

To apply the method of least squares to estimate the parameters $\beta_0, \beta_1, \ldots, \beta_k$, we rewrite the model in the form

$$Y|x_1, x_2, \ldots, x_k = \beta_0 + \beta_1 x_1 + \beta_2 x_2 + \cdots + \beta_k x_k + \mathrm{E}$$

where $Y|x_1, x_2, \ldots, x_k$ denotes the response variable when the predictor variables $X_1, X_2, \ldots, X_k$ assume the values $x_1, x_2, \ldots, x_k$, and E denotes the random difference between $Y|x_1, x_2, \ldots, x_k$ and its mean value. A random sample of size n consists of a set of n $(k + 1)$-tuples and takes the form

$$\{(x_{1i}, x_{2i}, \ldots, x_{ki}, Y|x_{1i}, x_{2i}, \ldots, x_{ki}) : i = 1, 2, 3, \ldots, n\}$$

where each of the first k members of each $(k + 1)$-tuple denotes a real number, and the last, a random variable. Dropping the conditional notation, we express the sample as

$$\{(x_{1i}, x_{2i}, \ldots, x_{ki}, Y_i) : i = 1, 2, \ldots, n\}$$

where

$$Y_i = \beta_0 + \beta_1 x_{1i} + \beta_2 x_{2i} + \cdots + \beta_k x_{ki} + \mathrm{E}_i$$

We again make the assumption that the random errors $\mathrm{E}_1, \mathrm{E}_2, \ldots, \mathrm{E}_n$ are independent with mean 0 and common variance σ^2.

The estimated curve of regression of Y on $X_1, X_2, \ldots, X_k$ is

$$\hat{y} = \hat{\mu}_{Y|x_1, x_2, \ldots, x_k} = b_0 + b_1 x_1 + b_2 x_2 + \cdots + b_k x_k$$

where $b_0, b_1, b_2, \ldots, b_k$ are the least-squares estimates for $\beta_0, \beta_1, \beta_2, \ldots, \beta_k$, respectively. To minimize the sum of the squares of the residuals, we minimize

$$\mathrm{SSE} = \sum_{i=1}^{n} e_i^2 + \sum_{i=1}^{n} [y_i - (b_0 + b_1 x_{1i} + b_2 x_{2i} + \cdots + b_k x_{ki})]^2$$

By taking the $k + 1$ partial derivatives

$$\frac{\partial \mathrm{SSE}}{\partial b_0}, \frac{\partial \mathrm{SSE}}{\partial b_1}, \frac{\partial \mathrm{SSE}}{\partial b_2}, \ldots, \frac{\partial \mathrm{SSE}}{\partial b_k}$$

and setting them each equal to 0, we obtain these normal equations:

Normal equations, multiple linear regression model

$$\begin{aligned}
b_0 n + b_1 \sum_{i=1}^{n} x_{1i} + b_2 \sum_{i=1}^{n} x_{2i} + \cdots + b_k \sum_{i=1}^{n} x_{ki} &= \sum_{i=1}^{n} y_i \\
b_0 \sum_{i=1}^{n} x_{1i} + b_1 \sum_{i=1}^{n} x_{1i}^2 + b_2 \sum_{i=1}^{n} x_{1i} x_{2i} + \cdots + b_k \sum_{i=1}^{n} x_{1i} x_{ki} &= \sum_{i=1}^{n} x_{1i} y_i \\
&\vdots \\
b_0 \sum_{i=1}^{n} x_{ki} + b_1 \sum_{i=1}^{n} x_{ki} x_{1i} + b_2 \sum_{i=1}^{n} x_{ki} x_{2i} + \cdots + b_k \sum_{i=1}^{n} x_{ki}^2 &= \sum_{i=1}^{n} x_{ki} y_i
\end{aligned}$$

(12.5)

These equations are solved simultaneously for $b_0, b_1, b_2, \ldots, b_k$. To illustrate, let us consider some hypothetical data.

Example 12.1.3. To develop an equation from which we can predict the gasoline mileage of an automobile based on its weight and the temperature at the time of operation, these data are gathered:

Car number	1	2	3	4	5	6	7	8	9	10
Miles per gallon (y)	17.9	16.5	16.4	16.8	18.8	15.5	17.5	16.4	15.9	18.3
Weight in tons (x_1)	1.35	1.90	1.70	1.80	1.30	2.05	1.60	1.80	1.85	1.40
Temperature in °F (x_2)	90	30	80	40	35	45	50	60	65	30

For these data $n = 10$ and

$$\sum_{i=1}^{10} x_{1i} = 16.75 \qquad \sum_{i=1}^{10} x_{1i}x_{2i} = 874.5 \qquad \sum_{i=1}^{10} x_{1i}y_i = 282.405$$

$$\sum_{i=1}^{10} x_{2i} = 525 \qquad \sum_{i=1}^{10} x_{2i}^2 = 31{,}475 \qquad \sum_{i=1}^{10} x_{2i}y_i = 8887.0$$

$$\sum_{i=1}^{10} x_{1i}^2 = 28.6375 \qquad \sum_{i=1}^{10} y_i = 170$$

The normal equations are

$$b_0 n + b_1 \sum_{i=1}^{n} x_{1i} + b_2 \sum_{i=1}^{n} x_{2i} = \sum_{i=1}^{n} y_i$$

$$b_0 \sum_{i=1}^{n} x_{1i} + b_1 \sum_{i=1}^{n} x_{1i}^2 + b_2 \sum_{i=1}^{n} x_{1i}x_{2i} = \sum_{i=1}^{n} x_{1i}y_i$$

$$b_0 \sum_{i=1}^{n} x_{2i} + b_1 \sum_{i=1}^{n} x_{2i}x_{1i} + b_2 \sum_{i=1}^{n} x_{2i}^2 = \sum_{i=1}^{n} x_{2i}y_i$$

Substituting, we obtain these equations:

$$10b_0 + 16.75b_1 + 525b_2 = 170$$

$$16.75b_0 + 28.6375b_1 + 874.5b_2 = 282.405$$

$$525b_0 + 874.5b_1 + 31{,}475b_2 = 8887$$

As you know, solving a system such as this by hand is time-consuming and monotonous. The use of a computer or at least a programmable calculator is becoming more and more appealing! The solution turns out to be $b_0 = 24.75$, $b_1 = -4.16$, and $b_2 = -.014897$.

We hope that you will agree that, in concept, the idea of estimating a polynomial or multiple linear regression model from a data set via least squares is not hard. We simply extend the ideas developed in the simple linear regression context to a more complex model.

Before closing this section, let us note that a third class of models can be developed by combining the polynomial and multiple linear regression models in a natural way. In particular, we can write a model which entails k distinct predictor variables $X_1, X_2, X_3, \ldots, X_k$ with one or more of these variables appearing to a power greater than 1 or with cross-product terms. Examples of such models are

$$\mu_{Y|x_1, x_2} = \beta_0 + \beta_1 x_1 + \beta_{11} x_1^2 + \beta_2 x_2$$

and

$$\mu_{Y|x_1, x_2} = \beta_0 + \beta_1 x_1 + \beta_2 x_2 + \beta_{12} x_1 x_2$$

As you can see, models of this sort can become extremely complicated very quickly. In practice, the experimenter hopes to obtain an adequate model without having to include many nonlinear or cross-product terms because the presence of these terms makes the practical interpretation of the model difficult. Mathematically, these models are no more difficult to handle than any other.

All the models mentioned in this section are special cases of the general linear model. In fact, the normal equations given by (12.5) are the normal equations for the general linear model. The procedures used for parameter estimation, prediction, and hypothesis testing are similar in all cases. In the next section we shall see that each of these models can be expressed in the same general matrix form. This greatly reduces the notational difficulties that exist and simplifies the equations involved in studying the model.

12.2 A MATRIX APPROACH TO LEAST SQUARES

It is evident from our work thus far that finding formulas for the least-squares estimators in a complex model is not easy. To overcome this problem, we turn to matrix algebra. In this section we shall:

1. express the general linear model in matrix form.
2. find a matrix expression for the normal equations for this model.
3. find a matrix expression for the least-squares estimates by solving the normal equations.
4. apply the results obtained to the polynomial and multiple linear regression models.

To begin, recall that the general linear model assumes the form

$$\mu_{Y|x_1, x_2, \ldots, x_k} = \beta_0 + \beta_1 x_1 + \beta_2 x_2 + \cdots + \beta_k x_k$$

This model can also be written in the form

$$Y_i = \beta_0 + \beta_1 x_{1i} + \beta_2 x_{2i} + \cdots + \beta_k x_{ki} + E_i \qquad i = 1, 2, \ldots, n$$

The matrix formulation of the model becomes fairly obvious by writing these equations in expanded form as shown:

$$
\begin{aligned}
Y_1 &= \beta_0 + \beta_1 x_{11} + \beta_2 x_{21} + \cdots + \beta_k x_{k1} + \mathrm{E}_1 \\
Y_2 &= \beta_0 + \beta_1 x_{12} + \beta_2 x_{22} + \cdots + \beta_k x_{k2} + \mathrm{E}_2 \\
Y_3 &= \beta_0 + \beta_1 x_{13} + \beta_2 x_{23} + \cdots + \beta_k x_{k3} + \mathrm{E}_3 \\
&\vdots \\
Y_n &= \beta_0 + \beta_1 x_{1n} + \beta_2 x_{2n} + \cdots + \beta_k x_{kn} + \mathrm{E}_k
\end{aligned}
\tag{12.6}
$$

We need to define three column vectors. These are

$$
\mathbf{Y} = \begin{bmatrix} Y_1 \\ Y_2 \\ \vdots \\ Y_n \end{bmatrix} \qquad \boldsymbol{\beta} = \begin{bmatrix} \beta_0 \\ \beta_1 \\ \beta_2 \\ \vdots \\ \beta_k \end{bmatrix} \qquad \mathbf{E} = \begin{bmatrix} \mathrm{E}_1 \\ \mathrm{E}_2 \\ \vdots \\ \mathrm{E}_n \end{bmatrix}
$$

Note that $\mathbf{Y}$ is the vector of responses, $\boldsymbol{\beta}$ is the vector of model parameters, and $\mathbf{E}$ is the vector of random errors. We also need to define an $n \times (k + 1)$ matrix X. The first member of each row of this matrix is 1. The remaining elements of the ith row for each i consists of the values assumed by the k predictor variables that give rise to the response Y_i. That is, the ith row takes the form

$$
1 \quad x_{1i} \quad x_{2i} \quad x_{3i} \quad \cdots \quad x_{ki}
$$

The entire X matrix is given by

$$
X = \begin{bmatrix}
1 & x_{11} & x_{21} & x_{31} & \cdots & x_{k1} \\
1 & x_{12} & x_{22} & x_{32} & \cdots & x_{k2} \\
1 & x_{13} & x_{23} & x_{33} & \cdots & x_{k3} \\
\vdots & & & & & \\
1 & x_{1n} & x_{2n} & x_{3n} & \cdots & x_{kn}
\end{bmatrix}
$$

We shall refer to this matrix as the *model specification matrix*. The reason for this name is that to change from one model to another, we simply change X. In this sense X determines or specifies the exact form of the model under study.

Note that since X is of dimension $n \times (k + 1)$ and $\boldsymbol{\beta}$ is of dimension $(k + 1) \times 1$, X and $\boldsymbol{\beta}$ are conformable. Their product $X\boldsymbol{\beta}$ is an $n \times 1$ vector. A simple matrix calculation should convince you that the system of equations given by (12.6) can be expressed in matrix form as

$$
\mathbf{Y} = x\boldsymbol{\beta} + \mathbf{E} \tag{12.7}
$$

These ideas are demonstrated in a less abstract setting by reconsidering a problem partially solved earlier (see Example 12.1.3).

Example 12.2.1. An equation is to be developed from which we can predict the gasoline mileage of an automobile based on its weight and the temperature at the time of operation. The model being estimated is

$$\mu_{Y|x_1, x_2} = \beta_0 + \beta_1 x_1 + \beta_2 x_2$$

These data are available:

Car number	1	2	3	4	5	6	7	8	9	10
Miles per gallon (y)	17.9	16.5	16.4	16.8	18.8	15.5	17.5	16.4	15.9	18.3
Weight in tons (x_1)	1.35	1.90	1.70	1.80	1.30	2.05	1.60	1.80	1.85	1.40
Temperature in °F (x_2)	90	30	80	40	35	45	50	60	65	30

The model for these data is

$$17.9 = \beta_0 + 1.35\beta_1 + 90\beta_2 + \varepsilon_1$$

$$16.5 = \beta_0 + 1.90\beta_1 + 30\beta_2 + \varepsilon_2$$

$$16.4 = \beta_0 + 1.70\beta_1 + 80\beta_2 + \varepsilon_3$$

$$\vdots$$

$$18.3 = \beta_0 + 1.40\beta_1 + 30\beta_2 + \varepsilon_{10}$$

In matrix form, these equations are expressed as

$$\mathbf{y} = X\boldsymbol{\beta} + \boldsymbol{\varepsilon}$$

where **y** denotes the vector of observed responses and $\boldsymbol{\varepsilon}$ denotes the vector of realizations on the random error vector **E**. In this case

$$\mathbf{y} = \begin{bmatrix} 17.9 \\ 16.5 \\ 16.4 \\ \vdots \\ 18.3 \end{bmatrix} \qquad \boldsymbol{\beta} = \begin{bmatrix} \beta_0 \\ \beta_1 \\ \beta_2 \end{bmatrix} \qquad \boldsymbol{\varepsilon} = \begin{bmatrix} \varepsilon_1 \\ \varepsilon_2 \\ \varepsilon_3 \\ \vdots \\ \varepsilon_{10} \end{bmatrix}$$

and

$$X = \begin{bmatrix} 1 & 1.35 & 90 \\ 1 & 1.90 & 30 \\ 1 & 1.70 & 80 \\ \vdots & & \\ 1 & 1.40 & 30 \end{bmatrix}$$

The Normal Equations

To find the matrix formulation of the normal equations, consider the matrix $X'X$, where X' denotes the transpose of the model specification matrix:

$$X'X = \begin{bmatrix} 1 & 1 & 1 & \cdots & 1 \\ x_{11} & x_{12} & x_{13} & \cdots & x_{1n} \\ x_{21} & x_{22} & x_{23} & \cdots & x_{2n} \\ \vdots & & & & \\ x_{k1} & x_{k2} & x_{k3} & \cdots & x_{kn} \end{bmatrix} \begin{bmatrix} 1 & x_{11} & x_{21} & \cdots & x_{k1} \\ 1 & x_{12} & x_{22} & \cdots & x_{k2} \\ 1 & x_{13} & x_{23} & \cdots & x_{k3} \\ \vdots & & & & \\ 1 & x_{1n} & x_{2n} & \cdots & x_{kn} \end{bmatrix}$$

$$= \begin{bmatrix} n & \sum_{i=1}^{n} x_{1i} & \sum_{i=1}^{n} x_{2i} & \cdots & \sum_{i=1}^{n} x_{ki} \\ \sum_{i=1}^{n} x_{1i} & \sum_{i=1}^{n} x_{1i}^2 & \sum_{i=1}^{n} x_{1i}x_{2i} & & \sum_{i=1}^{n} x_{1i}x_{ki} \\ \sum_{i=1}^{n} x_{2i} & \sum_{i=1}^{n} x_{1i}x_{2i} & \sum_{i=1}^{n} x_{2i}^2 & & \sum_{i=1}^{n} x_{ki}x_{2i} \\ \vdots & \vdots & \vdots & & \vdots \\ \sum_{i=1}^{n} x_{ki} & \sum_{i=1}^{n} x_{ki}x_{1i} & \sum_{i=1}^{n} x_{ki}x_{2i} & & \sum_{i=1}^{n} x_{ki}^2 \end{bmatrix}$$

Consider also the vector $X'\mathbf{y}$. This vector assumes the form

$$X'\mathbf{y} = \begin{bmatrix} 1 & 1 & 1 & \cdots & 1 \\ x_{11} & x_{12} & x_{13} & \cdots & x_{1n} \\ x_{21} & x_{22} & x_{23} & \cdots & x_{2n} \\ \vdots & & & & \\ x_{k1} & x_{k2} & x_{k3} & \cdots & x_{kn} \end{bmatrix} \begin{bmatrix} y_1 \\ y_2 \\ y_3 \\ \vdots \\ y_n \end{bmatrix}$$

$$= \begin{bmatrix} \sum_{i=1}^{n} y_i \\ \sum_{i=1}^{n} x_{1i}y_i \\ \sum_{i=1}^{n} x_{2i}y_i \\ \vdots \\ \sum_{i=1}^{n} x_{ki}y_i \end{bmatrix}$$

If we let **b** denote the vector of estimated model parameters, then

$$\mathbf{b} = \begin{bmatrix} b_0 \\ b_1 \\ b_2 \\ \vdots \\ b_k \end{bmatrix}$$

A quick matrix calculation should convince you that the normal equations given in (12.5) for the general linear model are given in matrix notation by

$$(X'X)\mathbf{b} = X'\mathbf{y}$$

To illustrate, let us find the normal equations for the data of our last example.

Example 12.2.2. The model specification matrix and vector of responses with which we are working are

$$X = \begin{bmatrix} 1 & 1.35 & 90 \\ 1 & 1.90 & 30 \\ 1 & 1.70 & 80 \\ 1 & 1.80 & 40 \\ 1 & 1.30 & 35 \\ 1 & 2.05 & 45 \\ 1 & 1.60 & 50 \\ 1 & 1.80 & 60 \\ 1 & 1.85 & 65 \\ 1 & 1.40 & 30 \end{bmatrix} \qquad \mathbf{y} = \begin{bmatrix} 17.9 \\ 16.5 \\ 16.4 \\ 16.8 \\ 18.8 \\ 15.5 \\ 17.5 \\ 16.4 \\ 15.9 \\ 18.3 \end{bmatrix}$$

$$X'X = \begin{bmatrix} 1 & 1 & 1 & 1 & 1 & 1 & 1 & 1 & 1 & 1 \\ 1.35 & 1.90 & 1.70 & 1.80 & 1.30 & 2.05 & 1.60 & 1.80 & 1.85 & 1.40 \\ 90 & 30 & 80 & 40 & 35 & 45 & 50 & 60 & 65 & 30 \end{bmatrix}$$

$$\times \begin{bmatrix} 1 & 1.35 & 90 \\ 1 & 1.90 & 30 \\ 1 & 1.70 & 80 \\ 1 & 1.80 & 40 \\ 1 & 1.30 & 35 \\ 1 & 2.05 & 45 \\ 1 & 1.60 & 50 \\ 1 & 1.80 & 60 \\ 1 & 1.85 & 65 \\ 1 & 1.40 & 30 \end{bmatrix}$$

$$= \begin{bmatrix} 10 & 16.75 & 525 \\ 16.75 & 28.6375 & 874.5 \\ 525 & 874.5 & 31{,}475 \end{bmatrix}$$

$$(X'X)\mathbf{b} = \begin{bmatrix} 10 & 16.75 & 525 \\ 16.75 & 28.6375 & 874.5 \\ 525 & 874.5 & 31{,}475 \end{bmatrix} \begin{bmatrix} b_0 \\ b_1 \\ b_2 \end{bmatrix}$$

$$= \begin{bmatrix} 10b_0 + 16.75b_1 + 525b_2 \\ 16.75b_0 + 28.6375b_1 + 874.5b_2 \\ 525b_0 + 874.5b_1 + 31{,}475b_2 \end{bmatrix}$$

Note that the entries in this vector constitute the left-side of the normal equations found earlier (see Example 12.1.3). The right-hand side of the system is given by $X'\mathbf{y}$. In this case,

$$X'\mathbf{y} = \begin{bmatrix} 1 & 1 & 1 & 1 & 1 & 1 & 1 & 1 & 1 & 1 \\ 1.35 & 1.90 & 1.70 & 1.80 & 1.30 & 2.05 & 1.60 & 1.80 & 1.85 & 1.40 \\ 90 & 30 & 80 & 40 & 35 & 45 & 50 & 60 & 65 & 30 \end{bmatrix} \times \begin{bmatrix} 17.9 \\ 16.5 \\ 16.4 \\ 16.8 \\ 18.8 \\ 15.5 \\ 17.5 \\ 16.4 \\ 15.9 \\ 18.3 \end{bmatrix} = \begin{bmatrix} 170 \\ 282.405 \\ 8887 \end{bmatrix}$$

Note that these values coincide with those of Example 12.1.3.

You should agree that even though the matrix approach to finding the normal equations entails some work, it is easier to remember that the normal equations are given by $(X'X)\mathbf{b} = X'\mathbf{y}$ than it is to remember the system of equations given in (12.5) in the last section!

Solving the Normal Equations

To find the matrix formulation for the least-squares estimates for $\beta_0, \beta_1, \beta_2, \ldots, \beta_k$, we solve the system

$$(X'X)\mathbf{b} = X'\mathbf{y}$$

We know that if the columns of X are linearly independent, that is, no column can be expressed as a linear combination of the others, then $X'X$ has an inverse. We denote this inverse by $(X'X)^{-1}$. To solve the normal equations for $\mathbf{b}$, we multiply both sides of the equation

$$(X'X)\mathbf{b} = X'\mathbf{y}$$

by $(X'X)^{-1}$ to obtain

$$\mathbf{b} = (X'X)^{-1}X'\mathbf{y}$$

Theoretically, to find the least-squares estimates for the model parameters, we simply compute

$$\hat{\boldsymbol{\beta}} = \mathbf{b} = (X'X)^{-1}X'\mathbf{y}$$

Again, this is easier said than done! It is no easy task to find the inverse of a matrix by hand except in the simplest cases. For this reason, in practice, we shall let the computer do the work for us. However, you should be aware of the fact that the computations are being performed via the matrix operations just described.

To illustrate, we find the least-squares estimates for β_0, β_1, and β_2 based on the mileage data given in Example 12.2.1.

Example 12.2.3. The matrix $X'X$ with which we are working is

$$X'X = \begin{bmatrix} 10 & 16.75 & 525 \\ 16.75 & 28.6375 & 874.5 \\ 525 & 874.5 & 31{,}475 \end{bmatrix}$$

We shall let the computer find $(X'X)^{-1}$ for us! (To see how this is done via SAS, refer to Computing Supplement XII at the end of this chapter.)

You can verify that the inverse of this matrix is, apart from round-off error,

$$(X'X)^{-1} = \begin{bmatrix} 6.070769 & -3.02588 & -.0171888 \\ -3.02588 & 1.738599 & .002166306 \\ -.0171888 & .002166306 & .0002582903 \end{bmatrix}$$

The vector of parameter estimates, apart from round-off error, is

$$\begin{aligned} \mathbf{b} &= (X'X)^{-1}X'\mathbf{y} \\ &= \begin{bmatrix} 6.070769 & -3.02588 & -.0171888 \\ -3.02588 & 1.738599 & .002166306 \\ -.0171888 & .002166306 & .0002582903 \end{bmatrix} \begin{bmatrix} 170 \\ 282.405 \\ 8887 \end{bmatrix} \\ &= \begin{bmatrix} 24.75 \\ -4.16 \\ -.014897 \end{bmatrix} \end{aligned}$$

The estimated model is

$$\hat{\mu}_{Y|x_1, x_2} = 24.75 - 4.16x_1 - .014897x_2$$

Based on this equation, we estimate the mileage for a car weighing 1.5 tons on a 70°F day to be $\hat{y} = 24.75 - 4.16(1.5) - .014897(70) = 17.47$ miles per gallon.

Recall that we developed a model earlier by which gas mileage could be predicted based *only* on the weight of the car. (See Example 11.3.3.) In our earlier model the estimates for the intercept and coefficient of the weight variable were 23.75 and -4.03, respectively. We should note here that the estimates for these parameters in our current model differ from those obtained previously. This usually happens when a new independent variable is introduced into an older model. We shall determine later whether or not the addition of the temperature variable improves our model.

Simple Linear Regression: Matrix Formulation

The simple linear regression model is a special case of the multiple regression model in which there is only one regressor. This regressor appears to the first power. The matrix techniques just developed can be applied to this model. In the next example we illustrate the matrix approach to simple linear regression by resolving the problem presented in Example 11.1.1.

Example 12.2.4. Consider Example 11.1.1 in which simple linear regression was used to examine the relationship between humidity, the regressor, and the extent of solvent evaporation in paint. These data are obtained:

(x) Relative humidity (%)	(y) Solvent evaporation (%wt)
35.3	11.0
29.7	11.1
30.8	12.5
58.8	8.4
61.4	9.3
71.3	8.7
74.4	6.4
76.7	8.5
70.7	7.8
57.5	9.1
46.4	8.2
28.9	12.2
28.1	11.9
39.1	9.6
46.8	10.9
48.5	9.6
59.3	10.1
70.0	8.1
70.0	6.8
74.4	8.9
72.1	7.7
58.1	8.5
44.6	8.9
33.4	10.4
28.6	11.1

The model for these data is

$$11.0 = \beta_0 + 35.3\beta_1 + \varepsilon_1$$
$$11.1 = \beta_0 + 29.7\beta_1 + \varepsilon_2$$
$$12.5 = \beta_0 + 30.8\beta_1 + \varepsilon_3$$
$$\vdots$$
$$11.1 = \beta_0 + 28.6\beta_1 + \varepsilon_{25}$$

In matrix form we can write this system of equations as

$$\mathbf{y} = X\boldsymbol{\beta} + \varepsilon$$

where $$\mathbf{y} = \begin{bmatrix} 11.0 \\ 11.1 \\ 12.5 \\ \vdots \\ 11.1 \end{bmatrix} \qquad \boldsymbol{\beta} = \begin{bmatrix} \beta_0 \\ \beta_1 \end{bmatrix} \qquad \boldsymbol{\varepsilon} = \begin{bmatrix} \varepsilon_1 \\ \varepsilon_2 \\ \varepsilon_3 \\ \vdots \\ \varepsilon_{25} \end{bmatrix}$$

The model specification matrix contains two columns. The first is a column of 1's and the second is a column that contains the numerical values of the regressor. Here

$$X = \begin{bmatrix} 1 & 35.3 \\ 1 & 29.7 \\ 1 & 30.8 \\ \vdots & \\ 1 & 28.6 \end{bmatrix}$$

You should verify for yourself that the matrix expression

$$\mathbf{y} = X\boldsymbol{\beta} + \boldsymbol{\varepsilon}$$

yields the system of algebraic equations presented earlier. Since X is of dimension 25×2, X' is a 2×25 matrix and $X'X$ is a 2×2 matrix. The general form of $X'X$ is

$$X'X = \begin{bmatrix} n & \Sigma\, x \\ \Sigma\, x & \Sigma\, x^2 \end{bmatrix}$$

For these data

$$X'X = \begin{bmatrix} 25 & 1314.9 \\ 1314.9 & 76308.53 \end{bmatrix}$$

The vector $X'\mathbf{y}$ is 2×1 and has the general form

$$X'\mathbf{y} = \begin{bmatrix} \Sigma\, y \\ \Sigma\, xy \end{bmatrix}$$

In this case

$$X'\mathbf{y} = \begin{bmatrix} 235.70 \\ 11824.44 \end{bmatrix}$$

The normal equations are

$$(X'X)\mathbf{b} = X'\mathbf{y}$$

where $\mathbf{b} = \begin{bmatrix} b_0 \\ b_1 \end{bmatrix}$. In this case the normal equations are

$$25b_0 + 1314.9b_1 = 235.70$$

$$1314.9b_0 + 76308.53b_1 = 11824.44$$

In Example 11.1.1 we found algebraically that the solution to this system of equations is

$$b_0 = 13.64 \qquad b_1 = -.08$$

To find the solution using matrices, we must invert the matrix $X'X$. Since this matrix is 2×2, the inverse can be found without the use of the computer. In Exercise 6 you are asked to show that for the simple linear regression model,

$$(X'X)^{-1} = \frac{1}{nS_{xx}} = \begin{bmatrix} \Sigma x^2 & -\Sigma x \\ -\Sigma x & n \end{bmatrix}$$

For these data $S_{xx} = \dfrac{n\Sigma x^2 - (\Sigma x)^2}{n} = 7150.05$ and $nS_{xx} = 178751.24$. Notice that nS_{xx} is always equal to the determinant of $X'X$. Here,

$$(X'X)^{-1} = \frac{1}{178751.24}\begin{bmatrix} 76308.53 & -1314.9 \\ -1314.9 & 25 \end{bmatrix}$$

$$= \begin{bmatrix} .42689 & -.00735 \\ -.00735 & .00014 \end{bmatrix}$$

You can verify for yourself that, apart from some round-off error, $(X'X)(X'X)^{-1}$ yields the identity matrix. The solution to the normal equations and the estimates for b_0 and b_1 are

$$\mathbf{b} = (X'X)^{-1}X'\mathbf{y}$$

or

$$\mathbf{b} = \begin{bmatrix} .42689 & -.00735 \\ -.00735 & .00014 \end{bmatrix}\begin{bmatrix} 235.70 \\ 11824.44 \end{bmatrix}$$

$$= \begin{bmatrix} 13.71 \\ -.077 \end{bmatrix}$$

Using the matrix approach, we have $b_0 = 13.71$ and $b_1 = -.077$. These values differ a little from those obtained earlier. The difference is due to some round-off error that occurs in forming $(X'X)^{-1}$. In doing simple linear regression on a calculator that does not have a built-in regression capability, we shall probably find that the algebraic formulas from Chap. 11 are easier than the matrix approach. However, when regression is done via the computer, the algorithm used to find b_0 and b_1 does entail setting up the matrices described here and computing the estimates for β_0 and β_1 by using matrix algebra.

Polynomial Model: Matrix Formulation

Since the polynomial model is a special case of the general linear model, to analyze such a model all we must do is to find the appropriate model specification matrix. The equations defining the model are

$$Y_1 = \beta_0 + \beta_1 x_1 + \beta_2 x_1^2 + \beta_3 x_1^3 + \cdots + \beta_p x_1^p + \mathrm{E}_1$$
$$Y_2 = \beta_0 + \beta_1 x_2 + \beta_2 x_2^2 + \beta_3 x_2^3 + \cdots + \beta_p x_2^p + \mathrm{E}_2$$
$$Y_3 = \beta_0 + \beta_1 x_3 + \beta_2 x_3^2 + \beta_3 x_3^3 + \cdots + \beta_p x_3^p + \mathrm{E}_3$$
$$\vdots$$
$$Y_n = \beta_0 + \beta_1 x_n + \beta_2 x_n^2 + \beta_3 x_n^3 + \cdots + \beta_p x_n^p + \mathrm{E}_n$$

From these equations it is easy to see that

$$X = \begin{bmatrix} 1 & x_1 & x_1^2 & \cdots & x_1^p \\ 1 & x_2 & x_2^2 & & x_2^p \\ 1 & x_3 & x_3^2 & & x_3^p \\ \vdots & \vdots & \vdots & & \vdots \\ 1 & x_n & x_n^2 & & x_n^p \end{bmatrix}$$

From this point on the analysis is identical to that of the general linear model.

Example 12.2.5. These data are available on X, the number of units of drug produced, and Y, the cost per unit of producing the drug. (See Example 12.1.1.)

x	5	5	10	10	15	15	20	20	25	25
y	14.0	12.5	7.0	5.0	2.1	1.8	6.2	4.9	13.2	14.6

The model specification matrix for a quadratic model is

$$X = \begin{bmatrix} 1 & 5 & 25 \\ 1 & 5 & 25 \\ 1 & 10 & 100 \\ 1 & 10 & 100 \\ 1 & 15 & 225 \\ 1 & 15 & 225 \\ 1 & 20 & 400 \\ 1 & 20 & 400 \\ 1 & 25 & 625 \\ 1 & 25 & 625 \end{bmatrix}$$

$$X'X = \begin{bmatrix} 10 & 150 & 2{,}750 \\ 150 & 2{,}750 & 56{,}250 \\ 2{,}750 & 56{,}250 & 1{,}223{,}750 \end{bmatrix}$$

$$X'\mathbf{y} = \begin{bmatrix} 81.3 \\ 1{,}228 \\ 24{,}555 \end{bmatrix}$$

Apart from round-off error

$$(X'X)^{-1} = \begin{bmatrix} 2.3 & -.33 & .01 \\ -.33 & .05342857 & -.00171429 \\ .01 & -.00171429 & .00005714286 \end{bmatrix}$$

(See Computing Supplement XIII.)

The least-squares estimates for β_0, β_1, and β_2 are

$$\mathbf{b} = (X'X)^{-1}X'\mathbf{y} = \begin{bmatrix} 27.3 \\ -3.313 \\ .111 \end{bmatrix}$$

The estimated model is

$$\hat{\mu}_{Y|x} = 27.3 - 3.313x + .111x^2$$

The predicted unit cost of producing 12 units of the drug is

$$\hat{y} = 27.3 - 3.313(12) + .111(12)^2 = 3.528$$

12.3 PROPERTIES OF THE LEAST-SQUARES ESTIMATORS

We now have a way to generate point estimates for the model parameters $\beta_0, \beta_1, \beta_2, \ldots, \beta_k$ in the general linear model. These estimates are denoted by $b_0, b_1, b_2, \ldots, b_k$, respectively, and are found via the matrix equation

$$\mathbf{b} = \begin{bmatrix} b_0 \\ b_1 \\ b_2 \\ \vdots \\ b_k \end{bmatrix} = (X'X)^{-1}X'\mathbf{y}$$

where $\mathbf{y}$ denotes the vector of observed values of the response vector $\mathbf{Y}$. The estimators for $\beta_0, \beta_1, \beta_2, \ldots, \beta_k$ are denoted by $B_0, B_1, B_2, \ldots, B_k$, respectively. The vector of parameter estimators is denoted by $\hat{\boldsymbol{\beta}}$. This vector is defined by

$$\hat{\boldsymbol{\beta}} = \begin{bmatrix} B_0 \\ B_1 \\ B_2 \\ \vdots \\ B_k \end{bmatrix} = (X'X)^{-1}X'\mathbf{Y}$$

As usual, we need to investigate the properties of these estimators. Before we begin, let us consider what is meant by the expected value of a vector of random variables.

Definition 12.3.1. Let $\mathbf{Y} = \begin{bmatrix} Y_1 \\ Y_2 \\ \vdots \\ Y_n \end{bmatrix}$ denote a vector of random variables. The expected value of this vector is denoted by $E[\mathbf{Y}]$ and is defined by

$$\mathrm{E}[\mathbf{Y}] = \begin{bmatrix} E[Y_1] \\ E[Y_2] \\ \vdots \\ E[Y_n] \end{bmatrix}$$

A simple example should convince you that the rules for expectation that were used in the case of a single random variable also hold when dealing with vectors and matrices.

Example 12.3.1. Let $\mathbf{Y} = \begin{bmatrix} Y_1 \\ Y_2 \end{bmatrix}$ be a random vector and let C be the 2×2 matrix.

$$C = \begin{bmatrix} 2 & 3 \\ 6 & 7 \end{bmatrix}$$

Since the entries in the matrix C are constants, the former rules for expectation suggest that

$$E[C\mathbf{Y}] = CE[\mathbf{Y}]$$

Let us show that this is true.

$$E[C\mathbf{Y}] = E\left[\begin{bmatrix} 2 & 3 \\ 6 & 7 \end{bmatrix}\begin{bmatrix} Y_1 \\ Y_2 \end{bmatrix}\right]$$

$$= E\begin{bmatrix} 2Y_1 + 3Y_2 \\ 6Y_1 + 7Y_2 \end{bmatrix}$$

By definition

$$E\begin{bmatrix} 2Y_1 + 3Y_2 \\ 6Y_1 + 7Y_2 \end{bmatrix} = \begin{bmatrix} E[2Y_1 + 3Y_2] \\ E[6Y_1 + 7Y_2] \end{bmatrix}$$

$$= \begin{bmatrix} 2E[Y_1] + 3E[Y_2] \\ 6E[Y_1] + 7E[Y_2] \end{bmatrix}$$

$$= \begin{bmatrix} 2 & 3 \\ 6 & 7 \end{bmatrix}\begin{bmatrix} E[Y_1] \\ E[Y_2] \end{bmatrix}$$

$$= CE[\mathbf{Y}]$$

For easy reference we list the matrix "rules for expectation." These rules are easy to verify and their proofs are left to the reader.

Matrix rules for expectation

Let $\mathbf{Y}$ and $\mathbf{Z}$ denote $n \times 1$ random vectors and let C denote an $m \times n$ matrix of constants. Then

1. $E[C] = C$
2. $E[C\mathbf{Y}] = CE[\mathbf{Y}]$
3. $E[\mathbf{Y} + \mathbf{Z}] = E[\mathbf{Y}] + E[\mathbf{Z}]$

Expected Value of $\hat{\beta}$

Recall that in the case of simple linear regression we were able to show that B_0 and B_1 are unbiased estimators for β_0 and β_1, respectively. The arguments given for this in Sec. 11.2 were fairly complex and algebraic in nature. Can we show that, in general, $E[\hat{\boldsymbol{\beta}}] = \boldsymbol{\beta}$? That is, can we show that $B_0, B_1, \ldots, B_k$ are unbiased estimators for $\beta_0, \beta_1, \ldots, \beta_k$ in the general linear model? As you should suspect, the answer is yes. The matrix rules for expectation just developed allow us to verify this quite easily.

To begin, recall that our general linear model in matrix notation is given by

$$\mathbf{Y} = X\boldsymbol{\beta} + \mathbf{E}$$

Recall also that we are assuming that the random errors $\mathrm{E}_1, \mathrm{E}_2, \mathrm{E}_3, \ldots, \mathrm{E}_n$ are independent random variables, each with mean 0 and variance σ^2. Thus $E[\mathbf{E}] = \mathbf{0}$ where $\mathbf{0}$ denotes the zero vector. Using the rules for expectation, we obtain

$$\begin{aligned} E[\mathbf{Y}] &= E[X\boldsymbol{\beta} + \mathbf{E}] \\ &= E[X\boldsymbol{\beta}] + E[\mathbf{E}] \\ &= X\boldsymbol{\beta} + \mathbf{0} = X\boldsymbol{\beta} \end{aligned}$$

The vector of least-squares estimators is given by

$$\hat{\boldsymbol{\beta}} = (X'X)^{-1}X'\mathbf{Y}$$

Once again, we use the rules for expectation to conclude that

$$\begin{aligned} E\left[\hat{\boldsymbol{\beta}}\right] &= E\left[(X'X)^{-1}X'\mathbf{Y}\right] \\ &= (X'X)^{-1}X'E[\mathbf{Y}] \\ &= (X'X)^{-1}X'X\boldsymbol{\beta} \\ &= \boldsymbol{\beta} \end{aligned}$$

This completes the argument that $\hat{\boldsymbol{\beta}}$ is an unbiased estimator for $\boldsymbol{\beta}$.

Estimation of σ^2 and Variance of $\hat{\beta}$

To determine the variances of the estimators $B_0, B_1, B_2, \ldots, B_k$, we need to define what we mean by the variance of a random vector.

Definition 12.3.2. Let $\mathbf{Y} = \begin{bmatrix} Y_1 \\ Y_2 \\ \vdots \\ Y_n \end{bmatrix}$ denote a vector of random variables. By $\text{Var}\,\mathbf{Y}$, we mean the matrix

$$\begin{bmatrix} \text{Var}\,Y_1 & \text{Cov}(Y_1, Y_2) & & \cdots & \text{Cov}(Y_1, Y_n) \\ \text{Cov}(Y_1, Y_2) & \text{Var}\,Y_2 & & & \text{Cov}(Y_2, Y_n) \\ \text{Cov}(Y_1, Y_3) & \text{Cov}(Y_2, Y_3) & \text{Var}\,Y_3 & & \text{Cov}(Y_3, Y_n) \\ \vdots & \vdots & & \ddots & \\ \text{Cov}(Y_1, Y_n) & \text{Cov}(Y_2, Y_n) & & & \text{Var}\,Y_n \end{bmatrix}$$

This matrix is called the *variance-covariance* matrix for the vector **Y**.

The name variance-covariance matrix is fitting since the elements along the main diagonal are the variances of the random variables $Y_1, Y_2, Y_3, \ldots, Y_n$; those off the main diagonal are the covariances between variable pairs (Y_i, Y_j), where $i \neq j$.

Although there are several matrix "rules for variance," we need only one. This rule is

$$\text{Var}\,C\mathbf{Y} = C\,\text{Var}\,\mathbf{Y}C'$$

where C is an $m \times n$ matrix of constants and **Y** is an $n \times 1$ vector of random variables. Let us illustrate this rule in a simple context.

Example 12.3.2. Let $\mathbf{Y} = \begin{bmatrix} Y_1 \\ Y_2 \end{bmatrix}$ be a random vector and let C be the 2×2 matrix.

$$C = \begin{bmatrix} 2 & 3 \\ 6 & 7 \end{bmatrix}$$

By definition $\text{Var}\,\mathbf{Y} = \begin{bmatrix} \text{Var}\,Y_1 & \text{Cov}(Y_1, Y_2) \\ \text{Cov}(Y_1, Y_2) & \text{Var}\,Y_2 \end{bmatrix}$

Applying the variance rule, we obtain

$$\text{Var}\,C\mathbf{Y} = C\,\text{Var}\,\mathbf{Y}C'$$

or

$$\text{Var}\,C\mathbf{Y} = \begin{bmatrix} 2 & 3 \\ 6 & 7 \end{bmatrix} \begin{bmatrix} \text{Var}\,Y_1 & \text{Cov}(Y_1, Y_2) \\ \text{Cov}(Y_1, Y_2) & \text{Var}\,Y_2 \end{bmatrix} \begin{bmatrix} 2 & 6 \\ 3 & 7 \end{bmatrix}$$

$$= \begin{bmatrix} 4\,\text{Var}\,Y_1 + 12\,\text{Cov}(Y_1, Y_2) + 9\,\text{Var}\,Y_2, & 12\,\text{Var}\,Y_1 + 32\,\text{Cov}(Y_1, Y_2) + 21\,\text{Var}\,Y_2 \\ 12\,\text{Var}\,Y_1 + 32\,\text{Cov}(Y_1, Y_2) + 21\,\text{Var}\,Y_2, & 36\,\text{Var}\,Y_1 + 84\,\text{Cov}(Y_1, Y_2) + 49\,\text{Var}\,Y_2 \end{bmatrix}$$

Note that this rule parallels our rule for a single random variable that requires that constants be squared when factoring.

We can now find a quick way to determine the variances of the least-squares estimators in the general linear model. We first note that in the context of the linear model we assume that $Y_1, Y_2, Y_3, \ldots, Y_n$ are independent with common variance σ^2. Since independence implies zero covariance, the variance-covariance matrix for the random vector $\mathbf{Y}$ is given by

$$\operatorname{Var}\mathbf{Y} = \begin{bmatrix} \sigma^2 & 0 & 0 & 0 & 0 \\ 0 & \sigma^2 & 0 & 0 & 0 \\ 0 & 0 & \sigma^2 & 0 & 0 \\ \vdots & \vdots & \vdots & \vdots & \vdots \\ 0 & 0 & 0 & \cdots & \sigma^2 \end{bmatrix}$$

This matrix can be rewritten as $\sigma^2 I$, where I is the $n \times n$ identity matrix, a matrix of 1's on the main diagonal with all other entries being 0. To find the variances of the least-squares estimators, recall that

$$\hat{\boldsymbol{\beta}} = (X'X)^{-1}X'\mathbf{Y}$$

Since the model specification matrix is of dimension $n \times (k + 1)$, $X'X$ and $(X'X)^{-1}$ are of dimension $(k + 1) \times (k + 1)$. The matrix $(X'X)^{-1}X'$ is a matrix of constants of dimension $(k + 1) \times n$. Using our matrix rule for variance with $C = (X'X)^{-1}X'$, we have

$$\begin{aligned} \operatorname{Var}\hat{\boldsymbol{\beta}} &= \operatorname{Var}\left[(X'X)^{-1}X'\mathbf{Y}\right] \\ &= (X'X)^{-1}X' \operatorname{Var}\mathbf{Y}\left[(X'X)^{-1}X'\right]' \end{aligned}$$

Rules for matrix algebra state that $(AB)' = B'A'$ and that $(A^{-1})' = (A')^{-1}$. Applying these rules here, we see that

$$\begin{aligned} \left[(X'X)^{-1}X'\right]' &= X\left[(X'X)^{-1}\right]' \\ &= X[(X'X)']^{-1} \\ &= X(X'X)^{-1} \end{aligned}$$

Substitution yields

$$\begin{aligned} \operatorname{Var}\hat{\boldsymbol{\beta}} &= (X'X)^{-1}X' \operatorname{Var}\mathbf{Y} X(X'X)^{-1} \\ &= (X'X)^{-1}X'\sigma^2 I X(X'X)^{-1} \\ &= \sigma^2(X'X)^{-1}(X'X)(X'X)^{-1} \\ &= \sigma^2(X'X)^{-1} \end{aligned}$$

Since σ^2 is unknown, we replace it by an appropriate estimator. As in the case of simple linear regression, to estimate σ^2 we use information concerning the variability of the data points about the fitted regression equation. That is, our estimator makes use of SSE, the sum of squares of the residuals. To obtain an unbiased estimator for σ^2, we divide SSE by $n - k - 1$. Thus our estimator is

$$\hat{\sigma}^2 = S^2 = \text{SSE}/(n - k - 1)$$

Note that in the case of simple linear regression, $k = 1$ and $\hat{\sigma}^2 = \text{SSE}/(n - 2)$. This coincides with the results obtained in Chap. 11.

To compute SSE, we again parallel the technique used in the simple linear regression context. In particular, we write SSE as the difference between two components whose sources are recognizable. Although the algebra is a bit messy, it can be shown that

$$\text{SSE} = \sum_{n=1}^{n} [Y_i - (B_0 + B_1 x_{1i} + B_2 x_{2i} + \cdots + B_k x_{ki})]^2$$

$$= \sum_{i=1}^{n} Y_i^2 - B_0 \sum_{i=1}^{n} Y_i - B_1 \sum_{i=1}^{n} x_{1i} Y_i - B_2 \sum_{i=1}^{n} x_{2i} Y_i - \cdots - B_k \sum_{i=1}^{n} x_{ki} Y_i$$

By adding and subtracting the term $(\sum_{i=1}^{n} Y_i)^2/n$, which is often called the *correction* factor, we obtain the expression

$$\text{SSE} = \left[\sum_{i=1}^{n} Y_i^2 - \left(\sum_{i=1}^{n} Y_i\right)^2 \Big/ n\right]$$

$$- \left[B_0 \sum_{i=1}^{n} Y_i + B_1 \sum_{i=1}^{n} x_{1i} Y_i + B_2 \sum_{i=1}^{n} x_{2i} Y_i + \cdots + B_k \sum_{i=1}^{n} x_{ki} Y_i - \left(\sum_{i=1}^{n} Y_i\right)^2 \Big/ n\right]$$

You should recognize the first component on the right as S_{yy}. We shall now refer to this term as the "corrected" total sum of squares. It measures the total variability in the data. The second component on the right is called the *regression* sum of squares. It is denoted by SSR and measures the variability in Y attributed to the linear association between the mean of Y and the predictor variables. Since

$$\text{SSE} = S_{yy} - \text{SSR}$$

it is easy to see that SSE is a measure of the random or unexplained variability in the response variable. That is, it helps to estimate σ^2. In Exercise 23 we outline a small example that will help you to understand the algebraic argument behind this derivation. To illustrate, we continue the gasoline mileage study begun earlier.

Example 12.3.3. In Example 12.2.3 we found that the estimated regression equation for predicting the gasoline mileage of a car based on its weight and the temperature at the time of operation is

$$\hat{\mu}_{Y|x_1,x_2} = 24.75 - 4.16x_1 - .014897x_2$$

Since

$$X'\mathbf{y} = \begin{bmatrix} 170 \\ 282.405 \\ 8887 \end{bmatrix} = \begin{bmatrix} \sum_{i=1}^{n} y_i \\ \sum_{i=1}^{n} x_{1i}y_i \\ \sum_{i=1}^{n} x_{2i}y_i \end{bmatrix}$$

we already have available most of the information needed to compute SSE. The only other term needed is $\sum_{i=1}^{n} y_i^2$. A quick computation yields a value of 2900.46 for this term. Substituting, we have

$$S_{yy} = \left[10\sum_{i=1}^{10} y_i^2 - \left(\sum_{i=1}^{10} y_i\right)^2\right]\Big/10 = \left[29004.6 - (170)^2\right]/10 = 10.46$$

$$\begin{aligned} \text{SSR} &= b_0\sum_{i=1}^{10} y_i + b_1\sum_{i=1}^{10} x_{1i}y_i + b_2\sum_{i=1}^{10} x_{2i}y_i - \left(\sum_{i=1}^{10} y_i\right)^2\Big/10 \\ &= 24.75(170) - 4.16(282.405) - .014897(8887) - (170)^2/10 \\ &= 10.31 \end{aligned}$$

By subtraction

$$\text{SSE} = S_{yy} - \text{SSR} = 10.46 - 10.31 = .15$$

Hence

$$\begin{aligned} \hat{\sigma}^2 = s^2 &= \text{SSE}/(n-k-1) \\ &= .15/(10-2-1) \\ &= .0214 \end{aligned}$$

In Example 12.2.3 we found that the matrix $(X'X)^{-1}$ for these data is

$$(X'X)^{-1} = \begin{bmatrix} 6.070769 & -3.02588 & -.0171888 \\ -3.02588 & 1.738599 & .002166306 \\ -.0171888 & .002166306 & .0002582903 \end{bmatrix}$$

Since $\text{Var}\,\hat{\boldsymbol{\beta}} = \sigma^2(X'X)^{-1}$, to find the estimates for the variances of B_0, B_1, and B_2, we multiply each number on the main diagonal of $(X'X)^{-1}$ by $\hat{\sigma}^2$. Thus

$$\widehat{\text{Var}\,B_0} = 6.070769(.0214) \doteq .1299$$

$$\widehat{\text{Var}\,B_1} = 1.738599(.0214) \doteq .0372$$

$$\widehat{\text{Var}\,B_2} = .0002582903(.0214) \doteq .000005$$

In practice, we let the computer do much of this work for us. However, to interpret the computer printout correctly, it is important to understand what is being done.

12.4 INTERVAL ESTIMATION

As in the past, it is helpful to be able to extend a point estimate for a parameter to an interval estimate so that its accuracy can be assessed. We consider three types of intervals here. These are

1. Confidence intervals on the parameters $\beta_0, \beta_1, \beta_2, \ldots, \beta_k$ of the general linear model
2. Confidence interval on $\mu_{Y|x_1, x_2, \ldots, x_k}$, the mean response for a given set of values of the predictor variables
3. Prediction interval on $Y|x_1, x_2, \ldots, x_k$, an individual response for a given set of values of the predictor variables

Confidence Interval on Coefficients

Recall that one way to express the general linear model is

$$Y_i = \beta_0 + \beta_1 x_{1i} + \beta_2 x_{2i} + \cdots + \beta_k x_{ki} + \mathrm{E}_i$$

where $\mathrm{E}_1, \mathrm{E}_2, \ldots, \mathrm{E}_n$ are assumed to be independent random variables, each with mean 0 and variance σ^2. We now make the additional assumption that these random variables are normally distributed. This in turn implies that we are assuming that the random variables $Y_1, Y_2, \ldots, Y_n$ are independent and normally distributed. In matrix form we know that the estimators $B_0, B_1, B_2, \ldots, B_k$ are given by

$$\hat{\boldsymbol{\beta}} = (X'X)^{-1}X'\mathbf{Y}$$

Since $(X'X)^{-1}X'$ is a matrix of constants, each component of the vector $(X'X)^{-1}X'\mathbf{Y}$ is a linear combination of the random variables $Y_1, Y_2, \ldots, Y_n$. Since any linear combination of independent normal random variables is also normal, it is easy to see that each of the estimators $B_0, B_1, \ldots, B_k$ is a normal random variable. We know that these estimators are unbiased for their respective parameters. The variance-covariance matrix for $\hat{\boldsymbol{\beta}}$ is $(X'X)^{-1}\sigma^2$. The variances of $B_0, B_1, B_2, \ldots, B_k$ are given by $c_{00}\sigma^2, c_{11}\sigma^2, \ldots, c_{kk}\sigma^2$, respectively, where c_{ii} denotes the element on the main diagonal in row $i+1$ of the matrix $(X'X)^{-1}$. The random variable

$$Z = \frac{\hat{\beta}_i - \beta_i}{\sigma\sqrt{c_{ii}}}$$

is standard normal. Since σ is unknown, we replace it with its estimator $S = \sqrt{\text{SSE}/(n-k-1)}$ to form the T_{n-k-1} random variable

$$T_{n-k-1} = \frac{\hat{\beta}_i - \beta_i}{S\sqrt{c_{ii}}}$$

This random variable has the same algebraic structure as many others encountered previously. Thus we know that the confidence bounds for β_i are given by

Confidence bounds for β_i, the ith model parameter in the general linear model

$\hat{\beta}_i \pm t_{\alpha/2}S\sqrt{c_{ii}}$, where the point $t_{\alpha/2}$ is the appropriate point based on the T_{n-k-1} distribution.

An example will demonstrate the use of these bounds.

Example 12.4.1. To predict the gasoline mileage of a car based on its weight and the temperature at the time of operation, we have developed the regression equation (see Example 12.2.3)

$$\hat{\mu}_{Y|x_1,x_2} = 24.75 - 4.16x_1 - .014897x_2$$

In Example 12.3.3 we found that $s^2 = .0214$ and hence that $s = \sqrt{s^2} = .1463$. The variance-covariance matrix is

$$(X'X)^{-1}\sigma^2 = \begin{bmatrix} 6.070769 & -3.02588 & -.0171888 \\ -3.02588 & 1.738599 & .002166306 \\ -.0171888 & .002166306 & .0002582903 \end{bmatrix}\sigma^2$$

A 95% confidence interval on β_0, the intercept for this model, is

$$\hat{\beta}_0 \pm t_{\alpha/2}s\sqrt{c_{00}}$$

Since $n = 10$ and $k = 2$, the number of degrees of freedom associated with the point $t_{.025}$ is $10 - 2 - 1 = 7$. The confidence interval is given by

$$24.75 \pm 2.365(.1463)\sqrt{6.070769}$$

or $24.75 \pm .853$. Since 0 does not lie in this interval, we have good evidence that $\beta_0 \neq 0$.

Confidence Interval on Estimated Mean

Although confidence intervals of the type just described can be formed, a more useful type of interval is that on the mean value of the response variable for a specific set of values of the predictor variables. We denote the values of interest by $x_{10}, x_{20}, \ldots, x_{k0}$ and note that these values are not necessarily those used to develop the regression equation. We know that an unbiased estimator for this mean is

$$\hat{\mu}_{Y|x_{10}, x_{20}, \ldots, x_{k0}} = \hat{\beta}_0 + \hat{\beta}_1 x_{10} + \hat{\beta}_2 x_{20} + \cdots + \hat{\beta}_k x_{k0}$$

To find the variance of this estimator, we write it in matrix form as

$$\hat{\mu}_{Y|x_{10}, x_{20}, \ldots, x_{k0}} = \mathbf{x}_0' \hat{\boldsymbol{\beta}}$$

where $\mathbf{x}_0' = [1 \quad x_{10} \quad x_{20} \quad \cdots \quad x_{k0}]$. Using our matrix rule for variance, we have

$$\begin{aligned}\operatorname{Var} \hat{\mu}_{Y|x_{10}, x_{20}, \ldots, x_{k0}} &= \operatorname{Var} \mathbf{x}_0' \hat{\boldsymbol{\beta}} \\ &= \mathbf{x}_0' \operatorname{Var} \hat{\boldsymbol{\beta}} \mathbf{x}_0 \\ &= \mathbf{x}_0' \sigma^2 (X'X)^{-1} \mathbf{x}_0 \\ &= \sigma^2 \mathbf{x}_0' (X'X)^{-1} \mathbf{x}_0\end{aligned}$$

Standardizing and replacing σ^2 by its unbiased estimator S^2, we obtain the T_{n-k-1} random variable

$$\frac{\hat{\mu}_{Y|x_{10}, x_{20}, \ldots, x_{k0}} - \mu_{Y|x_{10}, x_{20}, \ldots, x_{k0}}}{S\sqrt{\mathbf{x}_0'(X'X)^{-1}\mathbf{x}_0}}$$

It is easy to see that the bounds for a $100(1-\alpha)\%$ confidence interval on $\mu_{Y|x_{10}, x_{20}, \ldots, x_{k0}}$ are

Confidence bounds for $\mu_{Y|x_{10}, x_{20}, \ldots, x_{k0}}$, the mean response for a given set of values of the predictor variables

$$\hat{\mu}_{Y|x_{10}, x_{20}, \ldots, x_{k0}} \pm t_{\alpha/2} S \sqrt{\mathbf{x}_0'(X'X)^{-1}\mathbf{x}_0}$$

where the point $t_{\alpha/2}$ is the appropriate point based on the T_{n-k-1} distribution.

The next example illustrates the idea.

Example 12.4.2. In Example 12.2.3 we estimated the average gasoline mileage for a car weighing 1.5 tons being operated on a 70°F, day by

$$\hat{\mu}_{Y|x_{10}, x_{20}} = 24.75 - 4.16(1.5) - .014897(70) = 17.47$$

Let us now find a 95% confidence interval on this mean value. The vector $\mathbf{x}_0'$ required is given by

$$\mathbf{x}_0' = [1 \quad 1.5 \quad 70]$$

From previous work we know that

$$(X'X)^{-1} = \begin{bmatrix} 6.070769 & -3.02588 & -.0171888 \\ -3.02588 & 1.738599 & .002166306 \\ -.0171888 & .002166306 & .0002582903 \end{bmatrix}$$

and that $s = .1463$. A simple matrix calculation yields

$$\mathbf{x}_0'(X'X)^{-1}\mathbf{x}_0 \doteq .22$$

Based on the $T_{n-k-1} = T_{10-2-1} = T_7$ distribution, a 95% confidence interval in the average gasoline mileage when $x_1 = 1.5$ and $x_2 = 70$ is

$$\hat{\mu}_{Y|x_{10}, x_{20}} \pm t_{\alpha/2} S \sqrt{\mathbf{x}_0'(X'X)^{-1}\mathbf{x}_0}$$

$$17.47 \pm 2.365(.1463)\sqrt{.22}$$

or

$$17.47 \pm .16$$

We can be 95% confident that the average gasoline mileage of cars weighing 1.5 tons operated on a 70°F day lies between 17.31 and 17.63 miles per gallon (mi/gal).

Prediction Interval on Single Predicted Response

The confidence bounds for an individual response for a given set of values of the predictor variables are similar to those for the mean value. As in the case of simple linear regression, the only difference is that the variance of the estimator is a little larger. The bounds assume the form

Prediction bounds for $Y|x_{10}, x_{20}, \ldots, x_{k0}$, an individual response for a given set of values of the predictor variables

$$\hat{Y}|x_{10}, x_{20}, \ldots, x_{k0} \pm t_{\alpha/2} S \sqrt{1 + \mathbf{x}_0'(X'X)^{-1}\mathbf{x}_0}$$

where the point $t_{\alpha/2}$ is the appropriate point based on the T_{n-k-1} distribution.

To illustrate, we find a 95% prediction interval on the gas mileage obtained by a particular automobile weighing 1.5 tons when operated at 70°F.

Example 12.4.3. From our work in the previous example we know that $\hat{\mu}_{Y|x_{10}, x_{20}} = \hat{Y}|x_{10}, x_{20} = 17.47$, $s = .1463$, $\mathbf{x}_0'(X'X)^{-1}\mathbf{x}_0 = .22$. The desired 95% prediction interval is given by

$$\hat{Y}|x_{10}, x_{20} \pm t_{\alpha/2} s \sqrt{1 + \mathbf{x}_0'(X'X)^{-1}\mathbf{x}_0}$$

$$17.47 \pm 2.365(.1463)\sqrt{1 + .22}$$

or

$$17.47 \pm .38$$

We can be 95% confident that a given automobile weighing 1.5 tons will obtain between 17.09 and 17.85 mi/gal when operated on a 70°F day.

12.5 TESTING HYPOTHESES ABOUT MODEL PARAMETERS

In this section we consider three types of hypotheses concerning the parameters in the general linear model. These are:

1. Hypotheses concerning the value of a specific model parameter β_i
2. Hypotheses concerning the significance of the regression model as a whole
3. Hypotheses concerning the effectiveness of a subset of the original set of predictor variables

Testing a Single Predictor Variable

Occasionally an experimenter might suspect that a particular predictor variable is not really very useful. To decide whether or not this is the case, we test the null hypothesis that the coefficient for this variable is 0. That is, we test

$$H_0: \beta_i = 0$$

$$H_1: \beta_i \neq 0$$

The test statistic used is easy to derive. We know that the random variable

$$\frac{\hat{\beta}_i - \beta_i}{S\sqrt{c_{ii}}}$$

follows a T distribution with $n - k - 1$ degrees of freedom. If H_0 is true, then the statistic

$$\frac{\hat{\beta}_i - 0}{S\sqrt{c_{ii}}}$$

follows the T_{n-k-1} distribution. We reject H_0 for values of this statistic that are either too large or too small to have occurred by chance. If H_0 is rejected, then we have evidence that $\beta_i \neq 0$. In this case the predictor variable X_i is useful in predicting the value of the response. If H_0 is not rejected, then the predictor variable X_i is not needed in the model that contains the other predictor variables. Of course, null values other than 0 can be tested. However, 0 is the most commonly encountered value in practice. One-tailed tests can be conducted if desired.

Testing for Significant Regression

A more interesting hypothesis is the null hypothesis that the regression is "not significant." That is, we test the null hypothesis that the regression equation does not explain a sizable proportion of the variability in the response variable versus the alternative that it does explain a significant proportion of this variability. Mathematically, we are testing

$$H_0: \beta_1 = \beta_2 = \cdots = \beta_k = 0$$
$$H_1: \beta_i \neq 0 \text{ for at least one } i \qquad i = 1, 2, , \ldots, k$$

The groundwork for developing a logical test statistic has already been laid. We have shown that SSE, the residual sum of squares, can be expressed as

$$\text{SSE} = S_{yy} - \text{SSR}$$

Rewriting this expression, we see that

$$S_{yy} = \text{SSE} + \text{SSR}$$

That is, the total variability in the response, S_{yy}, can be partitioned into two components. These are SSE, the residual or unexplained variability about the regression line, and SSR, the variability in Y attributed to the linear association between the predictor variables and the mean of Y. If the regression is significant, then SSR should be large relative to SSE. Our test statistic makes use of this idea. In particular, we shall use the statistic

$$\frac{\text{SSR}/k}{\text{SSE}/(n-k-1)} = \frac{\text{SSR}/k}{S^2}$$

to test H_0. It can be shown that if H_0 is true, then this statistic follows an F distribution with k and $n - k - 1$ degrees of freedom. The test rejects for large values of the test statistic.

To illustrate, let us turn again to the multiple linear regression equation that we have developed to predict gasoline mileage based on the weight of the vehicle and the temperature at the time of operation. Let us see if this model explains a significant proportion of the variability that we observe in the response variable.

Example 12.5.1. From previous work we have this information (see Example 12.3.3):

$$n = 10 \qquad S_{yy} = 10.46 \qquad \text{SSR} = 10.31$$
$$k = 2 \qquad \text{SSE} = .15$$

To test

$$H_0: \beta_1 = \beta_2 = 0$$

$$H_1: \beta_i \neq 0 \text{ for some } i$$

we evaluate the $F_{k,n-k-1} = F_{2,10-2-1}$ statistic

$$\frac{\text{SSR}/k}{\text{SSE}/(n-k-1)}$$

For these data this statistic has value

$$\frac{10.31/2}{.15/7} = 240.56$$

Based on the $F_{2,7}$ distribution, we can reject H_0 with $P < .05$. We have good statistical evidence that β_1 and β_2 are not both 0.

Once we know that the regression is significant, a natural question to ask is, "What proportion of the total variability in Y is explained by our fitted regression model?" To answer this question, we parallel what was done in the case of simple linear regression. We define what is called the coefficient of *multiple determination*. This statistic, denoted by R^2, is defined by

$$R^2 = \frac{\text{SSR}}{S_{yy}}$$

When R^2 is multiplied by 100%, we get the percentage of the variation in Y explained by the fitted regression equation. Values of R^2 near 1 are taken as an indication that the model explains the data well. The square root of R^2 is called the *multiple correlation coefficient* between Y and the predictor variables. In our previous example

$$R^2 = \frac{\text{SSR}}{S_{yy}} = \frac{10.31}{10.46} = .9857$$

Our fitted model has explained 98.57% of the variation observed in Y.

Another interpretation of R^2 is possible. Note that if a model does a good job of explaining the variation in Y, then responses predicted by the model, $\hat{Y}$, should agree well with those actually observed; otherwise there will be substantial differences between the observed and predicted responses. This leads one to suspect that there is a relationship between R^2 and $\hat{\rho}$, the estimator for the Pearson coefficient of correlation between Y and $\hat{Y}$. In fact, it can be shown that $R^2 = \hat{\rho}^2$. Thus a strong linear association between Y and $\hat{Y}$ yields a large value of R^2 and vice versa. In SAS Supplement XII the actual responses and those predicted by the model are listed for the data of Example 12.2.3. These are:

y actual	$\hat{y}$ predicted
17.9	17.793
16.5	16.399
16.4	16.486
16.8	16.666
18.8	18.820
15.5	15.552
17.5	17.349
16.4	16.368
15.9	16.086
18.3	18.479

As you can see, the predicted responses agree closely with those actually observed. Thus $\hat{\rho}$ should lie close to 1 and $R^2 = \hat{\rho}^2$ should be large. A quick calculation gives $\hat{\rho} = .9932$ and $R^2 = \hat{\rho}^2 = .9864$. This agrees with the value found earlier, apart from round-off error.

To better understand the logic behind the F test for a significant regression, let us rewrite the test statistic in terms of R^2:

$$F_{k,n-k-1} = \frac{\mathrm{SSR}/k}{\mathrm{SSE}/(n-k-1)} = \frac{\dfrac{\mathrm{SSR}/k}{S_{yy}}}{\dfrac{\mathrm{SSE}/(n-k-1)}{S_{yy}}} = \frac{R^2/k}{(1-R^2)/(n-k-1)}$$

From this expression we see clearly that, apart from the constant multiple $(n - k - 1)/k$, the F statistic is the ratio of the explained to the unexplained variation in Y. It is natural that we say that the regression is significant only when the proportion of explained variation is large. This occurs only when the F ratio is large. For this reason, our F test is always to reject for values of F that are too large to have occurred by chance.

Testing a Subset of Predictor Variables

Even if we find the regression significant for a particular model, it is usually desirable to find the simplest model that fits the data well. Why use 10 predictor variables if 3 will suffice? A formal test that allows the experimenter to determine whether a subset of the original predictor variables is sufficient for purposes of prediction can be conducted. To see how this is done, let $X_1, X_2, X_3, \ldots, X_k$ denote the original predictor variables. The model being considered is given by

$$\mu_{Y|x_1, x_2, \ldots, x_k} = \beta_0 + \beta_1 x_1 + \beta_2 x_2 + \cdots + \beta_k x_k$$

This model is referred to as the *full model*. Assume that we propose to reduce the number of predictor variables by deleting all but m of them. Without loss of generality, we assume that the first m variables are to be retained. The new model, called the *reduced model*, is given by

$$\mu_{Y|x_1, x_2, \ldots, x_m} = \beta_0 + \beta_1 x_1 + \beta_2 x_2 + \cdots + \beta_m x_m$$

We want to choose between the reduced and the full models. We do so by testing

H_0: reduced model is appropriate

H_1: full model is needed

The method used to test H_0 is rather intuitive in nature. We first find the residual or error sum of squares for the full model in the usual way. We denote this sum of squares by SSE_f to indicate that this statistic is based on the full model, the model containing all k of the original predictor variables. We next find the residual sum of squares for the reduced model. This sum of squares is denoted by SSE_r to indicate that only a subset of the predictor variables is used in its computation. We know that for a given model the residual sum of squares reflects the variation in the response variable that is *not* explained by the model. If the predictor variables $X_{m+1}, X_{m+2}, \ldots, X_k$ are important, then deleting them from our model should result in a significant increase in the unexplained variation in Y. That is, SSE_r should become considerably larger than SSE_f. Our test statistic makes use of this idea. It is given by

$$F_{k-m, n-k-1} = \frac{(\text{SSE}_r - \text{SSE}_f)/(k-m)}{\text{SSE}_f/(n-k-1)}$$

Note that if H_0 is true, then the reduced model does as good a job of explaining the variability observed in the response variable as the full model. In this case SSE_r and SSE_f will not differ much in value, $\text{SSE}_r - \text{SSE}_f$ will be small, and the F ratio will be small in value. On the other hand, if H_0 is not true, then the reduced model is not appropriate. In this case SSE_r will be much larger than SSE_f, $\text{SSE}_r - \text{SSE}_f$ will be large, and the F ratio will be large in value. Logic dictates that we reject H_0 in favor of H_1 for values of the test statistic that are too large to have occurred by chance based on the $F_{k-m,n-k-1}$ distribution. Although this sounds complicated, a simple example should clarify things.

Example 12.5.2. At the moment we have two models proposed for predicting gasoline mileage of an automobile. One bases the prediction on both the weight of the car and the temperature at the time of operation; the other uses only the weight of the car in making predictions. The former is the full model, whereas the latter is the reduced model. Let us test

H_0: reduced model is appropriate

H_1: full model is needed

From past work (see Examples 11.3.3 and 12.3.3) we know that

$$\text{SSE}_r = 1.01 \qquad \text{SSE}_f = .15 \qquad n = 10$$

Since the full model entails two predictor variables while the reduced model entails only one, $k = 2$ and $m = 1$. The observed value of the test statistic is

$$F_{k-m,n-k-1} = \frac{(\text{SSE}_r - \text{SSE}_f)/(k-m)}{\text{SSE}_f/(n-k-1)}$$

$$= \frac{(1.01 - .15)/(2-1)}{.15/(10-2-1)}$$

$$= 40.13$$

Based on the $F_{1,7}$ distribution, H_0 can be rejected with $P < .05$. We conclude that adding the variable X_2, the temperature at which the automobile is operated, improves the original model.

12.6 USE OF INDICATOR OR "DUMMY" VARIABLES (OPTIONAL)

The previous sections dealt with multiple linear regression when the independent (predictor) variables are all *quantitative* such as height, temperature, time, or pressure. It is sometimes necessary to use *qualitative* or categorical variables. For example, variables such as sex, race, shift of work, and brand or type of product are qualitative; they have no natural scale of measurement. In such cases indicator or "dummy" variables are used in the model to account for the effect that the variable has on the response. Example 12.6.1 will illustrate the idea.

Example 12.6.1. Consider Example 11.1.1 in which simple linear regression was used to study the relationship between humidity X and the extent of solvent evaporation Y for a water-reducible paint. Suppose that the study is to be rerun by using two different brands of paint, A and B. It is believed that humidity generally affects both paints in the same way but that the responses for the two brands might be systematically different. Thus the regression model posed should contain two predictor variables x_1, humidity, and x_2, a variable that allows us to code the type of paint used. The model becomes

$$\mu_{Y|x_1, x_2} = \beta_0 + \beta_1 x_1 + \beta_2 x_2$$

where we let

$$x_2 = \begin{cases} 1 & \text{if type } A \text{ paint is used} \\ 0 & \text{if type } B \text{ paint is used} \end{cases}$$

The variable x_2 is called an *indicator* variable because it is used to indicate the presence or absence of paint A. To determine whether or not the qualitative variable, in addition to the humidity, is useful in predicting the value of Y, we test

$$H_0\colon \beta_2 = 0$$

$$H_1\colon \beta_2 \neq 0$$

This can be done using the T test or the equivalent F test presented in Sec. 12.5.

Consider an experiment involving one indicator variable. If $H_0\colon \beta_2 = 0$ is rejected, then there is evidence that the qualitative variable is important in the model. In this case we are actually dealing with two separate models. For example, in the paint experiment when $\beta_2 \neq 0$ and paint A is used, the model becomes

$$\mu_{Y|x_1, x_2} = \beta_0 + \beta_1 x_1 + \beta_2(1)$$

or

$$\mu_{Y|x_1, x_2} = (\beta_0 + \beta_2) + \beta_1 x_1$$

That is, the relationship between mean solvent evaporation and humidity is a straight line with slope β_1 and intercept $\beta_0 + \beta_2$. However, when paint B is used, the model is

$$\mu_{Y|x_1, x_2} = \beta_0 + \beta_1 x_1 + \beta_2(0)$$

or

$$\mu_{Y|x_1, x_2} = \beta_0 + \beta_1 x_1$$

This represents a straight line with slope β_1 and intercept β_0. Note that these models are linear with the *same* slope but *different* intercepts. Hence whenever we reject $H_0\colon \beta_2 = 0$ we are in effect concluding that we are dealing with two *parallel* regression lines with different intercepts. The estimated vertical distance between these two lines is the estimated difference in intercepts, namely, $\hat{\beta}_2$.

Example 12.6.2. These data are obtained for the study described in Example 12.6.1:

(x_1) **Relative humidity (%)**	(x_2) **Presence of paint** ***A***	y **Solvent evaporation (% wt.)**
35.3	1	11.2
29.6	1	11.0
31.0	1	12.6
58.0	1	8.3
62.0	1	10.1
72.1	1	9.6
74.0	1	6.1
77.0	1	8.7
71.1	1	8.1
57.0	1	9.0
46.4	1	8.2
29.6	1	13.0
28.0	1	11.7
39.1	0	6.7
46.8	0	7.7
48.5	0	6.8
59.3	0	7.0
70.0	0	5.2
70.0	0	4.0
74.4	0	5.7
72.1	0	4.9
58.1	0	5.5
44.6	0	6.1
33.4	0	7.5
28.6	0	8.0

The model specification matrix is given by

$$X = \begin{bmatrix} 1 & 35.3 & 1 \\ 1 & 29.6 & 1 \\ \vdots & \vdots & \vdots \\ 1 & 28.0 & 1 \\ 1 & 39.1 & 0 \\ 1 & 46.8 & 0 \\ \vdots & \vdots & \vdots \\ 1 & 28.6 & 0 \end{bmatrix}$$

With the help of the computer, it can be shown that

$$(X'X)^{-1} = \begin{bmatrix} .488429 & -.00753783 & -.0993029 \\ -.00753783 & .0001402605 & .0002971544 \\ -.0993029 & .0002971544 & .160886 \end{bmatrix}$$

and that

$$\hat{\beta}_0 = 10.3979 \qquad \hat{\beta}_2 = 3.3938 \qquad s = \hat{\sigma} = 1.0185$$

$$\hat{\beta}_1 = -.0770 \qquad \hat{\sigma}^2 = 1.0374$$

The T statistic used to test $H_0\colon \beta_2 = 0$ is

$$T_{n-k-1} = T_{25-2-1} = \frac{\hat{\beta}_2 - 0}{s\sqrt{c_{22}}}$$

$$= \frac{3.3938}{1.0185\sqrt{.160886}}$$

$$= 8.3074$$

Based on this statistic, H_0 can be rejected with $P < .0005$. It can be concluded that the type of paint used is an important factor in predicting the extent of solvent evaporation. The estimated model is

$$\hat{\mu}_{Y|x_1, x_2} = 10.3979 - .0770x_1 + 3.3938x_2$$

When paint A is used, the model is

$$\hat{\mu}_{Y|x_1, x_2} = 10.3979 - .0770x_1 + 3.3938(1)$$

or

$$\hat{\mu}_{Y|x_1, x_2} = 13.7917 - .0770x_1$$

The model for paint B is

$$\hat{\mu}_{Y|x_1, x_2} = 10.3979 - .0770x_1 + 3.3938(0)$$

or

$$\hat{\mu}_{Y|x_1, x_2} = 10.3979 - .0770x_1$$

Note that, as claimed, these estimated regression lines have the same slope, namely, $-.0770$, but different intercepts. The intercepts differ by $\hat{\beta}_2 = 3.3938$.

Since, when $H_0\colon \beta_2 = 0$ is rejected, two regression lines result, the natural question to ask is, "Why model them this way rather than simply fitting two separate regression lines?" The reason is that by pooling the data from the two groups, we obtain improved estimates of the common slope β_1 and the common variance σ^2.

A similar approach can be used for qualitative factors that have more than two levels. If, for example, we have three types of paint, A, B, and C, two indicator variables are required to code the type of paint being used. To illustrate, for three types of paint the model becomes

$$\mu_{Y|x_1, x_2, x_3} = \beta_0 + \beta_1 x_1 + \beta_2 x_2 + \beta_3 x_3$$

where x_1 denotes the humidity and x_2 and x_3 are given by

$$x_2 = 0 \text{ and } x_3 = 0 \qquad \text{if type } A \text{ is used}$$

$$x_2 = 1 \text{ and } x_3 = 0 \qquad \text{if type } B \text{ is used}$$

$$x_2 = 0 \text{ and } x_3 = 1 \qquad \text{if type } C \text{ is used}$$

In general, if a qualitative variable has l levels, then $l - 1$ indicator variables are needed to code the levels of the variable.

In the model just discussed it is assumed that the quantitative variable x_1 affects all levels of the qualitative variable in the same way. That is, it is assumed that one has good reason to believe that the regression lines for each level of the qualitative variable have the *same* slope with possibly different intercepts. When it is not known that this is the case, then a different model is needed so that a test for equality of slopes can be performed. In the case of one indicator variable x_2, with two levels, an appropriate model is

$$\mu_{Y|x_1, x_2} = \beta_0 + \beta_1 x_1 + \beta_2 x_2 + \beta_3 x_1 x_2$$

Note that when $x_2 = 1$, the model becomes

$$\mu_{Y|x_1, x_2} = \beta_0 + \beta_1 x_1 + \beta_2 + \beta_3 x_1$$

or

$$\mu_{Y|x_1, x_2} = (\beta_0 + \beta_2) + (\beta_1 + \beta_3) x_1$$

When $x_2 = 0$, the model reduces to

$$\mu_{Y|x_1, x_2} = \beta_0 + \beta_1 x_1$$

It should be clear that to test for equality of slopes, we test

$$H_0\colon \beta_3 = 0$$

$$H_1\colon \beta_3 \neq 0$$

via the T or F test described in Section 12.5.

As you can see, even though the use of indicator variables is a bit tricky, the analysis employs only those techniques already presented. Further reading on regression using indicator variables can be found in [12] and [39].

12.7 CRITERIA FOR VARIABLE SELECTION

As you can see, selecting the best model is not a trivial problem. In the case of polynomial regression the experimenter must decide on the degree of the polynomial to be used. In multiple linear regression he or she must determine which of the available predictor variables yields the simplest adequate model. Selecting a final model is, in many ways, an art rather than a science. Clearly, experience is valuable. However, there are several rather standard procedures that help in the model selection process. Most of these procedures are available in the standard statistical software packages. The use of these packages is straightforward, and this relieves us of the computational burden of regression analysis. However, it is important to understand what these procedures do. We summarize some of them here.

The basic problem is to find as simple a model as possible that has a "good fit." Since R^2 gives the proportion of the variability in the response that is explained by the fitted regression equation, we obviously desire R^2 to be large. However, most fitted models are used eventually for prediction purposes. Note that the width of the confidence intervals on β_i, $\mu_{Y|x_i, x_2, \ldots, x_k}$, and $Y|x_1, x_2, \ldots, x_k$ all depend in part on the statistic $S^2 = \text{SSE}/(n - k - 1)$. To get narrow confidence intervals and accurate estimates for these entities, we want S^2 to be small. This statistic is referred to as the *mean squared error*. We can always increase the value of R^2 by adding more terms to the model. However, the addition of unneeded variables may result in an increase in the mean squared error. Thus our real task is to balance these two measures of the goodness of fit of the model. We begin by considering some widely used methods for choosing an adequate model. Each of these methods is based on the statistic R^2.

Forward Selection Method

In the forward selection process, variables are added to the model one at a time until the addition of another variable does not significantly improve the model. That is, variables are added until we are unable to reject the reduced model.

Example 12.7.1. Assume that we have available three possible predictor variables X_1, X_2, and X_3. Suppose that our final model via forward selection contains only the variables X_3 and X_1 and that they entered the model in the order stated. These are the steps that are taken by the computer:

1. The three single-variable models

$$\mu_{Y|x_1} = \beta_0 + \beta_1 x_1$$

$$\mu_{Y|x_2} = \beta_0 + \beta_2 x_2$$

$$\mu_{Y|x_3} = \beta_0 + \beta_3 x_3$$

are fitted. The value of R^2 is found for each. The one with the highest R^2 is chosen and compared to the reduced model $\mu_Y = \beta_0$. In this case we test

$$H_0: \mu_Y = \beta_0 \qquad \text{(reduced model is appropriate)}$$

$$H_1: \mu_{Y|x_3} = \beta_0 + \beta_3 x_3 \qquad \text{(full model is needed)}$$

and H_0 is rejected. The variable X_3 is now included in our model.

2. The two two-variable models

$$\mu_{Y|x_3, x_1} = \beta_0 + \beta_1 x_1 + \beta_3 x_3$$

$$\mu_{Y|x_3, x_2} = \beta_0 + \beta_2 x_2 + \beta_3 x_3$$

are fitted. The value of R^2 is found for each. The one with the highest R^2 is chosen and compared to the reduced model $\mu_{Y|x_3} = \beta_0 + \beta_3 x_3$. In this case we test

$$H_0: \mu_{Y|x_3} = \beta_0 + \beta_3 x_3 \qquad \text{(reduced model is appropriate)}$$

$$H_1: \mu_{Y|x_3, x_1} = \beta_0 + \beta_1 x_1 + \beta_3 x_3 \qquad \text{(full model is needed)}$$

and H_0 is rejected. The variable X_1 is now included in our model.

3. The three-variable model

$$\mu_{Y|x_1,x_2,x_3} = \beta_0 + \beta_1 x_1 + \beta_2 x_2 + \beta_3 x_3$$

is fitted and we test

$$H_0: \mu_{Y|x_1,x_3} = \beta_0 + \beta_1 x_1 + \beta_3 x_3 \qquad \text{(reduced model is appropriate)}$$

$$H_1: \mu_{Y|x_1,x_2,x_3} = \beta_0 + \beta_1 x_1 + \beta_2 x_2 + \beta_3 x_3 \qquad \text{(full model is needed)}$$

In this case H_0 is not rejected. The variable X_2 does not appear to be needed in our model. The final model that we obtain is the two-variable model

$$\mu_{Y|x_1,x_3} = \beta_0 + \beta_1 x_1 + \beta_3 x_3$$

As you can see, doing this type of analysis by hand is impractical. The use of the computer makes the problem simple.

Backward Elimination Procedure

Another method of selecting a model is called *backward elimination*. In backward elimination one begins with the model that includes all the potential predictor variables. Variables are deleted from the model one at a time until the further deletion of a variable results in a rejection of the reduced model.

Example 12.7.2. Assume that we have three potential predictor variables and that via backward elimination we obtain a reduced model containing only the variable X_2. Assume that the variables X_1 and X_3 are deleted in the order mentioned. These are the steps that are taken:

1. The full model

$$\mu_{Y|x_1,x_2,x_3} = \beta_0 + \beta_1 x_1 + \beta_2 x_2 + \beta_3 x_3$$

is fitted. The value of R^2 is found.

2. The three two-variable models

$$\mu_{Y|x_1,x_2} = \beta_0 + \beta_1 x_1 + \beta_2 x_2$$

$$\mu_{Y|x_1,x_3} = \beta_0 + \beta_1 x_1 + \beta_3 x_3$$

$$\mu_{Y|x_2,x_3} = \beta_0 + \beta_2 x_2 + \beta_3 x_3$$

are fitted. The value of R^2 is found for each. The model with the largest R^2 is chosen and compared with the full model. In this case we test

$$H_0: \mu_{Y|x_2,x_3} = \beta_0 + \beta_2 x_2 + \beta_3 x_3 \qquad \text{(reduced model is adequate)}$$

$$H_1: \mu_{Y|x_1,x_2,x_3} = \beta_0 + \beta_1 x_1 + \beta_2 x_2 + \beta_3 x_3 \qquad \text{(full model is needed)}$$

and are unable to reject H_0. We delete the variable X_1 from the model since it appears that the reduced model is adequate.

3. The one-variable models

$$\mu_{Y|x_2} = \beta_0 + \beta_2 x_2$$

$$\mu_{Y|x_3} = \beta_0 + \beta_3 x_3$$

are fitted. In this case we test

$$H_0: \mu_{Y|x_2} = \beta_0 + \beta_2 x_2 \qquad \text{(reduced model is adequate)}$$

$$H_1: \mu_{Y|x_2, x_3} = \beta_0 + \beta_2 x_2 + \beta_3 x_3 \qquad \text{(full model is needed)}$$

and are unable to reject H_0. We delete the variable X_3 from the model since it appears to be unnecessary.

4. We now fit the model $\mu_Y = \beta_0$ and test

$$H_0: \mu_Y = \beta_0$$

$$H_1: \mu_{Y|x_2} = \beta_0 + \beta_2 x_2$$

In this case H_0 is rejected and we are left with the model that contains the one predictor variable X_2.

The third method of variable selection that is in widespread use is called *stepwise regression*.

Stepwise Method

Stepwise regression is a modified version of the forward selection process. In forward selection once a variable enters the model it stays. Unfortunately, it is possible for a variable entering at a later stage to render a previously selected variable unimportant because of the interrelationships of the variables. This usually occurs when the two predictor variables are themselves closely related. Forward selection does not consider this possibility. In stepwise regression each time a new variable is entered into the model, all the variables in the previous model are checked for continued importance.

It is hard to describe in general terms what is done in stepwise regression. However, an example should clarify matters.

Example 12.7.3. In a multiple linear regression model variables X_1 and X_3 are closely related, with variable X_1 being the best single predictor. Suppose that the final model contains the two variables X_2 and X_3, with variable X_2 entering on the second stage. The steps in the stepwise regression are

1. The three single-variable models

$$\mu_{Y|x_1} = \beta_0 + \beta_1 x_1$$

$$\mu_{Y|x_2} = \beta_0 + \beta_2 x_2$$

$$\mu_{Y|x_3} = \beta_3 + \beta_3 x_3$$

are fitted. The value of R^2 is computed for each and the model with the largest R^2 is compared to the model

$$\mu_Y = \beta_0$$

In this case we test

$$H_0\colon \mu_Y = \beta_0 \qquad \text{(reduced model is adequate)}$$

$$H_1\colon \mu_{Y|x_1} = \beta_0 + \beta_1 x_1 \qquad \text{(full model is needed)}$$

and reject H_0. The variable X_1 is inserted into the model.

2. The two-variable models

$$\mu_{Y|x_1, x_2} = \beta_0 + \beta_1 x_1 + \beta_2 x_2$$

$$\mu_{Y|x_1, x_3} = \beta_0 + \beta_1 x_1 + \beta_3 x_3$$

are fitted. The one with the largest R^2 is compared to our previous model. Here we test

$$H_0\colon \mu_{Y|x_1} = \beta_0 + \beta_1 x_1 \qquad \text{(reduced model is adequate)}$$

$$H_1\colon \mu_{Y|x_1, x_2} = \beta_0 + \beta_1 x_1 + \beta_2 x_2 \qquad \text{(full model is needed)}$$

and reject H_0. We also check to see if the variable X_1 is now needed. To do so, we test

$$H_0\colon \mu_{Y|x_2} = \beta_0 + \beta_2 x_2 \qquad \text{(reduced model is adequate)}$$

$$H_1\colon \mu_{Y|x_1, x_2} = \beta_0 + \beta_1 x_1 + \beta_2 x_2 \qquad \text{(full model is needed)}$$

and reject H_0. The variable X_2 alone is not sufficient. We still need X_1 in our model.

3. The model

$$\mu_{Y|x_1, x_2, x_3} = \beta_0 + \beta_1 x_1 + \beta_2 x_2 + \beta_3 x_3$$

is fitted. We test

$$H_0\colon \mu_{Y|x_1, x_2} = \beta_0 + \beta_1 x_1 + \beta_2 x_2 \qquad \text{(reduced model is adequate)}$$

$$H_1\colon \mu_{Y|x_1, x_2, x_3} = \beta_0 + \beta_1 x_1 + \beta_2 x_2 + \beta_3 x_3 \qquad \text{(full model is needed)}$$

and reject H_0. The variable X_3 is included in the model. To see if we still need variable X_2, we test

$$H_0\colon \mu_{Y|x_1, x_3} = \beta_0 + \beta_1 x_1 + \beta_3 x_3 \qquad \text{(reduced model is adequate)}$$

$$H_1\colon \mu_{Y|x_1, x_2, x_3} = \beta_0 + \beta_1 x_1 + \beta_2 x_2 + \beta_3 x_3 \qquad \text{(full model is needed)}$$

and reject H_0. This leaves X_2 in the model. To see if we still need variable X_1 in the model, we test

$$H_0\colon \mu_{Y|x_2, x_3} = \beta_0 + \beta_2 x_2 + \beta_3 x_3 \qquad \text{(reduced model is adequate)}$$

$$H_1\colon \mu_{Y|x_1, x_2, x_3} = \beta_0 + \beta_1 x_1 + \beta_2 x_2 + \beta_3 x_3 \qquad \text{(full model is needed)}$$

and are unable to reject H_0. At this point, X_1 is deleted from the model, leaving us with a prediction equation based on the two variables X_2 and X_3.

The three techniques just described have been used for many years and you will see references to them in the literature. These three techniques do not always lead to the same model, but they do allow the reseacher to find a reasonable linear combination of regressors without having to examine all possible combinations. This was a major advantage in the early years of computing when fitting a model was a time-consuming process. With the advent of efficient high-speed computing, it is now possible to examine all linear combinations of regressors. For a model with k regressors, the computer can fit 2^k models quickly. Some models may be quite good, whereas others are worthless. Recent research has concentrated on techniques that allow the reseacher to compare one model to another. In this way he or she can choose the model (or models) that appears to do the best job in describing the relationship between the regressor and the response. We now describe three of the statistics currently being used to make these judgments. Note that if no satisfactory linear combination of regressors is found, then other models can be tried.

Maximum R^2 Method

The Max R^2 procedure selects at each step j, $j = 1, 2, 3, \ldots, k$, the set of j variables that gives the largest R^2. Although R^2 will always increase as more variables are added, the mean squared error usually will first decrease and then increase as additional variables are selected. Typically, the experimenter selects the model corresponding to the smallest mean squared error.

Mallow's C_k Statistic

The Mallow's C_k statistic is based on the normalized expected total error of estimation, which is given by

$$\Gamma_k = \frac{E\left(\sum_{i=1}^{n} [Y_i - E(Y_i)]^2\right)}{\sigma^2} = \frac{\text{SSE}}{\sigma^2} + 2(k - 1) - n$$

where $k + 1$ is the total number of parameters in the model, including the intercept β_0. After substituting the sample estimator S^2 for σ^2, we see that the C_k statistic is

$$C_k = \frac{\text{SSE}}{S^2} + 2(k - 1) - n$$

A value of C_k near $k + 1$ suggests that the model bias is small. That is, there is no significant overfitting or underfitting of the model. Values of C_k near or below $k + 1$ are generally desirable.

PRESS Statistic

A somewhat different procedure for model selection is based on the PRESS (prediction sum of squares) statistic proposed by D. M. Allen. This statistic is used primarily to select a model for purposes of prediction. It is somewhat unnerving to realize that in the usual regression context we use each observation to develop an equation by which the value of the observation can be predicted. For example, we use information on the gasoline mileage of a car weighing 1.6 tons driven on a 50°F day to develop an equation by which we can predict the gasoline mileage of a car weighing 1.6 tons driven on a 50°F day! The PRESS statistic avoids this dilemma. For a specified model the PRESS statistic is formed by predicting each observation based on a model developed by using all the other observations. In short, the statistic is formed as follows:

1. All data points except the first are used to fit the model. The value of the first observation, y_1, is predicted from the fitted model. The PRESS residual $y_1 - \hat{y}_1$ is found.
2. All data points except the second are used to fit the model. The value of the second observation, y_2, is predicted from the new fitted model. The residual $y_2 - \hat{y}_2$ is formed.
3. This process is continued until each observation has been predicted from the others and the PRESS residual found.
4. The PRESS statistic is defined to be the sum of the squares of the PRESS residuals. That is, PRESS $= \sum_{i=1}^{n}(y_i - \hat{y}_i)^2$. The model chosen is that with a small value for PRESS.

It is evident that evaluating the PRESS statistic in this way entails a great deal of computation. Fortunately, a shortcut method is available and the statistic can be found via SAS.

The ideal model for prediction has small PRESS, small C_k, small mean squared error, and large R^2. Since it is almost too much to ask that one model have all these properties, the experimenter must use his or her own judgment to select the best model. Although all these criteria are useful, if forced to rate them in order of importance, we would rely upon PRESS, C_k, and mean squared error in that order.

In the next two examples we illustrate these criteria for two data sets. The first data set is unusual in that each of the criteria mentioned points to the same model. The second data set forces us to make some value judgments in selecting the model.

Example 12.7.4. It is known that in mammals the toxicity of various types of drugs, pesticides, and chemical carcinogens can be altered by inducing liver enzyme activity. A study to investigate this sort of phenomena in chickens was reported in "Organophosphate Detoxification Related by Induced Hepatic Microsomal Enzymes in Chickens," M. Ehrich, C. Larson, and J. Arnold, *American Journal of Veterinary Research*, vol. 45, 1983. Regression analysis was used to study the relationship between induced enzyme activity and detoxification of the insecticide malathion. Butylated hydroxytoluene (BHT) was the enzyme inducer used. Each number represents the percentage of activity relative to a control, an untreated

TABLE 12.1

BHT raw data

% Detoxification (y)	Enzyme 1 (x_1)	Enzyme 2 (x_2)	Enzyme 3 (x_3)	Enzyme 4 (x_4)	Enzyme 5 (x_5)
146.104	348.475	337.500	108.122	106.667	107.692
152.597	233.220	260.417	82.234	80.000	88.889
168.831	287.458	273.958	74.619	66.667	87.179
178.571	152.542	310.417	86.802	73.333	96.581
191.558	276.271	818.750	122.843	86.667	97.436
113.636	78.644	156.250	112.690	93.333	94.872
188.312	196.949	260.417	79.188	80.000	106.838
94.156	101.695	112.500	127.919	93.333	80.342
159.091	194.576	280.208	239.594	106.667	91.453
142.857	325.424	326.042	173.096	113.333	100.000

TABLE 12.2

Values of R^2 and C_k

Number of variables in model	R^2	C_k	Variables in the model
1	.037	151.5	x_3
1	.180	128.2	x_4
1	.186	127.2	x_1
1	.222	121.3	x_5
1	.408	90.9	x_2
2	.219	123.7	x_3, x_4
2	.227	122.5	x_1, x_3
2	.244	119.6	x_3, x_5
2	.283	113.3	x_1, x_5
2	.425	90.0	x_1, x_2
2	.455	85.1	x_2, x_3
2	.467	83.1	x_1, x_4
2	.488	79.8	x_2, x_5
2	.543	70.8	x_4, x_5
2	.574	65.8	x_2, x_4
3	.311	110.7	x_1, x_3, x_5
3	.473	84.2	x_1, x_2, x_3
3	.489	81.6	x_1, x_2, x_5
3	.523	76.0	x_2, x_3, x_5
3	.575	67.5	x_1, x_3, x_4
3	.594	64.5	x_2, x_3, x_4
3	.636	57.5	x_1, x_2, x_4
3	.642	56.6	x_1, x_4, x_5
3	.746	39.5	x_2, x_4, x_5
3	.833	25.3	x_3, x_4, x_5
4	.525	77.7	x_1, x_2, x_3, x_5
4	.691	50.6	x_1, x_2, x_3, x_4
4	.764	38.7	x_1, x_2, x_4, x_5
4	.927	12.0	x_2, x_3, x_4, x_5
4	.948	8.5	x_1, x_3, x_4, x_5
5	.976*	6.0*	x_1, x_2, x_3, x_4, x_5

TABLE 12.3
Estimated models ordered by PRESS

$\hat{\beta}_0$	$\hat{\beta}_1$	$\hat{\beta}_2$	$\hat{\beta}_3$	$\hat{\beta}_4$	$\hat{\beta}_5$	*MSE*	PRESS
54.079	.097	.034	.522	−2.655	2.559	54.00*	1995.1*
49.802	.131	.000	.588	−2.910	2.795	91.70	2003.6
16.103	.000	.000	.578	−2.782	3.351	245.39	3053.4
37.316	.000	.056	.472	−2.422	2.731	129.78	6343.8
43.403	.000	.000	.000	−1.187	2.281	577.35	7602.9
75.416	.122	.000	.000	−1.265	1.738	527.31	8306.2
230.841	.000	.000	.000	−.859	.000	905.74	9509.2
211.303	.188	.000	.000	−1.100	.000	672.15	9576.4
242.091	.215	.000	.317	−1.934	.000	625.16	10003.4
153.571	.000	.000	.000	.000	.000	981.64	10907.1
250.583	.000	.000	.187	−1.328	.000	985.45	11001.3
121.140	.148	.000	.000	.000	.000	899.15	12000.0
10.343	.000	.000	.000	.000	1.723	859.23	12051.7
167.842	.000	.000	−.118	.000	.000	1063.05	14569.6
11.771	.094	.000	.000	.000	1.273	905.29	15311.3
66.588	.000	.078	.000	−1.074	1.671	373.64	16149.6
5.672	.000	.000	−.092	.000	1.672	953.55	17375.7
135.775	.149	.000	−.124	.000	.000	975.73	18271.4
120.779	.000	.105	.000	.000	.000	654.19	20911.9
21.783	.000	.089	.000	.000	1.092	646.50	21241.3
30.820	.099	.000	−.103	.000	1.193	1014.61	21852.7
195.577	.000	.103	.000	−.825	.000	538.23	23842.5
136.482	.000	.106	−.134	.000	.000	687.35	28098.8
210.467	.000	.101	.134	−1.163	.000	598.37	30378.6
78.197	.057	.067	.000	−1.126	1.505	417.61	30380.0
42.773	.000	.091	−.116	.000	1.010	702.12	30904.6
193.349	.102	.078	.000	−.965	.000	535.74	41406.9
218.604	.135	.067	.233	−1.597	.000	546.09	41871.2
113.148	.052	.092	.000	.000	.000	725.29	49496.5
24.291	.015	.086	.000	.000	1.040	752.37	53154.6
128.842	.053	.094	−.134	.000	.000	775.61	60113.7
46.104	.019	.088	−.117	.000	.945	839.08	68821.9

chicken. The response variable is the percentage of detoxification of malathion. Five enzyme activities were measured and serve as the predictor variables. The data gathered are shown in Table 12.1. Table 12.2 gives the value of the R^2 and C_k statistics for all possible models. Table 12.3 shows the estimated regression coefficients, the mean squared error (MSE), and the PRESS statistic for each model. From Table 12.3 we see that the estimated model with all independent variables included

$$\hat{\mu}_{Y|x_1, x_2, x_3, x_4, x_5} = 54.079 + .097x_1 + .034x_2 + .522x_3 - 2.655x_4 + 2.559x_5$$

has the smallest mean squared error (54.00) and also the smallest PRESS statistic (1995.1). From Table 12.2 we see that the same model also has the largest R^2

TABLE 12.4
3-MC raw data

% Detoxification (y)	Enzyme 1 (x_1)	Enzyme 2 (x_2)	Enzyme 3 (x_3)	Enzyme 4 (x_4)	Enzyme 5 (x_5)
56.250	106.329	90.756	94.650	162.791	114.737
75.000	144.726	203.361	131.687	255.814	112.632
115.625	136.287	672.269	123.457	191.860	153.684
68.750	154.430	183.193	113.169	133.721	116.842
96.875	385.232	140.336	117.284	174.419	87.368
168.750	583.544	146.218	152.263	273.256	94.737
84.375	489.451	184.874	121.399	255.814	95.789
171.875	445.992	537.815	150.206	552.326	113.684
109.375	270.886	309.244	185.185	534.884	108.421
103.125	163.291	190.756	139.918	360.465	106.316

(.976) and the smallest value of C_k (6.00). Hence all our criteria suggest the same model, namely, the one containing all five predictor variables.

The situation just encountered makes the experimenter very confident in the model selected. Unfortunately, such a clear choice is rare. A more typical situation is demonstrated in the next example.

Example 12.7.5. An analysis similar to that described in the previous example was completed for the enzyme inducer 3-methylcholanthrene (3-MC). The complete data set is given in Table 12.4. Tables 12.5 and 12.6 give the values of R^2, C_k, the estimated model parameters, MSE, and PRESS statistics for these data. Let us now see which models are suggested by the various criteria. From Table 12.5 we see that, as expected, the five-variable model has the largest R^2. The smallest value of C_k is 1.4. This corresponds to the model containing only the two variables x_1 and x_2. From Table 12.6 we see that the smallest MSE is associated with the three-variable model containing the predictors x_1, x_2, and x_3. The smallest PRESS statistic corresponds to the model containing the variables x_1, x_4, and x_5. Summarizing, we find that our criteria suggest these models:

Criterion	Independent variables used
Largest R^2	x_1, x_2, x_3, x_4, x_5
C_k	x_1, x_2
MSE	x_1, x_2, x_3
PRESS	x_1, x_4, x_5

We have a problem! Which model do we choose? Let us assume that we want good predictive ability and as good a fit of the data as possible. Looking at Table 12.6 more closely, we see that the first five models listed differ only slightly in MSE. The second, fourth, and fifth of these models have reasonably close values of the PRESS statistic. Hence let us look further at models 2, 4, and 5. The

TABLE 12.5
Values of R^2 and C_k

Number of variables in model	R^2	C_k	Variables in the model
1	.003	13.2	x_5
1	.219	9.0	x_2
1	.347	6.6	x_4
1	.379	6.0	x_3
1	.448	4.6	x_1
2	.347	8.6	x_4, x_5
2	.380	8.0	x_3, x_5
2	.396	7.6	x_3, x_4
2	.421	7.2	x_2, x_4
2	.478	6.1	x_2, x_3
2	.593	3.8	x_2, x_5
2	.612	3.5	x_1, x_3
2	.616	3.4	x_1, x_4
2	.647	2.8	x_1, x_5
2	.719	1.4*	x_1, x_2
3	.397	9.6	x_3, x_4, x_5
3	.481	8.0	x_2, x_3, x_4
3	.603	5.7	x_2, x_4, x_5
3	.630	5.1	x_1, x_3, x_4
3	.642	4.9	x_2, x_3, x_5
3	.720	3.4	x_1, x_2, x_5
3	.765	2.5	x_1, x_2, x_4
3	.776	2.3	x_1, x_3, x_5
3	.779	2.3	x_1, x_4, x_5
3	.783	2.2	x_1, x_2, x_3
4	.652	6.7	x_2, x_3, x_4, x_5
4	.780	4.2	x_1, x_2, x_4, x_5
4	.784	4.2	x_1, x_2, x_3, x_4
4	.788	4.1	x_1, x_2, x_3, x_5
4	.790	4.1	x_1, x_3, x_4, x_5
5	.793*	6.0	x_1, x_2, x_3, x_4, x_5

characteristics of these models are given below:

Model	Variables	MSE	PRESS	R^2	C_k
2	x_1, x_4, x_5	507.00	7758.1	.779	2.3
4	x_1, x_2, x_4	537.89	9097.3	.765	2.5
5	x_1, x_2	552.80	8855.7	.719	1.4

We note that each of these models has a reasonably large R^2 relative to the maximum R^2 of .793. Model 5 has the smallest value of C_k but model 2 has the

TABLE 12.6
Estimated models ordered by MSE

$\hat{\beta}_0$	$\hat{\beta}_1$	$\hat{\beta}_2$	$\hat{\beta}_3$	$\hat{\beta}_4$	$\hat{\beta}_5$	*MSE*	PRESS
−16.46	.135	.090	.441	.000	.000	497.09*	11015.6
−101.92	.195	.000	.000	.100	1.103	507.00	7758.1*
−150.64	.190	.000	.595	.000	1.104	514.42	10557.8
22.34	.141	.087	.000	.065	.000	537.89	9097.3
30.64	.159	.107	.000	.000	.000	552.80	8855.7
−127.82	.189	.000	.307	.058	1.093	578.10	17167.8
−71.20	.158	057	.493	.000	.451	583.15	20719.5
−12.76	.135	.089	.395	.010	.000	595.42	19444.9
−79.07	.185	.018	.000	.093	.897	606.00	16131.7
54.13	.147	.122	.000	.000	−.215	641.93	16905.5
−94.18	.227	.000	.000	.000	1.211	693.16	10013.5
−89.57	.171	.033	.338	.040	.724	713.42	30456.1
37.34	.122	.000	.000	.112	.000	754.37	12085.2
−17.19	.116	.000	.668	.000	.000	761.60	15478.5
261.15	.000	.239	.000	.000	−1.989	799.87	8877.2
169.28	.000	.192	.416	.000	−1.545	821.08	15271.4
6.73	.116	.000	.346	.065	.000	847.91	27762.1
230.98	.000	.211	.000	.035	−1.742	910.08	11644.5
61.67	.150	.000	.000	.000	.000	948.16	12618.8
167.11	.000	.211	.639	−.055	−1.694	956.45	29312.3
−21.21	.000	.067	.815	.000	.000	1024.92	20604.9
−21.27	.000	.000	.950	.000	.000	1067.65	20324.6
60.25	.000	.000	.000	.155	.000	1122.04	14651.1
52.40	.000	.060	.000	.127	.000	1137.67	13480.9
2.34	.000	.000	.633	.064	.000	1186.27	37564.8
−12.36	.000	.064	.702	.024	.000	1190.57	41585.2
−29.07	.000	.000	.956	.000	.063	1218.50	24933.1
59.36	.000	.000	.000	.155	.008	1282.30	25800.4
79.36	.000	.096	.000	.000	.000	1341.90	20320.4
−4.57	.000	.000	.640	.064	.055	1382.52	48787.1
105.00	.000	.000	.000	.000	.000	1528.21	16980.1
117.81	.000	.000	.000	.000	−.116	1714.26	21392.9

smallest value of PRESS, the smallest MSE, and a small value of C_k. Practically speaking, each of these models would probably perform reasonably well. For predictive purposes, model 2 is our choice. This fitted model is given by

$$\hat{\mu}_{Y|x_1, x_2} = -101.92 + 0.195x_1 + 0.100x_4 + 1.103x_5$$

12.8 MODEL TRANSFORMATION AND CONCLUDING REMARKS

We have seen how to estimate a curve of regression when it is appropriate to assume that a linear relationship exists between x and the mean of Y. We have also seen how to use scattergrams of the data and residual plots to get a visual

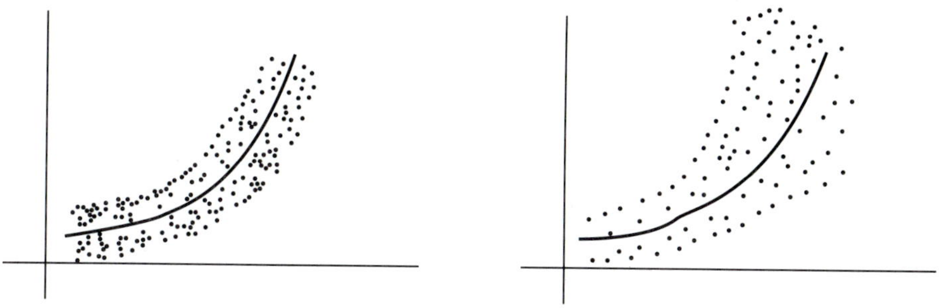

(*a*)

(*b*)

FIGURE 12.3
(*a*) A scattergram for which a polynomial model is appropriate; (*b*) a scattergram for which an exponential or a power model should be tried.

check of the validity of this assumption. If repeated observations are available at some regressor values, then we have learned how to test for lack of fit. What can we do if there is strong evidence that a linear regression is not appropriate? This question is not easy to answer since the approach taken depends on how the data deviates from linearity or violates the assumptions underlying the simple linear regression model. In this section we summarize a few useful "tricks of the trade."

To begin, consider the scattergrams shown in Fig. 12.3. In each case an "eyeball" curve of regression has been drawn through the data. What sort of equation would best describe these curves? This question is not easy to answer. In fact, there may be several different equations that work well. Notice that a fundamental difference exists in the scattergrams presented. In each case a curve rather than a straight line seems to fit the data. In Fig. 12.3(*a*) there is no apparent change in spread or variability as the value of x changes. Thus there is no suggestion that the assumption of equality of variance is violated. This is not the case in Fig. 12.3(*b*). Not only do the data suggest a curve rather than a straight line, but the variance does appear to be different at different values of x. Due to this difference in the data, the two cases are not handled in the same way. In the former case a polynomial model is probably appropriate. That is, we use the techniques presented in this chapter to fit a model of the form

$$\mu_{Y|x} = \beta_0 + \beta_1 x + \beta_2 x^2 + \cdots + \beta_p x^p$$

where p is a positive integer greater than 1.

In the latter case either an exponential model or a power model can be tried. These models are nonlinear models because they do not express the average response as a linear function of the parameters. They are explained in the following example.

Example 12.8.1 (Exponential model). The exponential model assumes the form

$$\mu_{Y|x} = \beta_0 e^{\beta_1 x} \qquad x > 0$$

or

$$y_i = \beta_0 e^{\beta_1 x_i} e_i$$

Notice that this model is different from those presented earlier in two respects. First, it is nonlinear. Second, the random error e_i is not added to the term $\beta_0 e^{\beta_1 x_i}$; rather, it is a multiplier. This fact accounts for the difference in variance for different values of x. Even though the model is nonlinear, it is intrinsically linear. To say that a model is intrinsically linear means that it can be tranformed or rewritten in an equivalent form that is linear. This process, called *linearization*, is accomplished in this case by taking natural logarithms to obtainy

$$\ln y_i = \ln \beta_0 e^{\beta_1 x_i} e_i$$

By the laws of logarithms the right-hand side of this equation can be written as $\ln \beta_0 + \beta_1 x_i + \ln e_i$. If we let $\ln y_i = y_i^*$, $\ln \beta_0 = \beta_0^*$, $\beta_1 = \beta_1^*$, and $\ln e_i = e_i^*$, then the transformed model becomes

$$y_i^* = \beta_0^* + \beta_1^* x_i + e_i^*$$

This model is a simple linear regression model. The method of least squares is used to estimate β_0^* and β_1^*. We are fitting the model by regression, the natural logarithm of the original response versus the regressor. Notice that $\beta_1^* = \beta_1$, the parameter in the exponent of the original model. Hence $\hat{\beta}_1 = \hat{\beta}_1^*$. However $\beta_0^* \neq \beta_0$. To estimate β_0, the coefficient in the original model, we use the relationship $\beta_0^* = \ln \beta_0$ or $\beta_0 = e^{\beta_0^*}$. Hence $\hat{\beta}_0 = e^{\hat{\beta}_0^*}$.

Keep in mind the fact that the scattergram shown in Fig. 12.3(*b*) is not the only pattern that suggests an exponential model. In fact, the pattern shown is one for which β_0 and β_1 are both positive. In Exercises 59 and 60 you are asked to plot scattergrams for other possible exponential models.

Example 12.8.2 (Power model). The power model is a model of the form

$$\mu_{Y|x} = \beta_0 x^{\beta_1} \qquad x > 0$$

or

$$y_i = \beta_0 x_i^{\beta_1} e_i$$

This model, like the exponential model, is intrinsically linear. The logarithmic transformation also linearizes this model as follows:

$$\ln y_i = \ln \beta_0 x_i^{\beta_1} e_i$$

or

$$\ln y_i = \ln \beta_0 + \beta_1 \ln x_i + \ln e_i$$

Here $y_i^* = \ln y_i$, $\beta_0^* = \ln \beta_0$, $\beta_1^* = \beta_1$, $x_i^* = \ln x_i$, and $e_i^* = \ln e_i$. The new model becomes

$$y_i^* = \beta_0^* + \beta_1^* x_i^* + e_i^*$$

The parameters are estimated by using least squares. Notice that we are regressing

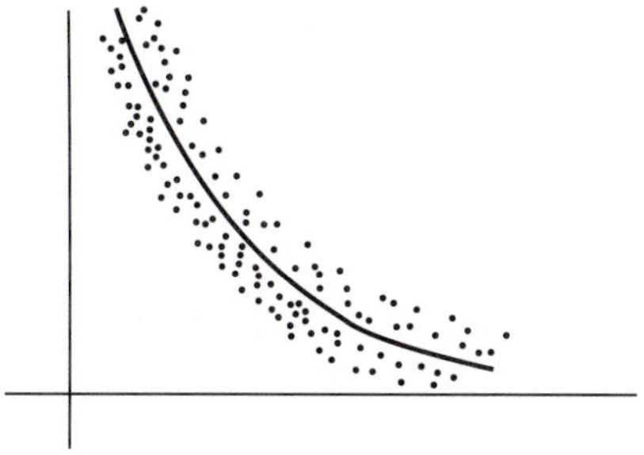

FIGURE 12.4
A scattergram for which a reciprocal with $\beta_1 > 0$ model is appropriate.

the natural logarithm of the original response versus the natural logarithm of the original regressor. Parameter estimates are $\hat{\beta}_1 = \hat{\beta}_1^*$ and $\hat{\beta}_0 = e^{\hat{\beta}_0^*}$.

Figure 12.3(*b*) illustrates a scattergram in which a power model with $\beta_0 > 0$ and $\beta_1 > 1$ might be appropriate. In Exercises 61 through 63 you are asked to explore other possible forms for scattergrams that might suggest a power model.

The next model is a linear model in which the regressor is a function of x rather than x itself. A scattergram for a data set for which this model might be appropriate is shown in Fig. 12.4.

Example 12.8.3 (Reciprocal model). The reciprocal model assumes the form

$$\mu_{Y|x} = \beta_0 + \beta_1(1/x) \qquad x > 0$$

or

$$y_i = \beta_0 + \beta_1(1/x) + e_i$$

The model is linear already since it does express the average response as a linear function of the parameters β_0 and β_1. In fitting the model by using least squares, we regress the response versus the reciprocal of the regressor rather than the regressor itself.

Before closing this section, we need to make an additional point. One of the basic assumptions of simple linear regression is that of equality of variance. If a scattergram suggests that this assumption is not valid, then using a logarithmic transformation often helps to stabilize the variance. That is, we let $y^* = \ln y$ and then regress y^* versus x. In this way we estimate the model

$$\mu_{Y^*|x} = \beta_0 + \beta_1 x$$

where $\mu_{Y^*|x}$ is the average value of the natural logarithm of the response for a given x value. Once estimates for β_0 and β_1 are obained, the fitted line of regression can be used to estimate y^*, the natural logarithm of the original response for a given value of x. The estimated response can be recovered by using the relationship

$$\hat{y} = e^{\hat{y}^*}$$

Remember that regression is an art. There may be several models that appear to fit a given set of data. The job of the researcher is to investigate these models to find the one that yields the best fit and gives the most reasonable explanation of the relationship between the regressor and the average response.

In this chapter we have only touched on the powerful statistical tool known as regression analysis. There are many aspects of this topic that have not been mentioned. For example, if predictor variables are correlated (they are linearly dependent), then we say that we have *multicollinearity*. When this happens, the least-squares estimators are ubiased but their variances can be very large. A procedure called *ridge regression* is often used in this situation. This procedure yields biased estimators, but the variance is usually reduced so that the mean squared error is relatively small. The procedure is discussed in detail in texts on regression analysis [39]. Another rather recent approach to regression analysis is called *robust regression*. This procedure is useful when the assumption of normality does not seem realistic or when outliers which greatly influence the usual least-squares estimators are present in the data. The procedure is still somewhat controversial [12].

Our best suggesion is this. If you are engaged in a serious research project that might involve regression, seek the help of a statistician who is knowledgeable in this area in the design stages of your study. You have learned enough about regression from this text to be able to converse with such a person; you have not learned enough to be able to use regression to its fullest. There are several excellent texts on the market that are devoted solely to the discussion of regression analysis, among them are [12] and [39].

CHAPTER SUMMARY

The simple linear regression model discussed in Chap. 11 was extended in this chapter. Extensions included the model for several linear independent variables, the polynomial model for a single independent variable, and combinations of both these cases. These models were then developed in matrix form, and the least-squares estimation procedure and properties of this procedure were presented. Methods of confidence interval estimation were given for these models for a single slope, the predicted mean, and a single predicted value. Hypothesis testing methods were also discussed for testing the significance of a single predictor variable, for testing for significant regression, and for a subset of predictor variables. We pointed out that these models are very useful in applications but typically require a computer for estimation of model parameters.

The multiple correlation coefficient and the coefficient of multiple determination was also defined and discussed.

In applications, deciding which predictor variables should be included in a selected model is not a trivial chore. Several of the more commonly used methods of variable selection were presented and discussed. These included forward selection, backward elimination, stepwise procedure, maximum R^2, Mallow's C_k, and the PRESS statistic.

We also introduced and defined important terms that you should know. These are:

Multiple linear regression
Model in matrix form
Predictor variable
Variable selection
Coefficient of determination
Power model
Exponential model
Reciprocal model
Polynomial regression
Least-squares estimators
Significant regression
Multiple correlation
Indicator variable
C_k
PRESS statistic

EXERCISES

Section 12.1

1. The simple linear regression model is a polynomial model of what degree? Verify that, in this case, the normal equations given in (12.3) reduce to those given in Chap. 11 for the simple linear regression model.
2. The simple linear regression model is also a multiple linear regression model with $k = 1$. Verify that, in this case, the normal equations given in (12.5) reduce to those given in Chap. 11 for the simple linear regression model.
3. Consider the model $\mu_{Y|x_1, x_2} = \beta_0 + \beta_1 x_1 + \beta_2 x_2$. These data are available:

x_1	x_2	y
0	8	9
2	9	8
4	8	7

(*a*) Find

$$\sum_{i=1}^{3} x_{1i} \qquad \sum_{i=1}^{3} x_{2i} \qquad \sum_{i=1}^{3} x_{1i}x_{2i} \qquad \sum_{i=1}^{n} x_{1i}y_i$$

$$\sum_{i=1}^{3} x_{1i}^2 \qquad \sum_{i=1}^{3} x_{2i}^2 \qquad \sum_{i=1}^{3} y_i \qquad \sum_{i=1}^{3} x_{2i}y_i$$

(*b*) Find the normal equations.

(*c*) Show that $b_0 = 9$, $b_1 = -.5$, and $b_2 = 0$ are solutions to the normal equations.

4. Consider the model

$$\mu_{Y|x_1, x_2} = \beta_0 + \beta_1 x_1 + \beta_{11} x_1^2 + \beta_2 x_2$$

Express this model in the general linear form of (12.1).

5. Consider the model

$$\mu_{Y|x_1,x_2} = \beta_0 + \beta_1 x_1 + \beta_2 x_2 + \beta_{12} x_1 x_2$$

Express this model in the general linear form of (12.1).

Section 12.2

6. Consider the simple linear regression model

$$\mu_{Y|x} = \beta_0 + \beta_1 x$$

Show that

$$(X'X)^{-1} = \frac{1}{nS_{xx}} \begin{bmatrix} \sum_{i=1}^{n} x_i^2 & -\sum_{i=1}^{n} x_i \\ -\sum_{i=1}^{n} x_i & n \end{bmatrix}$$

7. Consider the data of Exercise 7 of Chap. 11.
(*a*) Use the matrix approach to find the normal equations and compare your answer to that found in Exercise 8 of Chap. 11.
(*b*) Solve the normal equations by using matrix algebra and compare your answers to those found earlier.

8. Consider the data of Exercise 9 of Chap. 11. Estimate the regression line by using the matrix approach.

9. Consider the data of Exercise 10 of Chap. 11. Use the matrix approach to estimate the line of regression.

10. In simple linear regression, $\hat{y}$, the estimated response for a given value of x, can be written in matrix notation as follows:

$$\hat{y} = [b_0 \quad b_1]\begin{bmatrix} 1 \\ x \end{bmatrix}$$

(*a*) Verify that the above technique is valid.
(*b*) Use this method and the results of Exercise 7 to estimate the average energy consumed by a household for which the income is $50,000.
(*c*) Use this method and the results of Exercise 8 to estimate the pitch of a particular connector of length .03 inch.
(*d*) Use this method and the results of Exercise 9 to estimate the surface feet per minute that can be covered when a wheel is used at 3450 rpm (revolutions per minute).

11. In developing a simple linear regression model for predicting gasoline mileage, based on the weight of the car, these data are available:

x	1.35	1.90	1.70	1.80	1.30	2.05	1.60	1.80	1.85	1.40
y	17.9	16.5	16.4	16.8	18.8	15.5	17.5	16.4	15.9	18.3

(*a*) Find the model specification matrix.
(*b*) Find $X'X$.
(*c*) Find $X'\mathbf{y}$.
(*d*) Find the normal equations via matrix algebra.
(*e*) Find $(X'X)^{-1}$.
(*f*) Find the least-squares estimates for β_0 and β_1 by using matrix algebra. Compare these to the values found in Example 11.3.3

12. Consider these data for Exercises 12 and 13:

x	1.6	1.8	1.4	2.0	1.2	2.2	1.0	2.4	.8	2.6
y	12	6	13	5	10	1	20	1	24	0

(*a*) Find the model specification matrix for the quadratic model

$$\mu_{Y|x} = \beta_0 + \beta_1 x + \beta_2 x^2$$

(*b*) Find $X'X$.
(*c*) Find $X'\mathbf{y}$.
(*d*) Show that, apart from round-off error,

$$(X'X)^{-1} = \begin{bmatrix} 8.733 & -10.8182 & 3.0303 \\ -10.8182 & 13.9867 & -4.0246 \\ 3.0303 & -4.0246 & 1.1837 \end{bmatrix}$$

Hint: Simply show that $(X'X)(X'X)^{-1} = \mathrm{I}$.
(*e*) Find $\mathbf{b}$.

13. (*a*) Write the expression for the estimated model.
(*b*) Argue that in a model of this sort, for a specific value of x,

$$\hat{\mu}_{Y|x} = \hat{y} = [b_0 \quad b_1 \quad b_2] \begin{bmatrix} 1 \\ x \\ x^2 \end{bmatrix}$$

(*c*) Use the estimated model to predict the mean value of y when $x = 2.5$.

14. In Exercise 3 we considered the model $\mu_{Y|x_1, x_2} = \beta_0 + \beta_1 x_1 + \beta_2 x_2$ based on these data:

x_1	x_2	y
0	8	9
2	9	8
4	8	7

(*a*) Find the model specification matrix.
(*b*) Find $X'X$.
(*c*) Find $X'\mathbf{y}$.
(*d*) Find the normal equations and compare them to those found in Exercise 3.

(*e*) Show that

$$(X'X)^{-1} = \begin{bmatrix} \frac{1680}{16} & \frac{-4}{16} & \frac{-200}{16} \\ \frac{-4}{16} & \frac{2}{16} & 0 \\ \frac{-200}{16} & 0 & \frac{24}{16} \end{bmatrix}$$

(*f*) Verify that $\mathbf{b} = \begin{bmatrix} 9 \\ -.5 \\ 0 \end{bmatrix}$

15. Write the model specification matrix for the model

$$\mu_{Y|x_1, x_2} = \beta_0 + \beta_1 x_1 + \beta_{11} x_1^2 + \beta_2 x_2$$

based on a random sample of size 8.

16. Write the model specification matrix for the model

$$\mu_{Y|x_1, x_2} = \beta_0 + \beta_1 x_1 + \beta_2 x_2 + \beta_{12} x_1 x_2$$

based on a random sample of size 10.

17. Reconsider the data of Exercise 1 of Chap. 11. Can you suggest a model that might be appropriate for these data?

Section 12.3

18. Let $C = \begin{bmatrix} 2 & 8 \\ 3 & 6 \\ 1 & 0 \end{bmatrix}$. Let Y_1 and Y_2 be independent random variables with $E[Y_1] = 3$, $E[Y_2] = 9$, and $\text{Var } Y_1 = \text{Var } Y_2 = 16$.

(*a*) Find $E[C]$.

(*b*) Find $E[C\mathbf{Y}]$, where $\mathbf{Y} = \begin{bmatrix} Y_1 \\ Y_2 \end{bmatrix}$

(*c*) Find Var $\mathbf{Y}$.

(*d*) Find Var $C\mathbf{Y}$.

19. Let $C = \begin{bmatrix} 3 & 2 \\ 1 & 7 \end{bmatrix}$. Let Y_1 and Y_2 be independent random variables with $E[Y_1] = 5$, $E[Y_2] = 10$, $\text{Var } Y_1 = \text{Var } Y_2 = 6$.

(*a*) Find $E[C\mathbf{Y}]$ where $\mathbf{Y} = \begin{bmatrix} Y_1 \\ Y_2 \end{bmatrix}$.

(*b*) Find Var $\mathbf{Y}$.

(*c*) Find Var $C\mathbf{Y}$.

Consider the simple linear regression model for Exercises 20 through 22.

20. Find Var $\hat{\boldsymbol{\beta}}$. That is, find the variance-covariance matrix for this model. *Hint:* See Exercise 6.

21. Find Var $\hat{\beta}_0$ and Var $\hat{\beta}_1$ from the variance-covariance matrix. Compare your results to those given in Chap. 11, Sec. 2.

22. Are B_0 and B_1 uncorrelated? Explain, based on the variance-covariance matrix.

23. Consider the model

$$\mu_{Y|x_1, x_2} = \beta_0 + \beta_1 x_1 + \beta_2 x_2$$

By definition

$$\text{SSE} = \sum_{i=1}^{n} [Y_i - (B_0 + B_1 x_{1i} + B_2 x_{2i})]^2$$

(*a*) Square the term on the right and sum over i to obtain

$$\text{SSE} = \sum_{i=1}^{n} \Big[Y_i^2 - Y_i(B_0 + B_1 x_{1i} + B_2 x_{2i})$$

$$+ (B_0 + B_1 x_{1i} + B_2 x_{2i})^2 - Y_i(B_0 + B_1 x_{1i} + B_2 x_{2i}) \Big]$$

(*b*) Show that

$$\sum_{i=1}^{n} \Big[(B_0 + B_1 x_{1i} + B_2 x_{2i})^2 - Y_i(B_0 + B_1 x_{1i} + B_2 x_{2i}) \Big]$$

$$= B_0 \sum_{i=1}^{n} (B_0 + B_1 x_{1i} + B_2 x_{2i} - Y_i)$$

$$+ B_1 \sum_{i=1}^{n} \left(B_0 x_{1i} + B_1 x_{1i}^2 + B_2 x_{1i} x_{2i} - x_{1i} Y_i \right)$$

$$+ B_2 \sum_{i=1}^{n} \left(B_0 x_{2i} + B_1 x_{1i} x_{2i} + B_2 x_{2i}^2 - x_{2i} Y_i \right)$$

(*c*) Use the normal equations for the multiple linear regression model given in Sec. 12.1 to argue that each of the components on the right of the equation in part (*b*) is equal to 0.

(*d*) Show that

$$\text{SSE} = \sum_{i=1}^{n} Y_i^2 - B_0 \sum_{i=1}^{n} Y_i - B_1 \sum_{i=1}^{n} x_{1i} Y_i - B_2 \sum_{i=1}^{n} x_{2i} Y_i$$

thus partially verifying the computations used to find SSE.

24. For the simple linear regression model we found that

$$\text{SSE} = S_{yy} - B_1 S_{xy}$$

(See Sec. 11.2.) In this section we defined SSR for this model by

$$\text{SSR} = B_0 \sum_{i=1}^{n} Y_i + B_1 \sum_{i=1}^{n} x_{1i} Y_i - \left(\sum_{i=1}^{n} Y_i \right)^2 \Big/ n$$

Show that $B_1 S_{xy} = \text{SSR}$, thus verifying that the results obtained here coincide with those found earlier.

25. In Example 12.2.5, we developed a quadratic regression equation from which the unit cost of producing a drug can be predicted based on the number of units produced. Use the information given there to estimate Var B_0, Var B_1, and Var B_2.

26. For the quadratic model developed in Exercise 12, estimate Var B_0, Var B_1, and Var B_2.

Section 12.4

27. Use the data of Example 12.4.1 to find 95% confidence intervals on β_1 and β_2. Is there evidence that $\beta_1 \neq 0$? That $\beta_2 \neq 0$? Explain.

28. Use the data of Example 12.2.5 to find 95% confidence intervals on β_1 and β_2. Is there evidence that $\beta_1 \neq 0$? That $\beta_2 \neq 0$? Explain.

29. Use the information given in Example 12.3.3 to find a 90% confidence interval on the mean gasoline mileage obtained by cars weighing 1.5 tons when operated on a 40°F day. Find a 90% prediction interval on the gasoline mileage obtained by a specific automobile weighing 1.5 tons when operated on a 40°F day. Which interval is wider?

30. Use the information given in Example 12.2.5 to find a 95% confidence interval on the mean unit cost of producing 12 units of the given drug. Find a 95% prediction interval on the unit cost of producing a particular lot of 12 units of the drug.

31. Use the information from Example 12.2.4 to find a 95% confidence interval on the mean extent of solvent evaporation when the humidity at the time of spraying is 50%. Find a 95% prediction interval on the extent of solvent evaporation for a particular day on which the humidity is 50%.

The three basic structural elements of a data processing system are files, flows, and processes. Files are collections of permanent records in the system, flows are data interfaces between the system and the environment, and processes are functionally defined logical manipulations of data. An investigation of the cost of developing software as related to files, flows, and processes was investigated in "A Software Matrix for Cost Estimation and Efficiency Measurement in Data Processing System Development," *Journal of Systems Software*, Vol. 3, 1983. The following data are based on that study:

Cost (in units of 1000) (y)	Files (x_1)	Flows (x_2)	Processes (x_3)
22.6	4	44	18
15.0	2	33	15
78.1	20	80	80
28.0	6	24	21
80.5	6	227	50
24.5	3	20	18
20.5	4	41	13
147.6	16	187	137
4.2	4	19	15
48.2	6	50	21
20.5	5	48	17

Exercises 32 through 36 refer to these data.

32. Consider the model

$$\mu_{Y|x_1, x_2, x_3} = \beta_0 + \beta_1 x_1 + \beta_2 x_2 + \beta_3 x_3$$

(*a*) Find the model specification matrix.
(*b*) For these data

$$(X'X)^{-1} = \begin{bmatrix} .3197263 & -.0408268 & -.00202208 & .005305965 \\ -.0408268 & .0140738 & .0003717104 & -.00224159 \\ -.00202208 & .0003717104 & .00005188447 & -.000113861 \\ .005305965 & -.00224159 & -.000113861 & .0004938527 \end{bmatrix}$$

$$X'\mathbf{y} = \begin{bmatrix} 489.7 \\ 5259 \\ 59088.4 \\ 33845.2 \end{bmatrix}$$

$$s^2 = 98.37516961$$

Use this information to estimate the model.

33. Find a 95% confidence interval on $\mu_{Y|x_1, x_2, x_3}$ when $x_1 = 12$, $x_2 = 40$, and $x_3 = 20$.

34. Find a 95% prediction interval on the cost of a single system when $x_1 = 12$, $x_2 = 40$, and $x_3 = 20$.

35. Find a 90% confidence interval on β_0.

36. Find a 90% confidence interval on β_1.

Section 12.5

Consider the following data:

x	1	2	3	4	5	6	7
y	8	17	29	34	46	42	52

Use these data for Exercises 37 through 42.

37. Fit a regression curve of the form $\mu_{Y|x} = \beta_0 + \beta_1 x$.

38. Find the model specification matrix for a model of the form

$$\mu_{Y|x} = \beta_0 + \beta_1 x + \beta_2 x^2$$

39. For the model of Exercise 38,

$$(X'X)^{-1} = \begin{bmatrix} 2.428571 & -1.28571 & .1428571 \\ -1.28571 & .797619 & -.0952381 \\ .1428571 & -.0952381 & .01190476 \end{bmatrix}$$

$$X'\mathbf{y} = \begin{bmatrix} 228 \\ 1111 \\ 6091 \end{bmatrix}$$

$$s^2 = 12.27380952$$

Use this information to estimate the model.

40. Test an appropriate hypothesis to decide whether the quadratic regression curve significantly fits the data better than the linear regression curve.

41. Using the regression curve selected from Exercise 40, compute a 95% confidence interval on $\mu_{Y|x}$, when $x = 4.8$.

42. Using the regression curve selected from Exercise 40, compute a 95% prediction interval on an individual response when $x = 4.8$.

A research study was conducted on cracking of latex paint on wooden structures. The primary concern in the study is to investigate the effect of water permeability and fracture energy (energy to propagate a crack through paint film) on paint crack rating. The investigation yielded the following data:

Sample number	Crack rating (y)	Permeability (x_1)	Fracture energy (x_2)
1	2	2.1	4.31
2	9	8.4	22.11
3	5	5.1	11.40
4	10	14.5	24.15
5	3	4.4	6.21
6	3	6.2	5.65
7	8	12.5	9.71
8	7	7.0	12.00
9	8	17.2	14.25
10	5	7.1	8.63

Refer to these data for Exercises 43 through 49.

43. Plot y versus x_1 and y versus x_2.

44. Find the model specification matrix for the model

$$\mu_{Y|x_1, x_2} = \beta_0 + \beta_1 x_1 + \beta_2 x_2$$

45. For the model of Exercise 44,

$$(X'X)^{-1} = \begin{bmatrix} .529839 & -.02535552 & -.0182053 \\ -.0253552 & .00772376 & -.00337026 \\ -.0182053 & -.00337026 & .003942239 \end{bmatrix}$$

$$X'\mathbf{y} = \begin{bmatrix} 60 \\ 604.2 \\ 860.52 \end{bmatrix}$$

$$s^2 = .94334776$$

Use this information to estimate the model.

46. Find R^2 for the estimated model of Exercise 45.

47. Estimate ρ, the correlation between the observed and predicted responses for the model of Exercise 45.

48. Test for significant regression for the curve estimated in Exercise 45.

49. Test H_0: $\beta_2 = 0$. Do you think that x_2 is needed in a model that already contains the variable x_1? Explain.

50. Test for a significant regression effect in Exercise 32.

51. Using the data of Exercise 32, test H_0: $\beta_2 = \beta_3 = 0$.

Section 12.6

52. A study was conducted to study energy consumption versus household income and home ownership status. Let Y (in units of 10^8 Btu) denote energy consumption, x_1 (in units of \$1000/year) denote income, and $x_2 (x_2 = 1, 0)$ for ownership versus rental, respectively, denote ownership status. The following data were obtained:

Consumption (y)	**Income** (x_1)	**Ownership** (x_2)
1.8	20.0	0
4.7	25.0	1
3.0	30.5	0
5.8	35.1	1
4.8	40.0	0
7.1	48.2	1
5.0	55.1	0
8.0	60.5	1
7.0	74.9	0
9.9	80.3	1
9.0	88.4	0
11.3	90.1	1
9.2	95.2	0

(*a*) Assume that an appropriate model is

$$\mu_{Y|x_1, x_2} = \beta_0 + \beta_1 x_1 + \beta_2 x_2$$

Find the model specification matrix.

(*b*) For these data

$$(X'X)^{-1} = \begin{bmatrix} .5416711 & -.00690843 & -.151114 \\ -.00690843 & .0001196709 & .0001430353 \\ -.151114 & .0001430353 & .3096948 \end{bmatrix}$$

and

$$X'\mathbf{y} = \begin{bmatrix} 86.6 \\ 5751.14 \\ 46.8 \end{bmatrix}$$

Use this information to estimate the model.

(*c*) For the given data

$$\hat{\sigma}^2 = .13417107$$

Use this information to test H_0: $\beta_2 = 0$.

(*d*) State the estimated models that describe the relationship between energy consumption and income for homeowners; for renters.

(*e*) Find a point estimate for the difference in intercepts for the two models in part (*d*).

53. An engineer is investigating the recovery of heat lost to the environment in the form of exhaust gases for two types of furnaces. The experiment is designed to fix flow speed past heat pipes (in meters per second) and then to measure the recovery ratio.

Recovery (y)	**Flow speed** (x_1)	**Type furnace** (x_2)
.740	1	A
.745	1.5	A
.718	2	A
.678	2.5	A
.652	3	A
.627	3.5	A
.607	4	A
.507	4.5	A
.545	5	A
.402	1	B
.321	2	B
.255	2.5	B
.175	3	B
.115	4	B
.085	5	B

(*a*) Assume that it is not known whether or not flow speed affects each type of furnace in a similar way. The appropriate model is

$$\mu_{Y|x_1,x_2} = \beta_0 + \beta_1 x_1 + \beta_2 x_2 + \beta_3 x_1 x_2$$

where $x_2 = 1$ if furnace A is used and $x_2 = 0$ if furnace B is used. Find the model specification matrix.

(*b*) For these data

$$(X'X)^{-1} = \begin{bmatrix} 1 & -.285714 & -1 & .2857143 \\ -.285714 & .09795918 & .2857143 & -.0979592 \\ -1 & .2857143 & 1.711111 & -.485714 \\ .2857143 & -.0979592 & -.485714 & .1646259 \end{bmatrix}$$

and

$$X'\mathbf{y} = \begin{bmatrix} 7.172 \\ 19.665 \\ 5.819 \\ 16.5735 \end{bmatrix}$$

Use this information to estimate the model.

(*c*) Estimate the difference in slopes for the regression lines for the two furnaces.

(*d*) For the given data

$$\hat{\sigma}^2 = .0007562574$$

Use this information to test H_0: $\beta_3 = 0$. What practical conclusion can be drawn?

54. Consider the problem of Exercise 52. Suppose that in a future study we want to differentiate between types of rental property and home ownership so that x_2 assumes four levels. These are (i) owner of single family dwelling, (ii) owner of townhouse or condominium, (iii) renter of single family home, (iv) renter of condominium or apartment.

(*a*) How many indicator variables are needed to code ownership status?

(*b*) Assume that the slopes of the four regression lines are identical. Write an appropriate model.

(*c*) Let

$x_2 = x_3 = x_4 = 0$ for owners of a single family dwelling

$x_2 = x_3 = 0$ and $x_4 = 1$ for owners of a townhouse or condo

$x_2 = x_4 = 0$ and $x_3 = 1$ for renters of single family dwellings

$x_3 = x_4 = 0$ and $x_2 = 1$ for apartment renters

Write the model for each of these groups.

(*d*) What null hypothesis must be tested to test for equality of all four intercepts?

55. Suppose that we have a model with one quantitative variable, x_1, and one qualitative variable at three levels A, B, and C. Consider the model

$$\mu_{Y|x_1,x_2,x_3} = \beta_0 + \beta_1 x_1 + \beta_2 x_2 + \beta_3 x_3 + \beta_4 x_1 x_2 + \beta_5 x_1 x_3$$

where

$x_2 = x_3 = 0$ for level A

$x_2 = 1$ and $x_3 = 0$ for level B

$x_2 = 0$ and $x_3 = 1$ for level C

(*a*) Write the models for each of the three levels.

(*b*) What null hypothesis must be tested to test for equality of slopes among these three regression lines?

(*c*) What null hypothesis must be tested to test simultaneously for equality of slopes and intercepts? How many degrees of freedom are associated with the F ratio used to conduct this test?

Section 12.7

56. Assume that we have available four possible predictor variables x_1, x_2, x_3, x_4. Suppose that our final model via forward selection contains only the variables x_4 and x_2 and that they entered the model in the order stated. Outline the steps taken in developing this model. Follow the format given in Example 12.7.1.

57. Assume that we have four potential predictor variables and that via backward elimination we obtain a reduced model containing only the variables x_1 and x_2. Assume that the variables x_3 and x_4 are deleted in the order mentioned. Outline the steps taken in developing this model. Follow the format given in Example 12.7.2.

58. In a multiple linear regression model variables x_2 and x_4 are closely related, with variable x_4 being the best single predictor. Suppose that the final model contains the two variables x_2 and x_1, with variable x_1 entering on the second stage and x_2 entering on the third. Outline the steps used to develop this model via stepwise regression.

Section 12.8

59. Consider the model

$$\mu_{Y|x} = \beta_0 e^{\beta_1 x} \qquad x > 0$$

(a) Find the first derivative of the function $\beta_0 e^{\beta_1 x}$.
(b) If $\beta_0 > 0$ and $\beta_1 < 0$, is the regression curve increasing or decreasing?
(c) Find the second derivative of the function $\beta_0 e^{\beta_1 x}$.
(d) If $\beta_0 > 0$ and $\beta_1 < 0$, is the regression curve concave up or concave down?
(e) Sketch a scattergram for which it is reasonable to assume that an exponential model with $\beta_0 > 0$ and $\beta_1 < 0$ is appropriate.

60. Consider the model

$$\mu_{Y|x} = \beta_0 e^{\beta_1 x} \qquad x > 0$$

Sketch a scattergram for which this model is appropriate with $\beta_0 < 0$ and $\beta_1 > 0$, with $\beta_0 < 0$ and $\beta_1 < 0$.

61. Consider the model

$$\mu_{Y|x} = \beta_0 x^{\beta_1} \qquad x > 0$$

(a) Find the first derivative of the function $\beta_0 x^{\beta_1}$.
(b) Verify that if $\beta_0 > 0$ and $\beta_1 > 1$, then this function is increasing as shown in Fig. 12.3(b).
(c) Find the second derivative of the function $\beta_0 x^{\beta_1}$. Verify that if $\beta_0 > 0$ and $\beta_1 > 1$, then this function is concave up as shown in Fig. 12.3(b).

62. Consider the power model with $\beta_0 > 0$ and $0 < \beta_1 < 1$. Sketch a scattergram for which this model might be appropriate.

63. Consider the power model with $\beta_0 > 0$ and $\beta_1 < 0$. Sketch a scattergram for which this model might be appropriate.

64. Consider the model

$$\mu_{Y|x} = \beta_0 + \beta_1(1/x) \qquad x > 0$$

(a) Find the first derivative of the function $\beta_0 + \beta_1(1/x)$.
(b) Verify that if $\beta_1 > 0$, then this function is decreasing as shown in Fig. 12.4.
(c) Find the second derivative of the function $\beta_0 + \beta_1(1/x)$. Verify that if $\beta_1 > 0$, then this function is concave up as shown in Fig. 12.4.

65. Consider the reciprocal model with $\beta_1 < 0$. Sketch a scattergram for which this model might be appropriate.

REVIEW EXERCISES

A study was conducted on the effect of water temperature and time in solution on the amount of dye absorbed by a certain kind of fabric. A standard amount of dye (200 milligrams/mg) was added to a fixed amount of water. The three temperature levels used in the experiment were 105, 120, and 135°C. The fabric was left in the water 15, 30, or 60 minutes (min). For each of these temperature-time combinations the amount of dye left inside the fabric was measured. The experiment yielded the following data:

Dye in yarn (mg) (y)	**Time in solution (min)** (x_1)	**Temperature of H_2O (°C)** (x_2)
136	15	105
153	30	105
186	60	105
182	15	120
175	30	120
187	60	120
170	15	135
179	30	135
183	60	135

Problems 66 through 77 refer to these data.

66. Graph y versus x_1 and y versus x_2. Does there appear to be a relationship between time and/or temperature with the amount of dye left in fabric?

67. Estimate the curve of regression of Y on x_1.

68. Find and interpret the 95% confidence limits for $\mu_{Y|x_1=45}$.

69. Estimate the curve of regression of Y on x_2.

70. Sketch a 90% prediction band about a single predicted value of Y by using a few values of the regressor for the model of Exercise 69.

71. Find the model specification matrix for the model

$$\mu_{Y|x_1,x_2} = \beta_0 + \beta_1 x_1 + \beta_2 x_2$$

72. For the model of Exercise 71,

$$(X'X)^{-1} = \begin{bmatrix} 11.16667 & -.0111111 & -.0888889 \\ -.0111111 & .0003174603 & 0 \\ -.0888889 & 0 & .0007407407 \end{bmatrix}$$

$$X'\mathbf{y} = \begin{bmatrix} 1551 \\ 55890 \\ 186975 \end{bmatrix}$$

$$s^2 = 166.78571429$$

Use this information to estimate the model.

73. Consider the model containing only x_1 as the reduced model. Test

$$H_0\text{: reduced model is appropriate}$$

$$H_1\text{: full model is needed}$$

at the $\alpha = .05$ level. Which model do you prefer?

74. Find R^2 for the model chosen in Exercise 73. Find $\hat{\rho}$, the estimated correlation between the predicted and observed responses.

75. If a dye solution is prepared with the temperature at 125°C and the fabric left in the solution for 20 min, how much dye, on the average, would you predict to be left in the fabric?

76. Find the 95% confidence limits on your prediction from Exercise 75.

COMPUTING SUPPLEMENT

XII Multiple Regression

The SAS procedure PROC REG is used to handle a regression model. We illustrate its use first by analyzing the data of Examples 12.2.1 through 12.2.3.

Statement	*Function*
DATA MILEAGE;	names data set
INPUT Y X1 X2;	names variables
LABEL Y = MILEAGE;	
LABEL X1 = WEIGHT;	labels variables
LABEL X2 = TEMPERATURE;	
LINES;	signals beginning of data
17.9 1.35 90	
16.5 1.90 30	
⋮	data
18.3 1.40 30	allows us to predict the mileage
. 1.5 70	when $x_1 = 1.5$ and $x_2 = 70$
;	signals end of data
PROC REG;	calls for the regression procedure
MODEL Y = X1 X2/XPX I P;	identifies the model by naming the response variable on the left and the predictor variables on the right; asks for the X'X matrix and its inverse to be printed; asks for predicted values to be printed
TITLE1 A MULTIPLE;	titles output
TITLE2 LINEAR REGRESSION MODEL;	

On the printout the matrix $X'X$ is indicated by ①; $X'\mathbf{y}$ is shown by ②. The inverse of $X'X$ is given in ③ and ④ gives the vector of parameter estimates. The value predicted for Y when the car weighs 1.5 tons and is driven on a 70°F day is shown in ⑤.

A MULTIPLE
LINEAR REGRESSION MODEL

MODEL CROSSPRODUCTS X'X X'Y Y'Y

X'X	INTERCEPT	X1	X2		Y
INTERCEP	10	16.75	525	①	170
X1	16.75	28.6375	874.5		282.405
X2	525	874.5	31475		8887
Y	170	282.405	8887	②	2900.46

X′X INVERSE, B, SSE

INVERSE	INTERCEP	X1	X2		Y
INTERCEP	6.070769	−3.02588	−0.0171888		24.74887
X1	−3.02588	1.738599	0.002166306	③	−4.15933
X2	−0.0171888	0.002166306	0.0002582903		−0.014895
Y	24.74887	−4.15933	−0.014895	④	0.1403498

OBS	ACTUAL	PREDICT VALUE	RESIDUAL
1	17.900	17.793	0.106779
2	16.500	16.399	0.100712
3	16.400	16.486	−.086404
4	16.800	16.666	0.133729
5	18.800	18.820	−.020413
6	15.500	15.552	−.051962
7	17.500	17.349	0.150812
8	16.400	16.368	0.031629
9	15.900	16.086	−.185929
10	18.300	18.479	−.178955
11	.	17.467 ⑤	
SUM OF RESIDUALS			9.10383E − 14
SUM OF SQUARED RESIDUALS			0.1403498

XIII Polynomial Regression

PROC REG can also be used to analyze the polynomial model. We use the data of Example 12.2.5 to illustrate the idea.

Statement	*Function*
DATA DRUG;	names data set
INPUT X Y;	names variables
XSQ = X*X;	creates a new variable named *XSQ* whose value is the square of *X*
LABEL Y = COST;	
LABEL X = NUMBER PRODUCED;	labels variables
LINES;	signals that data follows
5 14.0	
5 12.5	data
⋮	
25 14.6	
12 .	allows us to predict *Y* when $x = 12$
;	signals end of data
PROC PLOT; PLOT Y*X;	asks for a scattergram
TITLE A POLYNOMIAL MODEL;	titles output
PROC REG;	asks for regression analysis

MODEL Y = X XSQ/XPX I P;	identifies the variable Y as the response variable; identifies X and XSQ as the predictor variables; asks for the $X'X$ matrix and its inverse to be printed; asks for predicted values to be printed

The printout on page 533 shows the scattergram of the data. Notice the nonlinear trend. The matrices $X'X$ and $X'\mathbf{y}$ are given by ① and ② respectively. $(X'X)^{-1}$ is shown in ③ and the parameter estimates are found in ④. The estimated cost of producing 12 units of product is given by ⑤.

A POLYNOMIAL MODEL

MODEL CROSSPRODUCTS X'X X'Y Y'Y

X'X	INTERCEPT	X	XSQ		Y
INTERCEPT	10	150	2750		81.3
X	150	2750	56250	①	1228
XSQ	2750	56250	1223750		24555
Y	81.3	1228	24555	②	883.75

X'X INVERSE, B, SSE

INVERSE	INTERCEP	X	XSQ		Y
INTERCEP	2.3	−0.33	0.01		27.3
X	−0.33	0.05342857	−0.00171429	③	−3.313
XSQ	0.01	−0.00171429	.00005714286		0.111
Y	27.3	−3.313	0.111	④	7.019

A POLYNOMIAL MODEL

DEP VARIABLE: Y COST

SOURCE	DF	SUM OF SQUARES	MEAN SQUARE	F VALUE	PROB > F
MODEL	2	215.762	107.881	107.589	0.0001
ERROR	7	7.019000	1.002714		
C TOTAL	9	222.781			

ROOT MSE	1.001356	R-SQUARE	0.9685
DEP MEAN	8.130000	ADJ R-SQ	0.9595
C.V.	12.3168		

VARIABLE	DF	PARAMETER ESTIMATE	STANDARD ERROR	T FOR H0: PARAMETER = 0	PROB > !T!
INTERCEP	1	27.300000	1.518632	17.977	0.0001
X	1	−3.313000	0.231460	−14.314	0.0001
XSQ	1	0.111000	0.007569542	14.664	0.0001

A POLYNOMIAL MODEL

PLOT OF Y*X   LEGEND: A = 1 OBS, B = 2 OBS, ETC.

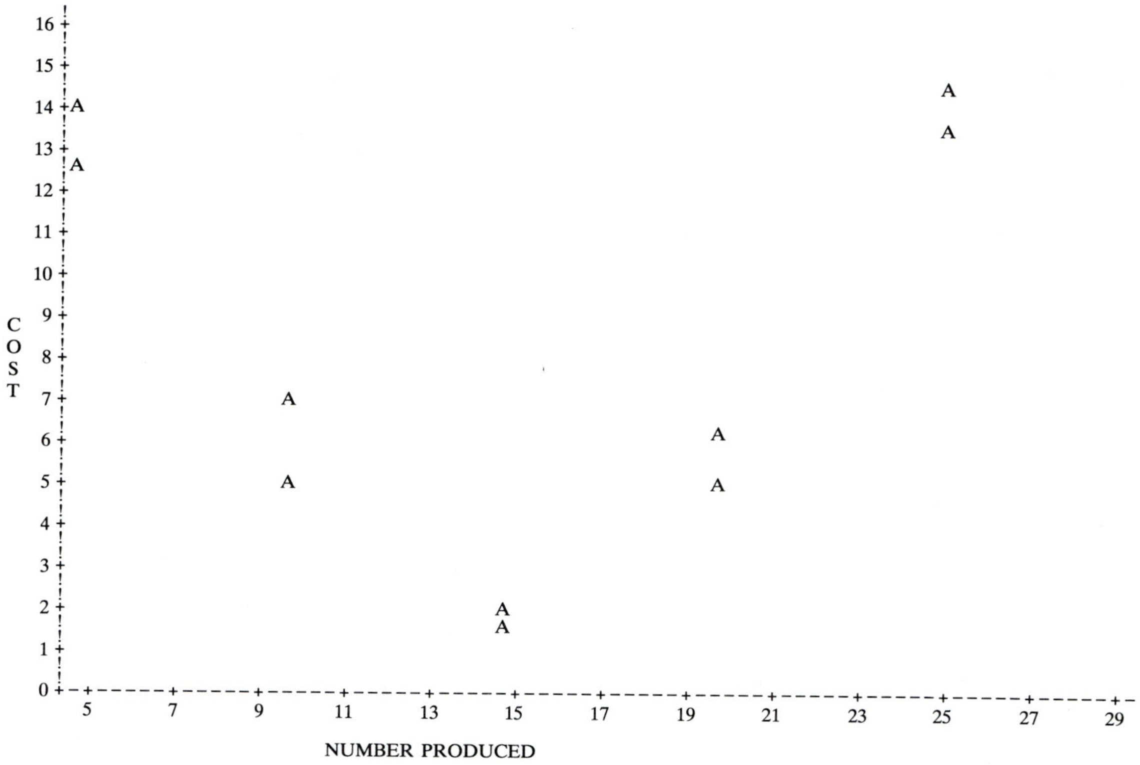

NOTE: 1 OBS HAD MISSING VALUES

OBS	ACTUAL	PREDICT VALUE	RESIDUAL
1	14.000	13.510	0.490000
2	12.500	13.510	−1.010
3	7.000	5.270	1.730
4	5.000	5.270	−.270000
5	2.100	2.580	−.480000
6	1.800	2.580	−.780000
7	6.200	5.440	0.760000
8	4.900	5.440	−.540000
9	13.200	13.850	−.650000
10	14.600	13.850	0.750000
11		3.528 ⑤	

SUM OF RESIDUALS	1.99840E − 14
SUM OF SQUARED RESIDUALS	7.019

CHAPTER 13

ANALYSIS OF VARIANCE

In Chap. 8 we discussed the problem of hypothesis testing on the mean of a single population. The problem was extended to testing the equality of two population means in Chap. 10. In the latter case, we were concerned primarily with comparing means based on independent samples drawn from normal populations. We used either the pooled T test or the Satterthwaite procedure. We also considered the paired T test, a method for comparing means based on paired data. In this chapter these problems are extended to that of comparing several population means via a statistical methodology called *analysis of variance* (*ANOVA*). This is a procedure in which the total variation in a measured response is partitioned into components which can be attributed to recognizable sources of variation. These individual components are useful in testing pertinent hypotheses.

In this chapter and in Chap. 14 we touch on an area of statistics called *experimental design*. Experimental design is a broad and important area of applied statistics that deals with the practical and theoretical aspects of designing experimental studies. There are three major phases of such a study. These are *problem formulation*, the *design* of the experiment, and the *analysis* of the data collected.

In phase 1 the researcher carefully states the problem to be solved. This process should include gathering all information currently known about the problem, a consideration of the point of view of others, a determination of the scope of the study, and a clear statement of the purpose of the study.

Phase 2, the design of the study, includes choosing the response variable(s) and trying to anticipate which other variables might have an influence on the response. Techniques for controlling or at least measuring the influence of these variables are determined. Cost, time, and other physical constraints are considered. Ultimately, decisions are made concerning the number of observations to be taken, the order of experimentation, and the method of randomization to be used. A statistical model is then formulated. Ideally, this model is one that describes the experimental design, is simple enough to be understood and analyzed by available analysis of variance techniques, and allows the questions posed in the formulation phase to be answered statistically.

If the experiment has been well designed, then phase 3, the analysis of the data collected, is not difficult since the proper analysis has been anticipated.

In this chapter we develop the ANOVA techniques for frequently encountered "single-factor" experimental designs. By "single-factor" we mean designs in which interest centers on a single primary factor that can influence the response. For example, in conducting an experiment to study the viscosity of a particular motor oil, we can think of the temperature of the oil as a factor; in a study of the speed with which a sorting algorithm is able to sort a random array the computer scientist could view the degree to which the array is out of sort at the outset as a factor. Multifactor experiments will be considered in Chap. 14.

13.1 ONE-WAY CLASSIFICATION FIXED-EFFECTS MODEL

Assume that we are interested in comparing the means of k populations. The experimental situation may be either of the following:

1. We have k populations, each identified by some common characteristic to be studied in the experiment. Independent random samples of sizes $n_1, n_2, \ldots, n_k$ are selected from each of the k populations, respectively. Differences observed in the measured response are attributed to basic differences among the k populations.
2. We have a collection of N homogeneous experimental units and wish to study the effects of k different treatments. These units are randomly divided into k subgroups of sizes $n_1, n_2, \ldots, n_k$ and each subgroup receives a different experimental treatment. The k subgroups are viewed as constituting independent random samples of size $n_1, n_2, \ldots, n_k$ drawn from k populations.

Although the above experimental situations are different, they are similar in that each results in independent random samples drawn from populations with means $\mu_1, \mu_2, \ldots, \mu_k$. Our interest is in testing the null hypothesis that the population means are equal. That is, we want to test

$$H_0\colon \mu_1 = \mu_2 = \cdots = \mu_k$$
$$H_1\colon \mu_i \neq \mu_j \qquad \text{for some } i \text{ and } j$$

(at least two of the means are not equal)

As you can see, this is an extension of the two sample problems based on independent samples studied in Chap. 10.

The model that we develop is called a *one-way classification fixed-effects* model. The term "one-way classification" refers to the fact that only one factor or attribute is being studied in the experiment. The factor is studied at k different *levels*. In the second experimental situation described we usually use the word "treatments" rather than factor levels. The term "fixed-effects" refers to the fact that the treatments or levels of the factor involved are specifically selected by the experimenter because they are of particular interest. They are not randomly selected from a larger group of possible treatments or levels. Random selection of treatments or levels leads to random-effects models discussed in Sec. 13.5. As you will see, the types of inferences we make depend on whether effects are fixed or random. Example 13.1.1 should make the meaning of these terms clear.

Example 13.1.1. A study is designed to investigate the sulfur content of the five major coal seams in a certain geographical region. Core samples are taken at randomly selected points within each seam and the measured response is the percentage of sulfur per core sample. We want to detect any differences that might exist in the average sulfur content for these five seams. Each seam constitutes a population. We want to compare population means by testing

$$H_0\colon \mu_1 = \mu_2 = \mu_3 = \mu_4 = \mu_5$$
$$H_1\colon \mu_i \neq \mu_j \qquad \text{for some } i \text{ and } j$$

based on independent samples drawn from these populations. The one factor under study is the coal seam involved. The factor is being studied at five levels. These levels are not selected at random. Rather, we have intentionally chosen to study the five major seams in the region. The design is a fixed-effects design. This study is an example of the first experimental situation described earlier. Figure 13.1 illustrates the idea.

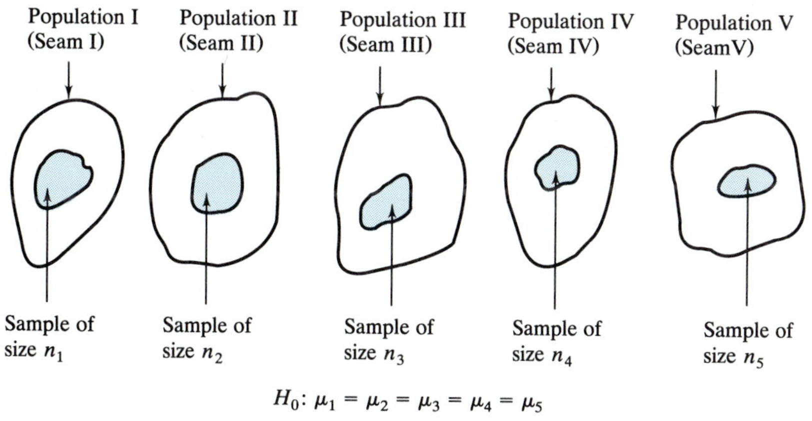

FIGURE 13.1
Random samples of sizes n_1, n_2, n_3, n_4, n_5 independently selected from the five major coal seams in a particular geographical region.

TABLE 13.1
Data layout one-way classification

	Treatment or factor level					
	1	2	3	$\cdots$	k	
	Y_{11}	Y_{21}	Y_{31}		Y_{k1}	
	Y_{12}	Y_{22}	Y_{32}		Y_{k2}	
	Y_{13}	Y_{23}	Y_{33}		Y_{k3}	
	$\vdots$	$\vdots$	$\vdots$		$\vdots$	
	Y_{1n_1}	Y_{2n_2}	Y_{3n_3}		Y_{kn_k}	
Total	$T_{1\cdot}$	$T_{2\cdot}$	$T_{3\cdot}$	$\cdots$	$T_{k\cdot}$	$T_{\cdot\cdot}$
Sample mean	$\bar{Y}_{1\cdot}$	$\bar{Y}_{2\cdot}$	$\bar{Y}_{3\cdot}$	$\cdots$	$\bar{Y}_{k\cdot}$	$\bar{Y}_{\cdot\cdot}$

Notationally, we let Y_{ij} denote the jth response for the ith treatment or factor level $i = 1, 2, \ldots, k$ and $j = 1, 2, \ldots, n_i$. Here n_i represents the size of the sample drawn from the ith population. The total number of observations for the k samples combined is $N = n_1 + n_2 + \cdots + n_k$. The data collected in a single-factor experiment as well as some important sample statistics are displayed conveniently as shown in Table 13.1. A dot in the following notation indicates the subscript over which summation is being conducted. Note that

$$T_{i\cdot} = \text{total of the } i\text{th treatment responses}$$

$$= \sum_{j=1}^{n_i} Y_{ij}$$

$$\bar{Y}_{i\cdot} = \text{sample mean for the } i\text{th treatment}$$

$$= T_{i\cdot}/n_i$$

$$T_{\cdot\cdot} = \text{total of all responses}$$

$$= \sum_{i=1}^{k} T_{i\cdot} = \sum_{i=1}^{k} \sum_{j=1}^{n_i} Y_{ij}$$

$$\bar{Y}_{\cdot\cdot} = \text{sample mean of all responses}$$

$$= T_{\cdot\cdot}/N$$

Example 13.1.2 illustrates the use of this notation.

Example 13.1.2. These data and summary statistics are obtained on the sulfur content of the five major coal seams in a particular geographical region:

Factor (coal seam)

1	2	3	4	5	
1.51	1.69	1.56	1.30	.73	
1.92	.64	1.22	.75	.80	
1.08	.90	1.32	1.26	.90	
2.04	1.41	1.39	.69	1.24	
2.14	1.01	1.33	.62	.82	
1.76	.84	1.54	.90	.72	
1.17	1.28	1.04	1.20	.57	
	1.59	2.25	.32	1.18	
		1.49		.54	
				1.30	
$T_{1\cdot} = 11.62$	$T_{2\cdot} = 9.36$	$T_{3\cdot} = 13.14$	$T_{4\cdot} = 7.04$	$T_{5\cdot} = 8.8$	$T_{\cdot\cdot} = 49.96$
$\bar{Y}_{1\cdot} = 1.66$	$\bar{Y}_{2\cdot} = 1.17$	$\bar{Y}_{3\cdot} = 1.46$	$\bar{Y}_{4\cdot} = .88$	$\bar{Y}_{5\cdot} = .88$	$\bar{Y}_{\cdot\cdot} = 1.189$

We know that the five sample means $\bar{Y}_{1\cdot}$, $\bar{Y}_{2\cdot}$, $\bar{Y}_{3\cdot}$, $\bar{Y}_{4\cdot}$, and $\bar{Y}_{5\cdot}$ are unbiased estimators for the population means μ_1, μ_2, μ_3, μ_4, and μ_5. By inspection we see that there are some differences among the sample means. The question to be answered is, "Are these differences extreme enough to conclude that there is a real difference in the average sulfur content among these five coal seams?" To answer this question, we need to develop an analytic method for testing H_0: $\mu_1 = \mu_2 = \mu_3 = \mu_4 = \mu_5$ based on these data.

The Model

To see how to test the null hypothesis of equal treatment means, we must devise a statistical model. To begin, note that each response can be expressed as

$$Y_{ij} = \mu_i + \mathrm{E}_{ij}$$

where μ_i denotes the theoretical mean of the ith population and E_{ij} represents the random difference between the jth observation taken from the ith population and the mean of that population. That is, $\mathrm{E}_{ij} = Y_{ij} - \mu_i$. An alternative way to write this model is obtained by letting $\alpha_i = \mu_i - \mu$, where

$$\mu = \sum_{i=1}^{k} n_i \mu_i / N$$

In a practical sense, μ represents an overall mean effect found by pooling the k

individual population means. Note that if the sample sizes are equal, then μ is just the average of the k population means. Since α_i is the difference between the overall mean μ and the mean of the ith population, α_i measures the effect of the ith treatment. Note that

$$\sum_{i=1}^{k} n_i\alpha_i = \sum_{i=1}^{k} n_i(\mu_i - \mu) = \sum_{i=1}^{k} n_i\mu_i - N\mu = 0$$

By substitution the one-way classification model with fixed effects can be expressed in any of the three ways given below:

One-way classification, fixed-effects model

$$Y_{ij} = \mu_i + E_{ij}$$
$$Y_{ij} = \mu + (\mu_i - \mu) + (Y_{ij} - \mu_i)$$
$$Y_{ij} = \mu + \alpha_i + E_{ij}$$

These models are demonstrated in Fig. 13.2.

The latter models express mathematically the idea that each response can be partitioned into three recognizable components as follows:

Response of jth experimental unit to ith treatment	=	overall mean response	+	deviation from overall mean due to the fact that unit received ith treatment	+	random deviation from ith population mean due to random influences

That is,

$$Y_{ij} \qquad = \mu \qquad + (\mu_i - \mu \text{ or } \alpha_i) \qquad + (Y_{ij} - \mu_i \text{ or } E_{ij})$$

The null hypothesis of equal treatment means can be expressed in an alternative form by noting that if $\mu_1 = \mu_2 = \cdots = \mu_k$, then

$$\mu = \sum_{i=1}^{k} n_i\mu_i/N = N\mu_i/N = \mu_i \qquad \text{for each } 1, 2, \ldots, k$$

and $\alpha_i = \mu_i - \mu = 0$ for each i. This implies that testing

$$H_0\colon \mu_1 = \mu_2 = \cdots = \mu_k$$

is equivalent to testing

$$H_0\colon \alpha_1 = \alpha_2 = \cdots = \alpha_k = 0$$

As you will see later, it is possible to express the one-way classification model in the general linear form discussed in Chap. 12. By rewriting the hypothesis of equal means in this alternative form, we can test H_0 by using regression techniques.

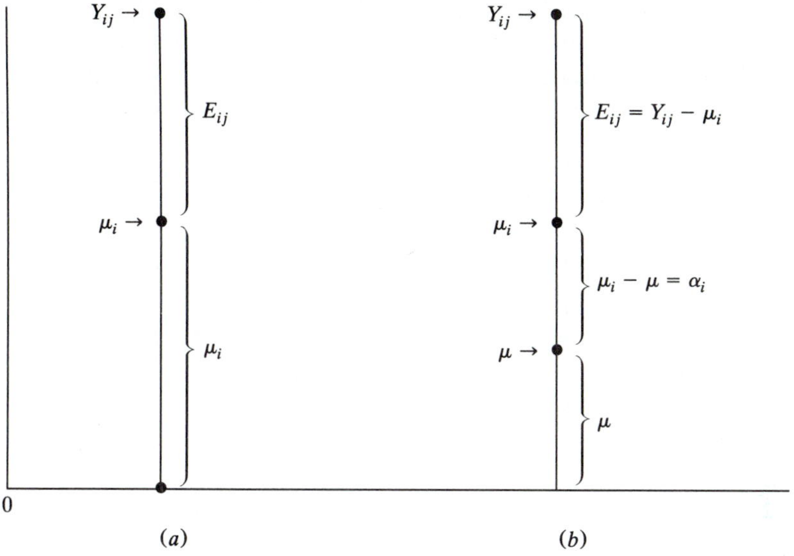

FIGURE 13.2
(a) $Y_{ij} = \mu_i + E_{ij}$; the jth observation in the ith sample is viewed as being partitional into two recognizable components. These are μ_i, the average for the ith treatment, and E_{ij}, a deviation from this average caused by random influences acting upon the jth experimental unit; (b) $Y_{ij} = \mu + (\mu_i - \mu) + (Y_{ij} - \mu_i)$ or $Y_{ij} = \mu + \alpha_i + E_{ij}$; the jth observation in the ith sample is partitioned into three recognizable components. These are μ, an overall average, $\mu_i - \mu$, an effect due to the fact that the ith treatment is involved, which could cause μ_i to differ from the overall average, and E_{ij}, the deviation due to random influences.

Testing H_0

To derive a test statistic, we must make some assumptions concerning the random differences E_{ij}. These assumptions are similar to those we made in the regression models considered earlier. In particular, we are assuming that the random differences E_{ij} are independent, normally distributed random variables, each with mean 0 and variance σ^2. In more easily understood terms we are assuming that:

1. The k samples represent independent samples drawn from k specific populations with unknown means $\mu_1, \mu_2, \ldots, \mu_k$.
2. Each of the k populations is normally distributed.
3. Each of the k populations has the same variance, σ^2.

When expressed in this form, it is easy to see that these assumptions parallel those made in Chap. 10 relative to the pooled T test for comparing two population means.

Analysis of variance has been defined as a procedure whereby the total variation in some measured response is subdivided into components that can be attributed to recognizable sources. Since $\mu, \mu_1, \mu_2, \ldots, \mu_k$ are theoretical population means, the model does this in only the theoretical sense. To partition an observation in a practical way, these theoretical means are replaced by their unbiased estimators $\bar{Y}.., \bar{Y}_1., \bar{Y}_2., \ldots, \bar{Y}_k.$, respectively. By replacing the theoretical means by their estimators in the model, we obtain the following identity:

$$Y_{ij} = \bar{Y}... + (\bar{Y}_i. - \bar{Y}..) + (Y_{ij} - \bar{Y}_i.)$$

Note that $\bar{Y}..$ is an estimator for μ, the overall pooled mean effect; $Y_i. - \bar{Y}..$ is an estimator for $\alpha_i = \mu_i - \mu$, the effect of the ith treatment; and $Y_{ij} - \bar{Y}_i.$ is an estimator for $E_{ij} = Y_{ij} - \mu_i$, the random error. The term $Y_{ij} - \bar{Y}_i.$ is usually called a *residual*. This identity is equivalent to

$$Y_{ij} - \bar{Y}.. = (\bar{Y}_i. - \bar{Y}..) + (Y_{ij} - \bar{Y}_i.)$$

If each side of the identity is squared and summed over all possible values of i and j, we get

$$\begin{aligned}\sum_{i=1}^{k}\sum_{j=1}^{n_i}(Y_{ij} - \bar{Y}..)^2 &= \sum_{i=1}^{k}\sum_{j=1}^{n_i}\left[(\bar{Y}_i. - \bar{Y}..) + (Y_{ij} - \bar{Y}_i.)\right]^2 \\ &= \sum_{i=1}^{k}\sum_{j=1}^{n_i}(\bar{Y}_i. - \bar{Y}..)^2 + 2\sum_{i=1}^{k}\sum_{j=1}^{n_i}(\bar{Y}_i. - \bar{Y}..)(Y_{ij} - \bar{Y}_i.) \\ &\quad + \sum_{i=1}^{k}\sum_{j=1}^{n_i}(Y_{ij} - \bar{Y}_i.)^2\end{aligned}$$

The middle term is 0 since

$$\sum_{j=1}^{n_i}(Y_{ij} - \bar{Y}_i.) = \sum_{j=1}^{n_i} Y_{ij} - n_i\bar{Y}_i. = 0$$

Noting that

$$\sum_{i=1}^{k}\sum_{j=1}^{n_i}(\bar{Y}_i. - \bar{Y}..)^2 = \sum_{i=1}^{k} n_i(\bar{Y}_i. - \bar{Y}..)^2$$

we get what is called the *sum of squares* identity for the one-way classification analysis of variance.

Sum of squares identity

$$\sum_{i=1}^{k}\sum_{j=1}^{n_i}(Y_{ij} - \bar{Y}..)^2 = \sum_{i=1}^{k} n_i(\bar{Y}_i. - \bar{Y}..)^2 + \sum_{i=1}^{k}\sum_{J=1}^{n_i}(Y_{ij} - \bar{Y}_i.)^2$$

Each of the components of this identity can be interpreted in a meaningful way. In particular,

$$\sum_{i=1}^{k}\sum_{j=1}^{n_i}(Y_{ij}-\bar{Y}..)^2 = \text{measure of the total variability in the data}$$

$$= \text{total sum of squares } (SS_{\text{Tot}})$$

$$\sum_{i=1}^{k} n_i(\bar{Y}_i.-\bar{Y}..)^2 = \text{measure of variability in data attributed to the fact that different factor levels or treatments are used}$$

$$= \text{treatment sum of squares } (SS_{\text{Tr}})$$

$$\sum_{i=1}^{k}\sum_{j=1}^{n_i}(Y_{ij}-\bar{Y}_i.)^2 = \text{measure of variability in data attributed to random fluctuation among subjects within the same factor level}$$

$$= \text{residual or error sum of squares } (SS_E)$$

Symbolically, the sum of squares identity can be written as

$$SS_{\text{Tot}} = SS_{\text{Tr}} + SS_E$$

If there are differences in the population means, then we expect most of the variation in the responses to be due to the fact that different treatments are being used. That is, we expect SS_{Tr} to be large relative to SS_E. The analysis of variance procedure uses this idea to test the null hypothesis of equal treatment means by comparing the between treatment variation (SS_{Tr}) to the within treatment variation (SS_E) via an appropriate F ratio.

To construct an appropriate F ratio, we must consider the expected values of the statistics SS_{Tr} and SS_E. To do so, we use the model assumption that the random errors E_{ij} are independent normally distributed random variables, each with mean 0 and variance σ^2. To begin, note that for each i,

$$\bar{Y}_i. = \sum_{j=1}^{n_i}(\mu + \alpha_i + \text{E}_{ij})/n_i$$

$$= \frac{n_i\mu + n_i\alpha_i + \sum_{j=1}^{n_i}\text{E}_{ij}}{n_i}$$

$$= \mu + \alpha_i + \bar{\text{E}}_i.$$

Also, since $\sum_{i=1}^{k} n_i\alpha_i = 0$,

$$\begin{aligned}
\bar{Y}.. &= \sum_{i=1}^{k}\sum_{j=1}^{n_i} Y_{ij}/N \\
&= \sum_{i=1}^{k}\sum_{j=1}^{n_i} (\mu + \alpha_i + \mathrm{E}_{ij})/N \\
&= \frac{N\mu + \sum_{i=1}^{k} n_i\alpha_i + \sum_{i=1}^{k}\sum_{j=1}^{n_i} \mathrm{E}_{ij}}{N} \\
&= \mu + \bar{\mathrm{E}}..
\end{aligned}$$

Substituting, we can rewrite SS_{Tr} as shown:

$$\begin{aligned}
SS_{\mathrm{Tr}} &= \sum_{i=1}^{k} n_i\left(\bar{Y}_i - \bar{Y}..\right)^2 \\
&= \sum_{i=1}^{k} n_i\left[\left(\mu + \alpha_i + \bar{\mathrm{E}}_i\right) - \left(\mu + \bar{\mathrm{E}}..\right)\right]^2 \\
&= \sum_{i=1}^{k} n_i\left(\alpha_i + \bar{\mathrm{E}}_i. - \bar{\mathrm{E}}..\right)^2 \\
&= \sum_{i=1}^{k} n_i\alpha_i^2 + 2\sum_{i=1}^{k} n_i\alpha_i\bar{\mathrm{E}}_i + \sum_{i=1}^{k} n_i\bar{\mathrm{E}}_i.^2 - N\bar{\mathrm{E}}..^2
\end{aligned}$$

Taking the expected value of each term, we get

$$E[SS_{\mathrm{Tr}}] = \sum_{i=1}^{k} n_i\alpha_i^2 + 2\sum_{i=1}^{k} n_i\alpha_i E\left[\bar{\mathrm{E}}_i\right] + \sum_{i=1}^{k} n_i E\left[\bar{\mathrm{E}}_i.^2\right] - NE[\bar{\mathrm{E}}..^2]$$

In Exercise 1 we outline the proof that $E[\bar{\mathrm{E}}_i.^2] = \sigma^2/n_i$ for each i. A similar argument shows that $E[\bar{\mathrm{E}}..^2] = \sigma^2/N$. It is easy to see that $E[\bar{\mathrm{E}}_i.] = 0$. Substituting, we see that

$$\begin{aligned}
E[SS_{\mathrm{Tr}}] &= \sum_{i=1}^{k} n_i\alpha_i^2 + \sum_{i=1}^{k} n_i\sigma^2/n_i - N\sigma^2/N \\
&= (k-1)\sigma^2 + \sum_{i=1}^{k} n_i\alpha_i^2
\end{aligned}$$

By dividing SS_{Tr} by $k - 1$, we obtain a statistic called the *treatment mean square*, which we denote by MS_{Tr}. That is,

$$MS_{\mathrm{Tr}} = SS_{\mathrm{Tr}}/(k-1)$$

It is easy to see that $E[MS_{\mathrm{Tr}}] = \sigma^2 + \sum_{i=1}^{k} n_i\alpha_i^2/(k-1)$. Recall that, in the regression context, the residual sum of squares helped to estimate σ^2. The same is true here. To obtain an unbiased estimator for σ^2 we divide the residual sum of squares SS_E by $N - k$. This estimator is called the *error mean square* and is denoted by MS_E. That is,

$$MS_E = SS_E/(N - k)$$

How can we use MS_{Tr} and MS_E to test H_0? To answer this question, we need only note that if H_0 is true, then $\alpha_1 = \alpha_2 = \cdots = \alpha_k = 0$, and hence $\sum_{i=1}^{k} n_i \alpha_i^2/(k - 1) = 0$. If H_0 is not true, then this term will be positive. Thus if H_0 is true, we would *expect* MS_{Tr} and MS_E to be close in value since both of them estimate σ^2; if H_0 is not true, we would *expect* MS_{Tr} to be somewhat larger than MS_E. This suggests the ratio

$$MS_{\text{Tr}}/MS_E$$

as a logical test statistic. If H_0 is true, its value is expected to be close to 1; otherwise it is expected to be larger than 1. This ratio can be used as a test statistic since if H_0 is true, it is known to have an F distribution with $k - 1$ and $N - k$ degrees of freedom. The test is always a right-tailed test, with rejection of H_0 occurring for values of the $F_{k-1, N-k}$ random variable that appear to be too large to have occurred by chance. Values of $F = MS_{\text{Tr}}/MS_E$ *can* be less than 1 since F is a random variable. Such an outcome can occur by chance alone or because the assumed linear model is incorrect.

Although in practice analysis of variance is usually performed via the computer, some computational shortcuts are available. We leave it as an exercise to show that

$$SS_{\text{Tot}} = \sum_{i=1}^{k} \sum_{j=1}^{n_i} Y_{ij}^2 - \frac{T_{..}^2}{N}$$

$$SS_{\text{Tr}} = \sum_{i=1}^{k} \frac{T_{i.}^2}{n_i} - \frac{T_{..}^2}{N}$$

$$SS_E = SS_{\text{Tot}} - SS_{\text{Tr}}$$

The theoretical ideas behind the analysis of variance procedure for the one-way classification fixed-effects model are summarized in Table 13.2. This type of table is called an *analysis of variance (ANOVA) table.*

To illustrate the use of the F ratio, we continue the analysis of the coal data begun in Example 13.1.2.

Example 13.1.3. We are testing

$$H_0: \mu_1 = \mu_2 = \mu_3 = \mu_4 = \mu_5$$

or

$$H_0: \alpha_1 = \alpha_2 = \alpha_3 = \alpha_4 = \alpha_5 = 0$$

TABLE 13.2
ANOVA table for the one-way classification design with fixed effects

Source of variation	Degrees of freedom (DF)	Sum of squares (SS)	Mean square (MS)	Expected mean square	F
Treatment or level	$k-1$	$\sum_{i=1}^{k} \frac{T_{i\cdot}^2}{n_i} - \frac{T_{\cdot\cdot}^2}{N}$ (SS_{Tr})	$\frac{SS_{\text{Tr}}}{k-1}$	$\sigma^2 + \sum_{i=1}^{k} \frac{n_i\alpha_i^2}{k-1}$	$\frac{MS_{\text{Tr}}}{MS_E}$
Error or residual	$N-k$	Subtraction (SS_E)	$\frac{SS_E}{N-k}$	σ^2	
Total	$N-1$	$\sum_{i=1}^{k}\sum_{j=1}^{n_i} Y_{ij}^2 - \frac{T_{\cdot\cdot}^2}{N}$			

based on our previous data. Recall that μ_i, $i = 1, 2, 3, 4, 5$, denote the mean sulfur content of the five major coal seams in a particular geographical region. We have these summary statistics available:

$$\begin{array}{llll} T_{1\cdot} = 11.62 & T_{4\cdot} = 7.04 & n_1 = 7 & n_4 = 8 \\ T_{2\cdot} = 9.36 & T_{5\cdot} = 8.8 & n_2 = 8 & n_5 = 10 \\ T_{3\cdot} = 13.14 & T_{\cdot\cdot} = 49.96 & n_3 = 9 & N = 42 \end{array}$$

The only new statistic needed is $\sum_{i=1}^{5}\sum_{j=1}^{n_i} Y_{ij}^2$. For the data given in Example 13.1.2 this statistic assumes the value 67.861. Substituting into the computational formulas, we obtain

$$SS_{\text{Tot}} = \sum_{i=1}^{5}\sum_{j=1}^{n_i} Y_{ij}^2 - T_{\cdot\cdot}^2/N = 67.861 - \frac{(49.96)^2}{42} = 8.432$$

$$SS_{\text{Tr}} = \sum_{i=1}^{5} \frac{T_{i\cdot}^2}{n_i} - \frac{T_{\cdot\cdot}^2}{N}$$

$$= \frac{(11.62)^2}{7} + \frac{(9.36)^2}{8} + \frac{(13.14)^2}{9} + \frac{(7.04)^2}{8} + \frac{(8.8)^2}{10} - \frac{(49.96)^2}{42}$$

$$= 3.935$$

$$SS_E = SS_{\text{Tot}} - SS_{\text{Tr}}$$

$$= 8.432 - 3.935 = 4.497$$

$$MS_{\text{Tr}} = \frac{SS_{\text{Tr}}}{k-1} = \frac{3.935}{4} = .984$$

$$MS_E = \frac{SS_E}{N-k} = \frac{4.497}{37} = .122$$

TABLE 13.3
ANOVA for coal seam data

Source of variation	Degrees of freedom (DF)	Sum of squares (SS)	Mean square (MS)	F
Treatments	4	3.935	.984	8.066
Error	37	4.497	.122	
Total	41	8.432		

The observed value of the $F_{k-1, N-k} = F_{4,37}$ test statistic is

$$F_{4,37} = \frac{MS_{\text{Tr}}}{MS_E} = \frac{.984}{.122} = 8.066$$

Since $f_{.05}(4, 37) \doteq 2.626$, we can reject H_0 with $P < .05$. We do have statistical evidence that at least two of the coal seams differ in mean sulfur content. The ANOVA table for these data is shown in Table 13.3.

Before concluding this section, recall that we assume that each of the k independent samples is drawn from normally distributed populations with equal variances σ^2. If the sample sizes are reasonably large, the test is quite robust to departures from normality in the sense that P values reported, although approximate, are fairly accurate. However, the test can be quite sensitive to departures from the assumption of equal variances. This is particularly true if the respective sample sizes differ considerably. When possible, it is advantageous to design experiments so that sample sizes are equal. A method for testing for equality of variances is given in the next section. If normality seems unreasonable and sample sizes are small or variances appear to differ, then a nonparametric method of analysis is appropriate. This method is discussed in Sec. 13.7.

13.2 COMPARING VARIANCES (OPTIONAL)

As indicated earlier, the F test for testing equality of means is sensitive to the violation of the assumption of equal variances. This is especially true when sample sizes differ greatly. Before performing an analysis of variance, we need to test the hypothesis

$$H_0: \sigma_1^2 = \sigma_2^2 = \cdots = \sigma_k^2$$

$$H_1: \sigma_i^2 \neq \sigma_j^2 \qquad \text{for some } i \text{ and } j$$

(at least two of the variances are not equal)

If H_0 is rejected, then either a nonparametric analysis should be used or else the data should be transformed in hopes of stabilizing the variances. A nonparametric alternative is discussed in Sec. 13.7; variance stabilization transformations are found in analysis of variance texts [14].

The most frequently used test for testing the null hypothesis of equal variances is called *Bartlett's test*. The statistic used in this test can be shown to follow an approximate chi-squared distribution with $k - 1$ degrees of freedom when sampling from normal populations.

To conduct Bartlett's test, we compute the sample variances $S_1^2, S_2^2, \ldots, S_k^2$ for each of the k samples. We also compute the error mean square, the pooled estimate of σ^2 under the assumption that H_0 is true. In this context it is convenient to compute MS_E directly from the individual sample variances by means of the formula

$$MS_E = S_p^2 = \sum_{i=1}^{k} \frac{(n_i - 1)S_i^2}{N - k}$$

It is left as an exercise to verify that this equation holds. (See Exercise 8.) We next form the statistic Q defined by

$$Q = (N - k)\log_{10} S_p^2 - \sum_{i=1}^{k} (n_i - 1)\log_{10} S_i^2$$

The observed value of this statistic is large when the sample variances S_i^2, $i = 1, 2, \ldots, k$ are quite different; it is near 0 when these sample variances are close in value. The Bartlett statistic is defined by

$$B = 2.3026Q/h$$

where

$$h = 1 + \frac{1}{3(k - 1)}\left(\sum_{i=1}^{k} \frac{1}{n_i - 1} - \frac{1}{N - k}\right)$$

An example should demonstrate the use of Bartlett's test.

Example 13.2.1. Let us return to the coal seam data given in Example 13.1.2. The sample variances and their logarithms must be found for each of the five factor levels. The results of the calculations are summarized below:

Coal seam	**Sample variance** (s_i^2)	$\text{Log}_{10}\, s_i^2$	**Sample size** (n_i)
1	.175	−.757	7
2	.144	−.842	8
3	.115	−.939	9
4	.123	−.910	8
5	.074	−1.131	10

The pooled estimate is

$$MS_E = s_p^2 = \sum_{i=1}^{k} \frac{(n_i - 1)s_i^2}{N-k}$$

$$= \frac{(7-1)(.175) + (8-1)(.144) + (9-1)(.115) + (8-1)(.123) + (10-1)(.074)}{42-5}$$

$$= .122$$

Note that this value agrees with the value obtained for MS_E in Example 13.1.3. By substitution

$$q = (N-k)\log_{10} s_p^2 - \sum_{i=1}^{k} (n_i - 1)\log_{10} s_i^2$$

$$= 37 \log_{10} .122 - [6(-.757) + 7(-.842) + 8(-.939) + 7(-.910) + 9(-1.131)]$$

$$= .692$$

and

$$h = 1 + \frac{1}{3(k-1)}\left[\sum_{i=1}^{k} \frac{1}{n_i - 1} - \frac{1}{N-k}\right]$$

$$= 1 + \frac{1}{3(4)}\left[\frac{1}{6} + \frac{1}{7} + \frac{1}{8} + \frac{1}{7} + \frac{1}{9} - \frac{1}{37}\right]$$

$$= 1.055$$

The observed value of the Bartlett statistic is

$$b = 2.3026q/h$$

$$= 2.3026(.692)/1.055$$

$$= 1.510$$

Based on the $X_{k-1}^2 = X_4^2$ distribution, the P value lies between .75 and .9. Since this value is large, we are unable to reject

$$H_0: \sigma_1^2 = \sigma_2^2 = \sigma_3^2 = \sigma_4^2 = \sigma_5^2$$

We have no reason to doubt that the assumption of equal population variances is not valid.

13.3 PAIRWISE COMPARISONS

In testing for equality of means in the one-way classification model, we either reject H_0 or fail to do so. If H_0 is rejected, we conclude that at least two of the population means differ in value. Unfortunately, the analysis of variance procedure does not tell us which of the k population means may be regarded as being different from the others.

There are many different techniques that have been suggested for comparing pairs of means. Several of these procedures do not require that the F test in the overall analysis of variance table be conducted prior to pairwise comparisons. However, this procedure is usually followed in practice. As pointed out at the end of this section, requiring a significant F test in the analysis of variance prior to conducting pairwise comparisons gives added protection against falsely declaring pairs of means significantly different.

Lentner and Bishop [31] present a good overview of most of the pairwise comparison procedures. Here we discuss two possibilities. These are the *Bonferroni T tests* and the *Duncan's multiple range test*.

Bonferroni T Tests

Consider a set of k population means. There are $\binom{k}{2} = k(k-1)/2$ possible pairs of means. Thus there are $k(k-1)/2$ possible tests of the form

$$H_0\colon \mu_i = \mu_j$$

$$H_1\colon \mu_i \neq \mu_j$$

that can be conducted. Since one of the model assumptions is that the population variances are equal, each of these hypotheses can be tested by using a pooled T test as described in Chap. 10. In such a test the test statistic is

$$T_{n_i+n_j-2} = \frac{\bar{Y}_{i\cdot} - \bar{Y}_{j\cdot}}{\sqrt{S_p^2\left(\dfrac{1}{n_i} + \dfrac{1}{n_j}\right)}}$$

where S_p^2 is the pooled estimator for the common population variance based on samples drawn from populations i and j. In this case, another estimator of σ^2 is available, namely, $\hat{\sigma}^2 = MS_E$. Since this estimator is based on all available data, the T test can be improved by using

$$T_{N-k} = \frac{\bar{Y}_{i\cdot} - \bar{Y}_{j\cdot}}{\sqrt{MS_E\left(\dfrac{1}{n_i} + \dfrac{1}{n_j}\right)}}$$

as the test statistic. The critical point for the two-tailed test at the α level of significance is

$$cp = t_{N-k,1-\alpha/2}\sqrt{MS_E\left(\frac{1}{n_i} + \frac{1}{n_j}\right)}$$

We reject H_0 whenever $|\bar{Y}_{i\cdot} - \bar{Y}_{j\cdot}|$ exceeds the computed critical point. Note that when sample sizes are unequal, it might be necessary to compute a different critical point for each test; if sample sizes are the same, then a single critical point will suffice. Performing $k(k-1)/2$ individual T tests is laborious but it can be done. However, this procedure has a more serious drawback that must be handled with care.

To understand the problem, suppose that we ask, "What is the probability of making at least one incorrect rejection and therefore drawing at least one incorrect conclusion?" If the $= k(k-1)/2$ tests are independent, then this probablity, called the *experimentwise error rate* and denoted by α', can be calculated as follows:

$$P[\text{at least one incorrect rejection}] = 1 - P[\text{no incorrect rejections}]$$

If each test is conducted at the α level of significance, then the probability of making an incorrect rejection in each case is α; the probability that a rejection is not incorrect is $1 - \alpha$. If the tests are independent, then the definition of independence guarantees that

$$P[\text{no incorrect rejections}] = (1 - \alpha)^c$$

Hence in the case of independent tests α' is given by $1 - (1 - \alpha)^c$. The problem in the case of multiple comparisons is that many tests are being conducted on the same set of data. For this reason, these tests are not independent. However, it can be shown that the above expression provides an upper bound for α' in this setting. In any hypothesis-testing setting in which more than one test is conducted we can say that

$$\text{Experimentwise error rate} = \alpha' \leq 1 - (1 - \alpha)^c$$

Here α is the common level of significance per test, called the *comparisonwise error rate*, and c is the number of tests run.

As an example for $k = 5$, there are $\binom{5}{2} = 10$ possible paired comparisons. If each test is conducted at the $\alpha = .05$ level, then an upper bound on the probability of making at least one incorrect rejection is $1 - (1 - .05)^{10} = .40$. It is easy to see that as k increases, the overall probability of error may become unacceptably high. To compensate for this problem, we suggest that only those tests of real interest to the researcher be conducted and that some reasonably small target upper bound, b, be chosen. We then conduct each T test at the b/c level of significance, where c denotes the actual number of tests run. It can be shown that if $\alpha = b/c$, then $\alpha' \leq b$. (See Exercise 14.) For example, if we want α' to be at most .10 and we run all possible tests for $k = 5$ groups, we would conduct each of our paired comparisons at the $\alpha = .10/10 = .01$ level of significance. The numerical value of the bound b is chosen at the discretion of the researcher. Its value is somewhat dependent on the number of tests being run. Notice that if b is small, then $\alpha = b/c$ is even smaller. If we try to force the experimentwise error rate to be too small, then α becomes so small that it becomes extremely difficult to reject $H_0\colon \mu_i = \mu_j$, even when this hypothesis should be rejected. That is, the price paid for setting too small a bound on α' is lack of power. For this reason, values of α' as high as .15 or .20 are not unreasonable.

The technique presented here is equivalent to what is called the *Fisher's least significant difference* (LSD) procedure run with $\alpha = b/c$. For an excellent discussion of LSD you are referred to [31]. Since the procedure is based on an inequality known as the Bonferroni inequality, it bears the same name. Bonferroni T tests are available on SAS and their use is illustrated in Computing Supplement XIV.

Duncan's Multiple Range Test

Duncan's multiple range test was developed by D. B. Duncan in the early 1950s. It was one of the first methods suggested for doing pairwise comparisons of means, and you will see references to this test in the literature.

To understand Duncan's idea, let us compare what is done to the Bonferroni technique just discussed. First, the Duncan procedure initially assumed that sample sizes are equal. That is, it is assumed that we have available a collection of k sample means $\bar{Y}_1, \bar{Y}_2, \bar{Y}_3, \ldots, \bar{Y}_k$, each based on a sample of size n. Notice that in this setting there is a common critical point cp, that applies to all Bonferroni type T tests conducted. We reject H_0: $\mu_i = \mu_j$ whenever $|\bar{Y}_{i.} - \bar{Y}_{j.}| > cp$. No attempt is made to account for the relative positions of $\bar{Y}_{i.}$ and $\bar{Y}_{j.}$ in the ordered list of sample means. The same critical point is used whether these means are side by side, or are separated by several other means, or are the extremes in the ordered list. In the Duncan procedure an attempt is made to adjust the critical point to account for positioning. Thus in the Duncan prodedure multiple critical points are used which lead to the name "multiple range" test. If two means are adjacent on the ordered list of sample means, then the critical point used is the same as that of an ordinary T test at the specified α level; however, if the sample means being compared are not adjacent, then this critical value is increased according to the span of the means in the ordered set of means. In other words, critical points are chosen so that sample means that lie close together are not required to exhibit as much difference as those that are more widely separated in order to declare the corresponding population means to be "significantly different." Duncan's contribution was to develop tables from which these adjusted critical points can be calculated.

The test was first developed under the assumption that sample sizes are equal. However, C. Y. Kramer adapted the procedure to the case of unequal sample sizes. The Duncan test is conducted as follows:

1. Linearly order the k sample means.
2. Find the value of the least significant "studentized range," r_p, for each $p = 2, 3, \ldots, k$. This value is given in Table XI of App. A for α levels of .1, .01, or .05. In this table γ denotes the number of degrees of freedom

associated with MS_E, the error mean square in the original analysis of variance.

3. For each $p = 2, 3, \ldots, k$ find the shortest or least significant range, SSR_p. This value is given by

$$\mathrm{SSR}_p = r_p\sqrt{\frac{MS_E}{n}} \qquad \text{if the sample sizes are all equal with value } n$$

$$\mathrm{SSR}_p = r_p\sqrt{MS_E} \qquad \text{if the sample sizes are unequal}$$

4. Consider any subset of p adjacent sample means. Let $|\bar{Y}_i. - \bar{Y}_j.|$ denote the range of the means in this subgroup. The population means, of span p, μ_i and μ_j are considered to be different if

$$|\bar{Y}_i. - \bar{Y}_j.| > \mathrm{SSR}_p \qquad \text{for equal sample sizes}$$

or

$$|\bar{Y}_i. - \bar{Y}_j.|\sqrt{\frac{2n_in_j}{n_i + n_j}} > \mathrm{SSR}_p \qquad \text{for unequal sample sizes}$$

5. Summarize your results by underlining any subset of adjacent sample means that are not considered to be significantly different at your chosen α level.

Although this sounds complicated, it is not! An example should make the idea clear.

Example 13.3.1. In Example 13.1.3 we rejected

$$H_0: \mu_1 = \mu_2 = \mu_3 = \mu_4 = \mu_5$$

and concluded that at least two of the coal seams sampled differ in mean sulfur content. To pinpoint the differences, we run the Duncan's multiple range test. The sample means in linear order are

$\bar{Y}_4.$	$\bar{Y}_5.$	$\bar{Y}_2.$	$\bar{Y}_3.$	$\bar{Y}_1.$
.88	.88	1.17	1.46	1.66

The values of r_p for span $p = 2, 3, 4, 5$ for an $\alpha = .01$ level test based on 37 degrees of freedom are found in Table XI of App. A. These values are

p	2	3	4	5
r_p	3.825	3.988	4.098	4.180

The error mean square found earlier is $MS_E = .122$. Since the sample sizes $n_1 = 7$, $n_2 = 8$, $n_3 = 9$, $n_4 = 8$, and $n_5 = 10$ are unequal, the shortest significant range for each p is given by

$$\mathrm{SSR}_p = r_p\sqrt{MS_E}$$

These values are

p	2	3	4	5
r_p	3.825	3.988	4.098	4.180
SSR_p	1.336	1.393	1.431	1.460

A pair of means μ_i and μ_j of span p are considered to differ if

$$|\bar{Y}_i. - \bar{Y}_j.|\sqrt{\frac{2n_in_j}{n_i + n_j}} > \text{SSR}_p$$

For five populations there are $\binom{5}{2} = 10$ possible comparisons. We first compare the largest to the smallest sample mean. In this case we compare $\bar{Y}_1.$ to $\bar{Y}_4.$. Since these means span the entire set of sample means, $p = 5$ and $\text{SSR}_p = 1.460$. The observed value of the test statistic is

$$|\bar{Y}_1. - \bar{Y}_4.|\sqrt{\frac{2n_1n_4}{n_1 + n_4}} = (1.66 - .88)\sqrt{\frac{2(7)(8)}{7 + 8}}$$

$$= 2.131$$

This value exceeds SSR_p, and hence we conclude that μ_1 and μ_4 are significantly different. The results of other comparisons are as follows:

Treatment pair	p	**Value of test statistic**	SSR_p	**Reject** $\mu_i = \mu_j$?
4-1	5	2.131	1.460	Yes
4-3	4	1.688	1.431	Yes
4-2	3	.820	1.393	No
		$\underline{\bar{Y}_4.\ \bar{Y}_5.\ \bar{Y}_2.}\ \bar{Y}_3.\ \bar{Y}_1.$		
4-5	2	0	1.336	No
5-1	4	2.238	1.431	Yes
5-3	3	1.785	1.393	Yes
5-2	2	.8646	1.336	No
2-1	3	1.339	1.393	No
		$\underline{\bar{Y}_4.\ \bar{Y}_5.\ \bar{Y}_2.}\ \bar{Y}_3.\ \bar{Y}_1.$		
2-3	2	.8440	1.336	No
3-1	2	.5612	1.336	No

In summary, we can conclude that

$$\mu_1 \neq \mu_4$$
$$\mu_1 \neq \mu_5$$
$$\mu_3 \neq \mu_4$$
$$\mu_3 \neq \mu_5$$

Each statement has probability .01 of being in error.

If sample sizes are all the same, then we might not have to make all $\binom{k}{2}$ comparisons. This is due to the fact that when sample sizes are equal, whenever the most extreme pair of means is found not to be significantly different then all means within the subgroup are assumed to be equal with no further testing required.

Duncan's test is widely used and available on SAS and other computer packages. However, it should be emphasized that the significance level given does *not* refer to α', the overall probability of making at least one incorrect rejection. Therefore a Bonferroni test run at the $\alpha' = .01$ or $.05$ level can give results that differ from that of a Duncan's test at the same level. In using SAS, you must be aware of this fact. This point is illustrated in Computing Supplement XIV. Keep in mind that in this supplement both procedures are run for purposes of comparison only. In an actual study only one procedure should be run. The choice of which one to run is yours.

The Bonferroni technique is analytically attractive since it is easy to calculate an upper bound for the experimentwise error rate. However, this technique tends to be very conservative, and therefore it may not detect differences in means when they exist. Duncan's procedure attempts to control the experimentwise error rate but yet not to be too conservative. The trade-off is that Duncan's procedure may yield large experimentwise error rates when the number of comparisons is relatively large.

An approach that is sometimes recommended is to use what can be called a *protected* Duncan's procedure. That is, do not use Duncan's technique unless the ANOVA F test is first significant at level α. This then assures the experimenter that if Duncan's technique is also used at level α, then the experimentwise error rate $\alpha' \leq \alpha$. Duncan's standard usage does not require a preliminary significant F test from the ANOVA.

13.4 TESTING CONTRASTS

In the last section we discussed methods for comparing pairs of means. There are situations in which other sorts of comparisons are needed. To illustrate, reconsider Example 13.1.2. The experimenter may believe that because of geological characteristics, coal seams 1 and 3 may be similar to each other, seams 4 and 5 may be similar, but that pairs $(1, 3)$ and $(4, 5)$ differ from each other. Hence it is of interest to test

$$H_0: \mu_1 + \mu_3 = \mu_4 + \mu_5$$

$$H_1: \mu_1 + \mu_3 \neq \mu_4 + \mu_5$$

Note that H_0 can be rewritten in the form $H_0: \mu_1 + \mu_3 - \mu_4 - \mu_5 = 0$. That is, it can be expressed as a linear combination of population means. It is, in fact, a special linear function called a *contrast*. This term is defined below:

Definition 13.4.1 (Contrast). Let $\mu_1, \mu_2, \ldots, \mu_k$ be the means of k populations. Any linear function of the form

$$L = \sum_{i=1}^{k} c_i \mu_i$$

such that $\sum_{i=1}^{k} c_i = 0$ is called a *contrast in treatment means.*

The function $\mu_1 + \mu_3 - \mu_4 - \mu_5$ is a contrast with coefficients $c_1 = c_3 = 1$, $c_2 = 0$, and $c_4 = c_5 = -1$.

To test a hypothesis of the form

$$H_0\colon \sum_{i=1}^{k} c_i \mu_i = 0$$

where $\sum_{i=1}^{k} c_i \mu_i$ is a contrast, we must develop a test statistic. To begin, we replace the population means with their unbiased estimators $\bar{Y}_{1\cdot}, \bar{Y}_{2\cdot}, \ldots, \bar{Y}_{k\cdot}$ to obtain the statistic

$$\hat{L} = \sum_{i=1}^{k} c_i \bar{Y}_{i\cdot}$$

Since it is assumed that we are drawing independent samples from k normal distributions with means $\mu_1, \mu_2, \ldots, \mu_k$ and common variance σ^2, the sample means $\bar{Y}_{1\cdot}, \bar{Y}_{2\cdot}, \ldots, \bar{Y}_{k\cdot}$ are each normally distributed with mean μ_i and variance σ^2/n_i, respectively. The statistic $\hat{L}$ is also normally distributed with mean $\sum_{i=1}^{k} c_i \mu_i$ and variance $\sigma^2 \sum_{i=1}^{k} (c_i^2/n_i)$. (See Exercise 41, Chap. 7.) If H_0 is true, then $\sum_{i=1}^{k} c_i \mu_i = 0$ and

$$\frac{\hat{L}}{\sqrt{\sigma^2 \sum_{i=1}^{k} \frac{c_i^2}{n_i}}}$$

follows the standard normal distribution. Exercise 76 of Chap. 4 guarantees that

$$Q_1 = \frac{\hat{L}^2}{\sigma^2 \sum_{i=1}^{k} \frac{c_i^2}{n_i}}$$

follows the chi-squared distribution with 1 degree of freedom. It can be shown that $Q_2 = SS_E/\sigma^2$ follows the chi-squared distribution with $N - k$ degrees of freedom and that Q_1 and Q_2 are independent. By Definition 10.2.1 the random variable

$$\frac{Q_1/1}{Q_2/(N-k)}$$

follows an F distribution with 1 and $N - k$ degrees of freedom. It is easy to verify that this statistic is given by

$$F_{1,N-k} = \left[\frac{\left(\sum_{i=1}^{k} c_i \bar{Y}_{i\cdot} \right)^2}{\sum_{i=1}^{k} \frac{c_i^2}{n_i}} \right] \Bigg/ MS_E$$

The numerator of this statistic is called a *contrast sum of squares* and is denoted by SS_L. To illustrate, let us test the null hypothesis posed earlier.

Example 13.4.1. We wish to test

$$H_0: \mu_1 + \mu_3 - \mu_4 - \mu_5 = 0$$
$$H_1: \mu_1 + \mu_3 - \mu_4 - \mu_5 \neq 0$$

based on the data given in Example 13.1.2. From the earlier analysis we know that

$\bar{y}_{1\cdot} = 1.66$	$\bar{y}_{2\cdot} = 1.17$	
$\bar{y}_{3\cdot} = 1.46$	$\bar{y}_{4\cdot} = .88$	$\bar{y}_{5\cdot} = .88$
$MS_E = .122$	$n_1 = 7$	$n_2 = 8$
$n_3 = 9$	$n_4 = 8$	$n_5 = 10$

The desired contrast sum of squares is

$$\frac{\left(\sum_{i=1}^{5} c_i \bar{y}_{i\cdot} \right)^2}{\sum_{i=1}^{5} \frac{c_i^2}{n_i}} = \frac{[1(1.66) + 1(1.46) + (-1)(.88) + (-1)(.88)]^2}{\frac{1^2}{7} + \frac{1^2}{9} + \frac{(-1)^2}{8} + \frac{(-1)^2}{10}}$$

$$= 3.81$$

The observed value of the F ratio used to test H_0 is

$$F_{1,37} = \frac{SS_L}{MS_E} = \frac{3.81}{.122} = 31.23$$

From Table IX of App. A we see that $P < .05$. Hence we can conclude that the aggregate mean for coal seams 1 and 3 is significantly different from that of seams 4 and 5.

It is possible to form a very special set of $k-1$ independent contrasts $L_1, L_2, \ldots, L_{k-1}$ that have the property that their contrast sum of squares sum to the treatment sum of squares for the entire ANOVA. That is,

$$SS_{L_1} + SS_{L_2} + \cdots + SS_{L_{k-1}} = SS_{\text{Tr}}$$

In this way the researcher can determine which contrasts contribute most heavily toward the rejection of the null hypothesis of equal treatment means. For this to occur the contrasts selected must be *orthogonal*. This term is defined in Definition 13.4.2.

Definition 13.4.2 (Orthogonal contrasts). Two contrasts

$$L_1 = \sum_{i=1}^{k} a_i\mu_i \qquad \text{and} \qquad L_2 = \sum_{i=1}^{k} b_i\mu_i$$

are orthogonal if

$$\sum_{i=1}^{k} \frac{a_i b_i}{n_i} = 0$$

Note that if sample sizes are equal, then L_1 and L_2 are orthogonal whenever $\sum_{i=1}^{k} a_i b_i = 0$. It is relatively easy to design orthogonal contrasts in this case. When sample sizes are unequal, the task becomes more difficult. An example will clarify this point.

Example 13.4.2. Consider a one-way classification model in which $k = 5$ and sample sizes are equal. The contrasts

$$L_1 = \mu_1 + \mu_2 - \mu_3 - \mu_5 \qquad \text{and} \qquad L_2 = \mu_1 + \mu_2 + \mu_3 - 4\mu_4 + \mu_5$$

are orthogonal since

$$\sum_{i=1}^{5} a_i b_i = 1(1) + 1(1) + (-1)(1) + 0(-4) + (-1)(1) = 0$$

Neither is orthogonal to $L_3 = \mu_1 - \mu_4$. However, if $n_1 = 20$, $n_2 = 22$, $n_3 = 24$, $n_4 = 21$, and $n_5 = 25$, then L_1 and L_2 are not orthogonal. In this case

$$\sum_{i=1}^{5} \frac{a_i b_i}{n_i} = \frac{1(1)}{20} + \frac{1(1)}{22} + \frac{(-1)(1)}{24} + \frac{0(-4)}{21} + \frac{(-1)(1)}{25} \neq 0$$

Orthogonal contrasts are used when the experimenter wants to partition the variation in response due to treatments, SS_{Tr}, into independent components. The $k-1$ contrasts chosen to do so can be selected rather arbitrarily. However, usually the researcher has certain contrasts that are of particular interest from a physical standpoint. These contrasts will be included among those tested; other contrasts that are chosen simply must be orthogonal to those of interest for the partitioning of SS_{Tr} into $k - 1$ single degrees of freedom contrast sums of squares to take place.

13.5 RANDOMIZED COMPLETE BLOCK DESIGN

The procedure discussed in this section is an extension of the paired T procedure discussed in Chap. 10 for comparing the means of two normal populations. The purpose of pairing is to control the effect of one extraneous variable, a variable not under study in the experiment, by pairing experimental units that are similar with respect to this variable. Each member of the pair receives a different treatment, and any differences in response are attributed to treatment effects since the effect of the extraneous variable has been controlled by pairing.

When we want to compare the means of k populations in the presence of an extraneous variable, a procedure known as *blocking* is used. A block is a collection of k experimental units that are as nearly alike as possible relative to the extraneous variable. Each treatment is randomly assigned to 1 unit within each block. Since the effect of the extraneous variable is controlled by matching like experimental units, any differences in response are attributed to treatment effects.

The experimental design presented here is called the *randomized complete block* design with fixed effects. The word "blocks" refers to the fact that experimental units have been matched relative to some extraneous variable; *randomized* refers to the fact that treatments are randomly assigned within blocks; and to say that the design is *complete* implies that each treatment is used exactly once within each block. The term "fixed effects" applies to both blocks and treatments. That is, it is assumed that neither blocks nor treatments are randomly chosen. Any inferences made apply only to the k treatments and b blocks actually used.

If blocking is done well in the sense that experimental units within blocks are relatively homogeneous and units in different blocks are relatively heterogeneous, then the randomized complete block design is usually more sensitive to differences in treatment means than is the one-way classification design. If blocking is not done well, then the reverse may be true.

The hypothesis of interest is the hypothesis of equal treatment means given by

$$H_0: \mu_{1\cdot} = \mu_{2\cdot} = \cdots = \mu_{k\cdot}$$

where $\mu_{i\cdot}$ denotes the mean of the ith treatment.

Notationally, Y_{ij} denotes the response for the ith treatment in the jth block for $i = 1, 2, \ldots, k$ and $j = 1, 2, \ldots, b$. Note that b denotes the number of blocks used in the experiment and the number of observations per treatment; k denotes the number of treatments being investigated and the number of observations per block; $N = kb$ denotes the total number of responses. The data collected in a randomized complete block experiment together with some

TABLE 13.4
Data layout randomized complete blocks

	Treatment						
Block	**1**	**2**	**3**	...	k	**Block total**	**Block mean**
1	Y_{11}	Y_{21}	Y_{31}		Y_{k1}	$T_{\cdot 1}$	$\bar{Y}_{\cdot 1}$
2	Y_{12}	Y_{22}	Y_{32}		Y_{k2}	$T_{\cdot 2}$	$\bar{Y}_{\cdot 2}$
3	Y_{13}	Y_{23}	Y_{33}		Y_{k3}	$T_{\cdot 3}$	$\bar{Y}_{\cdot 3}$
⋮	⋮	⋮	⋮		⋮	⋮	⋮
b	Y_{1b}	Y_{2b}	Y_{3b}		Y_{kb}	$T_{\cdot b}$	$\bar{Y}_{\cdot b}$
Treatment total	$T_{1\cdot}$	$T_{2\cdot}$	$T_{3\cdot}$		$T_{k\cdot}$	$T_{\cdot\cdot}$	
Treatment sample mean	$\bar{Y}_{1\cdot}$	$\bar{Y}_{2\cdot}$	$\bar{Y}_{3\cdot}$		$\bar{Y}_{k\cdot}$	$\bar{Y}_{\cdot\cdot}$	

important sample statistics are conveniently displayed as shown in Table 13.4. Note that

$$\begin{aligned}
T_{i\cdot} &= \text{total of all responses to } i\text{th treatment} \\
&= \sum_{j=1}^{b} Y_{ij} \\
\bar{Y}_{i\cdot} &= \text{sample mean for } i\text{th treatment} \\
&= T_{i\cdot}/b \\
T_{\cdot j} &= \text{total of all responses in } j\text{th block} \\
&= \sum_{i=1}^{k} Y_{ij} \\
\bar{Y}_{\cdot j} &= \text{sample mean for } j\text{th block} \\
&= T_{\cdot j}/k \\
T_{\cdot\cdot} &= \text{total of all responses} \\
&= \sum_{i=1}^{k} \sum_{j=1}^{b} Y_{ij} = \sum_{i=1}^{k} T_{i\cdot} = \sum_{j=1}^{b} T_{\cdot j} \\
\bar{Y}_{\cdot\cdot} &= \text{mean of all responses} \\
&= T_{\cdot\cdot}/N
\end{aligned}$$

Our first example illustrates these ideas.

Example 13.5.1. Officials of a small transit system with only five buses want to evaluate four types of tires with respect to wear. Each of the buses runs a different route so that terrain and driving conditions differ from bus to bus. To control the effect of this extraneous variable, a randomized complete block design is appropriate. Each bus constitutes a block and each tire type constitutes a treatment. One tire of each type is placed on each bus with the wheel positions being assigned randomly. The tires are run for 15,000 miles after which the tread wear, in millimeters, is measured. The data obtained along with pertinent summary statistics are given below:

Blocks (buses)	Treatment (tire type) 1	2	3	4	Block total	Block mean
1	9.1	17.1	20.8	11.8	$T._1 = 58.8$	$\bar{Y}._1 = 14.7$
2	13.4	20.3	28.3	16.0	$T._2 = 78.0$	$\bar{Y}._2 = 19.5$
3	15.6	24.6	23.7	16.2	$T._3 = 80.1$	$\bar{Y}._3 = 20.025$
4	11.0	18.2	21.4	14.1	$T._4 = 64.7$	$\bar{Y}._4 = 16.175$
5	12.7	19.8	25.1	15.8	$T._5 = 73.4$	$\bar{Y}._5 = 18.35$
Treatment total	$T_1. = 61.8$	$T_2. = 100$	$T_3. = 119.3$	$T_4. = 73.9$	$T.. = 355.0$	
Treatment mean	$\bar{Y}_1. = 12.36$	$\bar{Y}_2. = 20.0$	$\bar{Y}_3. = 23.86$	$\bar{Y}_4. = 14.78$	$\bar{Y}.. = 17.75$	

Note that, as expected, there appears to be substantial differences among block means. Also, some differences appear among treatment means. Are these differences extreme enough to allow us to conclude that there are differences in the average tread wear for these four tire types? To answer this question, we must develop a way to test

$$H_0: \mu_1. = \mu_2. = \mu_3. = \mu_4.$$

statistically based on these data.

The Model

To write the model for the randomized complete block design with fixed effects, we need the following notation:

$$\mu_{ij} = \text{mean for the } i\text{th treatment and } j\text{th block}$$

$$\mu_{i}. = \text{mean of } i\text{th treatment} = \sum_{j=1}^{b} \mu_{ij}/b$$

$$\mu._j = j\text{th block mean} = \sum_{i=1}^{k} \mu_{ij}/k$$

$$\mu = \text{overall mean} = \sum_{i=1}^{k} \sum_{j=1}^{b} \mu_{ij}/kb$$

$\tau_i = \mu_i. - \mu$ = effect due to the fact that the experimental unit received the ith treatment

$\beta_j = \mu._j - \mu$ = effect due to the fact that the experimental unit is in the jth block

$E_{ij} = Y_{ij} - \mu_{ij}$ = residual or random error

We can now express the model as shown below:

Model for randomized complete block design

$$Y_{ij} = \mu + \tau_i + \beta_j + E_{ij}$$

This model expresses symbolically the notion that each observation can be partitioned into four recognizable components: an overall mean effect μ, a treatment effect τ_i, a block effect β_j, and a random deviation attributed to unexplained sources E_{ij}. We make these model assumptions:

1. The $k \cdot b$ observations constitute independent random samples, each of size 1, from $k \cdot b$ populations with unknown means μ_{ij}.
2. Each of the $k \cdot b$ populations is normally distributed.
3. Each of the $k \cdot b$ populations has the same variance, σ^2.
4. Block and treatment effects are additive; that is, there is no interaction between blocks and treatments.

Assumptions 1 through 3 are identical to those made in the one-way classification model except that $k \cdot b$, rather than k, populations are under consideration. The fourth assumption is new and needs to be examined more closely. Briefly, to say that block and treatment effects are additive means that the treatments behave consistently across blocks and that the blocks behave consistently across treatments. Mathematically, this means that the difference in the mean values for any two treatments is the same in every block, and the difference in the means for any two blocks is the same for each treatment. If this is not the case, then we say that there is *interaction* between blocks and treatments.

The concept of additivity versus interaction is illustrated in Fig. 13.3. In each case we have graphed the theoretical means for three treatments A, B, and C in each of two blocks. When no interaction exists, the line segments joining any two means will be parallel across blocks. Such is the case in Fig. 13.3(a). Practically speaking, this means that it is possible to make general statements concerning the treatments without having to specify the block involved. For example, it is correct to say that the mean for treatment A is smaller than that of B and C. This statement holds for both blocks. In Fig. 13.3(b) the line segments are not parallel. This means that there is interaction between blocks and treatments. In other words, we must be very careful when making statements concerning the treatments because the block involved is also important. For example, it is no longer correct to say that treatment A has a smaller mean than B and C. This statement is true for block 1 but it is not true for block 2.

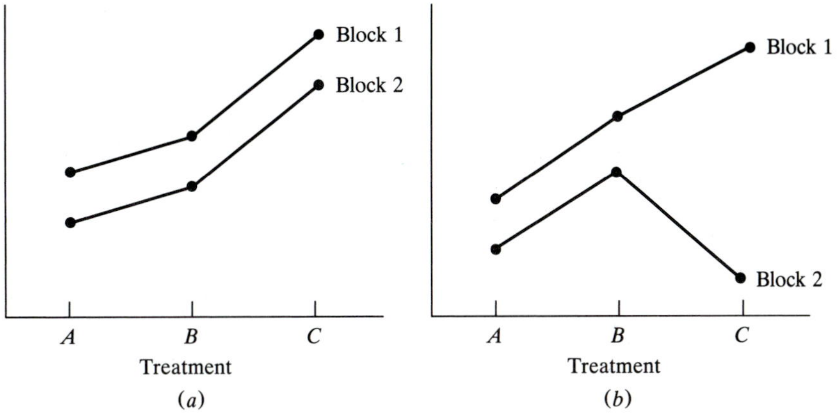

FIGURE 13.3
(a) Additive effects—no interaction (line segments are parallel); (b) interaction exists (line segments are not parallel).

Mathematically, additivity means that

$$\begin{aligned}\mu_{ij} &= \mu + \tau_i + \beta_j \\ &= \mu + (\mu_{i\cdot} - \mu) + (\mu_{\cdot j} - \mu)\end{aligned}$$

Substituting, we can rewrite the theoretical model as

$$\begin{aligned}\mu_{ij} - \mu &= \tau_i + \beta_j \\ &= (\mu_{i\cdot} - \mu) + (\mu_{\cdot j} - \mu) + \{\mu_{ij} - [\mu + (\mu_{i\cdot} - \mu) + (\mu_{\cdot j} - \mu)]\}\end{aligned}$$

Replacing parameters by their respective unbiased estimators, we have

$$\begin{aligned}Y_{ij} - \bar{Y}_{\cdot\cdot} = {} & (\bar{Y}_{i\cdot} - \bar{Y}_{\cdot\cdot}) + (\bar{Y}_{\cdot j} - \bar{Y}_{\cdot\cdot}) \\ & + \{Y_{ij} - [\bar{Y}_{\cdot\cdot} + (\bar{Y}_{i\cdot} - \bar{Y}_{\cdot\cdot}) + (\bar{Y}_{\cdot j} - \bar{Y}_{\cdot\cdot})]\}\end{aligned}$$

If each side of this identity is squared, summed over all possible values of i and j, and simplified, this sum-of-squares identity for the randomized complete block design results:

Sum of squares identity

$$\begin{aligned}\sum_{i=1}^{k}\sum_{j=1}^{b}(Y_{ij} - \bar{Y}_{\cdot\cdot})^2 = {} & \sum_{i=1}^{k} b(\bar{Y}_{i\cdot} - \bar{Y}_{\cdot\cdot})^2 + \sum_{j=1}^{b} k(\bar{Y}_{\cdot j} - \bar{Y}_{\cdot\cdot})^2 \\ & + \sum_{i=1}^{k}\sum_{j=1}^{b}(Y_{ij} - \bar{Y}_{i\cdot} - \bar{Y}_{\cdot j} + \bar{Y}_{\cdot\cdot})^2\end{aligned}$$

The practical interpretation for each component is similar to that of the one-way classification model. In particular,

$$\sum_{i=1}^{k}\sum_{j=1}^{b}(Y_{ij}-\bar{Y}_{..})^2 = \text{measure of total variability in data}$$
$$= \text{total sum of squares } (SS_{\text{Tot}})$$

$$\sum_{i=1}^{k} b(\bar{Y}_{i\cdot}-\bar{Y}_{..})^2 = \text{measure of variability in data attributable to use of different treatments}$$
$$= \text{treatment sum of squares } (SS_{\text{Tr}})$$

$$\sum_{j=1}^{b} k(\bar{Y}_{\cdot j}-\bar{Y}_{..})^2 = \text{measure of variability in data attributable to use of different blocks}$$
$$= \text{block sum of squares } (SS_{\text{Blks}})$$

$$\sum_{i=1}^{k}\sum_{j=1}^{b}(Y_{ij}-\bar{Y}_{i\cdot}-\bar{Y}_{\cdot j}+\bar{Y}_{..})^2 = \text{measure of variability in data due to random factors}$$
$$= \text{residual, or error sum of squares } (SS_E)$$

Symbolically, the sum of squares identity is

$$SS_{\text{Tot}} = SS_{\text{Tr}} + SS_{\text{Blks}} + SS_E$$

The hypothesis of equal treatment means can be stated in terms of the treatment effects τ_i. To see how this is done, note that if $\mu_{1\cdot} = \mu_{2\cdot} = \cdots = \mu_{k\cdot}$, then

$$\mu = \sum_{i=1}^{k}\sum_{j=1}^{b}\mu_{ij}/kb$$
$$= \sum_{i=1}^{k}\mu_{i\cdot}/k$$
$$= \mu_{i\cdot} \qquad \text{for each } i = 1, 2, \ldots, k$$

By definition $\tau_i = \mu_{i\cdot} - \mu$. Therefore if the treatment means are all the same, then their common value is μ and each treatment effect has value 0. The null hypothesis in our experiment

$$H_0\colon \mu_{1\cdot} = \mu_{2\cdot} = \cdots = \mu_{k\cdot}$$

is equivalent to

$$H_0\colon \tau_1 = \tau_2 = \cdots = \tau_k = 0$$

As in the case of the one-way classification model, this form for H_0 is useful if one wants to consider the randomized complete block design as a general linear model and to analyze the data via regression techniques.

Testing H_0

The test for this hypothesis is derived in a manner similar to that used for the one-way classification design. Utilizing the model assumptions and the rules for expectation, we can show that the expected mean square for treatments is given by

$$E[MS_{\text{Tr}}] = E[SS_{\text{Tr}}/(k-1)]$$

$$= \sigma^2 + \frac{b \sum_{i=1}^{k} \tau_i^2}{(k-1)}$$

To define the error mean square, we first note that the degrees of freedom associated with this statistic follow the usual pattern,

$$\begin{matrix}\text{Overall} \\ \text{sample size}\end{matrix} - 1 - \begin{matrix}\text{degrees of freedom associated} \\ \text{with other model components}\end{matrix}$$

or

$$kb - 1 - [(k-1) + (b-1)] = (k-1)(b-1)$$

As in the past, the error mean square is an unbiased estimator for σ^2. That is,

$$E[MS_E] = E[SS_E/(k-1)(b-1)] = \sigma^2$$

To test the null hypothesis,

$$H_0: \tau_1 = \tau_2 = \cdots = \tau_k = 0$$

we use the F ratio

$$F_{k-1,(k-1)(b-1)} = \frac{MS_{\text{Tr}}}{MS_E}$$

If H_0 is true, both the numerator and the denominator of the F statistic are estimators for σ^2 and the observed value of this statistic should lie close to 1; otherwise the numerator should be larger than the denominator, resulting in an F ratio larger than 1. Our test is to reject H_0 if the observed value of the test statistic is too large to have occurred by chance.

Table 13.5 summarizes the ideas developed in this section and gives some computational formulas for finding SS_{Tr} and SS_{Blks}.

Effectiveness of Blocking

Since blocking is designed to control the effect of an extraneous variable, the natural question to ask is, "Was blocking successful?" If so, then SS_{Blks} should account for a substantial portion of the total sum of squares. This in turn reduces SS_E, thereby increasing the value of the F ratio used to test for equality of treatment means and making it more likely that H_0 will be rejected. The power of the test will be improved. Note that the number of degrees of freedom for error in the one-way classification design is $N - k$; in the randomized

TABLE 13.5

ANOVA table for the fixed-effects randomized complete block design

Source of variation	Degrees of freedom (DF)	Sum of squares (SS)	Mean square (MS)	Expected mean square	F
Treatment	$k - 1$	$\sum_{i=1}^{k} \frac{T_{i\cdot}^2}{b} - \frac{T_{\cdot\cdot}^2}{kb}$	$SS_{\text{Tr}}/(k-1)$	$\sigma^2 + b\sum_{i=1}^{k} \frac{\tau_i^2}{(k-1)}$	$\frac{MS_{\text{Tr}}}{MS_E}$
Block	$b - 1$	$\sum_{j=1}^{b} \frac{T_{\cdot j}^2}{k} - \frac{T_{\cdot\cdot}^2}{kb}$	$SS_{\text{Blks}}/(b-1)$		
Error	$(k-1)(b-1)$	Subtraction	$SS_E/(k-1)(b-1)$	σ^2	
Total	$kb - 1$	$\sum_{i=1}^{k}\sum_{j=1}^{b} Y_{ij}^2 - \frac{T_{\cdot\cdot}^2}{N}$			

complete block design it is smaller than this, namely, $(k-1)(b-1) = (N-k) - (b-1)$. In Table IX of App. A observe that as the number of degrees of freedom associated with the denominator of an F ratio decrease, the tabled F value increases. The implication of this is that if blocking is done unnecessarily, we pay a price for this mistake. Namely, the number of degrees of freedom for error decreases, the critical point for testing H_0 increases in size, and it becomes harder to reject H_0. The power of the test will become smaller. It is clear that blocking can help when appropriate but that indiscriminate blocking should be avoided.

When an experiment is being conducted for the first time, intuition, based on knowledge of the subject matter, is the only guide in deciding whether to block. Once the initial experiment is conducted, an assessment of the effectiveness of blocking can be done so that future studies can be designed efficiently. It seems reasonable to suggest that if block means are equal, then blocking is unnecessary; otherwise blocking is useful. However, *there is no known valid way to test the null hypothesis of equal block means*. One approach used to investigate the effectiveness of blocking is to estimate the relative efficiency (RE) of the randomized complete block design as compared to the completely randomized design given in Sec. 13.1. The theoretical development of the notion of relative efficiency is beyond the scope of this text but can be found in [27] and [31]. Relative efficiency is a positive number that can be interpreted as the ratio of the number of observations per treatment needed for the two designs to be equivalent. For example, if $RE = 3$, then the completely randomized design requires 3 times as many observations as the randomized complete block design to produce a test with the same characteristics; blocking is desirable in this case. If $RE = .5$, then blocking is not desirable since the completely randomized design can accomplish the same thing as the randomized block design by using half as many observations. If $RE = 1$, then the designs are equivalent when sample sizes are identical.

Can we estimate the relative efficiency quickly from our original analysis of variance? Fortunately, there is an easy way to do so. It has been found [32] that there is a linear relationship between $\widehat{RE}$ and the ratio

$$MS_{\text{Blks}}/MS_E$$

where $\quad MS_{\text{Blks}} = \text{block mean square} = SS_{\text{Blks}}/(b-1)$

This relationship is given by

$$\widehat{RE} = c + (1-c)(MS_{\text{Blks}}/MS_E)$$

where $c = b(k-1)/(bk-1)$. It is easy to show that

$$\widehat{RE} = 1 \quad \text{if and only if } MS_{\text{Blks}}/MS_E = 1$$

$$\widehat{RE} < 1 \quad \text{if and only if } MS_{\text{Blks}}/MS_E < 1$$

$$\widehat{RE} > 1 \quad \text{if and only if } MS_{\text{Blks}}/MS_E > 1$$

Thus to make a judgment about whether blocking in a particular experiment helped or hurt, we can use the information available from our ANOVA to find the value of MS_{Blks}/MS_E. We then estimate the relative efficiency and decide from practical considerations such as the time, cost, and effort required to block whether blocking is worthwhile. As experience is gained, it will become unnecessary actually to compute $\widehat{RE}$. One need only consider the observed value of MS_{Blks}/MS_E. Values considerably larger than 1 indicate that blocking was beneficial; values near 1 indicate that blocking neither helped nor hurt; values somewhat less than 1 indicate that blocking was not helpful. In the latter case a completely randomized design is preferable in future studies.

It should be noted that some texts include a "test" for blocks based on the statistic MS_{Blks}/MS_E. However, due to the manner in which randomization is achieved in the randomized complete block design, the test is improper.

As an example, let us continue the analysis of the data given in Example 13.5.1.

Example 13.5.2. To compare four types of tires for use on buses, we test the null hypothesis that there are no differences with respect to average tread wear among these brands. That is, we test

$$H_0\colon \mu_{1\cdot} = \mu_{2\cdot} = \mu_{3\cdot} = \mu_{4\cdot}$$

We have computed these summary statistics:

$T_{1\cdot} = 61.8$	$T_{\cdot 1} = 58.8$	$T_{\cdot\cdot} = 355$
$T_{2\cdot} = 100$	$T_{\cdot 2} = 78.0$	
$T_{3\cdot} = 119.3$	$T_{\cdot 3} = 80.1$	
$T_{4\cdot} = 73.9$	$T_{\cdot 4} = 64.7$	
	$T_{\cdot 5} = 73.4$	

For the data given $\sum_{i=1}^{4}\sum_{j=1}^{5} Y_{ij}^2 = 6810.28$. Using the computational formulas given in Table 13.5, you should be able to verify most of the figures in the ANOVA

TABLE 13.6
ANOVA for tire wear data

Source of variation	Degrees of freedom (DF)	Sum of squares (SS)	Mean square (MS)	F
Treatments	3	401.338	133.779	61.340
Block	4	81.525	20.381	9.345*
Error	12	26.167	2.181	
Total	19	509.030		

table shown in Table 13.6. Since $f_{.05}(3, 12) = 3.49$ and $61.34 > 3.49$, we can reject

$$H_0\colon \mu_{1\cdot} = \mu_{2\cdot} = \mu_{3\cdot} = \mu_{4\cdot}$$

with $P < .05$. We have strong statistical evidence of differences in the mean tread wear among the four tire types. The starred value in the table is the observed value of the ratio MS_{Blks}/MS_E used to evaluate the effectiveness of blocking. Since this value exceeds 1, we can conclude that blocking was appropriate in this experiment. The estimated relative efficiency is

$$\widehat{RE} = c + (1 - c)(MS_{\text{Blks}}/MS_E)$$

where $c = b(k - 1)/(bk - 1) = 5(3)/(19) = .789$. In this case

$$\widehat{RE} = .789 + .211(9.345) = 2.76$$

Paired Comparisons

As in the one-way classification completely randomized design, paired comparisons can be made by performing $\binom{k}{2}$ Bonferroni type T tests with α carefully chosen so that α' is kept under control. In this case the T tests conducted are paired T tests. Keep in mind the fact that this method is feasible only when k is rather small. This is because large values of k force α to be *extremely* small, resulting in a test with very little power.

A Duncan's multiple range test is available also. The test is conducted as described earlier with

$$\text{SSR}_p = r_p\sqrt{\frac{MS_E}{b}}$$

Note that in this design sample sizes are equal.

We close this section by completing the analysis of the data given in Example 13.5.1.

Example 13.5.3. In the two previous examples we found significant differences in tread wear among the four types of tires under study and that blocking was appropriate.

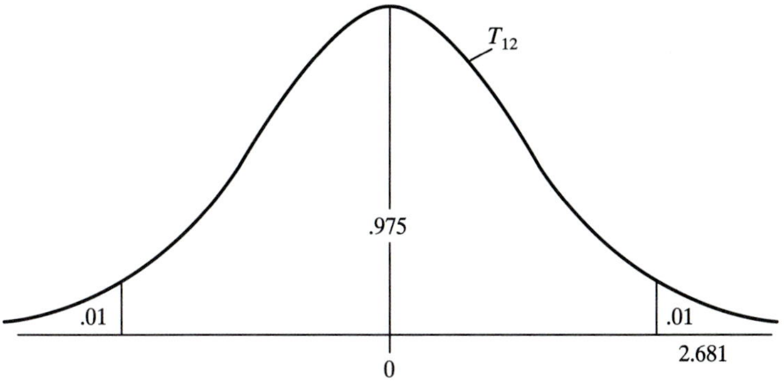

FIGURE 13.4
The T value used in finding the critical point for testing H_0: $\mu_i = \mu_j$ at $\alpha = .02$ level. The number of degrees of freedom is $(k - 1)(b - 1) = 12$.

The number of paired comparisons to be made is $\binom{4}{2} = 6$. We can use either Duncan's procedure or Bonferroni T tests. If Duncan's procedure is used, then α should be set at .01 so that the experimentwise error rate will be kept reasonably low. In this case let us conduct Bonferroni T tests with α' controlled at .12. To do so, we find that the comparisonwise error rate is $\alpha = .12/6 = .02$. Since sample sizes are equal, the common critical point for the Bonferroni T tests is

$$cp = t_{(k-1)(b-1),\,1-\alpha/2}\sqrt{MS_E[(1/n) + (1/n)]}$$

The point $t_{(k-1)(b-1),\,1-\alpha/2}$ is shown in Fig. 13.4. Its value is 2.681. The error mean square, 2.181, is found in Table 13.6. Since each treatment mean is based on a sample of size 4, $n = 4$ and the critical point is

$$cp = 2.681\sqrt{2.181[(1/4) + (1/4)]} \doteq 2.80$$

For two population means to be declared different their corresponding sample means must differ by at least 2.80. The ordered sample means with nonsignificant differences underlined are

$\bar{Y}_{1\cdot}$	$\bar{Y}_{4\cdot}$	$\bar{Y}_{2\cdot}$	$\bar{Y}_{3\cdot}$
12.36	14.78	20.0	23.86

All pairs of means are significantly different except 1 and 4. Since low tread wear is good, the experimenter can conclude that tire types 1 and 4 are both superior to types 2 and 3. He or she cannot say that one tire is clearly superior to all the others but can say that type 3 is clearly worse than the others based on these data.

13.6 LATIN SQUARES (OPTIONAL)

We have seen that the randomized complete block design is used to control the effect of one source of extraneous variation. In Example 13.5.1 four types of tires are compared. Since the buses on which the tires are to be used operate under different driving conditions, the buses are considered as blocks to account for this extraneous variable. In this example there is another recognizable source of variation that should be controlled. Namely, it is suspected that tire wear is affected by the tire's position on the bus. Front tires may wear faster than rear tires. It would be helpful to be able to design an experiment that takes this observation into account. This can be done by using what is called a *Latin square* design. In general, a Latin square is useful when the experimenter wants to control *two* extraneous sources of variation. Each of the extraneous variables is a blocking variable. Hypothesis tests for differences among levels of these variables are not appropriate. Thus a Latin square design is a design in which one factor is studied in the presence of two extraneous variables whose effects are controlled in the design of the experiment.

For a Latin square to be appropriate the number of levels at which each extraneous factor is studied must be the same as the number of population means being compared. In our example four types of tires are under study. Hence four buses are needed; on each bus the four natural tire positions are utilized. Tire types must be randomly assigned to buses and positions in such a way that each type of tire is used exactly once on each bus and exactly once in each position. Randomization is achieved by randomly selecting a design from existing tables of Latin square designs. These may be found in [15]. In our tire study a 4×4 design is needed. A typical layout is shown in Table 13.7. The letters A, B, C, and D refer to the type of tire employed.

For example, this design indicates that tire type B is to be used on bus I at tire position I; tire type C is to be used on bus I, position II, and so forth. Note that each tire type appears once on each bus and once at each position as desired. In general, when r population means are to be compared in the presence of two extraneous variables, an $r \times r$ design is required. Each treatment occurs exactly once in each row and once in each column of the design.

The theory behind the analysis of a Latin square design is an extension of that developed for a randomized complete block design. Notation becomes a little tricky. Here Y_{ijk} denotes the response to the treatment applied in row j and column k. For example, in the tire experiment the treatment applied in the first row and the first column is B, the second treatment. We indicate this by writing the response as Y_{211}. The entire data layout for the design given in Table 13.7 is shown in Table 13.8.

TABLE 13.7
A typical 4 × 4 Latin square design

Factor (tire position)	Factor (bus) I	II	III	IV
I	B	C	D	A
II	C	D	A	B
III	D	A	B	C
IV	A	B	C	D

TABLE 13.8
Data layout for the design of Table 13.7

Factor (tire position)	Factor (bus) I	II	III	IV
I	Y_{211}	Y_{312}	Y_{413}	Y_{114}
II	Y_{321}	Y_{422}	Y_{123}	Y_{224}
III	Y_{431}	Y_{132}	Y_{233}	Y_{334}
IV	Y_{141}	Y_{242}	Y_{343}	Y_{444}

The model for a Latin square design is given by

$$Y_{ijk} = \mu + \tau_i + \beta_j + \gamma_k + E_{ijk}$$

where τ_i is a treatment effect, β_j is a row effect, and γ_k is a column effect. In our example τ_i is an effect due to the type of tire being used, β_j is an effect associated with the position of the tire on the bus, and γ_k is an effect attributed to the bus itself. These model assumptions are made:

1. The r^2 observations constitute independent random samples, each of size 1, from r^2 populations, each with unknown means.
2. Each of the r^2 populations is normally distributed.
3. Each of the r^2 populations has the same variance, σ^2.
4. There is no interaction. That is, the only effects present are those due to the row, treatment, and column variables acting individually as indicated in the model.

The null hypothesis to be tested is

$$H_0: \mu_{1\cdot\cdot} = \mu_{2\cdot\cdot} = \cdots = \mu_{r\cdot\cdot} \qquad \text{(treatment means are identical)}$$

Necessary totals and sample means are

$$T_{i\cdot\cdot} = \text{total of all responses to treatment } i$$

$$\bar{Y}_{i\cdot\cdot} = \frac{T_{i\cdot\cdot}}{r} = \text{mean of all responses to treatment } i$$

$$T_{\cdot j\cdot} = \text{total of all responses in row } j$$

$$\bar{Y}_{\cdot j\cdot} = \frac{T_{\cdot j\cdot}}{r} \text{mean of all responses in row } j$$

$$T_{\cdot\cdot k} = \text{total of all responses in column } k$$

$$\bar{Y}_{\cdot\cdot k} = \frac{T_{\cdot\cdot k}}{r} = \text{mean of all responses in column } k$$

$$T_{\cdots} = \text{total of all responses}$$

$$\bar{Y}_{\cdots} = \frac{T_{\cdots}}{r^2} = \text{mean of all responses}$$

Symbolically, the sum of squares identity is

$$SS_{\text{Tot}} = SS_{\text{Tr}} + SS_{\text{Rows}} + SS_{\text{Col}} + SS_E$$

These sum of squares are defined and calculated in a manner analogous to those of the randomized complete block design. For example,

$$\text{Treatment sum of squares} = SS_{\text{Tr}} = r\sum_{i=1}^{r}\left(\bar{Y}_{i\cdot\cdot} - \bar{Y}_{\cdot\cdot\cdot}\right)^2 = \sum_{i=1}^{r}\frac{T_{i\cdot\cdot}^2}{r} - \frac{T_{\cdot\cdot\cdot}^2}{r^2}$$

You are asked to determine the form of the others as an exercise. The total number of degrees of freedom is given by total sample size minus 1 or $r^2 - 1$. These are partitioned as

$$DF(\text{total}) = DF(\text{treatments}) + DF(\text{rows}) + DF(\text{columns}) + DF(\text{error})$$

or

$$r^2 - 1 = (r-1) + (r-1) + (r-1) + (r-1)(r-2)$$

Dividing the sums of squares by their respective degrees of freedom yields the mean squares needed to test H_0. The test statistic is

$$F_{r-1,(r-1)(r-2)} = \frac{MS_{\text{Tr}}}{MS_E}$$

The complete ANOVA is shown in Table 13.9.

TABLE 13.9
ANOVA table for the fixed-effects Latin square design

Source of variation	Degrees of freedom (DF)	Sum of squares (SS)	Mean square (MS)	F
Treatment	$r-1$	$\sum_{i=1}^{r}\frac{T_{i\cdot\cdot}^2}{r} - \frac{T_{\cdot\cdot\cdot}^2}{r^2}$	$SS_{\text{Tr}/(r-1)}$	MS_{Tr}/MS_E
Rows	$r-1$	$\sum_{j=1}^{r}\frac{T_{\cdot j\cdot}^2}{r} - \frac{T_{\cdot\cdot\cdot}^2}{r^2}$	$SS_{\text{Rows}/(r-1)}$	
Columns	$r-1$	$\sum_{k=1}^{r}\frac{T_{\cdot\cdot k}^2}{r} - \frac{T_{\cdot\cdot\cdot}^2}{r^2}$	$SS_{\text{Col}/(r-1)}$	
Error	$(r-1)(r-2)$	Subtraction	$SS_{E/(r-1)(r-2)}$	
Total	r^2-1	$\sum\sum\sum_{\text{all possible } i,j,k} Y_{ijk}^2 - \frac{T_{\cdot\cdot\cdot}^2}{r^2}$		

Note that the only F ratio listed is that used to test for equality among treatment means. As was the case in the randomized complete block design, no valid *test* exists to determine the effectiveness of controlling the two extraneous sources of variation. However, the benefits of control in each case can be assessed by evaluating the ratios MS_{Rows}/MS_E and MS_{Col}/MS_E. Whenever these exceed 1, the design is helpful in accounting for extraneous sources of variation.

13.7 RANDOM-EFFECTS MODELS

One-Way Classification

The model discussed in Sec. 13.1 is called the *fixed-effects* model. This implies that the factor levels, or "treatments," are selected specifically by the experimenter because they are of particular interest. The purpose of the experiment is to make inferences about the means of the particular populations from which the samples are drawn. If, however, we want to make a broad generalization concerning a larger set of populations and not just the k populations from which we sample, then the appropriate model is called a *random-effects model.* In this case the k sampled populations are considered to be a random sample of populations drawn from the larger set. The hypothesis of interest is not that $\mu_1 = \mu_2 = \cdots = \mu_k$. Rather, we want to determine whether some variability exists among the population means of the larger set.

The random-effects model is written as follows:

$$Y_{ij} = \mu + T_i + \text{E}_{ij}$$

where μ = overall mean effect

μ_i = mean of the ith population selected for study

$T_i = \mu_i - \mu$ = effect of the ith treatment

$\text{E}_{ij} = Y_{ij} - \mu_i$ = residual or random error

The following *model assumptions* are made:

1. The k samples represent independent random samples from k populations randomly selected from a larger set of populations.
2. Each of the populations in the larger set is normally distributed, and therefore each of the sampled populations is also normal.
3. Each of the populations in the larger set has the same variance σ^2, and thus each of the k sampled populations also has variance σ^2.
4. $T_1, T_2, \ldots, T_k$ are independent normally distributed random variables, each with mean 0 and common variance σ^2_{Tr}.

The model itself and the first three model assumptions are similar to those of the fixed-effects model. However, an important difference between the two is expressed in assumption 4. In the fixed-effects model the treatments, or levels, used in the experiment are purposely chosen by the experimenter because they are of particular interest. If the experiment were replicated, or repeated, the same treatments would be used. That is, the same populations would be sampled each time and the k treatment effects $\alpha_i = \mu_i - \mu$ would not vary. This implies that in the fixed-effects model the k treatment effects are *unknown constants*. In the random-effects model this is not the case. Since the first step in a random-effects experiment is to select randomly k populations for study, those actually chosen will vary from replication to replication. Thus the k terms $T_i = \mu_i - \mu$ are not constants but, rather, are, in fact, *random variables* whose values for a given replication depend on the choice of the k populations to be studied. These random variables are assumed to be independent and normally distributed, each with mean 0 and common variance σ_{Tr}^2.

If the population means in the larger set are equal, then the treatment effects $T_i = \mu_i - \mu$ will not vary. That is, σ_{Tr}^2 will be 0. Thus in the random-effects model the hypothesis of equal means is expressed as

$$H_0\colon \sigma_{\text{Tr}}^2 = 0 \qquad \text{(no variability in treatment effects)}$$

$$H_1\colon \sigma_{\text{Tr}}^2 \neq 0$$

Even though the one-way classification random-effects model differs from the fixed-effects model, H_0 is tested in *exactly* the same way in each case. The difference in model assumptions concerning the nature of the treatment effects is reflected not in the way the data is handled but, rather, in the expected mean squares. These expectations are shown in Table 13.10. The term n_0, which appears in the expected mean square for treatments, is given by

$$n_0 = \frac{N^2 - \sum_{i=1}^{k} n_i^2}{N(k-1)}$$

TABLE 13.10

ANOVA table for the one-way classification design with random effects

Source of variation	**Degrees of freedom (DF)**	**Sum of squares (SS)**	**Mean square (MS)**	**Expected mean square**	F
Treatment or level	$k - 1$	$\sum_{i=1}^{k} \frac{T_{i\cdot}^2}{n_i} - \frac{T_{\cdot\cdot}^2}{N}$	$\frac{SS_{\text{Tr}}}{k-1}$	$\sigma^2 + n_0\sigma_{\text{Tr}}^2$	$\frac{MS_{\text{Tr}}}{MS_E}$
Residual or error	$N - k$	Subtraction	$\frac{SS_E}{N-k}$	σ^2	
Total	$N - 1$	$\sum_{i=1}^{k} \sum_{j=1}^{n_i} Y_{ij}^2 - \frac{T_{\cdot\cdot}^2}{N}$			

If sample sizes are equal, this reduces to n. You can see that if H_0 is true, then both MS_{Tr} and MS_E estimate σ^2. In this case the F ratio MS_{Tr}/MS_E should assume a value close to 1. Otherwise MS_{Tr} is expected to exceed σ^2 yielding an F ratio larger than 1. Just as in the fixed-effects model, we reject H_0 for values of the test statistic that are too large to have occurred by chance based on the $F_{k-1, N-k}$ distribution.

If H_0 is rejected, we do not perform multiple comparison tests as we did in the fixed-effects model. Rather, we estimate σ^2_{Tr}, the variability in the treatment effects. From the table of expected mean squares it is easy to see that an unbiased estimator for this parameter is given by

$$\hat{\sigma}^2_{Tr} = \frac{MS_{Tr} - MS_E}{n_0}$$

The next example illustrates these ideas.

Example 13.7.1. A utility company has a large stock of voltmeters that are used interchangeably by many employees. A study is conducted to detect differences among the average readings given by these voltmeters. If it appears that differences do exist, then all the meters in stock will be calibrated. A random sample of six meters is selected from stock and four readings are taken for each meter. The response variable is the difference between the meter reading and the known voltage being applied at the time of the reading. These data result:

Voltmeter

1	2	3	4	5	6
.18	−.15	−.25	1.95	−.90	1.10
−1.31	1.85	.77	1.03	−.50	1.21
.15	.63	1.65	.65	.25	.68
−.81	.45	1.24	1.25	−.88	.92

Since the six voltmeters used in the experiment represent a random sample of meters drawn from a larger population of meters, a one-way classification random-effects model is appropriate. The null and alternative hypotheses are

H_0: $\sigma^2_{Tr} = 0$ (there is no difference in treatment effects—all voltmeters in stock give the same average reading)

H_1: $\sigma^2_{Tr} \neq 0$

Treating the data exactly as in the fixed-effects model, we obtain the ANOVA shown in Table 13.11. If H_0 is true, the observed F ratio is expected to lie close to 1; otherwise it should be considerably larger than 1. Based on the $F_{5,18}$ distribution, we can reject H_0 with $P < .05$ ($f_{.05} = 2.773$ from Table IX of App. A). We do have statistical evidence of differences in average readings among the voltmeters in stock.

TABLE 13.11
ANOVA for voltmeter data

Source of variation	Degrees of freedom (DF)	Sum of squares (SS)	Mean square (MS)	F
Treatment	5	11.257	2.251	5.284
Error	18	7.669	.426	
Total	23	18.926		

We can estimate how much of the variability in meter readings is due to differences in meters and how much is due to random error. To do this, we estimate the variance components σ^2 and σ^2_{Tr}. From the ANOVA table unbiased estimates for these parameters are

$$\hat{\sigma}^2 = \text{MSE} = .426$$

$$\hat{\sigma}^2_{\text{Tr}} = \frac{MS_{\text{Tr}} - MS_E}{n_0}$$

$$= \frac{2.251 - .426}{4} = .456$$

The estimated total variability in meter readings is

$$\sigma^2_{\text{Tot}} = \hat{\sigma}^2 + \hat{\sigma}^2_{\text{Tr}} = .426 + .456 = .882$$

The proportion of total variability attributed to meter differences is

$$\frac{\hat{\sigma}^2_{\text{Tr}}}{\hat{\sigma}^2_{\text{Tot}}} = \frac{.456}{.882} = .517 \qquad \text{(51.7\% of total variability)}$$

Hence recalibration of meters seems to be called for since about 52% of total variability can be attributed to meter differences.

13.8 DESIGN MODELS IN MATRIX FORM

As you can see, as the design of an experiment becomes more complex, the computations needed to analyze the data become more cumbersome. This was also the case in the regression context; as more variables were added to the regression equation, the computations became so burdensome that we were forced to turn to matrix algebra to analyze the data efficiently. It is reasonable to assume that we might do the same here.

To see how to approach the problem, let us reconsider the one-way classification model with fixed effects. We know that this model can be expressed as

$$Y_{ij} = \mu + \alpha_i + \text{E}_{ij} \qquad i = 1, 2, \ldots, k; \qquad j = 1, 2, \ldots, n_i$$

In expanded form we have

$$
\begin{aligned}
Y_{11} &= \mu + \alpha_1 + \mathrm{E}_{11} \\
Y_{12} &= \mu + \alpha_1 + \mathrm{E}_{12} \\
&\vdots \\
Y_{1n_1} &= \mu + \alpha_1 + \mathrm{E}_{1n_1} \\
\hdashline
Y_{21} &= \mu + \alpha_2 + \mathrm{E}_{21} \\
Y_{22} &= \mu + \alpha_2 + \mathrm{E}_{22} \\
&\vdots \\
Y_{2n_2} &= \mu + \alpha_2 + \mathrm{E}_{2n_2} \\
\hdashline
&\vdots \\
\hdashline
Y_{k1} &= \mu + \alpha_k + \mathrm{E}_{k1} \\
Y_{k2} &= \mu + \alpha_k + \mathrm{E}_{k2} \\
&\vdots \\
Y_{kn_k} &= \mu + \alpha_k + \mathrm{E}_{kn_k}
\end{aligned}
$$

The vectors of responses, parameters, and random errors are given by

$$
\mathbf{Y} = \begin{bmatrix} Y_{11} \\ Y_{12} \\ Y_{1n_1} \\ Y_{21} \\ Y_{22} \\ Y_{2n_2} \\ \vdots \\ Y_{k1} \\ Y_{k2} \\ Y_{kn_k} \end{bmatrix}
\qquad
\boldsymbol{\alpha} = \begin{bmatrix} \mu \\ \alpha_1 \\ \alpha_2 \\ \vdots \\ \alpha_k \end{bmatrix}
\qquad
\mathbf{E} = \begin{bmatrix} \mathrm{E}_{11} \\ \mathrm{E}_{12} \\ \vdots \\ \mathrm{E}_{1n_1} \\ \mathrm{E}_{21} \\ \mathrm{E}_{22} \\ \vdots \\ \mathrm{E}_{2n_2} \\ \vdots \\ \mathrm{E}_{k1} \\ \mathrm{E}_{k2} \\ \vdots \\ \mathrm{E}_{kn_k} \end{bmatrix}
$$

respectively. The model specification matrix or design matrix is

$$X = \begin{bmatrix} 1 & 1 & 0 & & 0 \\ 1 & 1 & 0 & \cdots & 0 \\ \vdots & \vdots & \vdots & & \vdots \\ 1 & 1 & 0 & & 0 \\ 1 & 0 & 1 & \cdots & 0 \\ 1 & 0 & 1 & & 0 \\ \vdots & \vdots & \vdots & & \vdots \\ 1 & 0 & 1 & & 0 \\ \vdots & & & & \\ 1 & 0 & 0 & & 1 \\ 1 & 0 & 0 & \cdots & 1 \\ \vdots & \vdots & \vdots & & \vdots \\ 1 & 0 & 0 & & 1 \end{bmatrix}$$

A quick calculation will show that the system of equations defining the model can be expressed in matrix form as

$$\mathbf{Y} = X\boldsymbol{\alpha} + \mathbf{E}$$

This is exactly the matrix form for the general linear model derived earlier. That is, the *one-way classification model with fixed effects* is just a special case of the general linear model. It differs from the regression models studied earlier only in the format of the model specification matrix. In a regression model the first column is a column of 1's; the other k columns consist of the observed values of the independent variables $X_1, X_2, \ldots, X_k$. In a design model the first column is a column of 1's; the other k columns consist of blocks of 1's and 0's. In a regression model our primary null hypothesis is

$$H_0\colon \beta_1 = \beta_2 = \cdots = \beta_k = 0 \qquad \text{(regression is not significant)}$$

In general linear form we write the primary null hypothesis of no difference among treatment means in the form

$$H_0\colon \alpha_1 = \alpha_2 = \cdots = \alpha_k = 0 \qquad \text{(regression is not significant; that is, there are no differences among treatment effects)}$$

As you should suspect, the matrix analysis of a design model is almost identical to that of a regression model. The difference in approach is necessitated by the fact that in a design model the matrix $X'X$ has no inverse. We can no longer claim that the parameters $\mu, \alpha_1, \alpha_2, \ldots, \alpha_k$ are estimated by

$$\hat{\alpha} = (X'X)^{-1}X'\mathbf{Y}$$

There are various ways to remedy the situation. However, a full discussion of the problem is beyond the scope of this text. Excellent references are available if you wish to pursue the matter, [27] and [35].

Example 13.8.1. To illustrate in a simple context, consider these data:

Treatment		
1	**2**	**3**
1.3	2.7	1.6
1.5	1.1	3.2
2.1	1.6	
	2.0	

For these data,

$$\mathbf{Y} = \begin{bmatrix} 1.3 \\ 1.5 \\ 2.1 \\ 2.7 \\ 1.1 \\ 1.6 \\ 2.0 \\ 1.6 \\ 3.2 \end{bmatrix} \qquad \boldsymbol{\alpha} = \begin{bmatrix} \mu \\ \alpha_1 \\ \alpha_2 \\ \alpha_3 \end{bmatrix} \qquad \mathbf{E} = \begin{bmatrix} E_{11} \\ E_{12} \\ E_{13} \\ E_{21} \\ E_{22} \\ E_{23} \\ E_{24} \\ E_{31} \\ E_{32} \end{bmatrix}$$

The model specification matrix is

$$X = \begin{bmatrix} 1 & 1 & 0 & 0 \\ 1 & 1 & 0 & 0 \\ 1 & 1 & 0 & 0 \\ 1 & 0 & 1 & 0 \\ 1 & 0 & 1 & 0 \\ 1 & 0 & 1 & 0 \\ 1 & 0 & 1 & 0 \\ 1 & 0 & 0 & 1 \\ 1 & 0 & 0 & 1 \end{bmatrix}$$

$$X'X = \begin{bmatrix} 9 & 3 & 4 & 2 \\ 3 & 3 & 0 & 0 \\ 4 & 0 & 4 & 0 \\ 2 & 0 & 0 & 2 \end{bmatrix}$$

Note that $X'X$ has no inverse since the first column of this matrix is the sum of the last three columns.

The randomized complete block can be expressed in matrix form also. We leave this model to the reader to investigate.

13.9 ALTERNATIVE NONPARAMETRIC METHODS

As seen in Secs. 8.7 and 10.6, there are nonparametric analogs to the normal theory T tests for two independent samples and for paired samples when testing location differences. The multiple sample extensions of the normal theory two-independent-sample and paired-sample problems were seen in this chapter to be the one-way analysis of variance and the randomized complete block design. Both of these procedures assumed underlying normally distributed populations with rather restrictive assumptions on the population variances. Fortunately, nonparametric analogs are available for both of these procedures when the normal theory tests assumptions are not met. The nonparametric test for the one-way classification analysis of variance is the *Kruskal-Wallis test*. The *Friedman test* is the nonparametric alternative for the randomized complete block design. The two procedures are discussed in this section.

Kruskal-Wallis Test

Assume that k independent random samples of sizes $n_1, n_2, \ldots, n_k$ are drawn from continuously distributed populations. The Kruskal-Wallis procedure tests the hypothesis that each of the k samples has been drawn from identical populations. However, the test is particularly sensitive to location differences, and therefore the null hypothesis is usually stated in terms of equality of population medians. Thus the null hypothesis can be stated as

$$H_0\colon M_1 = M_2 = \cdots = M_k$$

$$H_1\colon \text{at least two population medians are not equal}$$

To perform the test, the $N = n_1 + n_2 + \cdots + n_k$ sample observations are pooled and ranked from the smallest to the largest, retaining group identity. As was done for the Wilcoxon tests, ties are assigned the average rank for their group. Let T_i, $i = 1, 2, \ldots, k$, denote the sum of the ranks associated with the observations from the ith population. The Kruskal-Wallis test statistic is given by

$$H = \frac{12}{N(N+1)} \sum_{i=1}^{k} n_i \left(\bar{T}_i - \frac{N+1}{2} \right)^2$$

where $\bar{T}_i = T_i/n_i$ denotes the average of the ranks assigned to the ith group. If the null hypothesis is true, it can be shown that $E(\bar{T}_i) = (N+1)/2$. Therefore the Kruskal-Wallis test statistic is a measure of the deviations of the observed average ranks for the k groups from the value expected if the null hypothesis is true. Large deviations lead to relatively large values of H, and hence to rejection of H_0. Although exact tables are available for small k and n_i, it has been shown that H approximately follows a chi-squared distribution with $k - 1$ degrees of freedom if all $n_i \geq 5$. Hence approximate critical values for H can be obtained from the chi-squared distribution given in Table IV of App. A. The reader can easily verify that an equivalent but computationally easier form for the test statistic H is

$$H = \left[\frac{12}{N(N+1)} \sum_{i=1}^{k} \frac{T_i^2}{n_i}\right] - 3(N+1)$$

Example 13.9.1. An experiment was conducted to compare the amount of pressure needed to compress three types of materials. Random samples of sizes 7, 7, and 10 were obtained for materials labeled A, B, and C, respectively. The pressure measurements and corresponding ranks (in parentheses) of the combined sample of 7 + 7 + 10 = 24 observations are given below:

Material A	Material B	Material C
207 (14)	194 (11)	288 (12.5)
150 (5)	146 (3)	269 (20)
197 (12)	175 (8)	288 (21.5)
173 (7)	186 (9)	358 (24)
147 (4)	223 (17)	229 (18)
144 (2)	143 (1)	249 (19)
192 (10)	170 (6)	346 (23)
		217 (16)
		203 (13)
		214 (15)

The rank sums are

$$T_1 = 14 + 5 + 12 + \cdots + 2 + 10 = 54$$

$$T_2 = 11 + 3 + 8 + \cdots + 1 + 6 = 55$$

$$T_3 = 21.5 + 20 + 21.5 + \cdots + 13 + 15 = 191$$

Calculating the Kruskal-Wallis test statistic, we obtain

$$H = \left[\frac{12}{N(N+1)} \sum_{i=1}^{3} \frac{T_i^2}{n_i}\right] - 3(N+1)$$

$$= \frac{12}{24(25)}\left(\frac{54^2}{7} + \frac{55^2}{7} + \frac{191^2}{10}\right) - 3(25)$$

$$= 14.94$$

From Table IV for $k - 1 = 3 - 1 = 2$ degrees of freedom we see that the critical value for a significance level of .005 is 10.6. Hence we conclude that the amount of pressure required to compress the materials is significantly different for at least two of the materials tested.

Like the Wilcoxon rank-sum test, the Kruskal-Wallis test is very robust relative to the usual normal theory F test. In addition, various multiple comparison procedures are available in standard texts on nonparametric statistics. The reader is referred to [23], [6], and [30].

Friedman Test

The Friedman test is the nonparametric analog to the randomized complete block design discussed in Sec. 13.5. As discussed in that section, our interest is in comparing the effects of k treatments (groups) when we are able to control for extraneous variation by blocking. We divide the experimental units into b blocks, each of size k, with elements within a block as nearly alike as possible with respect to the extraneous variable. The k treatments (groups) are then randomly assigned to elements within each block. To test the hypothesis of identical treatment effects (location differences), we use the ranking procedure as follows: Rank the observations *within each block* from 1 to k (the smallest to the largest), assigning tied scores the average group rank. Then the rank total, T_i, for each of the k treatments (groups) is computed. The Friedman test statistic is given by

$$S = \frac{12}{bk(k+1)} \sum_{i=1}^{k} \left[T_i - \frac{b(k+1)}{2} \right]^2$$
$$= \left[\frac{12}{bk(k+1)} \sum_{i=1}^{k} T_i^2 \right] - 3b(k+1)$$

When the null hypothesis is true, it can be shown that $E(T_i) = b(k+1)/2$. Therefore the Friedman statistic is a measure of deviations of the observed treatment rank totals from their expected value under H_0. The exact distribution of S has been tabulated for small values of b and k, but S is approximately distributed as a chi-squared distribution with $k - 1$ degrees of freedom. Using this approximation, we reject the null hypothesis for values of S exceeding critical values given in Table IV for specified significance levels.

Example 13.9.2. As implied in our sections on alternative nonparametric procedures, the data for some kinds of problems come originally in the form of ranks. For such data it should be clear that normality assumptions are not met, and hence normal theory tests are most likely not valid. In this example we wish to determine if there are significant differences in preferences for eight brands of computer terminals. Four judges are asked to rate the eight terminals, giving rank 1 to their first choice, rank 2 to their second choice, and so on. The experiment was conducted by using a randomized complete block design with the eight brands of terminals as treatments and the four judges as blocks. The resulting data are as follows:

	Terminal							
	1	**2**	**3**	**4**	**5**	**6**	**7**	**8**
Judge I	1	2	3	4	5	6	7	8
Judge II	4	1	3	2	5	6	8	7
Judge III	3	4	2	1	7	5	6	8
Judge IV	3	1	6	2	5	4	7	8
Rank totals	11	8	14	9	22	21	28	31

The hypothesis to be tested is

H_0: equal preference for terminals

H_1: significant difference in preference for at least two terminals

Calculating the Friedman test statistic, we obtain

$$S = \left[\frac{12}{(4)(8)(9)}(11^2 + 8^2 + 14^2 + \cdots + 31^2)\right] - 3(4)(9)$$

$$= 22.5$$

From Table IV the critical value for $k - 1 = 7$ degrees of freedom is 18.50 for $\alpha = 0.01$. Therefore we can conclude that there is a significant difference in terminal preferences for at least two terminals. Again, we mention that multiple comparison procedures are available in various nonparametric texts which enable one to determine which treatments differ significantly from one another.

CHAPTER SUMMARY

The fundamentals of analysis of variance were presented for some basic experimental design models. We considered the one-way classification, randomized complete block, and Latin square experiments. Multiple comparison methods were presented for fixed effects and variance component estimation was discussed for random effects. General analysis of variance (ANOVA) tables, useful for computational purposes and for selecting appropriate mean square error ratios for hypothesis testing, were given in the chapter. We attempted to be careful about assumptions necessary for the various models. A test for equality of variances was given and alternative nonparametric methods were discussed for possible use when the normal theory assumptions are questioned.

Finally, the design models were presented in matrix form to enable the student to see the relation to the general linear model.

We also introduced and defined important terms that you should know. These are:

One-way classification	Bonferroni inequality
Homogeneity of variances	Completely randomized design
Randomized complete block design	Multiple comparisons
Random effects	Fixed effects
Variance components	Nonparametric analysis of variance
Contrast	Latin square
Comparisonwise error rate	Bonferroni T tests
Experimentwise error rate	Duncan's multiple range test

EXERCISES

Section 13.1

1. Show that $E[\overline{\mathrm{E}}_{i\cdot}{}^2] = \sigma^2/n_i$ for each $i = 1, 2, \ldots, k$. *Hint:*

$$E[\overline{\mathrm{E}}_{i\cdot}{}^2] = E\left[\left(\sum_{j=1}^{n_i} \mathrm{E}_{ij} \Big/ n_i\right)^2\right]$$

$$= E\left[\left(\sum_{j=1}^{n_i} \mathrm{E}_{ij}\right)^2\right] \Big/ n_i^2$$

Argue that due to the independence of the terms E_{ij}, all cross-product terms $\mathrm{E}_{ij}\mathrm{E}_{if}$, $j \neq f$, have expectation 0. Argue also that $E[\mathrm{E}_{ij}^2] = \sigma^2$.

2. Show that

$$SS_{\mathrm{Tot}} = \sum_{i=1}^{k} \sum_{j=1}^{n_i} Y_{ij}^2 - T_{\cdot\cdot}{}^2/N$$

3. Show that

$$SS_{\mathrm{Tr}} = \sum_{i=1}^{k} T_{i\cdot}{}^2/n_i - T_{\cdot\cdot}{}^2/N$$

4. Experiments were conducted to study whether commercial processing of various foods changes the concentration of essential elements for human consumption. One such experiment was to study the concentration of zinc in green beans. A batch of green beans was divided into four groups. The four groups were then randomly assigned to be measured for zinc as follows: group 1 measured raw; group 2 measured before blanching; group 3 measured after blanching; and group 4 measured after the final processing step. Independent measurements were taken from the four groups (treatments), yielding the following observations:

Zinc concentration

Group 1	Group 2	Group 3	Group 4
2.23	3.71	2.53	5.46
2.20	4.67	2.87	5.19
2.44	3.45	2.83	5.51
2.11	2.73	2.33	4.82
2.30	2.58	2.19	6.63
1.72	1.85	1.80	2.39
1.78	1.81	1.75	2.09
2.36	2.32	1.83	2.27
2.91	2.50	1.97	2.39

(*a*) State the appropriate null and alternative hypotheses.
(*b*) Test your hypothesis for significance at the 5% level.
(*c*) Verbally state your conclusion.

5. It was known that a toxic material was dumped in a river leading into a large salt water commercial fishing area. Civil engineers studied the way the water carried the toxic material by measuring the amount of the material (in parts per million) found in oysters harvested at three different locations, ranging from the estuary out into the bay where the majority of commercial fishing was carried out. The resulting data are given below:

Site 1	Site 2	Site 3
15	19	22
26	15	26
20	10	24
20	26	26
29	11	15
28	20	17
21	13	24
26	15	
	18	

(*a*) Test whether there is a significant difference in the average parts per million of toxic material found in oysters harvested at the three sites. Use $\alpha = .05$.
(*b*) Would the means be significantly different at the .01 level of significance? (Use a computer package to find the P value.)
(*c*) Do your answers in parts (*a*) and (*b*) contradict each other? Justify your answer.

6. As landfills begin to fill up and land for such operations becomes more scarce, recycling will become more and more necessary. Some fast-food chains have become aware of this problem and are beginning to use biodegradable packaging or to recycle their packaging material. A study is conducted to investigate the amount of money received per year for recycled goods by various sized businesses. These data, in thousands of dollars, are obtained (based on figures in "New Ways to Take Out the Trash," Tom Andel, *Transportation and Distribution*, May 1993):

Company Size 101–250 employees		251–500 employees		Over 500 employees	
3.5	3.8	13.4	14.1	9.0	10.8
3.4	3.0	14.0	14.7	10.3	10.7
3.7	2.8	14.5	15.0	10.1	9.5
3.3	3.2	14.7	13.6	10.6	11.2
2.9	3.9	15.2	14.3	10.4	9.7
2.6	4.1	15.5	14.6	10.5	10.0
3.5	4.0	14.2	14.8	10.3	11.3
4.2	4.9	15.3	15.6	10.7	11.8

(*a*) Construct stem-and-leaf diagrams for each of these samples and comment on the reasonableness of the normality assumption.

(*b*) Estimate the variance of each of the populations from which these data are drawn and comment on the reasonableness of the assumption of equal variances. (Do not try to test anything.) What has been done on this study to guard against a possible violation of the assumption of equal variances?

(*c*) Test H_0: $\mu_1 = \mu_2 = \mu_3$ and report the P value of your test. What practical conclusion can be drawn?

7. Scroll speed is an important consideration in the development of color graphics cards. A study is conducted to compare the time, in seconds, required to scroll one screen of WORD documents using five different color graphics cards with 24-inch monitors. The test conducted was a standard Hydra Quick Draw performance test. These data are obtained (based on information found in "Gauging Video Speed," *MAC WORLD*, June 1993, p. 28):

Graphics cards

A	*B*	*C*	*D*	*E*
30.5	48.3	79.2	51.6	79.0
32.4	42.1	84.7	59.4	85.3
27.2	43.5	85.0	57.3	86.2
26.3	40.6	88.2	59.0	82.0
25.1	38.6	76.3	58.7	87.2
38.2	32.1	83.1	68.1	81.7
30.6	41.6	92.6	64.8	93.5
33.7	38.8	88.5	55.5	89.1

Use these data to test for equality of means. State your conclusion and report the P value of the test.

Section 13.2

8. Recall that MS_E was defined as the ratio $SS_E/(N - k)$. Show that

$$MS_E = S_p^2 = \frac{\sum_{i=1}^{k} (n_i - 1)S_i^2}{N - k}$$

for the one-way classification fixed-effects model.

9. In Exercise 4, use Bartlett's test to determine whether it is reasonable to assume homogeneity of variances for the four treatment groups. If not, what type of alternative procedure is available to analyze these data?

10. Test for equality of variances by using the data in Exercise 6. Be sure to state your conclusion and to report the P value of your test.

11. Test for equality of variances by using the data of Exercise 7.

Section 13.3

12. In each case, determine the number of paired comparisons possible.
(*a*) $k = 3$
(*b*) $k = 6$
(*c*) $k = 10$

13. Let $\alpha = .05$ and assume that all possible paired comparisons are to be made. For each value of k given in Exercise 12, use the Bonferroni procedure to find an upper bound for α', the overall probability of making at least one incorrect rejection. In each case, what level should be used to guarantee that $\alpha' \leq .10$?

14. Consider a situation in which c tests are to be run. Let b denote the target upper bound for α'. In each case, show that $1 - [1 - (b/c)]^c \leq b$, thus guaranteeing that the choice of $\alpha = b/c$ implies that $\alpha' \leq b$.
(*a*) $c = 2$
(*b*) $c = 3$

In each case, find the value of $1 - [1 - (b/c)]^c$ for $b = .10$.

15. (*Bonferroni inequality*.) The following inequality provides the basis for determining the comparisonwise error rate needed to control the experimentwise error rate at some specified level.

Let $A_1, A_2, \ldots, A_c$ be events. Then

$$P[A_1 \cap A_2 \cap \cdots \cap A_c] \geq 1 - [P[A_1'] + P[A_2'] + \cdots + P[A_c']]$$

(*a*) Prove this result for $c = 2$. *Hint*: Notice that by the general addition rule, $P[A_1' \cup A_2'] \leq P[A_1'] + P[A_2']$. Use the fact that $A_1 \cap A_2 = (A_1' \cup A_2')'$ to complete the proof.

(*b*) Let A_i denote the event that no incorrect rejection is made on the ith comparison, $i = 1, 2, \ldots, c$. In set notation, express the probability that no incorrect rejections are made. In terms of this probability, what is the probability that at least one incorrect rejection is made? If $\alpha = P[A_i']$ for $i = 1, 2, \ldots, c$, then use the Bonferroni inequality to show that $\alpha' \leq \alpha c$. Argue that if a target upper bound b is chosen and then c tests are seen at $\alpha = b/c$, then $\alpha' \leq b$.

16. Consider the study described in Exercise 6. If Duncan's multiple comparisons are run at $\alpha = .05$, what is an upper bound for the experimentwise error rate? Conduct these tests and discuss your conclusions.

17. Reconsider the data of Exercise 5.
(*a*) How many paired comparisons are possible?
(*b*) Suppose that these comparisons are to be made using Bonferroni T tests. If we want $\alpha' \leq .15$, at what α level should each test be conducted?
(*c*) Conduct each of the tests indicated in part (*b*) and discuss your results.

18. Use Bonferroni T tests with $\alpha' \leq .10$ to pinpoint differences among groups in Exercise 4.

19. Use Bonferroni T tests with $\alpha' \leq .20$ to pinpoint differences among graphics cards based on the data of Exercise 7.

20. Scientists concerned with treatment of tar sand wastewater studied three treatment methods for the removal of organic carbon. (Based on "Statistical Planning and Analysis for Treatments of Tar Sand Wastewater," W. R. Pirie, Technical Report, Virginia Tech University). The three treatment methods used were air flotation (A.F.), foam separation (F.S.), and ferric–chloride coagulation (F.C.C.). The organic carbon material measurements for the three treatments yielded the following data:

A.F.	F.S.	F.C.C.
34.6	38.8	26.7
35.1	39.0	26.7
35.3	40.1	27.0
35.8	40.9	27.1
36.1	41.0	27.5
36.5	43.2	28.1
36.8	44.9	28.1
37.2	46.9	28.7
37.4	51.6	30.7
37.7	53.6	31.2

Test for differences among means. Be sure to report the P value of your test and state your conclusion.

21. If you found significant differences in Exercise 20, determine which treatment methods differ from each other.

22. A study on the tensile strength of aluminum rods is conducted. Forty identical rods are randomly divided into four groups, each of size 10. Each group is subjected to a different heat treatment and the tensile strength, in thousands of pounds per square inch, of each rod is determined. The following data result:

Treatment			
1	2	3	4
18.9	18.3	21.3	15.9
20.0	19.2	21.5	16.0
20.5	17.8	19.9	17.2
20.6	18.4	20.2	17.5
19.3	18.8	21.9	17.9
19.5	18.6	21.8	16.8
21.0	19.9	23.0	17.7
22.1	17.5	22.5	18.1
20.8	16.9	21.7	17.4
20.7	18.0	21.9	19.0

(*a*) Construct a stem-and-leaf diagram for each data set. Does the assumption that each sample is drawn from a normal distribution appear reasonable?

(*b*) Test the null hypothesis of equal treatment means by using a one-way ANOVA.

(*c*) Compare all possible pairs of means by using Bonferroni T tests with an overall level of significance that is at most .05.

(*d*) Conduct a Duncan's multiple range test at the .05 level.

(*e*) Did the tests of parts (*c*) and (*d*) lead you to the same conclusions? Even though a level of significance of .05 is used in each case, these probabilities are not interpreted in the same way. Explain the difference between the two.

Section 13.4

23. Verify that $(Q_1/1)/(Q_2/(N-k))$ as defined in Sec. 13.4 can be expressed as SS_L/MS_E, where

$$SS_L = \frac{\left(\sum_{i=1}^{k} c_i \bar{Y}_{i\cdot}\right)^2}{\sum_{i=1}^{k} \frac{c_i^2}{n_i}}$$

as claimed.

24. In each case, decide whether the linear functions given are contrasts. If so, determine whether or not they are orthogonal.

(*a*) $L_1 = \mu_1 - (1/3)(\mu_2 + \mu_3 + \mu_4)$ $\quad n_1 = n_2 = n_3 = n_4 = n$
$L_2 = \mu_1 - \mu_2$

(*b*) $L_1 = \mu_1 - \mu_2$ $\quad n_1 = n_2 = n_3 = n_4 = n$
$L_2 = \mu_3 - \mu_4$

(*c*) $L_1 = \mu_1 - \mu_2$ $\quad n_1 = n_2 = 6$
$L_2 = \mu_3 - \mu_4$ $\quad n_3 = n_4 = 10$

(*d*) $L_1 = \mu_1 + 2\mu_2$ $\quad n_1 = n_2 = n_3 = n_4 = n$
$L_2 = \mu_1 - 2\mu_2 + \mu_3$

(*e*) $L_1 = \mu_1 - \mu_2 + \mu_3 - \mu_4$ $\quad n_1 = n_2 = n_3 = n_4 = n$
$L_2 = \mu_1 - \mu_3$

(*f*) $L_1 = \mu_1 - \mu_2 + \mu_3 - \mu_4$ $\quad n_1 = n_2 = 5$
$L_2 = \mu_1 - \mu_3$ $\quad n_3 = n_4 = 10$

25. Let $\bar{Y}_{1\cdot}$, $\bar{Y}_{2\cdot}$, $\bar{Y}_{3\cdot}$, and $\bar{Y}_{4\cdot}$ denote the sample means based on independent samples drawn from normal distributions with means μ_1, μ_2, μ_3, and μ_4, respectively. Find the contrast sum of squares for each of the contrasts of Exercise 24, parts (*a*), (*e*), and (*f*).

26. Consider the orthogonal contrasts of Exercise 24(*e*). Use the data of Exercise 22 to find the numerical values of SS_{L_1} and SS_{L_2}. Since $k = 4$, there exists a third contrast that is orthogonal to L_1 and L_2. What is the numerical value of SS_{L_3}?

27. Use the data of Exercise 22 to test

$$H_0\colon \mu_1 - \mu_2 + \mu_3 - \mu_4 = 0$$

and to test

$$H_0\colon \mu_1 - \mu_3 = 0$$

A chemical engineer is studying a newly developed polymer to be used in removing toxic wastes from water. Experiments are conducted at five different temperatures. The response noted is the percentage of impurities removed by the treatment. Use these data to answer Exercises 28 through 30:

Temperature				
I	II	III	IV	V
40	36	49	47	55
35	42	51	49	60
42	38	53	51	62
48	39	53	52	63
50	37	52	50	59
51	40	50	51	61

28. Test the null hypothesis of equal treatment means at the $\alpha = .05$ level.

29. Use Bonferroni T tests with $\alpha' = .25$ to compare all possible pairs of means. Report your results schematically by ordering the sample means from the smallest to the largest and underlining pairs that are not significantly different.

30. Write a contrast that can be used to compare the mean of population 5 to the aggregate of the other four means. Test the null hypothesis that this contrast assumes the value 0. Interpret your results in the context of this exercise.

Section 13.5

31. Each table given shows the theoretical treatment means for each of four treatments in each of three blocks. Graph these means in a manner similar to that of Fig. 13.3. In each case, decide whether or not there is interaction between blocks and treatments.

(*a*)

	Treatment			
Block	*A*	*B*	*C*	*D*
1	1	3	4	0
2	4	6	7	3
3	2	4	5	1

(*b*)

	Treatment			
Block	*A*	*B*	*C*	*D*
1	1	3	0	0
2	4	6	5	3
3	2	4	5	1

(*c*)

	Treatment			
Block	*A*	*B*	*C*	*D*
1	1	3	4	0
2	4	5	7	3
3	2	4	5	1

32. Show that $\sum_{i=1}^{k}\tau_i = 0$.

33. Explain the meaning of each of these relative efficiencies in assessing the effectiveness of blocking.
(*a*) $RE = 2$
(*b*) $RE = 10$
(*c*) $RE = .25$
(*d*) $RE = .10$
(*e*) $RE = 1$

A quality control engineer conducted an experiment to investigate the effect of experience on an assembly line in terms of the average time required to complete an assembly task. If experience is found to be a factor, a training program is planned for newer employees. The engineer randomly selected eight employees from groups who had completed 1, 2, 3, and 4 years of work experience, respectively. He set up the experiment as a randomized block design with tasks as blocks and years of experience as treatments. The resulting data are given below:

Time to complete assembly tasks

	Experience			
Task	**1 year**	**2 years**	**3 years**	**4 years**
1	40.3	34.2	28.8	26.6
2	25.4	25.4	29.2	21.1
3	28.2	28.0	24.6	23.2
4	41.6	24.9	29.1	27.0
5	28.8	39.2	34.8	27.1
6	38.7	29.5	26.6	27.3
7	29.4	29.0	36.0	34.2
8	37.7	25.6	25.6	25.2

Use this information to answer Exercises 34 through 39.

34. Write the appropriate statistical model for this experiment.

35. Test for any significant differences among years of experience for average assembly time. Use $\alpha = .05$.

36. Was blocking appropriate? Explain.

37. How many paired comparisons are possible? If these comparisons are made by means of Bonferroni paired T tests, what α level must be used to attain an overall level of significance that is at most .10?

38. Use Bonferroni T tests to determine which treatments (number of years of experience), if any, differ significantly with $\alpha' = 10$.

39. Do the data suggest that a training program might be productive?

A computer scientist is studying four different algorithms used to perform numerical integration. The speed (in seconds) with which a problem can be solved is measured. Each algorithm is to be used to solve each of 10 different problems and hence a problem serves as a block. Use these data to answer Exercises 40 through 43:

Problem	Algorithm I	II	III	IV
1	.10	.11	.16	.10
2	.11	.13	.17	.09
3	.13	.14	.17	.12
4	.08	.11	.13	.09
5	.15	.16	.19	.16
6	.17	.18	.23	.16
7	.05	.07	.11	.07
8	.13	.15	.18	.14
9	.12	.14	.15	.12
10	.16	.15	.20	.15

40. Test for differences among the mean times required to solve integration problems for these algorithms.

41. Estimate the relative efficiency and use it to discuss the desirability of blocking in a future study of this sort.

42. If the Bonferroni technique is used to compare all pairs of means, how many comparisons are possible? If each is done at the $\alpha = .01$ level, what is an upper bound for α', the overall probability of committing at least one Type I error?

43. Use Bonferroni T tests with $\alpha = .01$ to determine whether there appears to be a "fastest" algorithm.

Section 13.6

44. Give an example of a Latin square design with $r = 3$.

45. Give an example of a 4×4 Latin square that is different from that given in Table 13.7. Give the data layout for your design.

46. The design given in Table 13.12 is called a *knight's move* Latin square because one can get from each treatment type to another of the same type by making a knight's move as in the game of chess. The observed responses are given in parentheses in the table. Use these data to test H_0: $\mu_{1\cdot\cdot} = \mu_{2\cdot\cdot} = \mu_{3\cdot\cdot} = \mu_{4\cdot\cdot}$ at the $\alpha = .05$ level:

TABLE 13.12
A knight's move Latin square

	I	II	III	IV
I	*A* (3)	*B* (4.2)	*C* (.9)	*D* (5.3)
II	*C* (1.0)	*D* (5.6)	*A* (3.8)	*B* (4.3)
III	*B* (4.5)	*A* (3.5)	*D* (5.7)	*C* (1.2)
IV	*D* (5.5)	*C* (.8)	*B* (3.9)	*A* (3.7)

TABLE 13.13
Gasoline mileage obtained controlling for driver and automobile type (A = regular, B = extra, C = super)

	Factor (auto type)		
	4 cylinder	6 cylinder	8 cylinder
1	A (36.0)	B (33.0)	C (26.5)
2	B (36.5)	C (33.5)	A (25.0)
3	C (38.0)	A (32.5)	B (26.0)

47. Note that $SS_{\text{Tr}} = r\sum_{i=1}^{r}(\bar{Y}_{i}.. - \bar{Y}...)^2$. Express SS_{Tot}, SS_{Rows}, and SS_{Col} in a similar manner.

48. In a study to compare the average gasoline mileage of three different grades of a particular brand of gasoline, two extraneous sources of variation are recognized. These are the driver and the type of car used in the road trials. Three drivers and three new automobiles are obtained. Table 13.13 gives the mileage obtained over each test run. Test the null hypothesis of equal means among grades of gasoline at the .05 level.

49. A company is considering three word processors for use by their secretarial staff. All of them are acceptable, so the choice depends on which one is the easiest to learn. Since speed is affected by the type of article being produced and the learner, a Latin square design is used. The data obtained are given in Table 13.14. Test

$$H_0\colon \mu_{1}.. = \mu_{2}.. = \mu_{3}..$$

Section 13.7

50. A large number of laboratories are regularly used to measure the amount of toxic substances in various materials. There is concern that results not only vary due to normal measurement variability, but that there may be substantial variability due to different laboratory techniques. If true, this might raise a need for enforcing one "standard" procedure for all laboratories. To test this concern, four laboratories were randomly selected and asked to measure the amount of a certain chemical content in parts per million. Each laboratory was given six identical samples for

TABLE 13.14
Time required to learn various subroutines in a word processor (in hours)

	Factor (type of article)		
	I prose	II technical	III report
I	B (.5)	A (1.2)	C (1.0)
II	C (1.2)	B (1.0)	A (.95)
III	A (.8)	C (1.7)	B (.9)

testing. The resulting data are given below:

Laboratory

1	2	3	4
	(parts per million)		
53.2	51.0	47.4	51.0
54.5	40.5	46.2	51.5
52.8	50.8	46.0	48.8
49.3	51.5	45.3	49.2
50.4	52.4	48.2	48.3
53.8	49.9	47.1	49.8

(*a*) State the appropriate null and alternative hypotheses.
(*b*) Test the null hypothesis at the .05 level of significance.
(*c*) Estimate the proportion of total variance due to within laboratory variation (error mean square).
(*d*) Estimate the proportion of total variance due to between laboratory variation (variance due to treatments).
(*e*) Does it appear that standardizing laboratory techniques has merit?

Section 13.8

51. Consider the one-way classification model

$$Y_{ij} = \mu + \alpha_i + E_{ij} \qquad i = 1, 2, 3; \qquad j = 1, 2$$

For this model, find $\mathbf{Y}$, $\boldsymbol{\alpha}$, X, and $\mathbf{E}$. Find $X'X$ and show that this matrix has no inverse.

52. Consider the randomized complete block model

$$Y_{ij} = \mu + \alpha_i + \beta_j + E_{ij} \qquad i = 1, 2, 3; \qquad j = 1, 2, 3, 4$$

For this model, find $\mathbf{Y}$, $\boldsymbol{\alpha}$, X, and $\mathbf{E}$. Find $X'X$ and show that this matrix has no inverse.

Section 13.9

53. For a certain manufacturing plant, filters used to remove solid pollutants must be replaced as soon as they fail due to cracking or holes in the filter. An experiment was conducted to test five types of filters made from different fabrics. Six filters of each type were used under the same conditions, with the number of hours until failure recorded for each. The experiment yielded the following information:

Filter type (hours until failure)

1	2	3	4	5
261.1	221.9	201.4	600.9	160.6
186.2	188.7	146.1	301.2	135.0
239.1	167.6	96.8	608.9	455.1
243.3	224.9	173.9	283.3	402.3
296.8	178.8	280.8	193.3	457.9
270.5	147.9	100.3	159.4	559.6

(*a*) Use the Kruskal-Wallis test to determine whether there is significant evidence that the median time to failure among the filter types is different at the .05 level of significance.

(*b*) Repeat this test by using the appropriate normal theory test procedure.

54. Following a major accidental spill from a chemical manufacturing plant near a river, a study was conducted to determine whether certain species of fish caught from the river differ in terms of the amounts of the chemical absorbed. If differences are found, regulations on human consumption may be recommended. Samples from catches of three major species were measured in parts per million. The resulting data are given below:

Species

A	*B*	*C*
18.1	29.1	26.6
16.5	15.8	16.1
21.0	20.4	18.8
18.7	23.5	25.0
7.4	18.5	21.8
12.4	21.3	15.4
16.1	23.1	19.9
17.9	23.8	15.5
	20.1	21.1
	11.9	25.5

Test whether the median amounts of chemical absorbed by the three species of fish differ at the .05 level of significance.

55. A laboratory manager plans to purchase machines used to analyze blood samples. Five types of machines are being considered for purchase. After trial use, each of the eight technicians is asked to rank the machines in order of preference, with a rank of 1 being assigned to the machine most preferred. The respective rankings were as follows:

	Machine				
Technician	**I**	**II**	**III**	**IV**	**V**
A	1	3	4	2	5
B	4	5	1	2	3
C	4	1	3	5	2
D	4	1	5	2	3
E	1	3	2	5	4
F	1	2	3	4	5
G	5	1	3	2	4
H	5	1	4	3	2

Use the Friedman test to determine whether the group of technicians rate the machines differently at the .10 level of significance.

56. Four brands of tires are tested for tread wear. Since different cars may lead to different amounts of wear, cars are considered as blocks to reduce the effect of differences among cars. An experiment is conducted with cars considered as blocks and brands of tires randomly assigned to the four positions of tires on the cars. After a predetermined number of miles driven, the amount of tread wear (in milliliters) is measured for each tire. The resulting data are given below:

	Tire brand			
Car	*A*	*B*	*C*	*D*
1	8.9	6.6	5.6	4.2
2	7.2	6.9	7.3	6.9
3	3.1	6.2	7.2	4.1
4	7.1	8.3	6.3	5.8
5	6.7	6.4	5.9	9.4
6	5.3	6.7	8.0	7.9
7	2.4	5.5	6.1	3.1
8	5.7	9.2	9.6	4.2

(*a*) Rank the data appropriately for this experiment.

(*b*) If the null hypothesis is assumed to be true, what is the expected average rank total for brands of tires.

(*c*) Test the hypothesis that the median tread wear is equal for each brand of tire. Use $\alpha = .05$.

REVIEW EXERCISES

Carbon dioxide is known to have a critical effect on microbiological growth. Small amounts of CO_2 stimulate growth of some organisms, whereas high concentrations inhibit the growth of most. The latter effect is used commercially when perishable food products are stored. A study is conducted to investigate the effect of CO_2 on the growth rate of *Pseudomonas fragi*, a food-spoiling organism. Carbon dioxide is administered at five predetermined different atmospheric pressures. The response measured was the percentage change in cell mass after a 1-hour growing time. Ten cultures were used at each atmospheric pressure level, resulting in the following data:

Factor level (CO_2 pressure)				
0.0	**.083**	**.29**	**.50**	**.86**
62.6	50.9	45.5	29.5	24.9
59.6	44.3	41.1	22.8	17.2
64.5	47.5	29.8	19.2	7.8
59.3	49.5	38.3	20.6	10.5
58.6	48.5	40.2	29.2	17.8
64.6	50.4	38.5	24.1	22.1
50.9	35.2	30.2	22.6	22.6
56.2	49.9	27.0	32.7	16.8
52.3	42.6	40.0	24.4	15.9
62.8	41.6	33.9	29.6	8.8

Exercises 57 through 59 refer to these data.

57. State the assumptions required to test the null hypothesis

$$H_0: \mu_1 = \mu_2 = \cdots = \mu_5$$

58. (*a*) Numerically complete the appropriate analysis of variance (ANOVA) table.
(*b*) Is there sufficient evidence to reject H_0 at the $\alpha = .05$ level of significance?

59. Use Bartlett's test to test for homogeneity of variances. Does this test lend support to the assumptions given in Exercise 57?

60. Three treatments were randomly selected from a large population of possible treatments. Ten randomly selected observations were then obtained from each treatment selected.

(*a*) State an appropriate null hypothesis to be tested and list all assumptions necessary to make this test for the described experiment.

(*b*) The data yielded the following partial analysis of variance table:

ANOVA

Source	*DF*	*SS*	*MS*	*F*	*EMS*
Treatment	2	110.6			
Error	27				
Total	29	608.3			

Complete the ANOVA table and test your null hypothesis given in part (*a*) at the .05 level of significance.

(*c*) Estimate the proportions of total variability due to error and treatments, respectively.

61. An experiment is conducted to compare the energy requirements of three physical activities: running, walking, and bicycle riding. The variable of interest is the number of kilocalories expended per kilometer traveled. To control for possible metabolic differences, eight subjects were selected and then randomly assigned (in terms of order) each of the three tasks, with ample rest between tasks to eliminate fatigue. Each activity is monitored exactly once for each individual. The resulting data are given below:

	Task		
Individual	**Running**	**Walking**	**Bicycling**
1	1.4	1.1	0.7
2	1.5	1.2	0.8
3	1.8	1.3	0.7
4	1.7	1.3	0.8
5	1.6	0.7	0.1
6	1.5	1.2	0.7
7	1.7	1.1	0.4
8	2.0	1.3	0.6

(*a*) Test for possible differences in average kilocalories expended among the three tasks. State an approximate P value.

(*b*) Is there evidence that blocking (individuals) was effective in eliminating extraneous variability?

(*c*) Which energy tasks, if any, differ from each other at the .05 level of significance?

62. The OPEC oil embargo made it evident that fuel economy in automobiles needed to be improved. Newer lightweight materials were sought for use in automobile engines. Comparisons are made among test samples of steel, aluminum, and phenolic thermoset composites containing glass fibers. Two variables, density (g/cm^3) and tensile strength (ksi) are considered. These data are obtained (based on information found in "Phenolics Creep Up on Engine Applications," John Arimind and William Ayles, *Advanced Materials and Processes*, vol. 143, no. 6, June 1993; pp. 34–36):

Material (density)

Steel		Aluminum		Phenolics	
7.60	7.82	2.90	2.71	1.79	1.73
7.80	7.90	2.65	2.77	1.74	1.72
7.81	7.75	2.67	2.78	1.69	1.66
7.65	7.73	2.75	2.72	1.68	1.67
7.72	7.75	2.80	2.86	1.50	1.70
7.71	7.80	2.81	2.73	1.67	1.72
7.68	7.87	2.85	2.75	1.80	1.71
7.79	7.89	2.72	2.79	1.78	1.63
7.76	7.78	2.6	2.81	1.62	1.62

Material (tensile strength)

Steel		Aluminum		Phenolics	
60	100	17	40	8	13
73	122	25	19	17	12
87	86	35	29	15	10
98	112	38	22	9	11
175	77	27	26	10	16

Analyze each of these data sets by using whatever tests you believe are appropriate. Write a report to summarize your results. Explain your choice of test statistics.

63. Studies are conducted to investigate the use of slag in road pavement base, subbase, and surfacing. The composition of slag obtained from various sources is of interest. These data are obtained on the percentage of S_1O_2 in samples from five different sources: (Based on information from "Steel Plant Slag in Road Pavements," B. S. Heaton, *Australian Civil Engineering Transactions*, March 1993, pp. 49–53.

Source

Blast furnace slag	Steel slag	Portland cement	Fly ash	Natural basalt
35.1	16.0	20.1	58.3	45.9
34.7	15.6	22.0	57.6	46.3
34.8	17.2	23.1	55.0	44.5
33.2	16.2	19.7	60.1	45.2
33.6	16.3	19.5	61.2	44.5
36.8	14.7	16.2	58.2	49.6

Analyze these data and write a report summarizing your results. Be sure to defend your choice of test statistics.

COMPUTING SUPPLEMENT

XIV One-Way Classification with Multiple Comparisons

The SAS procedure PROC GLM can be used to analyze the one-way classification design. To illustrate, we use the coal seam data of Example 13.1.2.

Statement	*Function*
DATA COAL;	names data set
INPUT SEAM SULFUR;	names variables
LINES;	signals that data follows
1 1.51	
1 1.92	data
1 1.08	
⋮	
5 1.3	
;	signals end of data
PROC GLM;	calls for the general linear models procedure
CLASSES SEAM;	indicates that data are grouped according to the values of the variable SEAM
MODEL SULFUR = SEAM;	identifies variable sulfur as the response variable
MEANS SEAM/DUNCAN ALPHA = .01 BON ALPHA = .01;	asks for Duncan's test and Bonferroni T-tests to be run at the $\alpha = .01$ level
TITLE1 ONE WAY ANOVA;	
TITLE2 WITH PAIRED COMPARISONS;	titles output

On the printout the source called "model" ① corresponds to our source "treatments." The observed value of the F ratio used to test

$$H_0\colon \mu_1 = \mu_2 = \mu_3 = \mu_4$$

is shown in ② with its P value of .0001 given by ③. The results of the Duncan test are given by ④.

The results of the Bonferroni test are given by ⑤. Note that, as expected, these results differ from the Duncan results due to the fact that the α levels are not equivalent. Each comparison in the Duncan test is run at the .01 level; α', the overall probability of committing a type I error, is controlled at the .01 level in the Bonferroni test. Hence the Duncan test is able to detect a significant difference that was not picked up by the Bonferroni procedure. However, the overall level of significance for the Duncan test is not known.

ONE WAY ANOVA
WITH PAIRED COMPARISONS

GENERAL LINEAR MODELS PROCEDURE

DEPENDENT VARIABLE: SULFUR

SOURCE	DF	SUM OF SQUARES	MEAN SQUARE	F VALUE
MODEL ①	4	3.93539048	0.98384762	② 8.09
ERROR	37	4.49700000	0.12154054	PR > F
CORRECTED TOTAL	41	8.43239048		③ 0.0001

R-SQUARE	C.V.	ROOT MSE	SULFUR MEAN
0.466699	29.3081	0.34862665	1.18952381

SOURCE	DF	TYPE I SS	F VALUE	PR > F
SEAM	4	3.93539048	8.09	.0001

SOURCE	DF	TYPE III SS	F VALUE	PR > F
SEAM	4	3.93539048	8.09	0.0001

ONE WAY ANOVA
WITH PAIRED COMPARISONS
GENERAL LINEAR MODELS PROCEDURE

DUNCAN'S MULTIPLE RANGE TEST FOR VARIABLE: SULFUR
NOTE: THIS TEST CONTROLS THE TYPE I COMPARISONWISE ERROR RATE,
NOT THE EXPERIMENTWISE ERROR RATE

ALPHA = .01 DF = 37 MSE = 0.121541

WARNING: CELL SIZES ARE NOT EQUAL.
HARMONIC MEAN OF CELL SIZES = 8.27858

NUMBER OF MEANS	2	3	4	5
CRITICAL RANGE	0.465528	0.48537	0.499231	0.508566

MEANS WITH THE SAME LETTER ARE NOT SIGNIFICANTLY DIFFERENT.

DUNCAN	GROUPING		MEAN	N	SEAM
		A	1.6600	7	1
		A			
④	B	A	1.4600	9	3
	B				
	B	C	1.1700	8	2
		C			
		C	0.8800	10	5
		C			
		C	0.8800	8	4

```
                    ONE WAY ANOVA
                WITH PAIRED COMPARISONS

             GENERAL LINEAR MODELS PROCEDURE

BONFERRONI (DUNN) T TESTS FOR VARIABLE: SULFUR
NOTE: THIS TEST CONTROLS THE TYPE I EXPERIMENTWISE ERROR RATE
      BUT GENERALLY HAS A HIGHER TYPE II ERROR RATE THAN REGWQ

          ALPHA = .01     DF = 37     MSE = 0.121541
               CRITICAL VALUE OF T = 3.57367
          MINIMUM SIGNIFICANT DIFFERENCE = .61237

          WARNING: CELL SIZES ARE NOT EQUAL.
                   HARMONIC MEAN OF CELL SIZES = 8.27858

  MEANS WITH THE SAME LETTER ARE NOT SIGNIFICANTLY DIFFERENT.

 BON    GROUPING      MEAN       N      SEAM

              A      1.6600      7       1
              A
          B   A      1.4600      9       3
          B   A
 (5)      B   A      1.1700      8       2
          B
          B          0.8800     10       5
          B
          B          0.8800      8       4
```

XV Randomized Complete Blocks with Multiple Comparisons

The randomized complete block design can be analyzed by using the SAS procedure PROC ANOVA. The data of Examples 13.5.1 and 13.5.2 are used.

Statement	*Function*
DATA TIRE;	names data set
INPUT TRT BLK WEAR;	names variables
LINES;	signals that data follows
1 1 9.1	
1 2 13.4	
1 3 15.6	
1 4 11.0	data
1 5 12.7	
2 1 17.1	
⋮	
4 5 15.8	
;	signals end of data
PROC ANOVA;	calls for an analysis of variance to be conducted

Statement	*Function*
CLASSES TRT BLK;	indicates that the responses are grouped according to the values of the variables TRT and BLK
MODEL WEAR = TRT BLK;	identifies the variable WEAR as the response variable
MEANS TRT/DUNCAN ALPHA; = .01	asks for the Duncan procedure to be used to analyze treatment means (TRT) at the $\alpha = .01$ level
TITLE1 RANDOMIZED COMPLETE BLOCK DESIGN;	titles output
TITLE2 WITH MULTIPLE COMPARISONS;	

On the printout the treatment sum of squares, block sum of squares, and error sum of squares are given by ①, ②, and ③, respectively. The degrees of freedom for blocks, treatments, and error are shown in ④, ⑤, and ⑥, respectively. Block and treatment mean squares are not given, but the error mean square is shown in ⑦. The F ratio used to test H_0: $\mu_{1\cdot} = \mu_{2\cdot} = \mu_{3\cdot} = \mu_{4\cdot}$ is given by ⑧. Its corresponding P value is shown in ⑩. The results of the Duncan test are given in ⑪. The ratio $MS_{\text{BIKS}}/MS_{\text{E}}$ used to estimate relative efficiency is given by ⑨.

RANDOMIZED COMPLETE BLOCK DESIGN
WITH MULTIPLE COMPARISONS

ANALYSIS OF VARIANCE PROCEDURE

CLASS LEVEL INFORMATION

CLASS	LEVELS	VALUES
TRT	4	1 2 3 4
BLK	5	1 2 3 4 5

NUMBER OF OBSERVATIONS IN DATA SET = 20

RANDOMIZED COMPLETE BLOCK DESIGN
WITH MULTIPLE COMPARISONS

ANALYSIS OF VARIANCE PROCEDURE

DEPENDENT VARIABLE: WEAR

SOURCE	DF	SUM OF SQUARES	MEAN SQUARE	F VALUE
MODEL	7	482.86300000	68.98042857	31.63
ERROR	12 ⑥	26.16700000 ③	2.18058333 ⑦	PR > F
CORRECTED TOTAL	19	509.03000000		0.0001

R-SQUARE	C.V.	ROOT MSE	WEAR MEAN
0.948594	8.3193	1.47667983	17.75000000

SOURCE	DF	ANOVA SS	F VALUE	PR > F
TRT	3 ④	401.33800000 ①	61.35 ⑧	0.0001 ⑩
BLK	4 ⑤	81.52500000 ②	9.35 ⑨	

RANDOMIZED COMPLETE BLOCK DESIGN
WITH MULTIPLE COMPARISONS

ANALYSIS VARIANCE PROCEDURE

DUNCAN'S MULTIPLE RANGE TEST FOR VARIABLE: WEAR
NOTE: THIS TEST CONTROLS THE TYPE I COMPARISONWISE ERROR RATE, NOT THE EXPERIMENTWISE ERROR RATE.

ALPHA = 0.01 DF = 12 MSE = 2.18058

MEANS WITH THE SAME LETTER ARE NOT SIGNIFICANTLY DIFFERENT.

(11)

DUNCAN GROUPING	MEAN	N	TRT
A	23.860	5	3
B	20.000	5	2
C	14.780	5	4
C			
C	12.360	5	1

CHAPTER 14

FACTORIAL EXPERIMENTS

In many experiments two or more factors are being investigated. Neither factor is considered extraneous; each is of major concern to the experimenter. When this occurs, the experiment is called a *factorial experiment* to emphasize the fact that interest is centered on the effect of two or more factors on a measured response. We note that this is clearly a different experimental situation from that discussed in Sec. 4 of Chap. 13 for the randomized complete block and Latin square designs. For these designs other factors (blocks) were used only to control for extraneous variation in the experimental error. In Sec. 1 we discuss the *two-way classification completely random design* with fixed effects.

The values of the factors used in the experiment are called *levels*. Thus we deal with a model in which two factors, A and B, are studied with the levels of each factor being purposely, rather than randomly, selected by the experimenter. No matching of similar experimental units is done.

To demonstrate the meaning of a factorial experiment, consider the following illustration. Suppose that a certain performance characteristic of a microcomputer is to be investigated at three temperatures (50°F, 75°F, 100°F) and two humidity (25%, 75%) levels. There are $3 \times 2 = 6$ experimental conditions in this experiment. Suppose that we have 18 identical microcomputers for use in the experiment so that each of the 6 experimental conditions can be applied to 3 of the microcomputers. We use a random mechanism, such as a random digit table or random number generator, to assign 3 microcomputers to each of the 6 experimental conditions of temperature and humidity.

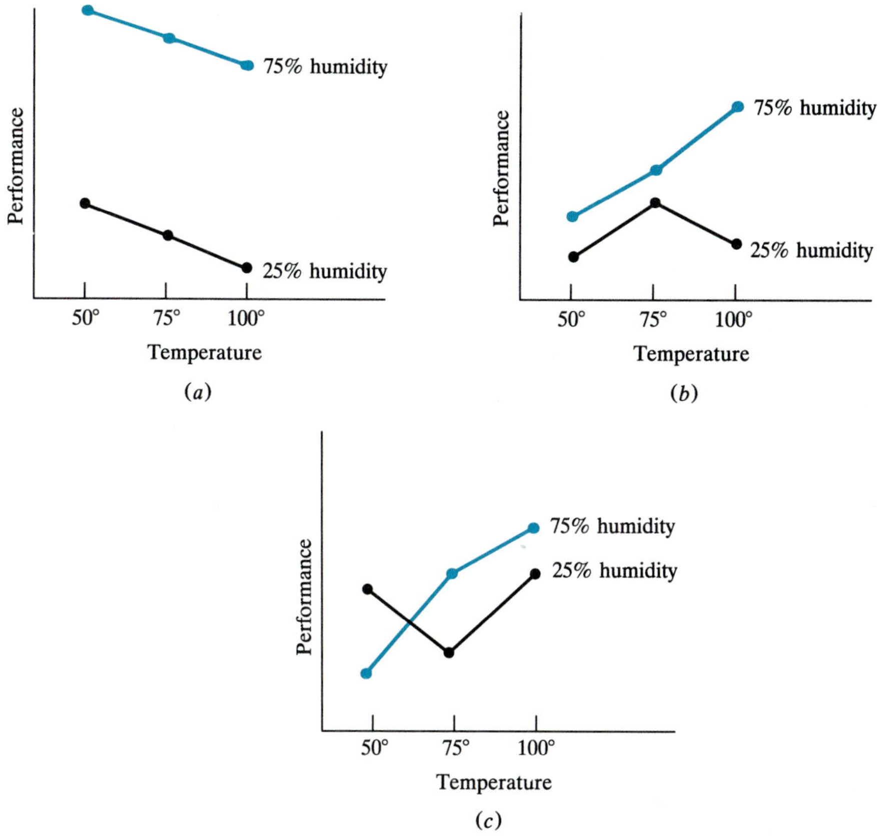

FIGURE 14.1
(*a*) No interaction between the factors temperature and humidity; (*b*) and (*c*) interaction exists between temperature and humidity.

The resulting design is called a *completely randomized design*; the experiment is a 3 × 2 factorial experiment. This design allows us to compare levels of each factor, as well as to compare treatment combinations in the same experiment. In addition, it allows us to detect interaction. Here interaction is a measure of parallelism. For example, if there is no interaction between temperature and humidity, then the graph of the means of the six treatment combinations is similar to that shown in Fig. 14.1(*a*). The means for the three temperature levels at 25% humidity fall on a line parallel to those at 75% humidity. When interaction exits, the lines will not be parallel, as in Figs. 14.1(*b*) and 14.1(*c*).

Care must be taken when interpreting the results of a two-factor experiment. Several meaningful tests can be conducted, but the first should be a test for interaction since the presence of interaction can mask differences in levels of main factors. A testing strategy will be suggested in Sec. 14.1.

14.1 TWO-FACTOR ANALYSIS OF VARIANCE

The data collected in a two-way classification design are conveniently displayed in the format shown in Table 14.1. Note that a denotes the number of levels of factor A used in the experiment, b denotes the number of levels of factor B, and $a \times b$ is the total number of treatment combinations, where a treatment combination is a level of factor A applied in conjunction with a level of factor B. We assume that there are $n \geq 1$ observations for *each* treatment combination. That is, we assume equal sample sizes. The analysis for unequal sample sizes is beyond the scope of this text. The total number of responses is $N = a \cdot b \cdot n$. We denote the response of the kth experimental unit to the ith level of factor A and jth level of factor B by Y_{ijk}. The following statistics are needed in analyzing the data (recall that a dot indicates the subscript over which summation is being conducted):

$$T_{ij\cdot} = \sum_{k=1}^{n} Y_{ijk} = \text{total of all responses to the } (i, j)\text{th treatment combination}$$

$$\bar{Y}_{ij\cdot} = T_{ij\cdot}/n = \text{sample mean for the } (i, j)\text{th treatment combination}$$

$$T_{i\cdot\cdot} = \sum_{j=1}^{b} T_{ij\cdot} = \text{total of all responses to the } i\text{th level of factor } A$$

$$\bar{Y}_{i\cdot\cdot} = T_{i\cdot\cdot}/bn = \text{sample mean for the } i\text{th level of factor } A$$

$$T_{\cdot j\cdot} = \sum_{i=1}^{a} T_{ij\cdot} = \text{total of all responses to the } j\text{th level of factor } B$$

$$\bar{Y}_{\cdot j\cdot} = T_{\cdot j\cdot}/an = \text{sample mean for the } j\text{th level of factor } B$$

$$T_{\cdots} = \sum_{i=1}^{a} T_{i\cdot\cdot} = \sum_{j=1}^{b} T_{\cdot j\cdot} = \sum_{i=1}^{a} \sum_{j=1}^{b} T_{ij\cdot} = \text{total of all responses}$$

$$\bar{Y}_{\cdots} = T_{\cdots}/abn = \text{sample mean for all responses}$$

The computation of these statistics is illustrated in the following example.

Example 14.1.1. A study is conducted to investigate the effect of temperature (factor A) and humidity (factor B) on the force required to separate an adhesive product from a certain material. The experimenter is interested in four specific temperature levels ($a = 4$) and two specific humidity levels ($b = 2$). These levels are not randomly selected, and thus the design is a two-factor design with fixed effects. Three measurements are taken at each of the $a \cdot b = 8$ treatment combinations. The resulting data and summary statistics are given in Table 14.2. The sample means for factor A appear to differ as do the sample means for factor B. The question to answer statistically is, "Are these differences extreme enough to allow us to conclude that there are real differences in the average responses obtained for these levels of both factors A and B?"

TABLE 14.1

Data layout two-way classification

Factor B level	Factor A level 1	2	3	$\cdots$	a	Totals (B)	Means (B)
1	Y_{111} Y_{112} $\vdots$ Y_{11n}	Y_{211} Y_{212} $\vdots$ Y_{21n}	Y_{311} Y_{312} $\vdots$ Y_{31n}	$\cdots$	Y_{a11} Y_{a12} $\vdots$ Y_{a1n}	$T_{\cdot 1 \cdot}$	$\bar{Y}_{\cdot 1 \cdot}$
2	Y_{121} Y_{122} $\vdots$ Y_{12n}	Y_{221} Y_{222} $\vdots$ Y_{22n}	Y_{321} Y_{322} $\vdots$ Y_{32n}		Y_{a21} Y_{a22} $\vdots$ Y_{a2n}	$T_{\cdot 2 \cdot}$	$\bar{Y}_{\cdot 2 \cdot}$
$\vdots$							
b	Y_{1b1} Y_{1b2} $\vdots$ Y_{1bn}	Y_{2b1} Y_{2b2} $\vdots$ Y_{2bn}	Y_{3b1} Y_{3b2} $\vdots$ Y_{3bn}		Y_{ab1} Y_{ab2} $\vdots$ Y_{abn}	$T_{\cdot b \cdot}$	$\bar{Y}_{\cdot b \cdot}$
Totals (A) Means (A)	$T_{1\cdot\cdot}$ $\bar{Y}_{1\cdot\cdot}$	$T_{2\cdot\cdot}$ $\bar{Y}_{2\cdot\cdot}$	$T_{3\cdot\cdot}$ $\bar{Y}_{3\cdot\cdot}$		$T_{a\cdot\cdot}$ $\bar{Y}_{a\cdot\cdot}$	$T_{\cdots}$ $\bar{Y}_{\cdots}$	

TABLE 14.2

		Factor A (temperature) 1	2	3	4	Factor B total	Factor B mean
Factor B (humidity)	1	40 36 43 $T_{11\cdot} = 119$	39 36 33 $T_{21\cdot} = 108$	32 34 29 $T_{31\cdot} = 95$	33 27 25 $T_{41\cdot} = 85$	$T_{\cdot 1 \cdot} = 407$	$\bar{Y}_{\cdot 1 \cdot} = 33.92$
	2	36 34 29 $T_{12\cdot} = 99$	32 26 25 $T_{22\cdot} = 83$	26 23 24 $T_{32\cdot} = 73$	20 22 18 $T_{42\cdot} = 60$	$T_{\cdot 2 \cdot} = 315$	$\bar{Y}_{\cdot 2 \cdot} = 26.25$
Factor A total		$T_{1\cdot\cdot} = 218$	$T_{2\cdot\cdot} = 191$	$T_{3\cdot\cdot} = 168$	$T_{4\cdot\cdot} = 145$	$T_{\cdots} = 722$	
Factor A mean		$\bar{Y}_{1\cdot\cdot} = 36.33$	$\bar{Y}_{2\cdot\cdot} = 31.83$	$\bar{Y}_{3\cdot\cdot} = 28$	$\bar{Y}_{4\cdot\cdot} = 24.17$	$\bar{Y}_{\cdots} = 30.08$	

These parameters and random variables are used to define the model:

$$\mu_{ij\cdot} = \text{mean for the } (i, j)\text{th treatment combination}$$

$$\mu_{i\cdot\cdot} = \text{mean for the } i\text{th level of factor } A = \sum_{j=1}^{b} \mu_{ij\cdot}/b$$

$$\mu_{\cdot j\cdot} = \text{mean for the } j\text{th level of factor } B = \sum_{i=1}^{a} \mu_{ij\cdot}/a$$

$$\mu = \text{overall mean} = \sum_{i=1}^{a} \sum_{j=1}^{b} \mu_{ij}/ab$$

$$\alpha_i = \mu_{i\cdot\cdot} - \mu = \text{effect due to the fact that the experimental unit received the } i\text{th level of factor } A$$

$$\beta_j = \mu_{\cdot j\cdot} - \mu = \text{effect due to the fact that the experimental unit received the } j\text{th level of factor } B$$

$$(\alpha\beta)_{ij} = \mu_{ij\cdot} - \mu_{i\cdot\cdot} - \mu_{\cdot j\cdot} + \mu = \text{effect of interaction between } i\text{th level of factor } A \text{ and } j\text{th level of factor } B$$

$$\mathrm{E}_{ijk} = Y_{ijk} - \mu_{ij\cdot} = \text{residual or random error}$$

Using this notation, we can express the model as

Model for the two-way classification design with fixed effects

$$Y_{ijk} = \mu + \alpha_i + \beta_j + (\alpha\beta)_{ij} + \mathrm{E}_{ijk}$$

The model expresses symbolically the idea that each observation can be partitioned into five components: an overall mean effect (μ), an effect due to factor A (α_i), an effect due to factor B (β_j), an effect due to interaction $(\alpha\beta)_{ij}$, and a random deviation due to unexplained sources (E_{ijk}). We make these *model assumptions*:

1. The observations for each treatment combination constitute independent random samples, each of size n from $a \cdot b$ populations with means $\mu_{ij\cdot}$.
2. Each of the $a \cdot b$ populations is normally distributed.
3. Each of the $a \cdot b$ populations has the same variance, σ^2.

Testing H_0

The sum of squares identity obtained by replacing each of the theoretical means μ, $\mu_{i\cdot\cdot}$, $\mu_{\cdot j\cdot}$, and $\mu_{ij\cdot}$ by their unbiased estimators $\bar{Y}\ldots$, $\bar{Y}_{i\cdot\cdot}$, $\bar{Y}_{\cdot j\cdot}$ and $\bar{Y}_{ij\cdot}$, respectively, squaring, and summing over i, j, and k is as follows:

Sum of squares identity

$$SS_{\text{Tot}} = SS_A + SS_B + SS_{AB} + SS_E$$

In this identity,

$$SS_{\text{Tot}} = \sum_{i=1}^{a} \sum_{j=1}^{b} \sum_{k=1}^{n} (Y_{ijk} - \bar{Y}_{\cdots})^2$$

= measure of total variability in data

$$SS_A = bn \sum_{i=1}^{a} (\bar{Y}_{i\cdot\cdot} - \bar{Y}_{\cdots})^2$$

= measure of variability in data attributable to the use of different levels of factor A

$$SS_B = an \sum_{j=1}^{b} (\bar{Y}_{\cdot j\cdot} - \bar{Y}_{\cdots})^2$$

= measure of variability in data attributable to the use of different levels of factor B

$$SS_{AB} = n \sum_{i=1}^{a} \sum_{j=1}^{b} (\bar{Y}_{ij\cdot} - \bar{Y}_{i\cdot\cdot} - \bar{Y}_{\cdot j\cdot} + \bar{Y}_{\cdots})^2$$

= measure of variability in data due to interaction between levels of factors A and B

$$SS_E = \sum_{i=1}^{a} \sum_{j=1}^{b} \sum_{k=1}^{n} (Y_{ijk} - \bar{Y}_{ij\cdot})^2$$

= measure of variability in data due to random or unexplained sources

The *first* null hypothesis to be tested is the null hypothesis of no interaction. Mathematically, this hypothesis is

$$H_0\colon (\alpha\beta)_{ij} = 0 \qquad i = 1, 2, \ldots, a; \quad j = 1, 2, \ldots, b$$

If this hypothesis is not rejected, then the analysis is continued by testing the null hypothesis of no difference among levels of factor A,

$$H_0^{\text{I}}\colon \mu_{1\cdot\cdot} = \mu_{2\cdot\cdot} = \cdots = \mu_{a\cdot\cdot}$$

or

$$H_0^{\text{I}} = \alpha_1 = \alpha_2 = \cdots = \alpha_a = 0$$

and the null hypothesis of no difference among levels of factor B,

$$H_0^{\mathrm{II}}: \mu_{\cdot 1 \cdot} = \mu_{\cdot 2 \cdot} = \cdots = \mu_{\cdot b \cdot}$$

or

$$H_0^{\mathrm{II}}: \beta_1 = \beta_2 = \cdots = \beta_b = 0$$

These tests are referred to as tests for "main" effects. If the null hypothesis of no interaction is rejected, then the levels of factor A do not behave consistently across the levels of factor B. Thus it is difficult to make generalizations concerning the behavior of factors A and B. To understand exactly what is taking place, we need to compare levels of factor A for each level of factor B and vice versa. This can be done graphically by constructing line graphs of the treatment means of the type illustrated in Fig. 14.1. It can be done analytically by performing $a + b$ one-way analyses and by testing hypotheses of the form

$$H_0^{\mathrm{III}}: \mu_{1j\cdot} = \mu_{2j\cdot} = \cdots = \mu_{aj\cdot} \qquad j = 1, 2, \ldots, b.$$

and

$$H_0^{\mathrm{IV}}: \mu_{i1\cdot} = \mu_{i2\cdot} = \cdots = \mu_{ib\cdot} \qquad i = 1, 2, \ldots, a$$

by using the methods of Chap. 13. For example, if interaction is detected when the data of Example 14.1.1 is analyzed, then we can continue the analysis by running $a + b = 4 + 2$ one-way analyses. We would test

$H_0^{\mathrm{III}}: \mu_{11\cdot} = \mu_{21\cdot} = \mu_{31\cdot} = \mu_{41\cdot}$ (There are no differences among temperature levels at humidity level 1)

$H_0^{\mathrm{III}}: \mu_{12\cdot} = \mu_{22\cdot} = \mu_{32\cdot} = \mu_{42}$ (There are no differences among temperature levels at humidity level 2)

$H_0^{\mathrm{IV}}: \mu_{11\cdot} = \mu_{12}$ (There is no difference in humidity levels at temperature level 1)

$H_0^{\mathrm{IV}}: \mu_{21\cdot} = \mu_{22\cdot}$ (There is no difference in humidity levels at temperature level 2)

$H_0^{\mathrm{IV}}: \mu_{31\cdot} = \mu_{32\cdot}$ (There is no difference in humidity levels at temperature level 3)

$H_0^{\mathrm{IV}}: \mu_{41\cdot} = \mu_{42\cdot}$ (There is no difference in humidity levels at temperature level 4)

In the presence of interactions we might simply check for differences among treatment combinations. That is, we run a one-way classification analysis of variance to test the null hypothesis of equal treatment combination means. This hypothesis is expressed mathematically as

$$H_0^{\mathrm{V}}: \mu_{11\cdot} = \mu_{12\cdot} = \cdots = \mu_{ab\cdot}$$

Figure 14.2 summarizes these suggestions.

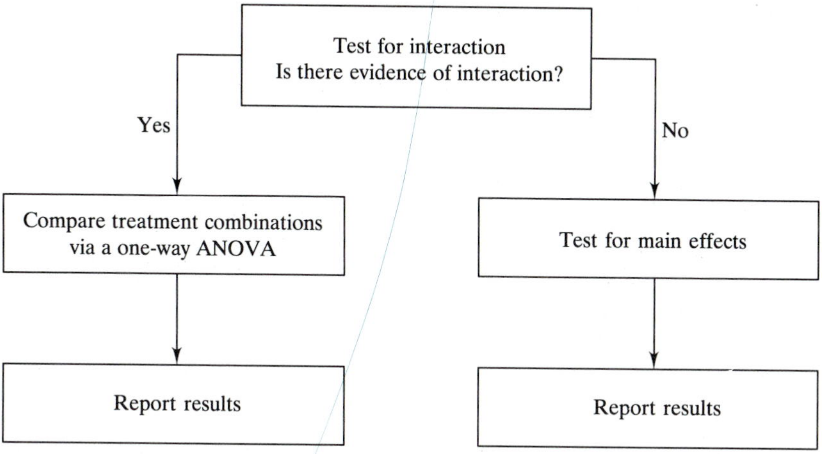

FIGURE 14.2
A suggested testing strategy in analyzing a two-factor experiment.

The computational formulas used to compute SS_A, SS_B, and SS_{Tot} are similar to those of previous models and are given by

$$SS_A = \sum_{i=1}^{a} \frac{T_{i\cdot\cdot}}{bn} - \frac{T_{\cdot\cdot\cdot}^2}{abn}$$

$$SS_B = \sum_{j=1}^{b} \frac{T_{\cdot j\cdot}^2}{an} - \frac{T_{\cdot\cdot\cdot}^2}{abn}$$

$$SS_{\text{Tot}} = \sum_{i=1}^{a} \sum_{j=1}^{b} \sum_{k=1}^{n} Y_{ijk}^2 - \frac{T_{\cdot\cdot\cdot}^2}{abn}$$

The interaction sum of squares is found by first computing what is called the treatment sum of squares. This is the usual treatment sum of squares that would be obtained if the $a \cdot b$ treatment combinations were analyzed as a one-way classification design. That is,

$$SS_{\text{Tr}} = \sum_{i=1}^{a} \sum_{j=1}^{b} \frac{T_{ij\cdot}^2}{n} - \frac{T_{\cdot\cdot\cdot}^2}{abn}$$

It can be shown that

$$SS_{\text{Tr}} = SS_A + SS_B + SS_{AB}$$

This allows us to compute the interaction sum of squares by subtraction with

$$SS_{AB} = SS_{\text{Tr}} - SS_A - SS_B$$

The error sum of squares is also obtained by subtraction with

$$SS_E = SS_{\text{Tot}} - SS_{\text{Tr}}$$

The analysis of variance table for this design is given in Table 14.3.

The first F ratio to consider in any experiment is

$$F_{(a-1)(b-1),\, ab(n-1)} = MS_{AB}/MS_E$$

This ratio is used to test the null hypothesis of no interaction. If this hypothesis is not rejected, then the F statistics

and

$$F_{a-1,\, ab(n-1)} = MS_A/MS_E$$
$$F_{b-1,\, ab(n-1)} = MS_B/MS_E$$

are used to test the null hypothesis of no differences among the means of levels of factors A and B, respectively. If the null hypothesis of no interaction is rejected, then the F statistic

$$F_{ab-1,\, ab(n-1)} = MS_{\text{Tr}}/MS_E$$

is used to test the null hypothesis of no difference among treatment combinations. The methods of Chap. 13 are used to conduct one-way analyses on a level-by-level basis. In each method rejection occurs for values of the F ratio that are too large to have occurred by chance.

Paired Comparisons

If significant main effects are detected, then Duncan's multiple range test or Bonferroni T tests are again applicable. To pinpoint differences among the levels of factor A, we use

$$\text{SSR}_p = r_p\sqrt{MS_E/bn}$$

To determine what differences exist among the levels of factor B, we use

$$\text{SSR}_p = r_p\sqrt{MS_E/an}$$

We illustrate these ideas by continuing the analysis of the data begun in Example 14.1.1.

TABLE 14.3

ANOVA table for the two-way classification design with fixed effects

Source of variation	Degrees of freedom (DF)	Sum of squares (SS)	Mean square (MS)	Expected mean square	F
Treatment	$ab - 1$	$\sum_{i=1}^{a} \sum_{j=1}^{b} \frac{T_{ij\cdot}^2}{n} - \frac{T_{\cdot\cdot\cdot}^2}{abn}$	$\frac{SS_{\mathrm{Tr}}}{ab - 1}$	$\sigma^2 + n \sum_{i=1}^{a} \sum_{j=1}^{b} (\mu_{ij} - \mu_{\cdot\cdot\cdot})^2/(ab - 1)$	MS_{Tr}/MS_E
A	$a - 1$	$\sum_{i=1}^{a} \frac{T_{i\cdot\cdot}^2}{ab} - \frac{T_{\cdot\cdot\cdot}^2}{abn}$	$\frac{SS_A}{a - 1}$	$\sigma^2 + nb \sum_{i=1}^{a} \alpha_i^2/(a - 1)$	MS_A/MS_E
B	$b - 1$	$\sum_{j=1}^{b} \frac{T_{\cdot j\cdot}^2}{an} - \frac{T_{\cdot\cdot\cdot}^2}{abn}$	$\frac{SS_B}{b - 1}$	$\sigma^2 + na \sum_{j=1}^{b} \beta_j^2/(b - 1)$	MS_B/MS_E
AB	$(a - 1)(b - 1)$	$SS_{\mathrm{Tr}} - SS_A - SS_B$	$\frac{SS_{AB}}{(a - 1)(b - 1)}$	$\sigma^2 + n \sum_{i=1}^{a} \sum_{j=1}^{b} (\alpha\beta)_{ij}^2/(a - 1)(b - 1)$	MS_{AB}/MS_E
Error	$ab(n - 1)$	$SS_{\mathrm{Tot}} - SS_{\mathrm{Tr}}$	$\frac{SS_E}{ab(n - 1)}$	σ^2	
Total	$abn - 1$	$\sum_{i=1}^{a} \sum_{j=1}^{b} \sum_{k=1}^{n} Y_{ijk}^2 - \frac{T_{\cdot\cdot\cdot}^2}{abn}$			

Example 14.1.2. In Example 14.1.1 we began a study of the effects of temperature (factor A) and humidity (factor B) on the force required to separate an adhesive product from a certain material. These summary statistics are available:

$$\begin{array}{lll} T_{1\cdot\cdot} = 218 & T_{\cdot 1\cdot} = 407 & T_{22\cdot} = 83 \\ T_{2\cdot\cdot} = 191 & T_{\cdot 2\cdot} = 315 & T_{31\cdot} = 95 \\ T_{3\cdot\cdot} = 168 & T_{11\cdot} = 119 & T_{32\cdot} = 73 \\ T_{4\cdot\cdot} = 145 & T_{12\cdot} = 99 & T_{41\cdot} = 85 \\ T_{\cdot\cdot\cdot} = 722 & T_{21\cdot} = 108 & T_{42\cdot} = 60 \\ n = 3 & N = 24 & \end{array}$$

For these data,

$$\sum_{i=1}^{4} \sum_{j=1}^{2} \sum_{k=1}^{3} Y_{ijk}^2 = 22{,}722$$

The required sums of squares are

$$SS_{\text{Tot}} = \sum_{i=1}^{a} \sum_{j=1}^{b} \sum_{k=1}^{n} Y_{ijk}^2 - \frac{T_{\cdot\cdot\cdot}^2}{abn} = 1001.83$$

$$SS_A = \sum_{i=1}^{a} \frac{T_{i\cdot\cdot}^2}{bn} - \frac{T_{\cdot\cdot\cdot}^2}{abn} = 488.83$$

$$SS_B = \sum_{j=1}^{b} \frac{T_{\cdot j\cdot}^2}{an} - \frac{T_{\cdot\cdot\cdot}^2}{abn} = 352.67$$

$$SS_{\text{Tr}} = \sum_{i=1}^{a} \sum_{j=1}^{b} \frac{T_{ij\cdot}^2}{n} - \frac{T_{\cdot\cdot\cdot}^2}{abn} = 844.50$$

$$SS_{AB} = SS_{\text{Tr}} - SS_A - SS_B = 3.00$$

$$SS_E = SS_{\text{Tot}} - SS_{\text{Tr}} = 157.33$$

The complete analysis of variance table is given in Table 14.4. Before testing for

TABLE 14.4
ANOVA for adhesive force data

Source of variation	Degrees of freedom (DF)	Sum of squares (SS)	Mean square (MS)	F
Treatment	7	844.50	120.64	12.270
A	3	488.83	162.94	16.580
B	1	352.67	352.67	35.880
AB	3	3.00	1.00	.102
Error	16	157.33	9.83	
Total	23	1001.83		

main effects, we first test for interaction. That is, we test

$$H_0\colon (\alpha\beta)_{ij} = 0 \qquad \text{for each } i \text{ and } j$$

The observed value of the $F_{(a-1)(b-1),\, ab(n-1)} = F_{3,16}$ test statistic is .102. This value is not significant even at the $\alpha = .1$ level ($f_{.1}(3, 16) = 2.46$). We do not reject H_0 for there is no statistical evidence of interaction between temperature and humidity. We continue the analysis by testing

$$H_0^{\mathrm{I}}\colon \mu_{1\cdot\cdot} = \mu_{2\cdot\cdot} = \mu_{3\cdot\cdot} = \mu_{4\cdot\cdot}$$

The observed value of the $F_{a-1,\, ab(n-1)} = F_{3,16}$ test statistic is 16.58. This value is significant at the $\alpha = .01$ level ($f_{.01}(3, 16) = 5.29$). We reject H_0^{I} and conclude that there are differences in mean values among the four temperature levels. We investigate these differences further via Duncan's multiple range test. The ordered sample means for the four temperature levels are

$\bar{Y}_{4\cdot\cdot}$	$\bar{Y}_{3\cdot\cdot}$	$\bar{Y}_{2\cdot\cdot}$	$\bar{Y}_{1\cdot\cdot}$
24.17	28.00	31.83	36.33

The required values of the shortest significant range, SSR_p, are found from Table XI of App. A with $\gamma = 16$ and $\alpha = .01$ by noting that

$$\mathrm{SSR}_p = r_p\sqrt{MS_E/bn} = r_p\sqrt{9.83/6} = 1.28r_p$$

p	**2**	**3**	**4**
r_p	4.131	4.309	4.425
SSR_p	5.288	5.156	5.664

For the $\binom{4}{2} = 6$ possible pairs of means we have

Pair	p	**Value of test statistic**	$\mathbf{SSR}_p$	**Reject** $\mu_{i\cdot\cdot} = \mu_{k\cdot\cdot}$?
1-4	4	12.16	5.664	Yes
1-3	3	8.33	5.156	Yes
1-2	2	4.50	5.288	No
2-4	3	7.66	5.156	Yes
2-3	2	3.83	5.288	No
3-4	2	3.83	5.288	No

In summary, we have

$$\bar{Y}_{4\cdot\cdot} \quad \bar{Y}_{3\cdot\cdot} \quad \bar{Y}_{2\cdot\cdot} \quad \bar{Y}_{1\cdot\cdot}$$

There is statistical evidence at the $\alpha = .01$ level that $\mu_{1\cdot\cdot} \neq \mu_{4\cdot\cdot}$, $\mu_{1\cdot\cdot} \neq \mu_{3\cdot\cdot}$, and $\mu_{2\cdot\cdot} \neq \mu_{4\cdot\cdot}$.

To complete the analysis of the data, we need to test

$$H_0^{\mathrm{II}}\colon \mu_{\cdot 1\cdot} = \mu_{\cdot 2\cdot}$$

Sample Size

A few remarks concerning sample sizes should be made. Note that for each of the models considered in the two-factor design the degrees of freedom for error is given by $ab(n-1)$. Hence if we have only one observation per cell, this term assumes the value 0. In this case MS_E is undefined and we cannot test for interaction. Even though interaction effects may be very important, we must assume no interaction. In this cases SS_E and SS_{AB} are pooled into one sum of squares with associated degrees of freedom $(a-1)(b-1)$ and are used $(SS_E + SS_{AB})/(a-1)(b-1)$ as our error mean square. It should be clear that if at all possible, the researcher should have at least two observations per treatment combination. Furthermore, if the number of observations per treatment combination differ, then the analysis is *not* simple. It does *not* simply parallel the ideas presented here with n being replaced by n_i. Be careful! If you are working with such a design, you should consult a professional statistician for help in analyzing your data and in interpreting your results.

14.2 EXTENSION TO THREE FACTORS (OPTIONAL)

By now it is probably clear that the analysis for a two-factor experiment is easy to extend to the case of three or more factors in concept. However, it should also be clear that the computations involved become very tedious. Hence such experiments are usually analyzed by computer in practice. In Exercises 16 through 23 we lead you through the development of the analysis of a fixed-effects experiment with factors A, B, and C. The design assumes the possible existence of both two and three-way interactions. Exercises 24 show a model in which it is assumed a priori that some interactions are negligible.

14.3 RANDOM AND MIXED MODEL FACTORIAL EXPERIMENTS

The two-factor design discussed in Sec. 14.1 assumes that both factors A and B are fixed. As in the one-way classification and randomized complete block designs, factors can be random. In the case of a two-factor design these experimental situations can arise:

1. Both factors A and B are fixed. This type of design is called the *fixed-effects* model.
2. Both factors A and B are random. This type of design is called the *random-effects* model.
3. One factor is fixed and the other is random. This type of design leads to a *mixed-effects* model.

Computations of degrees of freedom, sums of squares, and mean squares are the same in all cases. However, testing procedures vary from one model to another and are dependent on the expected mean squares. To perform each test, we inspect the expected mean squares and form an F ratio with the property that if H_0 is true, both numerator and denominator are estimating the same parameter; if H_0 is not true, then the numerator should exceed the denominator resulting in an inflated F statistic.

Random-Effects Model

We illustrate these ideas by considering the analysis of variance table for the random-effects model given in Table 14.5. We follow the same order of testing as in the fixed-effects model. We first test for interaction. That is, we test

$$H_0\colon \sigma_{AB}^2 = 0$$

By examining the expected mean squares in Table 14.5, we see that the appropriate test statistic is

$$F_{(a-1)(b-1),\, ab(n-1)} = MS_{AB}/MS_E$$

If H_0 is true, then both the numerator and the denominator of this statistic estimates σ^2; if H_0 is not true, then the numerator should exceed the denominator, resulting in an inflated F statistic. If H_0 is not rejected, then we test for differences in levels of factors A and B by testing

$$H_0^{\mathrm{I}}\colon \sigma_A^2 = 0$$

and

$$H_0^{\mathrm{II}}\colon \sigma_B^2 = 0$$

The appropriate test statistic for testing H_0^{I} is

$$F_{a-1,\, ab(n-1)} = MS_A/MS_{AB}$$

From the expected mean squares we see that if H_0^{I} is true, then both the numerator and the denominator estimate $\sigma^2 + n\sigma_{AB}^2$; if H_0^{I} is not true, then the numerator should exceed the denominator, resulting in a large value for the F statistic. Similar reasoning shows that to test H_0^{II}, we use the test statistic

$$F_{b-1,\, ab(n-1)} = MS_B/MS_{AB}$$

Note that for the first time we are testing a hypothesis by using an F ratio that does *not* use MS_E as the denominator. If the null hypothesis of no interaction is rejected, then we treat the experiment as a one-way classification random-effects design with $a \cdot b$ treatments. The null hypothesis

$$H_0^{\mathrm{III}}\colon \sigma_{\mathrm{Tr}}^2 = 0$$

TABLE 14.5

ANOVA table for the two-way classification design with random effects

Source of variation	Degrees of freedom (DF)	Sum of squares (SS)	Mean square (MS)	Expected mean square	F
Treatment	$ab - 1$	$\sum_{i=1}^{a}\sum_{j=1}^{b}\frac{T_{ij\cdot}^2}{n} - \frac{T_{\cdot\cdot\cdot}^2}{abn}$	$SS_{\mathrm{Tr}}/(ab - 1)$	$\sigma^2 + n\sigma_{\mathrm{Tr}}^2$	MS_{Tr}/MS_E
A	$a - 1$	$\sum_{i=1}^{a}\frac{T_{i\cdot\cdot}^2}{bn} - \frac{T_{\cdot\cdot\cdot}^2}{abn}$	$SS_A/(a - 1)$	$\sigma^2 + n\sigma_{AB}^2 + bn\sigma_A^2$	MS_A/MS_{AB}
B	$b - 1$	$\sum_{j=1}^{b}\frac{T_{\cdot j\cdot}^2}{an} - \frac{T_{\cdot\cdot\cdot}^2}{abn}$	$SS_B/(b - 1)$	$\sigma^2 + n\sigma_{AB}^2 + an\sigma_B^2$	MS_B/MS_{AB}
AB	$(a - 1)(b - 1)$	$SS_{\mathrm{Tr}} - SS_A - SS_B$	$SS_{AB}/(a - 1)(b - 1)$	$\sigma^2 + n\sigma_{AB}^2$	MS_{AB}/MS_E
Error	$ab(n - 1)$	$SS_{\mathrm{Tot}} - SS_{\mathrm{Tr}}$	$SS_E/ab(n - 1)$	σ^2	
Total	$abn - 1$	$\sum_{i=1}^{a}\sum_{j=1}^{b}\sum_{k=1}^{n}Y_{ijk}^2 - \frac{T_{\cdot\cdot\cdot}^2}{abn}$			

TABLE 14.6

ANOVA table for the two-way classification design with mixed effects: *A* fixed, *B* random

Source of variation	Degrees of freedom (DF)	Sum of squares (SS)	Mean square (MS)	Expected mean square	F
Treatment	$ab - 1$	$\sum_{i=1}^{a}\sum_{j=1}^{b}\frac{T_{ij\cdot}^2}{n} - \frac{T_{\cdot\cdot\cdot}^2}{abn}$	$SS_{\text{Tr}}/(ab-1)$	$\sigma^2 + n\sigma_{\text{Tr}}^2$	MS_{Tr}/MS_E
A	$a - 1$	$\sum_{i=1}^{a}\frac{T_{i\cdot\cdot}^2}{bn} - \frac{T_{\cdot\cdot\cdot}^2}{abn}$	$SS_A/(a-1)$	$\sigma^2 + n\sigma_{AB}^2 + \frac{nb\sum_{i=1}^{a}\alpha_i^2}{a-1}$	MS_A/MS_{AB}
B	$b - 1$	$\sum_{j=1}^{b}\frac{T_{\cdot j\cdot}^2}{an} - \frac{T_{\cdot\cdot\cdot}^2}{abn}$	$SS_B/(b-1)$	$\sigma^2 + na\sigma_B^2$	MS_B/MS_E
AB	$(a-1)(b-1)$	$SS_{\text{Tr}} - SS_A - SS_B$	$SS_{AB}/(a-1)(b-1)$	$\sigma^2 + n\sigma_{AB}^2$	MS_{AB}/MS_E
Error	$ab(n-1)$	$SS_{\text{Tot}} - SS_{\text{Tr}}$	$SS_E/ab(n-1)$	σ^2	
Total	$abn - 1$	$\sum_{i=1}^{a}\sum_{j=1}^{b}\sum_{k=1}^{n} Y_{ijk}^2 - \frac{T_{\cdot\cdot\cdot}^2}{abn}$			

is tested via the F ratio

$$F_{ab-1,\,ab(n-1)} = MS_{\mathrm{Tr}}/MS_E$$

Mixed-Effects Model

The analysis of a mixed-effects model is similar. The biggest difference that occurs is in the statements of the hypothesis to be tested. For example, in a mixed model in which the levels of factor A are fixed and those of factor B are random we test

$$H_0\colon \sigma_{AB}^2 = 0$$

$$H_0^{\mathrm{I}}\colon \mu_{1\cdot\cdot} = \mu_{2\cdot\cdot} = \cdots = \mu_{a\cdot\cdot}$$

$$H_0^{\mathrm{II}}\colon \sigma_B^2 = 0$$

$$H_0^{\mathrm{III}}\colon \sigma_{\mathrm{Tr}}^2 = 0$$

The complete analysis of variance table for this model is given in Table 14.6.

We leave the analysis of the mixed model in which levels of factor A are random and those of factor B are fixed as an exercise (see Exercise 25).

14.4 2^k FACTORIAL EXPERIMENTS

In this section we investigate a special type of factorial experiment. In particular, we consider experiments in which each of k factors is investigated at exactly two levels. This results in 2^k treatment combinations. For example, a three-factor experiment will have $2^3 = 8$ treatment combinations; a six-factor experiment will entail 64 different sets of experimental conditions. Experiments of this sort can be analyzed by using the techniques presented in Sec. 14.1. However, in this case a second method of analysis that makes use of the regression techniques of Chap. 12 is possible. This approach is especially useful in the design of experiments.

To begin, consider a two-factor experiment. Since each factor has only two levels, we can denote these levels as being either "high" or "low." These four treatment combinations are possible:

Factor	
A	B
low	low
low	high
high	low
high	high

Notationally, it is convenient to denote the high level of factor A by a and that of factor B by b. The low level of each is denoted by 1. This allows us to code each treatment combination. For example, the treatment low-low is denoted by $1 \cdot 1 = (1)$; the treatment low-high is given by $1 \cdot b = b$. Thus the four treatment combinations in a two-factor experiment are (1), b, a, and ab. This notation can be extended to a k-factor design. In general, the presence of a letter denotes the use of the high level of the corresponding factor; its absence denotes the use of the low level. For example, in a three-factor experiment the treatment combination (1) entails the use of the low level of all three factors; the combination ab uses the high levels of factors A and B and the low level of C.

To assist in the meaning of factor and interaction effects for this type of an experiment, suppose that we have n replications of each of the treatment combinations (1), a, b, and ab. Interpret the symbols (1), a, b, and ab to represent the totals of the n sample observations for each treatment combination. They are represented in the following graph:

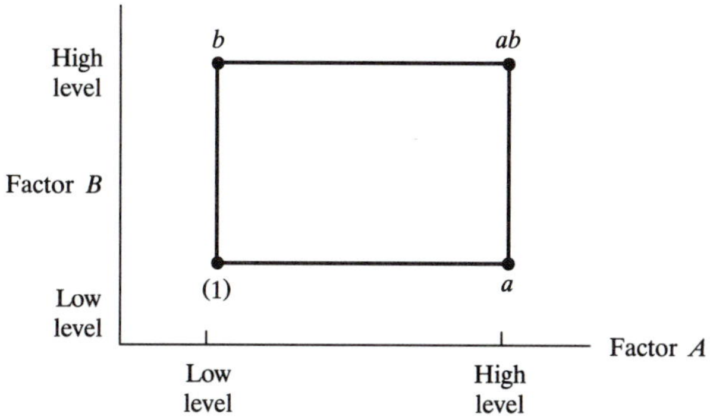

We define the *effect* of a factor as the change of a response when the level of the factor is changed. When factor B is at the low level, the effect of factor A is $a - (1)$. Similarly, when factor B is at the high level, the effect of A is $ab - b$. The *average effect of A* is then

$$A = 1/2\{[a - (1)] + [ab - b]\}$$
$$= 1/2[-(1) + a - b + ab]$$

Similarly, the *average effect of B* is

$$B = 1/2[-(1) - a + b + ab]$$

Now consider the interaction effect AB. At the low level of B the A effect is $a - (1)$, whereas at the high level of B the A effect is $ab - b$. When these two effects differ, it implies that there is an interaction between factors A and B. Thus the *interaction effect AB* is defined as the average difference between these two effects, which is

$$AB = 1/2\{[ab - b] - [a - (1)]\}$$
$$= 1/2[(1) - a - b + ab]$$

The regression model for the general two-factor design is

$$\mu_{Y|x_1, x_2} = \beta_0 + \beta_1 x_1 + \beta_2 x_2 + \beta_{12} x_1 x_2$$

Here x_1 denotes factor A, where we let $x_1 = -1$ if the low level of factor A is used and $x_1 = 1$ otherwise. The regressor x_2 represents factor B and is defined similarly. The cross-product term $x_1 x_2$ represents a possible interaction between factors A and B. Using this coding, we see that the entries of the model specification matrix are all either 1 or -1:

$$X = \begin{array}{c} \text{Intercept} \quad x_1 \quad x_2 \quad x_1x_2 \\ \left|\begin{array}{cccc} 1 & -1 & -1 & 1 \\ 1 & -1 & -1 & 1 \\ \vdots & \vdots & \vdots & \vdots \\ 1 & -1 & -1 & 1 \\ \hdashline 1 & 1 & -1 & -1 \\ 1 & 1 & -1 & -1 \\ \vdots & \vdots & \vdots & \vdots \\ 1 & 1 & -1 & -1 \\ \hdashline 1 & -1 & 1 & -1 \\ 1 & -1 & 1 & -1 \\ \vdots & \vdots & \vdots & \vdots \\ 1 & -1 & 1 & -1 \\ \hdashline 1 & 1 & 1 & 1 \\ 1 & 1 & 1 & 1 \\ \vdots & \vdots & \vdots & \vdots \\ 1 & 1 & 1 & 1 \end{array}\right| \end{array}$$

In this matrix each submatrix is of dimension $n \times 4$. Notice that the first submatrix represents the treatment combination (1), whereas the others represent a, b, and ab, respectively.

The matrix $X'X$ is always a diagonal matrix of the form

$$X'X = \begin{bmatrix} 2^2n & 0 & 0 & 0 \\ 0 & 2^2n & 0 & 0 \\ 0 & 0 & 2^2n & 0 \\ 0 & 0 & 0 & 2^2n \end{bmatrix}$$

where n is the number of observations per treatment combination. Its inverse is

$$(X'X)^{-1} = \begin{vmatrix} 2\frac{1}{2}n & 0 & 0 & 0 \\ 0 & 2\frac{1}{2}n & 0 & 0 \\ 0 & 0 & 2\frac{1}{2}n & 0 \\ 0 & 0 & 0 & 2\frac{1}{2}n \end{vmatrix}$$

The response vector is

$$\mathbf{y} = \begin{bmatrix} y_{111} \\ y_{112} \\ \vdots \\ y_{11n} \\ \hdashline y_{211} \\ y_{212} \\ \vdots \\ y_{21n} \\ \hdashline y_{121} \\ y_{122} \\ \vdots \\ y_{12n} \\ \hdashline y_{222} \\ \vdots \\ y_{22n} \end{bmatrix}$$

Notice that the first subvector of the response vector contains the responses for the treatment combination (1), whereas the others represent a, b, and ab, respectively. A quick calculation will show that the vector $X'\mathbf{y}$ is given by

$$X'\mathbf{y} = \begin{bmatrix} (1) + a + b + ab \\ a + ab - (1) - b \\ b + ab - (1) - a \\ (1) + ab - a - b \end{bmatrix}$$

where in this context (1), a, b, and ab denote the sum of the responses for the indicated treatment combination. An example will illustrate the idea.

Example 14.4.1. The *Mirogrex terrae-sanctae* is a commercial sardinelike fish found in the Sea of Galilee. A study is conducted to determine the effect of light and temperature on the growth of the ovary of these fish. Two photoperiods (14 hours of light–10 hours of dark and 9 hours of light–15 hours of dark) and two temperatures (16°C and 27°C) are used. In this way the experimenter can simulate both the summer and winter conditions in the region. These data are obtained:

Factor *B*	Factor *A* (photoperiod): 9 hours light (low)	Factor *A* (photoperiod): 14 hours light (high)
16° (low)	1.30, 2.88 (simulated winter) (1) = 1.30 + 2.88 = 4.18	1.01, 1.52 (unnatural) a = 1.01 + 1.52 = 2.53
27° (high)	.90, 1.06 (unnatural) b = .90 + 1.06 = 1.96	.83, .67 (simulated summer) ab = .83 + .67 = 1.50
		$T_{...} = (1) + a + b + ab = 10.17$

In matrix notation

$$\mathbf{y} = \begin{bmatrix} 1.30 \\ 2.88 \\ \hdashline 1.01 \\ 1.52 \\ \hdashline .90 \\ 1.06 \\ \hdashline .83 \\ .67 \end{bmatrix} \qquad X = \begin{bmatrix} 1 & -1 & -1 & 1 \\ 1 & -1 & -1 & 1 \\ \hdashline 1 & 1 & -1 & -1 \\ 1 & 1 & -1 & -1 \\ \hdashline 1 & -1 & 1 & -1 \\ 1 & -1 & 1 & -1 \\ \hdashline 1 & 1 & 1 & 1 \\ 1 & 1 & 1 & 1 \end{bmatrix}$$

Note that the entries of X are all either 1 or -1 as claimed. Furthermore, note that

$$X'X = \begin{bmatrix} 8 & 0 & 0 & 0 \\ 0 & 8 & 0 & 0 \\ 0 & 0 & 8 & 0 \\ 0 & 0 & 0 & 8 \end{bmatrix}$$

and

$$(X'X)^{-1} = \begin{bmatrix} 1/8 & 0 & 0 & 0 \\ 0 & 1/8 & 0 & 0 \\ 0 & 0 & 1/8 & 0 \\ 0 & 0 & 0 & 1/8 \end{bmatrix}$$

The vector $X'\mathbf{y}$ is given by

$$X'\mathbf{y} = \begin{bmatrix} 1 & 1 & 1 & 1 & 1 & 1 & 1 & 1 \\ -1 & -1 & 1 & 1 & -1 & -1 & 1 & 1 \\ -1 & -1 & -1 & -1 & 1 & 1 & 1 & 1 \\ 1 & 1 & -1 & -1 & -1 & -1 & 1 & 1 \end{bmatrix} \begin{bmatrix} 1.30 \\ 2.88 \\ 1.01 \\ 1.52 \\ .90 \\ 1.06 \\ .83 \\ .67 \end{bmatrix}$$

$$= \begin{bmatrix} 10.17 \\ (-4.18) + (2.53) - (1.96) + (1.50) \\ (-4.18) - (2.53) + (1.96) + (1.50) \\ (4.18) - (2.53) - (1.96) + 1.50 \end{bmatrix}$$

$$= \begin{bmatrix} (1) + a + b + ab \\ -(1) + a - b + ab \\ -(1) - a + b + ab \\ (1) - a - b + ab \end{bmatrix}$$

$$= \begin{bmatrix} (1) + a + b + ab \\ a + ab - (1) - b \\ b + ab - (1) - a \\ (1) + ab - a - b \end{bmatrix}$$

as claimed.

To test for interaction, we test

$$H_0\colon \beta_{12} = 0$$

Main effects are studied by testing

$$H_0^{\mathrm{I}}\colon \beta_1 = 0$$

and

$$H_0^{\mathrm{II}}\colon \beta_2 = 0$$

We already know how to conduct these tests by using the methods of Sec. 12.5. Recall that it has already been shown that, in any regression analysis,

$$S_{yy} = \mathrm{SSR} + \mathrm{SSE}$$

where S_{yy} denotes the corrected total sum of squares, SSR is the sum of squares due to regression, and SSE is the residual or error sum of squares. In matrix notation it is easy to see that

$$\mathrm{SSR} = \hat{\boldsymbol{\beta}}'X'\mathbf{y} - C$$

where C denotes the usual correction factor. In this case (see Exercise 31).

$$C = \frac{T^2\ldots}{N} = \frac{(1) + a + b + ab}{2^2 n}$$

It is known that

$$\hat{\boldsymbol{\beta}} = (X'X)^{-1}X'\mathbf{y}$$

Due to the special nature of $(X'X)^{-1}$ and $X'\mathbf{y}$ in this application, $\hat{\boldsymbol{\beta}}$ and SSR assume relatively simple forms. In particular,

$$\hat{\boldsymbol{\beta}} = \begin{bmatrix} \dfrac{(1) + a + b + ab}{2^2 n} \\ \dfrac{a + ab - (1) - b}{2^2 n} \\ \dfrac{b + ab - (1) - a}{2^2 n} \\ \dfrac{(1) + ab - a - b}{2^2 n} \end{bmatrix}$$

and
$$\text{SSR} = \frac{[(1) + a + b + ab]^2}{2^2 n} + \frac{[a + ab - (1) - b]^2}{2^2 n} + \frac{[b + ab - (1) - a]^2}{2^2 n} + \frac{[(1) + ab - a - b]^2}{2^2 n} - C$$

However, the term $(1) + a + b + ab = T\ldots$ so that the first term of the above expression is the correction factor. Hence SSR can be written as

$$\text{SSR} = \frac{[a + ab - (1) - b]^2}{2^2 n} + \frac{[b + ab - (1) - a]^2}{2^2 n} + \frac{[(1) + ab - a - b]^2}{2^2 n}$$

Using the techniques for testing a subset of predictor variables given in Sec. 12.5, we can show that the numerator of the F ratio used to test H_0: $\beta_{12} = 0$ is

$$\frac{[(1) + ab - a - b]^2}{2^2 n}$$

We denote this by SS_{AB} and write the test statistic as

$$F_{1,2^2n-3-1} = \frac{SS_{AB}}{\text{SSE}/(2^2 n - 3 - 1)}$$

(see Exercise 47).

It can also be verified that the test statistics for testing H_0^{I}: $\beta_1 = 0$ and H_0^{II}: $\beta_2 = 0$ are

$$F_{1,2^2n-3-1} = \frac{\dfrac{[a + ab - (1) - b]^2}{2^2n}}{\text{SSE}/(2^2n - 3 - 1)} = \frac{SS_A}{\text{SSE}/(2^2n - 3 - 1)}$$

$$F_{1,2^2n-3-1} = \frac{\dfrac{[b + ab - (1) - a]^2}{2^2n}}{\text{SSE}/(2^2n - 3 - 1)} = \frac{SS_B}{\text{SSE}/(2^2n - 3 - 1)}$$

Note that in this special case,

$$\text{SSR} = SS_A + SS_B + SS_{AB}$$

We illustrate by continuing the analysis of the data of Example 14.4.1.

Example 14.4.2. For the data of Example 14.4.1,

$$S_{yy} = (1.3^2 + 2.88^2 + 1.01^2 + \cdots + .67^2) - \frac{(10.17)^2}{8}$$

$$= 16.3863 - 12.9286 = 3.4577$$

$$SS_{AB} = \frac{[(1) + ab - a - b]^2}{8} = \frac{[4.18 + 1.50 - 2.53 - 1.96]^2}{8}$$

$$= .1770$$

$$SS_A = \frac{[a + ab - (1) - b]^2}{8} = [2.53 + 1.50 - 4.18 - 1.96]^2$$

$$= .5565$$

$$SS_B = \frac{[b + ab - (1) - a]^2}{8} = \frac{[1.96 + 1.50 - 4.18 - 2.53]^2}{8}$$

$$= 1.3203$$

$$\text{SSR} = SS_A + SS_B + SS_{AB}$$

$$= .5565 + 1.3203 + .1770$$

$$= 2.0538$$

$$\text{SSE} = S_{yy} - \text{SSR}$$

$$= 3.4577 - 2.0538$$

$$= 1.4039$$

The observed value of the F ratio used to test for interaction is

$$F_{1,4} = \frac{SS_{AB}}{\text{SSE}/4} = \frac{.1770}{1.4039/4} = .5043$$

The P value of this test exceeds .1, and hence $H_0\colon \beta_{12} = 0$ cannot be rejected. No interaction has been detected.

The two-factor design has been used as an illustration of the use of regression techniques in a factorial experiment. The ideas can be extended to any 2^k design. Fortunately, it is easy to anticipate the form of the numerator of any 1 degree of freedom F test needed without having to work directly with the appropriate reduced models. For example, in the two-factor case we construct a table that lists the algebraic signs of x_1, x_2, and x_1x_2 for each treatment combination. This table is shown in Table 14.7.

The term to be squared in the numerator of the F ratio is found by summing the treatment totals with the indicated coefficient found in the column corresponding to the effect being tested. For example, to test for interaction, we find the appropriate numerator by attaching the coefficients found in column AB to the treatment totals (1), a, b, and ab, summing these terms, squaring the sum, and dividing by $2^k n$. In this case we see that

$$SS_{AB} = \frac{[+(1) - a - b + ab]^2}{2^2 n}$$

$$SS_A = \frac{[-(1) + a - b + ab]^2}{2^2 n}$$

$$SS_B = \frac{[-(1) - a + b + ab]^2}{2^2 n}$$

As you can see, these values agree with those found earlier. In Exercises 32 and 33 you are asked to extend this idea to the three-factor model and to a restricted four-factor model, respectively.

TABLE 14.7
Coefficients of effects for 2^2 factorial

Treatment combination	Effect: $A(x_1)$	$B(x_2)$	$AB(x_1x_2)$
(1)	−	−	+
a	+	−	−
b	−	+	−
ab	+	+	+

Computational Techniques—Yates Method

If SAS or some other statistical package is available, then it can be used to analyze a 2^k factorial experiment easily. A sample SAS program is given in Computer Supplement XVII. If you must analyze the data by hand, then a technique developed by Yates simplifies the work. To use the technique, you must list the treatment combinations in a column in standard order. The standard order for a one-factor experiment is (1), a. When a second factor, b, is added, then multiply each of the previous combinations by b to obtain the four treatments (1), a, b, and ab. Similarly, if the factor c is added, you multiply the preceding terms by c to obtain the standard order (1), a, b, ab, c, ac, bc, and abc. The pattern continues for additional factors. The Yates method is outlined below and is illustrated in Example 14.4.3.

Yates Method

1. List the treatment combinations in standard order in a column.
2. Form a second column by listing the treatment totals in standard order. Pair the totals from top to bottom.
3. Label the next column 1 and form the first half of the entries by finding the totals of the pairs in the second column. Form the last half of the entires by subtracting the first member of each pair from the second. Pair the entries in this column.
4. Label the next column 2. Form this column from column 1 by the procedure described in part (iii).
5. Continue until you form k columns.
6. The first value in the kth column is the grand total of all responses. The remaining values when squared and divided by $2^k n$ give the numerator for the F ratio needed to test for the effect in question.

Example 14.4.3. A study is conducted on the effect of temperature, time in process, and rate of temperature rise on the amount of dye (in milligrams) left in the residue bath for a dying process. The experiment was run at two levels of temperature (120°C, 135°C), two levels of time in process (30 min, 60 min), and two levels of rate of temperature rise (R_1, R_2). The experiment is run as a 2^3 factorial with two replications, yielding the following data:

Temperature

		120°C		**135°C**	
		30 min	**60 min**	**30 min**	**60 min**
	R_1	19.9	17.4	25.0	19.5
Rate		18.6	16.8	22.8	18.3
	R_2	14.5	16.3	27.7	28.3
		16.1	14.6	18.0	26.2

Denoting the effects of temperature, time, and rate by A, B, and C, respectively, we compute the sum of squares by the Yates procedure as follows:

Treatment combination	Total ($n = 2$)	(1)	(2)	(3)	Effect	Sum of squares (SS)
(1)	38.5	86.3	158.3	320	Grand total	—
a	47.8	72.0	161.7	51.6	*A*	166.41
b	34.2	76.3	12.9	−5.2	*B*	1.69
ab	37.8	85.4	38.7	2.8	*AB*	0.49
c	30.6	9.3	−14.3	3.4	*C*	0.7225
ac	45.7	3.6	9.1	25.8	*AC*	41.6025
bc	30.9	15.1	−5.7	23.4	*BC*	34.2225
abc	54.5	23.6	8.5	14.2	*ABC*	12.6025
				Total = SSR = 257.74		

For these data the corrected total sum of squares is

$$\sum\sum\sum\sum y_{ijkl}^2 - \frac{T_{\cdots}^2}{16} = 6713.88 - \frac{320^2}{16}$$

$$= 313.88$$

Since $\text{SSE} = S_{yy} - \text{SSR}$,

$$\text{SSE} = 313.88 - 257.74 = 56.14$$

The complete analysis is shown in Table 14.8. From Table IX of App. A we see that the A effect (temperature) is significant at the .05 level and the AC (temperature–rate of rise) interaction is significant at the .05 level.

TABLE 14.8

Source	Degree of Freedom (DF)	Sum of squares (SS)	Mean square (MS)	F
Main effects				
A	1	166.41	166.41	23.71
B	1	1.69	1.69	0.24
C	1	0.7225	0.7225	0.10
Interactions				
AB	1	0.49	0.49	0.07
AC	1	41.6025	41.6025	5.93
BC	1	34.2225	34.2225	4.88
ABC	1	12.6025	12.6025	1.80
Error	8	56.14	7.0175	
Total	15	313.88		

14.5 2^k FACTORIAL EXPERIMENTS IN AN INCOMPLETE BLOCK DESIGN

In some experimental situations it is not possible to run every treatment combination under exactly the same experimental conditions. For example, the laboratory may not be equipped to handle all treatment combinations in the same day; some treatments must be run on one day and others later. This introduces the factor "day run" as a possible source of extraneous variation that should be taken into account. A chemical engineer might not have enough raw material from a single supplier to conduct experiments on all treatment combinations; three suppliers might be involved. Here the factor "supplier" is an extraneous variable to be controlled. A solution to the problem is to run the treatment combinations in subsets. Each subset is called an *incomplete block*. As you might suspect, there is a price paid for not running the entire experiment under identical conditions. This price is that some of the effects become "confounded" with or inseparable from block effects.

There are some restrictions on the number of incomplete blocks that can be used in a 2^k factorial experiment. In particular, we restrict this number to values of 2^p, where $p < k$. For example if $k = 2$, then p must be 1 and $2^p = 2$; the four treatment combinations can be divided into two sets of two treatments each. If $k = 3$, then $p = 1$ or $p = 2$ so that either two or four blocks can be used. To decide which treatment combinations are to be used in each of the 2^p incomplete blocks and to decide which effects are to be confounded by blocking, we must choose p "defining contrasts." A defining contrast is an effect that is not of interest to the researcher or one which is expected to be negligible. It will be confounded with the block effect, and hence it cannot be tested for significance later. Usually, higher-order interactions are chosen as defining contrasts.

Example 14.5.1. Suppose that we have a 2^4 factorial experiment to be run in four blocks so that $p = 2$. Two defining contrasts are needed. Suppose that the researcher has no interest in the interactions ABC and CD. These are chosen as defining contrasts. No inferences can be made concerning these effects later since these interactions will be confounded with block effects.

Once the defining contrasts have been selected, the treatment combinations assigned to each block can be determined. One scheme for doing so is outlined below. It is illustrated on a step-by-step basis by completing the problem posed in Example 14.5.1.

Forming 2^p Incomplete Blocks

1. Choose a value of p, where $p < k$. (In our example $p = 2$.)
2. Choose p defining contrasts. (We have chosen ABC and CD.)
3. Write each contrast in the form $A^{\gamma_{1j}}B^{\gamma_{2j}}C^{\gamma_{3j}}\ldots$, where $\gamma_{ij} = 1$ if the $i(jh)$ factor appears in the jth defining contrast and $\gamma_{ij} = 0$ otherwise. (Here $ABC = A^1B^1C^1D^0$ and $CD = A^0B^0C^1D^1$. That is, $\gamma_{11} = \gamma_{21} = \gamma_{31} = 1$ and $\gamma_{41} = 0$, whereas $\gamma_{12} = \gamma_{22} = 0$ and $\gamma_{32} = \gamma_{42} = 1$.)

4. Associated with the p defining contrasts are p functions $L_1, L_2, \ldots, L_p$ defined by

$$L_j = z_1\gamma_{1j} + z_2\gamma_{2j} + \cdots + z_k\gamma_{kj} \qquad j = 1, 2, \ldots, p$$

(In our example there are two such functions. These are

$$L_1 = z_1 + z_2 + z_3$$

$$L_2 = z_3 + z_4)$$

5. Evaluate each of these functions for each treatment combination by allowing $z_i = 1$ if the high level of the ith factor is used and $z_i = 0$ otherwise. (For example, for the treatment combination (1) in which the low level of each factor is used

$$L_1 = 0 + 0 + 0 = 0$$

$$L_2 = 0 + 0 = 0)$$

For the treatment combination *ab* in which the high levels of factors A and B and the low levels of C and D are used,

$$L_1 = 1 + 1 + 0 = 2$$

$$L_2 = 0 + 0 = 0$$

6. Reduce each of the values of L_j to 0 or 1 modulo 2. (To reduce L_j mod 2, we divide by 2 and report the remainder. Here both L_1 and L_2 reduce to 0 mod 2.)
7. Group all treatment combinations with identical values of $L_1, L_2, \ldots, L_p$ into a single block. The blocks containing treatment combination (1) is called the *principal block*. (The four blocks for our example have L_1 and L_2 mod 2 values of 0-0, 1-0, 0-1, and 1-1. They are given in Table 14.9.)

The analyses of a 2^k factorial experiment with blocking is similar to the factorial experiment without blocking. The sums of squares for effects that are not confounded with blocks are computed in the same manner as in the unconfounded case. We then compute a block sum of squares in the usual way:

$$SS_{\text{Blks}} = \sum_{\text{all blocks}} \frac{(\text{block total})^2}{\text{block size}} - \text{correction factor}$$

or by means of the Yates method. If the Yates method is used, then the block sum of squares is the total of all effects that are confounded with blocks. Which effects are these? In general, they are the p defining contrasts and their *generalized interactions*. The generalized interaction of two contrasts is defined as the product of the contrasts where the exponents of the product have been reduced modulo 2. For instance, in Example 14.5.1 the effects ABC and CD will be confounded with blocks. In addition, the generalized interaction $(ABC)(CD) = ABC^2D = ABD$ will also be confounded. The block sum of squares is given by $SS_{\text{Blks}} = SS_{ABC} + SS_{CD} + SS_{ABD}$. Example 14.5.2 illustrates the computations involved.

TABLE 14.9

Treatment combination	L_1	L_1(mod 2)	L_2	L_2(mod 2)	Block
(1)	0	0	0	0	1
a	1	1	0	0	2
b	1	1	0	0	2
c	1	1	1	1	4
d	0	0	1	1	3
ab	2	0	0	0	1
ac	2	0	1	1	3
ad	1	1	1	1	4
bc	2	0	1	1	3
bd	1	1	1	1	4
cd	1	1	2	0	2
abc	3	1	1	1	4
abd	2	0	1	1	3
acd	2	0	2	0	1
bcd	2	0	2	0	1
abcd	3	1	2	0	2

Block 1	Block 2	Block 2	Block 3	Block 4
(1)	*a*	*a*	*ac*	*c*
ab	*b*	*b*	*bc*	*ad*
acd	*cd*	*cd*	*d*	*bd*
bcd	*abcd*	*abcd*	*abd*	*abc*

Example 14.5.2. A study of the effect of four factors on the yield of a certain chemical is planned. A 2^4 factorial experiment is to be conducted. Two blocks will be used. In this case $p = 1$ and the defining contrast chosen is the four-factor interaction *ABCD*. All three-factor interactions are assumed to be negligible and are not included in the model. The resulting blocks along with the single response per treatment combination are shown in Table 14.10.

TABLE 14.10

Block 1	Block 2
(1) = 29.5	*a* = 39.8
ab = 38.2	*b* = 32.1
ac = 33.8	*c* = 27.3
bc = 25.3	*d* = 31.4
ad = 37.6	*abc* = 35.7
bd = 31.0	*abd* = 38.9
cd = 24.9	*acd* = 35.9
abcd = 34.1	*bcd* = 27.1

TABLE 14.11

Source	DF	Sum of squares (SS)	Mean square (MS)	F
Blocks	1	11.90		
Main effects				
A	1	267.32	267.32	789.14
B	1	.30	.30	
C	1	73.96	73.96	218.33
D	1	.04	.04	
Two-factor interactions				
AB	1	.42	.42	
AC	1	1.21	1.21	
AD	1	.09	.09	
BC	1	.16	.16	
BD	1	.01	.01	
CD	1	.02	.02	
Error	4	1.355	.75	

The analysis of variance table found using SAS is given in Table 14.11. In this table $SS_{\text{Blks}} = 11.90$ is a combination of the block effect and SS_{ABCD}; the error sum of squares is a combination of the three-factor interactions, SS_{ABC}, SS_{ABD}, SS_{ACD}, and SS_{BCD}.

14.6 FRACTIONAL FACTORIAL EXPERIMENTS

In most industrial experiments there are many variables that might influence a response. We have seen that in a complete factorial experiment as k increases, the number of treatment combinations becomes very large. For reasons of time or economics it might not be possible to collect data under all experimental conditions. In this section we consider a method of combating this problem.

To begin, note that as the number of factors under study in an experiment increases, so do the number of higher-order (three-way or above) interactions. It is often reasonable to assume that these are negligible. This reduces the number of model parameters or effects of real interest to the researcher and makes it possible to conduct the study by using fewer treatment combinations than would be required in a full 2^k model. Since such a study uses only a fraction of the 2^k possible treatment combinations, it is called a *fractional factorial* experiment. The fractions used are 1/2, 1/4, 1/8, and so forth.

A fractional factorial experiment basically is a randomized incomplete block experiment in which only one of the 2^p possible blocks is utilized. For this reason, most of the ideas needed to design these experiments were introduced in the last section.

We begin by choosing one or more higher-order interactions to use as defining contrasts. Since the effects of these and their generalized interactions are confounded with blocks, they should all be effects that are not of interest to the researcher. Another serious consequence of running only a fraction of the

2^k possible treatment combinations is that other effects also become confounded. If one is not careful, then effects that are of major interest to the researcher might become indistinguishable from one another. An example should clarify this idea.

Example 14.6.1. Suppose that we want to design a 1/2 fractional factorial experiment to study three factors. The defining contrast used to create two blocks is the three-factor interaction ABC. The resulting blocks are

Block 1	Block 2
(1)	a
ab	b
bc	c
ac	abc

Suppose that the treatment combinations listed in block 1 are to be used in the experiment. Table 14.12 is the table of coefficients for a 2^3 design.

The estimate for each of the model effects can be read from the columns of this table. The confounding referred to earlier can also be seen from this table. Note that if all eight treatment combinations are run, then the A effect is estimated by

$$\begin{aligned} -(1) + a - b - c + ab + ac - bc + abc \\ &= [a - b - c + abc] + [-(1) + ab + ac - bc] \\ &= R + S \end{aligned}$$

whereas the BC effect is estimated by

$$\begin{aligned} (1) + a - b - c - ab - ac + bc + abc \\ &= [a - b - c + abc] - [-(1) + ab + ac - bc] \\ &= R - S \end{aligned}$$

When all eight treatment combinations are run, these two estimates will be distinct with the difference in estimates being dependent on the magnitude of the term $[-(1) + ab + ac - bc]$. If we run only the treatment combinations listed in block 2, then it can be seen that the A and BC effects are both estimated by

TABLE 14.12

Coefficients for a 2^3 factorial experiment

Treatment combination	Block number	(x_1) A	(x_2) B	(x_3) C	(x_1x_2) AB	(x_1x_3) AC	(x_2x_3) BC
(1)	1	−	−	−	+	+	+
a	2	+	−	−	−	−	+
b	2	−	+	−	−	+	−
c	2	−	−	+	+	−	−
ab	1	+	+	−	+	−	−
ac	1	+	−	+	−	+	−
bc	1	−	+	+	−	−	+
abc	2	+	+	+	+	+	+

$R = a - b - c + abc$. The effects A and BC become indistinguishable; that is, they are confounded with one another.

When two effects are confounded, we say that one is the *alias* of the other or that one is aliased with the other. In Example 14.6.1 the A effect is aliased with the BC effect. This is denoted by $A \equiv BC$. An easy way to determine aliases is to find the generalized interactions between each effect and the defining contrasts. For instance, the aliases for effects B and C in Example 14.6.1 are

$$B(ABC) = AB^2C = AC$$

and

$$C(ABC) = ABC^2 = AB$$

We should emphasize that the 2^3 factorial design was used mainly as an illustration only. Fractional factorials are most useful when k is rather large, and we want to run one of several possible blocks. That is, we might run a $1/2$, $1/4$, $1/8$, or even smaller fraction of a very large factorial experiment. In general, for a 2^p fraction of a 2^k factorial experiment we construct p blocks after choosing $p - 1$ defining contrasts and then select exactly one of these p blocks for the actual experiment. When this is done, one must be aware of the resulting alias structure. This structure is crucial since it determines which effects can be tested from the experimental data. Often it will be necessary to consult a professional statistician for help in choosing an appropriate experimental design. Example 14.6.2 illustrates the problem of aliases in a 2^4 factorial experiment.

Example 14.6.2. Consider Example 14.5.1 in which $p = 2$ and $k = 4$. The two defining contrasts used to determine the four blocks given in Table 14.9 are ABC and CD. Suppose that main effect A is of primary interest. To determine the aliases for this effect, we multiply A by *each* of the defining contrasts and reduce the exponents modulo 2. These aliases are

$$A(ABC) \equiv A^2BC = BC$$

and

$$A(CD) \equiv ACD$$

The main effect A is aliased with two other effects, a two-factor and a three-factor interaction. If there are subject matter related reasons for believing that these interactions are negligible, then this does not create a problem; otherwise new defining contrasts must be found that do not confound effect A with other nonnegligible effects.

Example 14.6.3 illustrates the complete analysis of a fractional factorial experiment.

Example 14.6.3. An experiment was conducted to study the viscosity of star block copolymers. Five dependent variables were selected to determine their effect upon the response variable viscosity. The dependent variables were temperature, diblock concentration, frequency, casting solvents, and annealing time. Since experimentation was expensive, it was decided to use a $1/2$ fraction of a 2^5 factorial with the five-factor interaction confounded with blocks. The levels of the respective factors used in the experiment were as follows:

Factor	Level
A (temperature)	130°C, 230°C
B (diblock concentration)	10%, 100%
C (frequency)	.10, 100.0
D (casting solvent)	8.2, 9.3
E (annealing time)	0 hr, 2 hr

One of the two blocks obtained by confounding the $ABCDE$ interaction with blocks yielded the following responses on viscosity for the respective treatment combinations:

Factor level					Treatment	
A	B	C	D	E	combination	Response
0	0	0	0	0	(1)	62.7
1	1	0	0	0	*ab*	11.8
1	0	1	0	0	*ac*	1.8
1	0	0	1	0	*ad*	108.8
1	0	0	0	1	*ae*	26.4
0	1	1	0	0	*bc*	2.0
0	1	0	1	0	*bd*	36.4
0	1	0	0	1	*be*	21.7
0	0	1	1	0	*cd*	4.2
0	0	1	0	1	*ce*	2.3
0	0	0	1	1	*de*	117.9
1	1	1	1	0	*abcd*	1.2
1	1	1	0	1	*abce*	0.6
1	1	0	1	1	*abde*	1.1
1	0	1	1	1	*acde*	3.5
0	1	1	1	1	*bcde*	2.5

The reader can verify that the alias structure and respective sum of squares (using the Yates procedure) are as follows:

Source	Sum of squares (SS)
$A \equiv BCDE$	558.14
$B \equiv ACDE$	3915.63
$C \equiv ABDE$	8496.23
$D \equiv ABCE$	1337.73
$E \equiv ABCD$	174.90
$AB \equiv CDE$	.11
$AC \equiv BDE$	469.81
$AD \equiv BCE$	.18
$AE \equiv BCD$	1074.20
$BC \equiv ADE$	3579.03
$BD \equiv ACE$	1157.70
$BE \equiv ACD$	.23
$CD \equiv ABE$	1171.35
$CE \equiv ABD$	170.96
$DE \equiv ABC$	.18

TABLE 14.13

ANOVA table for $1/2$ fraction of 2^5 factorial

Source	Degrees of freedom (DF)	Sum of squares (SS)	Mean square (MS)	F
A	1	558.14	558.14	0.73
B	1	3915.63	3915.63	5.14
C	1	8496.23	8496.23	11.14
D	1	1337.73	1337.73	1.75
E	1	174.90	174.90	0.23
Error	10	7623.74	762.37	

Pooling the two-factor interactions into error, we obtain the ANOVA table for testing main effects, which is given in Table 14.13.

From Table IX of App. A we can conclude that factor B is significant at the .05 level and that factor C is significant at the .05 level.

In interpreting this example, we should add a warning. From the alias structure we see that the main effects are aliased with four-factor interactions and that two-factor interactions are aliased with three-factor interactions. Furthermore, to be able to test main effects, we pooled two-factor interactions into error. Otherwise we would have 0 degrees of freedom for error, and we have made some rather strong assumptions. To determine whether the assumptions are valid requires good judgment on behalf of the scientist and the statistician. If the assumptions are determined to be valid, this type of experiment gives very useful information at considerably reduced cost.

EXERCISES

Section 14.1

1. Show that in the two-way fixed-effects model

$$\sum_{i=1}^{a} \alpha_i = \sum_{j=1}^{b} \beta_j = \sum_{i=1}^{a} \sum_{j=1}^{b} (\alpha\beta)_{ij} = 0$$

2. Show that in the two-way classification fixed-effects model the null hypothesis

$$H_0^{\text{I}}: \mu_{1\cdot\cdot} = \mu_{2\cdot\cdot} = \cdots = \mu_{a\cdot\cdot}$$

is equivalent to H_0^{I}: $\alpha_i = 0$, $i = 1, 2, \ldots, a$, and that the null hypothesis

$$H_0^{\text{II}}: \mu_{\cdot 1 \cdot} = \mu_{\cdot 2 \cdot} = \cdots = \mu_{\cdot b \cdot}$$

is equivalent to H_0^{II}: $\beta_j = 0$, $j = 1, 2, \ldots, b$.

Ozonization as a secondary treatment for effluent, following absorption by ferrous chloride, was studied for three reaction times and three pH levels. The study yielded the following results for effluent decline:

Reaction time (minutes)	pH level	Effluent decline
20	7.0	23, 21, 22
	9.0	16, 18, 15
	10.5	14, 13, 16
40	7.0	20, 22, 19
	9.0	14, 13, 12
	10.5	12, 11, 10
60	7.0	21, 20, 19
	9.0	13, 12, 12
	10.5	11, 13, 12

Exercises 3 through 10 refer to the above experiment.

3. Assuming the two-way classification model with fixed effects, write the model for this experiment and explain what each parameter represents in the context of this experiment.

4. State the null hypothesis to be tested first.

5. (*a*) Plot the average of reaction time across pH levels.
(*b*) Plot the averages of pH level across reaction times.
(*c*) Do the plots in parts (*a*) and (*b*) suggest that interaction exists?

6. Test for significant interaction. State your appropriate P value.

7. Test for significant differences among reaction times. State the approximate P value.

8. Test for significant differences among pH levels at the 5% level.

9. Is it necessary to conduct one-way analyses on a level-by-level basis to be able to describe the nature of the differences in main effects? Explain.

10. Conduct paired comparisons to pinpoint differences in means for those main effects found to be significantly different.

Decomposition of leaf packs was studied for four environments and three time lengths of exposure. The measurements were in grams of weight loss of leaf packs. The study yielded the following results:

	Time		
Environment	1 month	2 months	3 months
E_1	1.09	1.35	1.60
	1.06	1.53	1.40
E_2	1.16	1.38	2.18
	1.03	1.35	1.77
E_3	1.01	1.63	1.66
	1.04	1.51	1.98
E_4	0.90	1.60	1.73
	1.03	1.72	1.76

Exercises 11 through 15 refer to the experiment just described.

11. Assuming the two-way classification model with fixed effects, write the model for this experiment and explain what each parameter represents in the context of this experiment.

12. (a) Plot the averages of time across environmental conditions.
(b) Plot the averages of environmental conditions across time.
(c) Do the plots in parts (a) and (b) suggest that interaction exists?

13. Test for significant interaction at the $\alpha = .10$ level.

14. Conduct any level-by-level one-way analyses deemed appropriate based on the results of Exercise 13.

15. Which combination of environment and time seems to yield the greatest decomposition of leaf packs?

Section 14.2

16. (*Three-factor ANOVA.*) In the three-factor design it is assumed that the $a \cdot b \cdot c$ treatment combinations represent random samples of size n drawn from $a \cdot b \cdot c$ populations, where each population is normally distributed with mean μ_{ijk} and variance σ^2. The model for the three-factor design is

$$Y_{ijkl} = \mu + \alpha_i + \beta_j + \gamma_k + (\alpha\beta)_{ij} + (\alpha\gamma)_{ik} + (\beta\gamma)_{jk} + (\alpha\beta\gamma)_{ijk} + E_{ijkl}$$

$$i = 1, 2, \ldots, a \qquad k = 1, 2, \ldots, c$$

$$j = 1, 2, \ldots, b \qquad l = 1, 2, \ldots, n$$

Here α_i, β_j, and γ_k represent main effects. The terms $(\alpha\beta)_{ij}$, $(\alpha\gamma)_{ik}$, and $(\beta\gamma)_{jk}$ represent interactions between factors A and B, A and C, and B and C, respectively. The interaction among factors A, B, and C is denoted by $(\alpha\beta\gamma)_{ijk}$. Let

$\mu_{ijk\cdot}$ = mean for the (ijk)th treatment combination

$\mu_{ij\cdot\cdot}$ = average of the population means for those populations receiving the ith level of A and the jth level of B

$$= \sum_{k=1}^{c} \frac{\mu_{ijk}}{c}$$

$\mu_{i\cdots}$ = average of the population means for those populations receiving the ith level of A

$$= \sum_{j=1}^{b} \sum_{k=1}^{c} \frac{\mu_{ijk\cdot}}{bc}$$

μ = average of the population means for all populations under consideration

$$= \sum_{i=1}^{a} \sum_{j=1}^{b} \sum_{k=1}^{c} \frac{\mu_{ijk\cdot}}{abc}$$

Other means of a similar nature are needed to define the terms of the model. They are given below. Express each verbally and in summation notation as an average of population means.

(a) $\mu_{i \cdot k \cdot}$
(b) $\mu_{\cdot jk \cdot}$
(c) $\mu_{\cdot j \cdot \cdot}$
(d) $\mu_{\cdot \cdot k \cdot}$

17. Main effects and interaction effects can be defined in a manner that parallels the definitions in the two-way model. For example, we define

$$\alpha_i = \mu_{i \cdot \cdot} - \mu$$

$$(\alpha\beta)_{ij} = \mu_{ij \cdot \cdot} - \mu_{i \cdot \cdot \cdot} - \mu_{\cdot j \cdot \cdot} + \mu$$

Define each of these model parameters in a similar manner.

(a) β_j
(b) γ_k
(c) $(\alpha\gamma)_{ik}$
(d) $(\beta\gamma)_{jk}$
(e) $(\alpha\beta\gamma)_{ijk}$

18. Using the definitions of Exercise 17, one can show that the main effects, as well as the two- and three-factor interaction terms, sum to 0. As an illustration, verify that $\sum_{i=1}^{a} \alpha_i = 0$.

19. Totals needed in the three-way ANOVA also parallel those seen in the two-way design. For example,

$$T_{i \cdot \cdot \cdot} = \text{sum of all responses from experimental units receiving level } i \text{ of } A$$

$$= \sum_{j=1}^{b} \sum_{k=1}^{c} \sum_{l=1}^{n} y_{ijkl}$$

$$T_{ij \cdot \cdot} = \text{sum of all responses from experimental units receiving level } i \text{ of } A \text{ and } j \text{ of } B$$

$$= \sum_{k=1}^{c} \sum_{l=1}^{n} y_{ijkl}$$

Define each of the following totals.

(a) $T_{\cdot j \cdot \cdot}$
(b) $T_{\cdot \cdot k \cdot}$
(c) $T_{i \cdot k \cdot}$
(d) $T_{\cdot jk \cdot}$
(e) $T_{ijk \cdot}$
(f) $T_{\cdot \cdot \cdot \cdot}$

20. Typical sums of squares needed in the three-way ANOVA are

$$SS_A = \sum_{i=1}^{a} \frac{T_{i \cdot \cdot \cdot}^2}{bcn} - \frac{T_{\cdot \cdot \cdot \cdot}^2}{abcn}$$

$$SS_{AB} = \sum_{i=1}^{a} \sum_{j=1}^{b} \frac{T_{ij \cdot \cdot}^2}{cn} - \sum_{i=1}^{a} \frac{T_{i \cdot \cdot \cdot}^2}{bcn} - \sum_{j=1}^{b} \frac{T_{\cdot j \cdot \cdot}^2}{acn} + \frac{T_{\cdot \cdot \cdot \cdot}^2}{abcn}$$

Define each of these sums of squares.

(*a*) SS_B
(*b*) SS_c
(*c*) SS_{AC}
(*d*) SS_{BC}
(*e*) SS_{ABC} [*Hint:* Look at the definition of $(\alpha\beta\gamma)_{ijk}$ given in Exercise 17.]
(*f*) SS_{Tot}

21. Write the sum of squares identity for the three-factor design.

22. Table 14.14 gives the ANOVA for the three-factor design. State the null hypothesis being tested by each of the given *F* ratios.

23. Find the ANOVA table for these data:

		Factor *A* 1		2		3	
	Factor *C*	1	2	1	2	1	2
Factor *B*	1	42 34 33	38 34 41	28 28 11	22 8 3	34 19 17	10 8 0
	2	47 59 43	38 41 32	52 33 29	33 15 24	22 3 17	16 10 1
	3	55 66 69	48 26 63	54 36 53	27 41 38	17 40 40	22 6 38
	4	64 60 94	58 51 67	63 55 40	41 40 64	33 34 40	29 24 20

TABLE 14.14
ANOVA table for three-fixed-factor factorial

Source of variation	Degrees of freedom (*DF*)	Sum of squares (*SS*)	Mean square (*MS*)	*F*
Main effect				
A	$a - 1$	SS_A	$SS_{A/a-1}$	MS_{A/MS_E}
B	$b - 1$	SS_B	$SS_{B/b-1}$	MS_{B/MS_E}
C	$c - 1$	SS_C	$SS_{C/c-1}$	MS_{C/MS_E}
Two-factor estimation				
AB	$(a-1)(b-1)$	SS_{AB}	$SS_{AB/(a-1)(b-1)}$	MS_{AB/MS_E}
AC	$(a-1)(c-1)$	SS_{AC}	$SS_{AC/(a-1)(c-1)}$	MS_{AC/MS_E}
BC	$(b-1)(c-1)$	SS_{BC}	$SS_{BC/(b-1)(c-1)}$	MS_{BC/MS_E}
Three-factor interaction				
ABC	$(a-1)(b-1)(c-1)$	SS_{ABC}	$SS_{ABC/(a-1)(b-1)(c-1)}$	MS_{ABC/MS_E}
Error	$abc(n-1)$	SS_E	$SS_{E/abc(n-1)}$	
Total	$abcn - 1$	SS_{Tot}		

TABLE 14.15
Reduced ANOVA for three fixed effects

Source of variation	Degrees of freedom (DF)	Sum of squares (SS)	Mean square (MS)	F
Main effect				
A	$a - 1$	SS_A	$SS_A/(a - 1)$	MS_A/MS_E
B	$b - 1$	SS_B	$SS_B/(b - 1)$	MS_B/MS_E
C	$c - 1$	SS_C	$SS_C/(c - 1)$	MS_C/MS_E
Interaction				
AB	$(a - 1)(b - 1)$	SS_{AB}	$SS_{AB}/(a - 1)(b - 1)$	MS_{AB}/MS_E
BC	$(b - 1)(c - 1)$	SS_{BC}	$SS_{BC}/(b - 1)(c - 1)$	MS_{BC}/MS_E
Error	DF_E = subtraction	SS_E = subtraction	SS_E/DF_E	
Total	$abcn - 1$	SS_{Tot}		

24. (*Restricted three-way model.*) Often there are a priori reasons to suspect that the full three-way model given in Exercise 16 is not needed. For example, suppose that it is believed that there is no three-way interaction and that there is no AC two-way interaction. In this case the model of interest is

$$Y_{ijkl} = \mu + \alpha_i + \beta_j + \gamma_k + (\alpha\beta)_{ij} + (\beta\gamma)_{jk} + E_{ijkl}$$

The ANOVA for such a model is identical to the full three-way analysis except for the error term. In this case the error sum of squares is given by

$$SS_E = SS_{\text{Tot}} - SS_A - SS_B - SS_C - SS_{AB} - SS_{BC}$$

This sum is called a "pooled" error term since it is found by combining or "pooling" variation usually attributed to the two-factor interaction AC and the three-factor interaction ABC with pure error. The degrees of freedom for error are also pooled from these three sources. Table 14.15 gives the ANOVA for this model.

(*a*) Suppose that $a = 4$, $b = 3$, $c = 3$, and $n = 2$. How many degrees of freedom are associated with SS_A, SS_B, SS_C, SS_{AB}, SS_{BC}, SS_{Tot}, and SS_E, respectively?

(*b*) Show that the degrees of freedom for error in general can be expressed as

$$(a - 1)(c - 1) + (a - 1)(b - 1)(c - 1) + abc(n - 1)$$

Section 14.3

25. In a mixed-effects model in which the levels of factor A are random and those of factor B are fixed, what null hypotheses are of interest? The analysis of variance table for such a model is given in Table 14.16. Based on the expected mean squares, determine the appropriate F ratios for testing each null hypothesis.

Section 14.4

26. How many treatment combinations are possible in an experiment with k factors, each at two levels, where $k = 4; 5; 7$?

TABLE 14.16

ANOVA table for the two-way classification design with mixed effects: *A* random, *B* fixed

Source of variation	Degrees of freedom (*DF*)	Sum of squares (*SS*)	Mean square (*MS*)	Expected mean square	*F*
Treatment	$ab - 1$	$\sum_{i=1}^{a}\sum_{j=1}^{b}\frac{T_{ij\cdot}^2}{n} - \frac{T_{\cdots}^2}{abn}$	$SS_{\text{Tr}}/(ab-1)$	$\sigma^2 + n\sigma_{\text{Tr}}^2$	?
A	$a - 1$	$\sum_{i=1}^{a}\frac{T_{i\cdots}^2}{bn} - \frac{T_{\cdots}^2}{abn}$	$SS_A/(a-1)$	$\sigma^2 + nb\sigma_A^2$	?
B	$b - 1$	$\sum_{j=1}^{b}\frac{T_{\cdot j\cdot}^2}{an} - \frac{T_{\cdots}^2}{abn}$	$SS_B/(b-1)$	$\sigma^2 + n\sigma_{AB}^2 + \frac{na\sum_{j=1}^{b}\beta_j^2}{(b-1)}$	?
AB	$(a-1)(b-1)$	$SS_{\text{Tr}} - SS_A - SS_B$	$SS_{AB}/(a-1)(b-1)$	$\sigma^2 + \sigma_{AB}^2$	?
Error	$ab(n-1)$	$SS_{\text{Tot}} - SS_{\text{Tr}}$	$SS_E/ab(n-1)$	σ^2	
Total	$abn - 1$	$\sum_{i=1}^{a}\sum_{j=1}^{b}\sum_{k=1}^{n}Y_{ijk}^2 - \frac{T_{\cdots}^2}{abn}$			

27. Assume that for a particular two-way model $\beta_0 = 12$, $\beta_1 = 3$, and $\beta_2 = 2$. Assume also that $\beta_{12} = 0$, which we have claimed can be interpreted as no interaction. Note that the theoretical mean for the treatment combination (1) is

$$\mu_{Y|x_1,x_2} = \beta_0 + \beta_1(-1) + \beta_2(-1)$$
$$= 12 - 3 - 2 = 7$$

(*a*) Find the theoretical means for combinations *a*, *b*, and *ab*, respectively.
(*b*) Label the remaining means on the graph of Fig. 14.3.
(*c*) Find the slopes of the line segments shown in Fig. 14.3. Argue that these line segments are parallel as expected when there is no interaction.

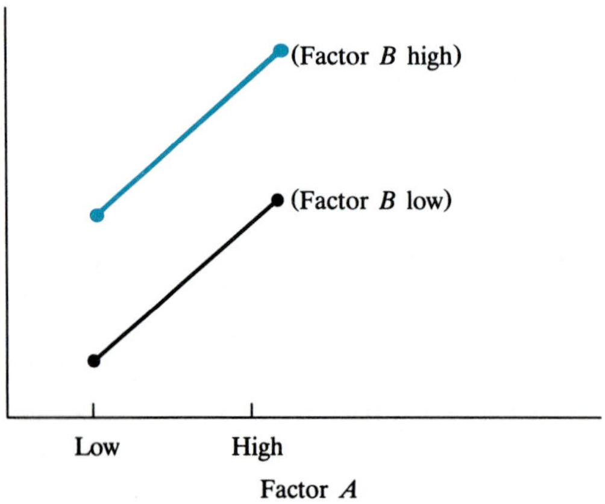

FIGURE 14.3
A graph of levels of factor *B* across levels of factor *A* when it is assumed that $\beta_{12} = 0$.

28. Assume that for a particular two-way model, $\beta_0 = 12$, $\beta_1 = 3$, and $\beta_2 = 2$. Assume also that $\beta_{12} = 1$, which we have claimed can be interpreted as the existence of interaction between factors A and B.

(*a*) Find the theoretical means for the treatment combinations (1), a, b, and ab.

(*b*) Construct a graph of those means similar to that of Fig. 14.3.

(*c*) Find the slopes of the line segments in your graph. Argue that these line segments are not parallel as expected when interaction exists.

29. Assume that for a particular two-way model, $\beta_0 = 12$, $\beta_1 = 3$, and $\beta_2 = 2$. Give a numerical value of β_{12} that causes the graphs of the means for the levels of factor B to cross.

30. In Sec. 12.3 it was shown that

$$\text{SSR} = \beta_0 \sum_{i=1}^{n} Y_i + \beta_1 \sum_{i=1}^{n} X_{2i}Y_i$$

$$+ \beta_2 \sum_{i=1}^{n} X_{2i}Y_i + \cdots + \beta_k \sum_{i=1}^{n} X_{ki}Y_i - C$$

where C is the usual correction factor. Verify that SSR can be expressed in matrix notation as

$$\text{SSR} = \hat{\boldsymbol{\beta}}'X'\mathbf{Y} - C$$

31. Complete the analysis of the data of Example 14.4.1 by testing for main effects.

32. The full 2^3 model is given by

$$\mu_{Y|x_1,x_2,x_3} = \beta_0 + \beta_1 x_1 + \beta_2 x_2 + \beta_3 x_3 + \beta_4 x_1 x_2 + \beta_5 x_1 x_3 + \beta_6 x_2 x_3 + \beta_7 x_1 x_2 x_3$$

(*a*) Complete the table of coefficients begun in Table 14.17.

(*b*) From Table 14.17 it can be seen that

$$SS_A = \frac{[(-1) + a - b - c + ab + ac - bc + abc]^2}{2^3 n}$$

Find SS_B, SS_C, SS_{AB}, SS_{AC}, SS_{BC}, and SS_{ABC}.

TABLE 14.17

Coefficients of effects for 2^3 factorial

	Effects						
Treatment combination	x_1 A	x_2 B	x_3 C	x_1x_2 AB	x_1x_3 AC	x_2x_3 BC	$x_1x_2x_3$ ABC
(1)	−	−	−	+	+	+	−
a	+	−	−	−	−	+	+
b	−	+	−				
c	−	−	+				
ab	+	+	−				
ac	+	−	+				
bc	−	+	+				
abc	+	+	+				

33. Consider the four-factor model

$$\mu_{Y|x_1, x_2, x_3, x_4} = \beta_0 + \beta_1 x_1 + \beta_2 x_2 + \beta_3 x_3 + \beta_4 x_4$$

in which it is assumed a priori that there are no interaction terms.
(*a*) How many treatment combinations are possible?
(*b*) Construct a table of coefficients of effects similar to Table 14.17 for this model.
(*c*) Find SS_A, SS_B, SS_C, and SS_D.

34. A four-factor factorial experiment was conducted. The experiment was replicated (Rep.) two times, yielding the following responses:

Treatment combination	**Rep. 1**	**Rep. 2**
(1)	29.8	30.3
a	34.6	35.2
b	30.7	29.5
ab	35.2	34.3
c	28.7	27.7
ac	31.5	32.0
bc	26.7	27.1
abc	32.0	31.5
d	30.2	29.6
ad	35.1	34.7
bd	31.1	30.8
abd	36.0	34.9
cd	27.0	27.8
acd	32.2	33.0
bcd	26.5	27.3
abcd	27.1	26.9

(*a*) Using the Yates procedure, derive the complete ANOVA table for testing all main effects and all interactions.
(*b*) Form the complete ANOVA table if three-factor and higher-order interactions are pooled into error.
(*c*) Form the ANOVA table if it is known a priori that interactions *ABCD*, *ADB*, *BCD*, and *CD* are negligible.

35. Prove that in a 2^k factorial design with $n = 1$ the maximum number of interaction terms that can be tested is $2^k - k - 2$. *Hint:* Note that a model with no interaction terms has $2^k - k - 1$ degrees of freedom for error. Interaction degrees of freedom are subtracted from this and we must guarantee that the resulting term is at least 1.

36. In a 2^4 factorial experiment with $n = 1$, at most how many interaction terms can be tested?

Section 14.5

37. In each of the following situations, list the effects that will be confounded with blocks:
(*a*) A 2^3 experiment with defining contrast *AB*
(*b*) A 2^3 experiment with defining contrasts *AB* and *ABC*
(*c*) A 2^4 experiment with defining contrasts *ABCD*, *ABC*, and *BCD*

38. In each case of Exercise 37, list any main effects that cannot be tested due to the choice of the defining contrasts selected.

39. Consider a 2^4 factorial experiment with factors A, B, C, and D.

(a) Using defining contrasts AB and CD, show that if the experiment is run in four blocks the following treatment combinations are assigned to each block:

Block 1	Block 2	Block 3	Block 4
(1)	a	c	ac
ab	b	abc	bc
cd	acd	d	ad
$abcd$	bcd	abd	bd

(b) What is the generalized interaction for this experiment? Which block is called the *principal block*?

(c) Assuming all unconfounded interactions are negligible, complete the skeleton ANOVA table (source of variation and degrees of freedom) for testing main effects.

40. A 2^4 factorial experiment is to be run. Since laboratory facilities are such that only eight treatment combinations (experimental runs) can be run in a single laboratory, the scientist decided to use two laboratories, with the separate laboratories treated as blocks. Using the $ABCD$ interaction as the defining contrast, we obtained the following blocks and treatment combination responses:

Block 1	Block 2
$(1) = 229.5$	$a = 39.2$
$ab = 38.2$	$b = 32.1$
$ac = 33.8$	$c = 27.3$
$bc = 25.3$	$d = 31.4$
$ad = 37.6$	$abc = 35.7$
$bd = 31.0$	$abd = 38.9$
$cd = 24.9$	$acd = 35.9$
$abcd = 34.1$	$bcd = 27.1$

(a) Complete the ANOVA table by directly computing sums of squares and by pooling three-factor interactions into error.

(b) Repeat part (a) except complete the ANOVA table by using the Yates procedure.

(c) Estimate the main effects A, B, C, and D. Which effects, if any, are significant at the 5% level of significance?

Section 14.6

41. Consider Example 14.6.1. Give the estimator for the B effect if all eight treatment combinations were run. Give the estimator for this effect if only the treatment combinations in block 2 are run. With what effect is the B effect aliased?

42. A 1/2 fraction of a 2^5 factorial experiment was run yielding the following responses for respective treatment combinations in the principal block:

(1) = 31.3	cd = 7.1
ab = 10.8	ce = 6.3
ac = 5.4	de = 5.6
ad = 59.4	$abcd$ = 5.6
ae = 18.2	$abce$ = 5.3
bc = 6.0	$abde$ = 4.2
bd = 23.2	$acde$ = 7.8
be = 15.7	$bcde$ = 7.2

(*a*) Write the complete alias structure for main effects and two-factor interactions.
(*b*) Assuming unconfounded interactions are negligible, test all main effects at the 5% level of significance.
(*c*) Would it be possible to test two-factor interactions in this example? Why?

43. An experiment with six factors *A*, *B*, *C*, *D*, *E*, and *F* was studied on a certain response variable. Due to the expense and size of the experiment, it was decided to use only two levels of each factor and to run a 1/4 fractional factorial.
(*a*) Construct the four blocks by using defining contrasts *ABDE* and *ABCF*.
(*b*) Give an abbreviated ANOVA table to test all main effects and two-factor interactions.
(*c*) Give the alias structure for all main effects.

44. Consider a 2^4 factorial experiment with factors *A*, *B*, *C*, and *D*.
(*a*) How many treatment combinations are necessary for this experiment if a full factorial is run with only one replicate?
(*b*) Write out an abbreviated ANOVA table (only source of variation and degrees of freedom) for part (*a*). What must be assumed in order to have a test for the experiment in part (*a*)?
(*c*) Using defining contrasts *ABD* and *ABC*, construct four incomplete blocks of treatment combinations for this 2^4 factorial experiment.
(*d*) Give the complete alias structure for part (*c*).
(*e*) Write an abbreviated ANOVA table for testing main effects in part (*d*), assuming all unconfounded interactions are negligible.

REVIEW EXERCISES

45. Consider a 2^5 factorial experiment. Using defining contrasts *ABCD* and *BCDE*, construct the four blocks by showing appropriate treatment combinations in each block.
(*a*) Which interaction sums of squares are confounded with blocks?
(*b*) Assuming all unconfounded three-, four-, and five-factor interactions are pooled into error, construct the skeleton ANOVA table for testing main effects and appropriate two-factor interaction effects.
(*c*) Without pooling unconfounded interactions into error, can any test of hypothesis be completed? Why?

46. In a regression model the number of degrees of freedom for error is $n - k - 1$, where k is the number of regressors in the model and n is the number of responses.

(*a*) Show that in order to conduct F tests on the model parameters, the number of responses must exceed the number of parameters in the model.

(*b*) In a 2^2 factorial experiment with $n = 1$, is it possible to test for interaction? Explain.

(*c*) In a 2^3 factorial experiment with $n = 1$, is it possible to test for all interactions? Explain.

(*d*) In a 2^3 factorial experiment with $n = 1$, is it possible to test for any interaction?

(*e*) In a 2^3 factorial experiment with $n = 1$, how many interactions can be tested?

47. The full 2^2 factorial model is given by

$$\mu_{Y|x_1, x_2} = \beta_0 + \beta_1 x_1 + \beta_2 x_2 + \beta_{12} x_1 x_2$$

In testing H_0: $\beta_{12} = 0$, we use the methods of Sec. 12.5. In particular, we test

$$H_0\text{: reduced model is appropriate}$$

$$H_1\text{: full model is needed}$$

(*a*) Write the reduced model when testing for interaction.

(*b*) Write the model specification matrix for the reduced model.

(*c*) Find $X'X$ for the reduced model.

(*d*) Let Y be partitioned so that the first n observations represent the responses to treatment combination (1). These are followed by responses from treatments a, b, and ab, respectively. Find a general expression for $X'Y$ for the reduced model.

(*e*) Find $\hat{\beta}$ for the reduced model.

(*f*) Find SSR for the reduced model.

(*g*) Find SSE_r and SSE_f.

(*h*) Find $\text{SSE}_r - \text{SSE}_f$. How many degrees of freedom are associated with this statistic? (See Sec. 12.5.)

(*i*) Verify that $\text{SSE}_r - \text{SSE}_f = SS_{AB}$, thus justifying the F test described in this section.

48. Consider an experiment with three factors in which it is assumed that there is no three-way interaction. Let $a = 5$, $b = 4$, and $c = 3$.

(*a*) Write the model for such an experiment.

(*b*) Find the abbreviated ANOVA table for the design.

49. Consider a 2^5 factorial experiment with factors A, B, C, D, and E.

(*a*) If $ABCDE$ is used as the defining contrast, construct the principal block for a 1/2 replicate.

(*b*) Give an abbreviated ANOVA table (source of variation and degrees of freedom) for testing main effects after pooling all unconfounded interactions into error.

50. Consider Example 14.6.2. Find the aliases for each of these effects:

(*a*) B; C; D.

(*b*) In this experiment, is it possible to test all main effects? Explain.

(*c*) Give an example of two defining contrasts that would guarantee that no main effects are aliased.

COMPUTING SUPPLEMENT

XVI Two-Way Classification with Multiple Comparisons

When sample sizes are equal, the two-way classification design can be analyzed by using PROC ANOVA. The program given runs a two-way analysis with multiple comparisons first. This analysis is used if the null hypothesis of no interaction is not rejected. It also runs a one-way analysis with multiple comparisons. This portion of the output is used only if the null hypothesis of no interaction is rejected. The data used is that of Examples 14.1.1 and 14.1.2.

Statement	*Function*
DATA ADHESIVE;	names data set
INPUT TEMP HUMIDITY CELL F1-F3;	names variables; CELL denotes the treatment
ARRAY OUT F1-F3;	combination
DO OVER OUT;	allows all data in single
F = OUT;	cell to be read without
OUTPUT;	having to repeat cell and
END;	factor level codes
LINES;	signals that data follows
1 1 1 40 36 43	
2 1 2 39 36 33	
3 1 3 32 34 29	
4 1 4 33 27 25	data
1 2 5 36 34 29	
2 2 6 32 26 25	
3 2 7 26 23 24	
4 2 8 20 22 18	
;	signals end of data
PROC ANOVA;	calls for an analysis of variance procedure
CLASSES TEMP HUMIDITY;	indicates that the dependen variable is grouped according to values of the two variables TEMP and HUMIDITY
MODEL F = TEMP HUMIDITY TEMP * HUMIDITY;	identifies the response variable as F; indicates a two-way classification model with an interaction term TEMP * HUMIDITY
MEANS TEMP/DUNCAN ALPHA = .01;	asks for multiple comparisons among temperature levels
MEANS HUMIDITY/DUNCAN ALPHA = .01;	asks for multiple comparisons among humidity levels

Statement	*Function*
TITLE1 TWO WAY CLASSIFICATION;	titles output
TITLE2 WITH MULTIPLE COMPARISONS;	
PROC ANOVA;	asks for another analysis of variance to be conducted in case interaction is detected
CLASSES CELL;	in this analysis data are grouped according to the value of the variable CELL
MODEL F = CELL;	identifies the response variable as F and the model as a one-way classification
MEANS CELL/DUNCAN ALPHA = .01;	asks for multiple comparisons among all means
TITLE1 ONE WAY ANALYSIS;	titles output
TITLE2 IF INTERACTION IS DETECTED;	

TWO WAY CLASSIFICATION
WITH MULTIPLE COMPARISONS

ANALYSIS OF VARIANCE PROCEDURE

CLASS LEVEL INFORMATION

CLASS	LEVELS	VALUES
TEMP	4	1 2 3 4
HUMIDITY	2	1 2

NUMBER OF OBSERVATIONS IN DATA SET = 24

TWO WAY CLASSIFICATION
WITH MULTIPLE COMPARISONS

ANALYSIS OF VARIANCE PROCEDURE

DEPENDENT VARIABLE: F

SOURCE	DF	SUM OF SQUARES	MEAN SQUARE	F VALUE
MODEL ①	7	844.50000000	120.64285714	12.27
ERROR	16	157.33333333	9.83333333	PR > F
CORRECTED TOTAL	23	1001.83333333		0.0001

R-SQUARE	C.V.	ROOT MSE	F MEAN
0.842955	10.4238	3.13581462	30.08333333

SOURCE	DF	ANOVA SS	F VALUE	PR > F
TEMP	3	488.83333333	16.57 ⑤	0.0001 ⑥
HUMIDITY ②	1	352.66666667	35.86 ⑧	0.0001 ⑨
TEMP * HUMIDITY	3	3.00000000	0.10 ③	0.9579 ④

TWO WAY CLASSIFICATION
WITH MULTIPLE COMPARISONS

ANALYSIS OF VARIANCE PROCEDURE

DUNCAN'S MULTIPLE RANGE TEST FOR VARIABLE: F
NOTE: THIS TEST CONTROLS THE TYPE I COMPARISONWISE ERROR RATE, NOT THE EXPERIMENTWISE ERROR RATE.

ALPHA = 0.01 DF = 16 MSE = 9.83333

MEANS WITH THE SAME LETTER ARE NOT SIGNIFICANTLY DIFFERENT.

DUNCAN	GROUPING	MEAN	N	TEMP
	A	36.333	6	1
	A			
B	A	31.833	6	2
⑦ B				
B	C	28.000	6	3
	C			
	C	24.167	6	4

TWO WAY CLASSIFICATION
WITH MULTIPLE COMPARISONS

ANALYSIS OF VARIANCE PROCEDURE

DUNCAN'S MULTIPLE RANGE TEST FOR VARIABLE: F
NOTE: THIS TEST CONTROLS THE TYPE I COMPARISONWISE ERROR RATE, NOT THE EXPERIMENTWISE ERROR RATE.

ALPHA = 0.01 DF = 16 MSE = 9.83333

MEANS WITH THE SAME LETTER ARE NOT SIGNIFICANTLY DIFFERENT.

DUNCAN	GROUPING	MEAN	N	HUMIDITY
	A	33.917	12	1
⑩				
	B	26.250	12	2

ONE WAY ANALYSIS
IF INTERACTION IS DETECTED

ANALYSIS OF VARIANCE PROCEDURE

CLASS LEVEL INFORMATION

CLASS	LEVELS	VALUES
CELL	8	1 2 3 4 5 6 7 8

NUMBER OF OBSERVATIONS IN DATA SET = 24

ONE WAY ANALYSIS
IF INTERACTION IS DETECTED

ANALYSIS OF VARIANCE PROCEDURE

DEPENDENT VARIABLE: F

⑪ SOURCE	DF	SUM OF SQUARES	MEAN SQUARE	F VALUE
MODEL	7	844.50000000	120.64285714	12.27
ERROR	16	157.33333333	9.83333333	PR > F
CORRECTED TOTAL	23	1001.83333333		0.0001

R-SQUARE	C.V.	ROOT MSE	F MEAN
0.842955	10.4238	3.13581462	30.08333333

SOURCE	DF	ANOVA SS	F VALUE	PR > F
CELL	7	844.50000000	12.27	0.0001

ONE WAY ANALYSIS
IF INTERACTION IS DETECTED

ANALYSIS OF VARIANCE PROCEDURE

DUNCAN'S MULTIPLE RANGE TEST FOR VARIABLE: F
NOTE: THIS TEST CONTROLS THE TYPE I COMPARISONWISE ERROR RATE, NOT THE EXPERIMENTWISE ERROR RATE.

ALPHA = 0.01 DF = 16 MSE = 9.83333

MEANS WITH THE SAME LETTER ARE NOT SIGNIFICANTLY DIFFERENT.

DUNCAN	GROUPING				MEAN	N	CELL
			A		39.667	3	1
			A				
	B		A		36.000	3	2
	B		A				
	B		A	C	33.000	3	5
	B		A	C			
	B	D	A	C	31.667	3	3
	B	D		C			
	B	D		C	28.333	3	4
		D		C			
		D	E	C	27.667	3	6
		D	E				
		D	E		24.333	3	7
			E				
			E		20.000	3	8

The explanation of this output follows. Note that in the two-way ANOVA the source referred to as "treatment" is called MODEL ① by SAS. The breakdown of the model, or treatment, sum of squares into factors A (temperature), B (humidity), and AB (interaction) is given by ② on the printout. To interpret the data, we look first at the F ratio used to test for interaction ③ and its P value ④. Since this P value is large (.9579), we do not reject the null hypothesis of no interaction. We continue the analysis by looking at the F ratio used to test the null hypothesis of no difference among temperature levels. This ratio is given by ⑤ and its P value is found at ⑥. Since this P value is small (.0001), we reject H_0 and conclude that there are differences among the temperature levels. These differences are pinpointed by the Duncan procedure shown in ⑦. To test for differences between humidity levels, we look at the F ratio given by ⑧ and its P value ⑨. Since this P value is also small (.0001), we reject the null hypothesis and conclude that there is a difference between these two humidity levels. The Duncan summary is shown in ⑩. If the null hypothesis of no interaction had been rejected, the one-way analysis of treatment combinations ⑪ would have been needed. In this case it is not pertinent.

XVII 2^k Factorial Experiments

Since 2^k factorial experiments can be expressed by using multiple regression models, they can be handled on SAS by using PROC GLM. We illustrate the technique by using the data of Example 14.4.3.

Statement	*Function*
DATA YATES;	names the data set
INPUT Y X1 X2 X3;	names the basic regressors
X4 = X1‘ * ’X2;	
X5 = X1‘ * ’X3;	creates coefficients for interaction terms
X6 = X2‘ * ’X3;	
X7 = X1‘ * ’X2‘ * ’X3;	
LINES;	signals that the data follows
19.9 −1 −1 −1	
18.6 −1 −1 −1	
17.4 −1 1 −1	data with low and high levels of each factor coded as −1 and 1 respectively
⋮	
26.2 1 1 1	
;	signals end of data
PROC GLM;	calls for the general linear models procedure
MODEL Y = X1 X2 X3 X4 X5 X6 X7;	identifies the model as a multiple regression model with seven regressors.

GENERAL LINEAR MODELS PROCEDURE

DEPENDENT VARIABLE: Y

SOURCE	DF	SUM OF SQUARES	MEAN SQUARE	F VALUE	Pr > F
MODEL	7	257.74000	36.82000	5.25	0.0164
ERROR	8	56.14000	7.01750 ②		
CORRECTED TOTAL	15	313.88000			

	R-SQUARE	C.V.	ROOT MSE	Y MEAN
	0.821142	13.24528	2.6491	20.000

SOURCE	DF	TYPE I SS ①	MEAN SQUARE	F VALUE ③	Pr > F ④
X1	1	166.41000	166.41000	23.71	0.0012
X2	1	1.69000	1.69000	0.24	0.6368
X3	1	0.72250	0.72250	0.10	0.7565
X4	1	0.49000	0.49000	0.07	0.7983
X5	1	41.60250	41.60250	5.93	0.0409
X6	1	34.22250	34.22250	4.88	0.0582
X7	1	12.60250	12.60250	1.80	0.2170

The sums of squares for each effect are given in column ①. The error mean square is at ②. The observed values of the one degree of freedom *F* ratios are in column ③ with their *P* values listed in column ④.

CHAPTER 15

CATEGORICAL DATA

In this chapter we are concerned with the analysis of data characterized by the fact that each observation in the data set can be classed as falling into exactly one of several mutually exclusive "cells" or categories. Interest centers on the number of observations falling into each category. The statistical problem is to determine whether the observed category frequencies tend to refute a stated hypothesis. We are concerned with three problems in particular. These are:

1. Testing to see whether a set of observations was drawn from a specified probability distribution
2. Testing for independence between two variables used for classification purposes
3. Comparing proportions

The statistical procedures used in much of the work to come are based on the *multinomial distribution*. We begin by describing this distribution.

15.1 MULTINOMIAL DISTRIBUTION

To develop the definition of a multinomial random variable, we need to consider first the idea of a multinomial trial.

> **Definition 15.1.1 (Multinomial trial).** A multinomial trial with parameters $p_1, p_2, \ldots, p_k$ is a trial that can result in exactly one of k possible outcomes. The probability that outcome i will occur on a given trial is p_i for $i = 1, 2, 3, \ldots, k$.

Note that since $p_1, p_2, \ldots, p_k$ are probabilities, each of them lies between 0 and 1 inclusive. Furthermore, since each trial results in exactly one of the k possible outcomes, these probabilities sum to 1.

Example 15.1.1. It is noted that 1% of the items coming off a production line are defective and nonsalvageable, 5% are defective but salvageable, and the rest are nondefective. One item is selected at random and classified. Since exactly one of three possible outcomes can result, this experiment can be viewed as constituting a single multinomial trial with parameters $p_1 = .01$, $p_2 = .05$, and $p_3 = .94$.

The multinomial random variable arises quite naturally whenever we observe a series of independent and identical multinomial trials. This multivariate random variable is defined as follows:

Definition 15.1.2 (Multinomial random variable). Let an experiment consist of n independent and identical multinomial trials with parameters $p_1, p_2, \ldots, p_k$. Let X_i denote the number of trials that result in outcome i for $i = 1, 2, \ldots, k$. The k-tuple $(X_1, X_2, \ldots, X_k)$ is called a *multinomial random variable* with parameters $n, p_1, p_2, \ldots, p_k$.

Example 15.1.2. Assume that a random sample of 100 items is selected from the production line described in Example 15.1.1. This experiment can be viewed as consisting of $n = 100$ independent multinomial trials, each with parameters $p_1 = .01$, $p_2 = .05$, and $p_3 = .94$. Let X_1 denote the number of defective and nonsalvageable items selected, X_2 the number of defective but salvageable items selected, and X_3 the number of nondefective items selected. The triple or 3-tuple (X_1, X_2, X_3) is a multinomial random variable with parameters 100, .01, .05, and .94. For example, if we observe 2 defective items that cannot be salvaged, 6 that are defective and can be salvaged, and 92 that are nondefective, then the multinomial random variable (X_1, X_2, X_3) assumes the observed value of $(2, 6, 92)$.

Although it is not hard to derive the density for a multinomial random variable, we shall not need to do so here. However, we do need to determine the expected value of each of the random variables $X_1, X_2, \ldots, X_k$. This is easy to do. Consider a single multinomial trial and any fixed outcome i. This trial either does or does not result in outcome i. If outcome i does occur, we consider the trial a success; otherwise it is a failure. The probability of success is p_i, the probability that outcome i will result on a given trial; the probability of failure is $1 - p_i$. Consider now a series of n independent and identical multinomial trials. Let X_i denote the number of trials that result in outcome i. Note that X_i also denotes the number of successes in n independent trials, each with probability of success p_i. Therefore X_i is a binomial random variable with parameters n and p_i. From our discussion in Chap. 3 we know that for each i, $E[X_i] = np_i$. This result plays an important role in analyzing categorical data.

Example 15.1.3. In a random sample of 100 items selected from the production line described in Example 15.1.1 the expected number of items falling into each category is

Expected number of defective and nonsalvageable items	$= E[X_1] = np_1 = 100(.01) = 1$
Expected number of defective but salvageable items	$= E[X_2] = np_2 = 100(.05) = 5$
Expected number of nondefective items	$= E[X_3] = np_3 = 100(.94) = 94$

When the sampling is complete, we observe that $x_1 = 2$, $x_2 = 6$, and $x_3 = 92$. These values do not coincide exactly with the expected values, but they do not seem to differ drastically from them. For this reason, the values do not lead us to suspect the accuracy of the stated category probabilities of .01, .05, and .94, respectively.

The previous example illustrates the basic idea of categorical data analysis. In analyzing count data, we compare the observed category frequencies to those expected under a stated null hypothesis. If these agree fairly well, we do not reject H_0; if there are substantial disagreements, we do reject H_0. In the sections that follow we develop the statistics needed to determine when the differences are extreme enough to warrant rejection of the stated hypothesis.

15.2 CHI-SQUARED GOODNESS OF FIT TESTS

The purpose of the chi-squared goodness of fit test is to test the null hypothesis that a given set of observations is drawn from, or "fits," a specified probability distribution. We consider the situation in which the hypothesized distribution is completely specified before the sampling is done. The procedure for handling this case is based on the next theorem, which is offered without proof.

Theorem 15.2.1. Let $(X_1, X_2, \ldots, X_k)$ be a multinomial random variable with parameters $n, p_1, p_2, \ldots, p_k$. For large n the random variable

$$\sum_{i=1}^{k} \frac{(X_i - np_i)^2}{np_i}$$

follows an approximate chi-squared distribution with $k - 1$ degrees of freedom.

We make two notational changes to make this random variable easier to remember. Since in the multinomial context X_i is the actual or "observed" number of trials resulting in outcome i or falling into category i, we denote X_i by O_i. Recall that np_i is the theoretical expected number of trials resulting in outcome i, and so we let $np_i = E[X_i] = E_i$. Thus Theorem 15.2.1 states that

$$\sum_{i=1}^{k} \frac{(O_i - E_i)^2}{E_i} = \sum_{i=1}^{k} \frac{[(\text{observed frequency}) - (\text{expected frequency})]^2}{\text{expected frequency}}$$

is, for large n, approximately chi-squared with $k - 1$ degrees of freedom, where k is the number of mutually exclusive categories involved. This naturally brings up the question, "How large is large?" There are various opinions as to the answer to this question. However, it is usually felt that n should be large enough that *no expected frequency is less than 1 and no more than 20% of the expected frequencies are less than 5*. If this condition is not met, either categories should be combined or redefined or the sample size should be increased so that the expected frequencies will be of adequate size.

This random variable serves quite logically as a test statistic for testing a null hypothesis that a given set of observations is drawn from a specified probability distribution. If H_0 is true, the value of p_i will be known for each i, and hence E_i can be computed easily. In effect, the above statistic compares the observed number of observations per category with the number expected under H_0. If these figures agree fairly well (there is a good fit), then the term $(O_i - E_i)^2$ will be small for each i, $\sum_{i=1}^{k}[(O_i - E_i)^2]/E_i$ will be small, and H_0 should not be rejected. If the observed and expected frequencies differ greatly, then $(O_i - E_i)^2$ will be large for some i, $\sum_{i=1}^{k}[(O_i - E_i)^2]/E_i$ will be large, and H_0 should be rejected. The use of this test statistic is illustrated in the next example.

Example 15.2.1. Computer systems crash for many reasons, among them software failure, hardware failure, operator error, and system overloading. It is thought that 10% of the crashes are due to software failure, 5% to hardware failure, 25% to operator error, 40% to system overloading, and the rest to other causes. Over an extended study period 150 crashes are observed and each is classified according to its probable cause. It is found that 13 are due to software failure, 10 to hardware failure, 42 to operator error, 65 to system overloading, and the rest to other causes. Do these data lead us to suspect the accuracy of the stated percentages? To answer this question, we test

$$H_0\colon p_1 = .10,\ p_2 = .05,\ p_3 = .25,\ p_4 = .40,\ p_5 = .20$$

$$H_1\colon p_i \text{ is not as stated for some } i = 1,2,3,4,5$$

If H_0 is true, then

$$E[X_1] = E_1 = np_1 = 150(.10) = 15$$
$$E[X_2] = E_2 = np_2 = 150(.05) = 7.5$$
$$E[X_3] = E_3 = np_3 = 150(.25) = 37.5$$
$$E[X_4] = E_4 = np_4 = 150(.40) = 60$$
$$E[X_5] = E_5 = np_5 = 150(.20) = 30$$

TABLE 15.1

Category	Software failure 1	Hardware failure 2	Operator error 3	System overloading 4	Other 5
Observed frequency, O_i	13	10	42	65	20
Expected frequency, E_i	15	7.5	37.5	60	30

The situation is summarized in Table 15.1. Note that the expected and observed frequencies do not agree exactly. The question to be answered is, "Do they differ enough to cause us to reject H_0?" Since $E_i > 5$ in each case, the test statistic

$$\sum_{i=1}^{5} \frac{(O_i - E_i)^2}{E_i}$$

follows an approximate $X^2_{k-1} = X^2_4$ distribution. The observed value of this statistic is

$$\frac{(13-15)^2}{15} + \frac{(10-7.5)^2}{7.5} + \frac{(42-37.5)^2}{37.5} + \frac{(65-60)^2}{60} + \frac{(20-30)^2}{30} = 5.39$$

Is this value large enough to cause us to reject H_0? From Table IV of App. A we see that the probability of observing a value of 5.39 or larger is .25. That is, if we reject H_0, the P value of the test is .25. Since this probability is large, we do not reject H_0. The data gathered are not sufficient to allow us to conclude that the stated percentages are incorrect.

15.3 TESTING FOR INDEPENDENCE

In this section we discuss a problem involving categorical data that is somewhat different from that considered in the last section. However, the idea behind the test procedure used is identical to that studied earlier. Namely, the test statistic compares observed category frequencies with those expected under the assumption that the stated null hypothesis is true, with rejection coming if these differ too much to have occurred by chance.

Here we consider experiments in which two random variables are being studied. The purpose of the study is to test these random variables for independence. For example, a highway engineer may be interested in seeing whether the extent of an injury is independent of the type of restraint being used by an accident victim; a manufacturer may want to see if the quality of an item produced is independent of the day of the week on which it was made; a cancer researcher may want to see whether the development of lung cancer is independent of exposure to airborne asbestos.

We illustrate the test for independence in the context of what are called 2×2 *contingency* tables. These tables arise in experiments in which each of the two random variables being considered is studied at two levels. This naturally defines 2×2 or 4 mutually exclusive "cells" or categories. The data analysis is based on an examination of the number of observations falling into each cell.

Example 15.3.1. A cancer researcher performs what is called a prospective study by selecting a large group of individuals at random and following their progress for a long period of time. At the end of the study period each individual is classified according to whether or not lung cancer is present and according to whether or not the individual has been exposed to an identifiable source of airborne asbestos. Let C denote the presence of lung cancer and let A denote the fact that the individual has been exposed to airborne asbestos.

These four mutually exclusive categories result:

$C \cap A$: has cancer and exposed to asbestos

$C \cap A'$: has cancer but not exposed to asbestos

$C' \cap A$: no cancer but exposed to asbestos

$C' \cap A'$: no cancer and not exposed to asbestos

Each individual in the study falls into exactly one of these cells.

Since we are concerned with the number of observations falling into each cell, we need a notational convention for the cell frequencies. We also need a notational convention to indicate the number of observations falling into each level of each of the two classification variables. We use the following:

n_{11} = number of observations falling into cell in row 1 and column 1

n_{12} = number of observations falling into cell in row 1 and column 2

n_{21} = number of observations falling into cell in row 2 and column 1

n_{22} = number of observations falling into cell in row 2 and column 2

$n_{1\cdot} = n_{11} + n_{12}$ = number of observations in row 1

$n_{2\cdot} = n_{21} + n_{22}$ = number of observations in row 2

$n_{\cdot 1} = n_{11} + n_{21}$ = number of observations in column 1

$n_{\cdot 2} = n_{12} + n_{22}$ = number of observations in column 2

n = total number of observations

This notational convention is illustrated in Example 15.3.2.

Example 15.3.2. When the study of Example 15.3.1 is completed it is found that 50 of the 5000 persons involved had developed lung cancer. Of these, 10 had been exposed to an identifiable source of airborne asbestos. A total of 500 persons in

TABLE 15.2

	A	A'	
C	$10 = n_{11}$	$40 = n_{12}$	$50 = n_{1\cdot}$
C'	$490 = n_{21}$	$4460 = n_{22}$	$4950 = n_{2\cdot}$
	$500 = n_{\cdot 1}$	$4500 = n_{\cdot 2}$	$5000 = n$

the study had been exposed to an identifiable airborne asbestos source. These data are summarized in Table 15.2. Note that $n_{\cdot 1}$ and $n_{\cdot 2}$ are column totals that appear along the margins of the 2×2 table. They are called *marginal column totals*. Similarly, $n_{1\cdot}$ and $n_{2\cdot}$ are called *marginal row totals*.

The general null hypothesis to be tested via a contingency table is that there is "no association" between the two classification variables. The alternative is that there is an association. The tables studied in this section are characterized by the fact that *only* the overall sample size n is fixed by the researcher. Prior to data collection all other entries, including the row and column marginal totals, are *free to vary*. In this sort of study the null hypothesis of "no association" is equivalent to a null hypothesis of independence.

To develop the general test, let A and B denote the classification variables. We want to test

H_0: A and B are independent

H_1: A and B are not independent

If H_0 is true, then knowledge of the classification level of an object relative to characteristic A has no bearing on its level relative to characteristic B. To express this idea mathematically, we use the table of probabilities given in Table 15.3. Note that p_{11} denotes the probability that a randomly selected object has characteristics A *and* B, $p_{1\cdot}$ denotes the probability that it has characteristic A, and $p_{\cdot 1}$ denotes the probability that it has characteristic B. Recall that A and B are independent if and only if

$$P[A \cap B] = P[A] \cdot P[B]$$

TABLE 15.3

	B	B'	
A	p_{11}	p_{12}	$p_{1\cdot}$
A'	p_{21}	p_{22}	$p_{2\cdot}$
	$p_{\cdot 1}$	$p_{\cdot 2}$	1

Thus the null hypothesis that A and B are independent can be expressed as

$$H_0: p_{11} = p_{1\cdot}p_{\cdot 1}$$

This implies that $p_{ij} = p_{i\cdot}p_{\cdot j}$ for $i = 1,2$ and $j = 1,2$. That is, A and B are independent if and only if the cell probability for any cell can be found by multiplying the corresponding row and column probabilities.

Since in a 2×2 table each observation falls into exactly one of four mutually exclusive categories, a random sample of size n can be viewed as constituting a series of n independent multinomial trials, each with parameters p_{11}, p_{12}, p_{21}, and p_{22}. Hence the set $(n_{11}, n_{12}, n_{21}, n_{22})$ of observed cell frequencies is a multinomial random variable with parameters $n, p_{11}, p_{12}, p_{21}$, and p_{22}. Thus the expected cell frequencies are given by

$$E_{ij} = np_{ij}$$

where p_{ij} is the probability of an observation falling into the (ij)th cell and n is the sample size. These probabilities are not known and must be estimated from the data under the assumption that the null hypothesis is true. How can this be done? Quite simply! Note, for instance, that if H_0 is true and characteristics A and B are independent, then

$$p_{11} = p_{1\cdot}p_{\cdot 1}$$

Since $p_{1\cdot}$ is the probability of an observation falling into row 1, it is logical to estimate $p_{1\cdot}$ by

$$\hat{p}_{1\cdot} = \frac{\text{number of elements in row 1}}{\text{sample size}} = \frac{n_{1\cdot}}{n}$$

Similarly, since $p_{\cdot 1}$ is the probability of an observation falling into column 1, we estimate $p_{\cdot 1}$ by

$$\hat{p}_{\cdot 1} = \frac{\text{number of elements in column 1}}{n} = \frac{n_{\cdot 1}}{n}$$

Thus

$$\hat{p}_{11} = \hat{p}_{1\cdot}\hat{p}_{\cdot 1} = \frac{n_{1\cdot}}{n}\frac{n_{\cdot 1}}{n}$$

This in turn implies that

$$\begin{aligned}\hat{E}_{11} = \hat{p}_{11}n &= \frac{n_{1\cdot}}{n}\frac{n_{\cdot 1}}{n}n \\ &= \frac{n_{1\cdot}n_{\cdot 1}}{n} = \frac{(\text{marginal row total})(\text{marginal column total})}{\text{sample size}}\end{aligned}$$

A similar argument holds for other cell expectations. Thus we conclude that for each i and j

$$\hat{E}_{ij} = \frac{n_{i\cdot}n_{\cdot j}}{n} = \frac{(\text{marginal row total})(\text{marginal column total})}{\text{sample size}}$$

Recall that for large samples

$$\sum_{i=1}^{2}\sum_{j=1}^{2}\frac{\left(O_{ij}-\hat{E}_{ij}\right)^2}{\hat{E}_{ij}}=\sum_{i=1}^{2}\sum_{j=1}^{2}\frac{\left(n_{ij}-\hat{E}_{ij}\right)^2}{\hat{E}_{ij}}$$

follows an approximate chi-squared distribution. The number of degrees of freedom is $k-1-m$, where m is the number of parameters estimated from the data used in computing the expected cell frequencies. Note that we actually need estimate only $p_{1\cdot}$ and $p_{\cdot 1}$ from the data since $p_{\cdot 2}=1-p_{\cdot 1}$ and $p_{2\cdot}=1-p_{1\cdot}$. Hence the number of degrees of freedom associated with the test statistic is

$$k-1-m=4-1-2=1$$

In this case, to satisfy the rule that no expected frequency should be less than 1 and no more than 20% should be less than 5, we must, in fact, have *no expected frequency less than* 5. If this rule cannot be satisfied, then the data should be analyzed by a procedure called Fisher's exact test [6]. Let us now complete the analysis of the data of Example 15.3.2.

Example 15.3.3. We want to see if there is evidence that the development of lung cancer (C) is not independent of the exposure of the individual to airborne asbestos (A). We shall test

$$H_0\colon C \text{ and } A \text{ are independent}$$

$$H_1\colon C \text{ and } A \text{ are not independent}$$

Using the data given in Table 15.2, we see that the expected cell frequencies under H_0 are given by

$$\hat{E}_{11}=\frac{n_{1\cdot}n_{\cdot 1}}{n}=\frac{50(500)}{5000}=5$$

$$\hat{E}_{12}=\frac{n_{1\cdot}n_{\cdot 2}}{n}=\frac{50(4500)}{5000}=45$$

$$\hat{E}_{21}=\frac{n_{2\cdot}n_{\cdot 1}}{n}=\frac{4950(500)}{5000}=495$$

$$\hat{E}_{22}=\frac{n_{2\cdot}n_{\cdot 2}}{n}=\frac{4950(4500)}{5000}=4455$$

The situation is summarized in Table 15.4. Note that there are some differences between what is expected if H_0 is true, listed in parentheses, and what is actually observed. Are those differences too large to have occurred strictly by chance? Note that no expected cell frequency is less than 5, as required. The observed value of

TABLE 15.4

	A	A'	
C	10 (5)	40 (45)	50
C'	490 (495)	4460 (4455)	4950
	500	4500	5000

the test statistic is given by

$$\sum_{i=1}^{2}\sum_{j=1}^{2}\frac{\left(n_{ij}-\hat{E}_{ij}\right)^2}{\hat{E}_{ij}} = \frac{(10-5)^2}{5}+\frac{(40-45)^2}{45}+\frac{(490-495)^2}{495}$$

$$+\frac{(4460-4455)^2}{4455}$$

$$= 5.61$$

The number of degrees of freedom associated with this chi-squared statistic is 1. Since $\chi^2_{.01} = 6.63$ and $\chi^2_{.025} = 5.02$, the P value of the test lies between .01 and .025. Since these probabilities are small, we reject H_0 and conclude that the development of lung cancer is associated with exposure to airborne asbestos. An inspection of Table 15.4 reveals that the number of cancers observed among those exposed to asbestos is higher than that expected if no association exists.

$r \times c$ Test for Independence

We have illustrated the test for independence in the case in which each variable is studied at two levels. This results in a 2×2 contingency table. In general, we may study one variable at r levels and the other at c levels, leading to what is called an $r \times c$ contingency table. The data layout and associated probabilities are shown in Tables 15.5(a) and 15.5(b), respectively.

TABLE 15.5(a)

	Variable B					
Variable A	**1**	**2**	**3**	$\cdots$	c	
1	n_{11}	n_{12}	n_{13}		n_{1c}	$n_{1\cdot}$
2	n_{21}	n_{22}	n_{23}		n_{2c}	$n_{2\cdot}$
3	n_{31}	n_{32}	n_{33}		n_{3c}	$n_{3\cdot}$
$\vdots$						
r	n_{r1}	n_{r2}	n_{r3}		n_{rc}	$n_{r\cdot}$
A	$n_{\cdot 1}$	$n_{\cdot 2}$	$n_{\cdot 3}$		$n_{\cdot c}$	n

TABLE 15.5(*b*)

	Variable B					
Variable A	**1**	**2**	**3**	$\cdots$	c	
1	p_{11}	p_{12}	p_{13}		p_{1c}	$p_{1\cdot}$
2	p_{21}	p_{22}	p_{23}		p_{2c}	$p_{2\cdot}$
3	p_{31}	p_{32}	p_{33}		p_{3c}	$p_{3\cdot}$
$\vdots$						
r	p_{r1}	p_{r2}	p_{r3}		p_{rc}	$p_{r\cdot}$
	$p_{\cdot 1}$	$p_{\cdot 2}$	$p_{\cdot 3}$		$p_{\cdot c}$	1

The null hypothesis of independence is stated mathematically as

$$H_0\colon p_{ij} = p_{i\cdot}p_{\cdot j} \qquad i = 1, 2, 3, \ldots, r$$
$$j = 1, 2, 3, \ldots, c$$

The alternative is that $p_{ij} \neq p_{i\cdot}p_{\cdot j}$ for at least one i and j.

The test statistic is

$$\sum_{i=1}^{r} \sum_{j=1}^{c} \frac{\left(n_{ij} - \hat{E}_{ij}\right)^2}{\hat{E}_{ij}}$$

$$\text{where } \hat{E}_{ij} = \frac{(\text{marginal row total})(\text{column row total})}{\text{sample size}}$$

The only question to answer is, "How many degrees of freedom are associated with this test statistic?" We must estimate the $r - 1$ probabilities $p_{1\cdot}$, $p_{2\cdot}, \ldots, p_{(r-1)\cdot}$ and the $c - 1$ probabilities $p_{\cdot 1}, p_{\cdot 2}, \ldots, p_{\cdot(c-1)}$ from the data in order to compute the expected cell frequencies. Recall that the number of degrees of freedom is given by $k - 1 - m$, where k is the number of cells in the table and m is the number of parameters estimated from the data used to compute expected frequencies. In this case the number of degrees of freedom is

$$\begin{aligned} k - 1 - m &= rc - 1 - (r - 1 + c - 1) \\ &= rc - r - c + 1 \\ &= (r - 1)(c - 1) \end{aligned}$$

These ideas are illustrated in the next example.

Example 15.3.4. To try to convince the public to use safety equipment in automobiles, a random sample of 1000 accidents is chosen from the records. Each accident is classed according to the type of safety restraint used by the occupants and the severity of the injuries received. The data given in Table 15.6 results.

TABLE 15.6

Extent of injury	Type of restraint: Seat belt only		Seat belt and harness		None		
None	75	(70.0)	60	(50.0)	65	(80.0)	200
Minor	160	(157.5)	115	(112.5)	175	(180.0)	450
Major	100	(105.0)	65	(75.0)	135	(120.0)	300
Death	15	(17.5)	10	(12.5)	25	(20.0)	50
	350		250		400		1000

Expected cell frequencies are given in parentheses. Do these data indicate an association between the type of restraint used and the extent of injury? The observed value of the $X^2_{(r-1)(c-1)} = X^2_{3\cdot 2} = X^2_6$ statistic is

$$\frac{(75-70)^2}{70} + \frac{(60-50)^2}{50} + \frac{(65-80)^2}{80} + \cdots + \frac{(25-20)^2}{20} = 10.96$$

The probability of observing a value of 10.96 or greater is between .05 ($\chi^2_{.95} = 12.6$) and .1 ($\chi^2_{.90} = 10.6$). That is, the P value of the test is between .05 and .1.

Since this probability is fairly small we shall reject the null hypothesis and conclude that the extent of an individual's injury is not independent of the type of safety restraint being used at the time of the accident.

Keep in mind the fact that when *only* the sample size is fixed by the experimenter and the remainder of the entries in the contingency table are free to vary, we are testing for independence. Other types of tests are considered in the next section.

15.4 COMPARING PROPORTIONS

In this section we consider the use of the chi-squared statistic in comparing proportions. Once again, we begin by describing an experiment that results in a 2×2 table. This example will demonstrate an important difference between the problems presented in this section and those considered earlier.

Example 15.4.1. A large number of people living in a particular community have been exposed over the last 10 years to radioactivity from an atomic waste storage dump. A study is to be run to find out whether there is any association between this exposure and the development of a particular blood disorder. To conduct the experiment, random samples will be chosen of 300 persons from the community who have been exposed to the hazard and 320 persons who have not been so exposed. Each subject will be screened to determine whether or not the blood disorder is present. This experiment generates a table of the form given in Table 15.7. Note that although this 2×2 table looks exactly like those studied earlier, there is a difference. In particular, the marginal row totals are *fixed* at 300 and 320 *prior* to conducting the field study. That is, these marginal totals as well as the overall sample size are predetermined by the experimenter. All other entries in the table are free to vary.

TABLE 15.7

	D (has disorder)	D' (does not have disorder)	
E (exposed)	n_{11}	n_{12}	$n_{1\cdot} = 300$ (fixed)
E' (not exposed)	n_{21}	n_{22}	$n_{2\cdot} = 320$ (fixed)
	$n_{\cdot 1}$	$n_{\cdot 2}$	$n = 620$ (fixed)

In experiments such as this where either the row or column totals, but not both, are fixed by the researcher the null hypothesis of "no association" is stated in terms of proportions. To see how this is done, let A and B denote the classification variables and assume that the marginal totals for the levels of A are fixed. Thus we essentially have two independent random samples: one from the population of objects with trait A and the other from the population of objects without trait A. We want to test

H_0: proportion of objects with trait B among those with trait A = proportion of objects with trait B among those without trait A

This implies that characteristic B is no more prevalent among those with characteristic A than among those without characteristic A. Hence there is no apparent association between A and B. In terms of our example we want to test

H_0: proportion of individuals with the blood disorder among those exposed to the hazard = proportion of individuals with the blood disorder among those not exposed to the hazard

In other words, there is no apparent association of the blood disorder with exposure to the hazardous waste material.

The null hypothesis can be expressed mathematically with the aid of Table 15.8. Since p_{11} denotes the proportion of objects with trait B among those with trait A and p_{21} denotes the proportion with trait B among those that do not have trait A, we are testing

$$H_0\colon p_{11} = p_{21}$$

Note that since $p_{12} = 1 - p_{11}$ and $p_{22} = 1 - p_{21}$, we are also testing $p_{12} = p_{22}$.

TABLE 15.8

	B	B'	
A	p_{11}	$p_{12} = 1 - p_{11}$	1 (fixed)
A'	p_{21}	$p_{22} = 1 - p_{21}$	1 (fixed)

It is convenient to state the null hypothesis as

$$H_0\colon p_{1j} = p_{2j} \qquad j = 1, 2$$

When the null hypothesis is expressed in this form, the test is referred to as a test of *homogeneity*.

To understand the logic behind the test, we need to look closer at the structure of a 2×2 table with the marginal row totals fixed. In particular, note that we can view the data in row 1 as constituting a random sample of size $n_{1\cdot}$ drawn from a binomial distribution with probability of success p_{11}. Here success is finding an object with trait B. Similarly, the data in row 2 constitutes a random sample of size $n_{2\cdot}$ from a binomial distribution with parameter p_{21}. Thus when we test

$$H_0\colon p_{11} = p_{21}$$

we are actually comparing two population proportions as we did in Chap. 9. In fact, the test statistic that we develop here is simply the square of the Z statistic used earlier! The important point to note here is that since the number of objects in each group with trait B is binomially distributed,

$$E_{11} = n_{1\cdot} p_{11}$$

and

$$E_{21} = n_{2\cdot} p_{21}$$

Thus to estimate E_{11} and E_{21} from the contingency table, we need to find only a logical way to estimate p_{11} and p_{21}. This is not hard to do. Note that if H_0 is true, $p_{11} = p_{21}$. We denote this common population proportion by p. Furthermore, if the proportion of objects with trait B is the same for both populations, then the overall proportion of objects in the two populations combined will also be p. A logical estimator for the overall proportion of objects with trait B is

$$\hat{p} = \frac{\text{number of objects in column 1}}{\text{overall sample size}} = \frac{n_{\cdot 1}}{n}$$

Since we are assuming that $p_{11} = p_{21} = p$, we can also use $\hat{p}$ as an estimator for p_{11} and p_{21}. Substituting, we see that the estimated expected cell frequencies under H_0 are

$$\hat{E}_{11} = n_{1\cdot}\hat{p}_{11} = n_{1\cdot}\frac{n_{\cdot 1}}{n}$$

$$= \frac{(\text{marginal row total})(\text{marginal column total})}{\text{sample size}}$$

$$\hat{E}_{21} = n_{2\cdot}\hat{p}_{21} = n_{2\cdot}\frac{n_{\cdot 1}}{n}$$

$$= \frac{(\text{marginal row total})(\text{marginal column total})}{\text{sample size}}$$

TABLE 15.9

	D (has disorder)	D′ (does not have disorder)	
E (exposed)	52 (48.39)	248 (251.61)	300
E′ (not exposed)	48 (51.61)	272 (268.39)	320
	100	520	620

We leave it to you to verify that

$$\hat{E}_{12} = \frac{n_1.n._2}{n} \qquad \text{and} \qquad \hat{E}_{22} = \frac{n_2.n._2}{n}$$

These expectations are exactly the same as those used in testing for independence. From this point on the test for homogeneity is identical to that for independence. We illustrate the idea by reconsidering the experiment described in Example 15.4.1.

Example 15.4.2. When the experiment of Example 15.4.1 is conducted, the data of Table 15.9 results. The expected cell frequencies are given in parentheses. The observed value of the X_1^2 statistic is

$$\frac{(52 - 48.39)^2}{48.39} + \frac{(248 - 251.61)^2}{251.61} + \frac{(48 - 51.61)^2}{51.61} + \frac{(272 - 268.39)^2}{268.39} = .62$$

This value is not significant even at the $\alpha = .25$ level ($\chi^2_{.25} = 1.32$). These data do not allow us to conclude that there is an association between this particular blood disorder and exposure to this source of radioactivity.

$r \times c$ Test for Homogeneity

As in the test for independence, we can test for homogeneity via an $r \times c$ table. In this case we are dealing with two variables, one of which is studied at r levels and the other at c levels. The marginal totals for exactly one of these variables is fixed by the researcher prior to data gathering. To illustrate the idea, consider Example 15.4.3.

Example 15.4.3. A study is to be conducted to consider the association between the sulfur dioxide (SO_2) level in the air and the mean number of chloroplasts per leaf cell of trees in the area. Three regions are selected for study. One is known to have a high SO_2 concentration, one to have a normal level of SO_2, and the third to have a low SO_2 level. Twenty trees are to be randomly selected from within each area and the mean number of chloroplasts per leaf cell is to be determined for each tree. On this basis each tree will be classified as having a low, normal, or high chloroplast count. This experiment generates a table of the form given in Table

TABLE 15.10

	Chloroplast level			
SO_2 level	**High**	**Normal**	**Low**	
High	n_{11}	n_{12}	n_{13}	$n_{1\cdot} = 20$ (fixed)
Normal	n_{21}	n_{22}	n_{23}	$n_{2\cdot} = 20$ (fixed)
Low	n_{31}	n_{32}	n_{33}	$n_{3\cdot} = 20$ (fixed)
	$n_{\cdot 1}$	$n_{\cdot 2}$	$n_{\cdot 3}$	n

15.10. Note that by fixing the row totals prior to experimentation, we are essentially selecting three independent random samples. One sample is selected from the population of trees exposed to a high SO_2 concentration, one is selected from the population of trees exposed to a normal SO_2 level, and the third is selected from the population of trees with a low SO_2 exposure.

To express the null hypothesis of "no association" when one set of marginal totals is fixed in an $r \times c$ table, let us assume that the row totals are fixed. Consider the probabilities shown in Table 15.11. Note that p_{ij} denotes the proportion of objects in the ith level relative to variable A that are in the jth level relative to variable B. The null hypothesis of no association essentially states that within each column, no row classification is more prevalent than any other. The alternative is that for some columns this is not the case. Statistically, this null hypothesis takes the form

$$H_0\colon p_{1j} = p_{2j} = p_{3j} = \cdots = p_{rj} \qquad j = 1, 2, 3, \ldots, c$$

We can think of this null hypothesis as testing to see whether or not r multinomial populations are identical. For instance, in our last example we are testing the null hypothesis of no association between chloroplast level and level of exposure to SO_2. We want to see whether or not the proportions of trees falling into each chloroplast level are identical regardless of the level of SO_2 to which the trees are exposed. The null hypothesis of homogeneity is tested in exactly the same way as the null hypothesis of independence.

TABLE 15.11

	Variable B					
Variable A	**1**	**2**	**3**	$\cdots$	c	
1	p_{11}	p_{12}	p_{13}		p_{1c}	1
2	p_{21}	p_{22}	p_{23}		p_{2c}	1
3	p_{31}	p_{32}	p_{33}		p_{3c}	1
$\vdots$						$\vdots$
r	p_{r1}	p_{r2}	p_{r3}		p_{rc}	1

TABLE 15.12

	Chloroplast level			
SO_2 level	**High**	**Normal**	**Low**	
High	3 (5)	4 (8.33)	13 (6.67)	20
Normal	5 (5)	10 (8.33)	5 (6.67)	20
Low	7 (5)	11 (8.33)	2 (6.67)	20
	15	25	20	60

Example 15.4.4. When the study described in Example 15.4.3 is conducted, the data shown in Table 15.12 are obtained. Again, expected cell frequencies are given in parentheses. The observed value of the $X^2_{(r-1)(c-1)} = X^2_4$ statistic is

$$\frac{(3-5)^2}{5} + \frac{(4-8.33)^2}{8.33} + \cdots + \frac{(2-6.67)^2}{6.67} = 14.74$$

Since $\chi^2_{.01} = 13.3$ and $\chi^2_{.005} = 14.9$, we can reject H_0 with $.005 < P < .01$. We have strong evidence that there is an association between the SO_2 concentration in the area and the chloroplast level in the leaf cells of the trees. To see the association more clearly, note that if there is no association, the proportion of trees with a low chloroplast count should be the same in each of the three regions. However, it is easy to see that this is not the case. Based on the data of Table 15.12, the estimated proportions of trees with a low chloroplast count are $13/20 = .65$, $5/20 = .25$, and $2/20 = .10$, respectively. These proportions suggest that a high SO_2 level tends to suppress the chloroplast count.

Comparing Proportions with Paired Data: McNemar's Test

Before ending our discussion of categorical data, let us consider one additional type of problem. Note that thus far we have been concerned with the problem of comparing proportions based on *independent* samples drawn from two or more populations. Occasionally there is a need to compare two proportions when the samples drawn are *not* independent. In this case neither the methods presented thus far in this section nor those discussed in Chap. 9 are applicable. However, a method of comparison based on a chi-squared statistic can be used. This technique, called *McNemar's test* is illustrated now.

Example 15.4.5. One problem that concerns the industrial engineer is that of the economical storage of small items that are distributed in less than case lot quantities. Two schemes for storing items are being studied. The first, called *alphameric placement*, stores items in strict alphameric order. The second, the selection density factor (SDF) method, uses a numerical factor computed for each

TABLE 15.13

	Alphameric		
SDF	Within 10 ft	10 ft or farther	
Within 10 ft	4	33*	37
10 ft or farther	2*	61	63
	6	94	100

stock item to determine its position relative to the distributor's workstation. We want to see whether the two schemes result in the same proportion of items being placed within 10 feet (ft) of the workstation. To decide, 100 items are classified first by alphameric placement and then via SDF. In this way each item generates a pair of observations. Although we have a sample of 100 observations from each population, the samples are *not* independent. Rather, they are matched. We record the data obtained in the format shown in Table 15.13. Note that if there is a difference in the proportions of objects placed within 10 ft of the workstation by two schemes, then this difference will be reflected in the cells in which the methods disagree on the placement of an item. Thus we are interested only in the starred cells of Table 15.13. Altogether there are 35 observations in these two cells. If the schemes place the same proportion of objects within 10 ft of the workstation, then we expect half or 17.5 of these observations to fall into each of the two cells. Just as before, we now compare the observed cell frequencies to those expected under the assumption that the proportions are the same via a chi-squared statistic with 1 degree of freedom. For these data we obtain

$$\frac{(2-17.5)^2}{17.5}+\frac{(33-17.5)^2}{17.5}=27.46$$

Since this value is significant even at the $\alpha = .005$ level ($\chi^2_{.005} = 7.88$), we can reject the null hypothesis. We do have evidence that the two proportions are not the same. An inspection of Table 15.13 shows that the SDF method tends to place a higher proportion of objects closer to the workstation than does the alphameric procedure. ("Storage Method Saves Space and Labor in Open Package Area Packing Operations," A. Davies, *Industrial Engineering*, June 1983, pp. 68–74.)

In this chapter we have covered some of the most frequently used tests for data of a categorical nature. We should say that there is a large literature on categorical data analysis. For further study we refer the reader to [6] and [15].

CHAPTER SUMMARY

In this chapter we discussed problems involving categorical data. These are data characterized by the fact that each observation can be classed as falling into exactly one of several mutually exclusive categories or cells. All the procedures presented involve comparing the actual number of observations in a cell to that expected if a specified null hypothesis were true. We reject H_0 if the differences observed are too large to have occurred by chance. The theory behind the test statistic used is based on the multinomial distribution. For this reason, we began our study of categorical data by considering this important multivariate distribution.

We first learned how to test to see if a data set was drawn from a specified distribution. We next considered experiments in which two random variables are being studied. The purpose of the study is to test these random variables for independence. We examined 2×2 tables in which each variable is studied at two classification levels in some detail. We later extended the ideas presented to include $r \times c$ tables. In testing for independence, we see that the only fixed entry in the contingency table is n, the overall sample size. All other entries are random variables. The null and alternative hypotheses assume these forms:

$$H_0\colon p_{ij} = p_{i\cdot}p_{\cdot j} \qquad \text{for all } i \text{ and } j$$

$$H_1\colon p_{ij} \neq p_{i\cdot}p_{\cdot j} \qquad \text{for some } i \text{ and } j$$

The test procedure used does require a fairly large sample. Our guideline for using the procedure requires that no expected cell frequency should be less than 1 and that no more than 20% should be less than 5.

We next considered tests of homogeneity. These are tests that compare two binomial populations via a 2×2 contingency table or r multinomial populations via an $r \times c$ contingency table. The null hypothesis takes this form:

$$H_0\colon p_{1j} = p_{2j} = \cdots = p_{rj} \qquad j = 1, 2, 3, \ldots, c$$

We saw that in such tables the marginal row totals as well as the overall sample size are fixed by the experimenter. However, despite these differences, the mathematical analysis is mechanically the same as that used in testing for independence.

The last test that we considered, called McNemar's test, is used to test for equality of two population proportions based on paired data.

We also introduced and defined important terms that you should know. These are:

Categorical data	Goodness of fit test
Cell	Marginal row total
Multinomial trial	Marginal column total
Multinomial random variable	McNemar's test

EXERCISES

Section 15.1

1. A study is run to determine whether the general public favors the construction of a dam for the generation of electricity and flood control. It is thought that 40% favor dam construction, 30% are neutral, 20% oppose the dam, and the rest have given the issue no thought. A random sample of 150 individuals in the affected area is

selected and interviewed. If the above figures accurately reflect public opinion, how many individuals are expected in each category? If in the sample 42 are in favor, 61 are neutral, 33 are opposed, and the rest have given the issue no thought, do you think, on an intuitive basis, that the proposed percentages are incorrect? Explain.

2. It is assumed that the labor pool for a particular industry consists of 40% white males, 30% white females, 5% black females, 15% black males, and 10% others. Ideally the work force should reflect these percentages. To see if this is the case, a random sample of 200 workers is selected and each worker is placed into exactly one of the above categories. If the work force reflects the labor pool, how many workers are expected in each category? When the sampling is complete, it is observed that there are 95 white males, 50 white females, 2 black females, 20 black males, and 33 others employed. Do these data lead you to suspect that the work force does not reflect the percentages in the labor pool very well? Explain.

3. A random digit generator should produce the digits 0 to 9 inclusive with equal probability. If such a generator is activated 100 times, how many of each digit is expected? If we observe 10 zeros, 8 ones, 9 twos, 11 threes, 12 fours, 7 fives, 10 sixes, 13 sevens, 9 eights, and the rest nines, do you think that there is reason to suspect that the generator does not produce the digits with equal frequency in the long run?

Section 15.2

4. Use the data of Exercise 1 to test

$$H_0: p_1 = .4, \qquad p_2 = .3, \qquad p_3 = .2, \qquad p_4 = .1$$

Is there evidence to support the contention that the stated probabilities are incorrect? Explain based on the P value of the test.

5. Use the data of Exercise 2 to test the null hypothesis that the work force reflects the percentages in the labor pool.

6. Use the data of Exercise 3 to test the null hypothesis that the random digit generator produces the digits 0 to 9 inclusive with equal frequency.

7. Select a random sample of 50 one-digit random numbers from Table III of App. A. Test the null hypothesis that the digits 0 to 9 inclusive occur with equal frequency in this table.

Section 15.3

8. A study is conducted to see if there is an association between age and willingness to use computerized banking systems. The data shown in Table 15.14 are obtained in a

TABLE 15.14

	Use computerized banking		
Age	Yes	No	
Under 40	150	75	
40 or over	150	125	
			500

TABLE 15.15

	Suspect make		
Fire	**Yes**	**No**	
Yes	9	31	
No	16	144	
			200

survey of 500 randomly selected customers of a bank that has been offering computerized banking for over a year. Is there evidence of an association between these two variables? Explain, based on the P value of your test and inspection of the table.

9. It is suspected that the tendency of an automobile to catch fire in a rear end collision is not independent of the make of the car. To support this contention, a random sample of 200 cars involved in rear end collisions is selected from past records. Each car is classified as to make and whether or not it is one of the cars suspected of being especially susceptible to fire under these circumstances. The data gathered is shown in Table 15.15. Is there evidence of an association between this make of car and the presence of fire when involved in a rear end collision? Explain.

10. A study is conducted to test for independence between air quality and air temperature. These data are obtained from records on 200 randomly selected days over the last few years. (See Table 15.16.) Do these data indicate an association between these variables? Explain, based on the P value of the test.

11. In a study of the association between color and the effectiveness of a graphical display 100 graphs are randomly selected from among current scientific journals. Each is classified as to whether or not color is used. Each is also rated as to its effectiveness in making its point. Resulting data are given in Table 15.17. Is there evidence that the effectiveness of a graphical display is not independent of color? Explain, based on the P value of the test.

12. It is suspected that there is an association between the day of the week on which an item is produced and the quality of the item. To support this contention, a random sample of 500 items is selected from stock and each item is classified as to the day

TABLE 15.16

	Air quality			
Temperature	**Poor**	**Fair**	**Good**	
Below average	1	3	24	
Average	12	28	76	
Above average	12	14	30	
				200

TABLE 15.17

	Color present		
Effective	**Yes**	**No**	
Excellent	7	4	
Good	10	19	
Fair	9	26	
Poor	4	21	
			100

TABLE 15.18

	Day produced					
Quality	**M**	**T**	**W**	**Th**	**F**	
Excellent	44	74	79	72	31	
Good	14	25	27	24	10	
Fair	15	20	20	23	9	
Poor	3	5	5	0	0	
						500

on which it was produced via its lot number. The item is also rated for quality. The data gathered are shown in Table 15.18.

(*a*) Our guideline on expected cell frequencies states that no more than 20% can be less than 5 and none can be less than 1. Is this criterion satisfied in this case?

(*b*) To satisfy the criterion, combine the quality of categories "Fair" and "Poor" to form a new table with three rows and five columns. Use this table to test for independence.

(*c*) Has an association between quality and day of production been established? Explain.

Section 15.4

13. A study is conducted to assess the effectiveness of a new computerized system of filling orders in a particular industry. Random samples of 100 customers served via the old system and 100 served via the new system are selected. Each customer is contacted to determine whether or not the order was filled satisfactorily within 2 weeks. Table 15.19 gives the results of the study. Test the null hypothesis that the proportion of satisfied customers among those served by the new system is the same as that among those served by the old system at the $\alpha = .05$ level.

14. Although many jobs in the airline industry entail stress, it is thought that air traffic controllers are particularly susceptible to stress-related disorders such as heart problems, high blood pressure, and ulcers. To support this contention, a random sample of 500 air traffic controllers is selected and surveyed. For comparative purposes a sample of 700 workers from other areas of the airline industry is also

TABLE 15.19

	Satisfied		
	Yes	No	
New	82	18	100
Old	70	30	100
			200

TABLE 15.20

	Stress-related disorder present		
	Yes	No	
Controllers	115	385	500
Others	125	575	700
			1200

selected and surveyed. The data obtained is presented in Table 15.20. Test the null hypothesis that the proportion of air traffic controllers with stress-related disorders is the same as that of other workers in the airline industry. Explain your results in a practical sense based on the P value of the test and inspection of Table 15.20.

15. A new method for etching semiconductors is being studied. The quality of the etch is to be compared to that obtained using two older techniques. The results of the study are given in Table 15.21. State the null hypothesis of homogeneity mathematically. Test this hypothesis at the $\alpha = .05$ level. Interpret your result in a practical sense.

16. A study of the salary gains by workers in research, development, and quality control is conducted. The data in Table 15.22 gives a breakdown of the percentage salary increases over last year of men and women working in these areas. The study is based on a sample of 300 men and 150 women randomly selected from among these workers. Raises were classified according to their integer value. For example, a raise of 5.75% is classified in the category 2–5%. Do these data tend to support the claim

TABLE 15.21

	Quality				
Method	Excellent	Good	Fair	Poor	
High pressure (old)	113	34	21	32	200
Reactive ion (old)	117	31	25	27	200
Magnetron (new)	130	40	20	10	200
					600

TABLE 15.22

	% increase					
	< 2%	2–5%	6–9%	10–13%	> 14%	
Male	50	47	103	76	24	300
Female	21	27	50	35	17	150
						450

TABLE 15.23

	This year		
Last year	**Insured**	**Uninsured**	
Insured	650	5	655
Uninsured	28	70	98
	678	75	753

that there is an association between the percentage increase in the salary of the worker and the worker's sex? Explain, based on the *P* value of your test. Interpret your result in a practical sense by inspecting the data of Table 15.22.

17. A recent study claims that an increasing proportion of engineering firms are purchasing liability insurance. This claim is based on a survey of 753 engineering firms. The status of each firm is recorded for the current and for the previous year. The data on which the claim is based are shown in Table 15.23. Do the data support the claim? Explain, based on the *P* value of McNemar's test.

18. A study is conducted of the association between the rate at which words are spoken and the ability of a "talking computer" to recognize commands that it is programmed to accept. A random sample of 50 commands is spoken at a rate under 60 words per minute and the response of the computer is noted. The same commands are repeated at a rate of 60 words per minute or faster and the response is again noted. The data gathered are shown in Table 15.24. Is there a difference in the proportion of commands accepted at the two speaking speeds? Explain, based on the *P* value of the McNemar test. (*Technology Review*, April 1983, p. 77.)

TABLE 15.24

	Rate under 60		
Rate 60 or over	**Command accepted**	**Command rejected**	
Command accepted	14	1	
Command rejected	28	7	
			50

TABLE 15.25

Previous year, tamper-resistant	Current year, tamper-resistant: Yes	No	
Yes	30	3	
No	52	15	
			100

19. A study of packaging of "over-the-counter" drugs is conducted. The purpose of the study is to determine whether or not the proportion of drugs in tamper-resistant packages is the same this year as last. A sample of 100 products is selected and the manner of packaging for each year is determined. Table 15.25 gives the results of the study. Is there evidence that the proportions differ? Explain.

*20. Show that when testing for homogeneity in a 2×2 table, $\hat{E}_{12} = (n_1.n._2)/n$ and $\hat{E}_{22} = (n_2.n._2)/n$. *Hint:* $\hat{E}_{12} = n_1. - \hat{E}_{11}$.

*21. Rework Exercise 13 using the pooled Z test statistic given in Chap. 9. Show that the square of this Z statistic is identical to the observed chi-squared value obtained in Exercise 13.

*22. Rework Exercise 14 using the pooled Z statistic given in Chap. 9. Show that the square of this Z statistic is identical to the observed chi-squared value obtained in Exercise 14.

REVIEW EXERCISES

23. The industrial robot is a programmable mechanism designed to do work in a limited space. Spray-painting robots are used in the automobile industry. Their main advantage is that they can work in areas with ventilation levels that would be unhealthy for human workers. Robots are highly efficient but not infallible, and they, like humans, occasionally produce a paint job with heavy edges or thin spots. In a study of these robots 50 car hoods painted by a robot are randomly selected and classified as to whether or not the paint job is flawed. A second sample of 50 hoods painted by a skilled painter is also studied. The resulting data are given in Table 15.26. Do these data support the contention that robots produce a smaller

TABLE 15.26

	Flawed: Yes	No	
Robot	2	48	50
Human	4	46	50
	6	94	100

TABLE 15.27

		Anaerobic bacteria present		
		Yes	No	
Aerobic bacteria present	Yes	322	286	(Random)
	No	81	137	(Random)
		(Random)	(Random)	826

proportion of flawed hoods than do humans? Explain, based on the P value of the appropriate test.

24. Anaerobic bacteria are microbes that cannot grow in the presence of oxygen. These organisms are now recognized as a major causative agent in infectious disease. In the past their presence was often overlooked in the clinical laboratory. New methods for detection have been devised recently. A study of 826 specimens yields the data of Table 15.27. (Based on "Aerobiology," R. Edmonds, *McGraw-Hill Yearbook of Science and Technology*, 1980, pp. 42–51.) Do these data indicate that the presence of anaerobic bacteria is not independent of the presence of aerobic bacteria? Explain, based on the P value of the appropriate test.

25. Scientists have suggested that animals use the earth's magnetic field as a clue to their orientation. An experiment to investigate this theory is conducted by using homing pigeons. A pair of coils is placed around each pigeon and a magnetic field that reverses the earth's field is applied. This could disorient the bird. Each day for 118 consecutive days a single bird is released. The bird's orientation and the type of day is noted. Do the data of Table 15.28 indicate that the bird's orientation is not independent of the cloud cover? Explain, based on the P value of the appropriate test. (*McGraw-Hill Yearbook of Science and Technology*, 1981, pp. 233–235.)

26. Two types of coatings are being compared for use as a rust preventive. Fifty pieces of pipe, each of the same type and size, are used in the equipment. Half of each pipe is coated with a .5-mil layer of compound A; the other half receives a .5-mil layer of compound B. Each pipe is then subjected to 1000 hours of salt fog. At the end of the experiment an impartial judge compares the two compounds for effectiveness in preventing rust. The data gathered are shown in Table 15.29. Is there a difference in the proportion of pipes that are deemed effective for the two compounds? Explain, based on the P value of McNemar's test.

TABLE 15.28

	Sunny		
Orient home	**Yes**	**No**	
Yes	79	5	
No	16	18	
			118

TABLE 15.29

	A effective		
B effective	Yes	No	
Yes	35	5	
No	8	2	
			50

27. A study of a computer-based haul truck dispatching system for open-pit mines is conducted. A simulation of truck availability is included in the study. Simulation is done in such a way that each truck should be deemed fully operable 50% of the time, partially operable 25% of the time, and inoperable the rest of the time.

(*a*) The condition of a particular truck is simulated 350 times. How many of these simulations are expected to result in the truck being classified as fully operable? Partially operable? Inoperable?

(*b*) When the simulation is complete, it is found that 168 trucks are classified as fully operable, 94 are classified as partially operable, and the rest are classified as inoperable. Do these data lead you to suspect that the simulator is not functioning properly? Explain.

28. In Bangladesh some flooding occurs yearly and abnormally high floods occur every 2 or 3 years. Poldering is a flood control technique by which the flood level is regulated inside an enclosed area by means of drainage regulators such as sluice gates and pumping stations. A study of the perceived effectiveness of one such polder is conducted. Samples were taken of 259 longtime residents of the area within the polder and 121 residents of the region outside the polder after a recent abnormal flood. Each person was asked to estimate the depth of the water in and around his or her residence. Data obtained are shown in Table 15.30. (Based on a discussion found in "Poldering vs. Compartmentalization: The Choice of Flood Control in Bangladesh," Harun Rasid and Azim Mallik, *Environmental Management*, vol. 17, no. 1, January 1993, pp. 59–71.)

(*a*) From the description of the sampling scheme, is the test for association a test of independence or of homogeneity?

(*b*) Has an association been detected between location and perceived severity of the flood in question? Explain, based on your statistical test.

TABLE 15.30

	Water depth (d) (in meters)			
($d < .5$)	Very shallow ($.5 \le d < 1$)	Shallow ($1 \le d < 1.5$)	Moderate $d \ge 1.5$	Deep
Inside polder	25	145	81	8
Outside polder	0	13	43	65

CHAPTER 16

STATISTICAL QUALITY CONTROL

Statistical quality control methods have been used in industry since at least the early 1940s. During the last several years interest in these methods and research in the development of more sophisticated methods has increased dramatically. Manufacturers in industrialized countries have realized that to compete favorably in the international marketplace, the quality and reliability of products produced must be competitive.

After World War II Japan began to place a major emphasis on industrial quality control. An American statistician, Edward Deming, received international recognition for his early efforts in assisting Japanese industry in implementing industrial quality control methods. Most world consumers now recognize the high quality of products, such as electronic equipment and automobiles, produced in Japan. The large success by Japan partly stimulated greater interest in areas of quality control in the United States as well as in other industrialized countries. In fact, statistical quality control procedures are becoming a vital part of the manufacturing process. These methods are of particular importance to engineers due to their key role in the creation of new products, operation of production processes, and design of industrial and public works facilities.

Until his untimely death in 1993, Deming surely was the most influential person to stimulate interest in quality control, especially in the United States and Japan. He emphasized his philosophy to industrial management with great success. Many other scientists have recently made major contributions to the rapidly expanding areas of quality control. The Japanese scientist G. Taguchi has had a major impact in the area. Some of his work is closely related to statistical experimental design.

Control charts are widely used to monitor a process (*process control*). The purpose of such a chart is to detect a situation in which the process is "out of control" in the sense that it exhibits evidence that there has been a change in the process due to assignable causes. Some of the first control charts were developed by Walter A. Shewhart (1918–1967), a physicist who worked most of his life for Bell Laboratories. He is best known for his work in the area of quality control via the use of statistical methods. He was instrumental in the development of the control charts presented in this chapter which bear his name. In modern quality control procedures, emphasis is placed on designing and monitoring production processes to meet or exceed specifications. However, when interest is in determining whether a batch of items received from a vendor actually meets specifications (*product control*), a procedure called *acceptance sampling* is useful. These ideas are discussed in Sec. 16.4. Some extensions and modifications to basic control charts and some ideas related to Taguchi's approach for designing (*process design*) to control for product variability are given in Sec. 16.6.

The methods presented in this chapter are mostly elementary, but they are widely used in industry and other areas of application such as environmental monitoring.

16.1 PROPERTIES OF CONTROL CHARTS

One of the primary tools used in process control is the simple but effective *Shewhart control chart*. It permits the early detection of a process that is unstable or out of control. A process becoming unstable means that the process distribution has changed with respect to location (such as the mean), variability (such as process standard deviation), or some other process characteristic. Many factors can cause a process to become unstable. Among these are such things as malfunctioning machinery, use of inferior materials, negligence or error on the part of operators, or environmental disturbances. Once a process has been deemed unstable or out of control, based on statistical considerations, it is the job of the quality control engineer to determine the cause and to correct the problem. Shewhart control charts can be used in conjunction with measurement data, count data, or attribute data. In this section we consider the general characteristics that should be possessed by any control chart. We then illustrate the ideas with the $\bar{X}$ Shewhart control chart. This is a chart used to monitor the mean value of the product being produced.

Before a control chart can be developed, we must ask, "What properties should the chart possess?" There are several properties and each is based on practical considerations:

1. Since most control charts are developed by statisticians and engineers working together but used in the workplace, they should be kept simple. Although most production workers are intelligent, they are not engineers and must be given a tool that they can understand and use accurately.

2. A control chart should be designed in such a way that it will allow the detection of an out-of-control situation quickly. For example, we do not want to produce a large quantity of an unacceptable product before we realize that a problem exists.
3. A control chart must have what is called a low *false alarm rate*. That is, we do not often want to call a process out of control when, in fact, there is nothing wrong. False alarms lead to costly and unnecessary downtime in the production process.
4. Sampling is time-consuming and can be expensive. This is especially true when product testing is destructive in that the product is destroyed in the testing process. For this reason, a control chart should work well with small samples.

The Shewhart control chart for means satisfies the above criteria and is widely used in industry. We shall therefore illustrate the use of control charts and develop some of the common language used with regard to control charts via this chart.

Monitoring Means

The control charts used to monitor the mean is centered at some target value μ_0. The lower control limit (LCL) and the upper control limit (UCL) represent the minimum and maximum values that the sample mean $\bar{X}$ can assume without raising an alarm. That is, if LCL $\leq$ UCL, then the process is assumed to be in control or operating correctly in the sense that the mean value appears to be close to target. Values of $\bar{X}$ that fall below the LCL or above the UCL signal that the process mean has shifted away from the target value. Ideally, the upper and lower control limits assume the form

$$\text{LCL} = \mu_0 - k\sigma_{\bar{x}} \qquad \text{UCL} = \mu_0 + k\sigma_{\bar{x}}$$

where μ_0 is the desired or target mean value, $\sigma_{\bar{x}}$ is the standard deviation of the sample mean, and k is some positive real number. In practice, k is usually 2 or 3, and we speak of a 2-sigma or a 3-sigma control chart. A Shewhart control chart for means is used in the following way. Samples, typically of size 4 or 5, are taken at fixed time intervals. The time interval chosen is at the discretion of the quality engineer; it could be every hour, every half hour, or whatever time interval is desired. The sample mean is computed and plotted on the control chart. As long as the $\bar{x}$ value obtained lies within the control limits, the process continues. Successive samples are drawn until $\bar{x}$ is obtained that lies outside of the central limits. When this occurs, we say that a "signal" has been observed and the process is stopped to search for the cause of the signal. In Fig. 16.1, the first six samples obtained lead us to believe that the process is in control; the $\bar{x}$ values all lie within the control limits. The seventh sample sends a signal that the process may not be producing a product with the desired mean value; the process is deemed out of control and the reason for the apparent shift in mean value is sought.

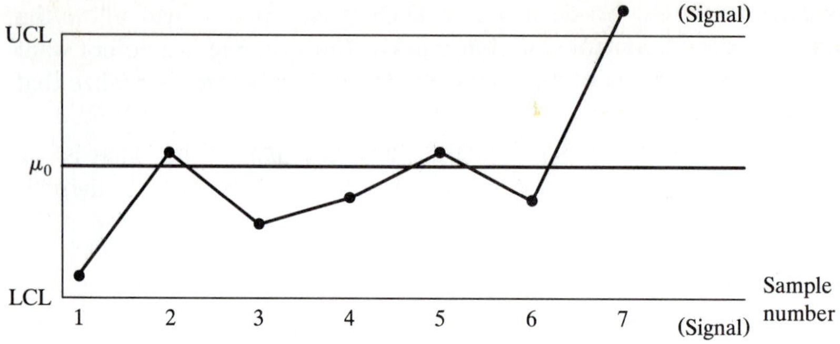

FIGURE 16.1

A typical Shewhart control chart for the mean. The desired target process mean is μ_0. The first six samples drawn reflect a process that appears to be "in control" with respect to the mean. On the seventh sample a "signal" is produced. The process is deemed to be "out of control."

To construct a 3-sigma control chart for the mean, we assume that X has a normal distribution with mean μ_0 and variance σ^2 when the process is in control. Let $\bar{X}$ denote the mean of a sample of size n items drawn at a given time. If the process is in control, then $\bar{X}$ is normally distributed with mean μ_0 and variance σ^2/n. The normal probability law studied in Chap. 3 tells us that a normal random variable will lie within 3 standard deviations of its mean approximately 99% of the time. That is,

$$P\left(\frac{-3\sigma}{\sqrt{n}} \le \bar{X} - \mu \le \frac{3\sigma}{\sqrt{n}}\right) \doteq .99$$

From the standard normal table it can be seen that the exact probability of this occurring is .9974. Isolating $\bar{X}$ in the middle of this inequality, we see that when the process is in control, $\bar{X}$ will fall in the interval $\mu \pm 3\sigma/\sqrt{n}$ with probability .9974. An observed value of $\bar{X}$ above $\mu_0 + 3\sigma/\sqrt{n}$ or below $\mu_0 - 3\sigma/\sqrt{n}$ is very unusual for a process that is in control. There are two explanations for observing such a value: (1) the process is in control and we simply obtain a very unusual sample or (2) the process is out of control. Since the probability that the former explanation is correct is so very small (.0026), we choose to believe the latter! That is, an observed sample mean outside the interval $\mu_0 \pm 3\sigma/\sqrt{n}$ leads us to declare the process out of control. This usually results in the process being stopped to locate the problem. We note that declaring the process out of control when there is really nothing wrong is equivalent to committing a Type I error in hypothesis testing. Since stopping the process is costly, we want to commit such an error only very infrequently.

Distribution of Run Length

Two important questions arise when control charts are used to monitor a process. One is, "How often will we make the wrong decision of declaring the process out of control (observing a value of the mean outside the control limits) when, in fact, the process is in control and we simply observed a rare random event?" The other question is, "How soon will we be able to detect the process being out of control (a true signal)?" For the simple case where we assume that the sample means are from a normal distribution with known mean and known variance, we can answer these questions by using the geometric distribution studied in Chap. 3.

Recall that the geometric distribution arises when a series of independent and identical trails are performed. It is assumed that each trial results in one of two possible outcomes, called *success* or *failure*. The probability of success is denoted by p, whereas the failure rate, $1 - p$, is denoted by q. The random variable Y is the number of trials needed to obtain the first success. It is known that the probability density for Y is given by

$$P[Y = y] = f(y) = pq^{y-1} \qquad y = 1, 2, 3, \ldots$$

and that the average value of Y is $1/p$.

To utilize the geometric distribution in the context of a control chart for the mean, let Y denote the number of samples needed to obtain the first signal. This random variable is called the run length. Notice that by defining Y in this manner, we are defining "success" as being the receipt of a signal or observing an X value that lies outside the control limits. Since $\bar{X}$ is a random variable, its value will vary from sample to sample even when the process is in control. Thus there is a small probability, p, of observing an unusual sample mean that creates a signal by chance even though $\mu = \mu_0$. Since samples are assumed to be independent, when the process is in control, p remains the same from sample to sample. Thus the random variable Y follows a geometric distribution with probability of success p. The value of p is inherent in the construction of the control chart. For a 2-sigma control chart the approximate probability that $\bar{X}$ will lie within the control limits when the process is in control is .95; the exact probability is .9544. The exact probability of receiving a false alarm is .0456. In this case $p = .056$. For a 3-sigma control chart,

$$P[\text{LCL} \le \bar{X} \le \text{UCL}] = .9974$$

and

$$P\left[\bar{X} < \text{LCL} \quad \text{or} \quad \bar{X} > \text{UCL}\right] = .0026$$

In this case $p = .0026$. Figure 16.2 illustrates this idea.

We now ask, "If a process is in control, on the average, how many samples will be taken in order to obtain the first false alarm?" That is, "If the process is in control, what is $E[Y]$ or the average run length?" Since Y is geometric, this question is easy to answer. For a 2-sigma control chart $E[Y] = 1/p = 1/.0456 = 21.929$. On the average, a false alarm will occur on about the 22nd sample. False alarms for a 3-sigma chart occur less frequently. In this case $E[Y] = 1/p = 1/.0026 = 384.6$. In other words, with a 3-sigma control chart we would expect about one false signal in every 385 samples.

Another interesting question to ask is, "If the process is out of control in the sense that the true mean has shifted off target, how many samples shall we have to take before a signal is received?" This question is harder to answer than

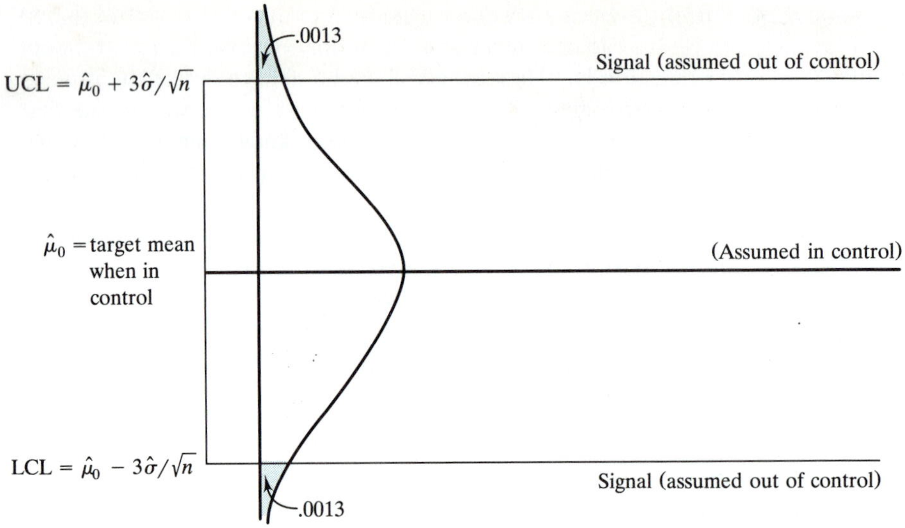

FIGURE 16.2
If the process is on control, then $\overline{X}$ is normally distributed with mean $\hat{\mu}_0$ and standard deviation $\hat{\sigma}/\sqrt{n}$. The probability that $\overline{X}$ will fall about UCL or below LCL is .0013 + .0013 = .0026.

the question concerning average run length because the probability of obtaining a signal is dependent on how far off target the process has become. A slight shift will be hard to find, whereas a dramatic change should be detected rather quickly. Example 16.1.1 illustrates this idea.

Example 16.1.1. Suppose that a manufacturer is producing bolts and it is known that the length of the bolts from this process follows a normal distribution with mean length $\mu = 0.5$ inches (in.) and standard error of the mean $\sigma/\sqrt{n} = 0.01$ in. Thus a 3-sigma control chart would have a centerline at $\mu = 0.5$ in., an upper control limit of UCL = 0.5 in. + 3(.01) in. = 0.53 in., and a lower control limit of LCL = 0.5 in. − 3(.01) in. = 0.47 in. If the process has shifted in such a way that, in fact, the average length of bolts being produced is 0.51 in., what is the probability that a given sample will signal this shift? What is the average number of samples that is required to detect this shift in mean? The probability of receiving a signal is

$$p = P[\overline{X} > 0.53 \quad \text{or} \quad \overline{X} < 0.47]$$

Standardization of the mean yields

$$P\left[\frac{\overline{X} - 0.51}{0.01} > \frac{0.53 - 0.51}{0.01}\right] = P[Z > 2.0] = .0228$$

and

$$P\left[\frac{\overline{X} - 0.51}{0.01} < \frac{0.47 - 0.51}{0.01}\right] = P[Z < -4.0] \doteq 0$$

The probability of detecting this rather small shift in the mean for an individual sample is $p = .0228 + 0 = .0228$. The average number of samples needed to detect the shift is $1/p = 43.86$. Hence it will take quite a while to detect this small shift in the average length of bolts produced. Suppose that the process has been disturbed so that the average has shifted to .56 in. How soon will this rather drastic change be detected? Calculations analogous to those just presented yield

$$P\left[Z > \frac{0.53 - 0.56}{.01}\right] = P[Z > -3] = .9987$$

and
$$P\left[Z < \frac{0.47 - 0.56}{.01}\right] = P[Z < -9.0] \doteq 0$$

In this case $p = .9987$. There is a very high probability that a change in the mean of this magnitude will be detected by a single sample. The expected run length is $1/.9987 \doteq 1.00$.

Questions of time can also be asked. If samples are taken at time intervals of size d, then the average time required to receive a signal is given by

$$\text{Average time} = E[Y] \cdot d$$

For instance, in the above example if samples are taken every hour, then, on the average a shift from the target value of 0.5–0.51 in. will take 43.86 hours to detect. If samples are taken every 1/2 hour, then this shift will be detected, on the average, in 43.86(.5) = 21.93 hours.

It should be pointed out that when a 2-sigma or a 3-sigma Shewhart control chart is used, a series of two-tailed hypothesis tests is conducted. At each sampling point we test

$$H_0: \mu = \mu_0 \qquad \text{(the process mean is in control)}$$

$$H_1: \mu \neq \mu_0 \qquad \text{(the process mean is out of control)}$$

The P value for each test is p, the probability that the test statistic $\bar{X}$ falls outside the control limits. If we receive a signal that turns out to be a false alarm, then we commit a Type I error. If we do not receive a signal when, in fact, the process mean has shifted off target, then a Type II error results. If the control chart is designed in such a way that the average run length is small for detecting some practical crucial shift in the process mean, then the tests conducted have high power for this change.

16.2 SHEWHART CONTROL CHARTS FOR MEASUREMENTS

When monitoring a process from which measurements such as length, diameter, and so on are observed, we almost always monitor stability of the process with respect to location and variability. In this section we consider two important control charts: the $\bar{X}$ chart and the R chart. The $\bar{X}$ chart monitors location (in terms of means) and the R chart monitors variability (in terms of range). The range chart should generally indicate control before the mean chart is constructed since the mean chart uses estimates of range chart parameters to determine control limits.

$\bar{X}$ Chart (Mean)

We have seen that the theoretical bounds for what is called a *3-sigma $\bar{X}$ chart* are

$$\mu_0 \pm \frac{3\sigma}{\sqrt{n}}$$

If the values of μ_0 and σ are known, then we can determine these bounds immediately. Unfortunately, as with most theoretical parameters, their exact values are seldom known in practice. They must be estimated experimentally. To do so, we set the process in motion and draw m samples of size n over a time period in which the process is assumed to be under control. Suggested guidelines are that $m = 20$ or more and $n = 4$ or 5. For each sample we compute $\bar{X}_j$, the sample mean. We estimate μ_0 by pooling the m sample means to obtain this estimator:

$$\hat{\mu}_0 = \frac{\sum_{j=1}^{m} \bar{X}_j}{m}$$

We could estimate σ by computing the sample deviation for each sample and by pooling the resulting estimates. In practice, this process is a bit cumbersome. For this reason, the sample ranges are used to estimate σ. For normal random variables it can be shown that the ratio of the expected value of the sample range R to the standard deviation σ is a constant that depends only on the sample size. This constant, denoted by d_2, is given by

$$d_2 = \frac{E[R]}{\sigma}$$

and hence

$$\sigma = \frac{E[R]}{d_2}$$

The appropriate value of d_2 is obtained from Table XII of App. A. The expected value of R is estimated from the m sample ranges $R_1, R_2, \ldots, R_m$ by averaging them. That is,

$$\widehat{E[R]} = \bar{R} = \frac{\sum_{j=1}^{m} R_j}{m}$$

Substitution yields this estimator for σ:

$$\hat{\sigma} = \frac{\widehat{E[R]}}{d_2} = \frac{\bar{R}}{d_2}$$

The estimated bounds for a 3-sigma $\bar{X}$ chart are

$$\hat{\mu}_0 \pm \frac{3\hat{\sigma}}{\sqrt{n}}$$

or

$$\hat{\mu}_0 \pm \frac{3\bar{R}}{d_2\sqrt{n}}$$

Example 16.2.1, illustrates the construction of an $\bar{X}$ chart.

Example 16.2.1. A new production line is designed to dispense 12 ounces (oz) of a drink into each can as it passes along the line. Regardless of the care taken, there will be some variability in the random variable X, the amount of drink dispensed per can. The process will be considered out of control if the mean amount of fill appears to differ considerably from the average fill obtained when the process is operating correctly or if the variability in fill appears to differ greatly from the variability obtained in a properly operating system. We use an $\bar{X}$ chart to monitor the mean of X. After calibration of the machinery and training of the assembly line personnel, five observations on X, the amount of drink dispensed per can, are taken each hour for a 24-hour period. For each sample of size 5 we compute the sample mean and the sample range. The data obtained are shown in Table 16.1. From these data we estimate μ_0 the centerline of the $\bar{X}$ chart by

$$\hat{\mu}_0 = \frac{\sum_{j=1}^{24} \bar{x}_j}{24} = \frac{(12.088 + 11.971 + \cdots + 12.007)}{24} = 11.987$$

The average sample range is given by

$$\bar{r} = \frac{\sum_{j=1}^{24} r_j}{24} = \frac{.133 + .268 + \cdots + .262}{24} = .254$$

TABLE 16.1
Liquid drink dispensed

Sample number	Weight (oz) per container					Mean $\bar{x}_j$	Range r_j
1	12.046	12.006	12.139	12.112	12.139	12.088	.133
2	12.091	12.118	11.850	11.931	11.863	11.971	.268
3	11.952	11.862	11.899	11.999	12.139	11.970	.277
4	11.821	11.989	11.866	12.104	12.028	11.962	.283
5	11.674	11.881	11.886	11.921	11.886	11.850	.247
6	12.020	12.016	12.227	12.004	11.887	12.031	.340
7	12.077	12.038	11.949	12.029	12.103	12.039	.154
8	11.867	11.971	12.016	11.866	11.124	11.969	.258
9	12.063	12.038	11.858	11.985	11.969	11.983	.205
10	12.042	12.059	12.086	12.024	11.915	12.025	.171
11	12.014	11.747	11.965	11.953	11.944	11.925	.267
12	11.949	11.894	11.951	12.076	12.023	11.979	.182
13	12.168	11.985	12.060	11.910	11.884	12.001	.284
14	11.974	11.964	12.183	12.054	11.794	11.994	.389
15	11.799	12.118	11.886	12.036	11.977	11.963	.319
16	12.021	11.993	12.061	11.969	11.814	11.972	.247
17	12.008	11.834	11.966	11.948	12.299	12.011	.465
18	12.128	11.986	11.911	12.019	11.980	12.005	.217
19	11.946	11.806	12.049	11.976	12.053	11.966	.247
20	11.956	12.066	11.911	11.937	12.040	11.982	.155
21	12.246	11.947	11.937	12.128	12.005	12.053	.309
22	11.947	12.000	11.984	11.838	12.038	11.961	.200
23	11.994	12.136	11.908	12.001	11.909	11.990	.228
24	12.124	11.862	11.904	12.073	12.072	12.007	.262
					Total	287.687	6.107
					Average	11.987	.254

Hence the standard deviation σ is estimated by

$$\hat{\sigma} = \frac{\bar{r}}{d_2}$$

The value $d_2 = 2.326$ is read from Table XII for $n = 5$, and hence

$$\hat{\sigma} = \frac{.254}{2.326 = .1092}$$

The 3-sigma bounds for the $\bar{X}$ chart are

$$\hat{\mu}_0 \pm \frac{3\hat{\sigma}}{\sqrt{n}}$$

or

$$11.987 \pm \frac{3(.1092)}{\sqrt{5}}$$

Hence the lower control limit (LCL) and upper control limit (UCL) are found to

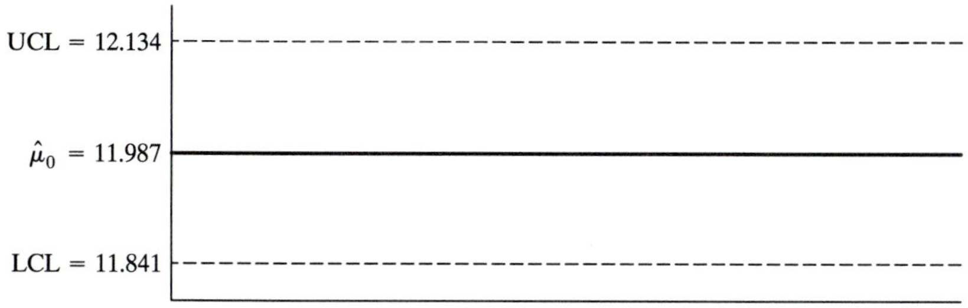

FIGURE 16.3
A 3-sigma $\overline{X}$ chart for controlling the number of ounces of drink contained in a can based on a sample of size 5.

be

$$\text{LCL} = 11.841$$

$$\text{UCL} = 12.134$$

The resulting $\overline{X}$ chart is shown in Fig. 16.3. This chart is used for future monitoring. When a future sample of size 5 is selected, its sample mean is plotted on the $\overline{X}$ chart. If it lies outside the control limits, the process is declared out of control. It is then the responsibility of the quality control engineer to locate and correct the problem. Note that occasionally there will be no problem to correct! On a few rare occasions the value of $\overline{X}$ will lie outside the control limits by chance even though the process is operating correctly.

One further comment should be made concerning the construction of an $\overline{X}$ chart. Once the bounds have been determined in the manner just illustrated, the values $\overline{X}_1, \overline{X}_2, \overline{X}_3, \ldots, \overline{X}_m$ used in the construction of the chart should be plotted on the chart. If these values all fall within the control limits, then the chart is complete and can be put into use. If one or more of these values falls outside the control limits, then these values should be deleted from the data set, μ_0 and σ should be reestimated based on the reduced data set, and new control limits should be computed. Note that the control limits found in our last example are 11.841 and 12.134. Since each of the 24 values for $\overline{X}_j$ listed in Table 16.1 fall between these limits, the $\overline{X}$ chart constructed is ready for use.

R Chart (Range)

Usually, the variability in a process is as important as the mean value. For example, suppose that we are producing bowling balls which are supposed to have an average diameter of 8.6 in. Is it enough to know that the production process is in control with respect to this mean value? Suppose that we observe a sample of five balls with these diameters coming off the line:

4.0	(about the size of a softball)
8.6	
8.7	
9.7	
12.0	(about the size of a basketball)

The sample mean for these data is $\bar{x} = 8.6$—right on target! The process clearly appears to be in control with respect to the mean, but is it a stable process? Obviously not. A bowling ball the size of a softball or a basketball is *not* acceptable. Control variability is as important as controlling the mean. In fact, one of Deming's major points in his method to total quality management is that variability must be reduced and controlled. We shall now introduce a Shewhart control chart that is used to monitor product variability. A chart for the standard deviation can be constructed analogous to the $\bar{X}$ chart. However, due to its simplicity, a control chart for the range is used more often. Such a chart is called a *Shewhart R control chart*. The theoretical bounds for the R chart are

$$\mu_R \pm 3\sigma_R$$

where μ_R denotes the mean value of the sample range R and σ_R denotes its standard deviation. We estimate μ_R by

$$\hat{\mu}_R = \bar{R} = \frac{\sum_{j=1}^{m} R_j}{m}$$

Although we shall not present the derivation, it can be shown that when sampling from a normal distribution, a good estimator for σ_R is

$$\hat{\sigma}_R = \frac{d_3 \bar{R}}{d_2}$$

where d_3 is also a constant whose value depends on the sample size. The values of d_3 are given in Table XII of App. A. Replacing μ_R and σ_R by their estimators, we see that the lower and upper control limits for the sample range are

$$\hat{\mu}_R \pm 3\hat{\sigma}_R$$

or

$$\bar{R} \pm 3\frac{d_3}{d_2}\bar{R}$$

Before illustrating the idea, we should point out one practical problem. The range of a distribution cannot be negative. However, occasionally the estimated lower bound for an R chart will be a negative number. When this occurs, the lower bound is taken to be zero.

Example 16.2.2. Let us construct an R chart based on the data of Table 16.1. We know already that

$$\hat{\mu}_R = \bar{r} = .254$$

The estimated mean of the sample range, when the process is in control, is .254 ounces. This is the centerline of the R chart. The values of d_2 and d_3 when $n = 5$ are found in Table XII. They are 2.326 and .864, respectively. The estimated control limits are

$$\bar{r} \pm 3\frac{d_3}{d_2}\bar{r}$$

$$.254 \pm 3\frac{(.864)(.254)}{2.326}$$

$$.254 \pm .283$$

Since the estimated lower control limit is $-.029$, we set the lower limit to 0. The UCL is .537. Note that since none of the sample ranges given in Table 16.1 falls above the UCL, the control chart is ready for use. This chart is shown in Fig. 16.4.

In practice, the $\bar{X}$ and R charts are nearly always used simultaneously. In this way an effort is made to control both the mean values and the variability of the product being produced. The next example illustrates this idea.

Example 16.2.3. The process described in Example 16.2.1 is monitored 5 times during the course of a day. The resulting data are shown in Table 16.2 and illustrated in Fig. 16.5. Note that the process goes out of control relative to location (the mean) at the fourth sampling period. At this time the engineer would usually look at the process and try to identify and correct the problem. The data suggest that the process is in control relative to variability.

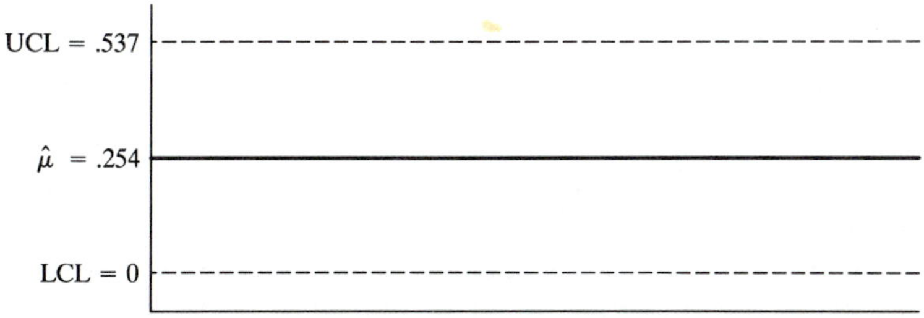

FIGURE 16.4
A 3-sigma R chart for controlling the variability in the number of ounces of drink contained in a can based on a sample of size 5.

TABLE 16.2

Sample number	Weight (oz) per container					Mean $\bar{x}_j$	Range r_j
1	12.016	12.088	11.792	11.971	12.118	11.997	.326
2	12.039	12.047	12.014	12.113	12.156	12.074	.142
3	11.998	12.053	12.058	12.077	12.049	12.047	.079
4	12.167	12.127	12.053	12.137	12.212	12.139	.159
5	12.048	12.048	11.931	12.083	12.045	12.031	.152

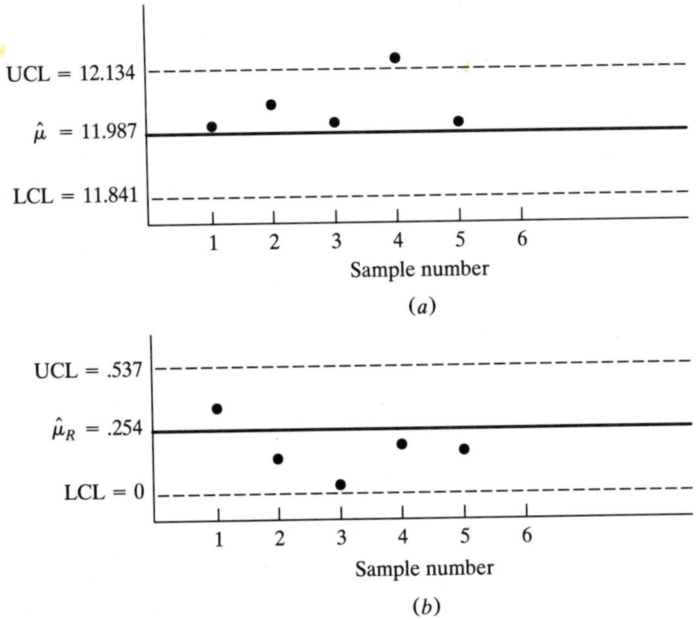

FIGURE 16.5
(*a*) An $\overline{X}$ chart for a sample of 5 days; the process is out of control relative to location on day 4; (*b*) an *R* chart for a sample of 5 days; the process is in control relative to variability each day.

16.3 SHEWHART CONTROL CHARTS FOR ATTRIBUTES

In the previous sections we considered Shewhart control charts used to monitor location (mean) and variability (range) of a continuous random variable. Data obtained were in the form of measurements. In this section we consider Shewhart control charts that entail the use of count data. In particular, we use *P* charts to monitor the proportion of defective items produced; we use *C* charts to monitor the average number of defects per the item produced. We shall present the standard 3-sigma control limits and refer to the number of defects in an item or the number of defective items as either "in control" or "out of control" based on these control limits.

P Chart (Proportion Defective)

The P chart is constructed in a manner similar to that used in constructing an $\overline{X}$ chart. Consider a sample of n items drawn from a process that is assumed to be in control. Since even under the best of circumstances a defective item will be produced occasionally, a certain proportion of items produced will fall into the defective range. Let Π denote this proportion and let X denote the number of defective items found in the sample. Since it is assumed that the quality of one item is not affected by others, the random variable X follows a binomial distribution with parameters n and Π. Notice that in this setting "success" is observing a defective item and the probability of success for a process that is in control is Π. This notation is different from that used in earlier chapters where the probability of success was denoted by p. The change is needed for two reasons. First, it is the notation commonly used in control chart literature for proportions. Second, weshall generate a series of estimates for Π and average these estimates. It will be convenient to denote the estimator for Π by P, successive estimates for Π by $p_1, p_2, p_3, \ldots, p_m$ and the average of these estimates by $\bar{p}$.

Based on our earlier study of sample proportions, it is known that the sample proportion P is an unbiased estimator for Π with variance $\Pi(1 - \Pi)/n$. That is, when the process is in control,

$$\mu_P = \Pi \qquad \sigma_P^2 = \frac{\Pi(1-\Pi)}{n} \qquad \sigma_P = \sqrt{\frac{\Pi(1-\Pi)}{n}}$$

In Chap. 4 we considered the normal approximation to the binomial distribution for large sample sizes. We shall utilize those results here in constructing control limits for P charts. Assuming a large sample size, the theoretical 3-sigma control limits for the P chart are

$$\mu_P \pm 3\sigma_p$$

As usual, we must estimate μ_p and σ_p from data obtained while the process is assumed to be functioning properly.

To estimate μ_p and σ_p, we obtain m random samples, each of size n. Let X_j represent the number of defective items in the jth sample. Then $P_j = X_j/n$ denotes the proportion of defective items in the jth sample. We estimate μ_p with the average value of these m sample proportions. That is,

$$\hat{\mu}_p = \overline{P} = \frac{\sum_{j=1}^{m} P_j}{m} = \frac{\sum_{j=1}^{m} X_j}{mn}$$

Note that $\hat{\mu}_p$ is just the total number of defectives found in the m samples combined divided by the total number of items examined. Since $\mu_p = \Pi$, $\hat{\mu}_p = \bar{P}$ is a pooled estimator for Π. It allows us to combine the m estimators $P_1, P_2, \ldots, P_m$ into a single unbiased estimator for Π. Since σ_p is a function of Π, this parameter can be estimated by

$$\hat{\sigma}_p = \sqrt{\frac{\bar{P}(1-\bar{P})}{n}}$$

The estimated limits for a 3-sigma control chart are

$$\hat{\mu}_p \pm 3\sigma_p$$

or

$$\bar{P} \pm 3\sqrt{\frac{\bar{P}(1-\bar{P})}{n}}$$

Since the proportion of defective items in a sample cannot be negative, the lower control limit (LCL) is set at 0 whenever $\bar{P} - 3\sqrt{\bar{P}(1-\bar{P})/n}$ is negative.

One other comment needs to be made. It seems a little strange that we would want to declare a process out of control when the proportion of defectives appears to be too small. However, in such situations it is sometimes necessary to run a check. Perhaps some change has occurred that results in a better production process than we had before; we would certainly want to discover the reason for this unexpected improvement. Perhaps we are getting too few defectives because of poor inspection techniques by our operators: we must uncover this sort of situation! Whether or not to stop the process when an observed proportion falls below the LCL is a judgment that must be made by the quality control engineer.

Example 16.3.1 demonstrates the construction and use of a P chart.

Example 16.3.1. An electronics firm produces computer memory chips. Statistical quality control methods are to be used to monitor the quality of the chips produced. A chip is classified as defective if any flaw is found that will make the chip unacceptable to the buyer. To set up a P chart to monitor the process, 300 chips are sampled on each of 20 consecutive workdays. The number and proportion of defective chips found each day are recorded in Table 16.3. From these data

$$\hat{\mu}_P = \bar{p} = \sum_{j=1}^{20} \frac{x_j}{mn} = \frac{206}{20(300)} = .0343$$

The estimated proportion of defective chips being produced is .0343. This value is

TABLE 16.3
Samples of memory chips

Work day	Number of defectives	Proportion defective (p)
1	16	.053
2	8	.027
3	1	.003
4	16	.053
5	9	.030
6	13	.043
7	10	.033
8	14	.047
9	11	.037
10	8	.027
11	6	.020
12	14	.047
13	13	.043
14	14	.047
15	4	.013
16	11	.037
17	4	.013
18	13	.043
19	9	.030
20	12	.040
	Total 206	

also the centerline of the P chart. The estimated standard deviation for P is

$$\hat{\sigma}_P = \sqrt{\frac{\bar{p}(1-\bar{p})}{n}} = \sqrt{\frac{(.0343)(.9657)}{300}}$$

$$= .0105$$

Substitution yields the UCL and LCL of

$$.0343 \pm 3(.0105)$$

$$.0343 \pm .0315$$

The LCL is .0028 and the UCL is .0658. The P chart with the 20 observations used in its construction is given in Fig. 16.6. Since none of the proportions used in the construction of the P chart lies outside the control limits, the chart is ready for use. If a future sample of 300 chips yields a sample proportion above .0658, then the process is considered to be unstable or out of control and the cause of the problem is investigated. If the sample proportion falls below .0028, then it is up to the quality control engineer to decide whether or not he or she thinks that the situation warrants investigation.

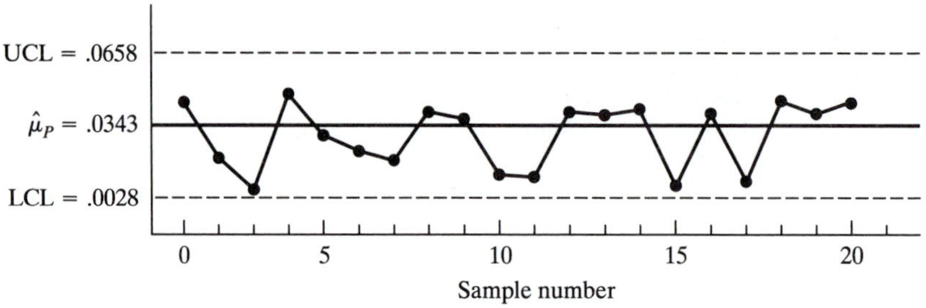

FIGURE 16.6
A 3-sigma P chart for controlling the proportion of defective computer chips produced by an electronics firm based on a sample of size 300; the process is in control on all 20 days used in constructing the chart.

C Charts (Average Number of Defects)

To construct a C chart, we let C denote the number of defects per item. If we think of an item as representing a continuous spacial "interval" of size $s = 1$ and a defect on the item as being a "discrete event," then C satisfies the description given in Chap. 3 of a Poisson random variable. The parameter k associated with C is $k = \lambda s = \lambda \cdot 1 = \lambda$, where λ denotes the average number of defects per item. Based on the Poisson properties derived earlier, it can be seen that

$$\mu_c = \lambda \qquad \sigma_c^2 = \lambda \qquad \sigma_c = \sqrt{\lambda}$$

Thus the theoretical control limits for a 3-sigma control chart are

$$\mu_c \pm 3\sigma_c$$

or

$$\lambda \pm 3\sqrt{\lambda}$$

To estimate λ, we take a sample of m items selected over a time period during which the process is assumed to be in control. Let C_j denote the number of defects found on the jth item. An unbiased estimator for λ is

$$\hat{\lambda} = \overline{C} = \frac{\sum_{j=1}^{m} C_j}{m}$$

Note that $\hat{\lambda}$ is the total number of defects found in the m items divided by the number of items sampled. The estimated limits for a 3-sigma control chart are

$$\hat{\lambda} \pm 3\sqrt{\hat{\lambda}}$$

or

$$\overline{C} \pm 3\sqrt{\overline{C}}$$

Once again, if the LCL is negative, it is set equal to 0.

TABLE 16.4 SAMPLES OF FABRIC

Sample number	Number of flaws	Sample number	Number of flaws
1	0	14	0
2	0	15	0
3	0	16	1
4	1	17	0
5	0	18	0
6	0	19	1
7	2	20	0
8	0	21	0
9	0	22	0
10	0	23	1
11	1	24	0
12	0	25	0
13	3		

To use a C chart of this sort in the future, a single item is sampled and the number of defects is determined. If this number falls outside the control limits, then the process is assumed to be out of control and the source of the problem is sought.

Example 16.3.2. In the making of synthetic fabrics, large rolls of cloth are produced. They are then inspected for flaws and graded as being either first quality, second quality, or unacceptable. In some factories a 100% inspection scheme is used. Each yard of cloth is visually examined for flaws in the inspection process. The quality engineer wants to develop a way to control for flaws in the production process rather than after the material has been packaged into rolls for shipment to the consumer. To construct a control chart for a particular loom, we obtain several 500-yard rolls of first-quality material produced by the loom. Twenty-five samples, each of length 1 linear yard, are selected from these rolls and the number of flaws per sample is recorded. In this case a single linear yard of material is considered to be an individual item. Data collected are shown in Table 16.4. In this case C denotes the number of flaws per linear yard in fabric judged to be of first quality and λ represents the average number of flaws per yard in this quality material. For these data,

$$\hat{\lambda} = \overline{C} = \frac{10}{25} = .4$$

The estimated control limits are

$$\overline{C} \pm 3\sqrt{\overline{C}}$$

or

$$.4 \equiv 3\sqrt{.4}$$

In this case the apparent LCL is negative. Since it is impossible to observe a negative number of defects, the LCL is taken to be 0. The working control limits are 0 to 2.3. In the future at randomly selected times, a yard of material will be examined as it comes from the loom. If more than two flaws are observed, the loom will be stopped and its settings checked.

Control charts for controlling the average number of defects per item can be constructed that utilize samples of more than one item. Charts of this type are useful when the items produced are small so that it is possible to examine many of them fairly quickly. Exercise 15 leads you through the derivation of the control limits for such a chart.

16.4 ACCEPTANCE SAMPLING

Although modern quality control techniques tend to emphasize process control so that defective items are not produced, another important area of statistical quality control is acceptance sampling. When a batch or lot of items has been received by the buyer, he or she must decide whether to accept the items. Usually, inspection of every item in the lot is impractical. This may be due to the time or cost required to do such an inspection; it may be due to the fact that inspection is destructive in the sense that inspecting an item thoroughly can be done only by cutting the item open or by testing it in some other way that renders it useless. Thus the decision to reject a lot must be made based on testing only a sample of items drawn from the lot. The sampling plans that we shall consider are called *attribute* plans. In these plans each item is classified as being either defective or acceptable. We make our decision as to whether or not to reject the lot based on the number of defectives found in the sample. As you will see, acceptance sampling is just an adaptation of classical hypothesis testing.

To begin, let us denote the number of items in the lot or batch by N. The true but unknown proportion of defective items in the lot is denoted by Π. We agree that the entire lot is acceptable if the proportion of defectives Π is less than or equal to some specified value Π_0. Since our job is to detect unacceptable lots, we want to test the hypothesis

$$H_0\colon \Pi \le \Pi_0 \qquad \text{(lot is acceptable)}$$

$$H_1\colon \Pi > \Pi_0 \qquad \text{(lot is unacceptable)}$$

Usually, to decide whether to reject H_0, we determine what is called an *acceptance number*, which we denote by c. If the number of defective items sampled exceeds c, we reject the lot; otherwise we accept it. As you know, two kinds of errors may be committed when testing a hypothesis. We might reject a lot that is, in fact, acceptable, thus committing a Type I error; we might fail to reject an unacceptable lot, thus committing a Type II error. Alpha, the probability of committing a Type I error in this context, is called the *producer's risk*. Beta, the probability of committing a Type II error, is called the *consumer's risk*.

As in the past, we shall be able to compute the value of α. In this case it will depend on the specific value of Π_0, the sample size n, and the lot size N.

Thus in a particular case we shall always know the risk to the producer. To see how to compute α, consider a lot of size N of which the proportion Π_0 is defective. Let $r = N\Pi_0$ denote the number of defective items. We select a random sample of size n from the lot and consider the random variable D, the number of defective items found in the sample. This random variable follows a hypergeometric distribution. From Sec. 3.6 we know that its probability density function is given by

$$f(d) = \frac{\binom{r}{d}\binom{N-r}{n-d}}{\binom{N}{n}}$$

where d is an integer lying between $\max[0, n - (N - r)]$ and $\min[n, r]$. For a preset acceptance number c the producer's risk is given by

$$\begin{aligned}\alpha &= P[\text{reject } H_0|\Pi = \Pi_0] \\ &= P[D > c|\Pi = \Pi_0] \\ &= \sum_{d>c} \frac{\binom{r}{d}\binom{N-r}{n-d}}{\binom{N}{n}}\end{aligned}$$

For relatively small samples this probability can be calculated directly. However, in practice we usually approximate it using either the binomial density or the Poisson density. In the binomial approximation the probability of "success," obtaining a defective part, is assumed to be r/N; in the Poisson approximation the parameter k is given by $k = nr/N$. These ideas are illustrated in the next example. We show you all three calculations. In practice, we would use the hypergeometric probability and would only turn to the approximations when the hypergeometric computations become too cumbersome to be practical.

Example 16.4.1. A construction firm receives a shipment of $N = 20$ steel rods to be used in the construction of a bridge. The lot must be checked to ensure that the breaking strength of the rods meets specifications. The lot will be rejected if it appears that more than 10% of the rods fail to meet specifications. We are testing

$$H_0: \Pi \leq .1 \qquad \text{(lot is acceptable)}$$

$$H_1: \Pi > .1 \qquad \text{(lot is unacceptable)}$$

We compute α under the assumption that the null value is correct. That is, we compute α under the assumption that the lot actually contains $r = N\Pi_0 = 20(.1) = 2$ defective rods. Since testing a rod requires that it be broken, we cannot test each rod. Let us assume that a sample of size $n = 5$ is selected for testing. Let us agree to reject the lot if more than one rod is found to be defective. In this way we are setting our acceptance number at $c = 1$. Note that D can assume only the values 0, 1, or 2. The producer's risk is given by

$$\begin{aligned}\alpha &= P[\text{reject } H_0|\Pi = .10]\\ &= P[D > 1|\Pi = .10]\\ &= P[D = 2|\Pi = .10]\\ &= \frac{\binom{2}{2}\binom{18}{5-2}}{\binom{20}{5}}\end{aligned}$$

Using the combination formula given in Chap. 1 to evaluate the terms shown above, we see that

$$\alpha = 816/15504 = .0526$$

That is, there is about a 5% chance that our sampling technique will lead us to reject an acceptable lot that contains only two defective items; there is about a 95% chance that we shall not reject such a lot. Since the numbers used in this example are small, the calculation based on the hypergeometric distribution is not difficult. For comparative purposes we approximate the value of α by using a binomial random variable X with $n = 5$ and $p = .1$. Since we want to find the probability associated with the right-tail region of the hypergeometric distribution, we approximate α by finding the probability associated with the right-tail region of the appropriate binomial distribution. In this case

$$\begin{aligned}\alpha = P[D = 2] &\doteq P[X \geq 2]\\ &= 1 - P[X < 2]\\ &= 1 - P[X \leq 1]\end{aligned}$$

From Table I of App. A, $\alpha \doteq 1 - .9185 = .0815$. We can also approximate α by using a Poisson random variable Y with parameter $k = nr/N = 5(2)/20 = .5$. From Table II of App. A,

$$\begin{aligned}\alpha = P[D = 2] &\doteq P[Y \geq 2]\\ &= 1 - P[Y < 2]\\ &= 1 - P[Y \leq 1]\\ &= 1 - .910\\ &= .09\end{aligned}$$

These approximations overestimate α, but considering the small numbers involved, they are not bad!

For a set sample size, a set lot size, and a set acceptance number, the probability of accepting a lot depends only on $\Pi = r/N$, the proportion of defectives actually in the lot. The hypergeometric distribution can be used to compute this probability for $r = 0, 1, 2, 3, \ldots, N$. The graph of this acceptance probability as a function of Π is called the *operating characteristic* or OC curve. In the next example we demonstrate how to construct and read an OC curve.

Example 16.4.2. Consider the problem described in Example 16.4.1 in which $N = 20$, $n = 5$, and $c = 1$. The probability of accepting this lot depends only on the proportion of defectives in the lot. We calculate the probability for various values of r and Π by using the equation

$$P\left[\text{accept lot}|\Pi = \frac{r}{N}\right] = \sum_{d \le 1} \frac{\binom{r}{d}\binom{N-r}{n-d}}{\binom{N}{n}}$$

For example, the probability of accepting a lot that contains no defective items is given by

$$\frac{\binom{0}{0}\binom{20}{5}}{\binom{20}{5}} = 1$$

The probability of accepting a lot that contains exactly one defective item is

$$\frac{\binom{1}{0}\binom{19}{5}}{\binom{20}{5}} + \frac{\binom{1}{1}\binom{19}{4}}{\binom{20}{5}} = \frac{11628 + 3876}{15504} = 1$$

We have already seen that the probability of accepting a lot which contains exactly two defective items is $1 - .0526 = .9474$. Similar calculations can be done for $r = 3, 4, 5, \ldots, 20$. The results of these calculations for selected values of r are shown in Table 16.5. Using Table 16.5, we can make a quick sketch of the OC curve for this sampling plan by plotting the lot proportion defective versus the probability of acceptance for these selected values and then by joining the points with a smooth curve. The resulting sketch is shown in Fig. 16.7. The producer's risk is found by projecting a vertical line up from the point $\Pi = \Pi_0$ until it intersects the OC curve. A horizontal line is then projected over to the vertical axis. It intersects this axis at the point $1 - \alpha$. The producer's risk (α) is the length of the line segment from this intersection point to 1 as shown in Fig. 16.7. The consumer's risk (β) for a specified alternative $\Pi_1 > \Pi_0$ can also be read from the OC curve. For example, suppose that we want to determine the probability of accepting a lot

TABLE 16.5

r	Π	**Probability of acceptance**
0	0	1
1	.05	1
2	.10	.9474
5	.25	.6339
10	.50	.1517
15	.75	.0049
20	1.00	0

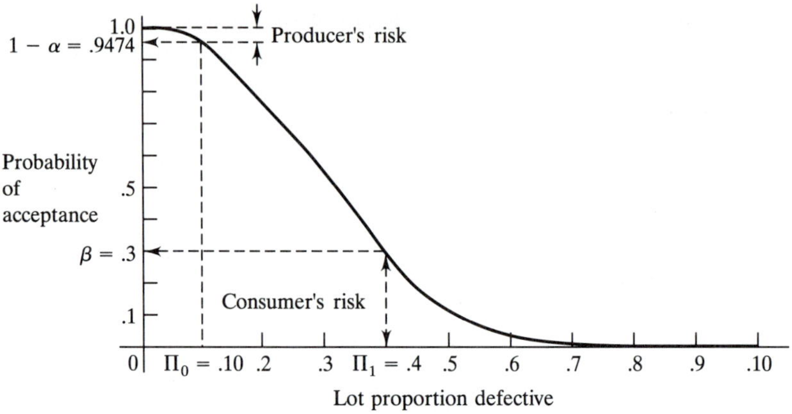

FIGURE 16.7
An OC curve with $N = 20$, $\Pi_0 = .10$, $c = 1$, $n = 5$; the producer's risk is $\alpha = .0526$; the consumer's risk when $\Pi = .4$, β, is approximately .3.

in which the true proportion of defectives is $\Pi_1 = .4$. We use the projection method to see that this probability is approximately .3 as shown in Fig. 16.7. Note that as the difference in Π_0 and Π_1 increases, β decreases. That is, as the proportion of defectives increases, we are less likely to accept an unacceptable lot.

As we have seen in earlier discussions on hypothesis testing, the typical approach is to specify a value for α and then to determine the appropriate rejection region. Here we would specify α and then determine the acceptance number that gives us this approximate α value. In this way we control the producer's risk. However, if samples are small, this might result in an unacceptably large risk to the consumer. In practice, efforts are made to obtain a balance between the producer's risk (α) and the consumer's risk (β). To do so, we specify a value $\Pi_1 > \Pi_0$ that represents to us a "barely acceptable" lot. For example, if we really want $\Pi \leq .10$, we might agree that a defective rate of .12, while not ideal, is at least barely acceptable. When N, Π_0, Π_1, α, and β are specified, it is possible to find a combination of n and C that meets the targets for α and β. That is, it is possible to find an OC curve such that at Π_0 the probability of accepting the lot is $1 - \alpha$ and at Π_1 the probability of accepting the lot is β. There are many sources available that give OC curves for specified values of N, n, α, and β. One of the most popular sources is Military Standard 105D [33].

16.5 TWO-STAGE ACCEPTANCE SAMPLING

Sometimes multiple-stage acceptance sampling plans are used. These plans can lead to smaller average sample sizes required to produce the same or similar

OC curves as those that result in single-stage sampling. This is important when sampling is expensive as in the case of destructive sampling. In this section we consider two-stage sampling in which lot sizes are large enough so that the binomial or normal distributions yield a good approximation to the hypergeometric distribution.

In a two-stage sampling scheme a single sample is drawn. If the number of defective items in the sample is large, the lot is rejected immediately and sampling ceases. If the number of defective items is very small, then the lot is accepted immediately and sampling also ceases. However, if the number of defective items is deemed to be moderate in size so that no clear decision is obvious, then a second sample is drawn. The decision to accept or reject the lot is made based on the total number of defective items in the two samples combined. The terms "very small," "large," and "moderate" are defined relative to the probability of obtaining various numbers of defective items.

The next example illustrates the computation of an OC chart in a two-stage sampling design. Recall that to compute an OC chart, we must find

$$P[\text{accept lot}|\Pi] = P[\text{accept lot on first or second sample}|\Pi]$$

That is, the OC chart is a graph of the probability of accepting a lot as a function of the true proportion of defectives in the lot.

Example 16.5.1. Consider the following two-stage sampling scheme. We draw a sample of size $n_1 = 50$ and decide to reject the lot if the number of defective items is four or more, to accept the lot if the number is 0 or 1, and to take a second sample otherwise. Figure 16.8(*a*) illustrates this first stage of sampling. If a second

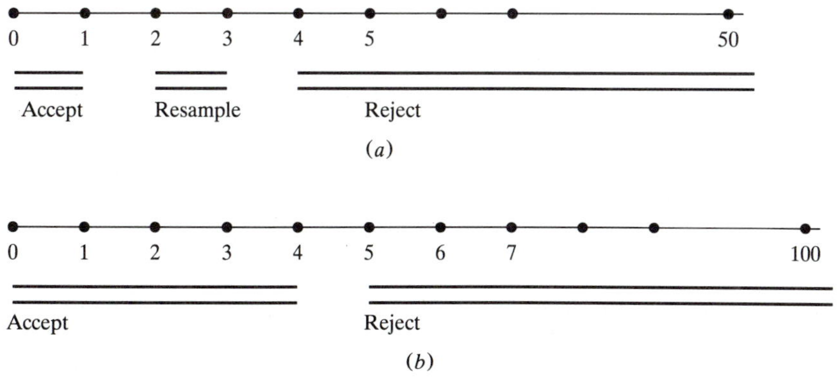

FIGURE 16.8
(*a*) A three-way decision rule is used in the first stage of a two-stage sampling scheme; (*b*) a two-way decision rule is used on the combined sample of size 100 in the second stage of sampling.

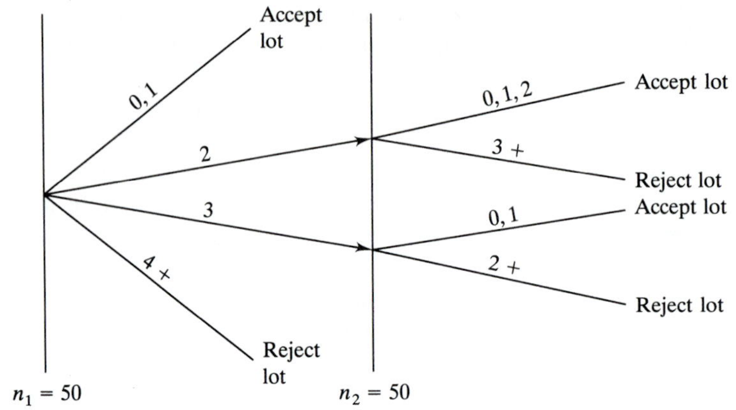

FIGURE 16.9
A two-stage lot acceptance plan.

sample of size $n_2 = 50$ is needed, then we reject the lot if the total number of defective items in the two samples combined is five or more; otherwise the lot is accepted. The second-stage acceptance rule is shown in Fig. 16.8(*b*). Notice that the rejection rule can change from four to five because the sample size has increased from 50 to 100. Figure 16.9 summarizes the entire sampling procedure. Let D denote the total number of defective items obtained while sampling. The probability of acceptance when the true proportion of defectives is Π is given by

$$P[\text{accept}] = P[\text{accept on first sample}] + P[\text{take a second sample and accept}]$$

By using the multiplication rule discussed in Sec. 2.3, we can express the latter probability as

$$P[\text{take a second sample and accept}] = P[\text{accept}|\text{take a second sample}] \times P[\text{take a second sample}]$$

In this example,

$$P[\text{accept}] = P[D \le 1|n_1 = 50, \Pi] + P[D = 2|n_1 = 50, \Pi]P[D \le 2|n_2 = 50, \Pi] + P[D = 3|n_1 = 50, \Pi]P[D \le 1|n_2 = 50, \Pi]$$

To illustrate, let us calculate the probability of acceptance when $\Pi = .10$. The probabilities can be found by using the binomial distribution with $n = 5$. Those probabilities given in Table 16.6 were found by using an extended binomial table that lists probabilities for $n = 50$ and $p = .01$ through $p = .10$. Since our binomial table does not list these values, we can either calculate the desired probabilities from the binomial density or approximate them by using the normal curve. The normal approximation technique is demonstrated below. In this approximation it is

TABLE 16.6

Π	Probability of acceptance
0	1
.01	.996
.02	.952
.03	.833
.04	.661
.05	.482
.06	.328
.07	.212
.08	.132
.09	.080
.10	.047

assumed that D is approximately normally distributed with $\mu = 50(.1) = 5$, $\sigma^2 = 50(.1)(.9) = 4.5$, and $\sigma = 2.12$:

$$P[D \leq 1|n_1{=}50, \Pi{=}.1] \doteq P\left[Z \leq \frac{1.5 - 5}{2.12}\right]$$

$$= P[Z \leq -1.65]$$

$$= .0495$$

$$P[D{=}2|n_1{=}50, \Pi{=}.1] \doteq P[-1.65 \leq Z \leq -1.18]$$

$$= .0695$$

$$P[D \leq 2|n_2{=}50, \Pi{=}.1] = .1190$$

$$p[D{=}3|n_1{=}50, \Pi{=}.1] = .1199$$

Substitution yields

$$P[\text{accept}] \doteq .0495 + .0695(.1190) + .1199(.0495)$$

$$= .0637$$

Notice that this approximation is fairly close to the binomial value given in Table 16.6.

The ideas illustrated here for two-stage acceptance sampling can be extended to multiple-stage sampling.

16.6 EXTENSIONS IN QUALITY CONTROL

Modifications of Control Charts

The basic concepts of quality control were presented in Secs. 16.1 and 16.5. More sophisticated methods are available which, although more complicated, are more sensitive to detecting changes in a process. The simpler Shewhart control charts are practical, simple, and widely used. However, the reader has most likely already realized that these charts lead to decisions which are based only on the most recent sample. A somewhat more complicated but also more efficient control chart utilizes past sample information as well as the present sample. These methods are called *CUSUM* (cumulative sum) control charts since they use functions of the cumulative sums of the previous and present samples. As the reader would guess, CUSUM control charts are more sensitive to small shifts in the process than Shewhart control charts. Furthermore, many processes monitor more than one variable simultaneously. Variables sampling plans are also available. Further reading about these procedures can be found in [13], [18], and [38].

The recent increase in the awareness of the importance of quality control has led to a renewed interest in research in quality control procedures. Some of the classical procedures have been improved to lead to considerably more sensitive sampling designs. For example, the control charts discussed in this chapter all assume that the time between samples and the size of the sample is constant. Intuition would suggest that it may be desirable to vary the time interval between samples or even the sample sizes between samples depending on the value of the most recent sample observation. For example, a sample value close to the target value suggests that the process is well in control, and hence we may be willing to wait longer before taking the next sample in order to save sampling costs. On the other hand, a current sample value that is some distance from the target value, but not outside the control limits, might suggest that the process is becoming unstable, and hence we may wish to take another sample (look again at the process) as soon as possible. These procedures have been formally investigated for Shewhart and CUSUM control charts and found to be much more sensitive than fixed interval procedures. The reader is referred to [1], [8], [42], and [43].

Parameter Design Procedures

Control charts are very useful to signal when a process goes out of control. However, the goal in production is to reduce quality variation and still be close to target (ideal values) of the product characteristics. In almost all production processes there are two types of variables that cause variation of product quality. These are *control factors* and *noise factors*. Control factors are variables that can be controlled by operators of the process. Noise factors are variables, such as environmental factors and manufacturing imperfection, which are difficult or very expensive to control. For example, suppose that transistor radios are being produced. These radios may be used on the beach in the hot sun or in a very cold climate. Here the producer has no real control on the variable, temperature. Thus it would be desirable to attempt to design the production process to control for variability in the product with regard to its intended use.

Taguchi has advanced the basic idea that values (or settings) of the controllable variables should be sought so that variation caused by noise (uncontrollable) variables is minimized. He developed an experimental design procedure called *parameter design* to reduce product variability in a cost-effective manner. These methods are related to both quality control and experimental design. The objective of these procedures is to identify optimal combinations of settings of control variables at which the quality characteristics are the least sensitive to uncontrollable (or at least very expensively controlled) noise variables. Briefly stated, parameter design procedures have the objective of minimizing product sensitivity to external conditions (for example, environmental), minimizing product variability due to use or deterioration, and minimizing product variation among produced units with the average value of the quality characteristic close to the target (ideal) value. For further reading about this approach to quality control, see [5], [24], [26], [29], [47], and [48].

As an engineer or computer scientist, you will find many references to quality control procedures in the literature of your field. You will also find that these procedures are widely used. This chapter is intended to introduce you to these ideas and to obtain a better understanding of why these methods are so important in modern industrial processes.

We also introduced and defined important terms that you should know. These are:

Statistical quality control	*P* charts
R charts	Acceptance sampling
C charts	Producer risk
Consumer risk	Control factors
CUSUM charts	Noise factors
Parameter design	Run length
Variable interval charts	Two-stage acceptance sampling
Unstable process	
$\overline{X}$ charts	OC curve

EXERCISES

Section 16.1

1. Consider the control chart constructed in Exercise 7. Find the probability of a false alarm and the average run length if the true mean has been shifted as described below (in each case assume that the variability has not been affected):

(*a*) μ has shifted up by .002.

(*b*) μ has shifted down by .01.

(*c*) In parts (*a*) and (*b*) on the average, how long will it take to detect the shifts described if samples are taken every half hour?

2. Consider a 3-sigma control chart for the data on 30-amp circuits given in Exercise 3. Find the probability of a signal and the average run length if the true mean has shifted from the target of 30.095 to 30.2. If it is absolutely essential that a shift of this magnitude be detected quickly, do you think that the chart at hand will do so if sampling is done every hour? Explain. If you want to detect the shift within a half hour on the average, how often should sampling be done?
3. In general, for a 3-sigma control chart, find the probability of detecting a downward shift of magnitude $2\sigma_{\bar{x}}$. That is, find the probability of obtaining a signal if the true mean has shifted from μ_0 to $\mu_0 + 2\sigma_{\bar{x}}$. Find the average run length in this setting.
4. A certain type of resistor is manufactured so that when opening properly, the value of the mean μ will be 5.0 and the variance will be $\sigma^2 = .0001$. In order to detect a change in μ, a sample of four resistors is taken hourly and tested. If the sample mean $\bar{X}$, for any sample, falls more than three standard errors from the mean (in either direction), the process is stopped so that the problem can be corrected.
   (*a*) Sketch the 3-sigma control chart for this process showing the centerline (target) value along with the UCL and LCL.
   (*b*) If $\mu = 5.0$, what is the probability that the sample mean $\bar{X}$ will fall above 5.015 or below 4.985?
   (*c*) If $\mu = 5.0$, what is the expected number of samples until the chart signals?
   (*d*) If $\mu = 5.0$, find the standard deviation of the number of samples until the control chart signals.
   (*e*) If $\mu = 5.01$, what is the expected number of samples until the control chart signals?
   (*f*) If samples are drawn every hour, how long, on the average, will it take to detect the shift from $\mu = 5.0$ to $\mu = 5.01$? How long will it take if samples are drawn every half hour? Every 15 minutes?
5. Consider a Shewhart control chart for means with LCL $= \mu_0 - 2\sigma_{\bar{x}}$ and UCL $= \mu_0 + 2\sigma_{\bar{x}}$. Assume that the random variable being measured follows a normal distribution with mean $\mu = 50$ and variance $\sigma^2 = 25$ when the process is in control. Assume that samples of size $n = 4$ are observed and that the sample means $\bar{X}$ are plotted on the control chart.
   (*a*) Sketch the control chart showing the target value and both control limits.
   (*b*) Find the false alarm rate.
   (*c*) If the process mean shifts from a target value of $\mu_0 = 50$ to $\mu = 45$, find the average number of samples required to detect the shift.
   (*d*) Repeat part (*c*) for a shift to $\mu = 40$.

Section 16.2

6. For each of the following sets of summary data, compute the lower control limits (LCL) and upper control limits (UCL) for a 3-sigma $\bar{X}$ chart ($\bar{x}$ and $\bar{r}$, each of which is based on m samples of size n over time when the process is assumed to be in control):
   (*a*) $\bar{x} = 24.5$, $\bar{r} = 2.4$, $n = 5$
   (*b*) $\bar{x} = 0.045$, $\bar{r} = .005$, $n = 10$
   (*c*) $\bar{x} = 8.65$, $\bar{r} = 2.15$, $n = 4$
7. A certain production process was designed for mass production of ball bearings 1/2 in. in diameter. The engineering specification calls for a standard deviation of 0.02 inch. A quality control engineer randomly sampled four ball bearings every

hour for 20 hours. The following data resulted:

Hour	Diameter of ball bearings (in.)
1	.481, .511, .463, .495
2	.507, .486, .491, .491
3	.495, .483, .480, .508
4	.511, .453, .539, .494
5	.479, .510, .554, .499
6	.488, .479, .497, .520
7	.511, .508, .501, .510
8	.507, .522, .521, .515
9	.494, .521, .503, .497
10	.522, .482, .524, .477
11	.509, .501, .530, .520
12	.518, .523, .470, .486
13	.490, .473, .480, .500
14	.464, .498, .473, .521
15	.464, .476, .513, .491
16	.504, .503, .545, .501
17	.515, .508, .490, .506
18	.473, .494,. 503, .473
19	.513, .508, .472, .497
20	.508, .512, .517, .505

(*a*) Compute the sample average and sample range for each of the 20 time periods and the overall average and range for the combined sample. Estimate the standard deviation based on the sample ranges.

(*b*) Calculate the UCL and LCL for a 3-sigma $\overline{X}$ chart and plot the 20 sample means. Does the production process appear to be in control with respect to the mean?

(*c*) Calculate the UCL and LCL for a 3-sigma $\overline{R}$ chart and plot the 20 sample ranges. Does the production process appear to be stable or in control with respect to variability?

8. An electronics firm manufactures circuits to produce 30 amps. The UCL and LCL on the mean of the circuits for samples of size 4 were computed to be LCL = 27.15 and UCL = 33.04. The UCL and LCL for the sample range were estimated to be LCL = 0 and UCL = 5.75. These control limits were used to monitor the circuits produced for 13 time periods, yielding the following data:

Time period	Sample amp measurements
1	30.9, 30.1, 32.9, 32.1
2	31.8, 32.4, 27.0, 28.6
3	29.0, 27.2, 28.0, 30.0
4	26.8, 29.8, 27.4, 32.1
5	26.8, 27.6, 31.4, 29.2
6	30.4, 30.4, 34.5, 30.1
7	31.6, 30.8, 29.0, 30.6
8	28.4, 29.4, 30.3, 28.5
9	31.5, 23.5, 37.0, 21.0
10	25.5, 33.4, 31.0, 20.1
11	34.5, 29.0, 32.1, 28.3
12	22.0, 29.5, 24.5, 23.5
13	29.5, 28.5, 32.5, 33.0

(*a*) Plot the sample means on the $\overline{X}$ chart for the 13 time periods. Discuss the quality of the process with respect to the mean.
(*b*) Plot the sample ranges on the R chart for the 13 time periods. Discuss the quality of the process with respect to variability.

9. When measurement data are used in quality control problems, why is it always important to use both the $\overline{X}$ chart and the $\overline{R}$ chart as opposed to using the $\overline{X}$ chart alone.

10. For each of the sets of summary data given in Exercise 6, construct 3-sigma R chart.

Section 16.3

11. A company manufacturing bolts plans a quality control program to monitor the breaking strength of the bolts. Random samples of 200 bolts are tested for breaking strength specifications on 20 consecutive working days. Any bolt that does not meet specifications is classified as defective. The 20 samples of size $n = 200$ yielded the following information:

Sample No.	1	2	3	4	5	6	7	8	9	10	11	12	13	14	15	16	17	18	19	20
No. of defectives	6	3	3	1	6	1	3	2	3	8	5	3	2	4	6	4	6	2	3	7

(*a*) Estimate the overall proportion defective and the standard deviation of your estimate.
(*b*) Calculate the lower and upper 3-sigma control limits for $\overline{P}$ and plot the 20 sample proportions used in constructing the control chart.
(*c*) If any of the plotted sample values is outside the control limits, how would you revise the control chart for future quality control monitoring?

12. A textile company wishes to implement a quality control program on a certain garment with respect to the number of defects found in the final product. A garment was sampled on 33 consecutive hours of production. The number of defects found per garment is given below:

Defects: 5, 1, 7, 1, 0, 2, 3, 4, 0, 3, 2, 4, 3, 4, 4, 1, 4, 2, 1, 3, 4, 3, 11, 3, 7, 8, 5, 6, 1, 2, 4, 7, 3

Compute the upper and lower 3-sigma control limits for monitoring the number of defects.

13. Consider a setting in which m samples of size n are selected. In this case m can be small but n should be 20 or more. Let C_{ij} denote the number of defects in the jth item of the ith sample. It is assumed that C_{ij} follows a Poisson distribution with parameter λ, where λ represents the average number of defects per item when the process is in control.
(*a*) Let $\overline{C}_i = \Sigma_{j=1}^{n} C_{ij}/n$, $i = 1, 2, \ldots, m$. Based on the Central Limit Theorem, what is the approximate distribution of $\overline{C}_i$?
(*b*) Let $\overline{C} = \Sigma_{i=1}^{m}/m$. What is the expected or average value of $\overline{C}$?
(*c*) To monitor the average number of defects per item in the future, a single sample of size n will be drawn. If the sample average number of defects falls outside the control limits, then the process is declared out of control. Argue that the 3-sigma control limits are given by

$$\overline{C} \pm \sqrt{\frac{\overline{C}}{n}}$$

(*d*) Explain why a chart of this sort might not be practical in the setting described in Example 16.4.2.

(*e*) Give an example of a setting in which a chart of this sort would be practical.

Section 16.4

14. Assume a lot of items to be inspected of size $N = 100$. Consider a sampling inspection plan which samples $n = 20$ items from the lot and rejects the lot if more than one defective item is found in the lot (acceptance number $c = 1$).

(*a*) Sketch the operating characteristic (OC) curve by calculating the exact probability of accepting the lot when the true proportion of defectives is assumed to be $\Pi = 0, .05, .10,$ and $.15$.

(*b*) If the acceptable quality level (Π_0) is .05 and the barely acceptable quality level (Π_1) is .15, calculate the producer's risk α and the consumer's risk β.

15. Repeat Exercise 14 except use the binomial approximation for the hypergeometric distribution. Comment on the validity of this approximation.

16. Repeat Exercise 14 except use the Poisson approximation for the hypergeometric distribution. Comment on the validity of this approximation.

Section 16.5

17. Consider a two-stage sampling acceptance plan. At the first stage a sample of size $n_1 = 10$ is taken from the lot. If zero or one defective is observed, the lot is accepted; and if three or more defectives are observed, the lot is rejected. If two defectives are observed, a second sample of size $n_2 = 10$ is taken. If the cumulative number of defectives (from both samples) is three or less, the lot is accepted. If four or more cumulative defectives are observed, the lot is rejected.

(*a*) Sketch a tree diagram for this acceptance sampling plan similar to that shown in Fig. 16.9.

(*b*) Use the binomial distribution to find the probability of accepting the lot if the true proportion of defectives is $\Pi = .2$.

(*c*) Calculate the acceptance probabilities for $\Pi = .1, .4,$ and $.6$ and then sketch the OC curve.

(*d*) If the true proportion of defectives is $\Pi = .2$, what is the probability that a lot will be rejected?

18. Consider the two-stage acceptance sampling plan defined in Exercise 17 for $n_1 = 10$ and $n_2 = 10$. For comparison consider a single-stage sampling plan for a sample size of $n = 20$ that accepts the lot for three or fewer defectives and rejects the lot for four or more defectives.

(*a*) If $\Pi = .2$, which sampling plan has a higher acceptance probability?

(*b*) If $\Pi = .4$, which sampling plan is more likely to reject the lot?

(*c*) Compare the two-stage versus the single-stage plans in terms of acceptance probabilities.

19. Use the normal approximation procedure to estimate the probabilities given in Table 16.6. If you have a calculator or computer software that allows you to compute binomial probabilities for $n_1 = n_2 = 50$, verify the binomial acceptance probabilities given in Table 16.6.

REFERENCES

1. Amin, R. W., M. R. Reynolds, Jr., and J. C. Arnold: "Cusum Charts with Variable Sampling Intervals," *Technometrics* Vol. 32, No. 4, pp. 371–384, 1990.
2. Beyer, William, ed.: *Handbook of Tables for Probability and Statistics*, 2d ed., CRC Press, Boca Raton, Fla., 1968.
3. Bishop, Y., S. Feinberg, and P. Holland: *Discrete Multivariate Analysis: Theory and Practice*, MIT Press, Cambridge, Mass., 1975.
4. Bowker, A., and G. Lieberman: *Engineering Statistics*, 2d ed., Prentice-Hall, Englewood Cliffs, N.J., 1972.
5. Box, G. E. P.: "Signal-to Noise Ratios, Performance Criteria, and Transformation," *Technometrics*, Vol. 30, pp. 1–17, 1988.
6. Bradley, J. V.: *Distribution Free Statistical Tests*, Prentice-Hall, Englewood Cliffs, N.J., 1968.
7. Burr, I. W.: *Engineering Statistics and Quality Control*, McGraw-Hill, New York, 1953.
8. Chengalur, I. N., J. C. Arnold, and M. R. Reynolds, Jr.: "Variable Sampling Intervals for Multiparameter Shewhart Charts," *Commun. Statist.—Theory Meth.*, Vol. 18, No. 5, pp. 1769–1792, 1989.
9. Conover, W. J.: *Practical Nonparametric Statistics*, Wiley, New York, 1971.
10. Daniel, W.: *Applied Nonparametric Statistics*, PWS-Kent, Boston, Mass., 1990.
11. Dodge, H., and H. Romig: *Sampling Inspection Tables: Single and Double Sampling*, 2d ed., Wiley, New York, 1959.
12. Draper, N., and H. Smith: *Applied Regression Analysis*, 2d ed., Wiley, New York, 1981.
13. Duncan, A. J.: *Quality Control and Industrial Statistics*, 4th ed., Irwin, Homewood, Ill., 1974.
14. Dunn, O., and V. Clark: *Applied Statistics: Analysis of Variance and Regression*, Wiley, New York, 1974.
15. Fisher, R., and F. Yates: *Statistical Tables for Biological Agriculture and Medical Research*, 6th ed., Oliver and Boyd Ltd., London, 1963.
16. Fleiss, J. L.: *Statistical Methods for Rates and Proportions*, 2d ed., Wiley, New York, 1981.
17. Gibbons, J., and J. Pratt: "*P* Values: Interpretation and Methodology," *Amer. Statist.*, Vol. 29, No. 29, 1, 1975.
18. Grant, E., and R. Leavenworth: *Statistical Quality Control*, 4th ed., McGraw-Hill, New York, 1972.
19. Graybill, F.: *Theory and Application of the Linear Model*, Duxbury Press, North Scituate, Mass., 1976.

20. Gupta, S.: "An Asymptotically Nonparametric Test for Symmetry," *Ann. Math. Statist.*, Vol. 38, pp. 849–866, 1967.
21. Hicks, C.: *Fundamental Concepts in the Design of Experiments*, Holt, Rinehart, and Winston, New York, 1965.
22. Hoel, P.: *Introduction to Mathematical Statistics*, 3rd ed., Wiley, New York, 1984.
23. Hollander, M., and D. Wolf: *Nonparametric Statistical Methods*, Wiley, New York, 1973.
24. Hunter, J. S.: "Statistical Design Applied to Production Design," *J. Qual. Tech.*, Vol. 17, pp. 210–221, 1985.
25. Iman, R.: "Graphs for Use with the Lilliefors Test for Normal and Exponential Distributions," *Amer. Statist.*, Vol. 36, No. 2, 1982.
26. Kackar, R. N.: "Off-line Quality Control, Parameter Design, and the Taguchi Method," *J. Qual. Tech.*, Vol. 17, pp. 176–221, 1985.
27. Kempthorne, O.: *Design and Analysis of Experiments*, Wiley, New York, 1952.
28. Kramer, C.: "Extension of Multiple Range Tests to Group Means with Unequal Numbers of Replications," *Biometrics*, Vol. 12, p. 307, 1956.
29. Leon, R. V., A. C. Shoemaker, and R. N. Kackar: "Performance Measures Independent of Adjustment," *Technometrics*, Vol. 29, pp. 253–286, 1987.
30. Lehmann, E.: *Nonparametric: Statistical Methods Based on Ranks*, Holden-Day, San Francisco, 1975.
31. Lentner, M., and T. Bishop: *Experimental Design and Analysis*, Valley Book Company, Blacksburg, Va., 1986.
32. Lentner, M., J. Arnold, and K. Hinkelmann: "How to Use the Ratio of MS(Blocks) and MS(Error) Correctly," *Amer. Statist.*, Vol. 43, no. 2, 1989.
33. *Military Standard 105D, Sampling Procedures and Tables for Inspection by Attributes*, Superintendent of Documents, Government Printing Office, Washington, D.C., 1963.
34. Miller, R.: *Simultaneous Statistical Inference*, McGraw-Hill, New York, 1966.
35. Milton, J., and R. Myers: *A First Course in the Theory of Linear Statistical Models*, Duxbury Press, Boston, 1991.
36. Milton, J., J. Corbet, and P. McTeer: *Introduction to Statistics*, D. C. Heath, Lexington, Mass., 1986.
37. Milton, J., and J. Tsokos: *Statistical Methods in the Biological and Health Sciences*, McGraw-Hill, New York, 1992.
38. Montgomery, D. C.: *Statistical Quality Control*, Wiley, New York, 1985.
39. Montgomery, D., and E. Peck: *Linear Regression Analysis*, Wiley, New York, 1982.
40. Mood, A., F. Graybill, and D. Boes: *Introduction to the Theory of Statistics*, McGraw-Hill, New York, 1974.
41. Olmsted, J.: *Real Variables*, Appleton-Century-Crofts, New York, 1956.
42. Reynolds, M. R., Jr., R. W. Amin, J. C. Arnold, and J. A. Nachlas: "$\bar{X}$-Charts with Variable Sampling Intervals," *Technometrics*, Vol. 30, pp. 181–192, 1988.
43. Reynolds, M. R., Jr., and J. C. Arnold: "Optimal One-Sided Shewhart Control Charts with Variable Sampling Intervals," *Seq. Anal.*, Vol. 8, pp. 51–57, 1989.
44. *SAS User's Guide*, SAS Institute Inc., Raleigh, N.C., 1982.
45. Snedecor, G., and W. Cochran: *Statistical Methods*, 7th ed., Iowa State University Press, Ames, 1980.
46. Stephens, M.: *Three Mile Island*, Random House, New York, 1981.
47. Taguchi, G.: *Introduction to Quality Engineering*, Asian Productivity Organization, Tokyo, 1986.
48. Taguchi, G., and Y. Wu: *Introduction to Off-line Quality Control: ehpl*, Central Japan Quality Control Association, Tokyo, 1980.
49. Tsokos, C.: *Probability Distributions: An Introduction to Probability Theory with Applications*, Duxbury Press, Belmont, Calif., 1972.
50. Tukey, J.: *Exploratory Data Analysis*, Addison-Wesley, Reading, Mass., 1977.
51. Wald, A., and J. Wolfowitz: "Sampling Inspection Plans for Continuous Production which Ensure a Prescribed Limit on the Outgoing Quality," *Ann. Math. Stat.*, Vol. 16, p. 30, 1945.

APPENDIX A

STATISTICAL TABLES

I. Cumulative binomial distribution
II. Poisson distribution function
III. A table of random digits
IV. Cumulative chi-squared distribution
V. Cumulative distribution: Standard normal
VI. *T* distribution
VII. Sample size for estimating the mean
VIII. Wilcoxon signed-rank test
IX. *F* distribution
X. Wilcoxon rank-sum test
XI. Least significant studentized ranges r_p
XII. Control chart constants

TABLE I

Cumulative binomial distribution

$$F_X(t) = P[X \le t] = \sum_{x \le t} \binom{n}{x} p^x (1-p)^{n-x}$$

		p										
n	*t*	0.1	0.2	0.25	0.3	0.4	0.5	0.6	0.7	0.75	0.8	0.9
5	0	0.5905	0.3277	0.2373	0.1681	0.0778	0.0312	0.0102	0.0024	0.0010	0.0003	0.0000
	1	0.9185	0.7373	0.6328	0.5282	0.3370	0.1875	0.0870	0.0308	0.0156	0.0067	0.0005
	2	0.9914	0.9421	0.8965	0.8369	0.6826	0.5000	0.3174	0.1631	0.1035	0.0579	0.0086
	3	0.9995	0.9933	0.9844	0.9692	0.9130	0.8125	0.6630	0.4718	0.3672	0.2627	0.0815
	4	1.0000	0.9997	0.9990	0.9976	0.9898	0.9688	0.9222	0.8319	0.7627	0.6723	0.4095
	5	1.0000	1.0000	1.0000	1.0000	1.0000	1.0000	1.0000	1.0000	1.0000	1.0000	1.0000
6	0	0.5314	0.2621	0.1780	0.1176	0.0467	0.0156	0.0041	0.0007	0.0002	0.0001	0.0000
	1	0.8857	0.6554	0.5339	0.4202	0.2333	0.1094	0.0410	0.0109	0.0046	0.0016	0.0001
	2	0.9841	0.9011	0.8306	0.7443	0.5443	0.3437	0.1792	0.0705	0.0376	0.0170	0.0013
	3	0.9987	0.9830	0.9624	0.9295	0.8208	0.6562	0.4557	0.2557	0.1694	0.0989	0.0159
	4	0.9999	0.9984	0.9954	0.9891	0.9590	0.8906	0.7667	0.5798	0.4661	0.3446	0.1143
	5	1.0000	0.9999	0.9998	0.9993	0.9959	0.9844	0.9533	0.8824	0.8220	0.7379	0.4686
	6	1.0000	1.0000	1.0000	1.0000	1.0000	1.0000	1.0000	1.0000	1.0000	1.0000	1.0000
7	0	0.4783	0.2097	0.1335	0.0824	0.0280	0.0078	0.0016	0.0002	0.0001	0.0000	0.0000
	1	0.8503	0.5767	0.4449	0.3294	0.1586	0.0625	0.0188	0.0038	0.0013	0.0004	0.0000
	2	0.9743	0.8520	0.7564	0.6471	0.4199	0.2266	0.0963	0.0288	0.0129	0.0047	0.0002
	3	0.9973	0.9667	0.9294	0.8740	0.7102	0.5000	0.2898	0.1260	0.0706	0.0333	0.0027
	4	0.9998	0.9953	0.9871	0.9712	0.9037	0.7734	0.5801	0.3529	0.2436	0.1480	0.0257
	5	1.0000	0.9996	0.9987	0.9962	0.9812	0.9375	0.8414	0.6706	0.5551	0.4233	0.1497
	6	1.0000	1.0000	0.9999	0.9998	0.9984	0.9922	0.9720	0.9176	0.8665	0.7903	0.5217
	7	1.0000	1.0000	1.0000	1.0000	1.0000	1.0000	1.0000	1.0000	1.0000	1.0000	1.0000
8	0	0.4305	0.1678	0.1001	0.0576	0.0168	0.0039	0.0007	0.0001	0.0000	0.0000	0.0000
	1	0.8131	0.5033	0.3671	0.2553	0.1064	0.0352	0.0085	0.0013	0.0004	0.0001	0.0000
	2	0.9619	0.7969	0.6785	0.5518	0.3154	0.1445	0.0498	0.0113	0.0042	0.0012	0.0000
	3	0.9950	0.9437	0.8862	0.8059	0.5941	0.3633	0.1737	0.0580	0.0273	0.0104	0.0004
	4	0.9996	0.9896	0.9727	0.9420	0.8263	0.6367	0.4059	0.1941	0.1138	0.0563	0.0050
	5	1.0000	0.9988	0.9958	0.9887	0.9502	0.8555	0.6846	0.4482	0.3215	0.2031	0.0381
	6	1.0000	0.9999	0.9996	0.9987	0.9915	0.9648	0.8936	0.7447	0.6329	0.4967	0.1869
	7	1.0000	1.0000	1.0000	0.9999	0.9993	0.9961	0.9832	0.9424	0.8999	0.8322	0.5695
	8	1.0000	1.0000	1.0000	1.0000	1.0000	1.0000	1.0000	1.0000	1.0000	1.0000	1.0000
9	0	0.3874	0.1342	0.0751	0.0404	0.0101	0.0020	0.0003	0.0000	0.0000	0.0000	0.0000
	1	0.7748	0.4362	0.3003	0.1960	0.0705	0.0195	0.0038	0.0004	0.0001	0.0000	0.0000
	2	0.9470	0.7382	0.6007	0.4628	0.2318	0.0898	0.0250	0.0043	0.0013	0.0003	0.0000
	3	0.9917	0.9144	0.8343	0.7297	0.4826	0.2539	0.0994	0.0253	0.0100	0.0031	0.0001
	4	0.9991	0.9804	0.9511	0.9012	0.7334	0.5000	0.2666	0.0988	0.0489	0.0196	0.0009
	5	0.9999	0.9969	0.9900	0.9747	0.9006	0.7461	0.5174	0.2703	0.1657	0.0856	0.0083
	6	1.0000	0.9997	0.9987	0.9957	0.9750	0.9102	0.7682	0.5372	0.3993	0.2618	0.0530
	7	1.0000	1.0000	0.9999	0.9996	0.9962	0.9805	0.9295	0.8040	0.6997	0.5638	0.2252
	8	1.0000	1.0000	1.0000	1.0000	0.9997	0.9980	0.9899	0.9596	0.9249	0.8658	0.6126
	9	1.0000	1.0000	1.0000	1.0000	1.0000	1.0000	1.0000	1.0000	1.0000	1.0000	1.0000

TABLE I
Cumulative binomial distribution (*continued*)

		p										
n	*t*	0.1	0.2	0.25	0.3	0.4	0.5	0.6	0.7	0.75	0.8	0.9
10	0	0.3487	0.1074	0.0563	0.0282	0.0060	0.0010	0.0001	0.0000	0.0000	0.0000	0.0000
	1	0.7361	0.3758	0.2440	0.1493	0.0464	0.0107	0.0017	0.0001	0.0000	0.0000	0.0000
	2	0.9298	0.6778	0.5256	0.3828	0.1673	0.0547	0.0123	0.0016	0.0004	0.0001	0.0000
	3	0.9872	0.8791	0.7759	0.6496	0.3823	0.1719	0.0548	0.0106	0.0035	0.0009	0.0000
	4	0.9984	0.9672	0.9219	0.8497	0.6331	0.3770	0.1662	0.0473	0.0197	0.0064	0.0001
	5	0.9999	0.9936	0.9803	0.9527	0.8338	0.6230	0.3669	0.1503	0.0781	0.0328	0.0016
	6	1.0000	0.9991	0.9965	0.9894	0.9452	0.8281	0.6177	0.3504	0.2241	0.1209	0.0128
	7	1.0000	0.9999	0.9996	0.9984	0.9877	0.9453	0.8327	0.6172	0.4744	0.3222	0.0702
	8	1.0000	1.0000	1.0000	0.9999	0.9983	0.9893	0.9536	0.8507	0.7560	0.6242	0.2639
	9	1.0000	1.0000	1.0000	1.0000	0.9999	0.9990	0.9940	0.9718	0.9437	0.8926	0.6513
	10	1.0000	1.0000	1.0000	1.0000	1.0000	1.0000	1.0000	1.0000	1.0000	1.0000	1.0000
11	0	0.3138	0.0859	0.0422	0.0198	0.0036	0.0005	0.0000	0.0000	0.0000	0.0000	0.0000
	1	0.6974	0.3221	0.1971	0.1130	0.0302	0.0059	0.0007	0.0000	0.0000	0.0000	0.0000
	2	0.9104	0.6174	0.4552	0.3127	0.1189	0.0327	0.0059	0.0006	0.0001	0.0000	0.0000
	3	0.9815	0.8389	0.7133	0.5696	0.2963	0.1133	0.0293	0.0043	0.0012	0.0002	0.0000
	4	0.9972	0.9496	0.8854	0.7897	0.5328	0.2744	0.0994	0.0216	0.0076	0.0020	0.0000
	5	0.9997	0.9883	0.9657	0.9218	0.7535	0.5000	0.2465	0.0782	0.0343	0.0117	0.0003
	6	1.0000	0.9980	0.9924	0.9784	0.9006	0.7256	0.4672	0.2103	0.1146	0.0504	0.0028
	7	1.0000	0.9998	0.9988	0.9957	0.9707	0.8867	0.7037	0.4304	0.2867	0.1611	0.0185
	8	1.0000	1.0000	0.9999	0.9994	0.9941	0.9673	0.8811	0.6873	0.5448	0.3826	0.0896
	9	1.0000	1.0000	1.0000	1.0000	0.9993	0.9941	0.9698	0.8870	0.8029	0.6779	0.3026
	10	1.0000	1.0000	1.0000	1.0000	1.0000	0.9995	0.9964	0.9802	0.9578	0.9141	0.6862
	11	1.0000	1.0000	1.0000	1.0000	1.0000	1.0000	1.0000	1.0000	1.0000	1.0000	1.0000
12	0	0.2824	0.0687	0.0317	0.0138	0.0022	0.0002	0.0000	0.0000	0.0000	0.0000	0.0000
	1	0.6590	0.2749	0.1584	0.0850	0.0196	0.0032	0.0003	0.0000	0.0000	0.0000	0.0000
	2	0.8891	0.5583	0.3907	0.2528	0.0834	0.0193	0.0028	0.0002	0.0000	0.0000	0.0000
	3	0.9744	0.7946	0.6488	0.4925	0.2253	0.0730	0.0153	0.0017	0.0004	0.0001	0.0000
	4	0.9957	0.9274	0.8424	0.7237	0.4382	0.1938	0.0573	0.0095	0.0028	0.0006	0.0000
	5	0.9995	0.9806	0.9456	0.8822	0.6652	0.3872	0.1582	0.0386	0.0143	0.0039	0.0001
	6	0.9999	0.9961	0.9857	0.9614	0.8418	0.6128	0.3348	0.1178	0.0544	0.0194	0.0005
	7	1.0000	0.9994	0.9972	0.9905	0.9427	0.8062	0.5618	0.2763	0.1576	0.0726	0.0043
	8	1.0000	0.9999	0.9996	0.9983	0.9847	0.9270	0.7747	0.5075	0.3512	0.2054	0.0256
	9	1.0000	1.0000	1.0000	0.9998	0.9972	0.9807	0.9166	0.7472	0.6093	0.4417	0.1109
	10	1.0000	1.0000	1.0000	1.0000	0.9997	0.9968	0.9804	0.9150	0.8416	0.7251	0.3410
	11	1.0000	1.0000	1.0000	1.0000	1.0000	0.9998	0.9978	0.9862	0.9683	0.9313	0.7176
	12	1.0000	1.0000	1.0000	1.0000	1.0000	1.0000	1.0000	1.0000	1.0000	1.0000	1.0000
13	0	0.2542	0.0550	0.0238	0.0097	0.0013	0.0001	0.0000	0.0000	0.0000	0.0000	0.0000
	1	0.6213	0.2336	0.1267	0.0637	0.0126	0.0017	0.0001	0.0000	0.0000	0.0000	0.0000
	2	0.8661	0.5017	0.3326	0.2025	0.0579	0.0112	0.0013	0.0001	0.0000	0.0000	0.0000
	3	0.9658	0.7473	0.5843	0.4206	0.1686	0.0461	0.0078	0.0007	0.0001	0.0000	0.0000
	4	0.9935	0.9009	0.7940	0.6543	0.3530	0.1334	0.0321	0.0040	0.0010	0.0002	0.0000
	5	0.9991	0.9700	0.9198	0.8346	0.5744	0.2905	0.0977	0.0182	0.0056	0.0012	0.0000
	6	0.9999	0.9930	0.9757	0.9376	0.7712	0.5000	0.2288	0.0624	0.0243	0.0070	0.0001
	7	1.0000	0.9988	0.9944	0.9818	0.9023	0.7095	0.4256	0.1654	0.0802	0.0300	0.0009
	8	1.0000	0.9998	0.9990	0.9960	0.9679	0.8666	0.6470	0.3457	0.2060	0.0991	0.0065

TABLE I

Cumulative binomial distribution (*continued*)

		p										
n	*t*	0.1	0.2	0.25	0.3	0.4	0.5	0.6	0.7	0.75	0.8	0.9
	9	1.0000	1.0000	0.9999	0.9993	0.9922	0.9539	0.8314	0.5794	0.4157	0.2527	0.0342
	10	1.0000	1.0000	1.0000	0.9999	0.9987	0.9888	0.9421	0.7975	0.6674	0.4983	0.1339
	11	1.0000	1.0000	1.0000	1.0000	0.9999	0.9983	0.9874	0.9363	0.8733	0.7664	0.3787
	12	1.0000	1.0000	1.0000	1.0000	1.0000	0.9999	0.9987	0.9903	0.9762	0.9450	0.7458
	13	1.0000	1.0000	1.0000	1.0000	1.0000	1.0000	1.0000	1.0000	1.0000	1.0000	1.0000
14	0	0.2288	0.0440	0.0178	0.0068	0.0008	0.0001	0.0000	0.0000	0.0000	0.0000	0.0000
	1	0.5846	0.1979	0.1010	0.0475	0.0081	0.0009	0.0001	0.0000	0.0000	0.0000	0.0000
	2	0.8416	0.4481	0.2811	0.1608	0.0398	0.0065	0.0006	0.0000	0.0000	0.0000	0.0000
	3	0.9559	0.6982	0.5213	0.3552	0.1243	0.0287	0.0039	0.0002	0.0000	0.0000	0.0000
	4	0.9908	0.8702	0.7415	0.5842	0.2793	0.0898	0.0175	0.0017	0.0003	0.0000	0.0000
	5	0.9985	0.9561	0.8883	0.7805	0.4859	0.2120	0.0583	0.0083	0.0022	0.0004	0.0000
	6	0.9998	0.9884	0.9617	0.9067	0.6925	0.3953	0.1501	0.0315	0.0103	0.0024	0.0000
	7	1.0000	0.9976	0.9897	0.9685	0.8499	0.6047	0.3075	0.0933	0.0383	0.0116	0.0002
	8	1.0000	0.9996	0.9978	0.9917	0.9417	0.7880	0.5141	0.2195	0.1117	0.0439	0.0015
	9	1.0000	1.0000	0.9997	0.9983	0.9825	0.9102	0.7207	0.4158	0.2585	0.1298	0.0092
	10	1.0000	1.0000	1.0000	0.9998	0.9961	0.9713	0.8757	0.6448	0.4787	0.3018	0.0441
	11	1.0000	1.0000	1.0000	1.0000	0.9994	0.9935	0.9602	0.8392	0.7189	0.5519	0.1584
	12	1.0000	1.0000	1.0000	1.0000	0.9999	0.9991	0.9919	0.9525	0.8990	0.8021	0.4154
	13	1.0000	1.0000	1.0000	1.0000	1.0000	0.9999	0.9992	0.9932	0.9822	0.9560	0.7712
	14	1.0000	1.0000	1.0000	1.0000	1.0000	1.0000	1.0000	1.0000	1.0000	1.0000	1.0000
15	0	0.2059	0.0352	0.0134	0.0047	0.0005	0.0000	0.0000	0.0000	0.0000	0.0000	0.0000
	1	0.5490	0.1671	0.0802	0.0353	0.0052	0.0005	0.0000	0.0000	0.0000	0.0000	0.0000
	2	0.8159	0.3980	0.2361	0.1268	0.0271	0.0037	0.0003	0.0000	0.0000	0.0000	0.0000
	3	0.9444	0.6482	0.4613	0.2969	0.0905	0.0176	0.0019	0.0001	0.0000	0.0000	0.0000
	4	0.9873	0.8358	0.6865	0.5155	0.2173	0.0592	0.0093	0.0007	0.0001	0.0000	0.0000
	5	0.9978	0.9389	0.8516	0.7216	0.4032	0.1509	0.0338	0.0037	0.0008	0.0001	0.0000
	6	0.9997	0.9819	0.9434	0.8689	0.6098	0.3036	0.0950	0.0152	0.0042	0.0008	0.0000
	7	1.0000	0.9958	0.9827	0.9500	0.7869	0.5000	0.2131	0.0500	0.0173	0.0042	0.0000
	8	1.0000	0.9992	0.9958	0.9848	0.9050	0.6964	0.3902	0.1311	0.0566	0.0181	0.0003
	9	1.0000	0.9999	0.9992	0.9963	0.9662	0.8491	0.5968	0.2784	0.1484	0.0611	0.0022
	10	1.0000	1.0000	0.9999	0.9993	0.9907	0.9408	0.7827	0.4845	0.3135	0.1642	0.0127
	11	1.0000	1.0000	1.0000	0.9999	0.9981	0.9824	0.9095	0.7031	0.5387	0.3518	0.0556
	12	1.0000	1.0000	1.0000	1.0000	0.9997	0.9963	0.9729	0.8732	0.7639	0.6020	0.1841
	13	1.0000	1.0000	1.0000	1.0000	1.0000	0.9995	0.9948	0.9647	0.9198	0.8329	0.4510
	14	1.0000	1.0000	1.0000	1.0000	1.0000	1.0000	0.9995	0.9953	0.9866	0.9648	0.7941
	15	1.0000	1.0000	1.0000	1.0000	1.0000	1.0000	1.0000	1.0000	1.0000	1.0000	1.0000
16	0	0.1853	0.0281	0.0100	0.0033	0.0003	0.0000	0.0000	0.0000	0.0000	0.0000	0.0000
	1	0.5147	0.1407	0.0635	0.0261	0.0033	0.0003	0.0000	0.0000	0.0000	0.0000	0.0000
	2	0.7892	0.3518	0.1971	0.0994	0.0183	0.0021	0.0001	0.0000	0.0000	0.0000	0.0000
	3	0.9316	0.5981	0.4050	0.2459	0.0651	0.0106	0.0009	0.0000	0.0000	0.0000	0.0000
	4	0.9830	0.7982	0.6302	0.4499	0.1666	0.0384	0.0049	0.0003	0.0000	0.0000	0.0000
	5	0.9967	0.9183	0.8103	0.6598	0.3288	0.1051	0.0191	0.0016	0.0003	0.0000	0.0000

TABLE I

Cumulative binomial distribution (*continued*)

							p					
n	*t*	0.1	0.2	0.25	0.3	0.4	0.5	0.6	0.7	0.75	0.8	0.9
	6	0.9995	0.9733	0.9204	0.8247	0.5272	0.2272	0.0583	0.0071	0.0016	0.0002	0.0000
	7	0.9999	0.9930	0.9729	0.9256	0.7161	0.4018	0.1423	0.0257	0.0075	0.0015	0.0000
	8	1.0000	0.9985	0.9925	0.9743	0.8577	0.5982	0.2839	0.0744	0.0271	0.0070	0.0001
	9	1.0000	0.9998	0.9984	0.9929	0.9417	0.7728	0.4728	0.1753	0.0796	0.0267	0.0005
	10	1.0000	1.0000	0.9997	0.9984	0.9809	0.8949	0.6712	0.3402	0.1897	0.0817	0.0033
	11	1.0000	1.0000	1.0000	0.9997	0.9951	0.9616	0.8334	0.5501	0.3698	0.2018	0.0170
	12	1.0000	1.0000	1.0000	1.0000	0.9991	0.9894	0.9349	0.7541	0.5950	0.4019	0.0684
	13	1.0000	1.0000	1.0000	1.0000	0.9999	0.9979	0.9817	0.9006	0.8029	0.6482	0.2108
	14	1.0000	1.0000	1.0000	1.0000	1.0000	0.9997	0.9967	0.9739	0.9365	0.8593	0.4853
	15	1.0000	1.0000	1.0000	1.0000	1.0000	1.0000	0.9997	0.9967	0.9900	0.9719	0.8147
	16	1.0000	1.0000	1.0000	1.0000	1.0000	1.0000	1.0000	1.0000	1.0000	1.0000	1.0000
17	0	0.1668	0.0225	0.0075	0.0023	0.0002	0.0000	0.0000	0.0000	0.0000	0.0000	0.0000
	1	0.4818	0.1182	0.0501	0.0193	0.0021	0.0001	0.0000	0.0000	0.0000	0.0000	0.0000
	2	0.7618	0.3096	0.1637	0.0774	0.0123	0.0012	0.0001	0.0000	0.0000	0.0000	0.0000
	3	0.9174	0.5489	0.3530	0.2019	0.0464	0.0064	0.0005	0.0000	0.0000	0.0000	0.0000
	4	0.9779	0.7582	0.5739	0.3887	0.1260	0.0245	0.0025	0.0001	0.0000	0.0000	0.0000
	5	0.9953	0.8943	0.7653	0.5968	0.2639	0.0717	0.0106	0.0007	0.0001	0.0000	0.0000
	6	0.9992	0.9623	0.8929	0.7752	0.4478	0.1662	0.0348	0.0032	0.0006	0.0001	0.0000
	7	0.9999	0.9891	0.9598	0.8954	0.6405	0.3145	0.0919	0.0127	0.0031	0.0005	0.0000
	8	1.0000	0.9974	0.9876	0.9597	0.8011	0.5000	0.1989	0.0403	0.0124	0.0026	0.0000
	9	1.0000	0.9995	0.9969	0.9873	0.9081	0.6855	0.3595	0.1046	0.0402	0.0109	0.0001
	10	1.0000	0.9999	0.9994	0.9968	0.9652	0.8338	0.5522	0.2248	0.1071	0.0377	0.0008
	11	1.0000	1.0000	0.9999	0.9993	0.9894	0.9283	0.7361	0.4032	0.2347	0.1057	0.0047
	12	1.0000	1.0000	1.0000	0.9999	0.9975	0.9755	0.8740	0.6113	0.4261	0.2418	0.0221
	13	1.0000	1.0000	1.0000	1.0000	0.9995	0.9936	0.9536	0.7981	0.6470	0.4511	0.0826
	14	1.0000	1.0000	1.0000	1.0000	0.9999	0.9988	0.9877	0.9226	0.8363	0.6904	0.2382
	15	1.0000	1.0000	1.0000	1.0000	1.0000	0.9999	0.9979	0.9807	0.9499	0.8818	0.5182
	16	1.0000	1.0000	1.0000	1.0000	1.0000	1.0000	0.9998	0.9977	0.9925	0.9775	0.8332
	17	1.0000	1.0000	1.0000	1.0000	1.0000	1.0000	1.0000	1.0000	1.0000	1.0000	1.0000
18	0	0.1501	0.0180	0.0056	0.0016	0.0001	0.0000	0.0000	0.0000	0.0000	0.0000	0.0000
	1	0.4503	0.0991	0.0395	0.0142	0.0013	0.0001	0.0000	0.0000	0.0000	0.0000	0.0000
	2	0.7338	0.2713	0.1353	0.0600	0.0082	0.0007	0.0000	0.0000	0.0000	0.0000	0.0000
	3	0.9018	0.5010	0.3057	0.1646	0.0328	0.0038	0.0002	0.0000	0.0000	0.0000	0.0000
	4	0.9718	0.7164	0.5187	0.3327	0.0942	0.0154	0.0013	0.0000	0.0000	0.0000	0.0000
	5	0.9936	0.8671	0.7175	0.5344	0.2088	0.0481	0.0058	0.0003	0.0000	0.0000	0.0000
	6	0.9988	0.9487	0.8610	0.7217	0.3743	0.1189	0.0203	0.0014	0.0002	0.0000	0.0000
	7	0.9998	0.9837	0.9431	0.8593	0.5634	0.2403	0.0576	0.0061	0.0012	0.0002	0.0000
	8	1.0000	0.9957	0.9807	0.9404	0.7368	0.4073	0.1347	0.0210	0.0054	0.0009	0.0000
	9	1.0000	0.9991	0.9946	0.9790	0.8653	0.5927	0.2632	0.0596	0.0193	0.0043	0.0000
	10	1.0000	0.9998	0.9988	0.9939	0.9424	0.7597	0.4366	0.1407	0.0569	0.0163	0.0002
	11	1.0000	1.0000	0.9998	0.9986	0.9797	0.8811	0.6257	0.2783	0.1390	0.0513	0.0012
	12	1.0000	1.0000	1.0000	0.9997	0.9942	0.9519	0.7912	0.4656	0.2825	0.1329	0.0064
	13	1.0000	1.0000	1.0000	1.0000	0.9987	0.9846	0.9058	0.6673	0.4813	0.2836	0.0282
	14	1.0000	1.0000	1.0000	1.0000	0.9998	0.9962	0.9672	0.8354	0.6943	0.4990	0.0982
	15	1.0000	1.0000	1.0000	1.0000	1.0000	0.9993	0.9918	0.9400	0.8647	0.7287	0.2662
	16	1.0000	1.0000	1.0000	1.0000	1.0000	0.9999	0.9987	0.9858	0.9605	0.9009	0.5497
	17	1.0000	1.0000	1.0000	1.0000	1.0000	1.0000	0.9999	0.9984	0.9944	0.9820	0.8499
	18	1.0000	1.0000	1.0000	1.0000	1.0000	1.0000	1.0000	1.0000	1.0000	1.0000	1.0000

TABLE I

Cumulative binomial distribution (*continued*)

		p										
n	*t*	0.1	0.2	0.25	0.3	0.4	0.5	0.6	0.7	0.75	0.8	0.9
19	0	0.1351	0.0144	0.0042	0.0011	0.0001	0.0000	0.0000	0.0000	0.0000	0.0000	0.0000
	1	0.4203	0.0829	0.0310	0.0104	0.0008	0.0000	0.0000	0.0000	0.0000	0.0000	0.0000
	2	0.7054	0.2369	0.1113	0.0462	0.0055	0.0004	0.0000	0.0000	0.0000	0.0000	0.0000
	3	0.8850	0.4551	0.2631	0.1332	0.0230	0.0022	0.0001	0.0000	0.0000	0.0000	0.0000
	4	0.9648	0.6733	0.4654	0.2822	0.0696	0.0096	0.0006	0.0000	0.0000	0.0000	0.0000
	5	0.9914	0.8369	0.6678	0.4739	0.1629	0.0318	0.0031	0.0001	0.0000	0.0000	0.0000
	6	0.9983	0.9324	0.8251	0.6655	0.3081	0.0835	0.0116	0.0006	0.0001	0.0000	0.0000
	7	0.9997	0.9767	0.9225	0.8180	0.4878	0.1796	0.0352	0.0028	0.0005	0.0000	0.0000
	8	1.0000	0.9933	0.9713	0.9161	0.6675	0.3238	0.0885	0.0105	0.0023	0.0003	0.0000
	9	1.0000	0.9984	0.9911	0.9674	0.8139	0.5000	0.0861	0.0326	0.0089	0.0016	0.0000
	10	1.0000	0.9997	0.9977	0.9895	0.9115	0.6762	0.3325	0.0839	0.0287	0.0067	0.0000
	11	1.0000	1.0000	0.9995	0.9972	0.9648	0.8204	0.5122	0.1820	0.0775	0.0233	0.0003
	12	1.0000	1.0000	0.9999	0.9994	0.9884	0.9165	0.6919	0.3345	0.1749	0.0676	0.0017
	13	1.0000	1.0000	1.0000	0.9999	0.9969	0.9682	0.8371	0.5261	0.3322	0.1631	0.0086
	14	1.0000	1.0000	1.0000	1.0000	0.9994	0.9904	0.9304	0.7178	0.5346	0.3267	0.0352
	15	1.0000	1.0000	1.0000	1.0000	0.9999	0.9978	0.9770	0.8668	0.7369	0.5449	0.1150
	16	1.0000	1.0000	1.0000	1.0000	1.0000	0.9996	0.9945	0.9538	0.8887	0.7631	0.2946
	17	1.0000	1.0000	1.0000	1.0000	1.0000	1.0000	0.9992	0.9896	0.9690	0.9171	0.5797
	18	1.0000	1.0000	1.0000	1.0000	1.0000	1.0000	0.9999	0.9989	0.9958	0.9856	0.8649
	19	1.0000	1.0000	1.0000	1.0000	1.0000	1.0000	1.0000	1.0000	1.0000	1.0000	1.0000
20	0	0.1216	0.0115	0.0032	0.0008	0.0000	0.0000	0.0000	0.0000	0.0000	0.0000	0.0000
	1	0.3917	0.0692	0.0243	0.0076	0.0005	0.0000	0.0000	0.0000	0.0000	0.0000	0.0000
	2	0.6769	0.2061	0.0913	0.0355	0.0036	0.0002	0.0000	0.0000	0.0000	0.0000	0.0000
	3	0.8670	0.4114	0.2252	0.1071	0.0160	0.0013	0.0000	0.0000	0.0000	0.0000	0.0000
	4	0.9568	0.6296	0.4148	0.2375	0.0510	0.0059	0.0003	0.0000	0.0000	0.0000	0.0000
	5	0.9887	0.8042	0.6172	0.4164	0.1256	0.0207	0.0016	0.0000	0.0000	0.0000	0.0000
	6	0.9976	0.9133	0.7858	0.6080	0.2500	0.0577	0.0065	0.0003	0.0000	0.0000	0.0000
	7	0.9996	0.9679	0.8982	0.7723	0.4159	0.1316	0.0210	0.0013	0.0002	0.0000	0.0000
	8	0.9999	0.9900	0.9591	0.8867	0.5956	0.2517	0.0565	0.0051	0.0009	0.0001	0.0000
	9	1.0000	0.9974	0.9861	0.9520	0.7553	0.4119	0.1275	0.0171	0.0039	0.0006	0.0000
	10	1.0000	0.9994	0.9961	0.9829	0.8725	0.5881	0.2447	0.0480	0.0139	0.0026	0.0000
	11	1.0000	0.9999	0.9991	0.9949	0.9435	0.7483	0.4044	0.1133	0.0409	0.0100	0.0001
	12	1.0000	1.0000	0.9998	0.9987	0.9790	0.8684	0.5841	0.2277	0.1018	0.0321	0.0004
	13	1.0000	1.0000	1.0000	0.9997	0.9935	0.9423	0.7500	0.3920	0.2142	0.0867	0.0024
	14	1.0000	1.0000	1.0000	1.0000	0.9984	0.9793	0.8744	0.5836	0.3828	0.1958	0.0113
	15	1.0000	1.0000	1.0000	1.0000	0.9997	0.9941	0.9490	0.7625	0.5852	0.3704	0.0432
	16	1.0000	1.0000	1.0000	1.0000	1.0000	0.9987	0.9840	0.8929	0.7748	0.5886	0.1330
	17	1.0000	1.0000	1.0000	1.0000	1.0000	0.9998	0.9964	0.9645	0.9087	0.7939	0.3231
	18	1.0000	1.0000	1.0000	1.0000	1.0000	1.0000	0.9995	0.9924	0.9757	0.9308	0.6083
	19	1.0000	1.0000	1.0000	1.0000	1.0000	1.0000	1.0000	0.9992	0.9968	0.9885	0.8784
	20	1.0000	1.0000	1.0000	1.0000	1.0000	1.0000	1.0000	1.0000	1.0000	1.0000	1.0000

TABLE II
Poisson distribution function

$$F_X(t) = P[X \le t] = \sum_{x \le t} e^{-\lambda s}(\lambda s)^x/x!$$

	λs															
t	0.5	1.0	2.0	3.0	4.0	5.0	6.0	7.0	8.0	9.0	10.0	11.0	12.0	13.0	14.0	15.0
0	0.607	0.368	0.135	0.050	0.018	0.007	0.002	0.001	0.000	0.000	0.000	0.000	0.000	0.000	0.000	0.000
1	0.910	0.736	0.406	0.199	0.092	0.040	0.017	0.007	0.003	0.001	0.000	0.000	0.000	0.000	0.000	0.000
2	0.986	0.920	0.677	0.423	0.238	0.125	0.062	0.030	0.014	0.006	0.003	0.001	0.001	0.000	0.000	0.000
3	0.998	0.981	0.857	0.647	0.433	0.265	0.151	0.082	0.042	0.021	0.010	0.005	0.002	0.001	0.000	0.000
4	1.000	0.996	0.947	0.815	0.629	0.440	0.285	0.173	0.100	0.055	0.029	0.015	0.008	0.004	0.002	0.001
5	1.000	0.999	0.983	0.916	0.785	0.616	0.446	0.301	0.191	0.116	0.067	0.038	0.020	0.011	0.006	0.003
6	1.000	1.000	0.995	0.966	0.889	0.762	0.606	0.450	0.313	0.207	0.130	0.079	0.046	0.026	0.014	0.008
7	1.000	1.000	0.999	0.988	0.949	0.867	0.744	0.599	0.453	0.324	0.220	0.143	0.090	0.054	0.032	0.018
8	1.000	1.000	1.000	0.996	0.979	0.932	0.847	0.729	0.593	0.456	0.333	0.232	0.155	0.100	0.062	0.037
9	1.000	1.000	1.000	0.999	0.992	0.968	0.916	0.830	0.717	0.587	0.458	0.341	0.242	0.166	0.109	0.070
10	1.000	1.000	1.000	1.000	0.997	0.986	0.957	0.901	0.816	0.706	0.583	0.460	0.347	0.252	0.176	0.118
11	1.000	1.000	1.000	1.000	0.999	0.995	0.980	0.947	0.888	0.803	0.697	0.579	0.462	0.353	0.260	0.185
12	1.000	1.000	1.000	1.000	1.000	0.998	0.991	0.973	0.936	0.876	0.792	0.689	0.576	0.463	0.358	0.268
13	1.000	1.000	1.000	1.000	1.000	0.999	0.996	0.987	0.966	0.926	0.864	0.781	0.682	0.573	0.464	0.363
14	1.000	1.000	1.000	1.000	1.000	1.000	0.999	0.994	0.983	0.959	0.917	0.854	0.772	0.675	0.570	0.466
15	1.000	1.000	1.000	1.000	1.000	1.000	0.999	0.998	0.992	0.978	0.951	0.907	0.844	0.764	0.669	0.568
16	1.000	1.000	1.000	1.000	1.000	1.000	1.000	0.999	0.996	0.989	0.973	0.944	0.899	0.835	0.756	0.664
17	1.000	1.000	1.000	1.000	1.000	1.000	1.000	1.000	0.998	0.995	0.986	0.968	0.937	0.890	0.827	0.749
18	1.000	1.000	1.000	1.000	1.000	1.000	1.000	1.000	0.999	0.998	0.993	0.982	0.963	0.930	0.883	0.819
19	1.000	1.000	1.000	1.000	1.000	1.000	1.000	1.000	1.000	0.999	0.997	0.991	0.979	0.957	0.923	0.875
20	1.000	1.000	1.000	1.000	1.000	1.000	1.000	1.000	1.000	1.000	0.998	0.995	0.988	0.975	0.952	0.917
21											0.999	0.998	0.994	0.986	0.971	0.947
22											1.000	0.999	0.997	0.992	0.983	0.967
23											1.000	1.000	0.999	0.996	0.991	0.981
24											1.000	1.000	0.999	0.998	0.995	0.989
25											1.000	1.000	1.000	0.999	0.997	0.994
26											1.000	1.000	1.000	1.000	0.999	0.997
27											1.000	1.000	1.000	1.000	0.999	0.998
28											1.000	1.000	1.000	1.000	1.000	0.999
29											1.000	1.000	1.000	1.000	1.000	1.000

TABLE III
A table of random digits

Line / Col.	1	2	3	4	5	6	7	8	9	10	11	12	13	14
1	10480	15011	01536	02011	81647	91646	69179	14194	62590	36207	20969	99570	91291	90700
2	22368	46573	25595	85393	30995	89198	27982	53402	93965	34095	52666	19174	39615	99505
3	24130	48360	22527	97265	76393	64809	15179	24830	49340	32081	30680	19655	63348	58629
4	42167	93093	06243	61680	07856	16376	39440	53537	71341	57004	00849	74917	97758	16379
5	37570	39975	81837	16656	06121	91782	60468	81305	49684	60672	14110	06927	01263	54613
6	77921	06907	11008	42751	27756	53498	18602	70659	90655	15053	21916	81825	44394	42880
7	99562	72905	56420	69994	98872	31016	71194	18738	44013	48840	63213	21069	10634	12952
8	96301	91977	05463	07972	18876	20922	94595	56869	69014	60045	18425	84903	42508	32307
9	89579	14342	63661	10281	17453	18103	57740	84378	25331	12566	58678	44947	05585	56941
10	85485	36857	43342	53988	53060	59533	38867	62300	08158	17983	16439	11458	18593	64952
11	28918	69578	88231	33276	70997	79936	56865	05859	90106	31595	01547	85590	91610	78188
12	63553	40961	48235	03427	49626	69445	18663	72695	52180	20847	12234	90511	33703	90322
13	09429	93969	52636	92737	88974	33488	36320	17617	30015	08272	84115	27156	30613	74952
14	10365	61129	87529	85689	48237	52267	67689	92294	01511	26358	85104	20285	29975	89868
15	07119	97336	71048	08178	77233	13916	47564	81056	97735	85977	29372	74461	28551	90707
16	51085	12765	51821	51259	77452	16308	60756	92144	49442	53900	70960	63990	75601	40719
17	02368	21382	52404	60268	89368	19885	55322	44819	01188	65255	64835	44919	05944	55157
18	01011	54092	33362	94904	31273	04146	18594	29852	71585	85030	51132	01915	92747	64951
19	52162	53916	46369	58586	23216	14513	83149	98736	23495	64350	94738	17752	35156	35749
20	07056	97628	33787	09998	42698	06691	76988	13602	51851	46104	88916	19509	25625	58104
21	48663	91245	85828	14346	09172	30168	90229	04734	59193	22178	30421	61666	99904	32812
22	54164	58492	22421	74103	47070	25306	76468	26384	58151	06646	21524	15227	96909	44592
23	32639	32363	05597	24200	13363	38005	94342	28728	35806	06912	17012	64161	18296	22851
24	29334	27001	87637	87308	58731	00256	45834	15398	46557	41135	10367	07684	36188	18510
25	02488	33062	28834	07351	19731	92420	60952	61280	50001	67658	32586	86679	50720	94953

26	81525	72295	04839	96423	24878	82651	66566	14778	76797	14780	13300	87074	79666	95725
27	29676	20591	68086	26432	46901	20849	89768	81536	86645	12659	92259	57102	80428	25280
28	00742	57392	39064	66432	84673	40027	32832	61362	98947	96067	64760	64584	96096	98253
29	05366	04213	25669	26422	44407	44048	37937	63904	45766	66134	75470	66520	34693	90449
30	91921	26418	64117	94305	26766	25940	39972	22209	71500	64568	91402	42416	07844	69618
31	00582	04711	87917	77341	42206	35126	74087	99547	81817	42607	43808	76655	62028	76630
32	00725	69884	62797	56170	86324	88072	76222	36086	84637	93161	76038	65855	77919	88006
33	69011	65797	95876	55293	18988	27354	26575	08625	40801	59920	29841	80150	12777	48501
34	25976	57948	29888	88604	67917	48708	18912	82271	65424	69774	33611	54262	85963	03547
35	09763	83473	73577	12908	30883	18317	28290	35797	05998	41688	34952	37888	38917	88050
36	91567	42595	27958	30134	04024	86385	29880	99730	55536	84855	29080	09250	79656	73211
37	17955	56349	90999	49127	20044	59931	06115	20542	18059	02008	73708	83517	36103	42791
38	46503	18584	18845	49618	02304	51038	20655	58727	28168	15475	56942	53389	20562	87338
39	92157	89634	94824	78171	84610	82834	09922	25417	44137	48413	25555	21246	35509	20468
40	14577	62765	35605	81263	39667	47358	56873	56307	61607	49518	89656	20103	77490	18062
41	98427	07523	33362	64270	01638	92477	66969	98420	04880	45585	46565	04102	46880	45709
42	34914	63976	88720	82765	34476	17032	87589	40836	32427	70002	70663	88863	77775	69348
43	70060	28277	39475	46473	23219	53416	94970	25832	69975	94884	19661	72828	00102	66794
44	53976	54914	06990	67245	68350	82948	11398	42878	80287	88267	47363	46634	06541	97809
45	76072	29515	40980	07391	58745	25774	22987	80059	39911	96189	41151	14222	60697	59583
46	90725	52210	83974	29992	65831	38857	50490	83765	55657	14361	31720	57375	56228	41546
47	64364	67412	33339	31926	14883	24413	59744	92351	97473	89286	35931	04110	23726	51900
48	08962	00358	31662	25388	61642	34072	81249	35648	56891	69352	48373	45578	78547	81788
49	95012	68379	93526	70765	10593	04542	76463	54328	02349	17247	28865	14777	62730	92277
50	15664	10493	20492	38391	91132	21999	59516	81652	27195	48223	46751	22923	32261	85653

From W. H. Beyer (ed.), in *CRC Handbook of Tables for Probability and Statistics*, 2d ed., 1968. Copyright CRC Press, Inc., Boca Raton, Fla.

TABLE IV
Cumulative chi-squared distribution

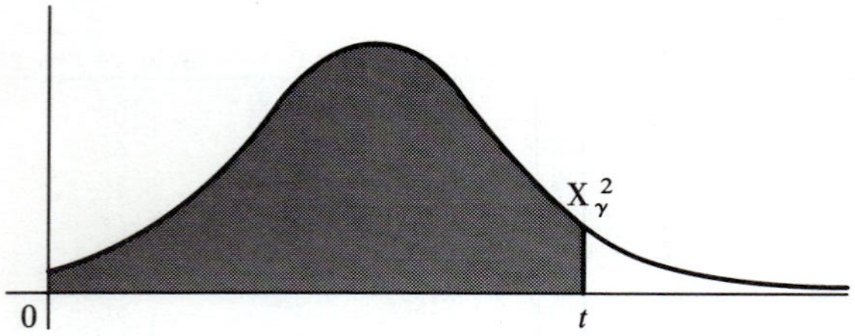

$P[X_\gamma^2 \le t]$

γ \ F	0.005	0.010	0.025	0.050	0.100	0.250	0.500	0.750	0.900	0.950	0.975	0.990	0.995
1	0.0000393	0.000157	0.000982	0.00393	0.0158	0.102	0.455	1.32	2.71	3.84	5.02	6.63	7.88
2	0.0100	0.0201	0.0506	0.103	0.211	0.575	1.39	2.77	4.61	5.99	7.38	9.21	10.6
3	0.0717	0.115	0.216	0.352	0.584	1.21	2.37	4.11	6.25	7.81	9.35	11.3	12.8
4	0.207	0.297	0.484	0.711	1.06	1.92	3.36	5.39	7.78	9.49	11.1	13.3	14.9
5	0.412	0.554	0.831	1.15	1.61	2.67	4.35	6.63	9.24	11.1	12.8	15.1	16.7
6	0.676	0.872	1.24	1.64	2.20	3.45	5.35	7.84	10.6	12.6	14.4	16.8	18.5
7	0.989	1.24	1.69	2.17	2.83	4.25	6.35	9.04	12.0	14.1	16.0	18.5	20.3
8	1.34	1.65	2.18	2.73	3.49	5.07	7.34	10.2	13.4	15.5	17.5	20.1	22.0
9	1.73	2.09	2.70	3.33	4.17	5.90	8.34	11.4	14.7	16.9	19.0	21.7	23.6
10	2.16	2.56	3.25	3.94	4.87	6.74	9.34	12.5	16.0	18.3	20.5	23.2	25.2
11	2.60	3.05	3.82	4.57	5.58	7.58	10.3	13.7	17.3	19.7	21.9	24.7	26.8
12	3.07	3.57	4.40	5.23	6.30	8.44	11.3	14.8	18.5	21.0	23.3	26.2	28.3

13	3.57	4.11	5.01	5.89	7.04	9.30	12.3	16.0	19.8	22.4	24.7	27.7	29.8
14	4.07	4.66	5.63	6.57	7.79	10.2	13.3	17.1	21.1	23.7	26.1	29.1	31.3
15	4.60	5.23	6.26	7.26	8.55	11.0	14.3	18.2	22.3	25.0	27.5	30.6	32.8
16	5.14	5.81	6.91	7.96	9.31	11.9	15.3	19.4	23.5	26.3	28.8	32.0	34.3
17	5.70	6.41	7.56	8.67	10.1	12.8	16.3	20.5	24.8	27.6	30.2	33.4	35.7
18	6.26	7.01	8.23	9.39	10.9	13.7	17.3	21.6	26.0	28.9	31.5	34.8	37.2
19	6.84	7.63	8.91	10.1	11.7	14.6	18.3	22.7	27.2	30.1	32.9	36.2	38.6
20	7.43	8.26	9.59	10.9	12.4	15.5	19.3	23.8	28.4	31.4	34.2	37.6	40.0
21	8.03	8.90	10.3	11.6	13.2	16.3	20.3	24.9	29.6	32.7	35.5	38.9	41.4
22	8.64	9.54	11.0	12.3	14.0	17.2	21.3	26.0	30.8	33.9	36.8	40.3	42.8
23	9.26	10.2	11.7	13.1	14.8	18.1	22.3	27.1	32.0	35.2	38.1	41.6	44.2
24	9.89	10.9	12.4	13.8	15.7	19.0	23.3	28.2	33.2	36.4	39.4	43.0	45.6
25	10.5	11.5	13.1	14.6	16.5	19.9	24.3	29.3	34.4	37.7	40.6	44.3	46.9
26	11.2	12.2	13.8	15.4	17.3	20.8	25.3	30.4	35.6	38.9	41.9	45.6	48.3
27	11.8	12.9	14.6	16.2	18.1	21.7	26.3	31.5	36.7	40.1	43.2	47.0	49.6
28	12.5	13.6	15.3	16.9	18.9	22.7	27.3	32.6	37.9	41.3	44.5	48.3	51.0
29	13.1	14.3	16.0	17.7	19.8	23.6	28.3	33.7	39.1	42.6	45.7	49.6	52.3
30	13.8	15.0	16.8	18.5	20.6	24.5	29.3	34.8	40.3	43.8	47.0	50.9	53.7

From Beyer, W. H. (ed.), in *CRC Handbook of Tables for Probability and Statistics*, 2d ed., 1968. Copyright CRC Press, Inc., Boca Raton, Fla.

TABLE V
Cumulative distribution: Standard normal

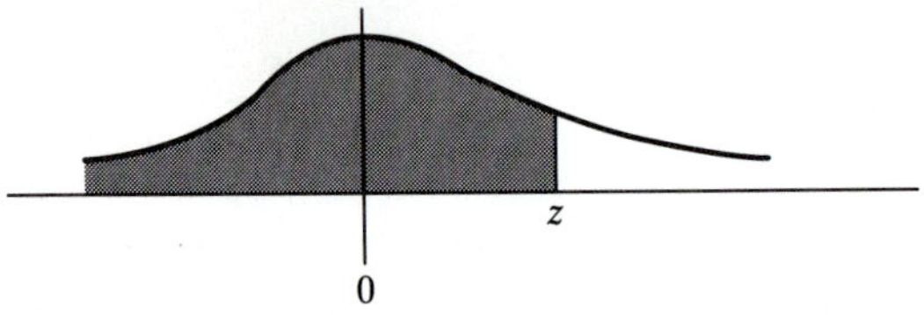

$F_Z(z) = P[Z \leq z]$

z	0.00	0.01	0.02	0.03	0.04	0.05	0.06	0.07	0.08	0.09
−3.4	0.0003	0.0003	0.0003	0.0003	0.0003	0.0003	0.0003	0.0003	0.0003	0.0002
−3.3	0.0005	0.0005	0.0005	0.0004	0.0004	0.0004	0.0004	0.0004	0.0004	0.0003
−3.2	0.0007	0.0007	0.0006	0.0006	0.0006	0.0006	0.0006	0.0005	0.0005	0.0005
−3.1	0.0010	0.0009	0.0009	0.0009	0.0008	0.0008	0.0008	0.0008	0.0007	0.0007
−3.0	0.0013	0.0013	0.0013	0.0012	0.0012	0.0011	0.0011	0.0011	0.0010	0.0010
−2.9	0.0019	0.0018	0.0018	0.0017	0.0016	0.0016	0.0015	0.0015	0.0014	0.0014
−2.8	0.0026	0.0025	0.0024	0.0023	0.0023	0.0022	0.0021	0.0021	0.0020	0.0019
−2.7	0.0035	0.0034	0.0033	0.0032	0.0031	0.0030	0.0029	0.0028	0.0027	0.0026
−2.6	0.0047	0.0045	0.0044	0.0043	0.0041	0.0040	0.0039	0.0038	0.0037	0.0036
−2.5	0.0062	0.0060	0.0059	0.0057	0.0055	0.0054	0.0052	0.0051	0.0049	0.0048
−2.4	0.0082	0.0080	0.0078	0.0075	0.0073	0.0071	0.0069	0.0068	0.0066	0.0064
−2.3	0.0107	0.0104	0.0102	0.0099	0.0096	0.0094	0.0091	0.0089	0.0087	0.0084
−2.2	0.0139	0.0136	0.0132	0.0129	0.0125	0.0122	0.0119	0.0116	0.0113	0.0110
−2.1	0.0179	0.0174	0.0170	0.0166	0.0162	0.0158	0.0154	0.0150	0.0146	0.0143
−2.0	0.0228	0.0222	0.0217	0.0212	0.0207	0.0202	0.0197	0.0192	0.0188	0.0183
−1.9	0.0287	0.0281	0.0274	0.0268	0.0262	0.0256	0.0250	0.0244	0.0239	0.0233
−1.8	0.0359	0.0351	0.0344	0.0336	0.0329	0.0322	0.0314	0.0307	0.0301	0.0294
−1.7	0.0446	0.0436	0.0427	0.0418	0.0409	0.0401	0.0392	0.0384	0.0375	0.0367
−1.6	0.0548	0.0537	0.0526	0.0516	0.0505	0.0495	0.0485	0.0475	0.0465	0.0455
−1.5	0.0668	0.0655	0.0643	0.0630	0.0618	0.0606	0.0594	0.0582	0.0571	0.0559
−1.4	0.0808	0.0793	0.0778	0.0764	0.0749	0.0735	0.0721	0.0708	0.0694	0.0681
−1.3	0.0968	0.0951	0.0934	0.0918	0.0901	0.0885	0.0869	0.0853	0.0838	0.0823
−1.2	0.1151	0.1131	0.1112	0.1093	0.1075	0.1056	0.1038	0.1020	0.1003	0.0985
−1.1	0.1357	0.1335	0.1314	0.1292	0.1271	0.1251	0.1230	0.1210	0.1190	0.1170
−1.0	0.1587	0.1562	0.1539	0.1515	0.1492	0.1469	0.1446	0.1423	0.1401	0.1379
−0.9	0.1841	0.1814	0.1788	0.1762	0.1736	0.1711	0.1685	0.1660	0.1635	0.1611
−0.8	0.2119	0.2090	0.2061	0.2033	0.2005	0.1977	0.1949	0.1921	0.1894	0.1867
−0.7	0.2420	0.2389	0.2358	0.2327	0.2296	0.2266	0.2236	0.2206	0.2177	0.2148
−0.6	0.2743	0.2709	0.2676	0.2643	0.2611	0.2578	0.2546	0.2514	0.2483	0.2451
−0.5	0.3085	0.3050	0.3015	0.2981	0.2946	0.2912	0.2877	0.2843	0.2810	0.2776
−0.4	0.3446	0.3409	0.3372	0.3336	0.3300	0.3264	0.3228	0.3192	0.3156	0.3121
−0.3	0.3821	0.3783	0.3745	0.3707	0.3669	0.3632	0.3594	0.3557	0.3520	0.3483
−0.2	0.4207	0.4168	0.4129	0.4090	0.4052	0.4013	0.3974	0.3936	0.3897	0.3859
−0.1	0.4602	0.4562	0.4522	0.4483	0.4443	0.4404	0.4364	0.4325	0.4286	0.4247
−0.0	0.5000	0.4960	0.4920	0.4880	0.4840	0.4801	0.4761	0.4721	0.4681	0.4641

TABLE V

Cumulative distribution: Standard normal (*continued*)

z	0.00	0.01	0.02	0.03	0.04	0.05	0.06	0.07	0.08	0.09
0.0	0.5000	0.5040	0.5080	0.5120	0.5160	0.5199	0.5239	0.5279	0.5319	0.5359
0.1	0.5398	0.5438	0.5478	0.5517	0.5557	0.5596	0.5636	0.5675	0.5714	0.5753
0.2	0.5793	0.5832	0.5871	0.5910	0.5948	0.5987	0.6026	0.6064	0.6103	0.6141
0.3	0.6179	0.6217	0.6255	0.6293	0.6331	0.6368	0.6406	0.6443	0.6480	0.6517
0.4	0.6554	0.6591	0.6628	0.6664	0.6700	0.6736	0.6772	0.6808	0.6844	0.6879
0.5	0.6915	0.6950	0.6985	0.7019	0.7054	0.7088	0.7123	0.7157	0.7190	0.7224
0.6	0.7257	0.7291	0.7324	0.7357	0.7389	0.7422	0.7454	0.7486	0.7517	0.7549
0.7	0.7580	0.7611	0.7642	0.7673	0.7704	0.7734	0.7764	0.7794	0.7823	0.7852
0.8	0.7881	0.7910	0.7939	0.7967	0.7995	0.8023	0.8051	0.8078	0.8106	0.8133
0.9	0.8159	0.8186	0.8212	0.8238	0.8264	0.8289	0.8315	0.8340	0.8365	0.8389
1.0	0.8413	0.8438	0.8461	0.8485	0.8508	0.8531	0.8554	0.8577	0.8599	0.8621
1.1	0.8643	0.8665	0.8686	0.8708	0.8729	0.8749	0.8770	0.8790	0.8810	0.8830
1.2	0.8849	0.8869	0.8888	0.8907	0.8925	0.8944	0.8962	0.8980	0.8997	0.9015
1.3	0.9032	0.9049	0.9066	0.9082	0.9099	0.9115	0.9131	0.9147	0.9162	0.9177
1.4	0.9192	0.9207	0.9222	0.9236	0.9251	0.9265	0.9279	0.9292	0.9306	0.9319
1.5	0.9332	0.9345	0.9357	0.9370	0.9382	0.9394	0.9406	0.9418	0.9429	0.9441
1.6	0.9452	0.9463	0.9474	0.9484	0.9495	0.9505	0.9515	0.9525	0.9535	0.9545
1.7	0.9554	0.9564	0.9573	0.9582	0.9591	0.9599	0.9608	0.9616	0.9625	0.9633
1.8	0.9641	0.9649	0.9656	0.9664	0.9671	0.9678	0.9686	0.9693	0.9699	0.9706
1.9	0.9713	0.9719	0.9726	0.9732	0.9738	0.9744	0.9750	0.9756	0.9761	0.9767
2.0	0.9772	0.9778	0.9783	0.9788	0.9793	0.9798	0.9803	0.9808	0.9812	0.9817
2.1	0.9821	0.9826	0.9830	0.9834	0.9838	0.9842	0.9846	0.9850	0.9854	0.9857
2.2	0.9861	0.9864	0.9868	0.9871	0.9875	0.9878	0.9881	0.9884	0.9887	0.9890
2.3	0.9893	0.9896	0.9898	0.9901	0.9904	0.9906	0.9909	0.9911	0.9913	0.9916
2.4	0.9918	0.9920	0.9922	0.9925	0.9927	0.9929	0.9931	0.9932	0.9934	0.9936
2.5	0.9938	0.9940	0.9941	0.9943	0.9945	0.9946	0.9948	0.9949	0.9951	0.9952
2.6	0.9953	0.9955	0.9956	0.9957	0.9959	0.9960	0.9961	0.9962	0.9963	0.9964
2.7	0.9965	0.9966	0.9967	0.9968	0.9969	0.9970	0.9971	0.9972	0.9973	0.9974
2.8	0.9974	0.9975	0.9976	0.9977	0.9977	0.9978	0.9979	0.9979	0.9980	0.9981
2.9	0.9981	0.9982	0.9982	0.9983	0.9984	0.9984	0.9985	0.9985	0.9986	0.9986
3.0	0.9987	0.9987	0.9987	0.9988	0.9988	0.9989	0.9989	0.9989	0.9990	0.9990
3.1	0.9990	0.9991	0.9991	0.9991	0.9992	0.9992	0.9992	0.9992	0.9993	0.9993
3.2	0.9993	0.9993	0.9994	0.9994	0.9994	0.9994	0.9994	0.9995	0.9995	0.9995
3.3	0.9995	0.9995	0.9995	0.9996	0.9996	0.9996	0.9996	0.9996	0.9996	0.9997
3.4	0.9997	0.9997	0.9997	0.9997	0.9997	0.9997	0.9997	0.9997	0.9997	0.9998

0.005376

TABLE VI

T distribution

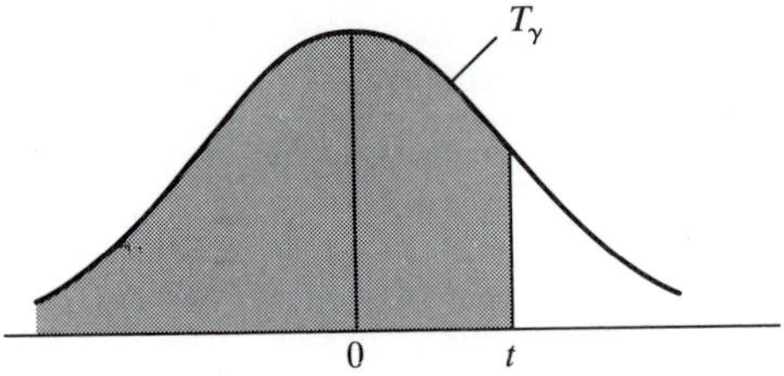

Column heading = cumulative probability

Row heading = degrees of freedom

Row ∞ = standard normal values

$P[T_\gamma \le t]$

γ	.6	.75	.9	.95	.975	.99	.995	.999	.9995
1	0.325	1.000	3.078	6.314	12.706	31.821	63.657	318.317	636.607
2	0.289	0.816	1.886	2.920	4.303	6.965	9.925	22.327	31.598
3	0.277	0.765	1.638	2.353	3.182	4.541	5.841	10.215	12.924
4	0.271	0.741	1.533	2.132	2.776	3.747	4.604	7.173	8.610
5	0.267	0.727	1.476	2.015	2.571	3.365	4.032	5.893	6.869
6	0.265	0.718	1.440	1.943	2.447	3.143	3.707	5.208	5.959
7	0.263	0.711	1.415	1.895	2.365	2.998	3.499	4.785	5.408
8	0.262	0.706	1.397	1.860	2.306	2.896	3.355	4.501	5.041
9	0.261	0.703	1.383	1.833	2.262	2.821	3.250	4.297	4.781
10	0.260	0.700	1.372	1.812	2.228	2.764	3.169	4.144	4.587
11	0.260	0.697	1.363	1.796	2.201	2.718	3.106	4.025	4.437
12	0.259	0.695	1.356	1.782	2.179	2.681	3.055	3.930	4.318
13	0.259	0.694	1.350	1.771	2.160	2.650	3.012	3.852	4.221
14	0.258	0.692	1.345	1.761	2.145	2.624	2.977	3.787	4.140
15	0.258	0.691	1.341	1.753	2.131	2.602	2.947	3.733	4.073
16	0.258	0.690	1.337	1.746	2.120	2.583	2.921	3.686	4.015
17	0.257	0.689	1.333	1.740	2.110	2.567	2.898	3.646	3.965
18	0.257	0.688	1.330	1.734	2.101	2.552	2.878	3.611	3.922
19	0.257	0.688	1.328	1.729	2.093	2.539	2.861	3.579	3.883
20	0.257	0.687	1.325	1.725	2.086	2.528	2.845	3.552	3.850
21	0.257	0.686	1.323	1.721	2.080	2.518	2.831	3.527	3.819
22	0.256	0.686	1.321	1.717	2.074	2.508	2.819	3.505	3.792
23	0.256	0.685	1.319	1.714	2.069	2.500	2.807	3.485	3.768
24	0.256	0.685	1.318	1.711	2.064	2.492	2.797	3.467	3.745
25	0.256	0.684	1.316	1.708	2.060	2.485	2.787	3.450	3.725
26	0.256	0.684	1.315	1.706	2.056	2.479	2.779	3.435	3.707
27	0.256	0.684	1.314	1.703	2.052	2.473	2.771	3.421	3.690
28	0.256	0.683	1.313	1.701	2.048	2.467	2.763	3.408	3.674
29	0.256	0.683	1.311	1.699	2.045	2.462	2.756	3.396	3.659
30	0.256	0.683	1.310	1.697	2.042	2.457	2.750	3.385	3.646
31	0.256	0.682	1.309	1.696	2.040	2.453	2.744	3.375	3.633
32	0.255	0.682	1.309	1.694	2.037	2.449	2.738	3.365	3.622
33	0.255	0.682	1.308	1.692	2.035	2.445	2.733	3.356	3.611
34	0.255	0.682	1.307	1.691	2.032	2.441	2.728	3.348	3.601
35	0.255	0.682	1.306	1.690	2.030	2.438	2.724	3.340	3.591
36	0.255	0.681	1.306	1.688	2.028	2.434	2.719	3.333	3.582
37	0.255	0.681	1.305	1.687	2.026	2.431	2.715	3.326	3.574
38	0.255	0.681	1.304	1.686	2.024	2.429	2.712	3.319	3.566
39	0.255	0.681	1.304	1.685	2.023	2.426	2.708	3.313	3.558
40	0.255	0.681	1.303	1.684	2.021	2.423	2.704	3.307	3.551
41	0.255	0.681	1.303	1.683	2.020	2.421	2.701	3.301	3.544
42	0.255	0.680	1.302	1.682	2.018	2.418	2.698	3.296	3.538
43	0.255	0.680	1.302	1.681	2.017	2.416	2.695	3.291	3.532
44	0.255	0.680	1.301	1.680	2.015	2.414	2.692	3.286	3.526

TABLE VI

T distribution (*continued*)

γ	.6	.75	.9	.95	.975	.99	.995	.999	.9995
45	0.255	0.680	1.301	1.679	2.014	2.412	2.690	3.281	3.520
46	0.255	0.680	1.300	1.679	2.013	2.410	2.687	3.277	3.515
47	0.255	0.680	1.300	1.678	2.012	2.408	2.685	3.273	3.510
48	0.255	0.680	1.299	1.677	2.011	2.407	2.682	3.269	3.505
49	0.255	0.680	1.299	1.677	2.010	2.405	2.680	3.265	3.500
50	0.255	0.679	1.299	1.676	2.009	2.403	2.678	3.261	3.496
51	0.255	0.679	1.298	1.675	2.008	2.402	2.676	3.258	3.492
52	0.255	0.679	1.298	1.675	2.007	2.400	2.674	3.255	3.488
53	0.255	0.679	1.298	1.674	2.006	2.399	2.672	3.251	3.484
54	0.255	0.679	1.297	1.674	2.005	2.397	2.670	3.248	3.480
55	0.255	0.679	1.297	1.673	2.004	2.396	2.668	3.245	3.476
56	0.255	0.679	1.297	1.673	2.003	2.395	2.667	3.242	3.473
57	0.255	0.679	1.297	1.672	2.002	2.394	2.665	3.239	3.470
58	0.255	0.679	1.296	1.672	2.002	2.392	2.663	3.237	3.466
59	0.254	0.679	1.296	1.671	2.001	2.391	2.662	3.234	3.463
60	0.254	0.679	1.296	1.671	2.000	2.390	2.660	3.232	3.460
61	0.254	0.679	1.296	1.670	2.000	2.389	2.659	3.229	3.457
62	0.254	0.678	1.295	1.670	1.999	2.388	2.658	3.227	3.455
63	0.254	0.678	1.295	1.669	1.998	2.387	2.656	3.225	3.452
64	0.254	0.678	1.295	1.669	1.998	2.386	2.655	3.223	3.449
65	0.254	0.678	1.295	1.669	1.997	2.385	2.654	3.221	3.447
66	0.254	0.678	1.295	1.668	1.997	2.384	2.652	3.218	3.444
67	0.254	0.678	1.294	1.668	1.996	2.383	2.651	3.217	3.442
68	0.254	0.678	1.294	1.668	1.995	2.382	2.650	3.215	3.440
69	0.254	0.678	1.294	1.667	1.995	2.382	2.649	3.213	3.437
70	0.254	0.678	1.294	1.667	1.994	2.381	2.648	3.211	3.435
71	0.254	0.678	1.294	1.667	1.994	2.380	2.647	3.209	3.433
72	0.254	0.678	1.293	1.666	1.993	2.379	2.646	3.207	3.431
73	0.254	0.678	1.293	1.666	1.993	2.379	2.645	3.206	3.429
74	0.254	0.678	1.293	1.666	1.993	2.378	2.644	3.204	3.427
75	0.254	0.678	1.293	1.665	1.992	2.377	2.643	3.203	3.425
76	0.254	0.678	1.293	1.665	1.992	2.376	2.642	3.201	3.423
77	0.254	0.678	1.293	1.665	1.991	2.376	2.641	3.200	3.422
78	0.254	0.678	1.292	1.665	1.991	2.375	2.640	3.198	3.420
79	0.254	0.678	1.292	1.664	1.990	2.375	2.640	3.197	3.418
80	0.254	0.678	1.292	1.664	1.990	2.374	2.639	3.195	3.416
81	0.254	0.678	1.292	1.664	1.990	2.373	2.638	3.194	3.415
82	0.254	0.677	1.292	1.664	1.989	2.373	2.637	3.193	3.413
83	0.254	0.677	1.292	1.663	1.989	2.372	2.636	3.191	3.412
84	0.254	0.677	1.292	1.663	1.989	2.372	2.636	3.190	3.410
85	0.254	0.677	1.292	1.663	1.988	2.371	2.635	3.189	3.409
86	0.254	0.677	1.291	1.663	1.988	2.371	2.634	3.188	3.407
87	0.254	0.677	1.291	1.663	1.988	2.370	2.634	3.187	3.406
88	0.254	0.677	1.291	1.662	1.987	2.369	2.633	3.186	3.405
89	0.254	0.677	1.291	1.662	1.987	2.369	2.632	3.184	3.403
90	0.254	0.677	1.291	1.662	1.987	2.369	2.632	3.183	3.402
91	0.254	0.677	1.291	1.662	1.986	2.368	2.631	3.182	3.401
92	0.254	0.677	1.291	1.662	1.986	2.368	2.630	3.181	3.400
93	0.254	0.677	1.291	1.661	1.986	2.367	2.630	3.180	3.398
94	0.254	0.677	1.291	1.661	1.986	2.367	2.629	3.179	3.397
95	0.254	0.677	1.291	1.661	1.985	2.366	2.629	3.178	3.396
96	0.254	0.677	1.290	1.661	1.985	2.366	2.628	3.177	3.395
97	0.254	0.677	1.290	1.661	1.985	2.365	2.627	3.176	3.394
98	0.254	0.677	1.290	1.661	1.984	2.365	2.627	3.176	3.393
99	0.254	0.677	1.290	1.660	1.984	2.365	2.626	3.175	3.392
100	0.254	0.677	1.290	1.660	1.984	2.364	2.626	3.174	3.391
∞	0.253	0.674	1.282	1.645	1.960	2.326	2.576	3.090	3.291

TABLE VII
Sample size for estimating the mean

	Level of t-test																				
Single-sided test	$\alpha = 0.005$					$\alpha = 0.01$					$\alpha = 0.025$					$\alpha = 0.05$					
Double-sided test	$\alpha = 0.01$					$\alpha = 0.02$					$\alpha = 0.05$					$\alpha = 0.1$					
$\beta =$	0.01	0.05	0.1	0.2	0.5	0.01	0.05	0.1	0.2	0.5	0.01	0.05	0.1	0.2	0.5	0.01	0.05	0.1	0.2	0.5	
0.05																					0.05
0.10																					0.10
0.15																				122	0.15
0.20										139					99					70	0.20
0.25					110					90				128	64			139	101	45	0.25
0.30				134	78				115	63			119	90	45		122	97	71	32	0.30
0.35			125	99	58			109	85	47		109	88	67	34		90	72	52	24	0.35
0.40		115	97	77	45		101	85	66	37	117	84	68	51	26	101	70	55	40	19	0.40
0.45		92	77	62	37	110	81	68	53	30	93	67	54	41	21	80	55	44	33	15	0.45
0.50	100	75	63	51	30	90	66	55	43	25	76	54	44	34	18	65	45	36	27	13	0.50
Value of $\Delta = (\mu - \mu_0)/\sigma$																					
0.55	83	63	53	42	26	75	55	46	36	21	63	45	37	28	15	54	38	30	22	11	0.55
0.60	71	53	45	36	22	63	47	39	31	18	53	38	32	24	13	46	32	26	19	9	0.60
0.65	61	46	39	31	20	55	41	34	27	16	46	33	27	21	12	39	28	22	17	8	0.65
0.70	53	40	34	28	17	47	35	30	24	14	40	29	24	19	10	34	24	19	15	8	0.70
0.75	47	36	30	25	16	42	31	27	21	13	35	26	21	16	9	30	21	17	13	7	0.75
0.80	41	32	27	22	14	37	28	24	19	12	31	22	19	15	9	27	19	15	12	6	0.80
0.85	37	29	24	20	13	33	25	21	17	11	28	21	17	13	8	24	17	14	11	6	0.85
0.90	34	26	22	18	12	29	23	19	16	10	25	19	16	12	7	21	15	13	10	5	0.90
0.95	31	24	20	17	11	27	21	18	14	9	23	17	14	11	7	19	14	11	9	5	0.95
1.00	28	22	19	16	10	25	19	16	13	9	21	16	13	10	6	18	13	11	8	5	1.00

1.1	24	19	16	14	9	21	16	14	12	8	18	13	11	9	6	15	11	9	7	1.1
1.2	21	16	14	12	8	18	14	12	10	7	15	12	10	8	5	13	10	8	6	1.2
1.3	18	15	13	11	8	16	13	11	9	6	14	10	9	7		11	8	7	6	1.3
1.4	16	13	12	10	7	14	11	10	9	6	12	9	8	7		10	8	7	5	1.4
1.5	15	12	11	9	7	13	10	9	8	6	11	8	7	6		9	7	6		1.5
1.6	13	11	10	8	6	12	10	9	7	5	10	8	7	6		8	6	6		1.6
1.7	12	10	9	8	6	11	9	8	7		9	7	6	5		8	6	5		1.7
1.8	12	10	9	8	6	10	8	7	7		8	7	6			7	6			1.8
1.9	11	9	8	7	6	10	8	7	6		8	6	6			7	5			1.9
2.0	10	8	8	7	5	9	7	7	6		7	6	5			6				2.0
Value of Δ $= (\mu - \mu_0)/\sigma$																				
2.1	10	8	7	7		8	7	6	6		7	6				6				2.1
2.2	9	8	7	6		8	7	6	5		7	6				6				2.2
2.3	9	7	7	6		8	6	6			6	5				5				2.3
2.4	8	7	7	6		7	6	6			6									2.4
2.5	8	7	6	6		7	6	6			6									2.5
3.0	7	6	6	5		6	5	5			5									3.0
3.5	6	5	5			5														3.5
4.0	6																			4.0

From Beyer, W. H. (ed.), in *CRC Handbook of Tables for Probability and Statistics*, 2d ed., 1968. Copyright CRC Press, Inc., Boca Raton, Fla.

TABLE VIII

Wilcoxon signed-rank test

$n = 5(1)50$

One-sided	Two-sided	$n = 5$	$n = 6$	$n = 7$	$n = 8$	$n = 9$	$n = 10$
$P = 0.05$	$P = 0.10$	1	2	4	6	8	11
$P = 0.025$	$P = 0.05$		1	2	4	6	8
$P = 0.01$	$P = 0.02$			0	2	3	5
$P = 0.005$	$P = 0.01$				0	2	3
One-sided	**Two-sided**	$n = 11$	$n = 12$	$n = 13$	$n = 14$	$n = 15$	$n = 16$
$P = 0.05$	$P = 0.10$	14	17	21	26	30	36
$P = 0.025$	$P = 0.05$	11	14	17	21	25	30
$P = 0.01$	$P = 0.02$	7	10	13	16	20	24
$P = 0.005$	$P = 0.01$	5	7	10	13	16	19
One-sided	**Two-sided**	$n = 17$	$n = 18$	$n = 19$	$n = 20$	$n = 21$	$n = 22$
$P = 0.05$	$P = 0.10$	41	47	54	60	68	75
$P = 0.025$	$P = 0.05$	35	40	46	52	59	66
$P = 0.01$	$P = 0.02$	28	33	38	43	49	56
$P = 0.005$	$P = 0.01$	23	28	32	37	43	49
One-sided	**Two-sided**	$n = 23$	$n = 24$	$n = 25$	$n = 26$	$n = 27$	$n = 28$
$P = 0.05$	$P = 0.10$	83	92	101	110	120	130
$P = 0.025$	$P = 0.05$	73	81	90	98	107	117
$P = 0.01$	$P = 0.02$	62	69	77	85	93	102
$P = 0.005$	$P = 0.01$	55	61	68	76	84	92
One-sided	**Two-sided**	$n = 29$	$n = 30$	$n = 31$	$n = 32$	$n = 33$	$n = 34$
$P = 0.05$	$P = 0.10$	141	152	163	175	188	201
$P = 0.025$	$P = 0.05$	127	137	148	159	171	183
$P = 0.01$	$P = 0.02$	111	120	130	141	151	162
$P = 0.005$	$P = 0.01$	100	109	118	128	138	149
One-sided	**Two-sided**	$n = 35$	$n = 36$	$n = 37$	$n = 38$	$n = 39$	
$P = 0.05$	$P = 0.10$	214	228	242	256	271	
$P = 0.025$	$P = 0.05$	195	208	222	235	250	
$P = 0.01$	$P = 0.02$	174	186	198	211	224	
$P = 0.005$	$P = 0.01$	160	171	183	195	208	
One-sided	**Two-sided**	$n = 40$	$n = 41$	$n = 42$	$n = 43$	$n = 44$	$n = 45$
$P = 0.05$	$P = 0.10$	287	303	319	336	353	371
$P = 0.025$	$P = 0.05$	264	279	295	311	327	344
$P = 0.01$	$P = 0.02$	238	252	267	281	297	313
$P = 0.005$	$P = 0.01$	221	234	248	262	277	292
One-sided	**Two-sided**	$n = 46$	$n = 47$	$n = 48$	$n = 49$	$n = 50$	
$P = 0.05$	$P = 0.10$	389	408	427	446	466	
$P = 0.025$	$P = 0.05$	361	379	397	415	434	
$P = 0.01$	$P = 0.02$	329	345	362	380	398	
$P = 0.005$	$P = 0.01$	307	323	339	356	373	

TABLE IX

F distribution

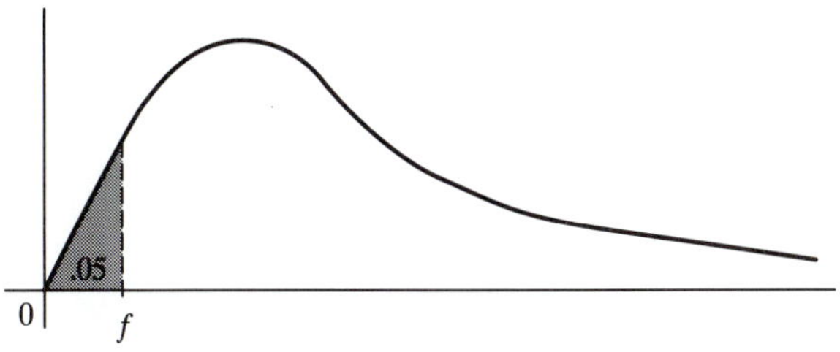

Column heading = numerator degrees of freedom

Row heading = denominator degrees of freedom

Points given as $f_{.95}$ points

For degrees of freedom > 120, use row or column 120

$$P[F_{\gamma_1, \gamma_2} \leq f] = 0.05$$

γ_1 \ γ_2	1	2	3	4	5	6	7	8	9
1	0.00620	0.05402	0.09874	0.12972	0.15133	0.16702	0.17884	0.18804	0.19541
2	0.00501	0.05263	0.10469	0.14400	0.17285	0.19443	0.21108	0.22427	0.23494
3	0.00464	0.05218	0.10779	0.15170	0.18486	0.21021	0.23005	0.24592	0.25888
4	0.00445	0.05196	0.10968	0.15653	0.19260	0.22057	0.24269	0.26056	0.27523
5	0.00434	0.05182	0.11094	0.15985	0.19800	0.22793	0.25179	0.27119	0.28722
6	0.00427	0.05174	0.11185	0.16226	0.20201	0.23343	0.25866	0.27928	0.29641
7	0.00422	0.05167	0.11253	0.16409	0.20509	0.23771	0.26406	0.28567	0.30369
8	0.00419	0.05163	0.11306	0.16553	0.20754	0.24115	0.26840	0.29085	0.30962
9	0.00416	0.05159	0.11348	0.16669	0.20954	0.24396	0.27197	0.29514	0.31457
10	0.00413	0.05156	0.11382	0.16766	0.21119	0.24631	0.27499	0.29876	0.31873
11	0.00412	0.05154	0.11411	0.16846	0.21258	0.24830	0.27754	0.30185	0.32231
12	0.00410	0.05152	0.11435	0.16915	0.21378	0.25000	0.27974	0.30451	0.32542
13	0.00409	0.05150	0.11455	0.16974	0.21481	0.25149	0.28166	0.30684	0.32813
14	0.00408	0.05148	0.11473	0.17026	0.21571	0.25278	0.28334	0.30889	0.33052
15	0.00407	0.05147	0.11490	0.17071	0.21651	0.25391	0.28484	0.31071	0.33266
16	0.00406	0.05146	0.11504	0.17111	0.21721	0.25495	0.28617	0.31233	0.33456
17	0.00405	0.05145	0.11517	0.17146	0.21784	0.25587	0.28737	0.31379	0.33627
18	0.00404	0.05144	0.11528	0.17179	0.21841	0.25669	0.28844	0.31512	0.33782
19	0.00404	0.05143	0.11538	0.17208	0.21892	0.25744	0.28942	0.31632	0.33923
20	0.00403	0.05143	0.11547	0.17234	0.21938	0.25812	0.29031	0.31742	0.34055
21	0.00403	0.05142	0.11555	0.17258	0.21981	0.25874	0.29112	0.31843	0.34171
22	0.00402	0.05141	0.11563	0.17279	0.22020	0.25931	0.29188	0.31936	0.34281
23	0.00402	0.05141	0.11570	0.17299	0.22056	0.25983	0.29257	0.32021	0.34382
24	0.00401	0.05140	0.11576	0.17318	0.22088	0.26032	0.29321	0.32100	0.34476
25	0.00401	0.05140	0.11582	0.17335	0.22118	0.26077	0.29380	0.32173	0.34564
26	0.00401	0.05140	0.11587	0.17350	0.22148	0.26118	0.29436	0.32240	0.34645
27	0.00401	0.05139	0.11592	0.17365	0.22174	0.26157	0.29487	0.32306	0.34721
28	0.00400	0.05139	0.11597	0.17379	0.22199	0.26194	0.29535	0.32365	0.34791
29	0.00400	0.05138	0.11601	0.17391	0.22222	0.26227	0.29580	0.32421	0.34858
30	0.00400	0.05138	0.11605	0.17403	0.22243	0.26259	0.29622	0.32473	0.34919
31	0.00400	0.05138	0.11609	0.17414	0.22263	0.26289	0.29662	0.32523	0.34978
32	0.00399	0.05138	0.11613	0.17425	0.22282	0.26317	0.29700	0.32569	0.35035
33	0.00399	0.05137	0.11616	0.17435	0.22300	0.26343	0.29735	0.32613	0.35087
34	0.00399	0.05137	0.11620	0.17444	0.22317	0.26367	0.29769	0.32655	0.35137
35	0.00399	0.05137	0.11623	0.17453	0.22333	0.26393	0.29800	0.32694	0.35183
36	0.00399	0.05137	0.11625	0.17461	0.22348	0.26415	0.29830	0.32732	0.35228
37	0.00399	0.05137	0.11628	0.17469	0.22362	0.26437	0.29859	0.32767	0.35270
38	0.00398	0.05136	0.11631	0.17477	0.22376	0.26457	0.29886	0.32801	0.35310
39	0.00398	0.05136	0.11633	0.17484	0.22388	0.26476	0.29911	0.32833	0.35348
40	0.00398	0.05136	0.11635	0.17491	0.22401	0.26495	0.29935	0.32864	0.35385
50	0.00397	0.05135	0.11654	0.17544	0.22500	0.26641	0.30130	0.33106	0.35677
60	0.00397	0.05134	0.11666	0.17580	0.22565	0.26739	0.30263	0.33274	0.35876
120	0.00395	0.05132	0.11696	0.17672	0.22735	0.26992	0.30605	0.33705	0.36396

TABLE IX
F distribution (*continued*)

γ_2 \ γ_1	10	11	12	13	14	15	16	17	18
1	0.20143	0.20642	0.21065	0.21428	0.21738	0.22010	0.22252	0.22465	0.22656
2	0.24373	0.25111	0.25738	0.26277	0.26746	0.27156	0.27520	0.27843	0.28133
3	0.26966	0.27875	0.28651	0.29319	0.29905	0.30419	0.30875	0.31281	0.31646
4	0.28752	0.29791	0.30683	0.31455	0.32131	0.32726	0.33257	0.33730	0.34156
5	0.30068	0.31212	0.32197	0.33053	0.33804	0.34467	0.35058	0.35586	0.36064
6	0.31083	0.32314	0.33376	0.34302	0.35115	0.35836	0.36479	0.37055	0.37576
7	0.31893	0.33197	0.34325	0.35309	0.36176	0.36946	0.37633	0.38251	0.38809
8	0.32555	0.33921	0.35105	0.36141	0.37055	0.37867	0.38593	0.39246	0.39837
9	0.33108	0.34527	0.35761	0.36840	0.37796	0.38645	0.39406	0.40091	0.40711
10	0.33577	0.35043	0.36319	0.37439	0.38428	0.39312	0.40104	0.40817	0.41464
11	0.33979	0.35487	0.36801	0.37956	0.38979	0.39891	0.40710	0.41450	0.42120
12	0.34329	0.35873	0.37221	0.38407	0.39459	0.40399	0.41243	0.42005	0.42697
13	0.34634	0.36213	0.37591	0.38806	0.39884	0.40846	0.41714	0.42498	0.43209
14	0.34907	0.36514	0.37920	0.39160	0.40262	0.41247	0.42133	0.42936	0.43667
15	0.35149	0.36783	0.38214	0.39477	0.40601	0.41607	0.42512	0.43333	0.44080
16	0.35366	0.37023	0.38478	0.39762	0.40906	0.41931	0.42854	0.43691	0.44452
17	0.35562	0.37242	0.38717	0.40021	0.41183	0.42224	0.43164	0.44016	0.44792
18	0.35739	0.37441	0.38934	0.40256	0.41435	0.42492	0.43447	0.44313	0.45100
19	0.35900	0.37621	0.39132	0.40471	0.41666	0.42738	0.43706	0.44585	0.45386
20	0.36047	0.37786	0.39314	0.40669	0.41878	0.42963	0.43944	0.44835	0.45648
21	0.36185	0.37938	0.39482	0.40849	0.42073	0.43172	0.44164	0.45066	0.45890
22	0.36310	0.38077	0.39636	0.41018	0.42254	0.43364	0.44368	0.45281	0.46114
23	0.36426	0.38209	0.39779	0.41174	0.42421	0.43543	0.44558	0.45481	0.46323
24	0.36533	0.38329	0.39911	0.41319	0.42576	0.43710	0.44734	0.45666	0.46518
25	0.36633	0.38441	0.40036	0.41454	0.42722	0.43865	0.44899	0.45840	0.46700
26	0.36726	0.38546	0.40152	0.41580	0.42858	0.44010	0.45053	0.46003	0.46871
27	0.36813	0.38644	0.40260	0.41696	0.42985	0.44145	0.45198	0.46155	0.47031
28	0.36893	0.38736	0.40362	0.41809	0.43105	0.44274	0.45334	0.46299	0.47182
29	0.36970	0.38822	0.40457	0.41913	0.43217	0.44394	0.45461	0.46434	0.47324
30	0.37042	0.38903	0.40547	0.42011	0.43324	0.44508	0.45581	0.46561	0.47457
31	0.37111	0.38980	0.40631	0.42103	0.43424	0.44615	0.45696	0.46682	0.47584
32	0.37175	0.39052	0.40712	0.42191	0.43518	0.44716	0.45804	0.46796	0.47704
33	0.37236	0.39120	0.40787	0.42274	0.43608	0.44813	0.45906	0.46903	0.47818
34	0.37293	0.39183	0.40859	0.42352	0.43693	0.44904	0.46003	0.47006	0.47926
35	0.37347	0.39245	0.40927	0.42427	0.43774	0.44990	0.46095	0.47104	0.48029
36	0.37398	0.39304	0.40992	0.42498	0.43850	0.45073	0.46184	0.47195	0.48126
37	0.37448	0.39360	0.41054	0.42565	0.43923	0.45151	0.46267	0.47286	0.48220
38	0.37495	0.39413	0.41113	0.42630	0.43993	0.45226	0.46347	0.47370	0.48309
39	0.37539	0.39463	0.41169	0.42691	0.44060	0.45297	0.46423	0.47451	0.48394
40	0.37581	0.39512	0.41220	0.42750	0.44123	0.45366	0.46496	0.47528	0.48476
50	0.37920	0.39895	0.41649	0.43219	0.44633	0.45914	0.47081	0.48148	0.49129
60	0.38151	0.40159	0.41943	0.43543	0.44985	0.46293	0.47488	0.48581	0.49587
120	0.38758	0.40849	0.42717	0.44396	0.45918	0.47302	0.48570	0.49735	0.50810

TABLE IX

***F* distribution** (*continued*)

γ_2 \ γ_1	19	20	21	22	23	24	25	26	27
1	0.22827	0.22982	0.23122	0.23250	0.23367	0.23475	0.23574	0.23666	0.23752
2	0.28394	0.28628	0.28845	0.29039	0.29222	0.29387	0.29540	0.29682	0.29814
3	0.31975	0.32274	0.32546	0.32796	0.33025	0.33236	0.33431	0.33612	0.33779
4	0.34541	0.34890	0.35209	0.35502	0.35770	0.36018	0.36247	0.36460	0.36659
5	0.36496	0.36888	0.37247	0.37576	0.37879	0.38158	0.38417	0.38658	0.38881
6	0.38045	0.38477	0.38869	0.39229	0.39562	0.39867	0.40154	0.40419	0.40665
7	0.39315	0.39777	0.40198	0.40588	0.40946	0.41277	0.41585	0.41870	0.42137
8	0.40375	0.40865	0.41314	0.41727	0.42109	0.42461	0.42789	0.43094	0.43378
9	0.41275	0.41791	0.42264	0.42698	0.43100	0.43473	0.43819	0.44140	0.44441
10	0.42052	0.42590	0.43084	0.43540	0.43961	0.44350	0.44713	0.45050	0.45366
11	0.42731	0.43289	0.43802	0.44275	0.44713	0.45119	0.45497	0.45849	0.46178
12	0.43328	0.43906	0.44437	0.44926	0.45378	0.45799	0.46190	0.46556	0.46898
13	0.43859	0.44454	0.45001	0.45506	0.45973	0.46406	0.46814	0.47186	0.47540
14	0.44334	0.44945	0.45507	0.46026	0.46506	0.46952	0.47367	0.47755	0.48120
15	0.44761	0.45387	0.45963	0.46495	0.46987	0.47446	0.47872	0.48270	0.48642
16	0.45149	0.45788	0.46377	0.46920	0.47423	0.47894	0.48330	0.48738	0.49121
17	0.45501	0.46153	0.46754	0.47309	0.47822	0.48302	0.48749	0.49166	0.49557
18	0.45826	0.46486	0.47099	0.47665	0.48189	0.48678	0.49133	0.49559	0.49958
19	0.46120	0.46794	0.47416	0.47992	0.48526	0.49023	0.49487	0.49920	0.50327
20	0.46393	0.47077	0.47709	0.48294	0.48837	0.49342	0.49814	0.50255	0.50669
21	0.46645	0.47340	0.47980	0.48573	0.49125	0.49638	0.50115	0.50566	0.50986
22	0.46879	0.47582	0.48231	0.48834	0.49391	0.49913	0.50399	0.50855	0.51281
23	0.47096	0.47808	0.48466	0.49076	0.49642	0.50170	0.50663	0.51123	0.51557
24	0.47299	0.48019	0.48685	0.49302	0.49875	0.50409	0.50909	0.51376	0.51815
25	0.47489	0.48216	0.48889	0.49513	0.50092	0.50634	0.51138	0.51613	0.52056
26	0.47668	0.48402	0.49081	0.49711	0.50298	0.50845	0.51356	0.51835	0.52283
27	0.47835	0.48576	0.49261	0.49898	0.50490	0.51043	0.51560	0.52043	0.52498
28	0.47993	0.48740	0.49432	0.50073	0.50671	0.51230	0.51752	0.52241	0.52699
29	0.48141	0.48895	0.49592	0.50240	0.50842	0.51406	0.51933	0.52427	0.52891
30	0.48281	0.49041	0.49745	0.50397	0.51005	0.51572	0.52105	0.52604	0.53072
31	0.48414	0.49180	0.49888	0.50546	0.51158	0.51734	0.52267	0.52770	0.53243
32	0.48540	0.49311	0.50025	0.50688	0.51305	0.51881	0.52425	0.52928	0.53406
33	0.48659	0.49435	0.50152	0.50822	0.51444	0.52024	0.52570	0.53078	0.53560
34	0.48772	0.49553	0.50277	0.50949	0.51576	0.52161	0.52708	0.53224	0.53706
35	0.48880	0.49665	0.50394	0.51070	0.51701	0.52290	0.52842	0.53363	0.53848
36	0.48982	0.49773	0.50505	0.51186	0.51821	0.52414	0.52969	0.53490	0.53982
37	0.49080	0.49875	0.50612	0.51296	0.51934	0.52531	0.53091	0.53616	0.54109
38	0.49174	0.49972	0.50713	0.51402	0.52043	0.52643	0.53207	0.53735	0.54232
39	0.49262	0.50066	0.50810	0.51502	0.52147	0.52752	0.53318	0.53850	0.54350
40	0.49348	0.50155	0.50903	0.51599	0.52247	0.52855	0.53424	0.53959	0.54462
50	0.50035	0.50874	0.51653	0.52378	0.53056	0.53690	0.54284	0.54845	0.55373
60	0.50516	0.51377	0.52177	0.52924	0.53623	0.54277	0.54893	0.55472	0.56017
120	0.51806	0.52732	0.53597	0.54406	0.55164	0.55877	0.56548	0.57182	0.57782

TABLE IX
***F* distribution** (*continued*)

γ_2 \ γ_1	28	29	30	31	32	33	34	35	36
1	0.23831	0.23905	0.23975	0.24040	0.24101	0.24159	0.24213	0.24264	0.24311
2	0.29935	0.30051	0.30158	0.30259	0.30353	0.30442	0.30526	0.30605	0.30680
3	0.33936	0.34083	0.34220	0.34348	0.34469	0.34583	0.34691	0.34792	0.34888
4	0.36845	0.37018	0.37180	0.37334	0.37475	0.37608	0.37736	0.37858	0.37970
5	0.39089	0.39287	0.39470	0.39641	0.39804	0.39957	0.40102	0.40239	0.40359
6	0.40895	0.41110	0.41313	0.41503	0.41681	0.41852	0.42011	0.42162	0.42304
7	0.42386	0.42619	0.42838	0.43044	0.43238	0.43422	0.43593	0.43759	0.43914
8	0.43644	0.43893	0.44126	0.44346	0.44553	0.44754	0.44939	0.45111	0.45278
9	0.44723	0.44986	0.45234	0.45466	0.45690	0.45896	0.46092	0.46279	0.46455
10	0.45661	0.45938	0.46198	0.46443	0.46674	0.46891	0.47100	0.47296	0.47482
11	0.46484	0.46773	0.47046	0.47302	0.47544	0.47772	0.47987	0.48194	0.48388
12	0.47217	0.47517	0.47800	0.58066	0.48317	0.48554	0.48779	0.48992	0.49198
13	0.47871	0.48181	0.48474	0.48748	0.49008	0.49255	0.49489	0.49710	0.49919
14	0.48459	0.48780	0.49082	0.49366	0.49634	0.49886	0.50129	0.50357	0.50574
15	0.48994	0.49323	0.49631	0.49925	0.50200	0.50462	0.50710	0.50944	0.51168
16	0.49479	0.49819	0.50135	0.50437	0.50717	0.50986	0.51241	0.51482	0.51711
17	0.49924	0.50269	0.50593	0.50902	0.51192	0.51465	0.51727	0.51975	0.52211
18	0.50332	0.50685	0.51017	0.51331	0.51630	0.51909	0.52176	0.52428	0.52670
19	0.50709	0.51068	0.51407	0.51728	0.52031	0.52316	0.52590	0.52848	0.53093
20	0.51057	0.51423	0.51768	0.52095	0.52404	0.52695	0.52973	0.53238	0.53488
21	0.51381	0.51752	0.52103	0.52435	0.52750	0.53048	0.53331	0.53600	0.53854
22	0.51682	0.52060	0.52420	0.52755	0.53074	0.53379	0.53663	0.53937	0.54196
23	0.51964	0.52347	0.52708	0.53053	0.53377	0.53684	0.53976	0.54255	0.54517
24	0.52228	0.52616	0.52982	0.53331	0.53660	0.53972	0.54268	0.54550	0.54818
25	0.52475	0.52869	0.53240	0.53595	0.53926	0.45242	0.54543	0.54828	0.55101
26	0.52705	0.53106	0.53482	0.53838	0.54177	0.54497	0.54801	0.55091	0.55367
27	0.52926	0.53330	0.53710	0.54070	0.54415	0.54737	0.55045	0.55338	0.55618
28	0.53133	0.53540	0.53926	0.54290	0.54635	0.54964	0.55276	0.55572	0.55855
29	0.53328	0.53740	0.54129	0.54498	0.54846	0.55177	0.55494	0.55794	0.56080
30	0.53513	0.53929	0.54322	0.54694	0.55047	0.55381	0.55701	0.56004	0.56293
31	0.53690	0.54107	0.54503	0.54879	0.55237	0.55575	0.55896	0.56203	0.56495
32	0.53854	0.54278	0.54679	0.55058	0.55418	0.55759	0.56083	0.56395	0.56688
33	0.54013	0.54440	0.54844	0.55227	0.55589	0.55934	0.56261	0.56573	0.56871
34	0.54164	0.54594	0.55002	0.55386	0.55752	0.56101	0.56431	0.56745	0.57045
35	0.54306	0.54741	0.55152	0.55541	0.55909	0.56260	0.56593	0.56910	0.57213
36	0.54445	0.54881	0.55295	0.55687	0.56058	0.56410	0.56747	0.57068	0.57373
37	0.54576	0.55015	0.55432	0.55825	0.56201	0.56556	0.56896	0.57218	0.57526
38	0.54702	0.55145	0.55563	0.55961	0.56338	0.56696	0.57035	0.57362	0.57672
39	0.54820	0.55268	0.55690	0.56089	0.56469	0.56830	0.57173	0.57499	0.57811
40	0.54937	0.55388	0.55810	0.56211	0.56594	0.56956	0.57303	0.57632	0.57947
50	0.55871	0.56342	0.56788	0.57212	0.57616	0.58000	0.58365	0.58712	0.59046
60	0.56534	0.57022	0.57485	0.57923	0.58344	0.58742	0.59122	0.59485	0.59831
120	0.58350	0.58888	0.59404	0.59890	0.60355	0.60799	0.61224	0.61630	0.62020

TABLE IX

***F* distribution** (*continued*)

γ_2 \ γ_1	37	38	39	40	50	60	120
1	0.24358	0.24400	0.24442	0.24481	0.24787	0.24992	0.25509
2	0.30751	0.30818	0.30882	0.30942	0.31421	0.31742	0.32554
3	0.34979	0.35065	0.35151	0.35227	0.35842	0.36257	0.37311
4	0.38080	0.38183	0.38280	0.38373	0.39106	0.39601	0.40862
5	0.40491	0.40607	0.40721	0.40825	0.41660	0.42225	0.43671
6	0.42441	0.42571	0.42692	0.42811	0.43735	0.44365	0.45977
7	0.44060	0.44203	0.44336	0.44464	0.45471	0.46157	0.47921
8	0.45437	0.45588	0.45730	0.45868	0.46950	0.47687	0.49593
9	0.46623	0.46783	0.46935	0.47080	0.48230	0.49017	0.51052
10	0.47659	0.47827	0.47988	0.48141	0.49355	0.50185	0.52343
11	0.48574	0.48750	0.48918	0.49078	0.50350	0.51224	0.53496
12	0.49388	0.49572	0.49747	0.49914	0.51242	0.52154	0.54535
13	0.50118	0.50310	0.50492	0.50665	0.52045	0.52993	0.55477
14	0.50780	0.50976	0.51167	0.51344	0.52772	0.53756	0.56337
15	0.51382	0.51584	0.51777	0.51961	0.53436	0.54453	0.57127
16	0.51931	0.52139	0.52338	0.52527	0.54045	0.55093	0.57855
17	0.52434	0.52647	0.52852	0.53047	0.54605	0.55683	0.58529
18	0.52899	0.53118	0.53326	0.53526	0.55123	0.56229	0.59157
19	0.53328	0.53552	0.53764	0.53969	0.55603	0.56736	0.59741
20	0.53728	0.53955	0.54173	0.54381	0.56050	0.57209	0.60288
21	0.54099	0.54331	0.54551	0.54765	0.56465	0.57651	0.60801
22	0.54445	0.54682	0.54908	0.55124	0.56857	0.58065	0.61286
23	0.54769	0.55010	0.55238	0.55459	0.57223	0.58453	0.61741
24	0.55072	0.55317	0.55551	0.55774	0.57568	0.58820	0.62172
25	0.55360	0.55608	0.55843	0.56071	0.57892	0.59165	0.62580
26	0.55630	0.55881	0.56121	0.56350	0.58199	0.59491	0.62967
27	0.55884	0.56138	0.56382	0.56614	0.58487	0.59799	0.63334
28	0.56125	0.56382	0.56628	0.56864	0.58762	0.60093	0.63684
29	0.56352	0.56613	0.56862	0.57101	0.59024	0.60372	0.64018
30	0.56569	0.56832	0.57084	0.57325	0.59271	0.60638	0.64336
31	0.56774	0.57040	0.57295	0.57539	0.59507	0.60890	0.64640
32	0.56969	0.57238	0.57496	0.57742	0.59731	0.61131	0.64931
33	0.57156	0.57427	0.57687	0.57936	0.59946	0.61361	0.65210
34	0.57333	0.57607	0.57869	0.58120	0.60151	0.61581	0.65478
35	0.57503	0.57779	0.58044	0.58297	0.60347	0.61791	0.65735
36	0.57664	0.57944	0.58210	0.58466	0.60534	0.61994	0.65982
37	0.57819	0.58099	0.58370	0.58628	0.60714	0.62188	0.66219
38	0.57968	0.58251	0.58523	0.58783	0.60887	0.62374	0.66447
39	0.58111	0.58396	0.58668	0.58931	0.61052	0.62552	0.66668
40	0.58248	0.58535	0.58809	0.59072	0.61211	0.62725	0.66880
50	0.59364	0.59668	0.59961	0.60241	0.62520	0.64143	0.68656
60	0.60163	0.60481	0.60785	0.61077	0.63466	0.65176	0.69978
120	0.62393	0.62751	0.63096	0.63426	0.66161	0.68154	0.73971

TABLE IX

***F* distribution** (*continued*)

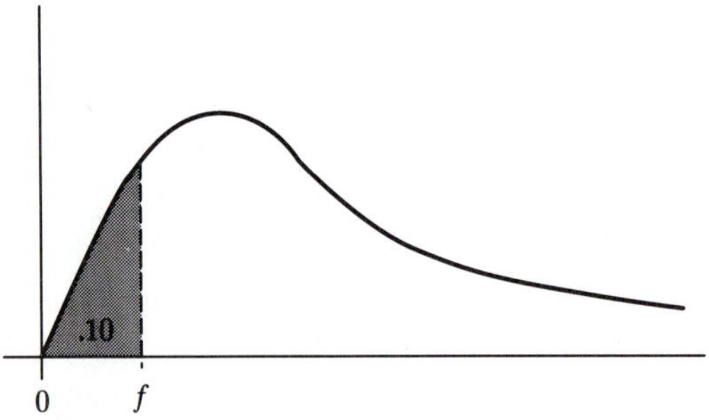

Column heading = numerator degrees of freedom

Row heading = denominator degrees of freedom

Points given as $f_{.90}$ points

For degrees of freedom > 120, use row or column 120

$P[F_{\gamma_1, \gamma_2} \le f] = .10$

γ_2 \ γ_1	1	2	3	4	5	6	7	8	9	10
1	0.025	0.117	0.181	0.220	0.246	0.265	0.279	0.289	0.298	0.304
2	0.020	0.111	0.183	0.231	0.265	0.289	0.307	0.321	0.333	0.342
3	0.019	0.109	0.186	0.239	0.276	0.304	0.325	0.342	0.356	0.367
4	0.018	0.108	0.187	0.243	0.284	0.314	0.338	0.356	0.371	0.384
5	0.017	0.108	0.188	0.247	0.290	0.322	0.347	0.367	0.383	0.397
6	0.017	0.107	0.189	0.249	0.294	0.327	0.354	0.375	0.392	0.406
7	0.017	0.107	0.190	0.251	0.297	0.332	0.359	0.381	0.399	0.414
8	0.017	0.107	0.190	0.253	0.299	0.335	0.363	0.386	0.405	0.421
9	0.017	0.107	0.191	0.254	0.302	0.338	0.367	0.390	0.410	0.426
10	0.017	0.106	0.191	0.255	0.303	0.340	0.370	0.394	0.414	0.431
11	0.017	0.106	0.191	0.256	0.305	0.343	0.373	0.397	0.417	0.434
12	0.016	0.106	0.192	0.257	0.306	0.344	0.375	0.400	0.420	0.438
13	0.016	0.106	0.192	0.257	0.307	0.346	0.377	0.402	0.423	0.441
14	0.016	0.106	0.192	0.258	0.308	0.347	0.378	0.404	0.425	0.443
15	0.016	0.106	0.192	0.258	0.309	0.348	0.380	0.406	0.427	0.446
16	0.016	0.106	0.192	0.259	0.310	0.349	0.381	0.407	0.429	0.448
17	0.016	0.106	0.193	0.259	0.310	0.350	0.382	0.409	0.431	0.450
18	0.016	0.106	0.193	0.260	0.311	0.351	0.384	0.410	0.432	0.451
19	0.016	0.106	0.193	0.260	0.311	0.352	0.385	0.411	0.434	0.453
20	0.016	0.106	0.193	0.260	0.312	0.353	0.385	0.412	0.435	0.454
21	0.016	0.106	0.193	0.260	0.312	0.353	0.386	0.413	0.436	0.456
22	0.016	0.106	0.193	0.261	0.313	0.354	0.387	0.414	0.437	0.457
23	0.016	0.106	0.193	0.261	0.313	0.354	0.388	0.415	0.438	0.458
24	0.016	0.106	0.193	0.261	0.313	0.355	0.388	0.416	0.439	0.459
25	0.016	0.106	0.193	0.261	0.314	0.355	0.389	0.417	0.440	0.460
26	0.016	0.106	0.193	0.261	0.314	0.356	0.389	0.417	0.441	0.461
27	0.016	0.106	0.193	0.262	0.314	0.356	0.390	0.418	0.442	0.462
28	0.016	0.106	0.193	0.262	0.315	0.356	0.390	0.419	0.442	0.463
29	0.016	0.106	0.193	0.262	0.315	0.357	0.391	0.419	0.443	0.463
30	0.016	0.106	0.193	0.262	0.315	0.357	0.391	0.420	0.444	0.464
31	0.016	0.106	0.194	0.262	0.315	0.357	0.392	0.420	0.444	0.465
32	0.016	0.106	0.194	0.262	0.315	0.358	0.392	0.421	0.445	0.465
33	0.016	0.106	0.194	0.262	0.316	0.358	0.392	0.421	0.445	0.466
34	0.016	0.106	0.194	0.262	0.316	0.358	0.393	0.421	0.446	0.466
35	0.016	0.106	0.194	0.262	0.316	0.359	0.393	0.422	0.446	0.467
36	0.016	0.106	0.194	0.263	0.316	0.359	0.393	0.422	0.446	0.467
37	0.016	0.106	0.194	0.263	0.316	0.359	0.394	0.423	0.447	0.468
38	0.016	0.106	0.194	0.263	0.316	0.359	0.394	0.423	0.447	0.468
39	0.016	0.106	0.194	0.263	0.317	0.359	0.394	0.423	0.448	0.469
40	0.016	0.106	0.194	0.263	0.317	0.360	0.394	0.423	0.448	0.469
50	0.016	0.106	0.194	0.263	0.318	0.361	0.396	0.426	0.451	0.472
60	0.016	0.106	0.194	0.264	0.318	0.362	0.398	0.428	0.453	0.475
120	0.016	0.105	0.194	0.265	0.320	0.365	0.401	0.432	0.458	0.480

TABLE IX

F distribution (*continued*)

γ_2 \ γ_1	11	12	13	14	15	16	17	18	19	20
1	0.310	0.315	0.319	0.322	0.325	0.328	0.330	0.333	0.334	0.336
2	0.350	0.356	0.362	0.367	0.371	0.375	0.378	0.381	0.384	0.386
3	0.376	0.384	0.391	0.396	0.402	0.406	0.410	0.414	0.417	0.420
4	0.394	0.403	0.411	0.418	0.423	0.429	0.433	0.437	0.441	0.445
5	0.408	0.418	0.426	0.433	0.440	0.446	0.451	0.455	0.460	0.463
6	0.419	0.429	0.438	0.446	0.453	0.459	0.465	0.470	0.474	0.478
7	0.427	0.438	0.448	0.456	0.463	0.470	0.476	0.481	0.486	0.490
8	0.434	0.446	0.456	0.464	0.472	0.479	0.485	0.491	0.496	0.500
9	0.440	0.452	0.462	0.471	0.479	0.487	0.493	0.499	0.504	0.509
10	0.445	0.457	0.468	0.477	0.486	0.493	0.500	0.506	0.511	0.516
11	0.449	0.462	0.473	0.482	0.491	0.499	0.506	0.512	0.518	0.523
12	0.453	0.466	0.477	0.487	0.496	0.504	0.511	0.517	0.523	0.528
13	0.456	0.469	0.481	0.491	0.500	0.508	0.515	0.522	0.528	0.533
14	0.459	0.472	0.484	0.494	0.504	0.512	0.519	0.526	0.532	0.538
15	0.461	0.475	0.487	0.498	0.507	0.515	0.523	0.530	0.536	0.542
16	0.464	0.478	0.490	0.500	0.510	0.519	0.526	0.533	0.540	0.546
17	0.466	0.480	0.492	0.503	0.513	0.522	0.529	0.537	0.543	0.549
18	0.468	0.482	0.494	0.505	0.515	0.524	0.532	0.539	0.546	0.552
19	0.469	0.484	0.496	0.508	0.518	0.527	0.535	0.542	0.549	0.555
20	0.471	0.486	0.498	0.510	0.520	0.529	0.537	0.544	0.551	0.557
21	0.472	0.487	0.500	0.511	0.522	0.531	0.539	0.547	0.554	0.560
22	0.474	0.489	0.502	0.513	0.523	0.533	0.541	0.549	0.556	0.562
23	0.475	0.490	0.503	0.515	0.525	0.534	0.543	0.551	0.558	0.564
24	0.476	0.491	0.504	0.516	0.527	0.536	0.545	0.552	0.559	0.566
25	0.477	0.492	0.506	0.517	0.528	0.538	0.546	0.554	0.561	0.568
26	0.478	0.493	0.507	0.519	0.529	0.539	0.548	0.556	0.563	0.570
27	0.479	0.494	0.508	0.520	0.531	0.540	0.549	0.557	0.564	0.571
28	0.480	0.495	0.509	0.521	0.532	0.542	0.550	0.558	0.566	0.573
29	0.481	0.496	0.510	0.522	0.533	0.543	0.552	0.560	0.567	0.574
30	0.482	0.497	0.511	0.523	0.534	0.544	0.553	0.561	0.568	0.575
31	0.482	0.498	0.512	0.524	0.535	0.545	0.554	0.562	0.570	0.577
32	0.483	0.499	0.513	0.525	0.536	0.546	0.555	0.563	0.571	0.578
33	0.484	0.499	0.513	0.526	0.537	0.547	0.556	0.564	0.572	0.579
34	0.484	0.500	0.514	0.526	0.538	0.548	0.557	0.565	0.573	0.580
35	0.485	0.501	0.515	0.527	0.538	0.549	0.558	0.566	0.574	0.581
36	0.485	0.501	0.515	0.528	0.539	0.549	0.559	0.567	0.575	0.582
37	0.486	0.502	0.516	0.529	0.540	0.550	0.559	0.568	0.576	0.583
38	0.486	0.502	0.517	0.529	0.541	0.551	0.560	0.569	0.577	0.584
39	0.487	0.503	0.517	0.530	0.541	0.552	0.561	0.569	0.577	0.585
40	0.487	0.503	0.518	0.530	0.542	0.552	0.562	0.570	0.578	0.585
50	0.491	0.508	0.522	0.535	0.547	0.558	0.567	0.576	0.584	0.592
60	0.494	0.510	0.525	0.538	0.550	0.561	0.571	0.580	0.589	0.596
120	0.500	0.518	0.533	0.547	0.560	0.571	0.582	0.591	0.600	0.609

TABLE IX

***F* distribution** (*continued*)

γ_2 \ γ_1	21	22	23	24	25	26	27	28	29	30
1	0.338	0.339	0.340	0.342	0.343	0.344	0.345	0.346	0.346	0.347
2	0.388	0.390	0.392	0.394	0.396	0.397	0.398	0.400	0.401	0.402
3	0.423	0.425	0.428	0.430	0.432	0.433	0.435	0.437	0.438	0.439
4	0.448	0.451	0.453	0.456	0.458	0.460	0.462	0.464	0.465	0.467
5	0.467	0.470	0.473	0.475	0.478	0.480	0.482	0.484	0.486	0.488
6	0.482	0.485	0.488	0.491	0.494	0.497	0.499	0.501	0.503	0.505
7	0.494	0.498	0.501	0.504	0.507	0.510	0.512	0.515	0.517	0.519
8	0.505	0.508	0.512	0.515	0.518	0.521	0.524	0.526	0.529	0.531
9	0.513	0.517	0.521	0.525	0.528	0.531	0.534	0.536	0.539	0.541
10	0.521	0.525	0.529	0.533	0.536	0.539	0.542	0.545	0.547	0.550
11	0.528	0.532	0.536	0.540	0.543	0.546	0.549	0.552	0.555	0.557
12	0.533	0.538	0.542	0.546	0.549	0.553	0.556	0.559	0.562	0.564
13	0.538	0.543	0.547	0.551	0.555	0.559	0.562	0.565	0.568	0.570
14	0.543	0.548	0.552	0.556	0.560	0.564	0.567	0.570	0.573	0.576
15	0.547	0.552	0.557	0.561	0.565	0.568	0.572	0.575	0.578	0.581
16	0.551	0.556	0.561	0.565	0.569	0.573	0.576	0.579	0.582	0.585
17	0.555	0.560	0.564	0.569	0.573	0.576	0.580	0.583	0.586	0.589
18	0.558	0.563	0.568	0.572	0.576	0.580	0.584	0.587	0.590	0.593
19	0.561	0.566	0.571	0.575	0.579	0.583	0.587	0.590	0.594	0.597
20	0.563	0.569	0.573	0.578	0.582	0.586	0.590	0.593	0.597	0.600
21	0.566	0.571	0.576	0.581	0.585	0.589	0.593	0.596	0.600	0.603
22	0.568	0.573	0.578	0.583	0.587	0.592	0.595	0.599	0.602	0.606
23	0.570	0.576	0.581	0.585	0.590	0.594	0.598	0.602	0.605	0.608
24	0.572	0.578	0.583	0.588	0.592	0.596	0.600	0.604	0.607	0.611
25	0.574	0.580	0.585	0.590	0.594	0.598	0.602	0.606	0.610	0.613
26	0.576	0.581	0.587	0.591	0.596	0.600	0.604	0.608	0.612	0.615
27	0.577	0.583	0.588	0.593	0.598	0.602	0.606	0.610	0.614	0.617
28	0.579	0.585	0.590	0.595	0.600	0.604	0.608	0.612	0.616	0.619
29	0.580	0.586	0.592	0.597	0.601	0.606	0.610	0.614	0.617	0.621
30	0.582	0.588	0.593	0.598	0.603	0.607	0.611	0.615	0.619	0.622
31	0.583	0.589	0.594	0.599	0.604	0.609	0.613	0.617	0.621	0.624
32	0.584	0.590	0.596	0.601	0.606	0.610	0.614	0.618	0.622	0.626
33	0.585	0.591	0.597	0.602	0.607	0.611	0.616	0.620	0.624	0.627
34	0.586	0.592	0.598	0.603	0.608	0.613	0.617	0.621	0.625	0.628
35	0.587	0.594	0.599	0.604	0.609	0.614	0.618	0.622	0.626	0.630
36	0.588	0.595	0.600	0.606	0.611	0.615	0.619	0.624	0.627	0.631
37	0.589	0.596	0.601	0.607	0.612	0.616	0.621	0.625	0.629	0.632
38	0.590	0.597	0.602	0.608	0.613	0.617	0.622	0.626	0.630	0.633
39	0.591	0.597	0.603	0.609	0.614	0.618	0.623	0.627	0.631	0.635
40	0.592	0.598	0.604	0.610	0.615	0.619	0.624	0.628	0.632	0.636
50	0.599	0.605	0.611	0.617	0.622	0.627	0.632	0.636	0.640	0.644
60	0.604	0.610	0.616	0.622	0.628	0.633	0.638	0.642	0.646	0.650
120	0.616	0.623	0.630	0.636	0.642	0.648	0.653	0.658	0.663	0.667

TABLE IX
F distribution (*continued*)

γ_2 \ γ_1	31	32	33	34	35	36	37	38	39	40
1	0.348	0.349	0.349	0.350	0.350	0.351	0.351	0.352	0.352	0.353
2	0.403	0.404	0.405	0.406	0.406	0.407	0.408	0.409	0.409	0.410
3	0.441	0.442	0.443	0.444	0.445	0.446	0.447	0.448	0.448	0.449
4	0.468	0.470	0.471	0.472	0.473	0.474	0.475	0.476	0.477	0.478
5	0.490	0.491	0.493	0.494	0.495	0.496	0.498	0.499	0.500	0.501
6	0.507	0.508	0.510	0.512	0.513	0.514	0.516	0.517	0.518	0.519
7	0.521	0.523	0.524	0.526	0.528	0.529	0.530	0.532	0.533	0.534
8	0.533	0.535	0.537	0.538	0.540	0.541	0.543	0.544	0.546	0.547
9	0.543	0.545	0.547	0.549	0.550	0.552	0.554	0.555	0.556	0.558
10	0.552	0.554	0.556	0.558	0.560	0.561	0.563	0.564	0.566	0.567
11	0.560	0.562	0.564	0.566	0.568	0.569	0.571	0.573	0.574	0.576
12	0.567	0.569	0.571	0.573	0.575	0.577	0.578	0.580	0.582	0.583
13	0.573	0.575	0.577	0.579	0.581	0.583	0.585	0.587	0.588	0.590
14	0.578	0.581	0.583	0.585	0.587	0.589	0.591	0.593	0.594	0.596
15	0.583	0.586	0.588	0.590	0.592	0.594	0.596	0.598	0.600	0.602
16	0.588	0.590	0.593	0.595	0.597	0.599	0.601	0.603	0.605	0.607
17	0.592	0.595	0.597	0.599	0.602	0.604	0.606	0.608	0.609	0.611
18	0.596	0.599	0.601	0.603	0.606	0.608	0.610	0.612	0.614	0.615
19	0.599	0.602	0.605	0.607	0.609	0.612	0.614	0.616	0.618	0.619
20	0.603	0.605	0.608	0.611	0.613	0.615	0.617	0.619	0.621	0.623
21	0.606	0.609	0.611	0.614	0.616	0.618	0.620	0.623	0.625	0.626
22	0.609	0.611	0.614	0.617	0.619	0.621	0.624	0.626	0.628	0.630
23	0.611	0.614	0.617	0.619	0.622	0.624	0.626	0.629	0.631	0.633
24	0.614	0.617	0.619	0.622	0.624	0.627	0.629	0.631	0.633	0.635
25	0.616	0.619	0.622	0.624	0.627	0.629	0.632	0.634	0.636	0.638
26	0.618	0.621	0.624	0.627	0.629	0.632	0.634	0.636	0.638	0.640
27	0.620	0.623	0.626	0.629	0.631	0.634	0.636	0.638	0.641	0.643
28	0.622	0.625	0.628	0.631	0.634	0.636	0.638	0.641	0.643	0.645
29	0.624	0.627	0.630	0.633	0.635	0.638	0.640	0.643	0.645	0.647
30	0.626	0.629	0.632	0.635	0.637	0.640	0.642	0.645	0.647	0.649
31	0.627	0.631	0.634	0.636	0.639	0.642	0.644	0.646	0.649	0.651
32	0.629	0.632	0.635	0.638	0.641	0.643	0.646	0.648	0.650	0.653
33	0.630	0.634	0.637	0.640	0.642	0.645	0.647	0.650	0.652	0.654
34	0.632	0.635	0.638	0.641	0.644	0.646	0.649	0.651	0.654	0.656
35	0.633	0.636	0.640	0.642	0.645	0.648	0.650	0.653	0.655	0.657
36	0.635	0.638	0.641	0.644	0.647	0.649	0.652	0.654	0.657	0.659
37	0.636	0.639	0.642	0.645	0.648	0.651	0.653	0.656	0.658	0.660
38	0.637	0.640	0.643	0.646	0.649	0.652	0.655	0.657	0.659	0.662
39	0.638	0.641	0.645	0.648	0.650	0.653	0.656	0.658	0.661	0.663
40	0.639	0.643	0.646	0.649	0.652	0.654	0.657	0.659	0.662	0.664
50	0.648	0.651	0.655	0.658	0.661	0.664	0.667	0.669	0.672	0.674
60	0.654	0.658	0.661	0.665	0.668	0.671	0.674	0.676	0.679	0.682
120	0.671	0.675	0.679	0.683	0.686	0.690	0.693	0.696	0.699	0.702

TABLE IX
***F* distribution (*continued*)**

γ_2 \ γ_1	50	60	120
1	0.356	0.358	0.364
2	0.415	0.418	0.426
3	0.455	0.459	0.469
4	0.485	0.490	0.502
5	0.509	0.514	0.527
6	0.528	0.533	0.548
7	0.543	0.550	0.566
8	0.557	0.563	0.581
9	0.568	0.575	0.594
10	0.578	0.586	0.605
11	0.587	0.595	0.615
12	0.595	0.603	0.625
13	0.602	0.611	0.633
14	0.609	0.618	0.640
15	0.615	0.624	0.647
16	0.620	0.629	0.654
17	0.625	0.635	0.659
18	0.630	0.639	0.665
19	0.634	0.644	0.670
20	0.638	0.648	0.675
21	0.641	0.652	0.679
22	0.645	0.655	0.683
23	0.648	0.659	0.687
24	0.651	0.662	0.691
25	0.654	0.665	0.694
26	0.657	0.668	0.698
27	0.659	0.670	0.701
28	0.661	0.673	0.704
29	0.664	0.675	0.707
30	0.666	0.678	0.710
31	0.668	0.680	0.712
32	0.670	0.682	0.715
33	0.672	0.684	0.717
34	0.674	0.686	0.719
35	0.675	0.688	0.721
36	0.677	0.689	0.724
37	0.678	0.691	0.726
38	0.680	0.693	0.728
39	0.681	0.694	0.729
40	0.683	0.696	0.731
50	0.694	0.708	0.746
60	0.702	0.717	0.757
120	0.725	0.742	0.791

F distribution (*continued*)

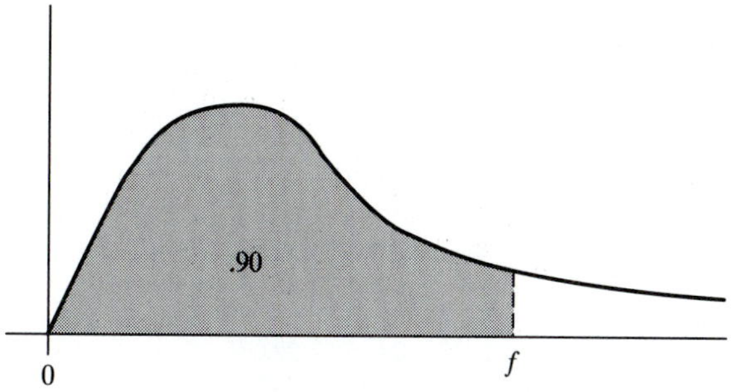

Column heading = numerator degrees of freedom

Row heading = denominator degrees of freedom

Points given as $f_{.10}$ points

For degrees of freedom > 120, use row or column 120

$$P[F_{\gamma_1, \gamma_2} \leq f] = .90$$

γ_2 \ γ_1	1	2	3	4	5	6	7	8	9	10
1	39.862	49.500	53.593	55.833	57.240	58.204	58.906	59.439	59.857	60.195
2	8.526	9.000	9.162	9.243	9.293	9.326	9.349	9.367	9.381	9.392
3	5.538	5.462	5.391	5.343	5.309	5.285	5.266	5.252	5.240	5.231
4	4.545	4.325	4.191	4.107	4.051	4.010	3.979	3.955	3.936	3.920
5	4.060	3.780	3.619	3.520	3.453	3.405	3.368	3.339	3.316	3.297
6	3.776	3.463	3.289	3.181	3.108	3.055	3.014	2.983	2.958	2.937
7	3.589	3.257	3.074	2.961	2.883	2.827	2.785	2.752	2.725	2.703
8	3.458	3.113	2.924	2.806	2.726	2.668	2.624	2.589	2.561	2.538
9	3.360	3.006	2.813	2.693	2.611	2.551	2.505	2.469	2.440	2.416
10	3.285	2.924	2.728	2.605	2.522	2.461	2.414	2.377	2.347	2.323
11	3.225	2.860	2.660	2.536	2.451	2.389	2.342	2.304	2.274	2.248
12	3.177	2.807	2.606	2.480	2.394	2.331	2.283	2.245	2.214	2.188
13	3.136	2.763	2.560	2.434	2.347	2.283	2.234	2.195	2.164	2.138
14	3.102	2.726	2.522	2.395	2.307	2.243	2.193	2.154	2.122	2.095
15	3.073	2.695	2.490	2.361	2.273	2.208	2.158	2.119	2.086	2.059
16	3.048	2.668	2.462	2.333	2.244	2.178	2.128	2.088	2.055	2.028
17	3.026	2.645	2.437	2.308	2.218	2.152	2.102	2.061	2.028	2.001
18	3.007	2.624	2.416	2.286	2.196	2.130	2.079	2.038	2.005	1.977
19	2.990	2.606	2.397	2.266	2.176	2.109	2.058	2.017	1.984	1.956
20	2.975	2.589	2.380	2.249	2.158	2.091	2.040	1.999	1.965	1.937
21	2.961	2.575	2.365	2.233	2.142	2.075	2.023	1.982	1.948	1.920
22	2.949	2.561	2.351	2.219	2.128	2.061	2.008	1.967	1.933	1.904
23	2.937	2.549	2.339	2.207	2.115	2.047	1.995	1.953	1.919	1.890
24	2.927	2.538	2.327	2.195	2.103	2.035	1.983	1.941	1.906	1.877
25	2.918	2.528	2.317	2.184	2.092	2.024	1.971	1.929	1.895	1.866
26	2.909	2.519	2.307	2.174	2.082	2.014	1.961	1.919	1.884	1.855
27	2.901	2.511	2.299	2.165	2.073	2.005	1.952	1.909	1.874	1.845
28	2.894	2.503	2.291	2.157	2.064	1.996	1.943	1.900	1.865	1.836
29	2.887	2.495	2.283	2.149	2.057	1.988	1.935	1.892	1.857	1.827
30	2.881	2.489	2.276	2.142	2.049	1.980	1.927	1.884	1.849	1.819
31	2.875	2.482	2.270	2.136	2.042	1.973	1.920	1.877	1.842	1.812
32	2.869	2.477	2.263	2.129	2.036	1.967	1.913	1.870	1.835	1.805
33	2.864	2.471	2.258	2.123	2.030	1.961	1.907	1.864	1.828	1.799
34	2.859	2.466	2.252	2.118	2.024	1.955	1.901	1.858	1.822	1.793
35	2.855	2.461	2.247	2.113	2.019	1.950	1.896	1.852	1.817	1.787
36	2.850	2.456	2.243	2.108	2.014	1.945	1.891	1.847	1.811	1.781
37	2.846	2.452	2.238	2.103	2.009	1.940	1.886	1.842	1.806	1.776
38	2.842	2.448	2.234	2.099	2.005	1.935	1.881	1.838	1.802	1.772
39	2.839	2.444	2.230	2.095	2.001	1.931	1.877	1.833	1.797	1.767
40	2.835	2.440	2.226	2.091	1.997	1.927	1.873	1.829	1.793	1.763
50	2.809	2.412	2.197	2.061	1.966	1.895	1.840	1.796	1.760	1.729
60	2.791	2.393	2.177	2.041	1.946	1.875	1.819	1.775	1.738	1.707
120	2.748	2.347	2.130	1.992	1.896	1.824	1.767	1.722	1.684	1.652

TABLE IX

***F* distribution** (*continued*)

γ_2 \ γ_1	11	12	13	14	15	16	17	18	19	20
1	60.473	60.705	60.903	61.072	61.220	61.350	61.464	61.566	61.658	61.740
2	9.401	9.408	9.414	9.420	9.425	9.429	9.432	9.435	9.438	9.441
3	5.223	5.216	5.210	5.205	5.200	5.196	5.193	5.190	5.187	5.185
4	3.907	3.896	3.886	3.878	3.870	3.864	3.858	3.853	3.849	3.844
5	3.282	3.268	3.257	3.247	3.238	3.230	3.223	3.217	3.212	3.207
6	2.920	2.905	2.892	2.881	2.871	2.863	2.855	2.848	2.842	2.836
7	2.684	2.668	2.654	2.643	2.632	2.623	2.615	2.607	2.601	2.595
8	2.519	2.502	2.488	2.475	2.464	2.455	2.446	2.438	2.431	2.425
9	2.396	2.379	2.364	2.351	2.340	2.329	2.320	2.312	2.305	2.298
10	2.302	2.284	2.269	2.255	2.244	2.233	2.224	2.215	2.208	2.201
11	2.227	2.209	2.193	2.179	2.167	2.156	2.147	2.138	2.130	2.123
12	2.166	2.147	2.131	2.117	2.105	2.094	2.084	2.075	2.067	2.060
13	2.116	2.097	2.080	2.066	2.053	2.042	2.032	2.023	2.014	2.007
14	2.073	2.054	2.037	2.022	2.010	1.998	1.988	1.979	1.970	1.962
15	2.037	2.017	2.000	1.985	1.972	1.961	1.950	1.941	1.932	1.924
16	2.005	1.985	1.968	1.953	1.940	1.928	1.917	1.908	1.899	1.891
17	1.978	1.958	1.940	1.925	1.912	1.900	1.889	1.879	1.870	1.862
18	1.954	1.933	1.916	1.900	1.887	1.875	1.864	1.854	1.845	1.837
19	1.932	1.912	1.894	1.878	1.865	1.852	1.841	1.831	1.822	1.814
20	1.913	1.892	1.875	1.859	1.845	1.833	1.821	1.811	1.802	1.794
21	1.896	1.875	1.857	1.841	1.827	1.815	1.803	1.793	1.784	1.776
22	1.880	1.859	1.841	1.825	1.811	1.798	1.787	1.777	1.768	1.759
23	1.866	1.845	1.827	1.811	1.796	1.784	1.772	1.762	1.753	1.744
24	1.853	1.832	1.814	1.797	1.783	1.770	1.759	1.748	1.739	1.730
25	1.841	1.820	1.802	1.785	1.771	1.758	1.746	1.736	1.726	1.718
26	1.830	1.809	1.790	1.774	1.760	1.747	1.735	1.724	1.715	1.706
27	1.820	1.799	1.780	1.764	1.749	1.736	1.724	1.714	1.704	1.695
28	1.811	1.790	1.771	1.754	1.740	1.726	1.715	1.704	1.694	1.685
29	1.802	1.781	1.762	1.745	1.731	1.717	1.705	1.695	1.685	1.676
30	1.794	1.773	1.754	1.737	1.722	1.709	1.697	1.686	1.676	1.667
31	1.787	1.765	1.746	7.729	1.714	1.701	1.689	1.678	1.668	1.659
32	1.780	1.758	1.739	1.722	1.707	1.694	1.682	1.671	1.661	1.652
33	1.773	1.751	1.732	1.715	1.700	1.687	1.675	1.664	1.654	1.645
34	1.767	1.745	1.726	1.709	1.694	1.680	1.668	1.657	1.647	1.638
35	1.761	1.739	1.720	1.703	1.688	1.674	1.662	1.651	1.641	1.632
36	1.756	1.734	1.715	1.697	1.682	1.669	1.656	1.645	1.635	1.626
37	1.751	1.729	1.709	1.692	1.677	1.663	1.651	1.640	1.630	1.620
38	1.746	1.724	1.704	1.687	1.672	1.658	1.646	1.635	1.624	1.615
39	1.741	1.719	1.700	1.682	1.667	1.653	1.641	1.630	1.619	1.610
40	1.737	1.715	1.695	1.678	1.662	1.649	1.636	1.625	1.615	1.605
50	1.703	1.680	1.660	1.643	1.627	1.613	1.600	1.588	1.578	1.568
60	1.680	1.657	1.637	1.619	1.603	1.589	1.576	1.564	1.553	1.543
120	1.625	1.601	1.580	1.562	1.545	1.530	1.516	1.504	1.493	1.482

TABLE IX

F distribution (*continued*)

γ_2 \ γ_1	21	22	23	24	25	26	27	28	29	30
1	61.815	61.882	61.945	62.002	62.054	62.103	62.148	62.189	62.228	62.265
2	9.444	9.446	9.448	9.450	9.451	9.453	9.454	9.456	9.457	9.458
3	5.182	5.180	5.178	5.176	5.175	5.173	5.172	5.170	5.169	1.168
4	3.841	3.837	3.834	3.831	3.828	3.826	3.824	3.821	3.819	3.817
5	3.202	3.198	3.194	3.191	3.187	3.184	3.181	3.179	3.176	3.174
6	2.831	2.827	2.822	2.818	2.815	2.811	2.808	2.805	2.803	2.800
7	2.589	2.584	2.580	2.575	2.571	2.568	2.564	2.561	2.558	2.555
8	2.419	2.414	2.409	2.404	2.400	2.396	2.392	2.389	2.386	2.383
9	2.292	2.287	2.282	2.277	2.272	2.268	2.265	2.261	2.258	2.255
10	2.194	2.189	2.183	2.178	2.174	2.170	2.166	2.162	2.159	2.155
11	2.117	2.111	2.105	2.100	2.095	2.091	2.087	2.083	2.080	2.076
12	2.053	2.047	2.041	2.036	2.031	2.027	2.022	2.019	2.015	2.011
13	2.000	1.994	1.988	1.983	1.978	1.973	1.969	1.965	1.961	1.958
14	1.955	1.949	1.943	1.938	1.933	1.928	1.923	1.919	1.916	1.912
15	1.917	1.911	1.905	1.899	1.894	1.889	1.885	1.880	1.876	1.873
16	1.884	1.877	1.871	1.866	1.860	1.855	1.851	1.847	1.843	1.839
17	1.855	1.848	1.842	1.836	1.831	1.826	1.821	1.817	1.813	1.809
18	1.829	1.823	1.816	1.810	1.805	1.800	1.795	1.791	1.787	1.783
19	1.807	1.800	1.793	1.787	1.782	1.777	1.772	1.767	1.763	1.759
20	1.786	1.779	1.773	1.767	1.761	1.756	1.751	1.746	1.742	1.738
21	1.768	1.761	1.754	1.748	1.742	1.737	1.732	1.728	1.723	1.719
22	1.751	1.744	1.737	1.731	1.726	1.720	1.715	1.711	1.706	1.702
23	1.736	1.729	1.722	1.716	1.710	1.705	1.700	1.695	1.691	1.686
24	1.722	1.715	1.708	1.702	1.696	1.691	1.686	1.681	1.676	1.672
25	1.710	1.702	1.695	1.689	1.683	1.678	1.672	1.668	1.663	1.659
26	1.698	1.690	1.684	1.677	1.671	1.666	1.660	1.656	1.651	1.647
27	1.687	1.680	1.673	1.666	1.660	1.655	1.649	1.645	1.640	1.636
28	1.677	1.669	1.662	1.656	1.650	1.644	1.639	1.634	1.630	1.625
29	1.668	1.660	1.653	1.647	1.640	1.635	1.630	1.625	1.620	1.616
30	1.659	1.651	1.644	1.638	1.632	1.626	1.621	1.616	1.611	1.606
31	1.651	1.643	1.636	1.630	1.623	1.618	1.612	1.607	1.602	1.598
32	1.643	1.636	1.628	1.622	1.616	1.610	1.604	1.599	1.595	1.590
33	1.636	1.628	1.621	1.615	1.608	1.603	1.597	1.592	1.587	1.583
34	1.630	1.622	1.614	1.608	1.601	1.596	1.590	1.585	1.580	1.576
35	1.623	1.615	1.608	1.601	1.595	1.589	1.584	1.579	1.574	1.569
36	1.617	1.609	1.602	1.595	1.589	1.583	1.578	1.572	1.567	1.563
37	1.612	1.604	1.596	1.590	1.583	1.577	1.572	1.567	1.562	1.557
38	1.606	1.598	1.591	1.584	1.578	1.572	1.566	1.561	1.556	1.551
39	1.601	1.593	1.586	1.579	1.573	1.567	1.561	1.556	1.551	1.546
40	1.596	1.588	1.581	1.574	1.568	1.562	1.556	1.551	1.546	1.541
50	1.559	1.551	1.543	1.536	1.529	1.523	1.517	1.512	1.507	1.502
60	1.534	1.526	1.518	1.511	1.504	1.498	1.492	1.486	1.481	1.476
120	1.472	1.463	1.455	1.447	1.440	1.433	1.427	1.421	1.415	1.409

TABLE IX

F distribution (*continued*)

γ_2 \ γ_1	31	32	33	34	35	36	37	38	39	40
1	62.298	62.330	62.361	62.389	62.415	62.441	62.464	62.487	62.508	62.528
2	9.459	9.460	9.461	9.462	9.463	9.463	9.464	9.465	9.466	9.466
3	5.167	5.166	5.165	5.164	5.163	5.163	5.162	5.161	5.160	5.160
4	3.816	3.814	3.812	3.811	3.810	3.808	3.807	3.806	3.805	3.804
5	3.172	3.170	3.168	3.166	3.165	3.163	3.161	3.160	3.159	3.157
6	2.798	2.795	2.793	2.791	2.789	2.787	2.786	2.784	2.783	2.781
7	2.553	2.550	2.548	2.546	2.544	2.542	2.540	2.538	2.537	2.535
8	2.380	2.378	2.375	2.373	2.371	2.369	2.367	2.365	2.363	2.361
9	2.252	2.249	2.247	2.244	2.242	2.240	2.238	2.236	2.234	2.232
10	2.152	2.150	2.147	2.144	2.142	2.140	2.138	2.135	2.134	2.132
11	2.073	2.070	2.067	2.065	2.062	2.060	2.058	2.056	2.054	2.052
12	2.008	2.005	2.002	2.000	1.997	1.995	1.992	1.990	1.988	1.986
13	1.954	1.951	1.948	1.945	1.943	1.940	1.938	1.936	1.934	1.931
14	1.909	1.905	1.902	1.899	1.897	1.894	1.892	1.889	1.887	1.885
15	1.869	1.866	1.863	1.860	1.857	1.855	1.852	1.850	1.848	1.845
16	1.835	1.832	1.829	1.826	1.823	1.820	1.818	1.815	1.813	1.811
17	1.805	1.802	1.799	1.796	1.793	1.790	1.788	1.785	1.783	1.781
18	1.779	1.776	1.772	1.769	1.766	1.764	1.761	1.758	1.756	1.754
19	1.756	1.752	1.749	1.746	1.743	1.740	1.737	1.735	1.732	1.730
20	1.734	1.731	1.728	1.724	1.721	1.718	1.716	1.713	1.711	1.708
21	1.715	1.712	1.708	1.705	1.702	1.699	1.696	1.694	1.691	1.689
22	1.698	1.695	1.691	1.688	1.685	1.682	1.679	1.676	1.674	1.671
23	1.683	1.679	1.675	1.672	1.669	1.666	1.663	1.660	1.658	1.655
24	1.668	1.664	1.661	1.658	1.654	1.651	1.648	1.646	1.643	1.641
25	1.655	1.651	1.648	1.644	1.641	1.638	1.635	1.632	1.630	1.627
26	1.643	1.639	1.635	1.632	1.629	1.626	1.623	1.620	1.617	1.615
27	1.632	1.628	1.624	1.621	1.617	1.614	1.611	1.608	1.606	1.603
28	1.621	1.617	1.614	1.610	1.607	1.604	1.601	1.598	1.595	1.592
29	1.611	1.607	1.604	1.600	1.597	1.594	1.591	1.588	1.585	1.583
30	1.602	1.598	1.595	1.591	1.588	1.585	1.582	1.579	1.576	1.573
31	1.594	1.590	1.586	1.583	1.579	1.576	1.573	1.570	1.567	1.565
32	1.586	1.582	1.578	1.575	1.571	1.568	1.565	1.562	1.559	1.556
33	1.578	1.574	1.571	1.567	1.564	1.560	1.557	1.554	1.551	1.549
34	1.571	1.567	1.564	1.560	1.556	1.553	1.550	1.547	1.544	1.541
35	1.565	1.561	1.557	1.553	1.550	1.546	1.543	1.540	1.537	1.535
36	1.559	1.554	1.551	1.547	1.543	1.540	1.537	1.534	1.531	1.528
37	1.553	1.548	1.545	1.541	1.537	1.534	1.531	1.528	1.525	1.522
38	1.547	1.543	1.539	1.535	1.532	1.528	1.525	1.522	1.519	1.516
39	1.542	1.538	1.534	1.530	1.526	1.523	1.520	1.517	1.514	1.511
40	1.537	1.532	1.528	1.525	1.521	1.518	1.515	1.511	1.508	1.506
50	1.497	1.493	1.489	1.485	1.481	1.477	1.474	1.471	1.468	1.465
60	1.471	1.466	1.462	1.458	1.454	1.450	1.447	1.444	1.440	1.437
120	1.404	1.399	1.395	1.390	1.386	1.382	1.378	1.375	1.371	1.368

TABLE IX
F distribution (*continued*)

γ_2 \ γ_1	50	60	120
1	62.688	62.793	63.060
2	9.471	9.475	9.483
3	5.155	5.151	5.143
4	3.795	3.790	3.775
5	3.147	3.140	3.123
6	2.770	2.762	2.742
7	2.523	2.514	2.493
8	2.348	2.339	2.316
9	2.218	2.208	2.184
10	2.117	2.107	2.082
11	2.036	2.026	2.000
12	1.970	1.960	1.932
13	1.915	1.904	1.876
14	1.869	1.857	1.828
15	1.828	1.817	1.787
16	1.793	1.782	1.751
17	1.763	1.751	1.791
18	1.736	1.723	1.691
19	1.711	1.699	1.666
20	1.690	1.677	1.643
21	1.670	1.657	1.623
22	1.652	1.639	1.604
23	1.636	1.622	1.587
24	1.621	1.607	1.571
25	1.607	1.593	1.557
26	1.594	1.581	1.544
27	1.583	1.569	1.531
28	1.572	1.558	1.520
29	1.562	1.547	1.509
30	1.552	1.538	1.499
31	1.543	1.529	1.489
32	1.535	1.520	1.481
33	1.527	1.512	1.472
34	1.520	1.505	1.464
35	1.513	1.497	1.457
36	1.506	1.491	1.450
37	1.500	1.484	1.443
38	1.494	1.478	1.437
39	1.488	1.473	1.431
40	1.483	1.467	1.425
50	1.441	1.424	1.379
60	1.413	1.395	1.348
120	1.340	1.320	1.265

TABLE IX

***F* distribution** (*continued*)

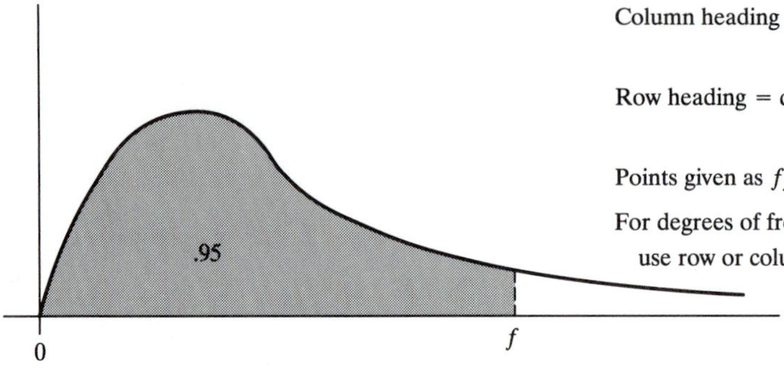

Column heading = numerator degrees of freedom

Row heading = denominator degrees of freedom

Points given as $f_{.05}$ points

For degrees of freedom > 120, use row or column 120

$P[F_{\gamma_1, \gamma_2} \leq f] = .95$

γ_2 \ γ_1	1	2	3	4	5	6	7	8
1	161.448	199.500	215.707	224.583	230.161	233.985	236.768	238.882
2	18.513	19.000	19.164	19.247	19.296	19.329	19.353	19.371
3	10.128	9.552	9.277	9.117	9.013	8.941	8.887	8.845
4	7.709	6.944	6.591	6.388	6.256	6.163	6.094	6.041
5	6.608	5.786	5.409	5.192	5.050	4.950	4.876	4.818
6	5.987	5.143	4.757	4.534	4.387	4.284	4.207	4.147
7	5.591	4.737	4.347	4.120	3.972	3.866	3.787	3.726
8	5.318	4.459	4.066	3.838	3.687	3.581	3.500	3.438
9	5.117	4.256	3.863	3.633	3.482	3.374	3.293	3.230
10	4.965	4.103	3.708	3.478	3.326	3.217	3.135	3.072
11	4.844	3.982	3.587	3.357	3.204	3.095	3.012	2.948
12	4.747	3.885	3.490	3.259	3.106	2.996	2.913	2.849
13	4.667	3.806	3.411	3.179	3.025	2.915	2.832	2.767
14	4.600	3.739	3.344	3.112	2.958	2.848	2.764	2.699
15	4.543	3.682	3.287	3.056	2.901	2.790	2.707	2.641
16	4.494	3.634	3.239	3.007	2.852	2.741	2.657	2.591
17	4.451	3.592	3.197	2.965	2.810	2.699	2.614	2.548
18	4.414	3.555	3.160	2.928	2.773	2.661	2.577	2.510
19	4.381	3.522	3.127	2.895	2.740	2.628	2.544	2.477
20	4.351	3.493	3.098	2.866	2.711	2.599	2.514	2.447
21	4.325	3.467	3.072	2.840	2.685	2.573	2.488	2.420
22	4.301	3.443	3.049	2.817	2.661	2.549	2.464	2.397
23	4.279	3.422	3.028	2.796	2.640	2.528	2.442	2.375
24	4.260	3.403	3.009	2.776	2.621	2.508	2.423	2.355
25	4.242	3.385	2.991	2.759	2.603	2.490	2.405	2.337
26	4.225	3.369	2.975	2.743	2.587	2.474	2.388	2.321
27	4.210	3.354	2.960	2.728	2.572	2.459	2.373	2.305
28	4.196	3.340	2.947	2.714	2.558	2.445	2.359	2.291
29	4.183	3.328	2.934	2.701	2.545	2.432	2.346	2.278
30	4.171	3.316	2.922	2.690	2.534	2.421	2.334	2.266
31	4.160	3.305	2.911	2.679	2.523	2.409	2.323	2.255
32	4.149	3.295	2.901	2.668	2.512	2.399	2.313	2.244
33	4.139	3.285	2.892	2.659	2.503	2.389	2.303	2.235
34	4.130	3.276	2.883	2.650	2.494	2.380	2.294	2.225
35	4.121	3.267	2.874	2.641	2.485	2.372	2.285	2.217
36	4.113	3.259	2.866	2.634	2.477	2.364	2.277	2.209
37	4.105	3.252	2.859	2.626	2.470	2.356	2.270	2.201
38	4.098	3.245	2.852	2.619	2.463	2.349	2.262	2.194
39	4.091	3.238	2.845	2.612	2.456	2.342	2.255	2.187
40	4.085	3.232	2.839	2.606	2.449	2.336	2.249	2.180
50	4.034	3.183	2.790	2.557	2.400	2.286	2.199	2.130
60	4.001	3.150	2.758	2.525	2.368	2.254	2.167	2.097
120	3.9301	3.072	2.6681	2.447	2.290	2.175	2.087	2.016

TABLE IX

F distribution (*continued*)

γ_2 \ γ_1	9	10	11	12	13	14	15	16
1	240.543	241.881	242.983	243.905	244.689	245.363	245.949	246.462
2	19.385	19.396	19.405	19.412	19.419	19.424	19.429	19.433
3	8.812	8.786	8.763	8.745	8.729	8.715	8.703	8.692
4	5.999	5.964	5.936	5.912	5.891	5.873	5.858	5.844
5	4.772	4.735	4.704	4.678	4.655	4.636	4.619	4.604
6	4.099	4.060	4.027	4.000	3.976	3.956	3.938	3.922
7	3.677	3.637	3.603	3.575	3.550	3.529	3.511	3.494
8	3.388	3.347	3.313	3.284	3.259	3.237	3.218	3.202
9	3.179	3.137	3.102	3.073	3.048	3.025	3.006	2.989
10	3.020	2.978	2.943	2.913	2.887	2.865	2.845	2.828
11	2.896	2.854	2.818	2.788	2.761	2.739	2.719	2.701
12	2.796	2.753	2.717	2.687	2.660	2.637	2.617	2.599
13	2.714	2.671	2.635	2.604	2.577	2.554	2.533	2.515
14	2.646	2.602	2.566	2.534	2.507	2.484	2.463	2.445
15	2.588	2.544	2.507	2.475	2.448	2.424	2.403	2.385
16	2.538	2.494	2.456	2.425	2.397	2.373	2.352	2.333
17	2.494	2.450	2.413	2.381	2.353	2.329	2.308	2.289
18	2.456	2.412	2.374	2.342	2.314	2.290	2.269	2.250
19	2.423	2.378	2.340	2.308	2.280	2.256	2.234	2.215
20	2.393	2.348	2.310	2.278	2.250	2.225	2.203	2.184
21	2.366	2.321	2.283	2.250	2.222	2.197	2.176	2.156
22	2.342	2.297	2.259	2.226	2.198	2.173	2.151	2.131
23	2.320	2.275	2.236	2.204	2.175	2.150	2.128	2.109
24	2.300	2.255	2.216	2.183	2.155	2.130	2.108	2.088
25	2.282	2.236	2.198	2.165	2.136	2.111	2.089	2.069
26	2.265	2.220	2.181	2.148	2.119	2.094	2.072	2.052
27	2.250	2.204	2.166	2.132	2.103	2.078	2.056	2.036
28	2.236	2.190	2.151	2.118	2.089	2.064	2.041	2.021
29	2.223	2.177	2.138	2.105	2.075	2.050	2.027	2.007
30	2.211	2.165	2.126	2.092	2.063	2.037	2.015	1.995
31	2.199	2.153	2.114	2.080	2.051	2.026	2.003	1.983
32	2.189	2.142	2.103	2.070	2.040	2.015	1.992	1.972
33	2.179	2.133	2.093	2.060	2.030	2.004	1.982	1.961
34	2.170	2.123	2.084	2.050	2.021	1.995	1.972	1.952
35	2.161	2.114	2.075	2.041	2.012	1.986	1.963	1.942
36	2.153	2.106	2.067	2.033	2.003	1.977	1.954	1.934
37	2.145	2.098	2.059	2.025	1.995	1.969	1.946	1.926
38	2.138	2.091	2.051	2.017	1.988	1.962	1.939	1.918
39	2.131	2.084	2.044	2.010	1.981	1.954	1.931	1.911
40	2.124	2.077	2.038	2.003	1.974	1.948	1.924	1.904
50	2.073	2.026	1.986	1.952	1.921	1.895	1.871	1.850
60	2.040	1.993	1.952	1.917	1.887	1.860	1.836	1.815
120	1.959	1.910	1.869	1.834	1.803	1.775	1.750	1.728

TABLE IX

***F* distribution (*continued*)**

γ_2 \ γ_1	17	18	19	20	21	22	23	24
1	246.917	247.322	247.685	248.012	248.308	248.577	248.824	249.051
2	19.437	19.440	19.443	19.446	19.448	19.450	19.452	19.454
3	8.683	8.675	8.667	8.660	8.654	8.648	8.643	8.639
4	5.832	5.821	5.811	5.803	5.795	5.787	5.781	5.774
5	4.590	4.579	4.568	4.558	4.549	4.541	4.534	4.527
6	3.908	3.896	3.884	3.874	3.865	3.856	3.849	3.841
7	3.480	3.467	3.455	3.445	3.435	3.426	3.418	3.411
8	3.187	3.173	3.161	3.150	3.140	3.131	3.123	3.115
9	2.974	2.960	2.948	2.936	2.926	2.917	2.908	2.900
10	2.812	2.798	2.785	2.774	2.764	2.754	2.745	2.737
11	2.685	2.671	2.658	2.646	2.636	2.626	2.617	2.609
12	2.583	2.568	2.555	2.544	2.533	2.523	2.514	2.505
13	2.499	2.484	2.471	2.459	2.448	2.438	2.429	2.420
14	2.428	2.413	2.400	2.388	2.377	2.367	2.357	2.349
15	2.368	2.353	2.340	2.328	2.316	2.306	2.297	2.288
16	2.317	2.302	2.288	2.276	2.264	2.254	2.244	2.235
17	2.272	2.257	2.243	2.230	2.219	2.208	2.199	2.190
18	2.233	2.217	2.203	2.191	2.179	2.168	2.159	2.150
19	2.198	2.182	2.168	2.156	2.144	2.133	2.123	2.114
20	2.167	2.151	2.137	2.124	2.112	2.102	2.092	2.082
21	2.139	2.123	2.109	2.096	2.084	2.073	2.063	2.054
22	2.114	2.098	2.084	2.071	2.059	2.048	2.038	2.028
23	2.091	2.075	2.061	2.048	2.036	2.025	2.014	2.005
24	2.070	2.054	2.040	2.027	2.015	2.003	1.993	1.984
25	2.051	2.035	2.021	2.007	1.995	1.984	1.974	1.964
26	2.034	2.018	2.003	1.990	1.978	1.966	1.956	1.946
27	2.018	2.002	1.987	1.974	1.961	1.950	1.940	1.930
28	2.003	1.987	1.972	1.959	1.946	1.935	1.924	1.915
29	1.989	1.973	1.958	1.945	1.932	1.921	1.910	1.901
30	1.976	1.960	1.945	1.932	1.919	1.908	1.897	1.887
31	1.965	1.948	1.933	1.920	1.907	1.896	1.885	1.875
32	1.953	1.937	1.922	1.908	1.896	1.884	1.873	1.864
33	1.943	1.926	1.911	1.898	1.885	1.873	1.863	1.853
34	1.933	1.917	1.902	1.888	1.875	1.863	1.853	1.843
35	1.924	1.907	1.892	1.878	1.866	1.854	1.843	1.833
36	1.915	1.899	1.883	1.870	1.857	1.845	1.834	1.824
37	1.907	1.890	1.875	1.861	1.848	1.837	1.826	1.816
38	1.899	1.883	1.867	1.853	1.841	1.829	1.818	1.808
39	1.892	1.875	1.860	1.846	1.833	1.821	1.810	1.800
40	1.885	1.868	1.853	1.839	1.826	1.814	1.803	1.793
50	1.831	1.814	1.798	1.784	1.771	1.759	1.748	1.737
60	1.796	1.778	1.763	1.748	1.735	1.722	1.711	1.700
120	1.709	1.690	1.674	1.659	1.645	1.632	1.620	1.608

TABLE IX

F distribution (*continued*)

γ_1 \ γ_1	25	26	27	28	29	30	31	32
1	249.258	249.451	249.629	249.795	249.949	250.093	250.228	250.355
2	19.456	19.457	19.459	19.460	19.461	19.462	19.463	19.464
3	8.634	8.630	8.626	8.623	8.620	8.617	8.614	8.611
4	5.769	5.764	5.759	5.754	5.750	5.746	5.742	5.739
5	4.521	4.515	4.510	4.505	4.500	4.496	4.492	4.488
6	3.835	3.829	3.823	3.818	3.813	3.808	3.804	3.800
7	3.404	3.397	3.391	3.386	3.381	3.376	3.371	3.367
8	3.108	3.102	3.095	3.090	3.084	3.079	3.075	3.070
9	2.893	2.886	2.880	2.874	2.869	2.864	2.859	2.854
10	2.730	2.723	2.716	2.710	2.705	2.700	2.695	2.690
11	2.601	2.594	2.588	2.582	2.576	2.570	2.565	2.561
12	2.498	2.491	2.484	2.478	2.472	2.466	2.461	2.456
13	2.412	2.405	2.398	2.392	2.386	2.380	2.375	2.370
14	2.341	2.333	2.326	2.320	2.314	2.308	2.303	2.298
15	2.280	2.272	2.265	2.259	2.253	2.247	2.241	2.236
16	2.227	2.220	2.212	2.206	2.200	2.194	2.188	2.183
17	2.181	2.174	2.167	2.160	2.154	2.148	2.142	2.137
18	2.141	2.134	2.126	2.119	2.113	2.107	2.102	2.096
19	2.106	2.098	2.090	2.084	2.077	2.071	2.066	2.060
20	2.074	2.066	2.059	2.052	2.045	2.039	2.033	2.028
21	2.045	2.037	2.030	2.023	2.016	2.010	2.004	1.999
22	2.020	2.012	2.004	1.997	1.990	1.984	1.978	1.973
23	1.996	1.988	1.981	1.973	1.967	1.961	1.955	1.949
24	1.975	1.967	1.959	1.952	1.945	1.939	1.933	1.927
25	1.955	1.947	1.939	1.932	1.926	1.919	1.913	1.908
26	1.938	1.929	1.921	1.914	1.907	1.901	1.895	1.889
27	1.921	1.913	1.905	1.898	1.891	1.884	1.878	1.872
28	1.906	1.897	1.889	1.882	1.875	1.869	1.863	1.857
29	1.891	1.883	1.875	1.868	1.861	1.854	1.848	1.842
30	1.878	1.870	1.862	1.854	1.847	1.841	1.835	1.829
31	1.866	1.857	1.849	1.842	1.835	1.828	1.822	1.816
32	1.854	1.846	1.838	1.830	1.823	1.817	1.810	1.804
33	1.844	1.835	1.827	1.819	1.812	1.806	1.799	1.793
34	1.833	1.825	1.817	1.809	1.802	1.795	1.789	1.783
35	1.824	1.815	1.807	1.799	1.792	1.786	1.779	1.773
36	1.815	1.806	1.798	1.790	1.783	1.776	1.770	1.764
37	1.806	1.798	1.789	1.782	1.775	1.768	1.761	1.755
38	1.798	1.790	1.781	1.774	1.766	1.760	1.753	1.747
39	1.791	1.782	1.774	1.766	1.759	1.752	1.745	1.739
40	1.783	1.775	1.766	1.759	1.751	1.744	1.738	1.732
50	1.727	1.718	1.710	1.702	1.694	1.687	1.680	1.674
60	1.690	1.681	1.672	1.664	1.656	1.649	1.642	1.636
120	1.598	1.588	1.579	1.570	1.562	1.554	1.547	1.540

TABLE IX

***F* distribution (*continued*)**

γ_2 \ γ_1	33	34	35	36	37	38	39	40
1	250.474	250.586	250.691	250.793	250.886	250.975	251.061	251.141
2	19.465	19.466	19.467	19.468	19.468	19.469	19.470	19.470
3	8.609	8.606	8.604	8.602	8.600	8.598	8.596	8.594
4	5.735	5.732	5.729	5.727	5.724	5.722	5.719	5.717
5	4.484	4.481	4.478	4.474	4.472	4.469	4.466	4.464
6	3.796	3.792	3.789	3.786	3.783	3.780	3.777	3.774
7	3.363	3.359	3.356	3.352	3.349	3.346	3.343	3.340
8	3.066	3.062	3.059	3.055	3.052	3.049	3.046	3.043
9	2.850	2.846	2.842	2.839	2.835	2.832	2.829	2.826
10	2.686	2.681	2.678	2.674	2.670	2.667	2.664	2.661
11	2.556	2.552	2.548	2.544	2.541	2.537	2.534	2.531
12	2.452	2.447	2.443	2.439	2.436	2.432	2.429	2.426
13	2.366	2.361	2.357	2.353	2.349	2.346	2.342	2.339
14	2.293	2.289	2.284	2.280	2.277	2.273	2.270	2.266
15	2.232	2.227	2.223	2.219	2.215	2.211	2.208	2.204
16	2.178	2.174	2.169	2.165	2.161	2.158	2.154	2.151
17	2.132	2.127	2.123	2.119	2.115	2.111	2.107	2.104
18	2.091	2.087	2.082	2.078	2.074	2.070	2.066	2.063
19	2.055	2.050	2.046	2.042	2.037	2.034	2.030	2.026
20	2.023	2.018	2.013	2.009	2.005	2.001	1.997	1.994
21	1.994	1.989	1.984	1.980	1.976	1.972	1.968	1.965
22	1.968	1.963	1.958	1.954	1.949	1.945	1.942	1.938
23	1.944	1.939	1.934	1.930	1.925	1.921	1.918	1.914
24	1.922	1.917	1.912	1.908	1.904	1.900	1.896	1.892
25	1.902	1.897	1.892	1.888	1.884	1.879	1.876	1.872
26	1.884	1.879	1.874	1.869	1.865	1.861	1.857	1.853
27	1.867	1.862	1.857	1.852	1.848	1.844	1.840	1.836
28	1.851	1.846	1.841	1.837	1.832	1.828	1.824	1.820
29	1.837	1.832	1.827	1.822	1.818	1.813	1.809	1.806
30	1.823	1.818	1.813	1.808	1.804	1.800	1.796	1.792
31	1.811	1.805	1.800	1.796	1.791	1.787	1.783	1.779
32	1.799	1.794	1.789	1.784	1.779	1.775	1.771	1.767
33	1.788	1.783	1.777	1.773	1.768	1.764	1.760	1.756
34	1.777	1.772	1.767	1.762	1.758	1.753	1.749	1.745
35	1.768	1.762	1.757	1.752	1.748	1.743	1.739	1.735
36	1.758	1.753	1.748	1.743	1.738	1.734	1.730	1.726
37	1.750	1.744	1.739	1.734	1.730	1.725	1.721	1.717
38	1.741	1.736	1.731	1.726	1.721	1.717	1.712	1.708
39	1.733	1.728	1.723	1.718	1.713	1.709	1.704	1.700
40	1.726	1.721	1.715	1.710	1.706	1.701	1.697	1.693
50	1.668	1.662	1.657	1.652	1.647	1.642	1.638	1.634
60	1.630	1.624	1.618	1.613	1.608	1.603	1.599	1.594
120	1.534	1.527	1.521	1.516	1.510	1.505	1.500	1.495

TABLE IX
***F* distribution (*continued*)**

γ_2 \ γ_1	50	60	120
1	251.773	252.191	253.252
2	19.475	19.478	19.487
3	8.581	8.572	8.549
4	5.699	5.688	5.658
5	4.444	4.431	4.398
6	3.754	3.740	3.705
7	3.319	3.304	3.267
8	3.020	3.005	2.967
9	2.803	2.787	2.748
10	2.637	2.621	2.580
11	2.507	2.490	2.448
12	2.401	2.384	2.341
13	2.314	2.297	2.252
14	2.241	2.223	2.178
15	2.178	2.160	2.114
16	2.124	2.106	2.059
17	2.077	2.058	2.011
18	2.035	2.017	1.968
19	1.999	1.980	1.930
20	1.966	1.946	1.896
21	1.936	1.916	1.866
22	1.909	1.889	1.838
23	1.885	1.865	1.813
24	1.863	1.842	1.790
25	1.842	1.822	1.768
26	1.823	1.803	1.749
27	1.806	1.785	1.731
28	1.790	1.769	1.714
29	1.775	1.754	1.698
30	1.761	1.740	1.683
31	1.748	1.726	1.670
32	1.736	1.714	1.657
33	1.724	1.702	1.645
34	1.713	1.691	1.633
35	1.703	1.681	1.623
36	1.694	1.671	1.612
37	1.685	1.662	1.603
38	1.676	1.653	1.594
39	1.668	1.645	1.585
40	1.660	1.637	1.577
50	1.599	1.576	1.511
60	1.559	1.534	1.467
120	1.457	1.429	1.352

TABLE X
Wilcoxon rank-sum test

$m = 3(1)245$ **and** $n = m(1)m + 25$

$P = .025$ **one-sided;** $P = .05$ **two-sided**

n	$m = 3$	$m = 4$	$m = 5$	$m = 6$	$m = 7$	$m = 8$	$m = 9$	$m = 10$	$m = 11$	$m = 12$	$m = 13$	$m = 14$
$n = m$	5,16	11,25	18,37	26,52	37,68	49,87	63,108	79,131	96,157	116,184	137,214	160,246
$n = m + 1$	6,18	12,28	19,41	28,56	39,73	51,93	66,114	82,138	100,164	120,192	141,223	165,255
$n = m + 2$	6,21	12,32	20,45	29,61	41,78	54,98	68,121	85,145	103,172	124,200	146,231	170,264
$n = m + 3$	7,23	13,35	21,49	31,65	43,83	56,104	71,127	88,152	107,179	128,208	150,240	174,274
$n = m + 4$	7,26	14,38	22,53	32,70	45,88	58,110	74,133	91,159	110,187	131,217	154,249	179,283
$n = m + 5$	8,28	15,41	24,56	34,74	46,94	61,115	77,139	94,116	114,194	135,225	159,257	184,292
$n = m + 6$	8,31	16,44	25,60	36,78	48,99	63,121	79,146	97,173	118,201	139,233	163,266	189,301
$n = m + 7$	9,33	17,47	26,64	37,83	50,104	65,127	82,152	101,179	121,209	143,241	168,274	194,310
$n = m + 8$	10,35	17,51	27,68	39,87	52,109	68,132	85,158	104,186	125,216	147,249	172,283	198,320
$n = m + 9$	10,38	18,54	29,71	41,91	54,114	70,138	88,164	107,193	128,224	151,257	176,292	203,329
$n = m + 10$	11,40	19,57	30,75	42,96	56,119	72,144	90,171	110,200	132,231	155,265	181,300	208,338
$n = m + 11$	11,43	20,60	31,79	44,100	58,124	75,149	93,177	113,207	135,239	159,273	185,309	213,347
$n = m + 12$	12,45	21,63	32,83	45,105	60,129	77,155	96,183	117,213	139,246	163,281	190,317	218,356
$n = m + 13$	12,48	22,66	33,87	47,109	62,134	80,160	99,189	120,220	143,253	167,289	194,326	222,366
$n = m + 14$	13,50	23,69	35,90	49,113	64,139	82,166	101,196	123,227	146,261	171,297	198,335	227,375
$n = m + 15$	13,53	24,72	36,94	50,118	66,144	84,172	104,202	126,234	150,268	175,305	203,343	232,384
$n = m + 16$	14,55	24,76	37,98	52,122	68,149	87,177	107,208	129,241	153,276	179,313	207,352	237,393
$n = m + 17$	14,58	25,79	38,102	53,127	70,154	89,183	110,214	132,248	157,283	183,321	212,360	242,402
$n = m + 18$	15,60	26,82	40,105	55,131	72,159	91,188	113,220	136,254	161,290	187,329	216,369	247,411
$n = m + 19$	15,63	27,85	41,109	57,135	74,164	94,194	115,227	139,261	164,298	191,337	221,377	252,420
$n = m + 20$	16,65	28,88	42,113	58,140	76,169	96,200	118,233	142,268	168,305	195,345	225,286	256,430
$n = m + 21$	16,68	29,91	43,117	60,144	78,174	99,205	121,239	145,275	171,313	199,353	229,395	261,439
$n = m + 22$	17,70	30,94	45,120	61,149	80,179	101,211	124,245	148,282	175,320	203,361	234,403	266,448
$n = m + 23$	17,73	31,97	46,124	63,153	82,184	103,217	127,251	152,288	179,327	207,369	238,412	271,457
$n = m + 24$	18,75	31,101	47,128	65,157	84,189	106,222	129,258	155,295	182,335	211,377	243,420	276,466
$n = m + 25$	18,78	32,104	48,132	66,162	86,194	108,228	132,264	158,302	186,342	216,384	247,429	281,475

n	$m = 15$	$m = 16$	$m = 17$	$m = 8$	$m = 19$	$m = 20$	$m = 21$	$m = 22$	$m = 23$	$m = 24$	$m = 25$
$n = m$	185,280	212,316	240,355	271,395	303,438	337,483	373,530	411,579	451,630	493,683	536,739
$n = m + 1$	190,290	271,327	246,366	277,407	310,450	345,495	381,543	419,593	460,644	502,698	546,754
$n = m + 2$	195,300	223,337	252,377	284,418	317,462	352,508	389,556	428,606	468,659	511,713	555,770
$n = m + 3$	201,309	229,347	258,388	290,430	324,474	359,521	397,569	436,620	477,673	520,728	565,785
$n = m + 4$	206,319	234,358	264,399	297,441	331,486	367,533	404,583	444,634	486,687	529,743	574,801
$n = m + 5$	211,329	240,368	271,409	303,453	338,498	374,546	412,596	452,648	494,702	538,758	584,816
$n = m + 6$	216,339	245,379	277,420	310,464	345,510	381,559	420,609	460,662	503,716	547,773	593,832
$n = m + 7$	221,349	251,389	283,431	316,476	351,523	389,571	428,622	469,675	512,730	556,788	603,847
$n = m + 8$	227,358	257,399	289,442	323,487	358,535	396,584	436,635	477,689	520,745	565,803	612,863
$n = m + 9$	232,368	262,410	295,453	329,499	365,547	403,597	443,649	485,703	529,759	575,817	622,878
$n = m + 10$	237,378	268,420	301,464	336,510	372,559	411,609	451,662	493,717	538,773	584,832	632,893
$n = m + 11$	242,388	274,430	307,475	342,522	379,571	418,622	459,675	502,730	546,788	593,847	641,909
$n = m + 12$	248,397	279,441	313,486	349,533	386,583	426,634	467,668	510,744	555,802	602,862	651,924
$n = m + 13$	253,407	285,451	319,497	355,545	393,595	433,647	475,701	518,758	564,816	611,877	660,940
$n = m + 14$	258,417	291,461	325,508	362,556	400,607	440,660	482,715	526,772	572,831	620,892	670,955
$n = m + 15$	263,427	296,472	331,519	368,568	407,619	448,672	490,728	535,785	581,845	629,907	679,971
$n = m + 16$	269,436	302,482	338,529	375,579	414,631	455,685	498,741	543,799	590,859	638,922	689,986
$n = m + 17$	274,446	308,492	344,540	381,591	421,643	463,697	506,754	551,813	599,873	648,936	699,1001
$n = m + 18$	279,456	314,502	350,551	388,602	428,655	470,710	514,767	560,826	607,888	657,951	708,1017
$n = m + 19$	284,466	319,513	356,562	395,613	435,667	477,723	522,780	568,840	616,902	666,966	718,1032
$n = m + 20$	290,475	325,523	362,573	401,625	442,679	485,735	530,793	576,854	625,916	675,981	727,1048
$n = m + 21$	295,485	331,533	368,584	408,636	449,691	492,748	537,807	584,868	633,931	684,996	737,1063
$n = m + 22$	300,495	336,544	374,595	414,648	456,703	500,760	545,820	593,881	642,945	693,1011	747,1078
$n = m + 23$	306,504	342,554	380,606	421,659	463,715	507,773	553,833	601,895	651,959	703,1025	756,1094
$n = m + 24$	311,514	348,564	387,616	427,671	470,727	515,785	561,846	609,909	660,973	712,1040	766,1109
$n = m + 25$	316,524	353,575	393,627	434,682	477,739	522,798	569,859	618,922	668,988	721,1055	775,1125

TABLE X
Wilcoxon rank-sum test *(continued)*

m = 3(1)25 **and** $n = m(1)m + 25$
P = .05 **one-sided**; P = .10 **two-sided**

n	$m = 3$	$m = 4$	$m = 5$	$m = 6$	$m = 7$	$m = 8$	$m = 9$	$m = 10$	$m = 11$	$m = 12$	$m = 13$	$m = 14$
$n = m$	6,15	12,24	19,36	28,50	39,66	52,84	66,105	83,127	101,152	121,179	143,208	167,239
$n = m + 1$	7,17	13,27	20,40	30,54	41,71	54,90	69,111	86,134	105,159	125,187	148,216	172,248
$n = m + 2$	7,20	14,30	22,43	32,58	43,76	57,95	72,117	89,141	109,166	129,195	152,255	177,257
$n = m + 3$	8,22	15,33	24,46	33,63	46,80	60,100	75,123	93,147	112,174	134,202	157,233	182,266
$n = m + 4$	9,24	16,36	25,50	35,67	48,85	62,106	78,129	96,154	116,181	138,210	162,241	187,275
$n = m + 5$	9,27	17,39	26,54	37,71	50,90	65,111	81,135	100,160	120,188	142,218	166,250	192,284
$n = m + 6$	10,29	18,42	27,58	39,75	52,95	67,117	84,141	103,167	124,195	147,225	171,258	197,293
$n = m + 7$	11,31	19,45	29,61	41,79	54,100	70,122	87,147	107,173	128,202	151,233	176,266	203,301
$n = m + 8$	11,34	20,48	30,65	42,84	57,104	73,127	90,153	110,180	132,209	155,241	181,274	208,310
$n = m + 9$	12,36	21,51	32,68	44,88	59,109	75,133	93,159	114,186	136,216	159,249	185,283	213,319
$n = m + 10$	13,38	22,54	33,72	46,92	61,114	78,138	96,165	117,193	139,224	164,256	190,291	218,328
$n = m + 11$	13,41	23,57	34,76	48,96	63,119	80,144	100,170	120,200	143,231	168,264	195,299	223,337
$n = m + 12$	14,43	24,60	36,79	50,100	65,124	83,149	103,176	124,206	147,238	172,272	199,308	228,346
$n = m + 13$	15,45	25,63	37,83	52,104	68,128	86,154	106,182	127,213	151,245	177,279	204,316	234,254
$n = m + 14$	15,48	26,66	39,86	53,109	70,133	88,160	109,188	131,219	155,252	181,287	209,324	239,363
$n = m + 15$	16,50	27,69	40,90	55,113	72,138	91,165	112,194	134,226	159,259	185,295	214,332	244,372
$n = m + 16$	17,52	28,72	42,93	57,117	74,143	94,170	115,200	138,232	163,266	190,302	218,341	249,381
$n = m + 17$	17,55	29,75	43,97	59,121	77,147	96,176	118,206	141,239	167,273	194,310	223,349	254,390
$n = m + 18$	18,57	30,78	44,101	61,125	79,152	99,181	121,212	145,245	171,280	198,318	228,357	260,398
$n = m + 19$	19,59	31,81	46,104	62,130	81,157	102,186	124,218	148,252	175,287	203,325	233,365	265,407
$n = m + 20$	19,62	32,84	47,108	64,134	83,162	104,192	127,224	152,258	178,295	207,333	237,374	270,416
$n = m + 21$	20,64	33,87	49,111	66,138	86,166	107,197	130,230	155,265	182,302	211,341	242,382	275,425
$n = m + 22$	21,66	34,90	50,115	68,142	88,171	109,203	133,236	159,271	186,309	216,348	247,390	280,434
$n = m + 23$	21,69	35,93	52,118	70,146	90,176	112,208	136,242	162,278	190,316	220,356	252,398	285,443
$n = m + 24$	22,71	37,95	53,122	72,150	92,181	115,213	139,248	166,284	194,323	224,364	257,406	291,451
$n = m + 25$	23,73	38,98	54,126	73,155	94,186	117,219	142,254	169,291	198,330	229,371	261,415	296,460

n	$m = 15$	$m = 16$	$m = 17$	$m = 18$	$m = 19$	$m = 20$	$m = 21$	$m = 22$	$m = 23$	$m = 24$	$m = 25$
$n = m$	192,273	220,308	249,346	280,386	314,427	349,471	386,517	424,566	465,616	508,668	552,723
$n = m + 1$	198,282	226,318	256,356	287,397	321,439	356,484	394,530	433,579	474,630	517,683	562,738
$n = m + 2$	203,292	232,328	262,367	294,408	328,451	364,496	402,543	442,592	483,644	527,697	572,753
$n = m + 3$	209,301	238,338	268,378	301,419	336,462	372,508	410,556	450,606	492,658	536,712	582,768
$n = m + 4$	215,310	244,348	275,388	308,430	343,474	380,520	418,569	459,619	501,672	546,726	592,783
$n = m + 5$	220,320	250,358	281,399	315,441	350,486	387,533	427,581	468,632	511,685	555,741	602,798
$n = m + 6$	226,329	256,368	288,409	322,452	358,497	395,545	435,594	476,646	520,699	565,755	612,813
$n = m + 7$	231,339	262,378	294,420	329,463	365,509	403,557	443,607	485,659	529,713	574,770	622,828
$n = m + 8$	237,348	268,388	301,430	336,474	372,521	411,569	451,620	494,672	538,727	584,784	632,843
$n = m + 9$	242,358	274,398	307,441	342,486	380,532	419,581	459,633	502,686	547,741	594,798	642,858
$n = m + 10$	248,367	280,408	314,451	349,497	387,544	426,594	468,645	511,699	556,755	603,813	652,873
$n = m + 11$	254,376	286,418	320,462	356,508	394,556	434,606	476,658	520,712	565,769	613,827	662,888
$n = m + 12$	259,386	292,428	327,472	363,519	402,567	442,618	484,671	528,726	574,783	622,842	672,903
$n = m + 13$	265,395	298,438	333,483	370,530	409,579	450,630	492,684	537,739	584,796	632,856	682,918
$n = m + 14$	270,405	304,448	340,493	377,541	416,591	458,642	501,696	546,752	593,810	642,870	692,933
$n = m + 15$	276,414	310,458	346,504	384,552	424,602	465,655	509,709	554,766	602,824	651,885	702,948
$n = m + 16$	282,423	316,468	353,514	391,563	431,614	473,667	517,722	563,779	611,838	661,899	712,963
$n = m + 17$	287,433	322,478	359,525	398,574	438,626	481,679	526,734	572,792	620,852	670,914	723,977
$n = m + 18$	293,442	328,488	366,535	405,585	446,637	489,691	534,747	581,805	629,866	680,928	733,992
$n = m + 19$	299,451	334,498	372,546	412,596	453,649	497,703	542,760	589,819	639,879	690,942	743,1007
$n = m + 20$	304,461	340,508	379,556	419,607	461,660	505,715	550,773	598,832	648,893	699,957	753,1022
$n = m + 21$	310,470	347,517	385,568	426,618	468,672	512,728	559,785	607,845	657,907	709,971	763,1037
$n = m + 22$	315,480	353,527	392,577	433,629	475,684	520,740	567,798	615,859	666,921	718,986	773,1052
$n = m + 23$	321,489	359,537	398,588	439,641	483,695	528,752	575,811	624,872	675,935	728,100	783,1067
$n + m + 24$	327,498	365,547	405,598	446,652	490,707	536,764	583,824	633,885	684,949	738,1014	793,1082
$n = m + 25$	332,508	371,557	411,609	453,663	498,718	544,776	592,836	642,898	694,962	747,1029	803,1097

From Beyer, W. H. (ed.), in *CRC Handbook of Tables for Probability and Statistics*, 2d ed., 1968. Copyright CRC Press, Inc., Boca Raton, Fla.

TABLE XI

Least significant studentized ranges r_p

Least significant studentized ranges r_p, $\alpha = 0.05$

r \ P	2	3	4	5	6
1	17.97	17.97	17.97	17.97	17.97
2	6.085	6.085	6.085	6.085	6.085
3	4.501	4.516	4.516	4.516	4.516
4	3.927	4.013	4.033	4.033	4.033
5	3.635	3.749	3.797	3.814	3.814
6	3.461	3.587	3.649	3.680	3.694
7	3.344	3.477	3.548	3.588	3.611
8	3.261	3.399	3.475	3.521	3.549
9	3.199	3.399	3.420	3.470	3.502
10	3.151	3.293	3.376	3.430	3.465
11	3.113	3.256	3.342	3.397	3.435
12	3.082	3.225	3.313	3.370	3.410
13	3.055	3.200	3.289	3.348	3.389
14	3.033	3.178	3.268	3.329	3.372
15	3.014	3.160	3.250	3.312	3.356
16	2.998	3.144	3.235	3.298	3.343
17	2.984	3.130	3.222	3.285	3.331
18	2.971	3.118	3.210	3.274	3.321
19	2.960	3.107	3.199	3.264	3.311
20	2.950	3.097	3.190	3.255	3.303
24	2.919	3.066	3.160	3.226	3.276
30	2.888	3.035	3.131	3.199	3.250
40	2.858	3.006	3.102	3.171	3.224
60	2.829	2.976	3.073	4.143	3.198
120	2.800	2.947	3.045	3.116	3.172
∞	2.772	2.918	3.017	3.089	3.146

Least significant studentized ranges r_p, $\alpha = 0.01$

r \ P	2	3	4	5	6
1	90.03	90.03	90.03	90.03	90.03
2	14.04	14.04	14.04	14.04	14.04
3	8.261	8.321	8.321	8.321	8.321
4	6.512	6.667	6.740	6.756	6.756
5	5.702	5.893	5.898	6.040	6.065
6	5.243	5.439	5.549	5.614	5.655
7	4.949	5.145	5.260	5.334	5.383
8	4.746	4.939	5.057	5.135	5.189
9	4.596	4.787	4.906	4.986	5.043
10	4.482	4.671	4.790	4.871	4.931
11	4.392	4.579	4.697	4.780	4.841
12	4.320	4.504	4.622	4.706	4.767
13	4.260	4.442	4.560	4.644	4.706
14	4.210	4.391	4.508	4.591	4.654
15	4.168	4.347	4.463	4.547	4.610
16	4.131	4.309	4.425	4.509	4.572
17	4.099	4.275	4.391	4.475	4.539
18	4.071	4.246	4.362	4.445	4.509
19	4.046	4.220	4.335	4.419	4.483
20	4.024	4.197	4.312	4.395	4.459
24	3.956	4.126	4.239	4.322	4.386
30	2.889	4.506	4.168	4.250	4.314
40	3.825	3.988	4.098	4.180	4.244
60	3.762	3.922	4.031	4.111	4.174
120	3.702	3.858	3.965	4.044	4.107
∞	3.643	3.796	3.900	3.978	4.040

Least significant studentized ranges r_p, $\alpha = 0.1$

r \ P	2	3	4	5	6
1	8.929	8.929	8.929	8.929	8.929
2	4.130	4.130	4.130	4.130	4.130
3	3.328	3.330	3.330	3.330	3.330
4	3.015	3.074	3.081	3.081	3.081
5	2.850	2.934	2.964	2.970	2.970
6	2.748	2.846	2.890	2.908	2.911
7	2.680	2.785	2.838	2.864	2.876
8	2.630	2.742	2.800	2.832	2.849
9	2.592	2.708	2.771	2.808	2.829
10	2.563	2.682	2.748	2.788	2.813
11	2.540	2.660	2.730	2.772	2.799
12	2.521	2.643	2.714	2.759	2.879
13	2.505	2.628	2.701	2.748	2.779
14	2.491	2.616	2.690	2.739	2.771
15	2.479	2.605	2.681	2.731	2.765
16	2.469	2.596	2.673	2.723	2.759
17	2.460	2.588	2.665	2.717	2.753
18	2.452	2.580	2.659	2.712	2.749
19	2.445	2.574	2.653	2.707	2.745
20	2.439	2.568	2.648	2.702	2.741
24	2.420	2.550	2.632	2.688	2.729
30	2.400	2.532	2.615	2.674	2.717
40	2.381	2.514	2.600	2.660	2.705
60	2.363	2.497	2.584	2.646	2.694
120	2.344	2.479	2.568	2.632	2.682
∞	2.326	2.462	2.552	2.619	2.670

Abridgment of H. L. Harter's "*Critical Values for Duncan's New Multiple Range Test*," *Biometrics*, Vol. 16, no. 4, 1960. With permission from the Biometric Society.

TABLE XII
Control chart constants

Number of observations in sample, n	d_2	d_3
2	1.128	0.853
3	1.693	0.888
4	2.059	0.880
5	2.326	0.864
6	2.534	0.848
7	2.704	0.833
8	2.847	0.820
9	2.970	0.808
10	3.078	0.797
11	3.173	0.787
12	3.258	0.778
13	3.336	0.770
14	3.407	0.762
15	3.472	0.755
16	3.532	0.749
17	3.588	0.743
18	3.640	0.738
19	3.689	0.733
20	3.735	0.729
21	3.778	0.724
22	3.819	0.720
23	3.858	0.716
24	3.895	0.712
25	3.931	0.709

With permission from *ASTM Manual on Quality Control of Materials*, American Society for Testing Materials, Philadelphia, Pa, 1951.

APPENDIX B

ANSWERS TO SELECTED PROBLEMS

Section 1.1

1. .3; relative frequency

3. .25; classical

Section 1.2

5. (*a*)

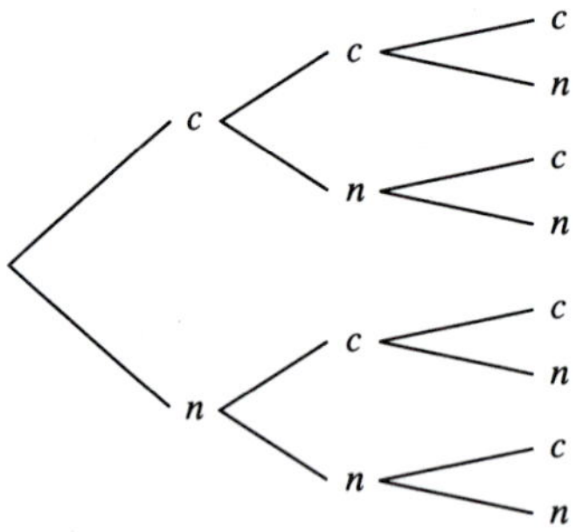

(*b*) $\{ccc, ccn, cnc, cnn, ncc, ncn, nnc, nnn\}$

(*c*) $A_1 = \{ccc, ccn, cnc, cnn, ncc, ncn, nnc\}$

$A_2 = \{ccc\}$

$A_3 = \{nnn\}$

(*d*) no; yes; yes; no

(*e*) no; the eight sample points are not equally likely

7. (*a*)

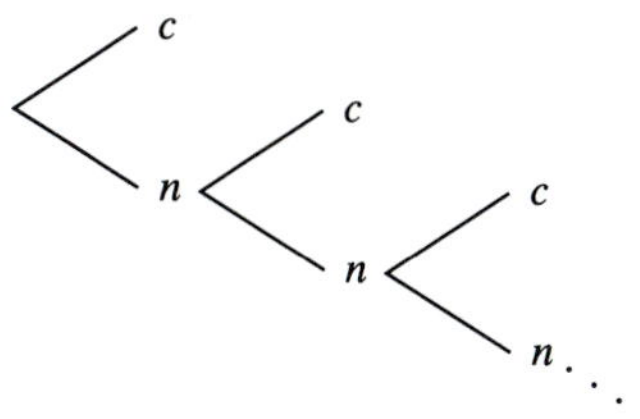

(*b*) no

(*c*) $\{c, nc, nnc, nnnc, \ldots\}$; the list cannot be completed

(*d*) $\{c, nc, nnc, nnnc\} = A$

(*e*) A_1 = contact is made on the first try = $\{c\}$

A_2 = contact is made on the second try = $\{nc\}$

Section 1.3

9. (*a*) 362,880
(*b*) 720
(*c*) 210
(*d*) 30
(*e*) 120
(*f*) 720

11. (*a*) 72
(*b*) 36

13. (*a*) 32
(*b*) 16
(*c*) 4
(*d*) 1

15. (*a*) $5! = 120$
(*b*) 48

17. (*a*) 126
(*b*) 56
(*c*) 56
(*d*) 1

19. 10

21. (*a*) 60,480
(*b*) 30,240

25. $12!/3!3!3!3! = 369{,}600$

27. (*a*) $\binom{3}{2} = 3$; $\{yyn, yny, nyy\}$; or $3!/2!1! = 3$

(*b*) $2^{10} = 1024$; $\binom{10}{7}$; or $10!/7!3! = 120$

Chapter 1 review exercises

28. 7; 15

29. (*a*) 44

(*b*) 11

(*c*) 11/44

30. $\binom{25}{10} = 3{,}268{,}760$; $\binom{5}{3}\binom{20}{7} = 775{,}200$

31. $\binom{6}{3}\binom{4}{0}\bigg/\binom{10}{3} = 1/6$; no, something that occurs by chance with probability 1/6 is not unusually rare

32. (*a*) $(26^5)10 = 118{,}813{,}760$

(*b*) $\left(\dfrac{5!}{3!2!}\right)5 = 50$

(*c*) 1/50

33. 2^{16}; $2^{16} - 1$

34. 15/28

35. (*a*)

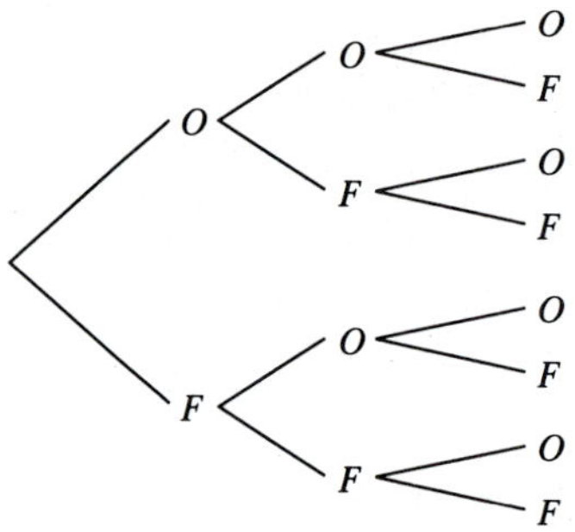

(*b*) $\{OOO, OOF, OFO, OFF, FOO, FOF, FFO, FFF\}$

(*c*) $A = \{OOO, OOF, OFO, OFF, FOO, FOF, FFO\}$

$B = \{OOO, OOF, OFO, OFF\}$

$C = \{FFF\}$

$D = \varnothing$

(*d*) no; yes; yes

(*e*) impossible event

(*f*) 1/8

36. (*a*)

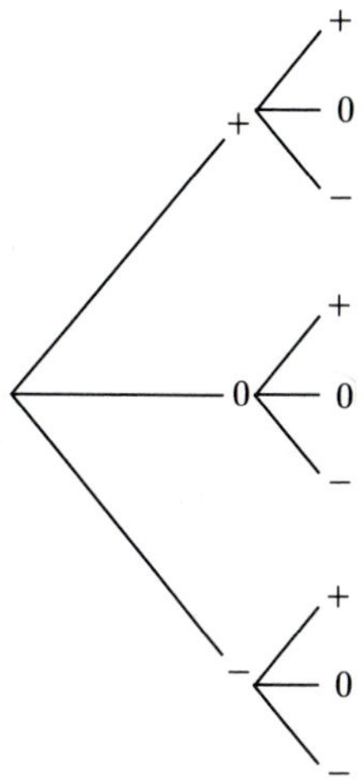

(*b*) $\{++, +0, +-, 0+, 00, 0-, -+, -0, --\}$
(*c*) $A = \{-+, -0, --\}$
$B = \{++, 00, --\}$
$C = \{+0, +-, 0-\}$
(*d*) no; yes
(*e*) The first item selected is not of inferior quality and both items are of the same quality ($A' \cap B = \{++, 00\}$); the first item selected is of inferior quality but the items are not of the same quality ($A \cap B' = \{-+, -0\}$); the first item selected is not of inferior quality and the two items are not of the same quality ($A' \cap B' = \{0+, 0-, +0, +-\}$); the first item selected is of inferior quality and the quality of the first does not exceed that of the second and the items are of the same quality ($A \cap C' \cap B = \{--\}$)
(*f*) The argument is invalid because the nine outcomes are not equally likely

37. (*a*)

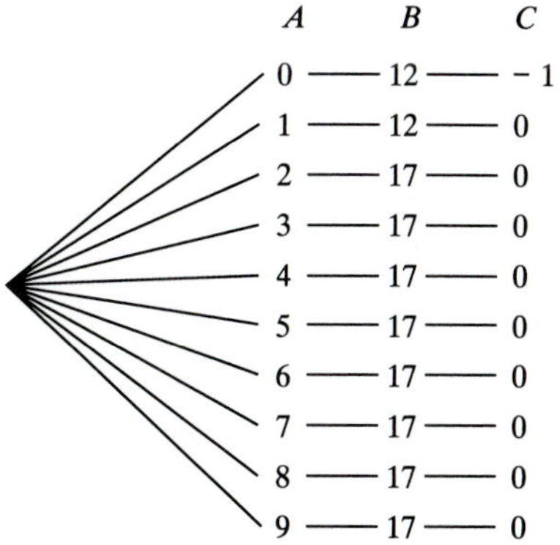

(*b*) {(0, 12, −1), (1, 12, 0), (2, 17, 0), (3, 17, 0), (4, 17, 0), (5, 17, 0), (6, 17, 0), (7, 17, 0), (8, 17, 0), (9, 17, 0)}
(*c*) yes
(*d*) 5/10
(*e*) 1/10
(*f*) 9/10
(*g*) 1

38. $180!/\underbrace{5!5!5!\cdots 5!}_{36\text{ terms}}$

Section 2.1

1. 12/13
3. .45; .13; .46
5. .58; .28; .1

Section 2.2

13. (*a*) 30/58
(*b*) 28/58
(*c*) Theorem 2.1.2
(*d*) 10/42
(*e*) no; exposure to the lethal dose should increase the probability of death

15. 5/35; 35/40

17. (*a*) 1/5
(*b*) 1/80
(*c*) .04
(*d*) 4/20
(*e*) .84

Section 2.3

19. no, $P[A_1 \cap A_2] = .2 \neq P[A_1]P[A_2]$
21. $(.39)^2 \doteq .15$
23. (*a*) .21
(*b*) 21/23
25. .085
27. .0144(.67) + .0012(.33) = .010044

Section 2.4

35. .85(.10)/[.85(.10) + .04(.90)] = .7025
37. .9999

Chapter 2 review exercises

38. (*a*) .85
(*b*) .15
(*c*) 5/20
(*d*) 5/10

39. (*a*) 1/2
(*b*) 1/8

40. .3529; .2353; .2647; .1471

41. .24; .6; .16

42. (*a*) .5
(*b*) .35
(*c*) .50
(*d*) 1/3
(*e*) 35/85

43. (*a*) .0008
(*b*) .0002
(*c*) .2

44. $(.99)^3(.01) = .00970299$; $.01 + .99(.01) + (.99)^2(.01) + (.99)^3(.01)$

Section 3.1

1. not discrete

3. discrete

5. not discrete

Section 3.2

7. (*a*) .01
(*b*)

x	0	1	2	3	4	5
$F(x)$	.7	.9	.95	.98	.99	1.00

(*c*) .98; .1
(*d*) .03

9. (*a*)

x	0	1	2	3
$f(x)$	$(.1)^3$	$3(.9)(.1)^2$	$3(.9)^2(.1)$	$(.9)^3$

(*b*) $k(x) = \dfrac{3!}{x!(3-x)!}$

(*c*)

x	0	1	2	3
$F(x)$	.001	.028	.271	1.00

(*d*) .999
(*e*) .028

11. (*b*) no
(*c*) yes
(*d*) 1; 0
(*e*) yes

13. right continuous, nondecreasing, $\lim_{x\to\infty} F(x) = 1$ and $\lim_{x\to-\infty} F(x) = 0$

Section 3.3

15. (*a*) 4.96; 26.34
(*b*) 1.7384; 1.3185
(*c*) holes per bit

17. $1/.7 = 10/7$; $1/p$

21. (*a*) 11
(*b*) -17
(*c*) 16
(*d*) 4
(*e*) 64
(*f*) 8
(*g*) 208
(*h*) 640
(*i*) 0; 1
(*j*) 0; 1
(*k*) $E\left[\frac{X-\mu}{\sigma}\right] = 0$ and $\text{Var}\left[\frac{X-\mu}{\sigma}\right] = 1$

***23.** (*e*) $163/60 = 2.7167$
(*f*) 2.7007
(*g*) $E[X_{100}] = 94.7953$

Section 3.4

25. (*a*) success is selecting an unacceptable lot; .05
(*b*) $f(x) = (.95)^{x-1}(.05)$
(*c*) $m_X(t) = \dfrac{.05e^t}{1 - .95e^t}$
(*d*) 20; 780; 380; 19.4936
(*e*) .9025
(*f*) .9025

27. $F(x) = 1 - (.95)^x$ for x, a positive integer; .142625

31. (*b*) $24/5$
(*c*) $(1/5)e^{4t} + (4/5)e^{5t}$
(*e*) $116/5$
(*g*) $4/25$; $2/5$

35. (*a*) $e - 1$
(*b*) $m_X(t) = \dfrac{(e-1)(e^{t-1})}{1 - e^{t-1}}$
(*c*) $E[X] = \dfrac{e^{-1}(e-1)}{(1-e^{-1})^2} = \dfrac{e}{e-1}$

Section 3.5

37. 2; .6778; yes, $P[X \geq 5] = .0328$

39. (*a*) $p = 1/4$

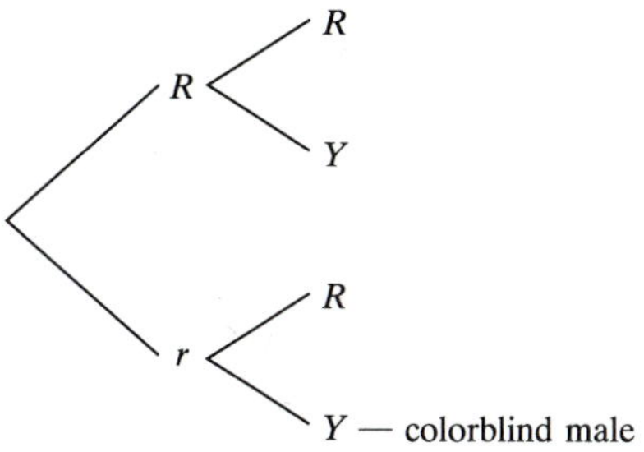

(*b*) 1.25; .1035

41. (*a*) .3758

(*b*) $(.3758)^5$

43. (*a*) $m_X(t) = (pe^t + q)^n$

45. (*a*) X = number of successes in one trial

$$f(x) = \binom{1}{x} p^x q^{1-x} \qquad x = 0, 1$$

(*b*) $f(x) = p^x q^{1-x} \qquad x = 0, 1$

(*c*) $m_X(t) = pe^t + q$

(*d*) p; pq

(*e*) .14

***53.**

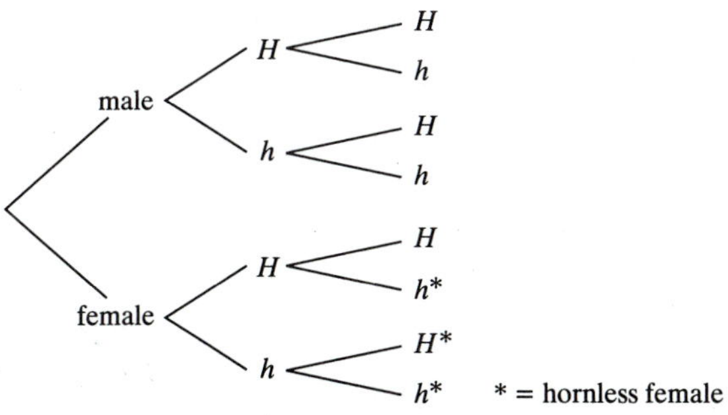

when $r = 2$ and $p = 3/8$,

$$E[X] = 16/3; \text{ no; } P[X \le 5] \doteq .6185$$

Section 3.7

55. $x = 0, 1, 2, 3$; $15/20$; $5\left(\frac{3}{20}\right)\left(\frac{17}{20}\right)\left(\frac{15}{19}\right) = .5033$

57. (*a*) $f(x) = \dfrac{\binom{3}{x}\binom{17}{5-x}}{\binom{20}{5}} \qquad x = 0, 1, 2, 3$

(*b*) .75; .5033
(*c*) .3991
(*d*) .6009

59. (*a*) $f(x) = \dfrac{\binom{600}{x}\binom{2400}{20-x}}{\binom{3000}{20}} \qquad x = 0, 1, 2, \ldots, 20$

(*b*) 4; $20\left(\frac{600}{3000}\right)\left(\frac{2400}{3000}\right)\left(\frac{2980}{2999}\right)$

(*c*) $\displaystyle\sum_{x=0}^{3} \frac{\binom{600}{x}\binom{2400}{20-x}}{\binom{3000}{20}}$

(*d*) .4114

Section 3.8

59. (*a*) 10
(*b*) 10
(*c*) $\sqrt{10}$
(*d*) $f(x) = \dfrac{e^{-10}10^x}{x!} \qquad x = 0, 1, 2, \ldots$
(*e*) .029
(*f*) .010
(*g*) .019
(*h*) .99
(*i*) .448

63. .010

65. yes, $P[X > 10] = .043$

67. yes, $P[X < 2] = .04$

71. .135; .865

Chapter 3 review exercises

75. .118; .882

76. (*a*) $5/4$; $15/16$
(*b*) $(3/4)^5$
(*c*) $(3/4)^{10}$

77. (*a*) X is binomial with $n = 100$ and $p = .1$; $E[X] = 10$
(*b*) Poisson
(*c*) $P[X \geq 17] \doteq .027$
78. $P[X \leq 6] \doteq .4662$
79. 5; 61/125; 64/125
80. 35/210; 105/210
81. $(999/1000)^{12,000} \doteq e^{-12}12^0/0! \doteq .000006$; .999994
82. yes; $P[X \geq 5] = .0127$
83. (*b*) 36/14; 98/14
(*c*) $m_X(t) = (e^t + 4e^{2t} + 9e^{3t})/14$
(*e*) 76/196; $\sqrt{76/196}$
84. $F(x) = 1 - \left(\frac{12}{13}\right)^x$ x a positive integer;

$$P[X \geq 3] = 1 - P[X \leq 2] = 1 - F(2) = \left(\frac{12}{13}\right)^2$$

85. (*a*) binomial, $n = 10$, $p = .8$
(*b*) Poisson, $k = 5$
(*c*) point binomial, $p = .3$ or binomial, $n = 1$, $p = .3$
(*d*) geometric, $p = .6$
(*e*) negative binomial, $n = 5$, $p = .3$
(*f*) Poisson, $k = 1$
86. (*a*) 8; 1.6; $\sqrt{1.6}$
(*b*) 5; 5; $\sqrt{5}$
(*c*) .3; .21; $\sqrt{.21}$
(*d*) 10/6; 10/9; $\sqrt{10/9}$
(*e*) 50/3; $5(.7)/(.3)^2 = 38.89$; $\sqrt{38.89}$
(*f*) 1; 1; 1
87. (*a*) $f(x) = (.99)^{x-1}(.01)$ $x = 1, 2, 3, \ldots$
(*b*) 100
(*c*) $F(x) = 1 - .99^x$ x a positive integer
(*d*) $P[X \leq 90] = 1 - (.99)^{90} \doteq .595$

Section 4.1

1. (*a*) 1/6
(*b*) 11/48
(*c*) 0
(*d*) 11/48
3. (*b*) $1 - e^{-.7} = .5034$; .4966; 0
(*c*) yes, $P[1 \leq X \leq 2] = e^{-.1} - e^{-.2} \doteq .086$
5. (*d*) .5
(*e*) equal; probabilities are constant or "uniform" over intervals of equal lengths
9. (*a*)

$$F(x) = \begin{cases} 0 & x < 2 \\ x^2/12 - 1/3 & 2 \leq x \leq 4 \\ 1 & x > 4 \end{cases}$$

(*c*) yes; yes; 0; 1; yes

(*d*) $\dfrac{dF(x)}{dx} = f(x)$

11. (*a*)

$$F(x) = \begin{cases} 0 & x < 0 \\ x/2\pi & 0 \le x < 2\pi \\ 1 & x \ge 2\pi \end{cases}$$

(*b*) yes; yes; 0; 1; yes

(*c*) $\dfrac{dF(x)}{dx} = f(x)$

13.

$$F(x) = \begin{cases} 0 & x < 25 \\ (\ln x - \ln 25)/\ln 2 & 25 \le x \le 50 \\ 1 & x > 50 \end{cases}$$

Section 4.2

15. (*a*) 56/18
(*b*) 10
(*c*) 104/324; $\sqrt{104/324}$

17. (*a*) $m_X(t) = (1 - 10t)^{-1} \qquad t < 1/10$
(*b*) 10 minutes
(*c*) 100; 10 minutes

19. π; $\pi^2/3$; $\pi/\sqrt{3}$

21. 10; 10; X

Section 4.3

25. (*a*) 2
(*b*) 5040
(*c*) 96
(*d*) 1

27. (*a*) 1; 2; 6; 24; 120
(*b*) $\Gamma(n) = (n - 1)! \qquad n > 1$
(*c*) yes, $\Gamma(1) = 1 = 0!$
(*d*) 14!

29. (*a*) $f(x) = (1/128)x^2e^{-x/4} \qquad x > 0$
(*b*) $m_X(t) = (1 - 4t)^{-3} \qquad t < 1/4$
(*c*) 12; 48; $\sqrt{48}$

33. $m_X(t) = (1 - \beta t)^{-1} \qquad t < 1/\beta$; β; β^2

37. (*a*) $e^{-.25} \doteq .779$
(*b*) $e^{-7/12}/e^{-4/12} = e^{-3/12}$; same as (*a*)

Section 4.4

39. (*a*) .9418
(*b*) .9418
(*c*) 0

(*d*) .0582
(*e*) .8543
(*f*) 1.28
(*g*) − 1.28
(*h*) 1.96
(*i*) 1.645

41. (*a*) .9544
(*b*) 1.24%
(*c*) 128 parsecs
(*d*) $m_X(t) = e^{5000t^2}$

43. (*a*) $f(z) = \frac{1}{\sqrt{2\pi}} e^{-(1/2)z^2} \qquad -\infty < z < \infty$

(*b*) $f'(z) = -\frac{1}{\sqrt{2\pi}} z e^{-(1/2)z^2}$

(*c*) $f''(z) = -\frac{1}{\sqrt{2\pi}} e^{-(1/2)z^2}[1 - z^2]$

Section 4.5

49. $P[-2\sigma < X - \mu < 2\sigma] = .95$

51. (*a*) no
(*b*) $P[|X - \mu| < .5\sigma] \geq -3$, $P[|X - \mu| < 1\sigma] \geq 0$,
$P[|X - \mu| < 2\sigma] \geq .75$, $P[|X - \mu| < 3\sigma] \geq .89$; $\quad k = 2$

Section 4.6

53. (*a*) yes; .9956
(*b*) .0668

55. (*a*) yes
(*b*) 37.5
(*c*) .0322
(*d*) .0392

57. .9484; .8413

Section 4.7

59. (*a*) random factors
(*b*) $R(t) = e^{-.02t}$
(*c*) $R(20) = .6703$
(*d*) $f(x) = .02e^{-.02x} \qquad x > 0$
(*e*) exponential
(*f*) 50; 2500
(*g*) .5488

65. (*a*) .9974; .96
(*b*) .8812
(*c*) .9996
(*d*) .8988
(*e*) III: .9964, V: .999999; .9537

67. 3

Section 4.8

71. $f_Y(y) = (2 - y)e^{-(2-y)} \qquad y \leq 2$

73. $f_F(x) = 5/54 \qquad 59 < x < 69.8$

Chapter 4 review exercises

82. (*a*) 1/18
(*b*) 0; 5.4
(*c*) 5.4; $\sqrt{5.4}$
(*d*) 35/54; 9/54; 26/54
(*e*)

$$F(x) = \begin{cases} 0 & x < -3 \\ (x^3 + 27)/54 & -3 \leq x \leq 3 \\ 1 & x > 3 \end{cases}$$

83. $\Gamma(11) = 10! = 3{,}628{,}800$

84. (*a*) $\alpha = 9$ and $\beta = 2$
(*b*) .01; .10; .725 (chi-squared with 18 degrees of freedom)

85. (*a*) $f(x) = \dfrac{1}{\sqrt{2\pi}\,(5)} \exp\left[-\dfrac{1}{2}\left(\dfrac{x - 15}{5}\right)^2\right]$
(*b*) .0082
(*c*) no; $P[10 \leq X \leq 20] \doteq .68$ by the normal probability rule
(*d*) yes; $P[X \geq 30] = .0013$

86. .7904; $e^{-6.25(6)}$; .042

87. (*a*) 24
(*b*) 122
(*c*) 152
(*d*) 67
(*e*) 4

89. (*a*) $f(x) = (2/5)x^{-4/5}e^{-2x^{1/5}}$
(*b*) 3.75 years
(*c*) $R(t) = e^{-2t^{1/5}}$
(*d*) .1005
(*e*) 2/5
(*f*) $\rho(t)$ is decreasing so that most failures are attributed to early burnout of defective modems

90. 8; .2912; no; $P[X \leq 5] \doteq .1788$

91. $f(x) = (3/2)x^2 + x, \qquad 0 \leq x \leq 1$

92. (*b*) 2; 6; 2

93. .4892

94. (*a*) normal, $\mu = 3$, $\sigma^2 = 16$
(*b*) gamma, $\alpha = 7$, $\beta = 3$
(*c*) chi-squared with 24 degrees of freedom
(*d*) uniform, $a = 1$, $b = 3$
(*e*) standard normal
(*f*) exponential, $\beta = 7$
(*g*) normal, $\mu = 3$, $\sigma^2 = 1$

95. (*a*) 3; 16
(*b*) 21; 63
(*c*) 24; 48
(*d*) 1; 4/2
(*e*) 0; 1
(*f*) 7; 49
(*g*) 3; 1

Section 5.1

1. (*a*) .005
(*b*) .03
(*c*) .98
(*d*) .045

3. (*b*) $f_X(x) = f_Y(y) = 1/n \qquad x, y = 1, 2, 3, \ldots, n$
(*c*) yes

5. (*a*) .4
(*b*) .429
(*d*) .121
(*e*) .22
(*f*) no

7. (*a*) .707
(*b*) .031
(*c*) .242

9. (*b*) .423
(*c*) .577
(*d*) .153
(*e*) $f_X(x) = 1 \qquad 0 < x < 1$
$f_Y(y) = -\ln y \qquad 0 < y < 1$
(*f*) .5
(*g*) .597
(*h*) no

11. (*a*) .005
(*b*) .0625
(*c*) .25
(*d*) $f_X(x) = .2 - .005x \qquad 20 < x < 40$
$f_Y(y) = .005y - .1 \qquad 20 < y < 40$
(*e*) .5625
(*f*) .25
(*g*) no

13. $f(x_1, x_2, \ldots, x_n) \geq 0$

$$\int_{-\infty}^{\infty}\int_{-\infty}^{\infty}\int_{-\infty}^{\infty}\cdots\int_{-\infty}^{\infty} f(x_1, x_2, \ldots, x_n)\, dx_1\, dx_2 \ldots dx_n = 1$$

$$P[a_1 \leq X_1 \leq b_1, a_2 \leq X_2 \leq b_2, \ldots, a_n \leq X_n \leq b_n]$$

$$= \int_{a_n}^{b_n}\int_{a_{n-1}}^{b_{n-1}}\cdots\int_{a_1}^{b_1} f(x_1, x_2, \ldots, x_n)\, dx_1\, dx_2 \ldots dx_n$$

Section 5.2

15. (*a*) negative
(*b*) 60/35; 80/35; 120/35; $-600/35^2$
17. 0
19. (*a*) 30.96; 28.99; 897.84; .3096
(*b*) 1.97; average difference in the inside and outside pressures
21. 1/2; 1/4; 1/6; 1/24
23. (*a*) positive
(*b*) 26.67; 33.33; 900; 11.09
(*c*) 6.66

Section 5.3

29. (*a*) negative; −1
(*b*) 120/35; 200/35; −1
31. 959.76; 841.67; 1.24; 1.25; .24
33. 1/3; 1/9; 1/12; 7/144; .655
37. 0

Section 5.4

39. (*a*) 31.99
(*b*) 28.5

41. (*a*) $\mu_{X|y} = \sum_{x=y}^{n} x\frac{1}{n-y+1} = \frac{n+y}{2}$; yes
(*b*) 7
(*c*) $\mu_{Y|x} = \sum_{y=1}^{x} \frac{y}{x} = \frac{x+1}{2}$; yes
(*d*) 5/2
43. (*a*) $\mu_{X|y} = (y+20)/2$; yes
(*b*) \$25.00
(*c*) $\mu_{Y|x} = (40+x)/2$; yes
(*d*) \$37.50

Section 5.5

45. (*a*) yes; $x = \frac{3u-v}{5}$, $y = \frac{2v-u}{5}$
(*b*) 1/5
47. $f_{UV}(u,v) = 1/10,\ 0 < u < 4,\ 0 < v < 7$
53. (*a*) $f_U(u) = \begin{cases} (1/6)(\ln 3 - \ln u/2) & 0 < u < 6 \\ 3/4 & 0 < u < 2/3 \end{cases}$
(*b*) $f_U(u) = 1/3u^2 \qquad 2/3 \le u < \infty$

Chapter 5 review exercises

54. (*b*) 8/18
(*c*) $P[Y \le 1/2 \text{ and } Y \le X \le Y + 1/4] = 11/96$

(*d*) $f_X(x) = 4x^3 \quad 0 < x < 1$; 4/5; 2/3
(*e*) $f_Y(y) = 4y(1 - y^2) \quad 0 < y < 1$; 8/15; 1/3
(*f*) no
(*g*) $f_{X|y} = 2x/(1 - y^2), \quad 0 < y < x < 1$
(*h*) $P[X \leq 3/4|y = 1/6] = 11/20$
(*i*) $\mu_{X|y} = 2(1 + y + y^2)/3(1 + y)$; no
(*j*) 43/63 (about 41 minutes after the system is activated)
(*k*) positive; .493

55. $f_X(x) = xe^{-x} \quad x > 0$; $f_Y(y) = ye^{-y} \quad y > 0$; yes; 0

56. (*a*) .356
(*b*) 3.279
(*c*)

x	0	1	2	3
$f_X(x)$	.210	.298	.277	.215

$\mu_X = 1.497$; $\alpha_X^2 = 1.1$

y	1	2	3	4
$f_Y(y)$	.267	.397	.302	.034

$\mu_y = 2.103$; $\sigma_y^2 = .694$

(*d*) .688
(*e*) .1308; .1497; yes

57. (*a*) $f_{XY}(x, y) = \left(\frac{e^{-5}5^x}{x!}\right)\left(\frac{e^{-3}3^y}{y!}\right) \quad x = 0, 1, 2, \ldots \quad y = 0, 1, 2, \ldots$
(*b*) .0572
(*c*) 0
(*d*) $f_{X|y} = \frac{e^{-5}5^x}{x!} \quad x = 0, 1, 2, \ldots$

Section 6.1

1. yes; set of all days from past, present, and future
3. no
5. yes; the 50,000 workers affected

Section 6.2

7. (*a*) 58.8
(*b*) 8.5
(*c*) 16.25
(*d*) 16.25 to 24.75
24.75 to 33.25
33.25 to 41.75
41.75 to 50.25
50.25 to 58.75
58.75 to 67.25
67.25 to 75.75

9. exponential

11. (*d*)

Category	Boundaries	Frequency	Relative frequency	Relative cumulative frequency
1	.45 to 1.25	7	7/50	7/50
2	1.25 to 2.05	15	15/50	22/50
3	2.05 to 2.85	15	15/50	37/50
4	2.85 to 3.65	8	8/50	45/50
5	3.65 to 4.45	3	3/50	48/50
6	4.45 to 5.25	2	2/50	50/50

the gamma distribution might be appropriate

13. (*a*) $P[X < p_{25/100}] \le 25/100$ and $P[X \le p_{25/100}] \ge 25/100$
(*b*) 8
(*c*) $-\ln .75 \doteq .288$

15. (*a*) approximately .044
(*b*) approximately 7.7
(*c*) approximately 2.2

Section 6.3

17. (*a*) group I 3; 3
group II 3; 2.5
(*b*) 4, 4
(*c*) group I 1.5; 1.2
group II 2.91; 1.7
(*d*) yes

19. .8; 1.07

21. 7.94; 1.221; 1.11

25. (*a*) yes
(*b*) 1.86
(*c*) 1.975

Section 6.4

27. (*b*) 1.34σ

29. (*a*)

1	1						
1	5	8					
2	4	3					
2	6	5	7	9			
3		2	0				
3	6	9	5	7	8	7	5
4	0	0	2				
4	7	9					
5	0	1					
5	6						
6	0						

(*b*) $\tilde{x} = 3.65$ $iqr = 1.45$ no outliers
$q_1 = 2.65$ $f_1 = .475$
$q_3 = 4.10$ $f_2 = 6.275$

(*c*)

2	0
4	
4	2 3
4	5 4
4	6 7 6
4	9 8
5	0 1 0
5	2 2 3 2 3
5	4 4
5	6
5	8 9
7	8

(*d*) $\tilde{x} = 5.05$ $iqr = .75$ $F_1 = 2.35$
$q_1 = 4.6$ $f_1 = 3.475$ $F_3 = 7.6$
$q_3 = 5.35$ $f_3 = 6.475$

2 is an extreme outlier; 7.8 is an extreme outlier

(*e*)

3	
3	9
4	2 4 3 1
4	5 7 9 8 7 6 8
5	0 1 1 2 0 4
5	9 6
6	1
6	

58 has a misplaced decimal and should read 5.8; $\bar{x} = 7.29$ (using bad point); $\bar{x} = 4.91$ (using corrected point)

Chapter 6 review exercises

26. (*a*) Poisson
(*b*) 1.9; 3.36; 1.8; 1
(*c*) no
(*d*) 6.24; 1.852; 1.36; 6.3

31. (*a*) range = 5.73

Category	Boundaries	Frequency	Relative frequency	Relative cumulative frequency
1	.235 to 1.195	2	2/50	2/50
2	1.195 to 2.155	18	18/50	20/50
3	2.155 to 3.115	13	13/50	33/50
4	3.115 to 4.075	7	7/50	40/50
5	4.075 to 5.035	7	7/50	47/50
6	5.035 to 5.995	3	3/50	50/50

(*b*) 2.788; 1.8594; 1.364
(*c*) approximately 2.644; about 16%

(*d*) p; 3/50; 3/50

(*e*) $p(1-p)$; (3/50)(47/50) = .0564; .0576; no; .0576

*(*f*) $\sigma_X \doteq .955$

32. (*a*)

1	1										
1	3										
1	5										
1											
1	9	8									
2	1	0									
2	2	2	2	3	3						
2	5	5	5	5	4	5	4	4	5	5	4
2	6	7	7	6	7	7	6				
2	9	8	8								
3	0										
3											
3											
3											
3											

(*b*) $\tilde{x} = 25$ $\quad iqr = 4$ $\quad F_1 = 10$

$q_1 = 22$ $\quad f_1 = 16$ $\quad F_3 = 38$

$q_3 = 26$ $\quad f_3 = 32$

the values 11, 13, 14, and 15 are flagged as mild outliers

(*c*) there is a left skew; gamma; no; outliers are flagged assuming normality

(*d*)

3									
3	3								
3	5	5							
3	6	7	7	6	6	6	7	7	
3	8	8	9	9	8	9	8	9	9
4	0	0	0	0	0	1	1		
4	2	3	3						
4	5	5							
4	6	6							
4									

yes

(*e*) $\tilde{x} = 3.9$ $\quad f_1 = 3.1$ $\quad$ no outliers

$q_1 = 3.7$ $\quad f_3 = 4.7$

$q_3 = 4.1$ $\quad F_1 = 2.5$

$iqr = .4$ $\quad F_3 = 5.3$

Section 7.1

1. 8; 5/20

3. (*a*) 5.25

(*b*) 5.25

(*c*) 5.25/9

5. (*b*) .44

7. (*a*) 9.7
(*b*) 3.57
(*c*) 19.4
(*d*) 1.9; no

9. (*b*) .817; no, the sample sizes are very different
(*d*) .708

13. (*a*)

16	0 3 2 0 2 3 1 2 4 3
16	5 7 8 7 9
17	0 2 3
17	6 9 8
18	0
18	5
19	
19	6

skewed right

(*b*) $\tilde{x} = 167$ $f_1 = 148$
$q_1 = 163$ $f_3 = 188$
$q_3 = 173$ $F_1 = 133$
$iqr = 10$ $F_3 = 203$

196 is flagged; the distribution is not bell-shaped, so the "outlier" is probably not unusual

(*c*) $\bar{x} = 169.0$ (169.04)
$s^2 = 81.54$

Section 7.2

15. .67

17. $\bar{X}$; $\bar{X}/s$

19. $M_2 - M_1^2$; no

21. $\hat{\beta} = \bar{x} = 169$; $\hat{\sigma} = M_2 - M_1^2 = 24/25(s^2) = 78.28$

23. $1/\bar{X}$

25. $\hat{\sigma}^2 = \dfrac{(1 - 1/\bar{x})}{(1/\bar{x})^2}$

27. (*a*)

31	9
32	0 1 2
32	5 6 7 8
33	0 0 1 3 4 2 4
33	9 8 5 7 8 9 8 6 7
34	0 3 1 1
34	9
25	0

(*b*) 333.8; 60.85
(*c*) 62.94

29. 333.8; 60.85

31. $\bar{X}$; no

33. (*a*) $\bar{X}/2$
(*b*) $\bar{X} = 2\hat{\beta}$; yes
(*c*) 35.25
(*d*) 70.5

35. .1

Section 7.3

37. (*a*) normal; $\mu = 2, \sigma^2 = 9$
(*b*) normal; $\mu = 0, \sigma^2 = 16$
(*c*) geometric, $p = .25$
(*d*) binomial, $n = 5, p = .5$
(*e*) Poisson, $k = 6$
(*f*) gamma, $\alpha = 5, \beta = 3$
(*g*) chi-squared, $\gamma = 16$
(*h*) exponential, $\beta = .5$

Section 7.4

47. (*a*) .643
(*b*) $.643 \pm .0039$
(*c*) shorter; $.643 \pm .0033$
(*d*) longer; $.643 \pm .0052$

49. (*a*) no, X is discrete
(*b*) 2.8
(*c*) $2.8 \pm .2036$; Central Limit Theorem
(*d*) no; 3.0 lies in this interval—it has not been ruled out as a possible value for μ

Chapter 7 review exercises

51. (*b*) $\dfrac{1 + \theta}{\theta + 2}$
(*c*) $(1 - 2\bar{X})/(\bar{X} - 1)$
(*d*) $-52/76 \doteq -.6842$
(*e*) $-1 - n/\ln \Pi x_i$
(*f*) $-.3836$; no

52. (*a*) $\theta/2$
(*b*) $2\bar{X}$; yes
(*c*) 2.06

53. (*a*) normal with mean μ and variance $4/16$
(*b*) 42.883
(*c*) $42.883 \pm .98$; no, 42.2 lies in the confidence interval

54. (*a*) \$3.698
(*b*) .0487
(*c*) .221; no
(*d*) .0438; no

55. (*a*) point binomial with probability of success p
(*b*) normal with mean p and variance $p(1 - p)/n$
(*c*) .05

56. (*a*) 7.1
(*b*) normal with mean μ and variance $25/36$
(*c*) 7.1 ± 1.9
(*d*) yes; 10 lies outside the 99% confidence interval

57. (*a*) $m_X(t) = e^{15t + 16(t^2/2)}$
$m_Y(t) = e^{1.5t + 25(t^2/2)}$

(*b*) $m_T(t) = e^{16.5t + 16.25(t^2/2)}$
(*c*) T is normally distributed with mean 16.5 minutes and variance 16.25
(*d*) .1922

58. (*a*) $m_Y(t) = (1 - 3t)^{-500}$
(*b*) gamma with $\alpha = 500$ and $\beta = 3$
(*c*) $m_{\bar{X}}(t) = m_Y(t/100) = [1 - (3/100)t]^{-500}$
(*d*) gamma with $\alpha = 500$ and $\beta = 3/100$
(*e*) .0681

59. (*a*) gamma with $\alpha = 2$ and $\beta = \theta$
(*b*) 2θ
(*c*) $\bar{X}/2$
(*d*) $\bar{X}/2$; no
(*e*) 1.55
(*f*) yes

60. (*a*) standard normal
(*b*) X_1^2
(*c*) X_{10}^2; $\bar{\gamma} \sim$ normal; $\mu = \dfrac{\lambda s}{n}$; $\sigma^2 = \dfrac{\lambda s}{n^2}$

62. (*a*)

x	1	2	3	4	5	6
$f(x)$	1/6	1/6	1/6	1/6	1/6	1/6

$$E[X] = 3.5$$

$$E[X^2] = 91/6$$

$$\sigma^2 = 35/12 = 2.92$$

(*b*) $E[\bar{X}] = 3.5$; Var $\bar{X} = 2.92/20 = .146$
(*c*) bell
(*d*) 3.5
(*e*) .146
(*f*) 24

Section 8.1

1. (*b*) .0129
(*c*) [.0082, .0234]
(*d*) [.091, .153]
(*e*) yes; .2 is not in the confidence interval

3. (*b*) 2.455
(*c*) [1.322, 5.699]
(*d*) [1.15, 2.39], reduce the confidence

***5.** (*b*) .00000375
(*c*) [0, .00000573]; [0, .00239]
(*d*) yes, σ appears to be at most .00239

7. (*a*) 122.94, 76.76
(*b*) [6.75, 8.48]

Section 8.2

9. (*a*) 1.86
(*b*) − 1.86
(*c*) − 2.179
(*d*) 2.179
(*e*) 1.645
(*f*) 1.645
(*g*) 1.708
(*h*) 2.060
(*i*) 1.753
(*j*) 1.325
(*k*) 1.746
(*l*) 1.310

11. (*a*) 1.2896; .0000123; .0035
(*b*) 1.2896 ± .0016
(*c*) no, 1.29 is contained in the confidence interval

13. (*a*) 2.35; .89
(*b*) 2.35 ± .45
(*c*) yes, we are 99% confident that the new mean time is at most 2.80 seconds

15. (*a*)

2	9
3	8 5 9
4	2 1 7 8
5	1 3 5 6 7 5
6	1 8 1 0 2 5 7 3
7	9 7 3 2
8	1 0
9	2
10	0

(*b*) $\tilde{x} = 605$ $\quad f_1 = 120$ $\quad$ no outliers
$q_1 = 480$ $\quad f_3 = 1080$
$q_3 = 720$
$iqr = 240$
(*c*) $\bar{x} = 602.3$; $s = 169.1$; 602.3 ± 85.1
(*d*) lower the confidence

***19.** (*b*) 385
(*c*) 153

Section 8.3

21. (*a*) H_0: $\mu \geq .08$
H_1: $\mu < .08$
(*b*) we shall conclude that the average percentage of metal in household wastes has been reduced when, in fact, it has not been reduced
(*c*) we shall be unable to detect the fact that the mean percentage of metal in household waste has been reduced
(*d*) we have a 5% chance of having committed a Type I error

23. (*a*) we shall conclude that the model is not credible when, in fact, it is a valid model
(*b*) we shall be unable to detect the fact that the proposed model is not credible

25. (*a*) $C = \{10, 11, 12, 13, 14, 15\}$, $\alpha = .0338$
(*b*) yes

27. (*a*) H_0: $p \leq .5$
H_1: $p > .5$
(*b*) 7.5
(*c*) .0592
(*d*) .7827; .4845; .1642; .0127
(*e*) .2173; .5155; .8358; .9873
(*f*) yes; Type I
(*g*) no; Type II

29. .0065; .0003; 0; 0; yes

Section 8.4

31. (*a*) H_0: $\mu \leq .05$
H_1: $\mu > .05$
(*b*) we shall assume that the percentage titanium exceeds 5% when, in fact, it does not; we shall be unable to detect a situation in which the percentage titanium exceeds 5%
(*c*) .1056; debatable, a P value of .1056 might be considered small by some and large by others

33. (*a*) H_0: $p \leq .15$
H_1: $p > .15$
(*b*) .1335; no, this probability is not unusually small; Type II

Section 8.5

35. (*a*) -1.711
(*b*) -1.282
(*c*) 2.093
(*d*) 2.602
(*e*) ± 1.729
(*f*) ± 2.045

37. (*a*) H_0: $\mu = .12$
H_1: $\mu > .12$
(*b*) 2.462
(*c*) yes, $t = 2.738$
(*d*) that X is at least approximately normal

39. (*a*) H_0: $\mu = 2.5$
H_1: $\mu < 2.5$
(*b*) $t = -3.5$; $.001 < P < .005$; yes, P seems to be small; at least approximate normality
(*c*) conclude that the mean noise level is below 2.5 db; we shall assume that the new product reduces noise when, in fact, it does not

41. (*a*) H_0: $\mu = 4.8$
H_1: $\mu < 4.8$
(*b*) $t = -2.828$; $.001 < P < .005$; yes

43. (*a*) H_1: $\mu < 5$
(*b*) $t = -3.47$; $.001 < P < .005$; reject H_0
(*c*) yes, because $P < .05$
(*d*) probably not since $\bar{x} = 4.28$; there is still a large accumulation there

Section 8.6

47. (*a*) unable to reject H_0
(*b*) $t = 1.154$, critical point $= \pm 2.145$; unable to reject H_0
(*c*) $\chi^2 = 17.81$, critical point $= 23.7$, unable to reject H_0
49. (*a*) unable to reject H_0
(*b*) $t = 1.22$, critical point $= 1.729$, unable to reject H_0
(*c*) $\chi^2 = 9.297$, critical point $= 11.7$, reject H_0; yes

Section 8.7

51. (*a*) yes, .0037
(*b*) yes, .0207
(*c*) yes, .0207
(*d*) yes, .0107
(*e*) no, .0547
(*f*) yes, .0074
(*g*) yes, .0352
53. no, $P = .3770$
55. (*a*) 110/4
(*b*) yes, $|W_-| = 8.5$, critical point $= 11$
57. (*a*) 100(101)/4
(*b*) 100(101)(201)/24
(*c*) H_0: $M = 2$
H_1: $M > 2$
(*d*) $P \doteq .0007$, yes

Chapter 8 review exercises

58. (*a*) 44
(*b*) no
(*c*) 3.10; .1213; .348, no
(*d*) $\chi^2_{.95} \doteq 33.65$; [.08998, .1766]; [.298, .418]
$\chi^2_{.05} \doteq 66.05$
(*e*) $3.10 \pm .082$
59. (*a*) H_0: $\mu = 3$
H_1: $\mu < .3$
(*c*) 13
(*d*) $t_{.95} = -1.753$; $t \doteq -3.71$; reject H_0; yes, the product should be marketed
60. (*a*) H_0: $p \le .5$
H_1: $p > .5$
(*b*) $C = \{15, 16, 17, 18, 19, 20\}$
(*c*) no; Type II
(*d*) .8744; .5836; .1958; .0113
(*e*) .1256; .4164; .8042; .9887
61. yes; $t = -2.69$; $.005 < P < .01$

62. $.10 < P < .25$

63. (*a*)

−.3	0
−.2	0 0 0 0
−.1	0 0 0 0 0 0 0 0
0	0 0 0 0 0 0 0 0 0 0 0
.1	0 0 0 0 0 0
.2	0 0 0 0
.3	0
.4	
.5	
.6	
.7	
.8	0

yes; .8 looks like an outlier

(*b*) $\tilde{x} = 0$ $f_1 = -.4$
$q_1 = -.1$ $f_3 = .4$
$q_3 = .1$ $F_1 = -.7$
$iqr = .2$ $F_3 = .7$

.8 is an extreme outlier

(*c*) $\bar{x} = -.008$; $s = .198$; $-.008 \pm .066$

64. (*a*) the number of trials is at most 3000; in a geometric setting there is no "a priori" number of trials
(*b*) 20
(*c*) exceed
(*d*) 59; $P[X \geq 59] = 0.51$; see Sec. 3.4
(*e*) no; Type II; we think that the system will not crash when, in fact, it will
(*f*) reject H_0 and conclude that the system will crash; Type I; we shall stop the system unnecessarily

65. (*b*) $8.1 \pm .8$
(*c*) [4.36, 10.56]
(*d*) 3.3 ± 10.7; [.80, 3.17]; [.90, 1.8]
(*e*) no
(*f*) no

66. (*a*) H_1: hollow arrows are faster than those made of solid aluminum
H_1: $\mu < 0$
(*b*) H_0: hollow arrows are no faster than those made of solid aluminum
H_0: $\mu = 0$
(*c*) $x = -20.41$; $s = .89$; negative sign means that the time with the hollow arrow was better than that with the solid arrow
(*d*) $t = -102.5$; $P < .0005$

Section 9.1

1. (*a*) .9
(*b*) $.9 \pm .069$
(*c*) 609

3. (*a*) $.6 \pm .025$
(*b*) $.9 \pm .019$

5. 1068

7. 6766

9. (*a*) $1 - 2\hat{p}$
(*b*) $1/2$
(*c*) -2
(*d*) $1/4$

Section 9.2

11. (*a*) H_0: $p = .99$
H_1: $p > .99$
(*b*) $z = .57$; no, $P = .2843$
(*c*) we are unable to show that the new network is compatible with more than 99% of the equipment already in use.

13. (*a*) H_0: $p = .6$
H_1: $p > .6$
(*b*) 1.645
(*c*) $z = .842$, no; Type II
(*d*) we shall not be able to show that more than 60% of the business offices in the United States have a mainframe when, in fact, this is true.

15. (*a*) ± 1.96
(*b*) $z = 1.38$, no; Type II

Section 9.3

17. (*a*) .356, .489, $-.133$
(*b*) $-.133 \pm .062$
(*c*) yes, 0 is not contained in the confidence interval

19. (*a*) .31, .40, $-.09$
(*b*) $-.09 \pm .079$
(*c*) yes, 0 is not contained in the confidence interval

23. 1527

25. 542

Section 9.4

27. (*a*) H_0: $p_1 = p_2$
H_1: $p_1 < p_2$
(*b*) -1.28
(*c*) .0015, .0025, $-.001$; $\hat{p} = .002$, $z = -.71$, no

29. (*a*) H_0: $p_1 = p_2$
H_1: $p_1 > p_2$
(*b*) $\hat{p} = .334$, $z = 12.35$, yes, $P \doteq 0$

Chapter 9 review exercises

33. (*a*) 533
(*b*) .21; $.21 \pm .034$

34. (*a*) H_0: $p = .5$
H_1: $p > .5$
(*b*) $z = .2828$; no, $P \doteq .39$

35. (*a*) H_0: $p = .5$
H_1: $p > .5$
(*b*) 1.645
(*c*) no; the observed value of the test statistic is 1.00

36. (*a*) $-.02$
(*b*) $-.02 \pm .047$
(*c*) no; 0 lies in the interval
(*d*) no; $n \doteq 2229$, using $\hat{p}_1 = .02$ and $\hat{p}_2 = .04$ as prior estimates

37. H_0: $p_1 = p_2$
H_1: $p_1 > p_2$
where p_1 denotes the proportion of customers reordering during the current year; $\hat{p} = .6$, $z = 1.87$, yes, $P = .0307$

38. (*a*) H_0: $p_1 - p_2 = .1$
H_1: $p_1 - p_2 > .1$
(*b*) 1.645
(*c*) $z = .25$, no, Type II

39. (*a*) $.2 \pm .078$
(*b*) $75{,}000 \pm 29{,}400$
(*c*) 1537

40. (*a*) .25
(*b*) $.25 \pm .003$
(*c*) [2468, 2532]

Section 10.1

1. 4.2; 2; 2.2

3. 8.008; 7.609; .399

Section 10.2

5. (*a*) .9
(*b*) 2.416
(*c*) .426
(*d*) .10
(*e*) .622
(*f*) 1.841
(*g*) .05
(*h*) 5.803
(*i*) .34890
(*j*) .05
(*k*) 1.659
(*l*) .52732

7. (*a*) $f = 5$, upper critical point = 2.725, reject H_0
(*b*) $f = 2$, upper critical point = 2.308, unable to reject H_0
(*c*) $f = 1.36$, upper critical point = 1.352, reject H_0

9. (*a*) H_0: $\sigma_1^2 = \sigma_2^2$
H_1: $\sigma_1^2 < \sigma_2^2$
(*b*) .41, $f = .92$, unable to reject H_0

11. (*a*) [.5604, 2.3644]
(*b*) no, 1 lies in the interval

Section 10.3

13. (*a*) $s_1^2 = .034$, $s_2^2 = .0525$, $f = 1.54$, upper critical point = 2.147, unable to reject H_0
(*b*) .0433
(*c*) $-.67 \pm .14$
(*d*) yes, the confidence interval does not contain 0

15. (*a*) $f = 1.23$, upper critical point = 2.086, unable to reject H_0
(*b*) 7.93
(*c*) 7.88 ± 3.18
(*d*) yes, 0 is not in the interval

Section 10.4

21. (*a*) H_0: $\mu_1 = \mu_2$ $\quad$ μ_1 = mean level for children
H_1: $\mu_1 > \mu_2$
(*b*) $f = 119.01$, upper critical point ($\alpha = .2$) = 1.348, do not pool
(*c*) $t = 20.00$, $\gamma = 123$, reject H_0; $P < .0005$; conclude that the average strontium level in children is higher than that in adults

23. (*a*) H_0: $\mu_1 = \mu_2$
H_1: $\mu_1 > \mu_2$ (premium > regular)
(*b*) $t = 1.583$, reject H_0, $.05 < P < .1$
(*c*) conclude that the average mileage using premium gasoline is higher than that using regular gasoline

25. (*a*) H_0: $\mu_1 = \mu_2$
H_1: $\mu_1 < \mu_2$ (acrylic < butyl)
(*b*) $f = .36$, lower critical point ($\alpha = .1$) = 1/2.4 = .41607, do not pool
(*c*) $t = -5.88$, $\gamma = 24$, reject H_0, $P < .0005$; conclude that the average tensile strength for butyl coating is higher than that for acrylic coatings

29. (*a*) $f = 3.32$, critical point ($\alpha = .10$) = 1.984, do not pool
(*b*) 2.8 ± 1.94, $\gamma = 37$
(*c*) yes, 0 is not in the interval

Section 10.5

31. (*b*) $-.412$, 1.14
(*c*) $-.412 \pm .723$, no, the confidence interval contains 0, so the results are inconclusive

33. $.04 \pm .585$; no, 0 lies in the interval

35. H_0: $\mu_X = \mu_Y$ $\quad$ $t = 3.04$, reject H_0; $.0005 < P < .005$
H_1: $\mu_X > \mu_Y$ $\quad$ X = travel lane

Section 10.6

37. H_0: $M_I = M_P$, $W_m = 101.5$, upper critical point = 98; reject H_0
H_1: $M_I < M_P$

39. $W_m = 20$, critical point = 22, reject H_0

41. (*a*) 6600
(*b*) 110000
(*c*) $z = -1.999$, $P = .0228$, yes

43. $|W_-| = 7$, can reject at $\alpha = .05$ level (critical point = 11); no, $P[Q_- \leq 3] = .1719$, the sign test ignores the magnitude of the differences involved

Chapter 10 review exercises

45. (*a*) H_0: $\mu_1 = \mu_2$ where μ_1 = mean temperature setting required by
H_1: $\mu_1 < \mu_2$ using the computerized system
H_0: $\sigma_1^2 = \sigma_2^2$ where σ_1^2 = variance in temperature settings required
H_1: $\sigma_1^2 < \sigma_2^2$ by using the computerized system
(*b*) $s_2^2/s_1^2 = 25$; reject H_0: $\sigma_1^2 = \sigma_2^2$; $P < .05$; do not pool; $\gamma \doteq 25$; $t = -8.73$; reject H_0: $\mu_1 = \mu_2$; $P < .0005$; yes, both claims are supported

46. (*a*) H_0: $\mu_1 - \mu_2 = 15$ where μ_1 = mean amount of dross obtained by using
H_1: $\mu_1 - \mu_2 > 15$ the old method
(*b*) $s_1^2/s_2^2 = 25$; reject H_0: $\sigma_1^2 = \sigma_2^2$; $P < .10$; do not pool; $\gamma \doteq 9$, $t = 4.96$; reject H_0; $P < .0005$; yes, it appears that the new process will be profitable

47. (*a*) yes
(*b*) 2.6 ± 1.0; yes, the confidence interval consists entirely of positive values

48. $s_1^2/s_2^2 = 1.14$; pool; 900 ± 58.2; yes, the confidence interval consists of positive values

49. yes, $t \doteq 7.56$; $P < .0005$

50. (*a*) $\bar{d} = -14.43$
(*b*) $-14.43 \pm .40$; $s_d = .72$
(*c*) yes; 0 is not contained in the confidence interval

51. (*a*)

A		R	
$\tilde{x} = 130$	$f_1 = 90.25$	$\tilde{x} = 253$	$f_1 = 188.5$
$q_1 = 119.5$	$f_2 = 168.25$	$q_1 = 238$	$f_3 = 320.5$
$q_3 = 139$	$F_1 = 61$	$q_3 = 271$	$F_1 = 139$
$iqr = 19.5$	$F_3 = 197.5$	$iqr = 33$	$F_3 = 370$

12.4 and 200 are extreme extreme outliers; 170 is a mild outlier

516 is an extreme outlier

(*b*) $f = 8.83$; upper critical point 1.908; reject at .20 level; conclude that $\sigma_A^2 \neq \sigma_R^2$
(*c*) $(267.4 - 133.9) \pm 2.093\sqrt{4448.89/17 + 503.83/19}$; $df \doteq 19$
(*d*) yes, the confidence interval is negative

52. (*a*)

15	1
16	2 9
17	2 5 6 4
18	9 8 1 3
19	8 5 7 9 8 2 3
20	0 6 1 5 4 1 2 7
21	2 1 8 1 4 3
22	6 1 1 4
23	0 6 3
24	1 9
25	0

nonsmokers

15	1 7 3 0 2 9
16	8 1 0 3
17	7 2 9
18	1 0 3
19	4 9 9
20	5
21	2 3 6
22	8 4 1
23	2 1 8 0
24	3 1 1 9 8
25	8 7 1 2 2 0 1

smokers

(*b*) yes; there is a difference in shape and also variability may be different
(*c*) use the Wilcoxon rank-sum test to test location

53.

bank

1	2
1	8 9 7 9 5 7
2	0 1 2 0 4 3 1 3 3
2	9 7 8 5 9 7 6 5 5
3	1 3 0 2 1 4
3	8

other

3	4
3	7 9 7
4	3 2 0 0 1
4	7 9 8
5	1 4 2 0 1 3 2 1 2 0
5	9 7 9 8
6	4 3 2
6	9

a two-sample t test should apply; $f = 2.19$; variances are unequal ($\alpha = .10$); $t = -13.17$; $df \doteq 50$

Section 11.2

7. (*b*) $\hat{\mu}_{Y|x} = 0.2177 + 0.0957x$
(*c*) 5.004 for both estimates
(*d*) increase $.1914 \times 10^8$ Btu
(*e*) decrease $.1914 \times 10^8$ Btu

9. (*b*) $\hat{\mu}_{Y|x} = -.1064 + 33.5178x$
(*d*) .8995
(*e*) .8995
(*f*) yes, yes, no

15. .1814

Section 11.3

17. significant ($P < .001$)
19. $t = 4.025$, significant at $\alpha = .01$
21. significant ($P < .001$)
23. $\hat{\mu}_{Y|x} = 14.491 + 1.498x$
25. $t = 1.197$, not significant
27. $\hat{\mu}_{Y|x=15} = 89.588$ (Regression not sign.)
29. (*a*) $\hat{\mu}_{Y|x} = 911.667 - 49.667x$
(*b*) $t = -19.431$, significant at $P < .0001$
(*c*) no
31. $\hat{\sigma}^2 = .0109$
33. (*b*) $\hat{\mu}_{Y|x} = 6.375 + 2.943x$
(*c*) 41.987
(*d*) $t = 6.378$, significant at .0002

Section 11.4

35. (*a*) $f = .392$, not significant at $\alpha = .05$
(*b*) yes
37. (*a*) $\hat{\mu}_{Y|x} = 2.1111 + .3167x$
(*b*) 13.194
41. $f = 4.725$, $P < .025$

Section 11.6

47. (*b*) $\hat{\rho} = .887$
49. $.657 \leq \rho \leq .966$
51. (*b*) $\hat{\rho} = .586$
(*c*) $-.068 \leq \rho \leq .888$
57. 96.9%
59. $R^2 \doteq 0$

Chapter 11 review exercises

60. (*b*) $\hat{\mu}_{Y|x} = 145.667 + 6.20x$
61. (*a*) no, $f = 27.91$, significant at $\alpha = .05$
(*b*) yes
62. $\hat{\mu}_{Y|x} = 99.383 - .0052x$
63. significant at $\alpha < .001$
64. $\hat{R}^2 = .967$
65. $99.1603 \leq \beta_0 \leq 99.6057$
66. $-.00602 \leq \beta_1 \leq -.00431$
67. $\hat{\mu}_{Y|x} = .8233 - .0589x$
68. $t = -9.264$, significant at $P < .0001$
69. (*a*) $\hat{\mu}_{Y|x=3.25} = .6318$; $(\hat{Y}|x = 3.25) = .6318$
(*b*) $.6121 \leq \mu_{Y|x=3.25} \leq .6516$
(*c*) $.5703 \leq (Y|x = 3.25) \leq .6933$
70. (*b*) $\hat{\rho} = -.959$
71. significant at $P < .0001$
72. $-.866 \leq \rho \leq -.988$
73. (*a*) $t = .889$, $P > .20$
(*b*) $t = 2.179$, $P < .05$
75. (*a*) $\hat{\rho} = .989$
(*b*) $.953 \leq \rho \leq .997$
(*c*) significant; yes
(*d*) 97.86%
A coefficient of determination of 97.86% implies that 97.86% of the variance in the dependent variable Y is explained by the linear regression model.

Section 12.2

11. (*a*) $X' = \begin{bmatrix} 1 & 1 & 1 & \cdots & 1 & 1 \\ 1.35 & 1.90 & 1.70 & & 1.85 & 1.40 \end{bmatrix}$

(*b*) $X'X = \begin{bmatrix} 10 & 16.75 \\ 16.75 & 28.6375 \end{bmatrix}$

(*c*) $X'\mathbf{y} = \begin{bmatrix} 170 \\ 282.405 \end{bmatrix}$

(*d*) $\begin{bmatrix} 10 & 16.75 \\ 16.75 & 28.6375 \end{bmatrix} \begin{bmatrix} b_0 \\ b_1 \end{bmatrix} = \begin{bmatrix} 170 \\ 282.405 \end{bmatrix}$

(*e*) $(X'X)^{-1} = \begin{bmatrix} 4.92688 & -2.88172 \\ -2.88172 & 1.72043 \end{bmatrix}$

(*f*) $\hat{\boldsymbol{\beta}} = \begin{bmatrix} 23.7576 \\ -4.0344 \end{bmatrix}$, results agree with Example 11.3.3

13. (*a*) $\hat{\mu}_{Y|x} = 42.921 - 28.625x + 4.640x^2$
(*b*) $\hat{\mu}_{Y|x=2.5} = .358$

15. $X' = \begin{bmatrix} 1 & 1 & 1 & \cdots & 1 & 1 \\ x_{11} & x_{12} & x_{13} & \cdots & x_{17} & x_{18} \\ x_{11}^2 & x_{12}^2 & x_{13}^2 & & x_{17}^2 & x_{18}^2 \\ x_{21} & x_{22} & x_{23} & \cdots & x_{27} & x_{28} \end{bmatrix}$

Section 12.3

19. (*a*) $E[C\mathbf{Y}] = \begin{bmatrix} 35 \\ 75 \end{bmatrix}$

(*b*) $\operatorname{Var} \mathbf{Y} = \begin{bmatrix} 6 & 0 \\ 0 & 6 \end{bmatrix}$

(*c*) $\operatorname{Var} C\mathbf{Y} = \begin{bmatrix} 72 & 102 \\ 102 & 300 \end{bmatrix}$

25. $s^2 = \dfrac{S_{yy} - \text{SSR}}{n - 2} = 1.002714$

$\operatorname{Var} \hat{\beta}_0 = (2.3)(1.002714) = 2.30624$
$\operatorname{Var} \hat{\beta}_1 = (.05342857)(1.002714) = .0535737$
$\operatorname{Var} \hat{\beta}_2 = (.00005714286)(1.002714) = .000057298$

Section 12.4

27. $-4.6162 \le \beta_1 \le -3.7038$; yes since CI excludes zero
$-.02046 \le \beta_2 \le -.00934$; yes since CI excludes zero

29. $17.6605 \le \mu_{Y|x_1=1.5,\ x_2=40} \le 18.1675$
$17.5383 \le Y|x_1 = 1.5,\ x_2 = 40 \le 18.2897$

31. $9.265 \le \mu_{Y|x=50} \le 10.007$
$7.766 \le Y|x = 50 \le 11.506$

33. $5.445 \le \mu_{Y|x_1=12,\ x_2=40,\ x_3=20} \le 47.301$

35. $-8.666 \le \beta_0 \le 12.590$

Section 12.5

37. $\hat{\mu}_{Y|x} = 4.143 + 7.107x$

39. $\hat{\mu}_{Y|x} = -4.571 + 12.917x - .726x^2$

41. $33.735 \le \mu_{Y|x=4.8} \le 42.780$

45. $\hat{\mu}_{Y|x_1, x_2} = .805 + .245x_1 + .264x_2$

47. $\hat{\rho} = .9517$

49. $f = 18.705$, $P \doteq .0035$, reject H_0; yes

51. $f = 32.2762$, $P \doteq .0003$, reject H_0

Section 12.6

53. (*a*)

$$X = \begin{bmatrix} 1 & 1 & 1 & 1 \\ 1 & 1.5 & 1 & 1.5 \\ 1 & 2 & 1 & 2 \\ 1 & 2.5 & 1 & 2.5 \\ 1 & 3 & 1 & 3 \\ 1 & 3.5 & 1 & 3.5 \\ 1 & 4 & 1 & 4 \\ 1 & 4.5 & 1 & 4.5 \\ 1 & 5 & 1 & 5 \\ 1 & 1 & 0 & 0 \\ 1 & 2 & 0 & 0 \\ 1 & 2.5 & 0 & 0 \\ 1 & 3 & 0 & 0 \\ 1 & 4 & 0 & 0 \\ 1 & 5 & 0 & 0 \end{bmatrix}$$

(*b*) $\hat{\mu}_{Y|x_1, x_2} = .470 - .084x_1 + .354x_2 + .025x_1x_2$
(*c*) $\hat{\beta}_3 = .0248$
(*d*) $f = 4.9506$, $P = .0479$, reject H_0 at $\alpha = .05$; regression lines for the two furnaces have different slopes

55. (*a*) A: $Y = \beta_0 + \beta_1 x_1 + E$
B: $Y = (\beta_0 + \beta_2) + (\beta_1 + \beta_4)x_1 + E$
C: $Y = (\beta_0 + \beta_3) + (\beta_1 + \beta_5)x_1 + E$
(*b*) H_0: $\beta_4 = \beta_5 = 0$
(*c*) H_0: $\beta_2 = \beta_3 = \beta_4 = \beta_5 = 0$; 4 and $n - 6$

Chapter 12 review exercises

67. $\hat{\mu}_{Y|x_1} = 154.5 + .5095x_1$
68. $164.163 \le \mu_{Y|x_1=45} \le 190.695$
69. $\hat{\mu}_{Y|x_1} = 96.333 + .6333x_2$
72. $\hat{\mu}_{Y|x_1, x_2} = 78.5 + .510x_1 + .633x_2$
73. Reduced model
74. $R^2 = 34.65$; $\hat{\rho} = .5886$
75. 164.69
76. $153.688 \le \mu_{Y|x_1, x_2} \le 182.027$

Section 13.1

5. (*a*) $f = 4.95$, $P = .0174$, significant at $\alpha = .05$
(*b*) no
(*c*) no

Section 13.2

7. $f_{4,35} = 216.34$, $p < .001$
9. $b = 23.187$ with 3 df, significant at $\alpha < .005$
not reasonable to assume homogeneity of variances; use a nonparametric test
11. $b = .354$, not significant

Section 13.3

13. (*a*) $\alpha' \le .1426, .0333$
(*b*) $\alpha' \le .5367, .0067$
(*c*) $\alpha' \le .9006, .0022$

17. (*a*) 3
(*b*) .05
(*c*) $\mu_1 \neq \mu_2, \mu_1 = \mu_3, \mu_2 = \mu_3$

21. $\mu_1 \neq \mu_2, \mu_1 \neq \mu_3, \mu_2 \neq \mu_3$ at $\alpha = .01$

Section 13.4

25. (*a*) $$SS_{L_1} = \frac{\left[\bar{Y}_1. - (1/3)\left(\bar{Y}_2. + \bar{Y}_3. + \bar{Y}_4.\right)\right]^2}{(1 + 1/9 + 1/9 + 1/9)/n}$$

$$SS_{L_2} = \frac{\left(\bar{Y}_1. - \bar{Y}_2.\right)^2}{(1 + 1)/n}$$

(*e*) $$SS_{L_1} = \frac{\left(\bar{Y}_1. - \bar{Y}_2. + \bar{Y}_3. - \bar{Y}_4.\right)^2}{(1 + 1 + 1 + 1)/n}$$

$$SS_{L_2} = \frac{\left(\bar{Y}_1. - \bar{Y}_3.\right)^2}{(1 + 1)/n}$$

(*f*) $$SS_{L_1} = \frac{\left(\bar{Y}_1. - \bar{Y}_2. + \bar{Y}_3. - \bar{Y}_4.\right)^2}{1/5 + 1/5 + 1/10 + 1/10}$$

$$SS_{L_2} = \frac{\left(\bar{Y}_1. - \bar{Y}_3.\right)^2}{1/5 + 1/10}$$

27. $f_{1,36} = 114.11$, $P < .05$, reject H_0
$f_{1,36} = 8.9246$, $P < .05$, reject H_0

29. 60 $\underline{51.333 \quad 50}$ 44.333 38.667

Section 13.5

33. (*a*) blocking is effective
(*b*) blocking is effective
(*c*) blocking is not effective
(*d*) blocking is not effective
(*e*) designs are equivalent

35. $f_{3,21} = 3.37$, significant at $\alpha = .05$

37. 6; .017

39. yes

41. $\widehat{RE} = 12.47$, blocking is highly desirable

Section 13.6

47. $$\text{SS}_{\text{Tot}} = \sum_i \sum_j \sum_k (Y_{ijk} - \bar{Y}_{\dots})^2$$

$$\text{SS}_{\text{Rows}} = r \sum_{j=1}^{r} (\bar{Y}_{.j.} - \bar{Y}_{\dots})^2$$

$$\text{SS}_{\text{Col}} = r \sum_{k=1}^{r} (\bar{Y}_{..k} - \bar{Y}_{\dots})^2$$

49. $f_{2,2} = 8.75$, $P \doteq .1026$, rejection of H_0 is debatable

Section 13.8

51.
$$\mathbf{Y} = \begin{bmatrix} Y_{11} \\ Y_{12} \\ Y_{13} \\ Y_{21} \\ Y_{22} \\ Y_{23} \end{bmatrix}, \quad \boldsymbol{\alpha} = \begin{bmatrix} \mu \\ \alpha_1 \\ \alpha_2 \\ \alpha_3 \end{bmatrix}, \quad X = \begin{bmatrix} 1 & 1 & 0 & 0 \\ 1 & 1 & 0 & 0 \\ 1 & 0 & 1 & 0 \\ 1 & 0 & 0 & 0 \\ 1 & 0 & 0 & 1 \\ 1 & 0 & 0 & 1 \end{bmatrix}, \quad \mathbf{E} = \begin{bmatrix} E_{11} \\ E_{12} \\ E_{13} \\ E_{21} \\ E_{22} \\ E_{23} \end{bmatrix},$$

$$X'X = \begin{bmatrix} 6 & 2 & 2 & 2 \\ 2 & 2 & 0 & 0 \\ 2 & 0 & 2 & 0 \\ 2 & 0 & 0 & 2 \end{bmatrix}$$

$X'X$ has no inverse since the first column is the sum of the last three columns.

Section 13.9

53. (*a*) $H = 9.05$, not significant at $\alpha = .05$ ($P \doteq .06$).
(*b*) $f = 3.30$, significant at $\alpha = .05$ ($P \doteq .026$)

55. $S = 3.4$ with 4 df; not significant

Chapter 13 review exercises

58. (*a*)

Source	Df	*SS*	*MS*
CO_2 level	4	11,274.32	2818.50
Error	45	1,248.04	27.73
Corrected total	49	12,522.36	

(*b*) yes, $f = 101.63$

59. $b = 1.07$, not significant; yes

60. (*b*)

Source	df	*SS*	*MS*	*F*	EMS
Treatment	2	110.6	55.3	3.0	$\sigma^2 + 10\sigma^2_{\mathrm{Tr}}$
Error	27	497.7	18.433		σ^2
Total	29	608.3			

$f = 3.0$ with 2 and 27 df; not significant at $\alpha = .05$

(*c*) due to error, 83.3%; due to treatments, 16.7%

61. (*a*) $f = 79.9$, significant at $\alpha < .0001$
(*b*) yes
(*c*) all three are significantly different

Section 14.1

3. $Y_{ijk} = \mu + \alpha_i + \beta_j + (\alpha\beta)_{ij} + E_{ijk}$
7. $P \leq .0001$, significant
9. no, interaction is not significant
11. $Y_{ijk} = \mu + \alpha_i + \beta_j + (\alpha\beta)_{ij} + E_{ijk}$
13. $P \doteq .064$, significant at $\alpha = .10$
15. E_2 and 3 months

Section 14.2

21. $SS_{\text{Tot}} = SS_A + SS_B + SS_C + SS_{AB} + SS_{AC} + SS_{BC} + SS_{ABC} + SS_E$
23.

Source of variation	Degrees of freedom (DF)	Sum of squares (SS)	Mean square (MS)	F	P value
Treatment	23	21100.99	917.43	8.36	.0001
A	2	10296.36	5148.18	46.88	.0001
B	3	7586.38	2528.79	23.03	.0001
C	1	1974.01	1974.01	17.98	.0001
AB	6	735.42	122.57	1.12	.3671
AC	2	6.03	3.01	.03	.9729
BC	3	25.49	8.50	.08	.9719
ABC	6	477.31	79.55	.72	.6320
Error	48	5270.67	109.81		
Total	71	26371.65			

Note: All SS, MS, and F statistics have been rounded to two places from SAS output.

Section 14.3

25. H_0: $\sigma^2_{AB} = 0$, $F = MS_{AB}/MS_E$
H_0^{I}: $\sigma^2_A = 0$, $F = MS_A/MS_E$
H_0^{II}: $\mu_{.1.} = \mu_{.2.} = \cdots = \mu_{.b.}$, $F = MS_B/MS_{AB}$
H_0^{III}: $\sigma^2_{\text{Tr}} = 0$, $F = MS_{\text{Tr}}/MS_E$

Section 14.4

27. (a) a: 13
b: 11
ab: 17
(c) B high: slope = 3
B low: slope = 3
29. Any value of β_{12} satisfying $\beta_{12} > 2$ or $\beta_{12} < -2$
31. A: H_0^{I}: $\beta_1 = 0$, $f_{1,4} = 1.59$; $p \doteq .2764$
B: H_0^{II}: $\beta_2 = 0$, $f_{1,4} = 3.76$, $P \doteq .1244$
33. (a) $2^4 = 16$

(*b*)

	Effects								
Treatment combination	x_1 A	x_2 B	x_3 C	x_4 D	**Treatment combination**	x_1 A	x_2 B	x_3 C	x_4 D
(1)	−	−	−	−	*bc*	−	+	+	−
a	+	−	−	−	*bd*	−	+	−	+
b	−	+	−	−	*cd*	−	−	+	+
c	−	−	+	−	*abc*	+	+	+	−
d	−	−	−	+	*abd*	+	+	−	+
ab	+	+	−	−	*acd*	+	−	+	+
ac	+	−	+	−	*bcd*	−	+	+	+
ad	+	−	−	+	*abcd*	+	+	+	+

(*c*)
$$SS_A = \frac{1}{2^4 n}[-(1) + a - b - c - d + ab + ac + ad - bc - bd - cd + abc + abd + acd - bcd + abcd]^2$$

$$SS_B = \frac{1}{2^4 n}[-(1) - a + b - c - d + ab - ac - ad + bc + bd - cd + abc + abd - acd + bcd + abcd]^2$$

$$SS_C = \frac{1}{2^4 n}[-(1) - a - b + c - d - ab + ac - ad + bc - bd + cd + abc - abd + acd + bcd + abcd]^2$$

$$SS_D = \frac{1}{2^4 n}[-(1) - a - b - c + d - ab - ac + ad - bc + bd + cd - abc + abd + acd + bcd + abcd]^2$$

Section 14.5

37. (*a*) *AB*
(*b*) *AB*, *ABC*, *C*
(*c*) *ABCD*, *ABC*, *BCD*, *D*, *A*, *AD*, *BC*

39. (*a*) $L_1 = z_1 + z_2$ and $L_2 = z_3 + z_4$

Treatment combination	L_1 **(mod 2)**	L_2 **(mod 2)**	**Block**
(1)	0	0	1
a	1	0	2
b	1	0	2
c	0	1	3
d	0	1	3
ab	0	0	1
ac	1	1	4
ad	1	1	4
bc	1	1	4
bd	1	1	4
cd	0	0	1
abc	0	1	3
abd	0	1	3
acd	1	0	2
bcd	1	0	2
abcd	0	0	1

(*b*) *ABCD*; block 1

(*c*)

Source	Degrees of freedom (*DF*)
Blocks	3
A	1
B	1
C	1
D	1
Error	8
Total	15

Section 14.6

41. $-(1) - a + b - c + ab - ac + bc + abc$; $-a + b - c + abc$; AC

43. (*a*)

Block 1	Block 2	Block 3	Block 4
(1)	*d*	*c*	*a*
ab	*e*	*f*	*b*
cf	*ac*	*ad*	*cd*
de	*af*	*ae*	*ce*
acd	*bc*	*bd*	*df*
ace	*bf*	*be*	*ef*
adf	*abd*	*abc*	*acf*
aef	*abe*	*abf*	*ade*
bcd	*cdf*	*cde*	*bcf*
bce	*cef*	*def*	*bde*
bdf	*acde*	*acdf*	*abcd*
bef	*adef*	*acef*	*abce*
abcf	*bcde*	*bcdf*	*abdf*
abde	*bdef*	*bcef*	*abef*
cdef	*abcdf*	*abcde*	*acdef*
abcdef	*abcef*	*abdef*	*bcdef*

(*b*)

Source	Degrees of freedom (*DF*)
A	1
B	1
C	1
D	1
E	1
F	1
Error	9
Total	15

(*c*)

$$\begin{aligned}
(1) &= ABDE = ABCF = CDEF \\
A &= BDE = BCF = ACDEF \\
B &= ADE = ACF = DEF \\
C &= ABCDE = ABF = CEF \\
D &= ABE = ABCDF = CDF \\
E &= ABD = ABCEF = CDE
\end{aligned}$$

Chapter 14 review exercises

45.

(1)	*a*	*e*	*b*
bc	*be*	*ab*	*c*
bd	*ce*	*ac*	*d*
cd	*de*	*ad*	*ae*
abe	*abc*	*bce*	*bcd*
ace	*abd*	*bde*	*abce*
ade	*acd*	*cde*	*abde*
abcde	*bcde*	*abcd*	*acde*

(*a*) *ABCD*, *BCDE*, *AE*

(*b*)

Source of variation	Degrees of freedom (*DF*)
Blocks	3
A	1
B	1
C	1
D	1
E	1
AB	1
AC	1
AD	1
BC	1
BD	1
BE	1
CD	1
CE	1
DE	1
Error	14
Total	31

(*c*) no, there would be no error term/no error degrees of freedom

46. (*b*) no, error degrees of freedom is 0

(*c*) no, error degrees of freedom is 0

(d) yes

(e) 3

47. (a) $\mu_{Y|x_1, x_2} = \beta_0 + \beta_1 x_1 + \beta_2 x_2$

(b) $$X = \begin{bmatrix} 1 & -1 & -1 \\ 1 & -1 & -1 \\ \cdot & \cdot & \cdot \\ \cdot & \cdot & \cdot \\ \cdot & \cdot & \cdot \\ \cdot & 1 & -1 \\ \cdot & 1 & -1 \\ \cdot & \cdot & \cdot \\ \cdot & \cdot & \cdot \\ \cdot & \cdot & \cdot \\ \cdot & -1 & 1 \\ \cdot & -1 & 1 \\ \cdot & \cdot & \cdot \\ \cdot & \cdot & \cdot \\ \cdot & \cdot & \cdot \\ \cdot & 1 & 1 \\ \cdot & 1 & 1 \\ \cdot & \cdot & \cdot \\ 1 & \cdot & \cdot \\ 1 & \cdot & \cdot \end{bmatrix}$$

(c) $$X'X = \begin{bmatrix} 2^2n & 0 & 0 \\ 0 & 2^2n & 0 \\ 0 & 0 & 2^2n \end{bmatrix}$$

(d) $$X'Y = \begin{bmatrix} (1) + a + b + ab \\ -(1) + a - b + ab \\ -(1) - a + b + ab \end{bmatrix}$$

(e) $$\hat{\beta} = \begin{bmatrix} [(1) + a + b + ab]/2^2n \\ [-(1) + a - b + ab]/2^2n \\ [-(1) - a + b + ab]/2^2n \end{bmatrix}$$

(f) $\text{SSR} = [-(1) + a - b + ab]^2/2^2n + [-(1) - a + b + ab]^2/2^2n$

(g) $$\text{SSE}_r = S_{yy} - [-(1) + a - b + ab]^2/2^2n \\ -[-(1) - a + b + ab]^2/2^2n$$

$$\text{SSE}_f = S_{yy} - [-(1) + a - b + ab]^2/2^2n \\ -[-(1) - a + b + ab]^2/2^2n - [(1) - a - b + ab]^2/2^2n$$

(h) $\text{SSE}_r - SS_f = [(1) - a - b + ab]^2/2^2n$

48. (a) $$Y_{ijkl} = \mu + \alpha_i + \beta_j + (\alpha\beta)_{ij} + \gamma_k \\ +(\alpha\beta)_{ij} + (\alpha\gamma)_{ik} + (\beta\gamma)_{jk} + E_{ijkl}$$

(*b*)

Source of variation	Degrees of freedom
A	4
B	3
C	2
AB	12
AC	8
BC	6
Error	24
Total	59

49. (*a*) (1), *ab*, *ac*, *ad*, *ae*, *bc*, *bd*, *be*, *cd*, *ce*, *de*, *abcd*, *abce*, *abde*, *acde*, *bcde*
(*b*)

Source of variation	Degrees of freedom
A	1
B	1
C	1
D	1
E	1
Error	10
Total	15

50. (*a*) $B \equiv AC \equiv BCD \equiv AD$
$C \equiv AB \equiv D \equiv ABCD$
$D \equiv ABCD \equiv C \equiv AB$
(*b*) no, the main effect C is aliased with the main effect D

Section 15.1

1. 60, 45, 30, 15; yes, these seem to differ quite a bit from the expected numbers
3. 10; questionable, the observed values are not drastically different from those expected

Section 15.2

5. $\chi^2 = 22.66$; reject H_0, $P < .005$ based on the X_4^2 distribution

Section 15.3

9. $\chi^2 = 4.57$; reject H_0, $.025 < P < .05$ based on X_1^2 distribution
11. $\chi^2 = 8.84$; reject H_0, $.025 < P < .05$ based on X_3^2 distribution

Section 15.4

13. $\chi^2 = 3.95$; reject H_0, critical point = 3.84
15. H_0: $p_{11} = p_{21} = p_{31}$, $p_{12} = p_{22} = p_{32}$, $p_{13} = p_{23} = p_{33}$, $p_{14} = p_{24} = p_{34}$; $\chi^2 = 14.72$; reject H_0, $.01 < P < .025$ based on the X_6^2 distribution

17. $\chi^2 = 16.03$; reject H_0; $P < .005$
19. $\chi^2 = 43.65$; reject H_0; $P < .005$

Chapter 15 review exercises

23. H_0: $p_{11} = p_{21}$; $\chi^2 = .709$; no, $.25 < P < .50$
24. yes; $\chi^2 \doteq 16.04$; $P < .005$
25. yes; $\chi^2 \doteq 34.05$; $P < .005$
26. no; $\chi^2 \doteq .69$; $.25 < P < .5$
27. (*a*) 175; 87.5; 87.5
(*b*) no; $\chi^2 \doteq .77$; $.50 < P < .75$

Section 16.1

2. $P = .0028$; ARL $\doteq$ 360 minutes; no; every 5 minutes
4. (*b*) .0026
(*c*) 384.6
(*d*) 384.1
(*e*) 6.30
(*f*) 6.30 hours; 3.15 hours; 1.575 hours

Section 16.2

6. (*a*) $\bar{X}$: LCL = 23.116, UCL = 25.884
$\bar{R}$: LCL = 0, UCL = 5.074
(*b*) $\bar{X}$: LCL = .04346, UCL = .04654
$\bar{R}$: LCL = .00112, UCL = .00888
(*c*) $\bar{X}$: LCL = 7.084, UCL = 10.216
$\bar{R}$: LCL = 0, UCL = 4.9067

Section 16.3

11. (*a*) $\bar{p} = .0195$, $\hat{\sigma}_P = .009777$
(*b*) LCL = 0, UCL = .0488
(*c*) All sample points are within control limits. If they were not one would delete points outside the limits and recompute limits with reduced data set.

Section 16.4

15. (*a*) $\Pi = 0$, $P_{acc} = 1$
$\Pi = .05$, $P_{acc} = .7358$
$\Pi = .1$, $P_{acc} = .3917$
$\Pi = .15$, $P_{acc} = .1756$
(*b*) producer risk = .2642
consumer risk = .1756

Section 16.5

17. (*b*) .6454
(*c*) .8623 for $p = .1$, .1666 for $\Pi = .4$, .0123 for $\Pi = .6$
(*d*) .3546

INDEX